Handbook of Air Pollution from Internal Combustion Engines

Pollutant Formation and Control

Handbook of Air Pollution from Internal Combustion Engines

Pollutant Formation and Control

Edited by

Eran Sher

ACADEMIC PRESS

Boston San Diego New York
London Sydney Tokyo Toronto

This book is printed on acid-free paper. ♾

ACADEMIC PRESS
525 B Street, Suite 1900, San Diego, CA 92101, USA
1300 Boylston Street, Chestnut Hill, MA 02167, USA
http://www.apnet.com

United Kingdom Edition published by
ACADEMIC PRESS LIMITED
24-28 Oval Road, London NW1 7DX
http://www.hbuk/co.uk/ap/

ISBN: 0-12-639855-0
Library of Congress Cataloging-in-Publication Data

Handbook of air pollution from internal combustion engines : pollutant formation and control / edited by Eran Sher.
p. cm.
Includes bibliographical references and index.
ISBN 0-12-639855-0 (alk. paper)
1. Motor vehicles—Motors—Exhaust gas—Environmental aspects.
2. Internal combustion engines—Environmental aspects. 3. Air-Pollution. I. Sher, Eran.
TD886.5.H36 1998
629.25′28—dc21 97-48256
CIP

Printed and bound in the United Kingdom
Transferred to Digital Print 2011

Dedication

I owe my roots to Professor Chaim Elata of the Ben-Gurion University, Beer-Sheva, Israel,
who taught me how to think.

I owe my stem to the late Professor Rowland S. Benson of UMIST Manchester, England,
who taught me how to observe.

I owe my foliage to Professor James C. Keck of MIT, Cambridge, Massachusetts, USA,
who taught me how to analyze.

Contents

List of Contributors

PART I

OVERVIEW

1. Motor Vehicle Emissions Control: Past Achievements, Future Prospects

Prof. John B. Heywood
Dept. of Mechanical Engineering
Massachusetts Institute of Technology
Cambridge, MA 02139
tel: 617-253-2243
fax: 617-253-5981
e-mail: jheywood@mit.edu

PART II

GLOBAL ASPECTS

2. Environmental Aspects of Air Pollution

Prof. Eran Sher
Dept. of Mechanical Engineering
Ben-Gurion University
Beer-Sheva 84 105
Israel
tel: 972-7-646-1394
fax: 972-7-647-2990
e-mail: sher@menix.bgu.ac.il

3. Health Aspects of Air Pollution

Prof. Rafael Carel
Soroka Medical Center

Beer-Sheva, Israel
tel: 972-7-6494-663
fax: 972-7-649-3934
e-mail: rcarel@post.tau.ac.il

4. Economic and Planning Aspects of Transportation Emission

Dr. Steven E. Plaut
Graduate School of Business Administration
University of Haifa
Haifa 31905, Israel
tel: 972-4824-0110
fax: 972-4824-9194
e-mail: rsec792@uvm.haifa.ac.il

Dr. Pnina O. Plaut
Faculty of Architecture and Town Planning
Technion, Haifa, Israel

PART III

SPARK-IGNITION ENGINES

5. Introductory Chapter: Overview and the Role of Engines with Optical Access

Dr. Richard Stone
Department of Engineering Science
University of Oxford
Oxford OX1 3PJ
United Kingdom
tel: 44-1865-273-000
fax: 44-1865-273-010
e-mail: richard.stone@eng.ox.ac.uk

6. Combustion-Related Emissions in SI Engines

Prof. Simone Hochgreb
Dept. of Mechanical Engineering
Massachusetts Institute of Technology
Cambridge, MA 02139
tel: 617-253-0972
fax: 617-253-9453
e-mail: simone@mit.edu

7. Pollution from Rotary Internal Combustion Engines

Dr. Mark Dulger
Department of Mechanical Engineering

Ben-Gurion University
Beer-Sheva 84 105
Israel
tel: 972-7-646-1353
fax: 972-7-647-2990
e-mail: markd@menix.bgu.ac.il

8. Control Technologies in Spark-Ignition Engines

Prof. Brian Milton
School of Mechanical and Manufacturing Engineering
The University of New South Wales
Barker Street, Gate 14
Kensington, Sydney 2052
Australia
tel: 61-2-385-4088
fax: 61-2-663-1222
e-mail: milton@unsw.edu.au

PART IV

COMPRESSION-IGNITION ENGINES

9. Introduction

Prof. Dr. Franz Pischinger
FEV Motorentechnik GmbH and Co. KG
Neuenhofstrasse 181
Aachen D-52078
Germany
tel: 49-241-5689-10
fax: 49-241-5689-224

10. Combustion-Related Emissions in CI Engines

Dr. Gary Hawley
School of Mechanical Engineering
University of Bath
Bath, United Kingdom
tel: 44-1225-826-860
fax: 44-1225-826-928
e-mail: ensjgh@bath.ac.uk

R. W. Horrocks
Advanced Diesel Engines
Research and Engineering Centre
Ford Motor Company, Ltd.
Laindon, United Kingdom

Frank Wallace and Chris Brace
School of Mechanical Engineering
University of Bath
Bath, United Kingdom

11. Control Technologies in Compression-Ignition Engines

Dr. Stephen Charlton
Director, Advanced Diesel Engine Technology
Cummins Engine Company, Inc.
Mail Code 50174
1900 McKinley Avenue
Columbus, Indiana 47201
tel: 812-377-8788
fax: 812-377-7226
e-mail: 103173.622@compuserve.com

PART V

TWO-STROKE ENGINES

12. Introductory Chapter: From a Simple Engine to an Electrically Controlled Gasdynamic System

Prof. Dr. Cornel Stan
College of Technology and Economics
Westsaxon Institute of Zwickau
Germany
tel: 49-375-536-1600
fax: 49-375-536-1193
e-mail: cornel.stan@fh-zwickau.de

13. Air Pollution from Small Two-Stroke Engines and Technologies to Control It

Prof. Yuji Ikeda and Tsuyoshi Nakjma
Department of Mechanical Engineering
Kobe University
Rokkodai, Nada
Kobe, Japan
tel: 81-78-803-1114
fax: 81-78-845-2736
e-mail: ikeda@mech.kobe-u.ac.jp

Prof. Eran Sher
Dept. of Mechanical Engineering
Ben-Gurion University

Beer-Sheva 84 105
Israel

14. Air Pollution from Large Two-Stroke Diesel Engines and Technologics to Control It

Dr. Svend Henningsen
MAN and B&W Diesel A/S
Copenhagen SV
DK-2450
Denmark
tel: 45-3385-1100
fax: 45-3385-1030
e-mail: svend.henningsen/copenhagen@manbw.dk

PART VI

FUELS

15. Introductory Chapter: Fuel Effects

Dr. David Blackmore
Research Centre
Shell Research, Ltd.
Thornton P.O. 1
Chester CH1 3SH
United Kingdom
tel: 44-151-373-5768
fax: 44-151-373-5674

16. Fuel Effects on Emissions

Prof. Yoram Zvirin
Department of Mechanical Engineering
Technion, Haifa 32000, Israel
tel: 972-4-292-070
fax: 972-4-324-533
e-mail: meryzvi@tx.technion.ac.il

Acknowledgments

The editor wishes to acknowledge the following organizations for their support and cooperation: Ford Motor Co. Ltd., Advanced Diesel Engines Research and Engineering Centre, UK; FEV Motorentechnik GmbH & Co KG, Aachen, Germany; Cummins Engine Company, Inc., Advanced Diesel Engine Technology Columbus, Indiana, USA; MAN and B&W Diesel A/S, Copenhagen Denmark; Research Centre, Shell Research Ltd., Chester UK; and the Pearlstone Center for Aeronautical Studies, Ben-Gurion University, Israel.

Academic Press and the editor would like to express their thanks to the following reviewers and other helpful persons for their invaluable comments and suggestions: David Blackmore, Shell Research Ltd., Chester UK; Mark Dulger, Ben-Gurion University, Israel; Elbert Hendricks, The Technical University of Denmark, Lyngby, Denmark; Itzik Henig, Ford Motor Co., UK; Simone Hochgreb, MIT, Cambridge, Massachusetts, USA; Uri Regev, Ben-Gurion University, Israel; Zvi Ruder, Academic Press, Boston, Massachusetts, USA; Roger Sierens, University of Gent, Gent, Belgium; Cornel Stan, Westsaxon Institute of Zwickau, Germany; Richard Stone, Oxford University, Oxford, UK; and Desmond Winterbone, UMIST, Manchester, UK.

The authors and editor wish to acknowledge the following publishers for their kind permission to reproduce figures from their publications: The Society of Automotive Engineers, American Society of Mechanical Engineers, The Institution of Mechanical Engineers, Gordon and Breach Science Publishers, The Combustion Institute, Elsevier Science Publishing Company, Edward Arnold Publishers, Macmillan Press, Automotive Matters International Ltd., and TNO Road-Vehicles Research Institute.

Special thanks are due to Elizabeth Voit of Academic Press, and to Ian Vinogradov and Ilai Sher for a careful preparation of some of the figures and illustrations in the handbook.

Eran Sher
Department of Mechanical Engineering
The Pearlstone Center for Aeronautical Studies
Ben-Gurion University of the Negev, Beer-Sheva, Israel

Handbook of Air Pollution from Internal Combustion Engines

Pollutant Formation and Control

PART I

Overview

CHAPTER 1

Motor Vehicle Emissions Control: Past Achievements, Future Prospects♦

John B. Heywood
Sun Jae Professor of Mechanical Engineering, Director, Sloan Automotive Laboratory, Massachusetts Institute of Technology, Massachusetts, USA

♦This chapter is based on the British Institution of Mechanical Engineers, Combustion Engine Group's Prestige Lecture, given by the author in London, May 21, 1996 and on the Institution's George Stephenson Centennial International Lecture given by the author in November 1997 in Hong Kong, Kuala Lumpur, Singapore, Australia, and New Zealand.

ISBN: 0-12-639855-0
$25.00

1.1 SYNOPSIS

Motor vehicles—cars, trucks, and buses—are a major source of air pollution. For 35 years we have been both learning about the problem and attempting to control vehicle emissions. In this introduction, we trace the history of our efforts to understand this important environmental issue and to find effective solutions, as well as look ahead to the future. While steady progress has been made and effective technology, such as engine controls, exhaust catalysts, and improved fuels, has been developed at the individual vehicle level, the full resolution of this problem still escapes us. Growth in vehicle use and the failure of the emission controls in a small but significant fraction of vehicles have offset a substantial part of the anticipated gains. Looking to the future, are prospects for effective emissions control better? Yes, improved fuel injection, sensors and controls, and catalyst technologies are being developed, more effective inspection programs are being implemented, and alternative fuels may play some role. However, growth in vehicle use will continue to present a major environmental challenge to both automotive engineers and regulators.

1.2 INTRODUCTION

In the 1950s through studies in Los Angeles, it became clear that emissions from automobiles were a major contributor to urban air pollution. This *smog,* formed in the atmosphere as a result of complex photochemistry between hydrocarbons—often called volatile organic compounds (HC or VOC), and oxides of nitrogen (NO_x)—on warm spring, summer, and fall days, results in high ambient levels of ozone and other oxidants. In addition, automobiles are the dominant source of carbon monoxide (CO) and of lead. It is not just cars: Light trucks, heavy trucks, and off-road vehicles also contribute significantly. So do stationary combustion systems. Even natural (i.e., biogenic) hydrocarbon emissions are important.

Starting in the late 1960s, vehicle emissions in the developed world have been regulated with increasing strictness. More recently, the fuels that the spark-ignition and diesel engines in these vehicles use (i.e., gasoline/petrol and diesel) have been or are about to be subject to more stringent constraints with the intent of further reducing emissions. This introduction traces the history of our efforts to understand this important environmental issue and to find effective solutions. We have made steady progress on improving urban air quality, yet the full resolution of the problem still eludes us. Looking at this problem of motor vehicles and air pollution from a broader perspective, there are several important questions. Just what is the problem? What have we done so far? Why is it proving to be such a difficult problem to solve, both fundamentally and in practice? What are the prospects for future improvements?

It has been my good fortune that the evolution of this problem and our attempts to resolve it have coincided with my own professional career. There is tremendous excitement and satisfaction in working on a new research problem with the opportunity to contribute to the development of technology that will help to resolve the problem. Over the past 30 years we have learned a great deal more about the internal combustion engine, the prime mover that is so ubiquitous and important to our modern lives. Whether it is a blessing or a curse is not the issue here: The internal combustion engine exists, is used worldwide in very large numbers, and that pattern will continue into the future. However, the internal combustion engine does need to become steadily more environmentally friendly.

1.3 MOTOR VEHICLES AND AIR POLLUTION

In the United States, cars, trucks, and off-road vehicles are currently estimated to be responsible for about 40 percent to 50 percent of the HC or VOC emissions, 50 percent of the NO_x emissions, and 80 percent to 90 percent of the CO emissions in urban areas. The relative contributions in other parts of the developed world such as in Europe and Japan are similar. A large fraction of these emissions still comes from cars and light trucks with spark-ignition engines, though the relative importance of NO_x and particulates from diesel engines is rising. Over the past decade (1982–1991) in the aggregate, CO and VOC emissions from mobile sources have decreased about 40 percent and NO_x emissions by 25 percent despite substantial growth in vehicle miles traveled. However, it is the changes in seasonal emissions—winter for CO and summer for VOC and NO_x—that matter, and significant differences exist from one urban area to another. It also has become clear that photochemical smog with its high ozone levels is now a large-scale regional problem transported by the prevailing winds, with ozone concentrations in rural areas often reaching about half the urban peaks. Air quality measurements in the United States show that urban ozone levels have decreased by about 12 percent over the 1984–1993 decade, and incidents when the ozone National Ambient Air Quality Standard is exceeded have decreased by 60 percent. Ambient carbon monoxide levels have decreased by about 40 percent over the same period. These improvements have come primarily from the engine technology changes that emissions regulations have demanded.

Auto emissions control has a long history. Exhaust emission standards for new cars were first set in 1968 (1965 in California), after which the standards for exhaust emissions became steadily stricter every couple of years until the early 1980s. Much more stringent standards for the 1990s and beyond have now been established, especially in the United States and Europe (Table 1.1). The strategy

Table 1.1

Future U.S. Light-Duty Vehicle Exhaust Emission Standards[1]

Standard type	NMOG	CO	NO_x	HCHO
Precontrol (1966)	10.6	84	4.1	
		Federal		
Tier I (1994)	0.25	3.4	0.4	
Tier II (2003)	0.125	1.7	0.2	
		California		
Conventional vehicles (1993)	0.25	3.4	0.4	
TLEVs (1994)	0.125	3.4	0.4	0.015
LEVs (1997)	0.04	1.7	0.2	0.015
Ultra LEVs (1997)	0.04	1.7	0.2	0.008

NMOG, nonmethane organic gas (sum of nonoxygenated and oxygenated HCs). Standards are for five years or 50,000 miles. Transitional low-emission vehicles (TLEVs). Low-emission vehicles (LEVs).

adopted to minimize smog was major reductions in unburned HC emissions with lesser reductions in NO_x. The strategy was chosen in part from our assessment of how the photochemical smog chemistry responds to changes in HCs and NO_x as well as from the technical feasibility of reducing HCs relative to NO_x. Emissions standards for engines in large vehicles (gasoline-fueled and diesel) have steadily become stricter too, though lagging in time.

Let us focus first on the emissions control issues of automobiles with gasoline-fueled spark-ignition (SI) engines. While diesel trucks are an important contributor to air pollution, and diesel cars are growing to be a significant fraction of new car sales in Europe due to high fuel prices and their higher efficiency, the spark-ignition engine still dominates the motor vehicle emissions problem. To provide some perspective on past and present emissions levels, Table 1.2 gives typical numbers for the fuel consumed, the engine emissions, and the vehicle exhaust emissions to the atmosphere per average mile of travel of precontrol and modern passenger cars. Unburned carbon-containing compounds in the exhaust are fuel HCs and partial

Table 1.2

Typical Automobile Fuel Consumption and Emissions

	HC*	CO grams/average mile	NO_x
Precontrol (1960s)	11	85	4
Current vehicle:			
Engine emissions	3	15	2
Tailpipe emissions	0.3	2	0.4

*Fuel consumption: 25 miles/US gallon = 120 g/mile.

Fig. 1.1.

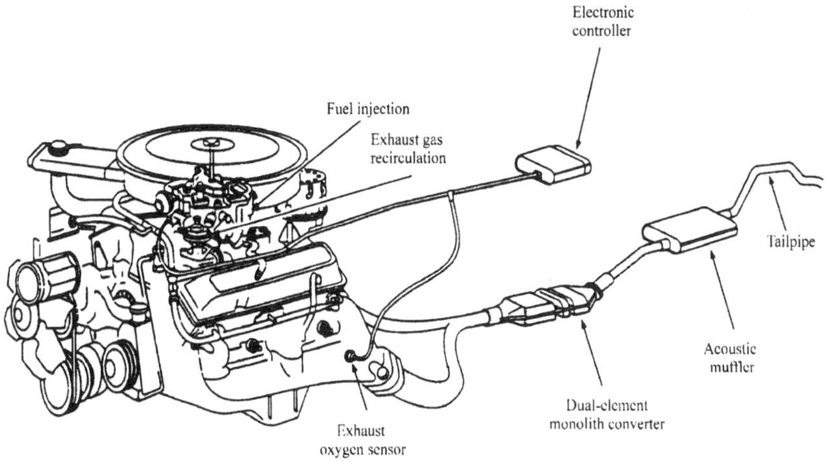

A modern automobile spark-ignition engine emission control system[2].

oxidation products that escape burning during the normal combustion events that occur in each cylinder of the spark-ignition engine. Carbon monoxide emissions are significant when the engine is operated under fuel-rich conditions, that is, when the air in the fuel-air mixture that enters the engine cylinder is insufficient to convert all the fuel carbon to CO_2. Rich mixtures are used as the engine approaches wide open throttle because they give the highest possible power from the engine. They also help with combustion stability during engine warm-up and, in older cars, at idle. Oxides of nitrogen are formed from nitrogen and oxygen in the high-temperature burned gases created during the combustion of the fuel-air mixture within the cylinder.

For the past 18 years, catalytic converters in engine exhaust systems have been used to achieve the large additional reductions in emissions required to meet mandated emissions standards (see Figure 1.1). In current new vehicles, a properly working catalyst reduces the emissions of each of the three pollutants—HCs, NO_x, and CO—that leave the engine's cylinders by a factor of about ten before the exhaust enters the atmosphere. However, it has taken two decades for the combined catalyst and engine technology to reach this point.

Evaporation of gasoline is an HC source comparable to exhaust HC. There are three categories of evaporative HC emissions from motor vehicle fuel systems: (1) diurnal emissions; (2) hot soak emissions; and (3) running losses, generally thought to occur in that order of importance. Diurnal emissions take place as the fuel tank of a parked vehicle draws air in at night as it cools down and expels air and gasoline vapor as it heats up during the day. This "diurnal breathing" of the fuel tank can produce evaporative HC emissions of as much as 50 g per day on hot days. Hot soak emissions occur just after the engine is shut down and the residual thermal energy of the engine heats the fuel system. Running losses can occur as gasoline vapors are expelled from the fuel tank while the car is driven and the fuel in the tank becomes hot. These losses can be high at high ambient temperatures or if the fuel system becomes particularly hot while running. Finally, gasoline vapor can escape from the fuel tank when a vehicle is filled at the service station. Evaporative HCs have been captured with carbon-containing canisters designed to absorb the gasoline vapors from these sources, as air is vented from the fuel system. The absorbed vapors are purged from the canister into the engine and burned during normal driving. While these evaporative controls have met the test requirements for two decades, many of these systems have not been nearly as effective at controlling evaporative emissions in the field.

It is the average emission rate from the total in-use vehicle fleet, as well as emissions from all other sources, that affect air quality. The average vehicle emission rate depends on the age distribution of the in-use vehicle fleet, the number of miles per year vehicles of a certain age are driven (new cars are driven more), the emissions from cars of a given age which depends on the rate of deterioration of emission controls and any tampering, and the reductions of emissions resulting from inspection and maintenance (IM) programs. Ambient temperature, average driving speed, and driving pattern also affect the average emission rate. Evaporative HC emissions can be converted to grams per mile and added to exhaust HC emissions to estimate total HC emissions.

Major efforts have and are being made to model these phenomena to provide quantitative input for evaluating air pollution reduction strategies. Figure 1.2 shows a typical output from such a calculation for the light-duty vehicle fleet. On a per car basis, progress looks encouraging. In the United States, today's average in-use

Fig. 1.2.

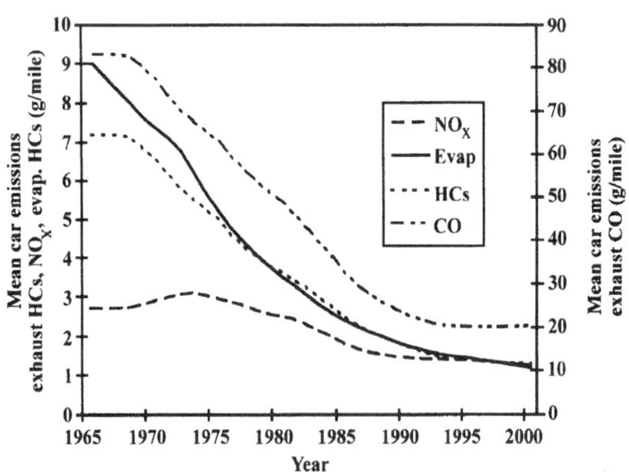

The exhaust HC, NO_X, CO, and evaporative (Evap) HC mean car emissions expressed in grams per vehicle mile traveled for the in-use U.S. light-duty vehicle fleet. The time period covered is from the late 1960s, when emissions controls were first introduced, to the year 2000. The curves show the effect on average predicted in-use fleet emissions of the introduction of cleaner new cars designed to meet the increasingly stringent federal emission standards.

car has about one-fifth the HC and CO emissions and one-half to one-third the NO_x emissions of a precontrol car of 25 years ago. However, the number of miles driven in major urban areas has gone up, and the emission rate is the product of grams per mile and miles driven. During this same 25-year period, the urban miles traveled in the United States per year increased by a factor of two, so part of this decrease in per car emissions (about one-quarter of the decrease in HCs and CO but some two-thirds of the decrease in NO_x) merely offsets this increase in mileage. The predicted future emission rates are based on the assumption that the future purchase of vehicles by consumers will follow the historical trends.

1.4 THE SCIENCE OF POLLUTANT FORMATION AND CONTROL

We now turn to the basic reasons why spark-ignition engines in cars are such a significant source of air pollutants. Engineers worldwide have learned a great deal about where these pollutants come from over the past 30 years. This knowledge has helped greatly in the development and design of effective emission control systems.

As mentioned previously the three pollutants of concern in the spark-ignition engine exhaust, CO, NO_x, and HC (or VOC), originate within the engine cylinder. Figure 1.3 illustrates the essential features of the processes involved.[3] Carbon monoxide is always present in the combustion products of close-to-stoichiometric fuel-air mixtures, that is, mixtures with just the right amount of air to fully oxidize the fuel. With excess air, CO levels are relatively low since almost complete oxidation of the fuel carbon occurs. With increasingly fuel-rich mixtures, the CO levels rise rapidly. As the burned gases inside the cylinder cool during expansion and exhaust, the CO oxidation chemistry becomes sufficiently slow so that CO levels freeze out well above equilibrium exhaust values. But the primary variable is whether the engine is lean, stoichiometric, or rich. Nevertheless, with the close-to-stoichiometric operation of modern spark-ignition engines, and with good exhaust catalyst systems, CO emissions can now be adequately controlled.

NO_x emissions also originate in the in-cylinder, high-temperature burned gases, when molecular collisions between nitrogen molecules and oxygen atoms become sufficiently vigorous to break the N-N bond. A nitric oxide molecule (NO) results and the N atom also formed rapidly finds oxygen to form another NO molecule. This air pollutant formation chemistry occurs in all combustion systems, making these significant NO_x sources too. But spark-ignition engines are an especially significant source because of the very high burned gas temperatures that result from the combination of compression due to piston motion and in-cylinder combustion inherent in the operation of the Otto cycle engine. The critical variables for NO formation are the maximum burned gas temperature and the relative concentration of oxygen. Today, under typical driving conditions with the engine at part load, in-cylinder NO control is achieved by recycling some

Fig. 1.3.

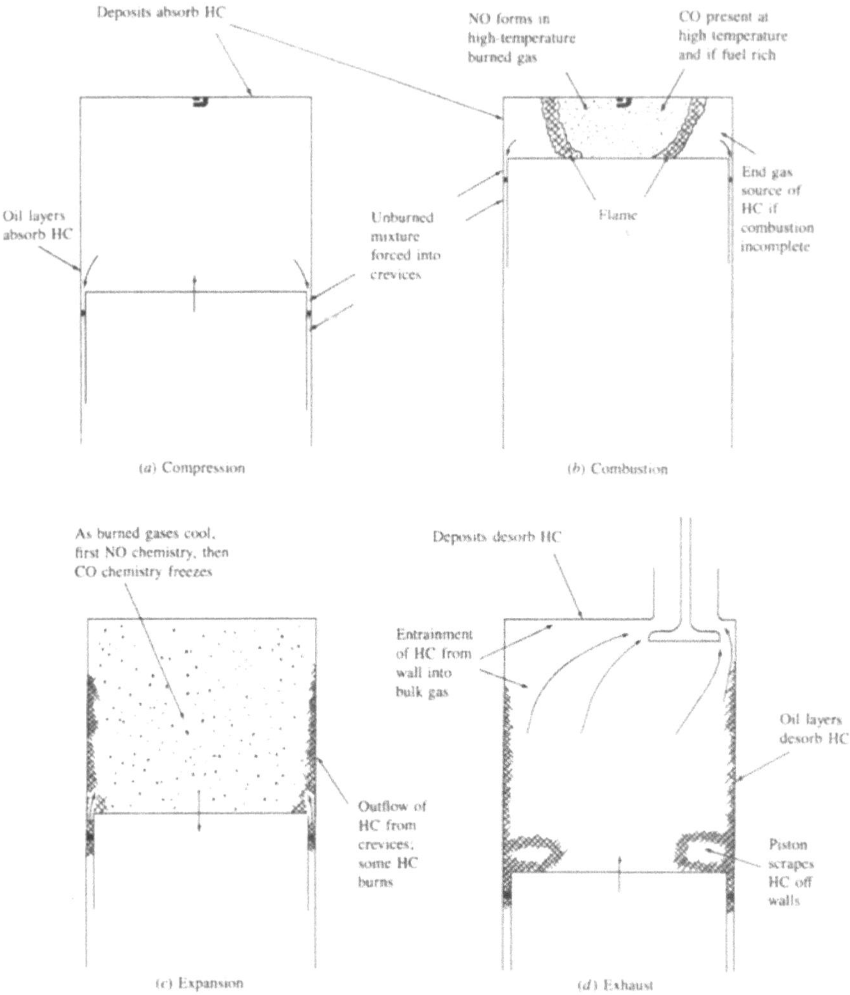

Summary of HC, CO, and NO pollutant formation mechanisms in the four-stroke cycle spark-ignition engine.

5 percent to as much as 20 percent of the engine's exhaust gas to the intake. This recycled exhaust dilutes the incoming fuel-air mixture (by effectively adding thermal capacity) so that after combustion, the burned gas temperatures are reduced by almost the same percentage as the amount of gas recycled.

The origin of the HC emissions is of special importance because California has set extremely stringent exhaust HC emission standards for the next few years, and the rest of the United States and Europe have largely followed California's standard-setting lead. The basic question with the HC emission problem is: Why doesn't *all* the fuel burn inside the engine? As Table 1.2 showed, 1.5 percent

Table 1.3

How Gasoline Compounds Escape Burning During Normal Combustion in the Four-Stroke SI Engine

1. Gasoline vapor-air mixture compressed into the combustion chamber crevice volumes.
2. Gasoline compounds absorbed in oil layers on the cylinder liner.
3. Gasoline absorbed by and/or contained within deposits on the cylinder head and piston crown.
4. Quench layers on the combustion chamber wall left as the flame extinguishes close to the walls.
5. Gasoline vapor-air mixture left unburned when the flame extinguishes prior to reaching the walls.
6. Liquid gasoline within the cylinder that does not evaporate and mix with sufficient air to burn prior to the end of combustion.
7. Leakage of unburned mixture through the (nominally) closed exhaust valve.

to 2 percent of the gasoline fuel escapes the engine unoxidized. The answer to this question is extremely complex, as yet imperfectly understood, impacts many critical engine processes (e.g., fuel injection and gasoline-air mixture preparation), and ends up showing that the HC emission problem is also a significant fuel economy problem too.[4] It is an important question, and it has been one of my major research interests for almost 30 years.

There are many mechanisms by which fuel or fuel-air mixture escapes burning during the normal engine flame propagation process that releases most of the fuel's chemical energy (see Table 1.3). Let us look at the largest of these: crevices or narrow volumes connected to the engine's combustion chamber, where fuel-air mixture can flow in but the flame cannot penetrate. Figure 1.4 shows the location

Fig. 1.4.

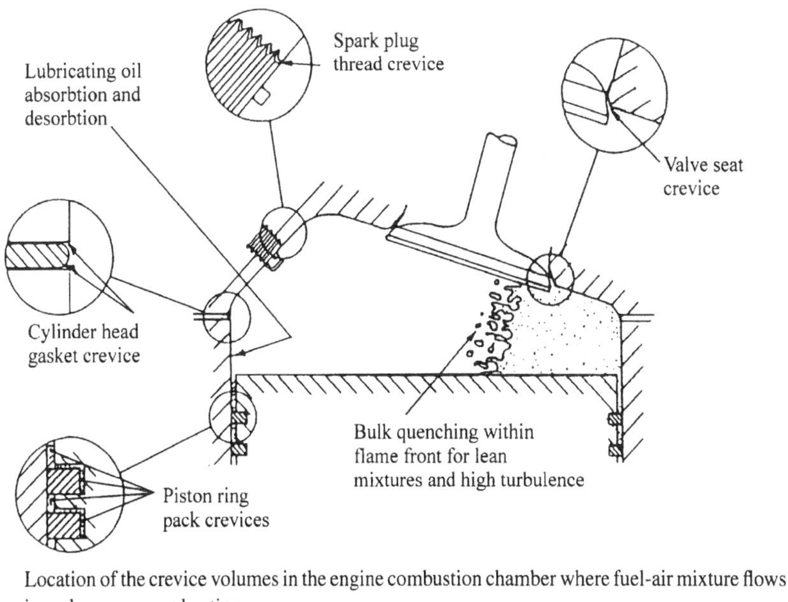

Location of the crevice volumes in the engine combustion chamber where fuel-air mixture flows in and escapes combustion.

of these crevice regions in more detail; the largest is the region between the piston, rings, and cylinder liner. At the end of the normal combustion process, some 4 percent or 5 percent of the fuel-air mixture, unburned, resides in these regions. The other HC sources listed in Table 1.3 together contain an additional 3 or 4 percent of the fuel. In total, recent estimates suggest that about 9 percent of the gasoline within the cylinder when the intake valve closes escapes burning during normal combustion.

How does this figure get reduced to the 1.5 percent to 2 percent of the fuel that leaves the engine as HC, as shown in Table 1.2? We now know that this is due to oxidation and retention of some of these hydrocarbons in the cylinder as shown in Figure 1.5. As the cylinder pressure falls during expansion, the HC stored in crevices, in oil layers on the liner, and in deposits on the cylinder head and piston,

Fig. 1.5.

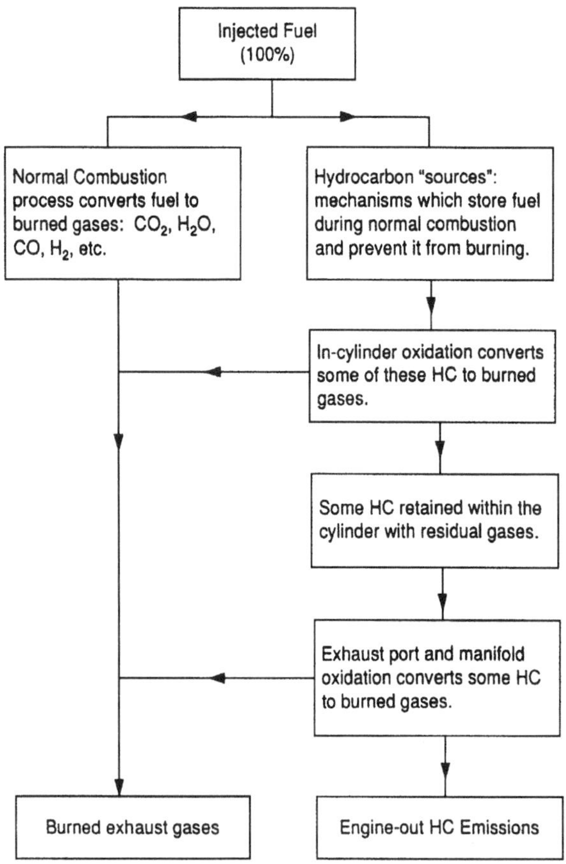

Schematic flowchart for the gasoline fuel that enters each cylinder during each cycle. Most of the fuel burns during the normal engine combustion process. A significant fraction escapes burning as the various mechanisms store fuel. These become the hydrocarbon sources which, when reduced by oxidation and retention in the cylinder, result in engine-out HC emissions.

flows or diffuses out into the hot burned gases in the cylinder and a significant fraction (about half) oxidizes. Then, during exhaust, only about two-thirds of the remaining hydrocarbons are exhausted from the cylinder, and about one-third of these hydrocarbons then oxidize in the exhaust port. Thus, the 1.8 percent of the fuel typically measured in the engine exhaust as HC corresponds to a much larger fraction of the fuel not burning when it should—during normal combustion. This unexpected "incompleteness" of combustion is a direct engine torque and fuel economy loss of significant magnitude.[4]

The engine controls implemented for HC emissions are in the design details that affect the various HC sources (e.g., smaller crevice volumes by raising the top ring), good control of the fuel injection process especially when the engine is started and warming up, and a combustion system that produces a fast flame propagation process with low variability from one cycle to the next.

The exhaust catalyst, however, has become the most important component in the total emission control system. Today, the so-called three-way catalyst (TWC) technology now removes a very large fraction of the three pollutants (CO, NO_x, and HC), that enter it. When fully warm, modern TWC catalysts, which must operate close to the stoichiometric air-fuel ratio, are over 95 percent effective. Close-to-stoichiometric operation is essential: CO and HC removal requires an oxidizing environment; NO removal requires a reducing environment. Stoichiometric exhaust gas fortunately provides both these environments together (see Figure 1.6 which shows how the catalyst efficiency—percent of entering pollutant removed—varies with the air-to-fuel ratio of the exhaust gases; 14.6 is the stoichiometric ratio). These catalysts (see Figure 1.7) currently use platinum and rhodium as the active catalyst materials, with ceria (CeO_2) to provide an oxygen-storage capability to continue CO and HC oxidation when the exhaust gas is slightly rich. The

Fig. 1.6.

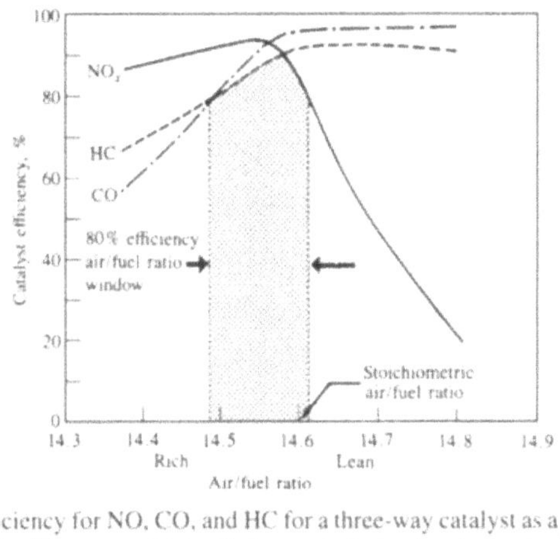

Conversion efficiency for NO, CO, and HC for a three-way catalyst as a function of the exhaust gas air-fuel ratio. The narrow high-efficiency air-fuel ratio window is clear.[3]

Fig. 1.7.

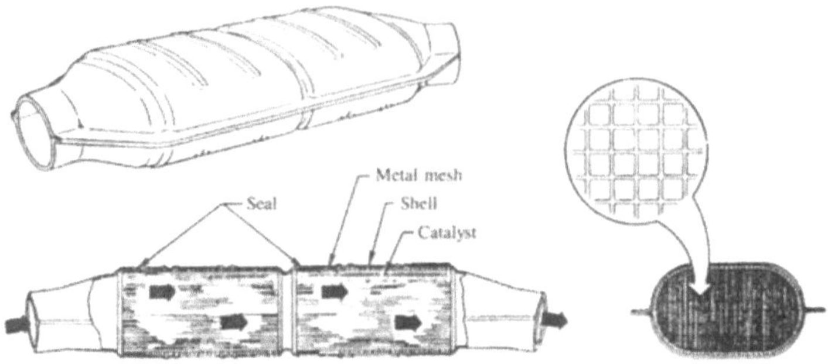

Catalytic converter used for automobile exhaust emission control.[3]

noble metals and ceria are dispersed on the surface of an alumina washcoat, which is bonded to a lightweight monolithic honeycomb support. This construction, through the appropriate length-diameter ratio of individual honeycomb channels and highly porous washcoat surface, exposes the exhaust gases to a very large surface area—some 100 m^2/g of washcoat material.

Precise operation of the engine at stoichiometric is not feasible without a feedback system. As shown in Figure 1.1, an oxygen sensor in the exhaust manifold is used to determine whether the exhaust gas is lean or rich, and the control system adjusts the fuel metering accordingly. The sensor used is an electrolytic cell, with a solid stabilized-zirconia electrolyte with high-surface-area platinum electrodes. Stabilized zirconia conducts current through oxygen ion transport, and the voltage generated across the cell is dependent on the oxygen partial pressure in the exhaust gases (which are chemically equilibrated by the platinum electrode) on one side of the cell, relative to that of air on the other side which acts as the reference. Thus, as the exhaust gas moves from lean through stoichiometric to rich, the exhaust gas oxygen partial pressure ratio across the cell decreases enormously and the voltage increases substantially. The cell acts like a switch. It proves advantageous for high catalyst efficiency for all three pollutants to oscillate the relative air-fuel ratio of the engine about stoichiometric with an amplitude of a few percent and frequency of about 1 Hz. For the catalyst and the oxygen sensor, critical durability issues are the levels of poisons (such as lead and sulfur) in the gasoline and the maximum temperature to which the catalyst is exposed.

With appropriate quality fuels, and in the absence of component failures or malfunctions, engine controls such as precise fuel metering to achieve a stoichiometric mixture during all modes of vehicle operation, fast combustion system, accurate control of spark timing and exhaust gas recycle rate, low thermal inertia exhaust system, durable and highly effective exhaust oxygen sensor and catalyst are proving remarkably successful at meeting the emission regulations. Problems result, however, when critical system components fail or malfunction; emission levels then rise substantially.

1.5 EFFECTIVENESS OF CURRENT EMISSION CONTROL TECHNOLOGY

Despite this remarkable progress in developing emission control technology, as explained previously, progress in reducing aggregate emissions from the total in-use vehicle fleet and hence in improving air quality has been much slower. (We do not belittle the progress that has been made—*any progress* is a real achievement.) Many reasons have been suggested for this slower than hoped for progress: Emissions test procedures underestimate in-use emissions; the extremes of vehicle use (e.g., many cold starts, lots of stop-and-go driving) add much more to the average than we now think; actual fuels are more variable and less clean than the specifications suggest; emissions at the extremes in temperature (really cold days for CO and really hot days for HC and NO_x) that determine when the air quality standards are exceeded are higher relative to standard tests than we expect; and tampering with and the malfunction or failure or incompetent repair of emission control system components result in a fraction of the vehicle fleet having very high emission levels. While all these factors do contribute to vehicle emissions being higher than we thought, it is the last of these that is the most important: This has become known as the high-emitter problem.

A valuable tool used to quantify the extent and impact of these high-emitting cars is remote sensing (Figure 1.8). Cars drive through a beam—IR for CO and HC, and UV for NO_x—across the roadway at the typical tailpipe exit height. The beam is absorbed by the CO, HC, and NO_x, and by CO_2, in proportion to the amount of these gases present. (The CO_2 absorption provides the calibration; its concentration in spark-ignition engine exhaust gases is a fixed quantity.) The license plate of the vehicle is recorded on a video camera so the vehicle registration can be accessed. The emissions of over 2 million cars in many different countries have been tested in this way. The results obtained (for HC, for example, in California, Figure 1.9) show a consistent pattern. The good news is that emission

Fig. 1.8.

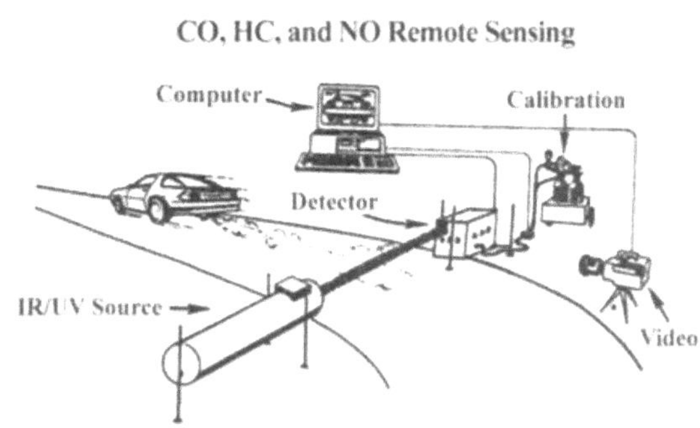

Remote sensing system for on-road emissions measurement.[5]

Fig. 1.9.

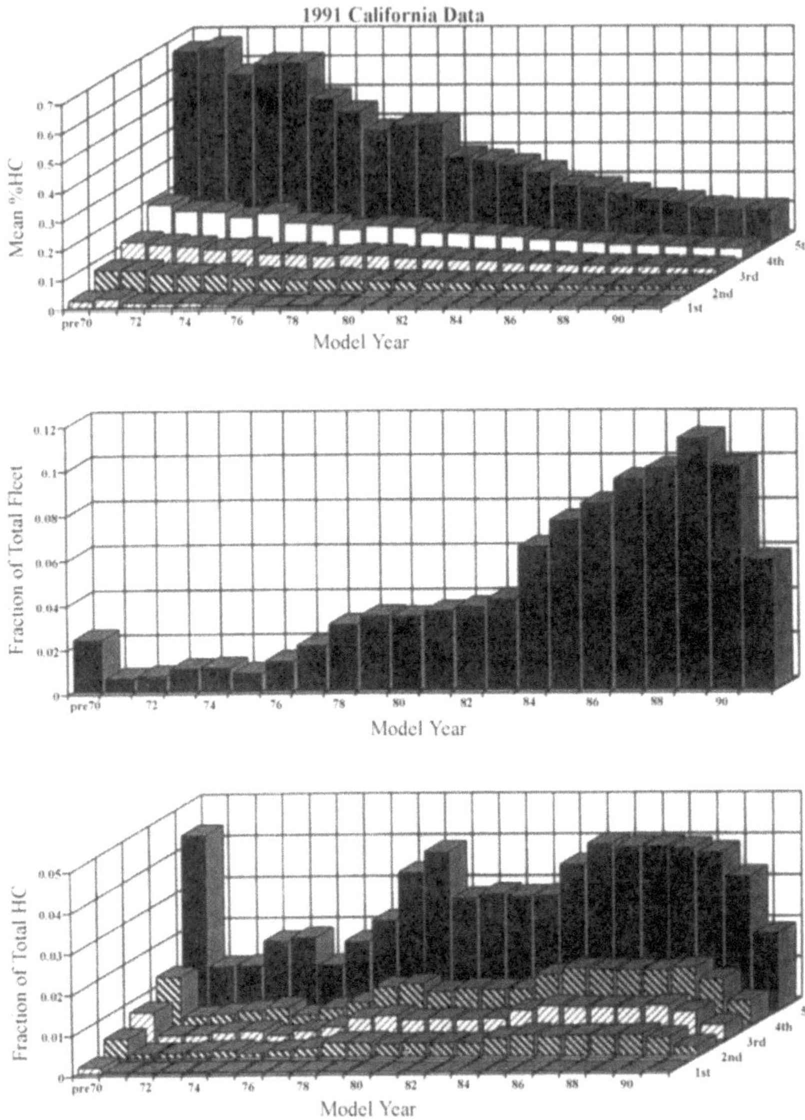

California data (1991) for 50,000 vehicles showing mean exhaust HC by model year and quintile (best to worst 20 percent), fleet distribution by model year, and percent of total HC by model year and quintile.[5]

levels have steadily come down as new and better emission control technology is employed. But for every model year of vehicles, newer through older, the worst-emitting 20 percent or quantile has much higher emissions than the rest of the distribution and, when weighted by its fraction of the total fleet, contributes more than half the emissions.[5]

It sounds too simple to say that half the total fleets' CO and HC emissions come from the worst 10 percent of the vehicles on the road, but it is indeed true! (The NO_x picture as yet is less clear.) And these high emitters are found in all model year vehicles; it is not just the old cars that are high emitters. Our emission control technology, developed with much hard work by engineers over the past 25 years, is very effective at reducing emissions from most cars on the road for much of their useful life, but its failure to achieve this control in a small fraction of the vehicle fleet offsets a substantial part of the reductions realized in the vast majority of the fleet. Our future efforts to reduce emissions must somehow deal with this reality.

1.6 DIRECT-INJECTION ENGINES, TWO-STROKES, AND DIESELS

Two other internal combustion engines are in widespread use besides the four-stroke cycle spark-ignition (SI) engine: the two-stroke gasoline engine and the diesel. In transportation, the two-stroke gasoline engine is used in the developing world for powering bicycles, small motor scooters, and motor bikes, because of its small size and weight, and low cost. The diesel dominates the truck engine market because its efficiency is substantially higher than that of the spark-ignition engine. In many countries the diesel has captured a significant share of the automobile engine market for similar reasons, especially in countries where fuel prices are high and where diesel fuel is taxed less than gasoline. While the fundamentals of the pollutant formation processes are similar in these other two engines, the details differ significantly and with the diesel there is a new problem—exhaust particulates.

The two-stroke cycle engine exhausts the burned gases from the cylinder largely by blowing in fresh air during approximately one-third of each crankshaft revolution as the crank moves through its bottom position. To make this scavenging process effective, a significant fraction of the fresh air flowing into the cylinder through the transfer ports in the bottom of the cylinder liner inevitably flows straight out of the exhaust ports (usually placed on the other side of the liner). With the simplest, small two-stroke SI engines that are carbureted, the gasoline is mixed with the air prior to entering the cylinder. So this short-circuiting of air directly through the cylinder results in a corresponding loss of fuel. This is a substantial fuel economy penalty (up to 25 percent), and results in very substantial hydrocarbon emissions. Thus, in cities with large numbers of motorized bicycles, motor scooters, motorcycles, and three-wheel taxis, the two-stroke cycle engine is an important source of emissions.

Fig. 1.10.

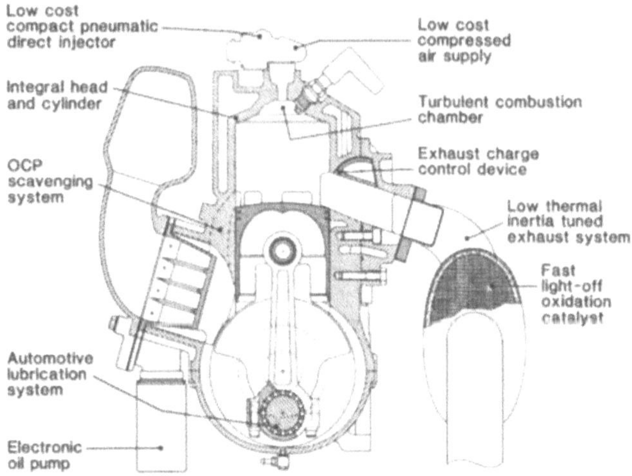

Features of the direct-injection two-stroke gasoline SI engine (courtesy Orbital Engine Co.).

Substantial development efforts over the past 15 years have explored the potential of using direct in-cylinder gasoline injection to avoid this loss of fuel during scavenging. These efforts have been targeted at the automotive, the marine, and the motorcycle sectors. Figure 1.10 shows one of the more promising direct-injection technologies developed by the Orbital Engine Company applied to a crankcase-scavenged, two-stroke cycle engine. The necessary emission control with this concept is achieved by direct injection of gasoline into the cylinder, with an air-assist injector that achieves good dispersion of the fuel with very small drop sizes, after the rising piston closes the exhaust ports. Additional scavenging control is achieved with an exhaust flow control device (shown in the figure), a low-thermal-inertia, tuned, exhaust system, and a close-coupled oxidation catalyst to achieve fast light-off for HC and CO control. NO_x control is achieved within the cylinder. The two-stroke scavenging process leaves significantly more burned gases inside the cylinder, mixed with the fresh air, than does the four-stroke gas-exchange process. This additional residual burned gas in the in-cylinder fuel-air mixture reduces peak burned gas temperatures and the NO formation rate significantly.

Whether or not this new direct-injection, two-stroke cycle technology will significantly penetrate the small engine/motorcycle market will depend on the cost of these fuel injection systems. Whether or not it becomes widely used in the automobile market will depend on the degree to which its durability and cost can be improved sufficiently to justify the development effort required to make this technology mass production feasible.

The diesel is the most efficient engine currently available and, consequently, is widely used in transportation (trucks, buses, railroads, and cars) when fuel economy is especially important. In the most efficient form of the diesel, the fuel

Fig. 1.11.

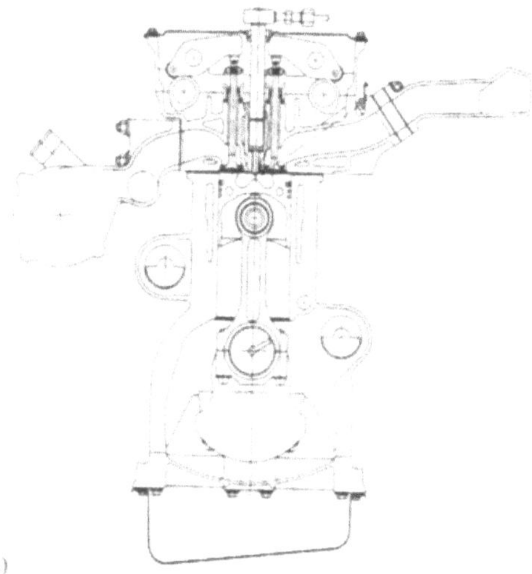
(a)

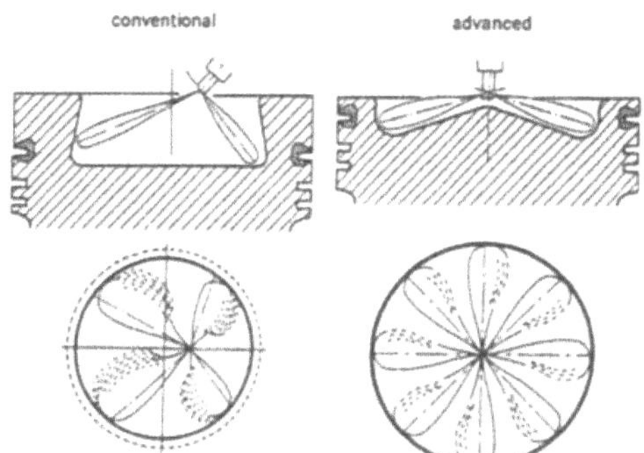

(b)

(a) Modern four-valve-per-cylinder turbocharged small high-speed direct-injection diesel (courtesy of Ford Motor Co.) (b) Fuel spray and bowl-in-piston combustion chamber characteristics of conventional (two valve, off axis inclined fuel nozzel, deep bowl, high swirl) and advanced (four valve, on-axis injector, shallow bowl, higher injection pressure-1600 bar, lower swirl) technology direct-injection diesel combustion systems (courtesy of Mercedes-Benz AG).

is injected with a high-pressure injection system into a combustion chamber or bowl in the top of the piston toward the end of the compression process, as shown in Figure 1.11. The injected liquid fuel atomizes, forms a spray, vaporizes, mixes with the high-temperature air, and spontaneously ignites shortly after injection. Once combustion starts, it continues as additional fuel mixes with air to form a combustible mixture. Diesel emissions of hydrocarbons and carbon monoxide are low because combustion is almost complete and the engine always operates lean, with excess air. NO_x emissions are high, however, because burned gas temperatures are high. The three-way catalyst technology employed to good effect in the standard gasoline engine cannot be used to reduce NO_x levels in the diesel exhaust because the exhaust gas is lean rather than stoichiometric. Also, the fuel-air mixing process during combustion produces soot particles in the highly rich regions of each fuel spray. Some of this soot survives the combustion process unburned and absorbs high molecular weight hydrocarbons from the oil and fuel and sulfur as sulfate in the exhaust to form particulates.

Substantial control of NO_x emissions, and especially particulate emissions, from diesels has been achieved by modifications to the combustion process. Use of fuel injection equipment with very high liquid fuel injection pressures (~2000 bar), and careful matching of the geometry of the bowl-in-piston combustion chamber, air motion, and spray geometry have significantly reduced soot formation by increasing fuel-air mixing rates. More careful control of lubricant behavior has reduced the high molecular weight hydrocarbon particulate component that is absorbed onto the soot. Use of low-sulfur fuels has reduced the sulfate component of the particulate. Oxidation catalysts in the diesel exhaust are increasingly being used to reduce further the soluble organic component of the particulate. NO_x reductions to date have been achieved by careful control of engine inlet air temperatures (e.g., turbocharged engines use an aftercooler to achieve low NO_x emissions), and with substantial injection retard to delay most of the combustion process to the early part of the expansion stroke. This latter strategy, of course, worsens fuel consumption by several percent.

While the diesel has made progress in reducing emissions (by about a factor of 3 to 4 for particulates, and a factor of 2 to 3 for NO_x), making this engine, which is the most efficient engine available, more environmentally friendly is an important task for engine developers and designers. Achieving substantially lower NO_x emissions is the major challenge. Some of this reduction could come from recycling exhaust gas, and a lesser amount from improvements in fuels. What is really needed is exhaust catalyst systems for NO_x reduction in the fuel-lean and low-temperature diesel exhaust environment. Lower levels of particulates will also be required.

In addition to these two-stroke and high-speed diesel engines, the direct-injection four-stroke spark-ignition engine is a potentially attractive new technology. Already in production in Japan,[6] this gasoline direct-injection engine offers improved fuel economy and is, therefore, one way to reduce vehicle CO_2 emissions. However, this engine's exhaust emissions are no better than those of the standard spark-ignition engine, and since it usually operates lean at light load, it requires a new catalyst technology to reduce NO_x.

1.7

FUTURE PROSPECTS

Our future efforts to reduce emissions from light-duty vehicles are focusing on three areas: improvements in fuels, stricter new car emission standards, and means for enforcing better in-use compliance. Let me comment on each of these.

Table 1.4

California Phase II Gasoline Specifications (Effective March 1, 1996)[1]

Measure	Flat limits*	Cap	1990 California average
Reid vapor pressure (psi)†	7.0		8.5
Sulfur (ppmw)	40	80	150
Benzene (percent by volume)	1.0	1.2	2.0
Aromatics (percent by volume)	25	30	32
Olefin (percent by volume)	6	10	9.9
Oxygen (percent by weight)	1.8 to 2.2	1.8 to 2.7	0
T_{90} (°F)	300	310	330
T_{50} (°F)	210	220	212

*Flat limits met by each gallon.

†Applies to summertime only and varies according to location.

Reformulated gasoline is being introduced in those areas in the United States where air pollution is most severe. Table 1.4 shows California's specifications for the second stage in its gasoline improvement program. The principal differences from standard gasoline are lower volatility, less sulfur, lower aromatic and olefin content, adding an oxygen-containing component, and a lower high end or maximum boiling point. The increase in cost is about 10 percent of the actual fuel cost (before taxes). The major emission benefits are reduced evaporative HC emissions, less reactive HC compounds in the exhaust, and improved catalyst performance from lower levels of sulfur—a catalyst poison. Most but not all of the required fuel changes are advantageous: The database available when these regulations were required by law was not complete enough to get it all right. The advantage of fuel changes is that they can impact the total car fleet within a short period of time (it takes only a few years to modify the fuel supply system), whereas changes in new vehicle technology take more than ten years to penetrate the fleet. A good case can be made for regulations that will reduce evaporative emissions via gasoline volatility controls, require low levels of sulfur, and reduce the upper boiling point of the fuel in regions with problems meeting ambient ozone standards. Overall, a reduction in exhaust and evaporative emissions in the 15 percent to 30 percent range is projected.[1]

Alternative fuels—natural gas, liquid petroleum gas, methanol, and ethanol—have been much in the news too. While the first two do achieve significantly lower emission levels, either the vehicle conversion cost or the fuel cost is too high at present for these alternatives to achieve any substantial market share.

Ever stricter new car emission standards are forcing the development of better emission control system technology, especially in catalysts, sensors, fuel injection technology, and engine controls. These are promising areas for technology improvements and offer hope of more effective and durable control of in-use vehicle emissions. California has set extremely stringent future exhaust emission standards, especially for exhaust HC (the ULEV requirements, see Table 1.1),

Fig. 1.12.

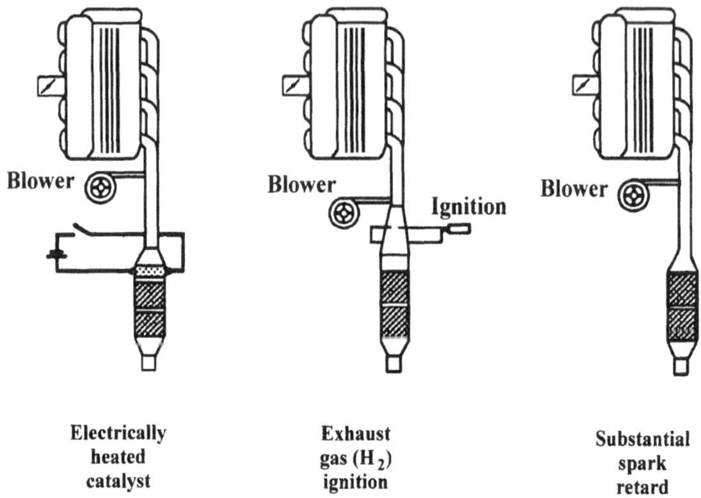

Methods likely to be used to heat the catalyst fast.

Three effective options for bringing the catalytic converter to its light-off temperature rapidly after engine start-up: electrically heated catalyst, left; exhaust gas ignition system, center; and substantial spark retard, right.

which are forcing the development of technologies that get the catalyst hot enough to be effective extremely quickly after a cold start. Examples of these technologies for four-stroke SI engines are electrically heated catalysts, exhaust gas ignition (which operates the engine very rich and then combusts the hydrogen produced in the exhaust with extra air to heat the catalyst), and use of excessive spark retard immediately after engine start-up so the very hot exhaust gases heat the catalyst directly (see Figure 1.12). A concern with these technologies (the first two, especially, are costly) is that they are "active" rather than "passive": That is, for a short period of time the engine and emission control system operate in a different nonstandard mode. Reliance upon such active systems, which must be carefully controlled, may not augur well for effective and durable in-use emission control behavior. This area is important; in the long run we *will* need steadily improving control technology to continue to offset the growth in vehicle use projected for the next several decades.

The third area I listed earlier is probably the most important for realizing significant reductions in *in-use* vehicle emissions in the next decade, and it is perhaps the most challenging because it involves people's behavior. Regulations are forcing the use of on-board diagnostics to monitor the behavior of critical engine/emission control system functions to ensure that they are not malfunctioning. Over time, an extensive set of functions must be continually evaluated (e.g., whether any misfires or noncombusting cycles occur, whether the catalyst still has substantial oxygen-storage capacity). Some of these requirements are relatively straightforward; many are extremely difficult to accomplish with the reliability and durability needed to avoid excessive warranty costs and customer complaints.

The other piece of the in-use emissions problem is implementing effective on-the-road emissions surveillance programs (using remote sensing) and emissions inspection programs (usually required annually), which separately or in combination identify high-emitting cars and force their repair. When operated well, honestly, and with the appropriate equipment, surveillance and inspection do effectively identify the cars that should be fixed. However, getting these cars fixed has proved more challenging. The vehicle service industry broadly lacks the diagnostic and repair competence to do this adequately, and often the cost of repair is beyond the financial resources of the vehicle owner and user. Making this critical aspect of achieving effective emission control work is a socioeconomic and, hence, political problem of substantial dimensions.

Will this engineering and regulation make the internal combustion engine with its petroleum-based fuel sufficiently environmentally friendly so that it remains our best choice? From the perspective of urban air pollution and the next couple of decades, my own judgment is that it will—but it will not be easy. However, making the alternatives (hybrids, fuel cells, battery electric vehicles) attractive and feasible will be even harder![7]

REFERENCES

1. Calvert, J.G., Heywood, J.B., Sawyer, R.F., and Seinfeld, J.H., "Achieving Acceptable Air Quality: Some Reflections on Controlling Vehicle Emissions," *Science*, Vol. 261, pp. 37–45, July 2, 1993.
2. Heck, R.M., and Farranto, R.J., *Catalytic Air Pollution Control*. New York: Van Nostrand, Reinhold, 1995.
3. Heywood, J.B., *Internal Combustion Engine Fundamentals*. New York: McGraw-Hill, 1988.
4. Cheng, W.K., Hamrin, D., Heywood, J.B., Hochgreb, S., Min, K., and Norris, M., "An Overview of Hydrocarbon Emissions Mechanisms in Spark-Ignition Engines," SAE paper 932708, *SAE Trans.*, Vol. 102, 1993.
5. Stedman, D.H., "Playing with Fire: Science and Politics of Air Pollution from Cars," University Lecture, University of Denver, 1995.
6. Iwamoto, Y., Noma, K., Nakayama, T., and Ando, H., "Development of Gasoline Direct Injection Engine," SAE paper 970541, 1997.
7. Amann, C.A., "Alternative Fuels and Power Systems in the Long Term," *Int. J. Vehicle Design*, Vol. 17, nos. 5/6 (Special Issue), 1996.

PART II

Global Aspects

CHAPTER 2

Environmental Aspects of Air Pollution

Eran Sher
Department of Mechanical Engineering, The Pearlstone Center for Aeronautical Engineering Studies, Ben-Gurion University of the Negev, Beer-Sheva, Israel

ISBN: 0-12-639855-0
$25.00

2.1 INTRODUCTION

An air pollutant is a substance or effect dwelling temporarily or permanently in the air, which adversely alters the environment by interfering with the health, the comfort, or the food chain, or by interfering with the property values of people. A polluting substance can be a solid, liquid, gas, or submolecular particle, and may originate from a natural or an anthropogenic source, or both. It is estimated that anthropogenic pollutants in the atmosphere have changed the composition of the global air less that 0.01 percent. It is, however, widely accepted among scientists that even such a small change can have a significant adverse effect on the climate, ecosystems, and species of the planet. This is true in particular when the effects of rain acidity, urban air composition, and solar ultraviolet (UV) radiation are considered. The world's primary air pollutants, their sources, and effects on human health are summarized in Table 2.1.

2.2 GLOBAL EFFECTS

2.2.1 The Stratospheric Ozone Layer

The ozone layer is a region of the atmosphere 15 to 30 km above the earth's surface that acts as a barrier to radiation by filtering out harmful ultraviolet rays from sunlight before they reach the surface of the planet, thus protecting the biosphere. The ozone layer is formed naturally in the upper atmosphere by the action of the sun's ultraviolet rays. In this layer most of the sun's ultraviolet radiation is absorbed by the ozone molecules, causing a rise in the temperature of the stratosphere and preventing vertical mixing so that the stratosphere forms a stable layer (Figure 2.1). Some ultraviolet light, however, does reach the ground. It is in a waveband from 290 to 320 nm, known as UV-B. It causes sunburn, some forms of skin cancer, and is associated with eye problems such as cataracts. In contrast to its harmful effects, the UV-B radiation is an important ingredient in the formation of vitamin D. Owing to the ozone layer, radiation with wavelengths in the band from 240 to 290 nm, known as UV-C, does not reach the ground at all. Radiation within these wavelengths destroys nucleic acids (RNA and DNA) and protein.

The ozone layer spans most of the stratosphere. It exists because oxygen filtering up from the top of the troposphere reacts under the influence of sunlight to form ozone. This photodissociation of oxygen is greatest above the equator and the tropics where solar radiation is strongest and most direct. From these regions, ozone is transported by winds within the stratosphere around the earth toward the polar regions to maintain the ozone layer.

Table 2.1

World's Primary Air Pollutants—Their Sources and Effects on Human Health

Pollutant	Natural source	Anthropogenic source	Known or suspected effect
Carbon monoxide (CO)	Unnoticeable	Fuel-rich and stoichiometric combustion mainly from motor vehicles	Reduces the oxygen-carrying capacity of the blood by combining with haemoglobin, thus deprives tissues of O_2
Nitrogen oxides (NO) and (NO_2)	Lightning and bacterial activity in soils	High-temperature combustion mainly from motor vehicles	Cause eye, throat, and lung irritation. Primary pollutants that produce photochemical smog and acid rain, destroy ozone at the stratosphere
Particulates	Forest fires, wind erosion, and volcanic eruptions	Coal, waste, and fossil burning	Breathing difficulties
Sulfur dioxide (SO_2)	Volcanic eruptions and decay	Coal combustion, ore smelters, petroleum refineries, and diesel engines	Causes eye, throat, and lung irritation. Primary pollutants that produce acid rain
Ozone (O_3)	Lightning and photochemical reactions in the troposphere	Product of photochemical reactions in photochemical smog	Causes eye, throat, and lung irritation, impairs lung function
Carbon dioxide (CO_2)	Animal respiration, decay, and release from oceans	Fossil-fuel and wood combustion	Partly responsible for the atmospheric greenhouse effect
Hydrocarbons other than methane (VOCs), i.e., volatile organic compounds	Biological processes	Incomplete combustion and volatiles	Primary pollutants that produce photochemical smog
Methane (CH_4)	Anaerobic decay, cud-chewing animals (cows, sheep, etc.), and oil wells	Natural-gas leak and combustion	Partly responsible for the atmospheric greenhouse effect
Chlorofluorocarbons (CFCs)	None	Used as solvent, aerosol propellant, and refrigerant	Destroy ozone at the stratosphere, thus reduce the ozone UV protective layer

Fig. 2.1.

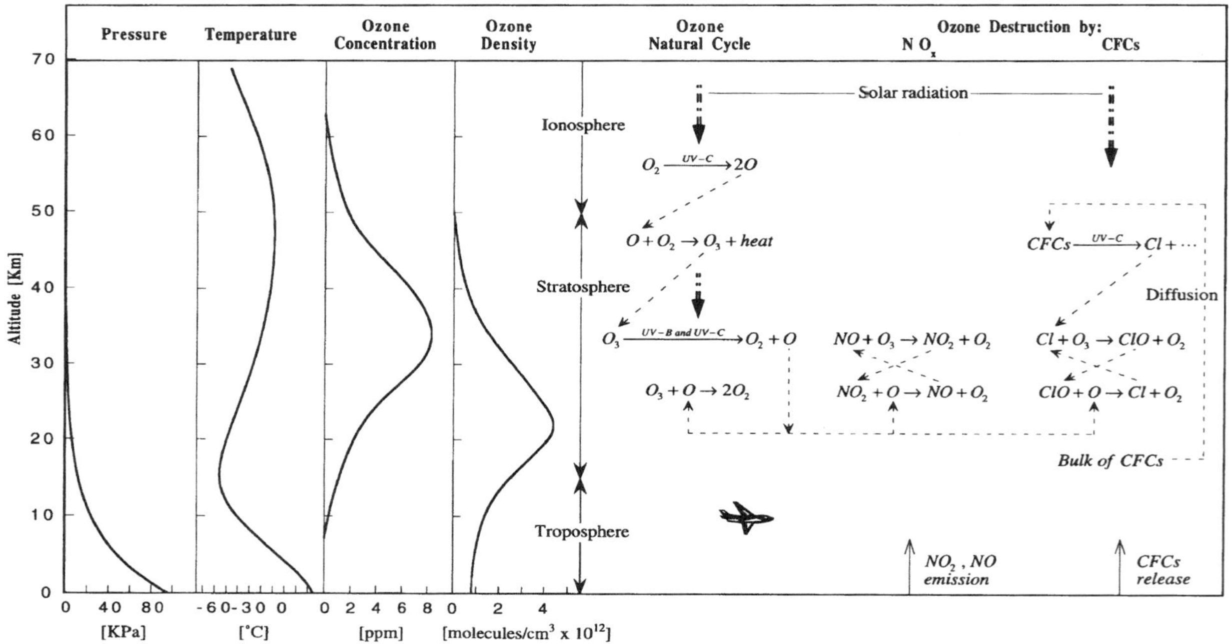

Profile of the earth's atmosphere and some important processes that determine the concentration of the ozone in the stratosphere.

In the whole stratosphere, at an altitude between 15 and 50 km, there are only about $5*10^9$ tons of ozone. If all the ozone could be brought down to sea level and spread evenly around the globe, the pressure of the atmosphere would squeeze it into a layer just 3 mm thick [1]. The total overhead amount of atmospheric ozone at any location is expressed in terms of Dobson units (DU). One such unit is equivalent to a 0.01-mm thickness of pure ozone at standard conditions. The normal amount of overhead ozone at temperate latitudes is about 350 DU. Ozone concentrations in the tropics usually average 250 DU, whereas those in subpolar regions are about 450 DU.

2.2.2 The Chemistry of the Ozone Layer

The chemistry of ozone depletion is driven by energy associated with light from the sun. Ozone in the stratosphere is constantly being formed, decomposed, and reformed during daylight hours by a series of reactions that proceed simultaneously. Ozone is produced in the stratosphere because there is adequate UV-C from sunlight to dissociate some O_2 molecules and so to produce oxygen atoms (this reaction absorbs and, therefore, contributes to the filtration of UV-C from sunlight):

$$O_2 \xrightarrow{\text{UV-C}} 2O \tag{2.1}$$

Most of the oxygen atoms collide with other O_2 molecules and form ozone:

$$O + O_2 \longrightarrow O_3 + \text{heat} \tag{2.2}$$

The ozone gas absorbs and, therefore, filters UV-C and partially filters UV-B from sunlight, and is destroyed temporarily in this process:

$$O_3 \xrightarrow{\text{UV-B and UV-C}} O_2 + O^* \tag{2.3}$$

where * denotes an excited state. The ozone is also destroyed when reacting with oxygen atoms, thus:

$$O_3 + O \longrightarrow 2O_2 \tag{2.4}$$

Ozone is not formed below the stratosphere due to a lack of the UV-C required to produce the O atoms necessary to form O_3; this type of sunlight has been absorbed by O_2 and O_3 in the stratosphere through reactions 2.1 and 2.3.

2.2.3 The Ozone Hole

Measurements of the overhead amounts of atmospheric ozone at the Antarctic Circle indicate that the total amounts of ozone each October have been gradually falling, with precipitous declines beginning in the mid-1970s. Many scientists believe that the changes in the size of the ozone hole are a naturally occurring

Table 2.2

The More Important Ozone-Depleting Chemicals [3]

	Chemical formula	Contribution to ozone depletion [%]	Lifetime in the atmosphere [years]	Main use
Dichlorodifluoromethane CFC-12	CCl_2F_2	45	111	Aerosols, air conditioning, foams, and refrigeration
Trichlorofluoromethane CFC-11	CCl_3F	26	74	Aerosols, foams, and refrigeration
Trifluoroethane CFC-113	CCl_2FCClF_2	12	90	Solvents
Carbon tetrachloride	CCl_4	8	67	Solvents
Methylchloroform	$C_2H_3Cl_3$	5	8	Solvents
Bromotrifluoromethane Halon 1301	$CBrF_3$	4	110	Fire extinguishers

phenomenon, and that many natural and human activities have amplified the event. For example, volcanic eruptions throw volcanic sulfuric acid high in the atmosphere, which can enhance the destructiveness of the chlorine chemicals that attack the ozone layer. It is therefore widely accepted that the hole increases each year because of human activities, especially the introduction of chlorofluorocarbons (CFCs) and hydrogen bromide (HBr) into the atmosphere. These substances were largely used (and still are being used today although to a lesser extent) as solvents, aerosol propellants, and refrigerants. The more important ozone-depleting chemicals are shown in Table 2.2. As the CFCs rise into the stratosphere, ultraviolet radiation is intense enough to split the CFC molecules and liberate chlorine atoms. The chlorine then attacks the ozone and detaches an oxygen atom from the ozone molecule to form chlorine monoxide (ClO), leaving behind an ordinary molecule of oxygen (O_2).When a ClO molecule collides with a free oxygen atom, the oxygens combine into an ordinary molecule of oxygen and free the chlorine to detach another oxygen molecule from another ozone molecule. Each chlorine atom can catalytically destroy many tens of thousands of ozone molecules in this manner. Past emissions of CFCs molecules will, therefore, remain in the atmosphere for decades with the potential to continue to harm the ozone layer. Thus, if a chloromethane molecule is considered as an example of a CFC species:

$$CH_3Cl(\text{chloromethane}) \xrightarrow{\text{UV-C}} Cl^{\bullet}(\text{atomic chlorine}) + CH_3^{\bullet} \tag{2.5}$$

$$Cl^{\bullet} + O_3 \longrightarrow ClO^{\bullet} + O_2 \tag{2.6}$$

$$ClO^{\bullet} + O \longrightarrow Cl^{\bullet} + O_2 \tag{2.7}$$

and the net effect is:

$$O_3 + O \longrightarrow 2O_2 \tag{2.8}$$

It is important to note that oxides of nitrogen act in a similar catalytic way to convert ozone and oxygen atoms into molecular oxygen:

$$NO^{\bullet} + O_3 \longrightarrow NO_2^{\bullet} + O_2 \tag{2.9}$$

and,

$$NO_2^{\bullet} + O \longrightarrow NO^{\bullet} + O_2 \tag{2.10}$$

with the same net effect as previously (equation 2.8).

By the mid-1980s the overall loss in ozone at some altitudes over Antarctica during springtime amounted to more than 50 percent of the total overhead amount—a *hole* in the ozone layer had been defined. A similar phenomenon now appears each spring over Antarctica and lasts for several months. (The period from September to November corresponds to the spring season at the South Pole.) In 1993, the ozone concentration dropped to a record low of 90 DU in early October. The average amount of ozone loss in the stratosphere is estimated to be 5 percent per decade (an average depletion rate of 0.51 percent per year). The worldwide loss of ozone has become a major environmental concern, since it results in less protection to life on the earth's surface from the harmful ultraviolet solar radiation.

It was not clear for several years after its discovery whether the hole was due to natural meteorological forces or due to chemical mechanisms involving air pollutants. Today it is commonly accepted that the hole indeed does occur as a result of chlorine and nitric oxide pollution. Furthermore, it is believed that the hole will continue to reappear each spring for the next several decades, and that a corresponding hole may appear one day above the Arctic (North Pole) region.

2.2.4 Global Warming

In 1896 Arrhenius coined the term *greenhouse effect* and predicted that the burning of fossil fuels would increase the amount of carbon dioxide in the atmosphere and lead to a global warming of the world's climate. Today there is no broad consensus among scientists concerning the increase in global temperatures. Some studies have indicated that the average temperature of the world has risen by 0.5°C since 1600. Other studies have noted temperature increases between 0.3°C and 0.6°C in the past 100 years. It is also not clear whether the temperature rise is due to the earth's natural climate cycle or a result of the increase in some specific gases from human activity.

Some scientists, however, believe that the greenhouse effect is leading to the most rapid climate change in the history of civilization, and this will have immense effects for all life on earth. One theory, for example, states that runaway greenhouse gases will cause a rise in temperature of 5°C over the next 100 years. Such a major increase could melt glacial ice sheets, raising sea levels by 50 cm or more, and inundating coastal cities. In addition, cloud cover would increase and affect flora and fauna. Another study predicts that the global average temperature will increase by 1°C by the year 2030 and will change the agricultural patterns around the world. It is, however, important to note that due to natural fluctuations in local climate, it is difficult to determine the global warming effect in quantitative terms. Furthermore, though there is an agreement that atmospheric concentration of carbon dioxide, which is considered as a major greenhouse gas (Section 2.2.5), is increasing, there is no solid evidence that the increase is due to human activities.

2.2.5 Greenhouse Gases

In the earth's natural warming cycle, the earth is blanketed by an atmosphere of nitrogen (78.1 percent by volume), oxygen (20.9 percent), and argon (0.9 percent), with smaller amounts of trace gases, such as carbon dioxide (0.035 percent), methane (0.0001 percent), and others. Some of the trace gases, the *greenhouse gases*, capture some of the heat that is reflected back from the earth (about 30 percent of the sun's energy is reflected back into space). As these gases are transparent to solar radiation and are opaque to longwave reradiation from the earth's surface, the shortwave incoming radiation is transmitted, heat is trapped, and the underlying surface is thereby warmed.

The absorption of light by a molecule occurs most efficiently when the frequency of the light matches exactly its vibration frequency. The ability of a molecule to absorb infrared light over a range of frequencies around the vibration frequency, rather than at a single frequency, occurs because it is not only the energy associated with vibration that changes when an infrared photon is absorbed, but also an energy associated with the rotation of the molecule about its internal axis. Basically, in order for infrared light to be absorbed by a molecule, the molecule must have a dipole moment. In homonuclear diatomic molecules like O_2 and N_2, the centers of charge coincide and the molecules have no dipole moments. In nonhomonuclear molecules, such as in CO_2, during vibration, the contraction of one CO bond occurs when the other is expanding, so that during the motion, the centers of charge do not necessarily coincide; consequently, infrared light can be absorbed.

The major greenhouse gases include carbon dioxide (CO_2), methane (CH_4), nitrous oxide (N_2O), water (H_2O), chlorofluorocarbons (CFC-11 and CFC-12), ozone, and other small trace gases. Some components are much more damaging than others; a molecule of methane, for example, is 23 times more powerful in greenhouse terms than a molecule of carbon dioxide, and CFCs are estimated to

Table 2.3

Greenhouse Gases, Based on [1–5]

Gas		Current inventory	Rate of increase per year in last 20 years	Relative warming effectiveness per mole	Relative contribution to the greenhouse effect
Carbon dioxide	CO_2	356 ppm	+0.4%	1	50%
Methane	CH_4	1.74 ppm	+0.6%	23	20%
Nitrous oxide "laughing gas"	N_2O	0.31 ppm	+0.25%	270	5%
CFC-11	$CFCl_3$	0.26 ppb	+2.1%	14,000	5%
CFC-12	CF_2Cl_2	0.47 ppb	+2.3%	19,500	10%
Tropospheric ozone	O_3	0.03 ppm	−0.51%	-	10%

be more than 10,000 times more powerful than carbon dioxide. It is estimated that carbon dioxide is the greatest single contributor to the greenhouse effect (50 percent), CFCs 15 percent, methane 20 percent, ozone 10 percent, and nitrous oxide 5 percent (Table 2.3). The main sources of the greenhouse gases are listed in Table 2.1. The nitrous oxide is a byproduct of the biological denitrification and nitrification processes in aerobic environments (released by the oceans and soils of tropical regions). Anthropogenic release of carbon dioxide (mainly due to combustion of fossil fuels) amounts to about 4 percent of the enormous amounts produced by nature.

2.3 REGIONAL EFFECTS

2.3.1 Photochemical Smog

Photochemical smog is a brownish-gray haze caused by the action of solar ultraviolet radiation on atmosphere polluted with hydrocarbons and oxides of nitrogen. It contains anthropogenic air pollutants, mainly ozone, nitric acid, and organic compounds, which are trapped near the ground by temperature inversion. These pollutants and also some others can affect human health and cause damage to plants. Photochemical smog often has an unpleasant odor due to some of its gaseous components. The term *smog* (not to be confused with *photochemical smog*) is commonly used to represent a near-ground haze made of a combination of smoke and fog rather than ozone, nitric acid, and organic compounds.

Photochemical smog appears to be initiated by nitrogen oxides that are emitted into the air as pollutants mainly from internal combustion engines (Figure 2.2).

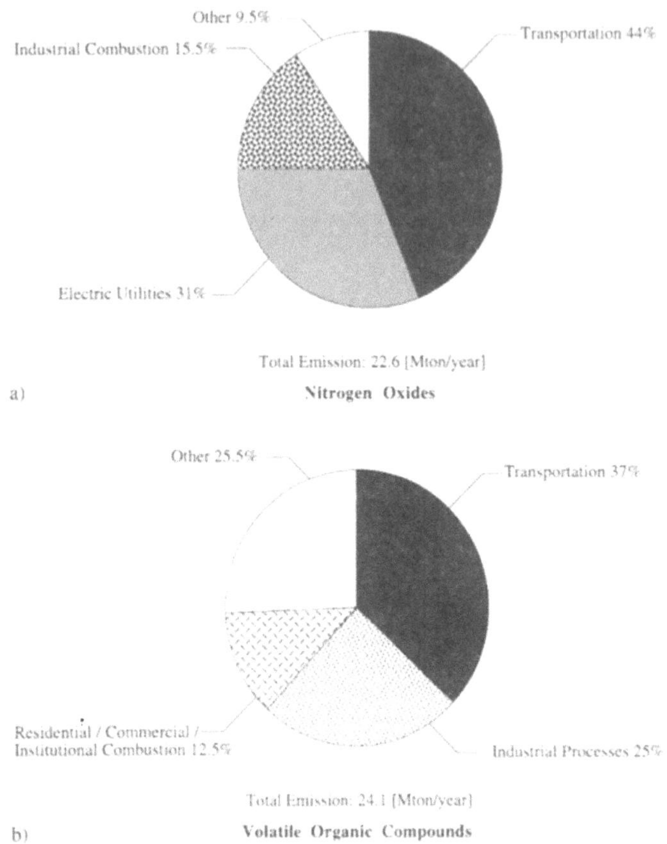

Primary pollutants from various anthropogenic sectors that are transformed into the final products of photochemical smog in the United States and Canada [5].

Absorbing the visible or ultraviolet energy of sunlight, it forms nitric oxide (NO) to free atoms of oxygen (O), which then combine with molecular oxygen (O_2) to form ozone (O_3). In the presence of hydrocarbons (other than methane), certain other organic compounds, and sunlight, various chemical reactions take place to form photochemical smog.

Unburned hydrocarbons are present in urban air as a result of uncompleted combustion and the evaporation of solvents and liquid fuels. Substances including hydrocarbons and their derivatives that readily vaporize are termed *volatile organic compounds* (VOCs). The most reactive VOCs in urban air are hydrocarbons that contain a C=C bond, since they can add free radicals. A simplified set of some of the reactions involved in photochemical smog formation is as follows [4]:

$$2NO^{\bullet} + O_2 \longrightarrow 2NO_2^{\bullet} \tag{2.11}$$

$$NO_2^{\bullet} \xrightarrow{\text{UV-A}} NO^{\bullet} + O \tag{2.12}$$

$$O + O_2 \longrightarrow O_3 \tag{2.13}$$

$$NO^{\bullet} + O_3 \longrightarrow NO_2^{\bullet} + O_2 \tag{2.14}$$

$$HC + NO^{\bullet} + O_2 \longrightarrow NO_2^{\bullet} + PAN \tag{2.15}$$

Or as a general global reaction [2]:

$$VOCs + NO \xrightarrow{\text{Sunlight}} O_3 + HNO_3 + \text{Organics} \tag{2.16}$$

In order that a city be subjected to photochemical smog, there must be a substantial source (usually vehicular traffic) in order to emit sufficient NO, hydrocarbon, and other VOCs into air. Warmth, ample sunlight, and a relatively little movement of air so that the reactants are not diluted are also required.

The major undesirable components of photochemical smog are nitrogen dioxide (NO_2), ozone (O_3), PAN (peroxyacetylnitrate), and chemical compounds that contain the –CHO group (aldehydes). PAN and aldehydes can cause eye irritation and plant damage if their concentrations are sufficiently high.

2.3.2 Ozone at Ground Level

Ozone is a form of oxygen that has three atoms in each molecule (trioxygen, O_3). Ozone is a colorless gas with a chlorinelike odor, soluble in cold water and in alkalis. The molecule is bent with considerable double-bond character, with a melting point of $-192.7°C$ and boiling point of $-11.9°C$. It is a powerful oxidizer agent that is widely used in bleaching, sterilizing water, and purifying air. In industry, ozone gas is made by passing oxygen through a silent electric discharge. In nature, it is produced near earth when lightning occurs and in the stratosphere by the action of high-energy ultraviolet radiation on oxygen (equations 2.1 and 2.2). Its presence in the stratosphere acts as a screen for ultraviolet radiation, which forms the lifesaving ozone layer. Near ground level it is a harmful pollutant.

Ozone occurs in small quantities in the earth's lower stratosphere, around areas where certain pollutants are prevalent, especially in city photochemical smog (Section 2.3.1). Surface ozone forms when certain volatile organic compounds, oxygen, and nitrogen oxides chemically react in the presence of sunlight, especially during hot weather. Surface ozone can reduce the yield of agricultural crops and damage forests and other vegetation. Ozone in combination with sulfur dioxide can have a more severe effect on human health than either pollutant can separately.

The odor threshold for ozone in the most sensitive individuals [6] is 0.01 ppm, but it is only recognized by most people at 0.05 ppm. At a concentration of 0.1 ppm of ozone (or oxidants), more than 5 percent of individuals will have symptoms of eye irritation. Individuals with obstructive lung diseases such as asthma or emphysema, when exposed to an ambient atmosphere containing 0.1–0.15 ppm of ozone, show increased breathing resistance and increased O_2 consumption [6]. Ozone and other oxidants presumably produce their irritant action as a result of their chemical reactivity at the point of contact. Experiments have shown that exposure to 0.2 ppm of ozone for 3 hours reduces visual acuity, increases peripheral vision, decreases night vision, and alters the balance of the muscles controlling the position of the eye. The industrial exposure limit for ozone is 0.1 ppm [6]. It is important to note that at an altitude of 25 km, the concentration of ozone is raised to 16 ppm by the direct action of the sunlight. Unless some means is used to decompose the ozone, the concentration inside pressurized aircraft at this altitude reaches 0.3–0.4 ppm.

2.3.3 Particulates

Particulates are fine solids (dust or soot) or liquid particles (mist or fog) suspended in air and that are individually invisible to the naked eye. Solid particulates are usually made of carbon or mineral base material that absorbs (dissolves within the particle) and adsorbs (sticks to the surface of the particle) different types of substances. Particulates enter the atmosphere from both fixed and mobile sources. Fixed sources include factories, electrical power plants, ore smelters, and farms, while mobile sources include all forms of transportation that burn fossil fuels. Mobile sources account for nearly 17 percent of the particulates emitted to the atmosphere in the United States [4]. Fuel combustion from stationary sources and industrial processes accounts for about 40 percent of the particulates' emissions in the United States. Natural sources include winds eroding dust from cultivated farm fields and smoke from forest fires and volcanic ash. One of the main sources of carbon-based particulates (soot) is the exhaust from diesel engines. Soot oxidation catalysts and soot filters are now being used to reduce this kind of pollution. The solid particulates often provide extended surfaces due to their irregularities and, therefore, other pollutants can be carried along. Pollutants may have greater effect on health when in combination (smoke particles and sulfur dioxide, for example) than separately.

Particles whose diameters are less than 2.5 μm are called fine particulates. Since the rate at which particles settle increases with the square of their diameter (Stoke's law), the rate at which fine particles settle is fairly low and they usually remain airborne for days. Large particles are of less concern to human health than small ones not only because they settle out quickly, but also because they are efficiently filtered by the respiratory system. Plots of daily mortality in cities versus their particular matter with diameters less than 10-μm values (PM_{10} index) for the same or the previous day are consistently linear [2]. An increase by 100 $\mu m/m^3$ in the PM_{10} sindex of a city (American standards for respirable particulates is 150 $\mu m/m^3$ averaged over 24 hours), is estimated [2] to increases its mortality by 6 percent–17 percent.

2.3.4 Acid Rain

Most rainfall is slightly acidic due to carbonic acid from the carbon dioxide content of the atmosphere. Acid rain in the pollution sense is produced by the conversion of the primary pollutants sulfur dioxide (SO_2) and nitrogen oxides (NO and NO_2) to sulfuric acid (H_2SO_4) and nitric acid (HNO_3), respectively, to generate an acid solution with a pH lower than 5. In fact, the term *acid rain* covers a variety of phenomena, including acid precipitation (rainfall or snow) and acid mist. It is important to note, however, that carbon dioxide in the atmosphere dissolves in rain reducing its pH from 7.0 to 5.6, while naturally occurring oxides of sulfur and nitrogen are responsible for unpolluted rain having a pH of about 5.0. Unpolluted rain is, therefore, mildly acidic due to the presence of dissolved atmospheric carbon dioxide in it:

$$CO_2(g) + H_2O(l) \Leftrightarrow H_2CO_3(l) \tag{2.17}$$

and

$$H_2CO_3(l) \Leftrightarrow H^+ + HCO_3^- \tag{2.18}$$

Only rain that is appreciably more acidic than natural (unpolluted) rain acidity (i.e., with a pH of less than 5) is considered to be acid rain.

Acid rain is precipitated far downstream from the source of the primary pollutants [2], namely sulfur dioxide, SO_2, and nitrogen oxides, NO_x. Nitrogen oxides are mainly attributable to vehicle emissions (Figure 2.2), while most of the sulfur dioxide is produced by volcanoes and by the oxidation of sulfur gases produced by the decomposition of plants. (A breakdown of the anthropogenic source of sulfur dioxide is shown in Figure 2.3.) The main anthropogenic source is the combustion of solid fuels and crude oil, and by the petroleum industry when oil is refined and natural gas is cleaned before delivery. It is worthwhile to note that since sulfur dioxide absorbs UV-B, a significant concentration of SO_2 in the air will reduce the amount of UV-B reaching ground level.

The primary pollutants themselves, SO_2 and NO_x, do not make rainwater particularly acidic. However, over a period of days they are converted into sulfuric acid, H_2SO_4, and nitric acid, HNO_3, both of which are very soluble in water and are strong acids.

The conversion of SO_2 to H_2SO_4 initiates with:

$$SO_2^\bullet(g) + OH^\bullet(g) \longrightarrow HSO_3(g) \tag{2.19}$$

Fig. 2.3.

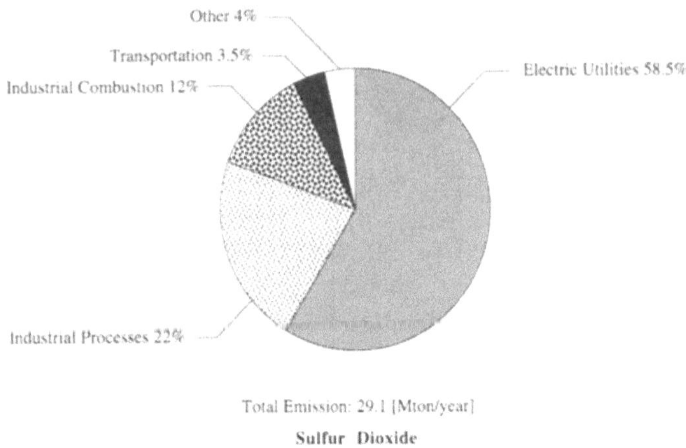

Sulfur dioxide pollution from various anthropogenic sectors in the United States and Canada [5].

The overall mechanism for the conversion of SO_2 to H_2SO_4 is as follows [2]:

$$SO_2(g) + NO(g) + O_2(g) \xrightarrow{H_2O} NO_2(g) + H_2SO_4(l) \qquad (2.20)$$

The conversion of NO_2 to HNO_3 proceeds through:

$$NO_2^\bullet(g) + OH^\bullet(g) \longrightarrow HNO_3(g) \qquad (2.21)$$

Since the initial steps involve an OH radical, the rate of producing H is much faster in the presence of photochemical smog atmosphere. In general, because reaction 2.20 is very slow, only a minority of sulfur dioxide is oxidized in air, and the majority is removed by dry deposition.

2.3.5 Air Quality

Air quality usually refers to the concentration in air of one or more pollutants. For many pollutants, air quality is expressed as an average concentration over certain period of time (e.g., $\mu g/m^3$ for a period of 24 hours). The concentrations of air pollutants that cannot legally be exceeded during a given time period in a specified location are determined by the air quality standards.

The objective of air pollution control is to prevent adverse responses by all receptor categories exposed to the atmosphere—human, animal, vegetable, and material. These adverse responses have characteristic response times which may be seconds, minutes, hours, days, months, or years. For there to be no adverse responses, the pollutant concentration in the air must be lower than the concentration level at which these responses occur. Air quality indexes have been devised for categorizing the air quality measurements of several individual

Table 2.4

U.S. Pollutant Standards Index (PSI) and pollutant levels [7].

Pollutant Standards Index (PSI)	Air quality description	Suspended particulates for a period of 24 hrs	Sulfur dioxide for a period of 24 hrs	Carbon monoxide for a period of 8 hrs	Ozone for a period of 1 hr	Nitrogen dioxide for a period of 1 hr
		PM_{10}-$\mu g/m^3$	SO_2-$\mu g/m^3$	CO-mg/m^3	O_3-$\mu g/m^3$	NO_2-$\mu g/m^3$
0–49	Good	0–50	0–79	0–4.9	0–119	—
50–99	Moderate	50–150	80–364	5.0–9.9	120–235	—
100–199	Unhealthy	150–380	365–799	10.0–16.9	235–400	—
200–299	Very unhealthy	380–420	800–1599	17.0–33.9	400–800	1130–2259
300–399	Hazardous	420–500	1600–2099	34.0–45.9	800–1000	2260–2999
>400	Hazardous	>500	>2100	>46.0	>1000	>3000

pollutants by one composite number. The index used by the U.S. Environmental Protection Agency is called the Pollutant Standards Index (PSI). Current PSI values and air pollution levels are given in Table 2.4. The term *air pollution index* is sometimes used to describe the air quality in semiquantitative terms; an arbitrary function of the concentration of several pollutants is used to scale the severity of air pollution.

REFERENCES

1. Gribbin, J., "The Ozone Layer," *New Scientist*, Inside Science No. 9, May 5, 1988.
2. Baird, C., *Environmental Chemistry*. Place: W.H. Freeman and Co., 1995.
3. Lean, G., and Hinrichsen, *Atlas of the Environment*, Place: Helicon, 1992.
4. Pepper, I.L., Gerba, C.P., and Brusseau, M.L., *Pollution Science*. New York: Academic Press, 1996.
5. Placet, M., and Streets, D.G., in "The Causes and Effects of Acidic Deposition," Vol. 2, The National Precipitation Assessment Program, Washington, D.C.: Government Printing Office, 20402, 1987.
6. Robert, H.D., and Robertson, W.O., *Handbook of Poisoning*. Place: Appleton & Lang, 1987.
7. Bouble, R.W., Fox, D.L., Turner, D.B., and Srern, A.C., *Fundamentals of Air Pollution*. New York: Academic Press, 1994.

CHAPTER 3

Health Aspects of Air Pollution

Rafael S. Carel
Division of Community Medicine, Faculty of Health Sciences, Soroka Medical Center, Beer-Sheva, Israel

ISBN: 0-12-639855-0
$25.00

3.1 ANATOMY AND PHYSIOLOGY OF THE RESPIRATORY SYSTEM

Pulmonary anatomy and physiology are discussed insofar as is necessary for an understanding of the noxious effects of the relevant environmental pollutants. For more detailed information about pulmonary medicine, the reader is referred to the following textbooks: Baum and Wolinsky [1], Cotes and Steel [2], and Isselbacher, Braunwald et al. [3]. The major function of the lungs is the exchange of gases with the environment. The overall function of the respiratory system is to supply oxygen to the tissues of the body and to remove carbon dioxide from the body. This gas exchange is accomplished by a complex system of organs, involving many anatomical structures and physiological mechanisms. The respiratory passages and the lung parenchyma are the organ systems most intimately exposed to the environment. A normal sedentary adult inhales 3 to 5 liters of air per minute (4500 to 7000 liters per day) to accomplish the physiologically required gas exchange. During exercise or strenuous work, the rate of inhalation may increase as much as 10 to 20 times that of rest.

Ambient air is seldom free from pollutants. Thus, various noxious agents have access to the lung, which makes it the body organ most frequently affected by airborne environmental pollutants. Some of these toxic agents also affect other body systems, such as the cardiovascular system, the central nervous system, the kidneys, and the bone marrow.

Respirable agents may be of different types: pathogens (viruses, bacteria, and fungi), allergens, gases (e.g., HC, CO, CO_2, SO_2, and radon), particulates (e.g., asbestos fibers, silica, and cotton dust), smoke (e.g., cigarettes, traffic, and industrial), or other chemical agents. Airborne contaminants may exist in a variety of physical states: gases, aerosols, fumes, smoke, mist, or dust (particles).

3.1.1 Upper Airways

The function of the upper airways is to transport air into and out of the lungs. This conducting system is composed of the nose, the mouth, the pharynx, the larynx, the trachea, and the bronchial tree. These anatomical structures are illustrated in Figure 3.1. During inspiration, ambient air enters the respiratory system through the nostrils and passes backward almost horizontally through

Fig. 3.1.

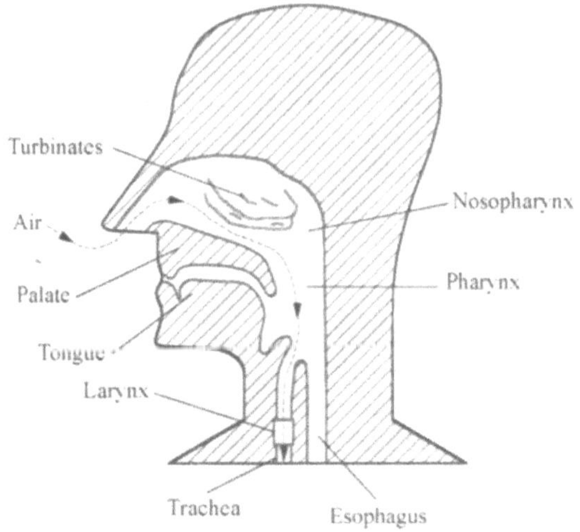

The upper airways.

the nasal cavity toward the pharynx. The convoluted turbinate bones, which have a large surface area lined with ciliated epithelium, partially fill the nasal cavity. The marked convolution of the turbinate bones serves to bring the incoming air into close contact with the large surface area of nasal epithelium. This epithelium is supported by a highly vascular tissue. During its passage through the nose, the air is warmed to body temperature and humidified to saturation (with water vapor). The mechanisms by which inhaled particles and gases are taken up by the upper airways are described later. From the posterior nasal passages, the incoming air moves downward through the pharynx to the larynx. The larynx is a boxlike structure that leads into the trachea. In its lower part, the larynx narrows markedly as a result of the presence of the vocal cords. During passage of the air through the vocal cords, the flow becomes turbulent as a result of sharp variations in the diameter of the conducting tube.

3.1.2 Bronchial Tree

The trachea, at its lower end, divides into the right and left main bronchi, each of which supplies one lung. Functionally, however, the two lungs work as a single organ and will thus be considered as such. Each main bronchus branches further as many as 20 to 25 times to form the bronchial tree (Figure 3.2). This branching is usually dichotomous, doubling the number of bronchi in each generation of branching. The conducting airways gradually decrease in diameter with each branching, but the total cross-sectional area in each step is somewhat greater than

Fig. 3.2.

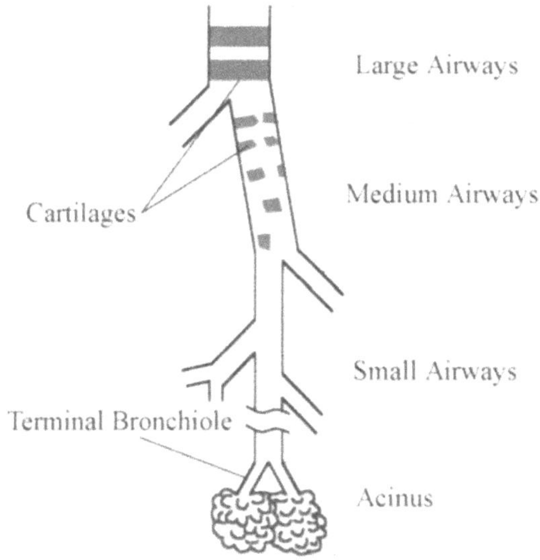

Schematic presentation of the bronchial tree.

that of the previous branching level. At the level of the small airways (branching generations 20 and beyond), the diameter is 2 mm or less.

The walls of the large and medium-size airways contain cartilage plates, which support the conducting tubes and prevent their collapse under external pressure. In the large airways, the cartilage plates form almost a complete ring. The frequency and size of the cartilage plates and extent of the circumference that they occupy decrease with each successive branching of the bronchi. Cartilage is absent below the level of the medium airways.

The walls of the large and medium airways are surrounded by a layer of smooth muscle (Figure 3.3). When contracting, these muscles can markedly decrease the diameter of the airways and increase the resistance to airflow. Resistance to the flow of a gas in a conducting tube is inversely proportional to the fourth power of the radius of the tube (Poiseuille-Hagen formula). Thus, stimuli that cause contraction of the muscles of the airways may cause a marked reduction in airflow and ventilation (bronchoconstriction).

The small airways are peripheral (subsegmental) bronchi devoid of cartilage. Their walls consist of fibrous tissue and bundles of smooth muscle. The small airways are conducting structures in which no gas exchange takes place. The basic unit of the lung—the pulmonary segment—is anatomically and functionally a discrete unit with its own bronchus and blood supply.

A schematic representation of the inner (epithelial) surface of a bronchus is shown in Figure 3.4. The bronchial epithelium consists primarily of a layer of columnar ciliated cells. Goblet (mucus-secreting) cells and mucus glands, which are spread along the entire length of the airways, continuously secrete mucus into

Fig. 3.3.

Cross-sectional presentation of a medium-sized bronchus.

Fig. 3.4.

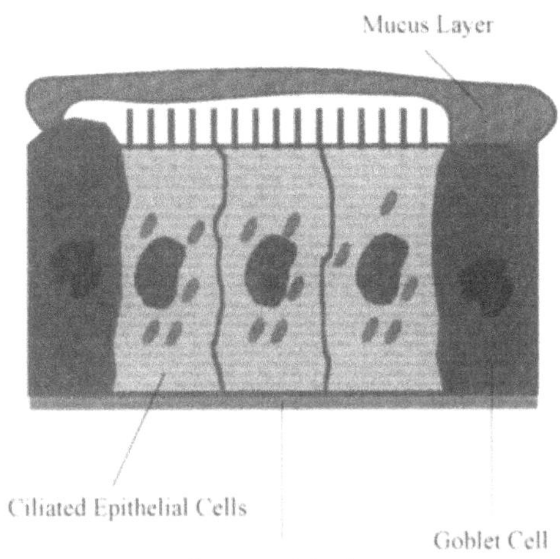

Schematic presentation of mucociliary escalator.

the lumen of the bronchial tree. The secreted mucus forms a continuous layer that lines the entire internal surface of the bronchi (mucus sheet). The mucus-secreting structures are almost completely absent from the level of respiratory bronchioles and distally (smaller airways).

Structures from the respiratory terminal bronchioles and beyond constitute the lung parenchyma. This is the section of the lung where gas exchange takes place. One respiratory bronchiole serves a group of alveoli joined together very much like a cluster of grapes; this structure is known as an acinus.

3.1.3 Acinus

A schematic representation of an acinus with its constituent alveoli is given in Figure 3.5. The respiratory bronchioles end in alveolar ducts that, in turn, terminate in two or more air sacs. The cluster of alveoli around an air sac forms an acinus, which constitutes the major part of the aerating surface. An acinus contains up to

Fig. 3.5.

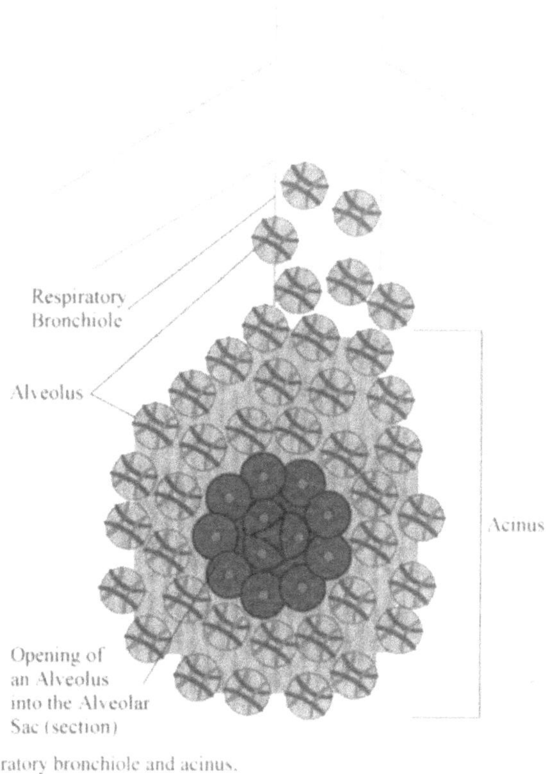

The respiratory bronchiole and acinus.

Fig. 3.6.

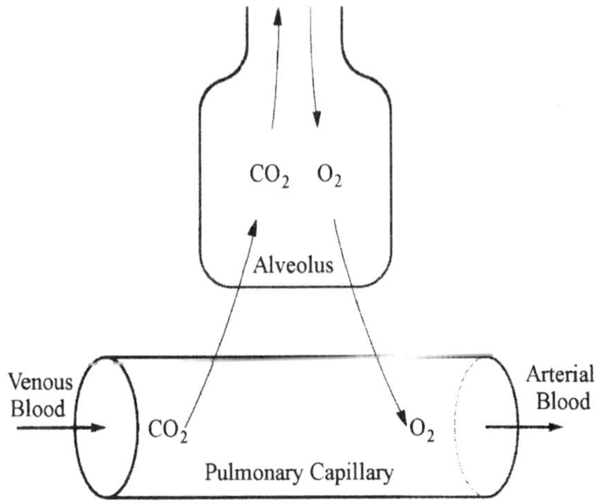

Function of the alveolus.

about 4000 alveoli. The diameter of an alveolus is about 75 to 300 mm and there are about $1 \cdot 10^6$ terminal bronchioles in the lung, serving about $3 \cdot 10^8$ alveoli.

The alveolus (Figure 3.6) is the terminal part of the respiratory system, where the gas exchange process takes place. The alveoli are tiny hollow sacs with an opening toward the alveolar sacs and respiratory bronchioles. Each alveolus is lined with a continuous but very attenuated layer of epithelium and the inner surface of the alveolus is covered with a film of a lipoprotein material that acts as a surfactant.

More than 95 percent of the wall of an alveolus is composed of a thin layer of protoplasm comprising alveolar epithelial cells (type I cells). The honeycomblike structure of lung parenchyma is formed mainly by the walls of contiguous alveoli. The alveolar wall forms the respiratory surface of the lung, where the gas exchange takes place.

Type II alveolar cells are almost cuboidal cells that bulge into the lumen of the alveolus. Their function is to secrete the surfactant layer that lines the internal surface of an alveolus.

The main function of the surfactant is to reduce the surface tension of the liquid that lines the inner surface of the alveoli. The surfactant alters the surface tension in response to changes in the alveolar area (radius): The surface tension is reduced when the alveolar radius becomes smaller. If the alveoli were lined with waterlike fluids, the surface tension would be sufficiently high to cause the emptying of smaller-diameter alveoli into larger ones, according to the Laplace relationship $P = 2T/r$, where P = alveolar pressure, T = surface tension, and r = radius of the alveolus. The equation indicates that the smaller the value of r is, the greater the value of P. The surfactant lining has the property that during inflation and deflation of the alveolus, the surface tension of the surfactant lining changes in such a way as to maintain the intra-alveolar pressure (P) at a constant value.

The overall gas exchange surface area of the alveoli in an adult is about 60 m^2. The alveolar wall also acts as a barrier between the external world (inspired air) and the interior of the body (blood and tissues). Blood capillaries are to be found in close proximity to the alveolar wall, so as to enable gas exchange to take place strictly by diffusion (Figure 3.6). The term *gas exchange tissue* or *lung parenchyma* thus refers to that part of the lung composed mainly of millions of alveoli and the rich network of capillaries intimately associated with them.

3.1.4 Pulmonary (Alveolo-Capillary) Membrane

The gas exchange process takes place through a membrane composed of alveolar epithelium, alveolar and capillary basement membrane, and capillary endothelium (Figure 3.7). The capillary basement membrane is a continuous layer of tissue attached to the layer of alveolar epithelial cells. The average thickness of this alveolo-capillary membrane is 0.2 to 0.5 μm.

Fig. 3.7.

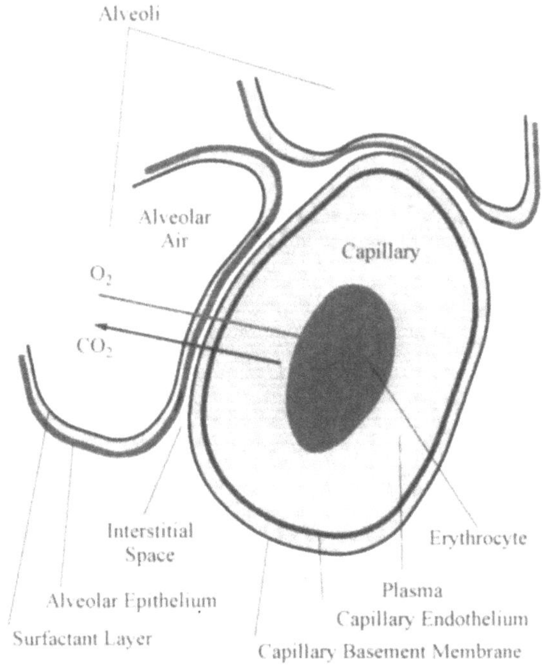

The respiratory membrane.

At any time the amount of blood present in the distensible capillary bed of the lung is about 60 cm^3. This relatively small amount of blood comes into contact with the large surface area of the pulmonary membrane. Diffusion involves the movement of molecular oxygen from the alveolar air, through the alveolar epithelium, the basement membrane, the capillary endothelium, and the capillary plasma into the erythrocytes (red blood cells). Carbon dioxide moves in the opposite direction.

The interstitial tissue of the lung parenchyma is composed of ground substance, collagen fibers, and cellular components (fibroblasts, mast cells, phagocytes, lymphocytes, and plasma cells). This tissue fills the space between the alveoli and around the blood vessels and lymphatics. The interstitium plays a major role in certain types of lung disease.

3.1.5 Pulmonary Ventilation

The term *pulmonary ventilation* refers to the mechanical component of respiration by which air is moved into and out of the lung. With each inspiration, about 200–250 ml of new air is supplied to the expanding alveoli. This ambient air, rich in oxygen, mixes with air already present in the alveoli. During expiration, air with a relatively high concentration of carbon dioxide derived from the blood is mechanically forced out of the alveoli to the external atmosphere.

The inspiratory and expiratory muscles are responsible for the continuous process of renewal of alveolar air. The main inspiratory muscles are the diaphragm and the external intercostal muscles. During inspiration, the volume of the thoracic cage increases, and air moves into the lung according to the gas pressure gradient between the alveoli and the atmosphere. During inspiration, the alveolar pressure becomes subatmospheric. When the pressure difference between the atmosphere and the alveoli, as a result of expansion of the thoracic case, reaches 3 mm Hg, air is drawn into the alveoli and causes them to expand. (Note: Normal atmospheric pressure at sea level is 760 mm Hg.)

The main expiratory muscles are the abdominal muscles and the internal intercostal muscles. When these contract, the volume of the thoracic cage decreases, the intra-alveolar gas pressure rises above atmospheric pressure (by about 3 mm Hg), and air is forced out of the alveoli into the atmosphere. In each normal expiration, the amount of air exhaled is similar to the amount of air brought into the lung in each inspiration.

When air is inspired, it is rapidly humidified by moisture from the lining of the nose and the upper airways. At normal body temperature (37°C), the partial pressure of water vapor is 47 mm Hg. Mixing of the incoming ambient air with water vapor dilutes the air so the partial pressure of the inspired gases in the airways is slightly lower than the partial pressure in the atmosphere. Thus, alveolar air contains less oxygen than atmospheric air. The partial pressure of oxygen in dry atmospheric air at sea level is 159 mm Hg, while that of alveolar oxygen is only 104 mm Hg (the balance is water vapor).

3.1.6 Pulmonary Gas Exchange

Alveolar gas is a mixture of inspired air, fully saturated with water vapor, and carbon dioxide that diffuses from the blood. At the end of each expiration about 2000–2500 ml of air remain in the lung. With each cycle of breathing, about 15 percent of alveolar air is renewed, thus maintaining the necessary gas gradient between alveolar air and the blood.

Atmospheric air at sea level is composed almost entirely of nitrogen (78.6 percent), oxygen (20.8 percent), water vapor (0.5 percent), and carbon dioxide (0.04 percent). Alveolar air contains much more water vapor (6.2 percent) and carbon dioxide (5.3 percent), resulting in a dilution of nitrogen to 74.9 percent and oxygen to 13.6 percent. Oxygen diffuses continuously from the alveolar air through the alveolo-capillary membrane into the blood. Similarly, carbon dioxide diffuses from the blood into the alveolar air. This process of diffusion reflects the differences in partial pressure of the relevant gases across the alveolo-capillary membrane (alveolo-arterial oxygen gradient and capillary alveolar carbon dioxide gradient). The amount of oxygen transported into the blood and carbon dioxide transported from the tissues would be grossly inadequate if it were not that oxygen combines with hemoglobin of the blood and that carbon dioxide enters into a series of reversible chemical reactions that convert it into other compounds. Hemoglobin increases the oxygen-carrying capacity of the blood by about 60 to 70 times, and the reactions of carbon dioxide increase the blood carbon dioxide content about 17 times, compared with what would be transported if oxygen and carbon dioxide were merely dissolved in the plasma (Henry's law).

The pulmonary membrane is composed of those pulmonary surface areas thin enough to allow gas diffusion into and out of the capillary blood. As illustrated in Figure 3.7, the membrane contains several layers. Gas exchange (diffusion) across the thin alveolo-capillary membrane occurs rapidly and completely. Ordinarily, venous blood is fully oxygenated during a single pass along the alveolo-capillary membrane, and the partial pressure of carbon dioxide in the blood is reduced to about 40 mm Hg. This process lasts less than 1 second.

The rate of diffusion of a gas across a membrane is determined by several physicochemical factors. It is proportional to the difference in the partial pressure of the gas across the membrane (pressure difference), the available surface area for diffusion, and the solubility of the gas in the membrane. It is inversely proportional to the membrane thickness (diffusion distance) and to the square root of the molecular weight of the gas. Thus, the greater the pressure difference of the gases on each side of the membrane, the higher the rate of diffusion; and the thinner the membrane, the higher the rate of diffusion. The larger the area of the pulmonary membrane, the greater the volume of gas that can diffuse in a given time; and the greater the solubility of the gas in the membrane, the larger the volume of gas that will diffuse. Gases with low molecular weights will diffuse more rapidly than those with higher molecular weights, in inverse proportion to the square root of their molecular weights.

3.2 DEFENSE MECHANISMS OF THE LUNG

During rest, a normal adult inhales about 5 liters of air per minute to accomplish the physiological requirements of gas exchange. During heavy exercise or work, the ventilation rate may increase up to five times that at rest.

Various defense mechanisms protect the respiratory system (and the body) from noxious elements carried into the body with the inhaled ambient air. In this chapter, we will describe primarily those defense mechanisms related to protection from toxic inhalants, while other mechanisms, such as those involved in viral or bacterial infection, will not be considered. There are broadly three types of defense mechanisms—physical, physiological and immunological—each being effective against certain characteristics of the penetrating injurious inhalants.

3.2.1 Physical Mechanisms

Particulates are filtered out and removed from the respiratory system by several physical mechanisms.

Deposition. This refers to the process that leads to contact between an inhaled particle and the mucus layer that lines the epithelium of the airways. This may occur by inertial impaction, sedimentation, or diffusion, depending on the size (*effective diameter*) of the particle, its density, airflow velocity, and breathing pattern.

Inertial Impaction. Particles with an aerodynamic diameter in the range of 7–10 μm or more are removed primarily by means of this mechanism. As the air passes over the large surface area of the nasal septum and turbinates, the inhaled particles impact on these structures as a result of the momentum, which is proportional to their velocity. The inhaled air (including particulates) has a relatively high velocity on entering the nose due to the relatively small cross-sectional area of the nares. Thus, a large proportion of particles over 7 μm in diameter are eliminated from the incoming air by impaction at this level. Inertial impaction is also an important reason for the deposition of particles in other parts of the airways system, particularly where there is a change in the direction of flow of the incoming air, such as the bronchial bifurcation sites. The turbulent motion that is created increases the probability of impact of a particle with the walls of the airways. Most of the larger inhaled particles are removed by this mechanism, since their relatively large mass and high velocity and, hence, their relatively high momentum increase the chance of impaction.

Sedimentation. Sedimentation of particles in the airways occurs when gravitational forces acting on the inhaled particles exceed the forces of flow and buoyancy. In this situation the particles will settle on the moist lining of the airways and will not travel further. This mechanism is effective mainly in the more peripheral airways where airflow velocity is low. Sedimentation is the process by which particles with an aerodynamic diameter in the range of 0.5 to 0.7 μm are deposited. The settling velocity of such particles is governed by Stoke's law.

Diffusion. For particles less than about 0.3–0.5 μm in diameter, the main mechanism of deposition is diffusion. In addition, many of the particles of this size are expelled in exhaled air. They are deposited mainly in the alveolar region.

Solubility. Water-soluble particles and gases are readily absorbed by the aqueous lining of the mucosa of the upper airways. Gases that are highly water soluble, such as SO_2, are almost completely extracted from the inhaled air by the nose and the pharynx in resting healthy subjects. Gases that are less water soluble, such as NO_2 or O_3, are much less completely removed in the upper airways, and they can penetrate more deeply into the respiratory tract, where they may exert their noxious effects.

3.2.2 Physiological Mechanisms

Airway Reflexes

Certain respiratory reflexes, such as coughing or sneezing, help to clear the airways from inhaled foreign agents and so protect the lungs. The physiological role of the cough reflex is to expel excess secretions or inhaled irritants from the airways. Sensory receptors (nerve endings) all along the respiratory tract are readily stimulated by inhalants. This stimulation initiates the cough reflex, which is an explosive release of intrathoracic air. The cough maneuver expectorates mucus and foreign inhalants out of the airways. The sneezing reflex is an explosive exhalation of air through the nose.

Bronchoconstriction

Certain irritants, when inhaled, stimulate specific receptors that initiate the bronchoconstriction reflex, resulting in a general constriction of the muscles in the walls of the bronchi (via the vagus nerve). This leads to a decrease in the diameter of the bronchi and increased airway resistance, and fewer particles may thus reach the lung parenchyma. This reflex has only a limited value as a defense mechanism and should be viewed rather as an untoward response of the respiratory system to certain stimuli.

Antioxidant Activity

Certain inhalants (environmental pollutants), such as NO_2, O_3, and quinones, are oxidizing agents and as such may cause injury to the lung tissue. Secretions of antioxidants by the tracheobronchial tree primarily prevent the formation of hydroxy radicals.

Mucociliary Clearance

The mucus that lines the airways forms a continuous "mucus blanket" covering the inner surface of the bronchi (Figure 3.4). This blanket is continuously propelled upward, from the small airways through the larger ones up to the larynx. This action is accomplished by the coordinated beating of the cilia of the epithelial cells that line the respiratory tract. The continuous upward movement of the mucus layer is known as the *mucociliary escalator*. Inhaled agents deposited on this mucus lining are carried upward until they reach the larynx, where they are swallowed. Mucociliary clearance is an important defense mechanism by which the greater bulk of the inhalants deposited along the airways is cleared from the lungs.

3.2.3 Immunological Repair Mechanisms

Various minute injuries caused by pollutants are continuously repaired by cellular and humoral responses. Such mechanisms should be distinguished from disease responses, since they constitute part of the ongoing defense mechanisms of the lung; for example, damaged bronchial epithelium is rapidly repaired by proliferation and differentiation of cells. This is a complex mechanism that will not be further discussed here. Inhalation of noxious particles and gases may also initiate a cellular response in the lung, mainly in the lower respiratory tract. Inflammatory cells that reside in the lung parenchyma, or are derived from the blood, migrate to the injured zone and initiate the host response. These are complex mechanisms that involve cellular, humoral, and immunological defense reactions. Recruited macrophages can remove foreign particles by phagocytosis, the process by which the host cell attacks and destroys or removes a foreign particle from the lung. In addition, stimulated macrophages release various mediators with chemotactic (attraction of cells by chemical substances) activity that recruit other inflammatory cells and thus amplify the defense response. A detailed description of these mechanisms is beyond the scope of the present discussion.

3.2.4 Inhalants Deposition, Clearance and Injury—General Considerations

The lungs are continuously exposed to various types of air pollutants, of which pollutants from internal combustion engines are only one. All pollutants generated by internal combustion engines also occur in the environment as a result of their production by other industrial processes. In practice, people are often exposed to a combination of pollutants released into the atmosphere from a variety of sources. The harmful effects of such pollutants may be synergistic in particular, those effects associated with cigarette smoking (voluntary or involuntary) and those of industrial origin.

The effects of pollutants in people with preexisting diseases or those with certain predispositions are to produce the responses (whether acute or chronic) earlier, in a more prominent way or at lower concentrations than would be expected in healthy (nonsmoking) adults. Similarly, certain susceptible groups, such as the very young or the very old, may show signs and symptoms of injury sooner than the general population.

As may be expected, the effects on health of certain inhalants differ in relation to their dose (the level of exposure or concentration of the agent in the ambient air), the duration and character of the exposure (continuous or intermittent), the combination of pollutants involved, the physicochemical properties of the agents, and various other conditions such as climate, geographic area, and the microenvironment (indoors versus outdoors). Special attention should be paid to occupational exposures and conditions, as well as to the pattern of breathing of the worker and the physical demands of the job. The use of protective devices (respirators) in certain occupations can reduce the dose of inhalants that reach the lung and hence prevent damage. However, quite often, such protective devices are not used or are used only intermittently. All these factors may affect the manifestations of the disease in a particular person. Similarly, the retention of an inhaled agent in the body, its clearance, and its biological effects depend on many factors, some related to the agent and others to the host. The physical properties of the inhalant, such as size, density, solubility, and physical state (particulate, mist, vapor, or gas), will to a great extent determine the deposition site in the bronchial tree or lung parenchyma. As a rule, smaller particles penetrate more deeply and reside for longer in the lung. The chemical and biological properties of the agent will affect the type and extent of the damage and the host response (local or general). Some inhalants are quite inert, while others may be strongly alkaline or acid. Some agents primarily cause a local reaction, while others react with body fluids or cells and have systemic effects. Different agents may cause markedly different biological effects such as inflammation, immunological reactions, fibrosis, or cancer.

It therefore follows that the overall host response is affected by many factors and involves complex relationships between the host and the environment. Determinants of the response to inhaled pollutants include individual susceptibility, age, sex, and previous morbidities. In particular, the efficiency of the pulmonary defense mechanisms is severely compromised in smokers. Smoking not only directly affects the lung, but also enhances the deleterious effect of other environmental pollutants. The marked ill effects of smoking on human health, and particularly on the respiratory system, cannot be overemphasized. Even though most pulmonary defense mechanisms are very effective, they cannot completely prevent the penetration and retention of injurious agents in the lung, especially when the ambient air is heavily polluted. Thus, control of these agents at their source should, as a rule, be the method of choice to prevent their deleterious effects on human, animal, and plant life.

3.3

VENTILATORY FUNCTION TESTS

Pulmonary ventilatory function tests provide quantitative information about lung physiology. The device used to record the volume and flow of air into and out of the lung is known as a spirometer. Spirometry is relatively easy to perform and noninvasive and is thus frequently used for the evaluation of pulmonary damage in exposed persons. It provides a measure of dynamic lung volumes, based on the forced vital capacity (FVC) maneuver (Figure 3.8a), as described

Fig. 3.8.

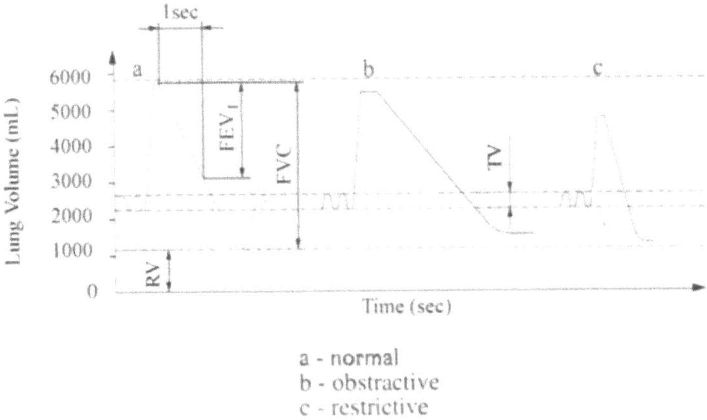

A spirogram.

next. Vital capacity (VC) is the volume of gas exhaled from the lung during a maximal expiratory effort after a maximal inspiration. FVC is a VC maneuver performed rapidly and completely immediately after taking a full inspiration. The amount of gas left in the lung at the end of a forceful exhalation is the residual volume (RV). RV cannot be determined by simple spirometry and is measured by other techniques. Total lung capacity (TLC) is the volume of gas present in the lung at the end of maximal inspiration; that is, TLC = VC + RV.

The forced expiratory volume in 1 sec (FEV) is the volume of air exhaled in the first second during the performance of the FVC maneuver. Tidal volume (TV) is the volume of air that passes into and out of the lung with each respiration during quiet breathing. The TV at rest is about 300–400 ml, and it may increase markedly during exercise and strenuous work. The term *minute ventilation* is used to describe TV multiplied by respiratory rate (RR).

Since lung volume is a function of body size (height) and is also influenced by age and sex, prediction formulas have been developed for the parameters described previously. These formulas provide expected or reference values that make it possible to adjust or standardize the actual measured parameter. Thus, spirometric measurements are expressed as a percentage of expected values (observed value/predicted value multiplied by 100). This method normalizes the measured parameters with respect to height, age, and sex and makes the interpretation of the measurements more uniform. The normal (predicted) values are derived from studies of large populations of healthy individuals.

The clinical and epidemiological value of measuring ventilatory lung function depends on several factors: well-maintained and calibrated equipment, proper execution of the procedure by well-trained technicians, standardized testing techniques and the use of appropriate prediction (reference) values, quality control procedures, and proper instruction of examinees. These factors may markedly influence the quality and reproducibility of the test results and their meaningful interpretation. Ventilatory lung function testing provides information on lung volumes and rates of flow. Other tests are used to provide information on other physiological functions of the lung, such as pulmonary diffusion capacity.

Spirometry is particularly useful as a measure of obstruction to airflow. Thus, persons with obstructive lung disease, in which reduction of the diameter of the bronchi is a major element, will show reduced FEV and reduced FEV-FVC ratio but preserved FVC (i.e., pure obstruction to airflow, see Figure 3.8b). Conditions that produce pure restrictive disease (stiff lung) are characterized spirometrically by a proportional decrease of FEV and FVC with a normal (preserved) FEV-FVC ratio (Figure 3.8c). Often, the results of pulmonary function tests show a mixed picture of obstruction and restriction. In addition, spirometry can be used to assess the reversibility of airflow obstruction (asthmalike disease) by performing the test before and after the administration of bronchodilating drugs.

3.4 PRINCIPLES OF INHALATION INJURIES

Exposure to air pollutants can cause or exacerbate various lung and systemic diseases. Each pollutant may have different noxious functional or pathological effects. A description of the unique pattern of toxicological effects of each of the major pollutants exhausted by internal combustion engines will be presented in order to facilitate an understanding of specific pulmonary reactions and indicate possible prevention measures. However, seldom will we encounter an isolated pollutant. As a rule, exposure is to multiple pollutants and their additive or synergistic combined effect should be considered (total environmental exposure).

3.4.1 Carbon Monoxide Intoxication

Carbon monoxide (CO), the most commonly encountered noxious gas, is produced in large amounts by industrial plants and gasoline engines. It is formed during incomplete combustion of organic (carbon-containing) materials. CO is an odorless gas, lighter than air, which is rapidly absorbed through the lungs and binds to the hemoglobin in the red blood cells to form carboxyhemoglobin. The affinity of hemoglobin for CO is about 200 to 240 times greater than that for oxygen. Thus, the oxygen-carrying capacity of hemoglobin is decreased proportionally to the concentration of CO in the inspired (ambient) air. The CO-hemoglobin complex is far more stable than O_2-hemoglobin (oxyhemoglobin). Thus, exposure to CO results in tissue hypoxia (a condition in which there is a reduced supply of oxygen to body tissues) because less oxygen is transported from the lung to the tissues. The damage caused by CO intoxication is primarily due to a decrease in the release of oxygen to tissues and cells (cellular anoxia = lack of O_2 needed for metabolic processes in cells). Tissues with a high metabolic rate (high O_2 demand), that is, the brain and other parts of the nervous system, are affected earlier by CO intoxication than tissues with lower metabolic rates.

Because CO is odorless, there are no early warnings of exposure. The clinical manifestations of CO poisoning are easily recognized and are related to the amount of carboxyhemoglobin formed and the amount of oxygen available to tissues. The rate of CO transfer from inspired air into the blood is primarily affected by the tension (partial pressure) difference between CO content in the ambient air and that in alveolar gas. CO is removed from the lung and rapidly taken up by the blood. Because of its high affinity to hemoglobin, the rate of carboxyhemoglobin formation reflects the ambient concentration of CO. Since ambient air contains 21 percent of oxygen by volume, exposure of a person to air containing as little as 0.1 percent of CO will result, at equilibrium, in 50 percent of the hemoglobin in the blood being converted to carboxyhemoglobin. When exposure to CO is discontinued, there is a gradual dissociation of the CO-hemoglobin complex, and CO is removed via the lung.

The early clinical signs and symptoms of CO poisoning are shortness of breath, tachypnea (rapid breathing), headache, dizziness, impaired judgment (confusion), and lack of motor coordination. All these signs are the result of brain and nervous system hypoxia. Nausea, vomiting, and diarrhea may appear later as may cardiac arrhythmias, pulmonary edema, and finally loss of consciousness and death. The symptoms may occur after only a few breaths of air contaminated with carbon monoxide. If CO concentration in the ambient air is high enough, rapid loss of consciousness and death may occur within a short time. As a rule, symptoms of CO intoxication may begin when carboxyhemoglobin saturation reaches 20 percent, and unconsciousness occurs at 60 percent saturation.

Treatment includes rapid removal of the affected person from further exposure and the administration of 100 percent oxygen to expedite the dissociation of carboxyhemoglobin into hemoglobin that can combine with O_2 and hence to correct the tissue hypoxia.

Normal urban dwellers may have up to 0.5 percent carboxyhemoglobin, but smokers may have up to 5 percent to 10 percent of carboxyhemoglobin. Thus, heavy smokers are more susceptible to CO intoxication than nonsmokers.

Prevention of CO intoxication is mainly by control of CO at its source by reducing the release of CO into the atmosphere. In addition, in areas with a high potential for exposure, ensuring a clean work environment by maintenance of efficient exhaust systems or catalytic oxygenation of CO to CO_2 is essential.

3.4.2 Carbon Dioxide

Carbon dioxide (CO_2) is an odorless gas, heavier than air, that is produced during the burning of organic matter and by several industrial processes. It is also an end-product of certain metabolic processes in the body.

Inhalation of CO_2 leads rapidly to hyperventilation (extremely rapid and deep breathing), sweating, and headache. The initial symptoms may be followed by loss of consciousness and death. The hyperventilation reflex is triggered by specific receptors for hypoxemia (lack of oxygen in the blood) and hypercarbia (excess of CO_2 in the blood).

The symptoms of CO_2 intoxication result from the displacement of oxygen from the inspired air by high concentrations of CO_2. This may lead to asphyxia (loss of consciousness as a result of too little oxygen and too much CO_2 in the blood). Treatment involves rapid removal from exposure, which is usually in a confined space.

3.4.3 Ozone

Ozone (O_3) is detrimental to health because it is a strong oxidant. O_3 oxidizes substances that contain double-bonded carbon or sulfur linkages. It is produced in nature by photochemical reactions involving ultraviolet (UV) radiation and nitrogen oxides. O_3 (like some other irritant substances such as SO_2, chlorine, and inert respirable dust) causes nonspecific bronchoconstriction by increasing the reactivity of the airways. These effects are augmented in persons with asthma or chronic bronchitis particularly in the presence of cigarette smoke.

Exposure to low concentrations of O_3 (0.3 to 0.9 ppm) will produce symptoms of cough, dryness of the throat, and chest discomfort. Pulmonary function tests will demonstrate obstructive impairment and increased reactivity. These effects are transient and will subside when exposure is discontinued (transient reduction in ventilatory lung function). The clinical effects are the result of irritation of the mucus membrane of the airways. Signs and symptoms of eye irritation may also be present.

It is thought that long-term exposure to ozone may contribute to the development of emphysema in the afflicted persons. At higher concentrations O_3 may cause pulmonary edema.

3.4.4 Nitrogen Oxides

Nitrogen oxides are by products of several industrial processes and usually exist as a mixture. The oxidant gas nitrogen dioxide, NO_2, is an efficient ultraviolet light absorber that leads to the formation of ozone (NO_2 photolytic cycle).

$$NO_2 \xrightarrow{UV} NO + O$$

$$O + O_2 \rightarrow O_3$$

NO_2 is a reddish-brown gas with an irritating odor. Because NO_2 is not very soluble, it is a deep lung irritant that may reach the bronchioles and alveoli. In low concentrations, it may produce only mild upper respiratory tract symptoms. The inhaled gas dissolves in the aqueous lining medium of the nasal and pharyngeal mucosa where it rapidly forms nitrous and nitric acid, as is shown in the following equation:

$$NO_2 + H_2O \rightarrow HNO_3 \rightarrow H^+ + NO_3^-$$

In acute exposure, the formation of an acid causes immediate irritation to the mucosa of the airways. Similar signs of mucosal irritation may also appear in the eyes (conjunctivitis). In the lungs the initial response will be cough, dyspnea, and a bronchospasm reaction.

As sufficient quantities of the gas reach the lung parenchyma, it may cause acute bronchiolitis and damage to the alveoli. Inhalation of higher concentrations may result in pulmonary edema and death may occur.

The acute phase of the pulmonary reaction may be followed, several weeks later, by a delayed reaction of bronchiolitis obliterans. This is a more chronic state of inflammation, progressing to bronchiolar fibrosis and emphysema. Such a progressive process may lead to respiratory failure due to massive pulmonary fibrosis. In some cases, acute relapses of pulmonary edema and pneumonia may occur but in other cases, complete recovery follows the acute stage. An association between NO_2 exposure and repeated childhood respiratory infections has been demonstrated. This increased susceptibility to respiratory infection is only rarely identified as being related to an earlier exposure to NO_2. Exposure to low concentrations of NO_2 may increase airway reactivity (airways resistance) in normal subjects without producing overt clinical symptoms. In asthmatics, such exposure may initiate asthmatic attacks.

Extrapulmonary Effects of NO_2 (Methemoglobinemia)

The heme iron of hemoglobin in the red blood cell is normally maintained in the reduced ferrous (Fe^{++}) state, in which it fulfills its oxygen transport function. The unique form of the hemoglobin complex has the remarkable property of being able to combine directly with oxygen, while the iron is retained in the reduced form. Oxyhemoglobin is carried in the red blood cells to all body tissues, where oxygen is delivered to the cells to supply their needs (according to oxygen pressure gradient). This process may be disrupted by a variety of nitrogen-containing substances that oxidize the iron in the heme moiety to its ferric (Fe^{+3}) state. This valance change may result in the formation of methemoglobin, an almost black pigment, which cannot combine reversibly with oxygen, resulting in a reduction in the supply of oxygen to the tissues (anemic hypoxia).

In addition to the delivery of oxygen to the tissues, the formation of methemoglobin from hemoglobin leads to denaturation of the hemoglobin and hemolysis (the destruction of the red corpuscles with liberation of hemoglobin into the surrounding fluid).

Clinical manifestations of methemoglobulinemia result from tissue hypoxia. When the methemogloblin level exceeds that of 10 percent of the total hemoglobin level, the affected person will show cyanosis (bluish coloration of the skin caused by lack of oxygen in the blood). At higher concentrations of methemoglobin (over 35 percent), symptoms of shortness of breath, weakness, and headache may appear. All these symptoms are due to the reduced capacity of the blood to carry oxygen to the tissues. The severity of symptoms is proportional to the fraction of hemoglobin that has been transformed to methemoglobin.

Usually exposure to nitrogen oxides is gradual and signs and symptoms develop insidiously. The major sign of this state is the development of cyanosis, particularly at the periphery, that is, lips, fingers, and toes (acrocyanosis). Since the symptoms (fatigue, headache, palpitations) are not pathognomonic and since methemoglobulinemia may be caused by a variety of agents and circumstances, a thorough medical history (occupational and environmental) should be taken, investigating different possible exposure situations. Clearly establishing the causative agent is of prime importance in order to prevent further exposure of the afflicted individual and other persons potentially similarly exposed.

Biological monitoring of persons potentially exposed to nitrogen oxides is feasible, since the level of methemoglobin in the blood can be determined in the laboratory.

Treatment of affected persons includes removal from further exposure, administration of 100 percent oxygen, and when the methemoglobin level is over 20 percent and the person is markedly symptomatic, administration of methylene blue.

3.4.5 Sulfur Dioxide Intoxication

Sulfur dioxide (SO_2) is released into the atmosphere during many industrial processes. It is the principal and most ubiquitous urban air pollutant. SO_2 is a highly irritating gas, and at levels as low as 0.3 to 1.0 ppm it can be identified by either smell or taste. Because of its marked solubility in water, it is almost 100 percent absorbed in the upper airways in quiet breathing. Even during exercise it is unlikely to penetrate beyond the large bronchi. Sulfur dioxide, as dissolved gas, reacts rapidly with the aqueous medium of the mucous membranes of the body to form sulfurous acid. Local irritation and response of membranes such as those of the eyes, nose, mouth, and upper respiratory tract appear quickly.

When exposure is mild (6 to 12 ppm), the respiratory tract reaction will be that of immediate cough and reflex bronchoconstriction. Epistaxis (nosebleed) may result from damage to the nasal mucosa. Cough and expectoration are also related to the direct stimulating effect of SO_2 on the mucous glands in the respiratory tract. This leads to an increase in the secretary capacity of these glands (and goblet cells) such that mucus is discharged faster and in larger quantities into the lumen of the bronchi (hypersecretion). With time, hypertrophy of the mucous glands can be identified and is accompanied by a chronic productive cough. (Similar phenomena can result from chronic irritation by other pollutants such as NO_2 and O_3.)

An intense exposure to SO_2 may lead to laryngeal spasm or pulmonary edema, both of which conditions could be fatal. Chronic low-level repeated exposures to SO_2 may exacerbate chronic bronchitis, increase airway resistance (obstruction) in asthmatics, and result in ulceration of the nasal septum and destruction of dental enamel. Exacerbation of states of chronic bronchitis have been shown to be clearly related to periods of heavy pollution with SO_2 in urban areas. Similarly, respiratory tract illnesses in childhood have been found to be associated with chronic exposure to even modestly elevated levels of SO_2. People with a state of chronic bronchitis quite often have increased susceptibility to intercurrent infections.

Sulfate ions in different substances vary widely in their irritant potency. The sulfate ion itself is not an irritant. The most potent agent is sulfuric acid (H_2SO_4), followed by ferric sulfate, zinc sulfate, and ammonium sulfate. Sulfur dioxide may be absorbed by particulates in the ambient air and thus exert a synergistic effect with that of the particle in question.

3.5 AIRBORNE POLLUTANTS CAUSING CANCER AND OTHER DISEASES

Identification and assessment of environmentally caused disease in a particular person is important so that he or she may be protected from further damage by removal from exposure. It is also of great importance in the primary prevention of similar diseases in other exposed, less susceptible people. Such assessment requires quantitative information concerning type and extent of exposures. Since many diseases become clinically evident only after chronic exposure over many years, current environmental measurements are of limited value. Moreover, the actual dose absorbed in the body is influenced by many factors such that even when the level of the pollutant in the environment is known, it is hard to evaluate the actual amount received by a particular individual. Since simultaneous exposure to various pollutants frequently occurs, it is hard to determine the exact role of each one of them in the development of the adverse effect in individuals or groups. The approach to reducing the adverse effects of inhalants is mainly by regulation and control of the quality of outdoor air.

As mentioned earlier, the lung is a primary gateway of entry into the body for various toxic agents, although many of them will exert their adverse effects in other organ systems. Metal fumes of cadmium affect the kidneys, while those of mercury affect primarily the central nervous system and kidneys. Carbon disulfide affects the cardiovascular and nervous systems. Benzene is known to harm bone marrow, and vinyl chloride the liver. A detailed toxicological presentation of these processes is not a part of the current work, but the examples given earlier illustrate the significant role of the lung as a port of entry to many pollutants with broad biological effects.

The lung is the organ system most intimately exposed to environmental putative agents; some of them are confirmed human carcinogens and others are suspected carcinogens. Among the air pollutants of concern here, none has been identified as a human carcinogen. Airborne agents that are known to cause lung cancer in man are bis(chloromethyl) ether (BCME), acrylonitrile, arsenic compounds, chromium, iron oxide, mustard gas, asbestos, vinyl chloride, nickel, uranium, and the output of coke ovens (exposure to polycyclic hydrocarbons such as benzo(a)pyrene). The association between cigarette smoking and lung cancer is well established and will not be discussed here.

REFERENCES

1. Baum, G.L., and Wolinsky E. (eds.), *Textbook of Pulmonary Diseases*, 5th ed. Boston: Little, Brown & Co., 1992.
2. Cotes, J.E., and Steel, J., *Work-Related Lung Disorders*. Oxford: Blackwell Scientific Publications, 1990.
3. Isselbacher, K.J., Braunwald, E., et al. (eds.), *Harrison's Principles of Internal Medicine*, 13th ed. New York: McGraw-Hill, Inc., 1994.

CHAPTER 4

Economic and Planning Aspects of Transportation Emission

Pnina O. Plaut
Faculty of Architecture and Town Planning, Technion, Israel Institute of Technology, Haifa, Israel

Steven E. Plaut
Graduate School of Business Administration, University of Haifa, Haifa, Israel

ISBN: 0-12-639855-0
$25.00

4.1 INTRODUCTION

Emissions from transportation vehicles are a major source of pollution, especially in urban areas. They are major sources of both local urban pollutants, such as oxides of nitrogen, hydrocarbons, and carbon monoxide, and global greenhouse emissions, like CO_2. (They are a major source of noise pollution as well.)

Vehicle use creates not only emission of pollution, but also problems of congestion, especially in urban areas. While congestion is not the focus of this paper, many of the comments that follow could be applied to congestion and traffic problems as well, not only to vehicle emissions of pollutants. Policy alternatives that reduce vehicle emissions also frequently reduce congestion problems, and so produce two sets of social benefits.

Automobiles dominate the transportation and travel activities in industrialized countries (except Japan). Automobiles are also the main source of vehicle emissions and are the main component of travel-related energy usage in all high-usage OECD countries (Schipper et al. [5]). They have been the subject of concern by international conferences on air quality and global warming, such as the Rio Framework Convention on Climate Change and the United Nations Conference on Environment and Development.

The contribution of vehicle emissions to pollution is illustrated in Table 4.1, with numbers taken from several selected studies. As can be seen, for all categories of pollutants investigated in these studies, the share of vehicle pollutants is quite large.

In many ways, vehicle emissions represent a policy problem similar to industrial emissions of pollutants. However, in some ways they are different, namely, in the inability to apply the direct pricing methods favored by economists and planners. The approach of economists and city planners to vehicle emissions is addressed in this chapter.

All pollution emissions problems involve two separate sets of issues: technical engineering issues for controlling pollution and policy/incentive issues. Vehicle emissions can be reduced through four principle methods: (1) fuel switching and alternative fuels (Sperling, Setiawan, and Hungerford [6]); (2) exhaust-emission controls; (3) reduced vehicle fuel intensity, that is, reducing fuel consumption per distance traveled; and (4) reduced vehicle use per capita. The first three involve technical and engineering solutions. The last one involves alterations in travel behavior.

Table 4.1

Selected Estimates of Proportion of Pollution Originating in Vehicle Emissions, from Previous Research Papers, by Country/City

	United States	United Kingdom	Japan	Sweden	Santiago, Chile
Volatile Organic Compunds—VOC's	27%	41%		40%	45%
Carbon Monoxide	50%	90%			78%
Nitrogen Oxides	29%	41%		80%	69%
Particulate Matter	17%				
All Greenhouse Gases	22%				
Black Smoke		46%			
Carbon Dioxide		19%	22%	46%	
Sulfur Dioxide				23%	

Sources: For the US, Environmental Protection Agency (1) and DeCicco (2); for Japan, Ono (3); for Sweden, data from Statistics Sweden; for Santiago, Hall et al. (4).

The policy/incentives issues can operate upon any of these four areas. They can address the problem of getting polluters to adopt engineering solutions available in order to reduce emissions. If there is no incentive or requirement to do so, even the most wonderful and impressive engineering solution will not be applied or adopted.

Economics and planning both address the policy problems of implementing emissions reduction. While command-and-control forms of pollution control are possible, where a regulator simply dictates abatement of emissions, such methods are generally *not* favored by economists. This is because they are generally considered inefficient and/or ineffective.

Instead, planners and economists generally prefer policy methods that operate through incentive structures. This is in spite of the fact that much real-world pollution abatement policy operates through administrative commands, which function primarily through regulation, but contain incentive elements in the form of penalties and subsidies.

Pollution will be reduced when it becomes profitable for polluters to reduce it. It can be made worthwhile through a variety of carrot-and-stick methods. In addition, any pollution abatement program will involve enforcement and monitoring problems, but it is often thought that those operating through the creation of an incentive structure have more manageable enforcement and monitoring (Lee [7]). So instead of a governmental authority issuing cease-and-desist orders to potential polluters, preferred methods involve "making it worth their while" to reduce pollution, such as through making reward payments for abatement or levying a fine or tax or charge for emissions. For further discussion of the role of economic incentives in environmental policy, see Barde [8], Baumol and Oates [9], Downing and White [10], Folmer, Gaebel, and Opschoor [11], Hahn [12], Markandyal and Richardson [13], Mills [14], Opschoor and Vos [15], Stavins [16], and Tietenberg [17, 18, 19, 20].

4.2

THE NOTION OF OPTIMAL POLLUTION ABATEMENT AND CONTROL

It sounds strange to most ears to hear economists and planners discuss the "optimal amount" of air pollution. Most people take it as axiomatic that the "optimal" amount must be zero, and that the goal of public policy must be to get as close to zero as possible. This is fallacious. Optimal pollution would be zero only when the costs to society from an incremental reduction in pollution are also zero. Since lowered pollution generally must be accompanied by a loss in some other commodity or service that members of society value, a trade-off problem arises. Do we want a bit less pollution with a bit less of those commodities and services we value, or not?

Zero pollution would be optimal only if the costs for reducing pollution were also zero. In general, these costs are not zero, and are indeed increasing with the level of environmental cleanliness. Hence, the marginal costs of any increment in pollution reduction become more and more expensive, the more pollution has already been abated.

While incremental costs for pollution abatement rise, the incremental benefits generally decline as pollution approaches zero. The sacrifice that people will be willing to make—for example, in terms of their living standards—in order to eliminate N tons of some pollutant from their environment will depend on the total amount of pollution they have already eliminated, or the residual level of pollution. If they live in a highly polluted town, the sacrifice they are willing to make to eliminate N tons of the pollutant is likely to be quite large. In other words, the marginal benefits from such pollution reduction are large. If they are living in a near-pristine rural environment where there is little pollution to begin with, the sacrifice they are willing to make to eliminate the same N tons of pollutants is likely to be much lower.

These basic notions are illustrated in Figure 4.1. The marginal benefits curve is downward sloping and indicates the value (in terms of what the representative consumer would be willing to pay or sacrifice) in order to eliminate a certain amount of pollution. It slopes downward because environmental "cleanliness," like every other consumer good or service, involves diminishing benefits or "saturation." As the environment becomes cleaner and cleaner, people will tend to prefer other consumption goods or public services to any further pollution abatement.

On the other hand, the marginal costs of pollution abatement are presumed to increase, for the same reason that marginal costs in all or nearly all other economic activities increase. At any point in time, any reduction in some pollutant by N tons

Fig. 4.1.

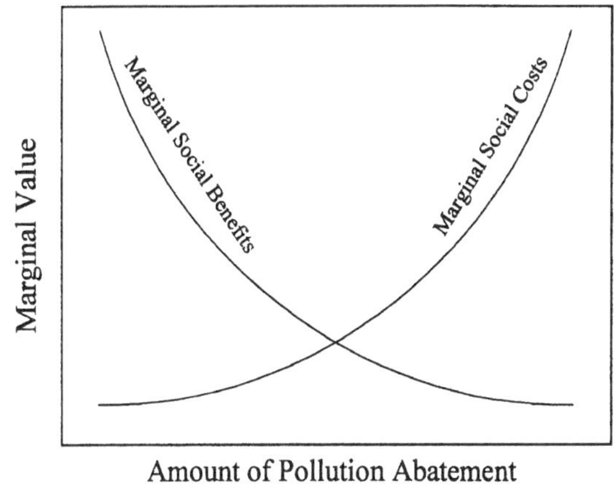

Cost-benefits trade-off in pollution abatement.

will be more expensive, the more it has *already* been eliminated. This is because the cheaper and easier steps or technologies for abatement are presumed to have been already utilized, leaving only the more expensive.

Optimal pollution or, if you prefer, optimal environmental cleanliness is defined theoretically by the intersection between the marginal benefits and marginal costs curves. This implies that any level of pollution abatement to the left of the intersection is too little, leaving "too much pollution." But any point to the right involves "too little pollution."

The danger of "too little pollution" is not a mere academic fantasy and can occur in the real world. For example, many public utilities operate under a cost-plus pricing environment, where they price electricity so that the price covers all costs plus some markup. Among the costs to be passed on to consumers automatically under such a cost-plus system would be costs of pollution abatement, such as at power generators. In the absence of some other regulatory constraint, the utility can continue to invest in pollution reduction well beyond the point of optimal abatement. In other words, under cost-plus pricing, there is nothing to prevent overinvestment in pollution reduction and, hence, "insufficient pollution" emanating from public utilities. There are reasons to suspect that such excessive pollution abatement really occurs in some utilities.

Thus far, the definition of *optimal pollution* closely resembles the definition of efficiency and optimality in all other areas of economics and for all other markets, goods, and services. There are several problems with applying ordinary economic tools and means for achieving efficiency to problems of vehicle pollution abatement.

1. The "public goods" nature of the abatement problem means that in most contexts, abatement must be decided through public policy and not by individual consumers. Environmental cleanliness is a public good in the sense that it is

enjoyed as a collective good. There exists a "free-rider problem" because once the environment is clean its cleanliness cannot be withheld from those who do not pay for it; and this means it cannot be priced or marketed as an ordinary good.

If we allowed all individuals to decide whether and how much of their pollutants to reduce, far too little would be abated. Each polluter would think to himself or herself, "Since I am very small compared with overall society and my pollutants are negligible relative to total pollutant levels, why should I bother to invest resources in abating my pollutant?"

When all individuals think in this way, little if any pollution is abated. This is known in economics as a market failure problem, and it means that individuals optimizing individually do *not* produce a social optimum. There are many other sets of circumstances in which market failures can occur, and where public policy measures are required to correct market behavior and incentives. This is the focus of public finance within economics.

Another way to explain this point is to note that in most markets, people can assess the marginal benefits from any form of consumption *individually*. Consumers can choose for themselves individual levels of consumption. The social marginal benefits curve is merely an aggregate of these individual consumer demand curves, revealed through observing how consumers' behavior changes as prices change. For public goods or for those services requiring provision through some governmental collective mechanism, there exists no economic method for translating individual preferences into socially preferred levels of consumption. This is particularly the case in problems of pollution control. Instead, some governmental or political mechanism must operate.

This will be true even when marketlike imitation methods for reducing pollution emissions are used, such as emissions taxes or indirect pricing policies. While these price mechanisms create individual incentives for polluters to expend costs and efforts in order to reduce pollution, the rate or level of such taxes must be determined collectively, not individually by consumers. This is much like the observation that income tax rates or traffic ticket fines must be determined collectively rather than allowing individuals to select their own individual rates. The earliest proposals for such taxes include Baumol and Oates [9], Mills [14], and Roberts and Spence [21]. For recent discussion of emissions taxes, see Goulder [22], Hahn [23], Lee [7], Markandy and Richardson [13], Parry [24], Shapiro and Warhit [25], Tahvonen [26], and Tietenberg [17, 27].

2. In many cases, the costs and benefits of pollution abatement are not out-of-pocket cash costs, but rather opportunity costs at the societywide macro level, involving the forgoing of other private or public goods and services. For example, suppose that all forms of power generation involve some pollution; the question of whether to produce more or less power is the same as whether to have higher or lower consumption of electricity, and also whether to have slightly higher rates of mortality and morbidity or not. If a particular form of vehicle operation (or an industrial process) creates inevitable pollution, then there may be no way to

reduce that pollution other than through curtailing travel activity, such as through mandatory trip reduction.

A direct trade-off will take place between the consumption of whatever it is that produces the pollutant emissions and environmental cleanliness. If there are no cost-effective ways to eliminate automobile emissions, then society faces a trade-off between curtailing automobile travel and use versus suffering more pollution. Other sorts of noncash trade-offs from pollution abatement may be a reduction in travel speed or safety, reducing real consumption levels, inconvenience, and discomfort.

3. Pollution and emission problems involve severe externalities, where there are spillovers of costs and disutilities to other producers and consumers. In economics, an *externality* is a situation in which one individual's behavior causes costs (or benefits) to spill over onto other people. For example, if I operate a discotheque in my apartment, I spill noise pollution costs on to my neighbors. If I pollute my part of the river, I spill losses on to all the fishermen downstream who will catch fewer fish. If my office is a fire hazard, I spill fire risks on to neighboring offices. And so on.

In classical economics it is well known that the existence of any form of externalities may mean that market operations and solutions are suboptimal, and public policy intervention may be called for to correct matters. While externalities have been postulated theoretically for many problems in public policy, there are few areas where they can credibly be shown to exist in significant levels, and the main such area is pollution emission. In some special cases, market participants may resolve externality problems on their own (this is known as the Coase theorem) through side payments, eliminating the inefficiency (for example, if the fishermen "bribe" the polluter not to dump pollutants in the stream, if the neighbors "bribe" the person or office not to engage in the behavior spilling risks or nuisances on to them). But where externalities are widescale and involve large numbers of people who are affected by the spillover of costs or harms, it is generally agreed that the Coase side-payment resolution is not feasible. Hence, some form of public policy intervention is required to avoid the excessive harms caused by the externalities. This will generally be the case for vehicle emissions.

4. Unlike most consumer goods, pollution abatement is a goal that often cannot be isolated from assorted other public policy goals. Consumption of cinema films or toys or tomatoes generally does not involve complex trade-offs and conflicts with other important competing social goals. Pollution control does. This is especially true for pollution emissions from transportation vehicles. We want transportation vehicles to emit as little pollution as possible, but we also want a well-developed transportation sector in which people and goods can move about quickly, safely, and cheaply and reach a large number of destinations. These goals are not fully consistent and require choices and trade-offs.

Hence, pollution abatement can only be addressed within the framework of complex multicentered or multiple-goal public policy discussion. Pollution control is just one of many goals and considerations in transportation policy.

Other considerations might include costs and speed of commuting and transporting goods, the impacts upon different social or interest groups, the impact upon land prices and conservation of land, safety, size, and shape of urban areas, population distribution, and so on. Often any change in policy to reduce vehicle pollution may necessitate negative impacts on one or more of these many other policy goals. On the other hand, abatement policies could also have *positive* impact on some other policy goals, such as the congestion problems already noted. Hence, pollution abatement cannot be addressed in isolation from these other policy goals.

All of the preceding considerations mean that the identification of the "right" level of pollution is difficult and complex. It does not change the fundamental observations that not all pollution abatement is desirable nor justifiable, and that some form of public policy intervention is generally necessary in order to achieve or approximate optimal pollution control.

4.3

ALTERNATIVE SETS OF ABATEMENT POLICIES FOR MOBILE-SOURCE EMISSIONS

Arguably, the most important idea in all of economics is the centrality of the role of incentives. Economists and public policy analysts believe people are rational in the sense that they respond (at least in aggregate) to incentives in all things. This would include pollution and emissions problems. Thus, most economists would argue that the best way to reduce emissions and pollution is to create incentives for their reduction. As noted in the previous section, ideally such incentives should be just enough to reduce pollution emissions to their optimal level, and no further.

In general there are two prototypes of policy methods that may be chosen to reduce pollution, with a wide variety of specific policy forms to be found under the headings of each prototype:

A. Administrative Methods

B. Pricing Methods

Pricing methods themselves may be either direct/Pigouvian or indirect. As we will see, direct/Pigouvian charges are not feasible for vehicle emissions, leaving administrative and indirect pricing policies as the feasible sets of alternatives. While not a relevant option for vehicle emissions, nevertheless, we will include direct/Pigouvian policies in the following discussion, as they illustrate the concerns of economists and planners in resolving emissions problems.

Administrative methods are dictates from policy makers and regulators that define acceptable maximum levels of pollution emission for polluters. They are directly legislated or mandated pollution levels, such as a maximal emissions level for each factory or each vehicle. An administrative dictate could be the requirement that a given fleet of transportation vehicles not emit a particular pollutant in quantities beyond some maximum amount. It differs from pricing methods in that polluters are left with no choice as to their level of pollution; as we will see, under pricing methods polluters' incentives are "doctored" by the policy maker but individuals ultimately decide on their level of pollution on their own. Administrative methods will require some enforcement or monitoring system to see to it that such regulation is obeyed (Lee [7]).

Under pricing methods, the policy maker or regulator does not impose a formal polluting quota upon producers and operators of vehicles or factories. Each polluter still can decide individually which level of pollutants to emit. Rather, a fine or tax is imposed upon polluting or pollution-related activities or upon each unit of a pollutant emitted for the purposes of altering incentives. Once incentives are altered, policy relies on individual polluters to reduce their emissions in order to save costs to themselves. For example, for a factory's pollution a charge could be imposed upon emissions of certain hydrocarbons.

The basic logic of pricing methods is that they motivate polluters to internalize the social costs of their emissions; that is, individual polluters must pay out of pocket for those emissions that in fact impose harms or costs on society. The pricing method makes individual polluters view their own out-of-pocket polluting costs and the harm they impose on society as one and the same. Therefore, they seek to reduce their harmful spillovers to society, not out of altruism and social conscience but rather in order to make greater profits.

Pricing methods operate through individual incentives. External costs imposed on society become transformed by the pricing method into internal private costs borne by the polluter. Hence, individual incentives are created to expend costs and efforts to reduce the emission.

It should be noted that both administrative methods and pricing methods require that the public policy makers or regulators be capable of measuring emissions from the vehicles (or factories) in question in a reasonable, reliable, and cost-effective way. Otherwise, neither can be enforced. Theoretically, any pollution level achievable through administrative dictate is achievable through judicious pricing as well. The two sets, however, differ in some ways; in particular they are not fully equivalent under conditions of uncertainty, in which case technical matters (such as the slopes of supply and demand curves) alter the relative effectiveness of the two sets of policies (see Weitzman [28]).

Having noted this, it should be emphasized that economists and policy analysts generally believe that pricing methods of emission abatement, especially *direct* pricing methods, are *superior* to administrative methods and operate more efficiently (for example, Swaney [29]). The reason for this is that pricing methods require only that policy makers assess the social damage from unit emissions of pollutants, rather than try to impose some global or individual cap to emissions themselves. Policy makers rely on incentives to get the job done rather than the prescience of bureaucrats. If each ton of hydrocarbons emitted inflicts X dollars of harm upon society in the view of policy makers, then in theory all vehicles or factories emitting such hydrocarbons could be charged an emissions tax or charge of X, and the incentives would do the rest. Unfortunately, this is not a feasible solution for vehicle emissions. For discussion of the uses of economic instruments in environmental policy, see Baumol and Oates [9], Markandya and Richardson [13], Mills [14], Opschoor and Vos [15], Stavins [16], and Tietenberg [17].

Under administrative dictates, policy makers have difficulties in setting individual maximal emission standards for individual polluters. If they set global emission standards that are reasonable and socially efficient, they still face a problem because each "quota" of pollution for individual polluters will be arbitrary and probably seen as unfair and preferential, making it politically unpopular as well. Under pricing methods, in a sense polluters select their own level of emission, based on the costs and benefits of pollution abatement, as well as the method for abatement. The benefits to individuals consist of avoiding the direct or indirect emissions tax. Thus, at the level of the individual polluter, emissions are abated just as long as the marginal costs of such abatement are less than the marginal benefits from avoiding the charge. In other words, the efficiency mechanism described in Figure 4.1 in the previous section operates at the level of the individual polluter.

Pollution charges under the pricing methods are also far more flexible and adapt more readily to changing circumstances than do administrative dictates. Direct or indirect pricing of emissions retains a marketlike flexibility. Under administrative controls, there are no responses at all or at best very slow responses to changes in the costs and social benefits from emission abatement. For example, administrative controls will respond poorly to the need to reassess control levels due to a technological change reducing costs of abatement, a seasonal change in climate, a change in the overall level of pollution in a specific environment, and so on. Under pricing methods, emissions levels are determined in a marketlike system where pollution costs are internalized by polluters, and so resource efficiency and flexibility of output are preserved as in ordinary market environments. As costs of abatement change, the supply and demand for the vehicles or products producing the emission shift automatically, with no need to wait for slow political or bureaucratic processes to catch up.

4.3.1 Pigouvian Pricing Mechanisms for Controlling Emissions

Pigouvian or direct pricing mechanisms are the pollution abatement methods preferred by economists. Unfortunately, they are not feasible for vehicle emissions. Nevertheless, it is worthwhile identifying the advantages of these methods. As we will see, because they cannot be applied to vehicle emissions, abatement policies for those emissions might be considered to be within the realm of second-best policies. In order to understand the considerations that make such policies second best, it is helpful to understand the theoretical advantages of what are generally considered first-best abatement policies.

Pigouvian taxes are taxes or charges levied directly upon any externality. They were first proposed by Pigou [30]. The idea is to force polluters or producers of any other externality to internalize the costs they impose on society by being charged a fee equal to the (negative) value of the externality. Because the Pigouvian charge is direct and proportional to the level of the externality activity, it is generally regarded as the most efficient method for controlling the externality. Polluters will strive to minimize the payout for this charge in exactly the same way that they try to minimize any other costs of production or consumption. If the charge is directly related to the quantity of emissions and adequately enforced, the only way to avoid its payment is to avoid emission in the first place. As long as the costs of abatement are lower than the emission charge or tax itself, polluters have incentive to spend those costs to reduce the pollution. This is true of both vehicle users and industry in theory.

Because the direct Pigouvian charge is levied on the emissions themselves, there is no interference other microdecisions by the polluter. Indeed, the full scope of the creativity and imagination of the polluter will be focused by the incentives upon abatement itself. Regulators do not need to dictate *how* to reduce emissions, only that their reduction is desirable and profitable. If reducing fuel consumption or if using public transit is the best way to do so, there are now incentives in place to reduce such consumption; but if there are *better* or *cheaper* ways to do so, then these will be selected. Decisions regarding which method should be utilized will be made at the level of individuals, not collectively by government bureaucrats and politicians.

Under direct pricing, different polluters may select *different* methods for achieving abatement, unlike the "one method fits all" abatement systems dictated under administrative policies. By concentrating the minds of individual vehicle users and other polluters on emissions abatement, the ingenuity of these same people is unleashed in a way that would be impossible under administrative command and control systems.

Theoretically, direct pricing can be set in such a way that the conditions for achieving optimal abatement, as discussed in Part II, are automatically met. These conditions are achieved simply by setting the pollution tax or charge equal to that level where the marginal social benefits curve equals the marginal social costs curve in Figure 4.1. At the optimum charge, emissions should be eliminated as long as the costs of abatement are less than the social value of abatement. When costs of abatement exceed this level, it is undesirable for the abatement to take place and so it will not. Polluters can be motivated to achieve exactly the optimal levels of abatement by setting the emissions charge equal to this level. Rational polluters will then take steps to avoid emissions as long as the costs of these steps are less than the emissions charge. (For recent analyses of the efficiency of Pigouvian taxes in different contexts, see Barthold [31], Carlton and Loury [32], Folmer et al. [11], Goulder [22], Johansson [33], Lee [7], Mills [13], Parry [24], Roberts and Spence [21], Tahvonen [26], Tietenberg [17, 18, 19, 20], Wirl [34], and Xepapadeas [35, 36].) These steps will include all choices that result in lower emissions, including such remote choices as using public transit more intensively or keeping vehicles tuned.

A variant on the notion of Pigouvian direct charges that has attracted considerable attention in recent years has been the idea of marketable emissions permits. Here the regulator distributes or auctions permits that allow polluters to emit a certain level of pollutant, and these permits then become transferable and marketable. Potential polluters can buy and sell them. The permits are then priced by the market, rather than by the regulator. In reality, they closely resemble Pigouvian taxes, except that the tax level is market determined rather than regulator selected.

Assertions of the optimality of Pigouvian taxes in environmental contexts under at least some circumstances can be found in Baumol and Oates [9], Carlton and Loury [32], Downing and White [10], Hahn [12], and Xepapadeas [34, 35]. While marketable and tradable emissions permits have recently captured the imagination of environmental economists, these too can be shown (Baumol and Oates [9]) to be equivalent to direct emissions charges, at least under certainty, and so will be regarded here as a variation on the theme of direct environmental policy alternatives. The principal difference is that under traded emissions permits, the market in effect sets the emissions tax, not policy makers, who set the quantity of emissions permitted. Tradable emissions permits are discussed in Atkinson and Tietenberg [18], Carson [37], Downing and White [10], Hahn and Noll [38], Maleug [39], Montgomery [40], Roberts and Spence [21], and Tietenberg [17]. There have been extensive experiments with such pollution emissions license marketing in the United States and some other countries. (For discussion, see Carson [37], Downing and White [10], Hahn [12], Kort [41], and Ledyard and Szakaly-Moore [42].)

Under direct pricing methods, the prices of emission, that is, the taxes or charges on emission, are themselves market determined. Policy makers can determine what the permissible maximum global emission of some pollutant may be in their jurisdiction and then auction off "polluting licenses." Market demand will then determine what the proper price should be for any "license to pollute." Similarly, policy makers could allot such licenses according to some formula believed to be fair, and then allow secondary trading in the same licenses. The recipient of a license or "pollution quota" could then sell it, for example, to someone else for whom the costs of pollution abatement are lower (and, hence, the license is more valuable). (For analyses of trading in marketable emissions permits, see Atkinson and Tietenberg [43], Baumol and Oates [9], Hahn [12], Hahn and Noll [38], Kort [41], Ledyard and Szakaly-Moore [42], Maleug [39], Montgomery [40], Shapiro and Warhit [25], and Tietenberg [17, 18].)

It should be noted that direct methods, including Pigouvian charges and marketable emissions permits, are all but irrelevant for vehicle emissions because of the impracticality of direct emissions monitoring and charges for vehicles, at least using current technology. (There have been some experiments with tradable permits within the framework of CAFE fuel efficiency standards by vehicle manufacturers.) But there are no practical ways to impose emissions taxes on vehicle users. This means that vehicle emissions control and abatement must be achieved through either administrative command-and-control methods or indirect pricing methods. *Both* of these alternatives are generally considered by economists and planners to be second best and inferior to Pigouvian and direct methods. The best of these second-best policies would be those that approximate as closely as possible the externality-correcting incentive structures of direct or Pigouvian policies.

4.4 ADMINISTRATIVE METHODS OF POLLUTION EMISSIONS CONTROL

Administrative controls are all those regulatory policies that work primarily through some mechanism other than a pricing apparatus. They work through governmental dictate and bureaucratic control rather than through incentives for individual polluters. Generally, administrative controls involve command-and-control types of regulations, for example, where policy makers dictate a maximum amount of emission for each vehicle or each polluter or a maximum number of trips or a minimal fuel efficiency standard.

The distinction between an administrative control and a pricing control mechanism is not always clear. For example, suppose an administrative control policy assigns a maximum level of emission to some polluter or a minimal fuel efficiency standard for some vehicle, but the consequence of violation of this limit is that the polluter pays some known fine or penalty. In this case, the administrative method may create incentives that operate very much like a pricing system, at least with respect to violations exceeding the assigned maximum/minimum targets. Technically, an emissions allotment plus a penalty for exceeding the limit resembles a pricing system known in economics as two-part pricing, where pricing is nonlinear. For example, electricity may be priced so that the initial kilowatt consumption is at a low rate, whereas additional consumption is at a higher rate.

Administrative controls may also resemble pricing methods whenever the allotments under the controls can themselves be bought and sold and exchanged among potential polluters. If a potential polluter is prohibited from polluting altogether (at least beyond some limit) unless he or she purchases part of the emissions allotment from some other polluter, then the price of those allotments once they are commercialized is the shadow price for emission. In this way the administrative control system may become a pricing system.

Let us consider several examples of administrative emissions abatement programs and identify their advantages and disadvantages and damages from such administrative controls.

4.4.1 CAFE Standards

Perhaps the best such example is the U.S. program establishing the Corporate Fuel Economy Standards (or CAFE) for fleets of automobiles, initiated in 1975. Under this federal mandate, fuel economy goals were assigned to each manufacturer of automobiles, not to individual consumers or users. CAFE standards were adopted in the midst of the "energy crisis" in the mid-1970s. At first they were thought to serve as a way to reduce energy demand. Later when energy prices dropped, they were retained, in part as a way to control emissions from vehicles by increasing fuel economy (see DeCicco [2] and Dellucchi, Green, and Wang [44]). As a means to lower energy demand, they were not particularly effective (Green [45] and Kleit [46]). By the mid-1980s the standards were increasingly unpopular among economists (Becker [47], Crandall [48], Crandall and Nivola [49], Krupnick et al. [50]).

The program to establish CAFE standards seems to have done everything that is wrong from the point of view of efficient policy making.

First, it established mileage standards as a proxy for pollution emission standards, an inefficient way to reduce pollution. Why not establish polluting standards directly or price emissions directly?

Second, it established mileage efficiency standards for the fleets of vehicles produced by car manufacturers rather than for individual vehicle owners and drivers. Public policy should always address the "marginal," the incentive of the individual to buy one more unit, to produce one more unit, to drive one more mile. CAFE standards apply to an arbitrary average of vehicles manufactured by each manufacturer. Under the CAFE system, some drivers buy cars with extremely low mileage and so presumably pollute a lot, whereas others buy fuel-efficient cars and pollute less. The CAFE system imposes supposedly "efficient" mileage standards upon *averages* of drivers, but not upon individual drivers, and so creates a haphazard and arbitrary system of individual incentives.

Third, even if the system consisted of a matrix of *individual* mileage efficiency standards for different vehicle classes, the CAFE system still would operate through administrative *dictate* rather than through pricing. It would be far more efficient to force individual vehicle users to pay a charge or tax for vehicle emission, based on quantity of emission. This would create an *individual* incentive for emission reduction. And if it is true, as conjectured, that emission is directly related to vehicle mileage efficiency, it would create a demand for mileage-efficient transportation vehicles. It would also create incentive to use public transportation extensively and reduce trips. By contrast, CAFE vehicle standards create no such direct incentives.

Fourth, CAFE standards have numerous harmful side effects because they ignore the fact that automobile consumption is part of a complex multigoal consumer and policy problem. The most well-documented side effect is vehicle safety. In order to meet CAFE standards of fuel efficiency, vehicle weight and mass were sharply reduced, by about 500 pounds per vehicle on average. However, it turns out that vehicle mass is closely correlated with the incidence of death and injury in accidents. Indeed, it may be the single factor *most highly correlated* with survivability in accidents (Crandall and Nivola [49] and Khazoom [53]). This connection was first noted by Robert Crandall of the Brookings Institution and John

Graham of Harvard's School of Public Health. They estimated that the CAFE standards caused an additional 2200 to 3900 accident deaths in the United States each year. Following their findings, a Washington, DC Court of Appeals in February 1992 temporarily overturned the CAFE standards altogether. Ironically, the CAFE standards imposed serious risks of death and injury on precisely those users of vehicles with the best fuel efficiency and lowest emission levels. Can there be a better illustration of the iron law of unintended consequences?

4.4.2 Transportation Demand Management Programs

These policy tools have been used in major cities in the United States and other OECD countries. The aim of transportation demand management (TDM) programs is to reduce congestion and mobile emissions through reducing or restricting travel demand, especially automobile travel demand to central urban locations. TDM includes strategies for affecting travel behavior such as the adoption of flexible work schedules and hours to avoid peak hours, the encouraging of telecommuting and teleworking (to avoid commuting altogether or to reduce miles traveled), encouraging carpooling, and subsidizing mass transit.

The main reason for the increasing use of TDM is concern over air quality and the negative impact thereon of metropolitan traffic congestion, especially during peak hours, although traffic congestion has also been an important consideration. TDM generally focuses on the journey to work and typically seeks to reduce drive-alone commuting through incentives for using transit and ridesharing. A particularly popular idea in recent years is the development of light-rail public transit and other forms of mass transit as a solution to pollution-cum-congestion problems. Advocates of these see them as long-term solutions that will result in large-scale changes in travel behavior.

Two examples of TDM will be discussed here, mandatory trip reduction and telecommuting.

4.4.3 Mandatory Trip Reduction

Mandatory trip reduction is an extreme form of the general set of policies known as transportation demand management (TDM). Such mandatory trip reduction programs are now in effect in the Los Angeles metropolitan area. In most cases, such programs are rationalized on grounds of their environmental intentions, namely, reducing vehicle emissions in metropolitan areas. They operate through issuing governmental dictates of trip quotas and rations by place of employment.

Severe air quality problems in southern California have induced the South Coast Air Quality Management District (the air pollution control agency for the Los Angeles area) to enact Regulation XV, which went into effect July 1, 1988. This requires all employers of at least 100 employees to develop, file, and implement a trip reduction program to achieve specified ridesharing goals, as measured by average vehicle ridership (AVR). This is measured by taking the number of employees reporting for work between 6:00 and 10:00 A.M. divided by the number of vehicles driven by them. Employers may choose their own internal incentive schemes to produce these results, such as parking preferences, transit pass subsidies, promotional activities, and rideshare matching procedure. (For a review of the early results of implementing Regulation XV, see Giuliano, Hwang and Wachs [52].) Fines for employers who failed to implement their own plans were introduced.

The program reportedly only produced a very modest mean increase in the mean AVR of 2.7 percent, mainly through increased carpooling. Other program features, including parking strategy incentives, seemed to have little effect upon AVR. Any significantly larger increase would clearly require much greater incentives. There have been no attempts to estimate the impact of the program on air quality in Los Angeles or to estimate what level of AVR would be required to achieve significant air quality improvement. Other than ridesharing, indirect methods used under Regulation XV for relieving vehicle congestion and emissions involved subsidized vanpools, employer-provided or subsidized transit passes, and telecommuting projects. The Los Angeles municipality gave a van to any employer that agreed to fill it with commuters.

4.4.4 Telecommuting and Teleworking Projects

Another area in which administrative programs seeking to reduce vehicle emissions (and traffic congestion) are operating with increasing frequency is telecommunications. Like mandatory trip reduction policies, these programs could be seen to belong to the area of travel demand management (TDM). These projects would include pilot programs designed to stimulate and encourage the use of teleworking, teleshopping, telebanking, teleconferencing, and so on. In all cases, telecommunications are supposed to serve as substitutes for travel.

Of these, telecommuting and teleworking have been the most popular and are considered the most promising. They can be considered substitutes for staggered work hours and arrangements. Some have argued that telecommuting could eliminate travel altogether (Lee and Meyburg [53]) or at least significantly alter travel behavior (Mokhtarian [54], Nilles [55], and Salomon [56]). However, there are other viewpoints, including some who argue that enhanced use of telecommunications will produce *increased* travel and use of motor vehicles (Salomon [56] and Plaut [57]). Telecommuting can take place either from home or from special telecenters that could be established in residential neighborhoods, reducing commuting and eliminating some traffic and vehicle emissions in central business locations.

Telecommuting pilot projects have been implemented and are now in operation in many places, including California, Washington state, the Netherlands, and Japan. Some air quality regulations, such as in mandatory trip regulation programs, grant "credit" for home-based and telecenter-based telecommuting. For example, Rule 2202 of the South Coast (California) Air Quality Management

District (AQMD 1996) states that vehicle trip emission credits will be granted for vehicle miles traveled (VMT) reduction programs, which may include telecommuting. The application of this and similar transport control measures are motivated by their conjecture that they produce nontrivial air quality benefits, but there has been virtually no empirical confirmation to date of such benefits.

Recently the first attempts at quantifying the impact of home-based and center-based telecommuting upon transportation use and on pollution emissions have begun (Henderson and Mokhtarian [58]). From a small sample of workers in the Puget Sound Telecommuting Demonstration Project, it appears that the number of vehicle miles traveled (VMT) was significantly reduced as a result of telecommuting (from 63.25 miles per person-day to 29.31 on telecommuting days). However, the number of personal vehicle trips did *not* change significantly. The drop in VMT translated into a 49 percent drop in emissions of oxides of nitrogen and a 53 percent drop in particulate matter emissions on telecommuting days compared with commuting days. Emissions from engine cold start (especially emissions of total organic gas and carbon monoxide), of course, are eliminated only in home telecommuting, not in center telecommuting. It should be emphasized that there are serious questions as to how much of the entire labor force can potentially utilize such telecommuting. If the percentage is negligible, so will be the impact on emissions and congestion. Debate over the pollution benefits of telecommunications use continues.

In addition to CAFE standards, direct assignments or allotments of trip quotas, and administering of telecommuting pilot projects, other administrative controls are commonly used by policy makers to control emissions of pollutants. Among these are:

1. Traffic management, especially in downtown urban areas.
2. Motor emissions standards and regulations. These are regulations requiring vehicle inspection and emissions standards, and are probably the most widely used abatement policy for nonvehicle emissions.
3. Land use policy.
4. Vehicle specifications, such as with respect to weight, size, power, and so on.
5. Targets for use of methanol and alternative fuels (Sperling, Setiawan, and Hungerford [6]).
6. Center-city parking policies.

As noted earlier, administrative methods contain severe practical disadvantages because they substitute bureaucratic command and control for market mechanisms. Pricing methods will be all the more preferable within the framework of complex multigoaled policy problems, such as vehicle emissions. Yet in spite of all the advantages of pricing methods, policy makers have traditionally preferred administrative command-and-control methods of abatement, even in multicentered policy settings. (See Baumol and Oates [9], Stavins [16], and Swaney [29] for further discussion.)

4.5

INDIRECT PRICING MECHANISMS

4.5.1 Theoretical Considerations

Indirect pricing of emissions would consist of levying a charge or tax against some variable that is closely correlated with the amount of the emissions, indeed, the closer the better, but *not* upon the quantity of emissions itself. If, for example, vehicle emissions were closely correlated with the quantity of fuel consumed, an indirect tax on emissions could be effectively created through fuel taxes. Similarly, pollution is presumably *negatively* correlated with the proportion of all travel that takes place on public transit. Hence, a *subsidy* for travel on public transit could be regarded as an indirect pollution pricing policy method.

One problem with indirect pricing is that if there is some way that the polluter can reduce or avoid the charge or tax that has been levied without at the same time reducing emissions, he or she will do so. Indirect policies can involve either a tax or a subsidy. The latter case will be relevant when the indirect activity is believed to *reduce* net emissions.

Formally, if Q is the amount of emissions and F is some activity correlated with emissions, then an indirect abatement policy can be modeled as:

$$Q = \zeta\{F\} + \varepsilon,$$

where ζ is some function, not necessarily linear. It is presumed that the first partial ζ' is positive, and ε is random noise and is distributed with mean 0 and variance σ^2. An indirect tax will be a tax on F, which reduces F. An indirect subsidy would be a subsidy that increases F, if F is negatively correlated with emissions Q.

In the special case where σ^2 is zero and ζ' is 1, then the indirect emissions tax is a perfect proxy for a direct emissions tax. More generally, an indirect tax (or subsidy) is generally considered less effective and less satisfactory than a direct emissions tax because its incentive effect is less unambiguous and less powerful.

Also, the greater is σ^2, the less closely will the tax burden or extra expense borne by the polluter be correlated with the quantity of the emissions, which is what policy is trying to control. If σ^2 is significant, polluters know that there is a chance that they will pay a low tax even if they pollute quite a lot, yet they may be charged a high tax while effectively reducing their emissions. Hence, the motivation to expend costs and efforts to reduce emissions through capital investments to reduce pollution or using a transportation vehicle less is reduced. At the extreme, where σ^2 is very large, there is no incentive value at all to the charge because Q no longer depends in any meaningful way upon F.

A second problem arises when ζ is not linear. Under a direct emissions charge, all units of a pollutant are in a sense created equal. Each is taxed at the same rate. Under a nonlinear relation between emissions and the indirect activity, the shadow price being imposed upon the polluter for each unit of the emissions varies from unit to unit and perhaps from polluter to polluter. The result is likely to be insufficient or excessive pollution abatement, depending in part on whether the function ζ is concave or convex. Incentives for abatement will also vary at the margin from polluter to polluter. Such arbitrariness across individual vehicle operators and owners (or across individual industries) could make such indirect pricing unpopular because it is viewed as unfair and politically controversial.

While direct emissions charges are generally regarded by economists and planners as the better choice in terms of the efficiency of the set of incentives created, in many situations they are simply not feasible or impractical. This may be because direct monitoring of the emissions is too expensive, because there exist no technologies for monitoring, because the direct pricing is politically unpopular for some reason, or because evasion of the monitoring by polluters is too easy.

This problem is particularly relevant in transportation vehicle emissions. One might imagine some sort of tamper-free monitor that someday might be installed in a car's engine that measures pollutants directly and reliably, but no such technology exists at the moment. Accordingly, only indirect pricing methods are feasible today.

This means that pollution emissions pricing for transportation vehicles may belong unavoidably to the realm of the second best, that is, policies that approximate the policy goal and the incentive structure as closely as possible, given imperfect and limited methods for achieving it. Formally, suppose that the social costs or damages from each unit emitted of some pollutant have been determined in some judicious manner (such as by elected legislators) to be equal to C. A direct emissions charge will be socially optimal if it imposes a charge of C on the emissions themselves.

An indirect charge, such as on the fuel consumed, also creates an incremental effect related to each unit of emission, but an indirect effect. Other things equal, the polluter is charged a higher tax when pollution increases, but the charge is levied against some variable other than the emissions themselves. That activity upon which the tax or charge is levied is presumed to be correlated with emissions. The tax creates an incentive to avoid doing that activity which is taxed, and because that activity is correlated with emissions, there is indirect incentive to avoid the emissions.

For example, again consider a fuel charge. Fuel consumption is correlated with vehicle emissions: The more fuel that is consumed, the greater the output of pollutants. By taxing fuel, its consumption will be reduced. Vehicle users will have incentive to purchase less fuel, and so may drive less and may use public transportation more and may keep their engines better tuned. The indirect result is the production of fewer emissions.

Even though administratively charged as an add-on cost to the price of fuel, one can formally define the value of the fuel charge as an indirect emissions tax equal to:

$$\Psi = \mathrm{dCosts}/\mathrm{d}Q = [\partial\mathrm{Costs}/\partial F] \cdot [\partial F/\partial Q] = [\partial\mathrm{Costs}/\partial F]/\zeta' \qquad (4.1)$$

where costs refer to total activity costs to the polluter.

Ψ is the effective indirect emissions charge, implicit and embedded within the fuel tax. Ψ may be stochastic, and will be so whenever either of the two terms on the right-hand side of (4.1) is stochastic. The last term on the right-hand side will be stochastic whenever $\sigma^2 > 0$, that is, where ε introduces noise into the relationship between fuel consumption and the amount of emissions. Such randomness could also result if the emissions from a unit of fuel depend on other factors and decisions taken by the polluter, not included in equation (4.1) and unaffected by the fuel charge. In the case of a simple fuel charge, then the other term $[\partial\, \mathrm{Costs}/\partial F]$ reduces to that charge itself.

Ideally, Ψ should be set equal to C, the socially optimal emissions charge, the charge that creates the optimal incentive for pollution abatement. But this may not be possible because of the stochastic nature of the relationship between F and costs.

In sum, once direct pricing of emissions is ruled out, indirect pricing alternatives must be compared and evaluated, using an efficiency criterion for judgment and ranking in order to identify the best of the second-best alternatives. To date, few formal methodologies for achieving such rankings have been developed, and this is an important area for research.

4.5.2 Examples of Indirect Pricing Methods

We do not propose to resolve the debate here over the question of which indirect pricing method is best of the second best. Instead, let us simply review which indirect pricing policy alternatives are commonly discussed in the transportation literature or by policy makers. Among these are car and fuel taxes, road user charges, parking pricing, vehicle miles of travel (VMT) fees, and smog fees. The goals of these are to internalize externality costs and to raise the effective user costs of automobile travel. It is quite common to supplement any of these with subsidization of public transit, in order to encourage the use of alternatives to private automobiles.

Among indirect policy methods used to achieve motor vehicle emissions abatement, perhaps the most common is the fuel charge. Fuel charges are taxes on fuel, and are commonly used in the United States, Europe, and other countries. While thought to reduce energy consumption, they are generally advocated as an environmental measure. They are part of the set of policies designed to raise the effective cost of automobile travel, including parking fees, automobile purchase taxes, and road-use charges. All have an environmental impact to the extent that they serve to dissuade travelers from using private automobiles and switching to public transit, or reducing travel altogether. Fuel charges may also serve as incentive for improved fuel economy.

There has been some research on the environmental impact of fuel use and fuel charges. Schipper et al. [5] analyze how fuel prices and other factors are linked with automobile fuel consumption, travel, and fuel economy. Based on data from Europe, the United States, and Japan, it was found that there is a clear relationship between fuel prices and both on-road fleet fuel intensity and new-car test fuel intensity. *Fuel intensity* is defined as the ratio of fuel consumed to distance traveled. It is measured in liters per 100 kilometers (in most European countries) or gallons per mile. They also found that there is a clear relationship between fuel prices in the countries examined and:

a. Fuel use per car;
b. Fuel use per capita;
c. Fuel use per kilometer.

Other econometric research has reached similar conclusions. (For a review of this research, see Sterner [59].)

Fuel prices are often considered an important causal factor in determining transit ridership. Tolls and road pricing also seem to affect both car and transit travel (Lind and Lindkvist [60] and TØI [61]), as do parking policies and costs (Jansson [62] and Nielsen [63]). Automobile taxation affects car ownership, and also the selection of automobile characteristics, such as size and engine power, which have an effect on emissions (Schipper et al. [5]).

There have been several plans implemented, such as *feebates*, which are under discussion in California and designed to stimulate the purchase of less fuel-intensive cars through reducing their taxation or through their subsidization. Such programs have been implemented in Denmark and Norway.

Virley [64] examined the impact on CO_2 emissions in the United Kingdom of increases in fuel prices and found that real fuel prices would have to double between 1990 and 2000 in order to force the level of CO_2 emissions back to their 1990 levels. It was found that most of the abatement in the emissions would come from improved fuel economy and not from reduced mileage traveled.

Another form of indirect pricing of emissions is the VMT fee. These are flat fees paid by the driver for each mile driven, regardless of where or how that mile contributed to pollution or congestion. VMT fees have been examined in southern California (Cameron [65]). VMT fees are easier to monitor and implement than smog fees or user charges, which are hour specific and location specific. User charges vary by the level of congestion and pollution; they are considered to be first and foremost congestion-reduction techniques, and only secondarily emissions abatement tools.

Parking pricing, especially in central business districts of cities, has often been suggested as an indirect emissions abatement policy. It is thought to raise the real costs of automobile commuting and so can induce greater use of public transit or ridesharing. In the United States it is thought to offset the effects of employer-subsidized parking (Cameron [65]). One survey in southern California discovered that over 90 percent of employers provide free parking for their employees and that the effect of free parking on travel demand is greater than would be the effect of granting free gasoline (Wilson et al. [66]). In order to neutralize this, a parking cash-out policy is being implemented in some parts of California, where employers are required to offer a choice of a cash grant in lieu of free or subsidized parking. In Los Angeles, a city ordinance requires payment of $15 per month as a transit subsidy to employees who forgo free parking privileges. President Clinton has also included this idea as a central feature of his Climate Action Plan.

Other examples of indirect pricing policies designed to affect emissions would include:

1. Highway user charges and tolls.
2. Charges and taxes on parking fees, especially in central business districts of urban areas.
3. Congestion fees.
4. Subsidization of public transport (Feitelson [67] and Site and Filippi [68]).

4.6 CONCLUSIONS

Vehicle emissions are a very large part of overall air quality problems, especially in urban areas. They are a major source for a number of pollutants and greenhouse gases. The methodologies generally most preferred by economists and planners for emissions control and abatement, direct Pigouvian charges and marketable emissions permits, are not applicable to vehicle emissions, at least under current technology. This makes the subject of vehicle emissions and pollution both an important and an interesting policy issue.

Once direct emissions pricing methods are ruled out, remaining policy methods are administrative command-and-control regulations and indirect pricing methods. Both sets of policies are commonly used by policy makers in developed countries. Both sets of policies have problems, including the iron rule of unintended consequences. Very little research has been done in developing tools for ranking and comparing the efficiencies of these methods. Mobile-source emissions policy must be formulated within the framework of a broad multigoaled transportation planning policy, including land use planning, and must recognize the existence of complex trade-offs among the competing policy goals.

REFERENCES

1. Environmental Protection Agency, *National Air Pollutant Emissions Estimates: 1940–1990, Office of Air Quality*, North Carolina (1991).
2. DeCicco, J.M., "Projected Fuel Savings and Emissions Reductions from Light Vehicle Fuel Economy Standards," *Transportation Research A* **29(3)**, May, 205–228 (1995).
3. Ono, Y., "Japan's Sustainable Transport Policy Towards the 21st Century," *Transport Policy* **1(1)**, 32–42 (1993).
4. Hall, S., Zegras, C., and Rojas, H.M., "Transportation and Energy in Santiago, Chile," *Transport Policy* **1(4)**, 233–243 (1994).
5. Schipper, L., Steiner, R., Figueroa, M.J., and Dolan, K., "Fuel Prices and Economy," *Transport Policy* **1(1)**, 6–20 (1993).
6. Sperling, D., Setiawan, W., and Hungerford, D., "The Target Market for Methanol Fuels," *Transportation Research A* **29(1)**, January, 33–45 (1995).
7. Lee, D.R., "The Economics of Enforcing Pollution Taxation," *Journal of Environmental Economics and Management* **11(2)**, June (1984).
8. Barde, J., "Environmental Policy and Policy Instruments," in Folmer, H., Gaebel, H.L., and Opschoor, H., *Principles of Environmental and Resource Economics*, Aldershot: Edgar Alder (1995).
9. Baumol, W.J., and Oates, W.E., *The Theory of Environmental Policy*. Cambridge: Cambridge University Press, 1988.
10. Downing, P.B., and White, L.J., "Innovation in Pollution Control," *Journal of Environmental Economics and Management* **13(1)**, January, 18–29 (1986).
11. Folmer, H., Gaebel, H.L., and Opschoor, H., *Principles of Environmental and Resource Economics*. Aldershot: Edgar Alder, 1995.
12. Hahn, R.W., "Economic Prescriptions for Environmental Problems: How the Patient Followed the Doctor's Orders," *Journal of Economic Perspectives* **3(2)**, Spring, 95–114 (1989).
13. Markandya, A., and Richardson, J., *Environmental Economics*. London: Earthscan Publications, 1992.
14. Mills, E.S., *The Economics of Environmental Quality*. New York: Norton, 1978.
15. Opschoor, J.B., and Vos, H.B., *The Application of Economic Instruments for Environmental Protection in OECD Countries*. Paris: OECD, 1989.
16. Stavins, R.N., "Harnessing Market Forces to Protect the Environment," *Environment* **31**, 28–35 (1989).
17. Tietenberg, T.H., "Economic Instruments for Environmental Regulation," *Oxford Review of Economic Policy* **6**, (1990).
18. Tietenberg, T.H., *Innovation in Environmental Policy*. Aldershot: Edward Elgar, 1992.
19. Tietenberg, T.H., *Environmental and Natural Resources Economics*. New York: Harper Collins, 1992.
20. Tietenberg, T.H., *Economics and Environmental Policy*. Aldershot: Edward Elgar, 1994.
21. Roberts, M.J., and Spence, M., "Effluent Charges and Licenses Under Uncertainty," *Journal of Public Economics* **5**, (1976).
22. Goulder, L.H., "Energy Taxes: Efficiency Effects and Environmental Implications," *National Bureau of Economic Research Working Paper 4582*, December (1993).
23. Hahn, R.W., "Marketable Permits: Lessons from Theory and Practice," *Ecology Law Quarterly* **16**, 361–406 (1989).
24. Parry, I.W.H., "Optimal Pollution Taxes and Endogenous Technological Progress," *Resource and Energy Economics* **17(1)**, May, 69–85 (1995).
25. Shapiro, M., and Warhit, E., "Marketable Permits: The Case of Chloroflurocarbons," *Natural Resource Journal* **23**, 577–91 (1983).
26. Tahvonen, O., "International CO2 Taxation and the Dynamics of Fossil Fuel Markets," *Interna-*

tional Taxation and Public Finance **2(2)**, August, 261–78 (1995).
27. Tietenberg, T.H., *Emissions Trading: An Exercise in Reforming Pollution Policy*. Washington, DC: Resources for the Future, 1985.
28. Weitzman, M.L., "Prices vs. Quantities," *Review of Economic Studies* **41**, 50–65 (1974).
29. Swaney, J.A., "Market vs. Command and Control Environmental Policies," *Journal of Economic Issues* **26**, June, 623–33 (1992).
30. Pigou, A.C., *The Economics of Welfare*. London: Macmillan, 1920.
31. Barthold, T.A., "Issues in the Design of Environmental Excise Taxes," *Journal of Economic Perspectives* **8(1)**, Winter, 133–151 (1994).
32. Carlton, D., and Loury, G., "The Limitations of Pigouvian Taxes as a Long-Run Remedy for Externalities," *Quarterly Journal of Economics* **95(5)**, 559–66 (1980).
33. Johansson, B., "Strategies for Reducing Emissions of Air Pollutants from the Swedish Transportation Sector," *Transportation Research A* **29(5)**, September, 371–385 (1995).
34. Wirl, F., "Energy Pricing When Externalities are Taxed," *Resource and Energy Economics* **15(3)**, September, 255–70 (1993).
35. Xepapadeas, A.P., "Observability and Choice of Instrument Mix in the Control of Externalities," *Journal of Public Economics* **56(3)**, March, 485–98 (1995).
36. Xepapadeas, A.P., "Optimal Taxes for Pollution Regulation: Dynamic, Spatial and Stochastic Characteristics," *Natural Resource Modeling* **6(2)**, Spring, 139–70 (1992).
37. Carson, T.N., "An Experimental Investigation of the Seller Incentives in the EPA's Emissions Trading Auction," *American Economic Review* **85(4)**, September, 905–922 (1995).
38. Hahn, R.W., and Noll, R., "Designing a Market for Tradeable Emmissions Permits," in W.A. Magat, ed., *Reform of Environmental Regulation*. Cambridge: Ballinger (1982).
39. Maleug, D.A., "Welfare Consequences of Emission Credit Trading Programs," *Journal of Environmental Economics and Management* **18(1)**, January, 66–77 (1990).
40. Montgomery, W.D., "Markets in Licenses and Efficient Pollution Control," *Journal of Economic Theory* **5**, 395–418 (1972).
41. Kort, P.M., "The Effects of Marketable Pollution Permits on the Firm's Optimal Investment Policies," *Central European Journal for Operations Research and Economics* **3(2)**, 139–55 (1995).
42. Ledyard, J.O., and Szakaly-Moore, K., "Designing Organizations for Trading Pollution Rights," *Journal of Economic Behavior and Organization* **25(2)**, October, 167–76 (1994).
43. Atkinson, S., and Tietenberg, T.H., "Market Failure in Incentive-Based Regulation: The Case of Emissions Trading," *Journal of Environmental Economics and Management* **21(1)**, January, 17–31 (1991).
44. Dellucchi, M.A., Green, D.L., and Wang, M.Q., "Motor-Vehicle Fuel Economy: The Forgotten Hydrocarbon Control Strategy?" *Transportation Research A* **28(3)**, May, 223–244 (1994).
45. Green, D.L., "CAFE or Price? An Analysis of the Effect of Federal Fuel Economy Regulation and Gasoline Price on New Car MPG, 1978–89," *Energy Journal* **11(3)**, July, 37–57 (1990).
46. Kleit, A.N., "The Effect of Annual Changes in Automobile Fuel Economy Standards," *Journal of Regulatory Economics* **2(2)**, June, 151–72 (1990).
47. Becker, G.S., "Fuel Efficiency Standards: An Idea Whose Time Has Passed," *Business Week*, September 2, 11 (1985).
48. Crandall, R.W., "Policy Watch: Corporate Average Fuel Economy Standards," *Journal of Economic Perspectives* **6**, Spring, 171–80 (1992).
49. Crandall, R.W., and Nivola, P.S., "The Extra Mile: Rethinking Energy Policy for Automobile Transportation," Policy Paper, Washington, DC: Brookings Institution (1995).
50. Krupnick, A.J., Walls, M.A., and Collins, C.T., "Global Warming and Urban Smog: The Cost Effectiveness of CAFE Standards and Alternative Fuels," *Energy Journal* **14(4)**, 75–97 (1993).
51. Khazoom, D.J., "Fuel Efficiency and Automobile Safety: Single-Vehicle Highway Fatalities for Passenger Cars," *Energy Journal* **15(4)**, 49–101 (1994).
52. Giuliano, G., Hwang, K., and Wachs, M., "Mandatory Trip Reduction in Southern California: First Year Results," *Working Paper #98*, University of California Transportation Center (1992).
53. Lee, A.M., and Meyburg, A.H., "Resource Implications of Electronic Message Transfer in Letter

Post Industry," *Transportation Research Record* **812**, 59–64 (1981).
54. Mokhtarian, P.L., "Telecommuting and Travel: State of the Practice, State of the Art," *Transportation* **18**, 319–42 (1991).
55. Nilles, J.M., "Traffic Reduction by Telecommuting: A Status Review and Selected Bibliography," *Transportation Research A* **22A**, 301–307 (1988).
56. Salomon, I., "Telecommunications and Travel Relationships: A Review," *Transportation Research* **20A(3)**, 223–38 (1986).
57. Plaut, P.O., "Transportation-Communications Relationships in Industry," *Transportation Research A* **31A (6)**, 419–29 (1997).
58. Henderson, D.K., and Mokhtarian, P.L., "Impacts of Center-Based Telecommuting on Travel and Emissions: Analysis of the Puget Sound Demonstration Project," *Transportation Research D* **1(1)**, 29–45 (1996).
59. Sterner, T., *The Pricing of and Demand for Gasoline*, TFB Rapport 9, Stolkholm: Swedish Transport Research Board, 1990.
60. Lind, F., and Lindqvist, E., *Afgifter i Trafiksystem*. Stolkholm: Transek, 1989.
61. TØI, *Reisevaner i Oslo-Omradet: endringer i reisvaner i Oslo og Akershus fra 1977 til 1990*, TØI Rapport 0096/1991, Oslo: Transportokonomisk Institutt, 1991.
62. Jansson, J.O., *Parkerings—och trafikpolitik for humanare innerstadsmiljo*. Stolkholm: Byggforskiningsradet, 1987.
63. Nielsen, H., "Redegorselse: erfaringer med parkeringsafdifter," Copenhagen: *Stadsingeniorens Direktorat*, Vejkontoret (internal memo) (1991).
64. Virley, S., "The Effect of Fuel Price Increases on Road Transport CO_2 Emissions," *Transport Policy* **1(1)**, 43–48 (1993).
65. Cameron, M., "A Consumer Surplus Analysis of Market-Based Demand Management in Southern California," *Transport Policy* **1(4)**, 213–220 (1994).
66. Wilson, D., Schipper, L., Bartlett, S., and Tyler, S., "Residential Energy Use Policies and Programs in Five OECD Countries." Berkeley: Lawrence Berkeley Laboratory (1989).
67. Feitelson, E., "The Potential of Rail as an Environmental Solution: Setting the Agenda," *Transportation Research A* **28(3)**, 209–21 (1994).
68. Site, P.D., and Filippi, F., "Bus Service Optimization and Car Pricing Policies to Save Fuel in Urban Areas," *Transportation Research A* **29(5)**, September, 345–358 (1995).

PART III

Spark-Ignition Engines

CHAPTER 5

Introductory Chapter: Overview and the Role of Engines with Optical Access

Richard Stone
Department of Engineering Science, University of Oxford, Oxford, United Kingdom

ISBN: 0-12-639855-0
$25.00

5.1 INTRODUCTION

This chapter provides an overview of the factors that are driving spark-ignition (SI) engine research with an overview of the subsequent chapters in the spark-ignition engine section. Illustrations are presented of the use of engines with optical access for studying the mixture preparation and combustion processes in spark-ignition engines.

It will be argued here that engines with optical access provide additional information that is invaluable when it comes to designing engines (and their control systems) that are capable of meeting the increasingly strict emissions legislation.

5.1.1 The Impact of Emissions Legislation on Spark-Ignition Engine Development

The parallel demands for reducing both fuel consumption and emissions continue to provide spark-ignition engineers and scientists with a challenge.

Emissions legislation has a major impact on engine development strategy. It is much easier for legislators to decree ever more stringent controls on emissions from spark-ignition engines than it is to implement other strategies, such as improved traffic management and alternatives to personal transport in congested cities. It is also much easier to impose legislation on new vehicles than it is to reduce emissions from the existing vehicle population. Concern about the environment is not likely to decrease, and as research continues, new concerns arise.

Originally with hydrocarbon emissions only the total quantity was of interest. Now that there is a better understanding of the environmental chemistry by which hydrocarbon species form ozone, emissions legislation is being developed that takes the relative reactivity of the different hydrocarbon species into account. Currently particulate emissions are not legislated for in SI engines. However, there are emissions of submicron particles, and these are of a particular health concern because of their ability to travel far into the lungs.

Of great current interest is the potential for direct-injection spark-ignition (DISI) engines to achieve the specific output of gasoline engines, yet with fuel economy that is comparable to diesel engines. Will these engines live up to current expectations, or will the outcome be the same as two-stroke engines for passenger cars? Mitsubishi has already launched such a DISI engine [1, 2], and Toyota [3] and Ricardo [4] are but two other organizations active in developing DISI systems (Mercedes, Ford, and GM among others are also known to be developing these engines). DISI engines operate at stoichiometric near full load, with early injection (during induction) so as to obtain a nominally homogeneous mixture. In contrast, at part load they operate with injection during the compression stroke. This enables the mixture to be stratified, so that a flammable mixture is formed in the region of the spark plug, yet the overall air-fuel ratio is weak (and the three-way catalyst operates in an oxidation mode). However, in order to keep the engine-out NO_x emissions low, it is necessary to be very careful in the way the mixture is stratified. It is far from clear yet whether or not DISI engines will be able to satisfy increasingly stringent legislation (such as EURO IV), but the potential fuel economy benefits are such that engine manufacturers cannot risk ignoring this technology. The Mitsubishi engine has a swept volume of 1800 cm^3, and it may be very difficult to make this technology work in smaller displacement engines. Even if DISI engines are possible in larger engine sizes, the higher cost of the fueling system still has to be justified.

Future fuels for spark-ignition engines are also likely to become more diverse than at present, and this is in part a consequence of emissions legislation. Only natural gas fueled vehicles can currently satisfy the U.S. EZEV(Equivalent Zero Emissions Vehicle) legislation. Other causes of diversity are the use of alcohol-based fuels derived from renewable sources. (When methanol is derived from hydrocarbon sources, then the high energy cost associated with this limits the use of methanol to special cases.) Fuel cell technology has also been developed enormously, and these can be expected to compete with heat engine powered vehicles in specialist markets (where a high premium is placed on low emissions). The requirement of hydrogen for fuel cells has hitherto led to bulky on-board storage or reformation of methanol. However, recent work (for example, by Johnson Matthey developing its hot-spot technology) is expected to result in a hydrogen output of more than 1 kW/litre-of-reformer [5]. This is a substantial reduction in bulk. The ready availability of on-board hydrogen offers considerable potential for SI engines to operate with very lean mixtures. Hydrogen can also be used as a supplement to more conventional fuels, thereby extending their lean limit and increasing burning velocities.

In the case of spark-ignition engines, there is scope for using heat from the exhaust gases as an energy input to the fuel reformation process, thereby giving two potential benefits. First, the calorific value of the fuel is increased, and second, the hydrogen in the reformed fuel can be used to extend the lean mixture limit of gasoline [6].

Apart from the need to conserve fossil fuel reserves, concern about global warming is also leading to legislation for reduced fuel consumption. In this context, it is usually referred to as a carbon dioxide tax, since there is a fuel dependency on the carbon dioxide emissions per unit calorific value of the fuel. Although there are no grounds for complacency, it is interesting to note that as atmospheric modeling has become more refined, predictions of global warming have been reduced. Recent estimates [7] suggest that global warming will be between 0.2 and 0.4 K/decade with no change in current fuel usage patterns. This is a level which is quite difficult to distinguish from natural changes in the earth's temperature caused by phenomena such as variations in sun spot activity. The difficulty of interpreting observed climate changes adds importance to the need to model climate change. However, climate modeling is extraordinarily difficult for a number of reasons. First, the oceans act as huge thermal stores, and they also have the ability to absorb carbon dioxide. Second, the most significant of the greenhouse gases is water vapor, and it is thus necessary to model the hydrological cycle, in other words, how much water is in the atmosphere, and in what form (clouds affect the incoming and outgoing infrared radiation, and the water in clouds can be either water drops or ice crystals of varying size). Climate modeling is clearly an immensely difficult problem.

5.1.2 Overview of Part III

The chapters in Part III of this handbook provide a sound basis on which to understand the origins and means of control of emissions from spark-ignition engines. Simone Hochgreb provides an authoritative treatment of the sources of the major emissions in spark-ignition engines, namely, nitric oxides, carbon monoxide, and unburnt hydrocarbons. This chapter contains a critical review of the mechanisms by which these pollutants are formed, and the treatment of the material is characterized by full details of submodels that can be used for predictive modeling.

The chapter by Mark Dulger on Wankel engines provides a useful reminder that these engines are still being developed for a number of niche markets, such as military applications, motorbikes, automobiles (Mazda), and electrical generators, where its high specific output and few moving parts are an advantage. A Wankel engine has recently attained flight approval by the CAA, and research is continuing in many countries to improve such engines. Mark Dulger discusses the combustion and emission characteristics of Wankel engines, and this includes the contribution of rotor seal leakage to hydrocarbon emissions formation. It has also been shown that leakage past the apex tip seal affects the flow field in the combustion chamber [8]. Rotor seal leakage and wear are major considerations in Wankel engines, and indeed there have been proposals to eliminate the seals altogether.

The chapter by Brian Milton on emission control for spark-ignition engines is very broad in its content; in particular, there is a comprehensive treatment of the topic of non-steady-state operation (both short-term transients such as speed and load change, and longer-term transients like engine warm-up). As would be expected there is a discussion of catalyst technology and the prospects for improvement, such as fast light-off and an ability to reduce NO in an oxidizing environment. Another important topic that is covered is the mixture preparation process, which of course has a particular impact on non-steady-state operation. Among other topics included in this chapter are the means of reducing fuel consumption or carbon dioxide emissions, for example, direct-injection stratified charge engines, and the use of variable valve timing.

In conclusion, these chapters provide an excellent basis for researchers to become familiar with existing knowledge, and to identify the issues where further work is needed.

5.2

ENGINES WITH OPTICAL ACCESS

Engines with optical access through the piston crown are widely used in studies of flow, mixture preparation, and combustion. A typical arrangement is shown in Figure 5.1. Above the conventional piston crown is a slotted tubular extension that supports the window in the piston crown. The piston slots enable an elliptical mirror to be fixed, so that an image of the combustion chamber can be viewed through the opposite slot. The piston crown window is typically 75 percent of the bore diameter. The piston rings in the combustion chamber are often based on graphite and fiber-reinforced polymers, so as to eliminate the need for any lubrication. Such rings can run on cast-iron cylinder liners or indeed an annular window. The lower piston will have a conventional oil control ring to avoid oil contamination of the mirror and piston window.

Fig. 5.1.

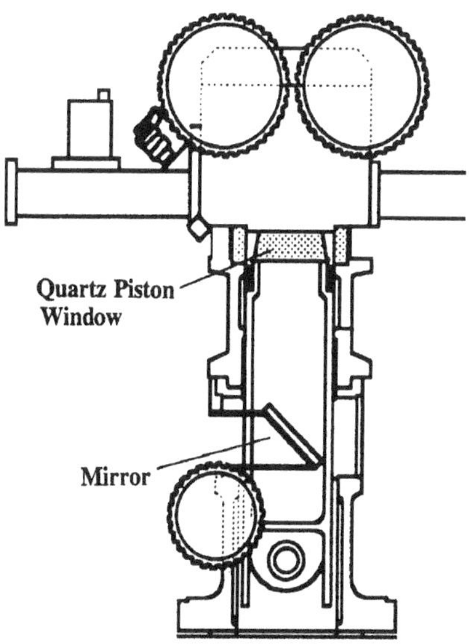

Typical arrangement for optical access through the piston, and an annulus below the cylinder head.

The windows are usually made of fused silica (often referred to as quartz) because of its high strength and low coefficient of thermal expansion, which lead to a good resistance to thermal shock. However, fused silica is very brittle, so great care is needed in the installation and support of the fused silica windows. Fully annealed copper gaskets are a widely used method in firing engines, while silicone sealant is sometimes used in nonfiring applications. The choice of window material is governed by the radiation wavelengths that are of interest. Sapphire is used when ultraviolet radiation is being studied, while germanium or silicon is used when infrared radiation needs to be transmitted [10].

Engines have been built with full or partial optical access in the bore by means of annular fused silica windows. When the optical access is the full length of the stroke, then operation is usually confined to nonfiring or skip-firing tests. An annular window causes substantial optical distortion, but Reeves et al. [11] demonstrate how it is possible to correct for this. This particular study used particle image velocimetry (PIV) for whole flow field studies. Flow measurement techniques will not be discussed further here, but a comprehensive overview has been published by Adrian [12]. The alternative to correcting for optical distortion is to eliminate it, either by the use of plane windows set into the bore, or by means of an engine that has a rectangular piston [13].

5.3 HIGH-SPEED PHOTOGRAPHY

For high-speed photography a choice has to be made between conventional film cameras and digital (CCD) cameras. It will be argued here that when it is necessary to have a large number of consecutive images, it is still necessary to use a film-based system. The discussion that follows here is based on color photography, since the spectral information is often of great significance.

5.3.1 High-Speed Cine Photography

Rotating prism cine cameras have been used for many decades. The film runs continuously through the camera, and the rotating prism causes the image to move with the film for a limited period, during which time a rotating shutter is also open. Typically, rotating prism cine cameras use 16 mm cine film with framing rates of up to 10,000 pictures/second (pps) at full frame (7.5 $\times$ 10 mm), with 40 frames/foot. The option for using half-frame at double the framing rate is usually not very useful, since this elongates the image. Time is obviously required for the camera to accelerate to the desired framing rate, and at 6000 pps this may use the first 70 feet of a 100- or 400-foot reel of film. The advantage of a rotating prism cine camera is that the filming is continuous for several consecutive cycles. For example, at 6000 pps it is possible to obtain 12 consecutive combustion cycles of a

Fig. 5.2.

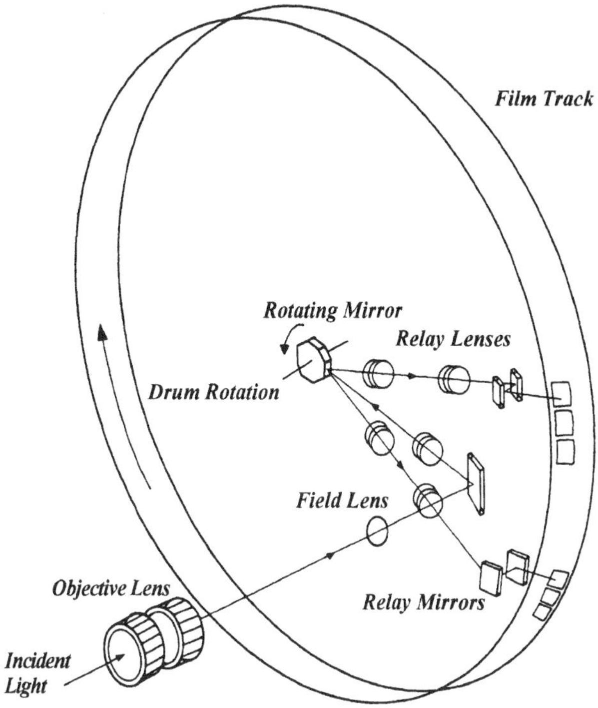

The optical arrangement of a Cordin rotating drum camera.

four-stroke engine running at 1000 rpm (with 1 frame/°ca), and this can be useful in studying cycle-by-cycle variations in SI engine combustion.

Rotating drum cameras are an alternative to rotating prism cine cameras, but with a limited number of images on each film. For example, the Cordin 350 rotating drum camera uses a standard length 35 mm film, and records 224 frames in two columns (each 7.5 × 10 mm) at up to 35,000 pps. The use of a standard 35 mm film means that specialized processing is not required. Figure 5.2 illustrates the optical arrangement. The film is carried on the inside of the rotating drum, and the rotating mirror reflects the light through the relay lenses at regular intervals. The relay lenses focus the light image on the moving film, creating regularly spaced discrete images. The obvious disadvantage of rotating drum cameras is that they cannot record consecutive combustion cycles.

The choice of the film and its processing are also important, and the same arguments apply to both type of cameras. Color negative and reversal films are both readily available with a maximum sensitivity of about 400 ASA. The color reversal film has the advantage that the processed film can be projected to give true color images, but the exposure tolerance is much smaller than with color negative film. Once color negative film has been processed, then a print has to be made; this film then has the original colors in it. If the original negative film has been underexposed, then this can be corrected for when the print is made. The negative colors can also be reversed to give the original colors when the film is being transferred to video or being input to a computer.

The resolution of the image will depend on both the camera optics and the film, but it is invariably the grain size of the film which limits the resolution. The resolving power is expressed as line pairs/mm (lp/mm—the finest spacing at which black and white lines can be distinguished in the image). The camera's optical system resolution is likely to be about 40 lp/mm, and the resolution of the film should be no worse, but this does depend on its type and processing. In particular, when an underexposed film is forced (the developer stage of the process is extended, so as to double or quadruple the sensitivity), then there is an increase in the grain size, and a corresponding reduction in the resolution. On a 7.5×10 mm image, a resolution of 40 lp/mm would be the equivalent of 300 by 400 pixels. It should also be noted that the maximum effective aperture of the camera is likely to be determined by the internal optics of the camera, and not the maximum aperture of the lens.

5.3.2 Digital Photographic Systems

A typical three-color charge coupled device (CCD) video camera has a maximum framing rate of about 30 pps and 750×480 pixel resolution, but a shutter that can give an exposure of 0.1 ms or shorter. Such cameras are also very sensitive, with an ASA equivalent of about 4000. These CCD cameras usually have an 8-bit resolution for each of the three color channels, so that each image will require about 1 MB of storage. The output from a CCD video camera can be stored on videotape or logged to a (PC) workstation via a frame-grabber card; such a system is illustrated in Figure 5.3 [14].

Fig. 5.3.

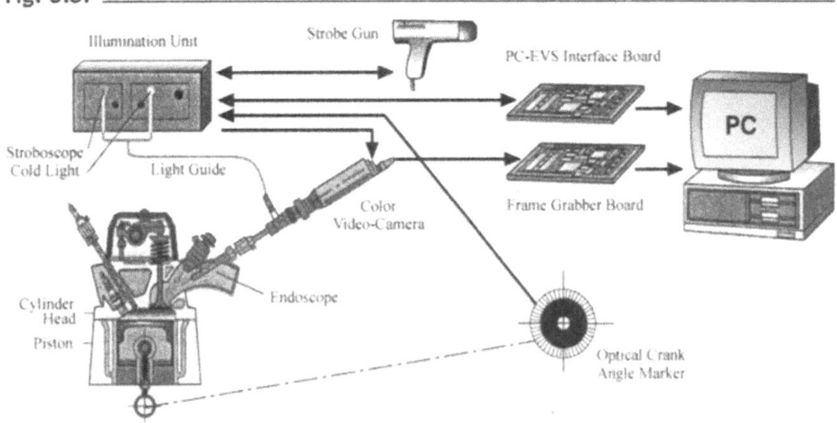

Example of a CCD color video camera-based combustion video system [14], courtesy of AVL.

Table 5.1

Comparison of Color Imaging Technologies; the Data Are Intended to be Indicative of Typical Systems. (Note: The Image Converter System is Monochrome)

Imaging System	Framing Rate (images/second)	Sensitivity (ASA)	Capacity (frames)	Resolution (pixels)
High-Speed Cine	10,000	1,000	10,000	300 × 400
Standard Video	30	4,000	25	480 × 750
High-Speed Video	1,000	100	5,000	384 × 512
Image Converter	1,000,000	>10,000	20	512 × 752

A 200 MHz Pentium PC with appropriate hardware is typically capable of grabbing 25 images/s. The number of images that can be stored is a function of the hard disk (or other storage media) capacity, but for a large numbers of images, it is probably most convenient to record the images on videotape for subsequent selective frame grabbing. When a videotape system is being used, it is best to use a professional system in which the three color channels are recorded separately.

High-speed CCD systems are at least an order of magnitude more expensive, and are currently only capable of recording about 1000 pps, with a 512 × 384 pixel resolution. High-speed CCD systems have dedicated digital memory, and typically 1000 to 5000 images can be stored. However, the sensitivity of the camera is low and if image intensifiers are used, then the output is restricted to a monochrome image. Image converter cameras are also limited to monochrome, and combine very high framing rates (over 10^6 images/second) with high sensitivity (equivalent to over 10,000 ASA), but with a very limited number of images (20 or so). Table 5.1 summarizes typical performance figures for different imaging systems.

Digital CCD systems invariably have an 8-bit output for each of the three color channels (red, green, and blue), while standard video cameras usually have analog outputs (although some professional cameras are available with digital outputs). The analog outputs are processed by frame grabber cards which have 8-bit resolution for each of the three colors. Conventional film has the equivalent of an analog output, but when the image analysis is digital, then the resolution will almost certainly be 8 bits, since this is determined by the scanner and CCD cameras/frame grabber resolution. The spatial resolution of frame grabbers varies but is usually linked to the monitor resolution, such as SVGA with a 1024 × 678 pixel resolution. It should also be noted that the aspect ratios of the imaging/display systems differ; for example, high-speed video has an 0.75 aspect ratio, while for standard video it is 0.64.

Once an image has been digitized, a wide range of software is available for image analysis. Typically this can be used for detecting the flame front position by setting threshold levels, with outputs available of the flame front coordinates, area, and perimeter. The perimeter on successive images might then be superimposed, with false coloring of the areas between successive perimeters. Other software options that are useful include particle tracking, since there are frequently fuel droplets or other particles present.

Clearly, the high-speed CCD systems are not currently fast enough for IC engine studies, so much use has been made of CCD video cameras, with one image being recorded from each engine cycle (25 pps corresponding to 3000 rpm for a four-stroke engine). By externally gating the camera, it is possible to image mixture preparation or combustion in one of two ways:

a. Looking at cycle-by-cycle variations in the spray (probably small) and the flame kernel, by grabbing every 720°ca.
b. Grabbing every $(720 + x)^\circ$crankangle to look at repetitive phenomenon (such as spray development) with an increment of x°crankangle.

AVL provides such a system commercially [14], which includes a facility to assemble images from representative cycles when there are cycle-by-cycle variations, thereby recreating a representative combustion sequence.

5.4 FLAME FRONT DETECTION

The simplest method of flame front detection is by direct observation, as discussed in the previous section. However, the flame front is three-dimensional and the image is only two-dimensional, so the image from direct observation will be recorded as the outer extremity of any part of the flame that falls within the field of view. A further complication is the depth of focus, which only results in objects within a certain distance range being within focus—an out-of-focus image of part of the flame front will compromise the in-focus part of the image, thereby hindering detection of the flame front. Nonetheless, direct observation is very useful, since in addition to recording the flame front, there is likely to be information on droplet/particle trajectories and the combustion of fuel that is still in the liquid state.

There is, of course, some difficulty in defining the flame front. The chemical reactions that comprise combustion lead to radiation over a wide range of frequencies (not all visible). Reacting species and heat both diffuse, so the flame front is a zone of finite width in which there is also a high-temperature gradient (at least 1000 K/mm). The flame front can thus be detected by the presence of a temperature gradient, or the emission/absorption of radiation at a defined wavelength by specific species.

The temperature gradient at the flame front leads to a density gradient, which in turn causes a change in the refractive index. The change in refractive index can be utilized by a schlieren system to visualize the flame front, albeit as a monochrome image. Comprehensive details of how the schlieren system works and the necessary optical arrangements can be found in many texts, for example, Merzkirch [15]. What is required is either a through-the-chamber optical path, or a mirror inside

Fig. 5.4.

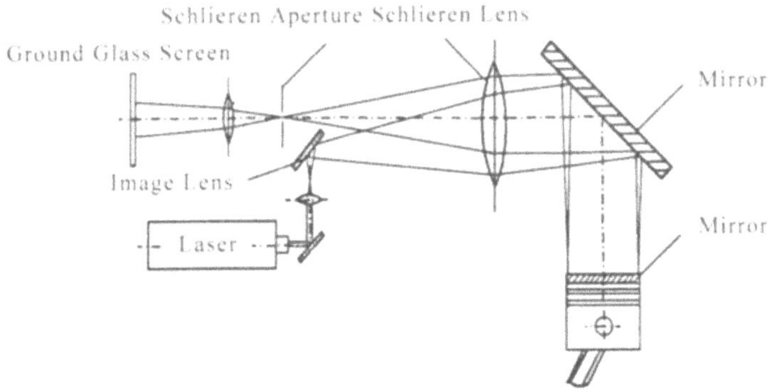

Optical arrangement for a schlieren system that uses a single window, but a piston crown mounted window, adapted from [16].

Fig. 5.5.

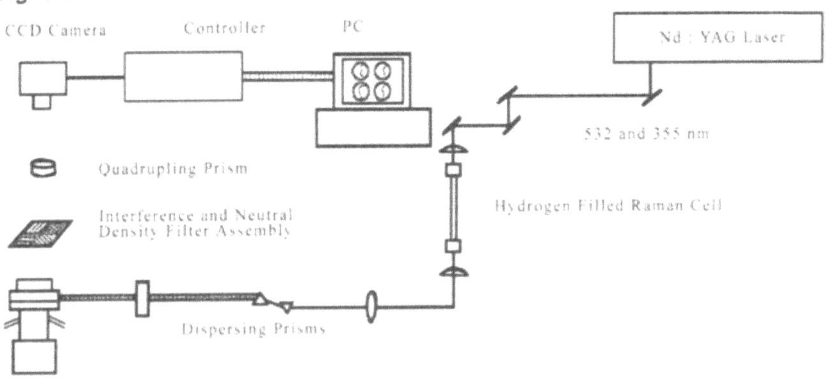

A multiple laser sheet imaging system that uses Mie scattering, adapted from [17].

the combustion chamber. An example of this (with details of the schlieren optical arrangement) is shown in Figure 5.4.

There are a number of techniques for providing unambiguous two-dimensional flame images, and in general the principle is to use a cylindrical lens to form a sheet of laser light, with viewing normal to the sheet, to detect laser light absorption, or some other optical interaction.

One technique is to use Mie scattering, in which the laser light is reflected from particles that are larger than the wavelength of the light. More light is reflected from the unburnt gas because its density (and the concentration of the particles) is higher. Typically submicron titanium dioxide particles are used, produced by the hydrolysis of titanium tetrachloride. Hicks et al. [17] employed a system with four sheets of laser light (see Figure 5.5), each at different frequencies, so that by means of filters and a CCD camera, four simultaneous images were obtained from different planes. This system was limited to one set of images from a combustion event, but a single sheet of light was used in conjunction with a copper vapor laser and a rotating prism camera, to obtain up to 10,000 images/s. The four simultaneous planar images provide useful information about the flame shape in the third dimension. Burnt gas islands ahead of the flame front in one plane can be seen to be connected to the burnt gas in another plane.

Fiber optics enable two-dimensional imaging systems to be devised that do not require special engines. Witze [18] has devised a system in which fiber-optic cables are distributed around the circumference of a spark plug, with their axes parallel to the spark plug body. This enables studies to be made of the early stages of flame propagation, but there is obviously some ambiguity about the axial positioning of the measurements.

A tomographic technique has been implemented by Philipp et al. [19], using fiber optics embedded in a 2.5-mm thick cylinder head gasket; see Figure 5.6. Up to 150 channels are used, with measurements being taken at a minimum of 0.1°ca increments for up to 32,000 events. Good agreement has been demonstrated between the tomographic reconstruction and conventional flame front images, in an engine with through-the-piston optical access [19].

Fig. 5.6.

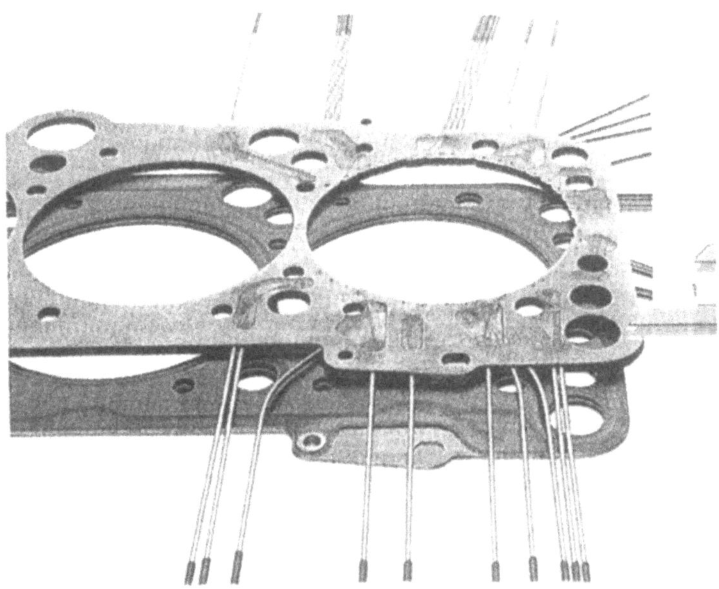

Optical fibers

A cylinder head gasket with fiber-optic sensors used for tomographic reconstruction of the flame front position, courtesy of AVL.

5.5 MIXTURE PREPARATION AND COMBUSTION DIAGNOSTICS

The following material is restricted to laser-based techniques because of their noninvasive nature. In an overview article such as this, it is not possible to give full details of the underlying physics of the various techniques, nor indeed the experimental details that can be found in papers. The intention here is to illustrate the types of measurement that can be made, and further details can be found in the associated papers. For details of the underlying physics there are useful monographs, such as Chigier [20], Taylor [21], and Eckbreth [22].

5.5.1 Fuel Sprays and Vapor Measurements

5.5.1.1 Fuel Concentration Measurements

Laser-induced fluorescence (LIF) can be used to detect the presence of both fuel vapor and liquid. Production gasoline and diesel fuels both fluoresce when subject to UV light, but the fluorescence is only due to some components within the fuels. These components tend to be the high boiling point constituents, and they will thus not be representative of the whole fuel. Since the particular components and their fluorescence are not likely to be characterized, then a more elegant approach is to use single-component reference fuels (or blends) with dopants that will fluoresce. A useful review of these techniques has recently been presented by Felton [23].

Dopants such as 3-pentanone and valeraldehyde will fluoresce in both the liquid and vapor phases with an almost identical spectra. However, by the use of exciplex (excited-state complex) dopants it is possible to obtain different spectra in the liquid and vapor phases. A fluorescent excited molecule (M^*, such as fluorobenzene) can be reacted with a second molecule (G, such as diethyl-methyl-amine) to form a second emitting molecule M-G^* (the exciplex). The exciplex (M-G^*) has an emission that is shifted toward the longer wavelengths (compared to M^*), and the relative concentrations of the dopants can be used to make the exciplex (M-G^*) the dominant emitter in the liquid phase, and the dissociated species (M^*) the dominant emitter in the vapor phase [23].

Great care is needed in the interpretation of fluorescence spectra because of the dependence on temperature and pressure, and the presence of molecules (such as oxygen) that can quench the vapor-phase fluorescence. A typical experimental configuration would utilize passband filters (centered on the wavelength of interest) and intensified CCD cameras interfaced to frame grabber hardware. Figure 5.7 shows the optical arrangement for planar laser-induced fluorescence (PLIF) in a firing spark-ignition engine [24]. Great care is needed with the calibration in order

Fig. 5.7.

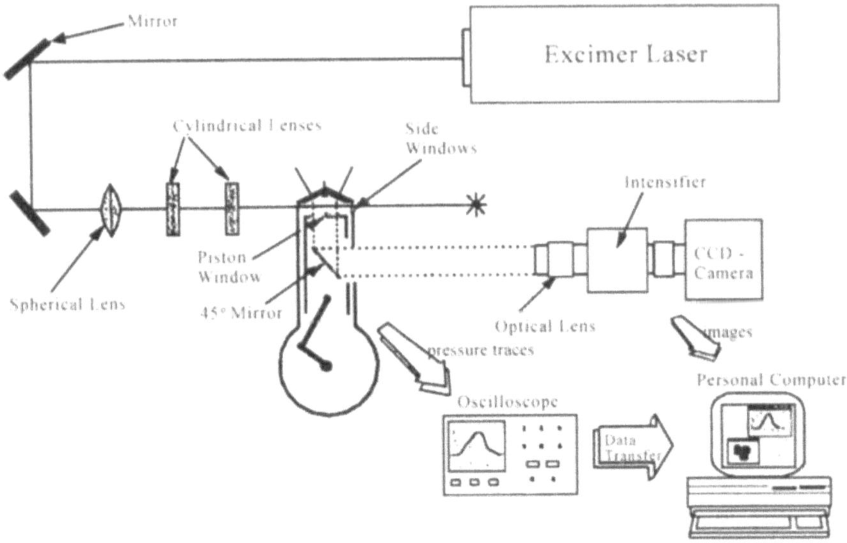

The optical arrangement for planar laser-induced fluorescence (PLIF) in a firing spark-ignition engine, adapted from [24].

to allow for effects such as spatial variations in the laser intensity and the collection efficiency of the detection system, and temporal laser intensity fluctuations. It is thus necessary to generate calibration images (when the engine cylinder is filled with vapor at a known concentration), and to record background level images. Corrections then have to be made to the data from the firing engine to correct for pressure and temperature effects so as to then produce two-dimensional plots of fuel concentration.

5.5.1.2 Droplet Size and Velocity Measurements

Droplet velocity measurements can be made by Laser Doppler Anemometry (LDA). Interference fringes are generated by coherent laser beams of equal intensity (see Figure 5.8), and a particle traveling through the interference fringes scatters intensity-modulated light which is detected by the photomultiplier. The rate of intensity variation is proportional to the velocity component normal to the fringes. If three noncolinear (and preferably orthogonal) interference fringes of different wavelengths are generated within the same measurement volume, then all three velocity components can be determined.

Phase Doppler Anemometry (PDA) [26] uses the same principle as LDA, but makes use of three photodetectors at different angles to the measurement volume. Each detector receives the same modulated signal as in LDA, but with a time delay. The time delay is a phase shift which is proportional to the particle diameter. Thus, simultaneous measurements can be made of the droplet sizes and

Fig. 5.8.

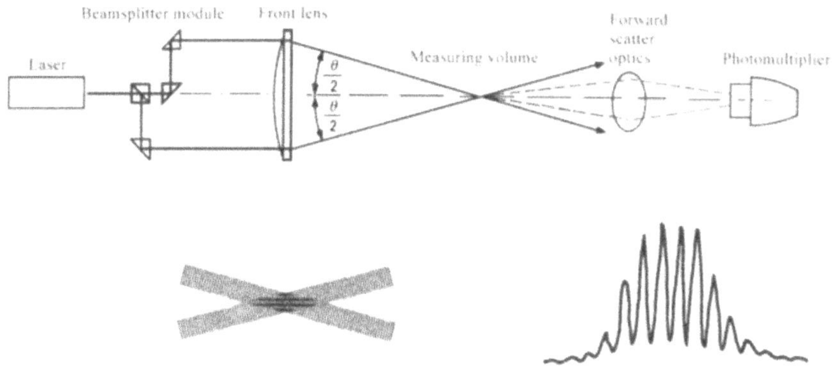

Optical arrangement for laser Doppler Anemometry (LDA), showing the interference fringe pattern, collection of light in the forward scatter mode, and a typical output from the photo multiplier [25].

velocities. Examples of such measurements are included in Figure 5.9, along with air velocity measurements from tests on a motored engine. The measurements are on a diametral plane, 10 mm below the cylinder head, 75°ca after the start of the induction stroke. The shaded area represents the piston position, and for the droplet size distributions the Sauter Mean Diameter (the total volume of the droplets divided by their area) is greater than the arithmetic mean diameter. Similar measurements have been reported for the Honda Vtec engine using a steady flow rig, with data presented for the axial and radial directions [28].

5.5.2 Temperature and Combustion Species Measurements

Laser-based diagnostic techniques lead to either incoherent or coherent signals coming from the region being probed. LIF (see Section 5.5.1.1) is an example with an incoherent signal. This means that the signal is radiated in all directions, and in order to obtain a reasonable signal level, the signal has to be collected from a large collection angle. However, optical access is comparatively simple, since a single window can be used for the laser probe beams, and for collecting the signal by means of back scatter. The poor signal-to-noise ratio that this implies also means that such measurements are prone to interference. However, the measurements can be two-dimensional when the excitation is by a laser light sheet.

Methods generating coherent (laserlike) signals tend to have stronger signals but require line-of-sight access and tend to have only point or line measurements. (Point measurements are made when the probe volume is defined by the intersection of simple focused beams, whereas line or two-dimensional measurements are made by the intersection of light sheets.) However, the resulting spectra are complex, with a nonlinear dependence on the phenomenon (temperature or species concentration) being measured.

Fig. 5.9.

Air axial velocities	Droplet axial velocities Ensemble window of 15°	Droplet data rate Ensemble window of 15°	Droplet size distribution Ensemble window of 15°

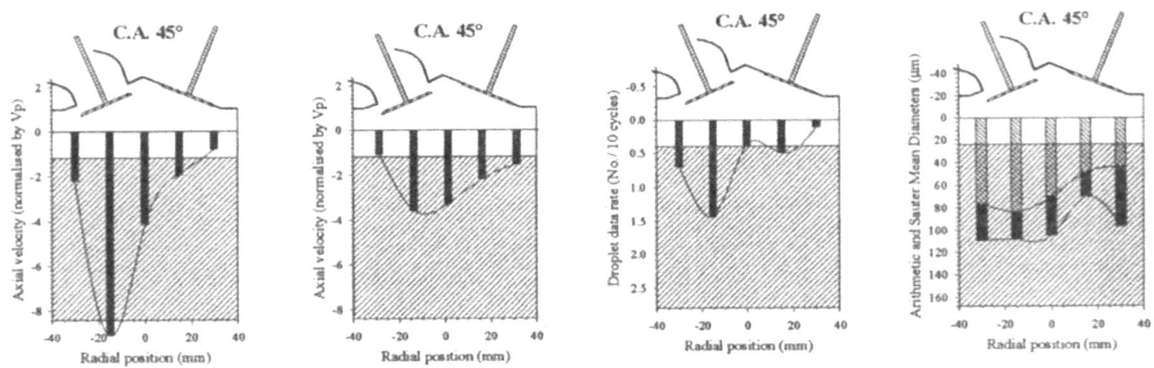

Air axial velocity measurements, and fuel droplet axial velocity and size measurements in a motored spark-ignition engine [27].

5.5.2.1 Incoherent Techniques

Laser-induced fluorescence (LIF) has already been discussed in the context of fuel concentration measurements, and because the fluorescence spectra has a temperature dependency, it can also be used for temperature measurements.

Elastic and inelastic scattering of the laser light source can both be used for temperature measurements, and inelastic scattering can be used for species measurements. Elastic (Rayleigh) scattering is when there is no energy exchange between the incident light and the target molecules, and the scattered light is unshifted from its initial frequency—it is thus not possible to identify individual species concentrations. With inelastic (Raman or LIF) scattering there is energy interchange between the incident photons and the molecule, leading to radiation shifted from the incident frequency by the characteristic vibrational frequencies of the molecule being excited. There will also be elastically scattered radiation at the incident frequency (the Rayleigh scattering). The use of "inelastic scattering" is perhaps confusing for a process in which energy is absorbed and then reradiated, but this is the accepted convention. The vibrational frequencies of the molecule are species and temperature dependent, and thus Raman scattering generates information on both species concentrations and temperature. The radiation downshifted in frequency is known as the *Stokes* spectrum, while the radiation upshifted in frequency is known as the *anti-Stokes* spectrum. The Raman spectra (Stokes and anti-Stokes) are specific to particular species and are linearly dependent on their concentration. The major disadvantage is the weak signal level.

5.5.2.2 Coherent Techniques

The most important coherent techniques are probably coherent anti-Stokes Raman spectroscopy (CARS) and degenerate four-wave mixing (DFWM). Both techniques can measure temperature and concentrations, but with CARS being most appropriate for major species, and DFWM being most appropriate for minor species.

Figure 5.10 shows a CARS beam arrangement. The crossing of the pump laser beams (ω_1) defines where the measurements are to be made. A laser beam at a different frequency (ω_2) known as the probe or Stokes beam, mixes with the pump beams to generate the CARS beam at a frequency of $(2\omega_1 - \omega_2)$. The frequency of the Stokes beam has to fall within the Stokes spectra in order to generate the anti-Stokes signal. The CARS spectra is generated by either using a broadband Stokes beam, which excites all the resonances simultaneously, or by scanning the Stokes beam over the relevant frequency range. The broadband (or multiplex) approach enables a time-resolved measurement to be made from a single laser pulse. However, the reduced intensity at each frequency means a lower signal-to-noise ratio. Nonetheless, the coherent (i.e., laserlike) nature of the signal beam and its strength (orders of magnitude greater than Raman scattering) facilitates the collection and detection of the signal, albeit with the need for line-of-sight optical access.

Fig. 5.10.

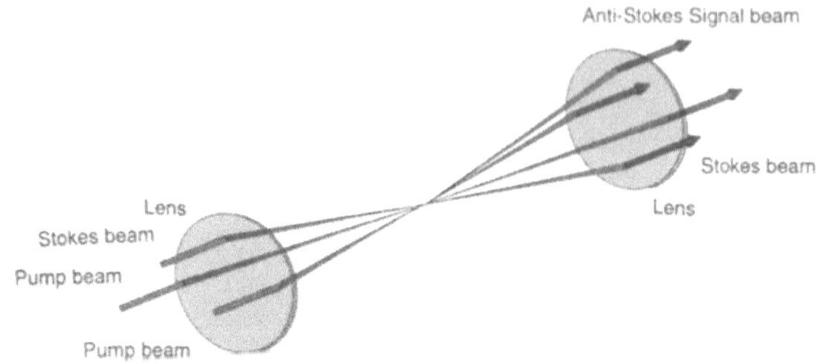

The optical arrangement for CARS (coherent anti-Stokes Raman spectroscopy).

Fig. 5.11.

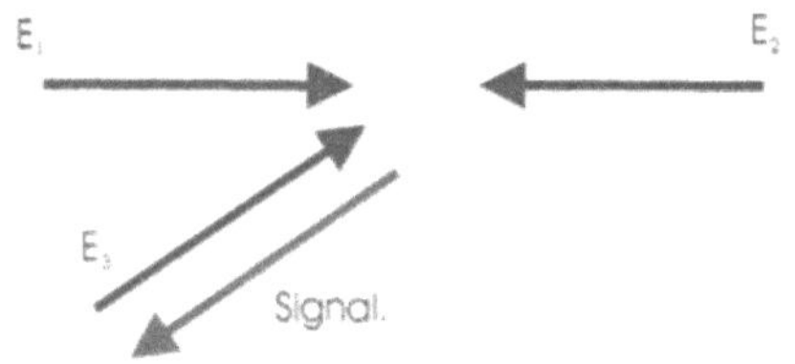

The counterpropagation arrangement for degenerate four-wave mixing (DFWM).

CARS requires phase matching of the pump and Stokes beams, and this dictates the angular separation that must be used between the pump and Stokes beams. When the pump and Stokes beams are generated from a single laser, a suitable arrangement is the crossed-beam arrangement of Figure 5.10 known as BOXCARS. As previously mentioned, CARS is best suited to major species, so for temperature measurements in internal combustion engines the nitrogen spectra is usually used. Typically, a YAG laser is frequency doubled to 532 nm to provide the pump beams, and part of this is used to optically pump a dye laser operating with a 3–4-nm bandwidth in the region of 607 nm to generate the Stokes beam. The interaction with the pump beam generates the anti-Stokes spectra in the region of 474 nm.

The advantage of degenerate four-wave mixing (DFWM) is that it can be used to detect the presence of minor species such as OH and NO. The optical requirements for DFWM are very similar to CARS, the difference being that the pump and Stokes probe beams are all at the same frequency (thereby simplifying the phase matching requirements), which is chosen to correspond with a resonance of the species being detected. The simplest arrangement for DFWM is the counterpropagating arrangement shown in Figure 5.11. The two pump beams (E_1 and E_2) are in-line and opposed, and the probe beam (E_3) generates a counterpropagating signal beam. However, signal collection requires care, since the same frequency components are present in the probe and signal beams. Another geometry, folded BOXCARS (shown in Figure 5.12), is also quite easy to define and leads to a

Fig. 5.12.

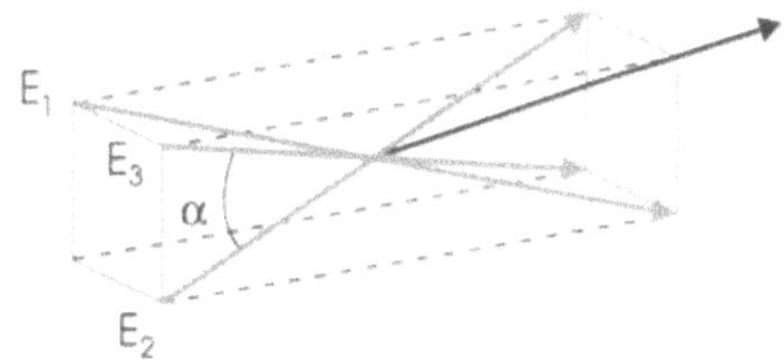

The folded BOXCARS arrangement for degenerate four-wave mixing (DFWM).

Fig. 5.13.

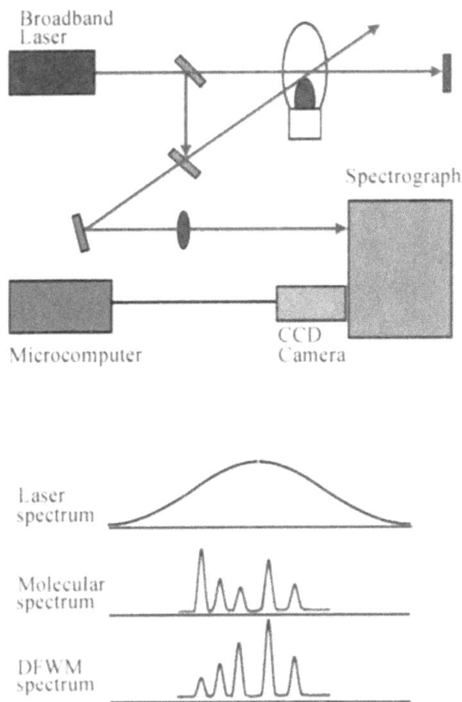

The use of a broadband laser with the folded BOXCARS beam arrangement for multiplex DFWM.

signal that is not counterpropagating, and thus it is easier to collect and analyze. The geometric requirement is for the pump beams and probe beam to lie on the diagonals of a rectangular box; the signal beam generated by any interaction then aligns with the fourth diagonal. The folded BOXCARS arrangement is also used with CARS, but because the pump and probe beams are at different frequencies, instead of the box being rectangular, it is a parallelepiped.

Figure 5.13 illustrates the use of a broadband laser so as to excite all resonances, thereby producing what is known as multiplex DFWM. The broadband laser encompasses all of the frequencies of interest, and the resulting spectrum depends on the molecular spectrum and the intensity of the laser spectrum at each frequency. DFWM is geometrically simpler to implement than CARS but has not been so widely used as it is a more recent technique. DFWM is more susceptible to interference effects that make the signal analysis more difficult and subject to greater uncertainties. In the case of broadband DFWM, the signal-to-noise ratio is compromised by scatter from the pump beams. DFWM can also be extended to two-dimensional imaging.

5.6 SOME APPLICATIONS OF ENGINES WITH OPTICAL ACCESS

Ever more demanding emissions legislation places an increased importance on combustion performance, especially in the period prior to catalyst light-off. Particularly troublesome can be the hydrocarbon emissions that are the consequence of ensuring a flammable air-fuel mixture following a cold start. In a warmed-up engine, a major source of unburned hydrocarbons can be outgassing from crevices as the pressure in the cylinder is falling: These have been studied by LIF.

5.6.1 Natural Light Photography

The photographs shown in Figure 5.14 are from experiments in which high-speed cine photography was used to film combustion before the engine was fully warmed up [29]. The fuel used here (toluene) has a comparatively high boiling point (110°C), and the fuel injection occurred during the compression stroke. Consequently, there was a tendency for the fuel to pool behind the inlet valve, and then fall from the inlet valve onto the piston crown, which would not have reached its full operating temperature. The engine was operating at 1000 rpm with a wide open throttle and a stoichiometric mixture that was ignited 15°btdc. The 0 percent to 10 percent mass fraction burn period occupied 12°ca, and the 0 percent to 90 percent mass fraction burn period occupied 26°ca, yet Figure 5.14 shows evidence of combustion continuing well into the expansion stroke. This is a result of the evaporation of liquid fuel from the piston crown, which is then burned by diffusion-controlled combustion with a luminous yellow flame.

Recent evidence of the liquid fuel behavior in SI engines has been provided by Meyer and Heywood [30] who have measured the droplet size distribution during warm-up (using phase Doppler anemometry), for injection when the inlet valve is both open and closed. With injection when the inlet valve was closed (closed-valve injection—CVI), the Sauter mean diameter was about 35 μm, compared to about 20 μm when injecting with the inlet valve open (open-valve injection—OVI).

Fig. 5.14.

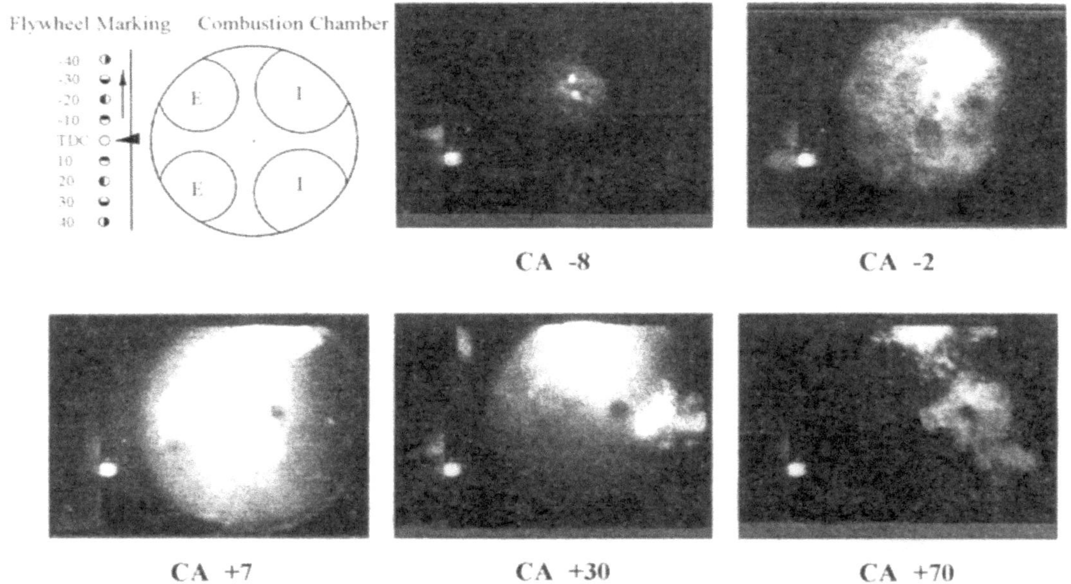

Flame front images (the numbers refer to crankangle degrees relative to TDC, with ignition at −15) with a diagram to show the image orientation (the piston window is 75 percent of the bore diameter).

Fig. 5.15.

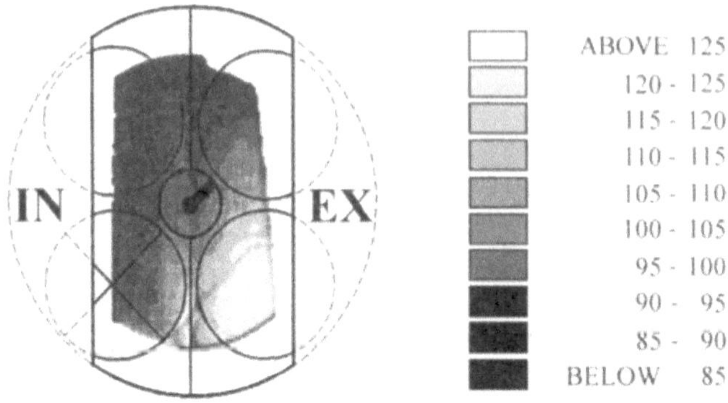

Measurements made by planar laser-induced fluorescence (PLIF) of air-fuel ratio; the mean air-fuel ratio has been set to a level of 100. Adapted from [24].

In-cylinder liquid fuel films have recently been observed by Witze and Green [31] using flood-illuminated laser-induced fluorescence, and this showed that CVI led to more extensive fuel films than OVI. Witze and Green [31] also used direct video imaging to observe pool fires, and noted that these pool fires are a likely source of particulate emissions.

5.6.2 Laser-Induced Fluorescence (LIF)

LIF is now being widely used for measuring in-cylinder hydrocarbon concentrations [24, 31, 32, 33, 34, 35], and nitric oxide (NO) levels [24, 34, 36, 37]. Figure 5.15 provides an example of air-fuel measurements made by LIF [24, 34] in an engine with a deactivated inlet valve (marked with a cross) so as to obtain charge stratification at part load. At full load both the inlet valves are used, and the engine works as a conventional homogeneous charge SI engine. The image shown here is an average of 200 consecutive cycles, but the images from individual cycles can also be studied and correlated with the burn rate derived from the pressure history. It has been shown that the fuel concentration in a small region close to the spark plug has a dominating effect on the subsequent combustion history of lean mixtures. By using LIF measurements of the NO from the residual gases, it has been demonstrated that the residual gas is not mixed homogeneously with the air, and that there are significant cyclic variations in the residual gas concentration. These fluctuations are thought to be large enough to contribute significantly to the fluctuations in mixture concentration [24].

5.6.3 Direct-Injection Spark-Ignition Engines

During development of the Mitsubishi direct-injection spark-ignition engine, a number of techniques have been used to elucidate the air motion, mixture preparation, and combustion [2]. Laser sheet illumination was used with a seeded airflow and high-speed cine photography in order to visualize the bulk flow. The mixture preparation was imaged with shadow and schlieren photography. The shadow photography was tuned so as to identify the liquid component of the fuel, while the schlieren system was sensitive to the presence of fuel in both the liquid and vapor states. During combustion, spectral analysis of the flame radiation was used to identify the different types of combustion. With early injection there was confirmation of premixed lean or stoichiometric combustion from luminescence attributed to OH and CH, and CO-O recombination. In contrast at light load, the intention is to achieve stratification by means of late fuel injection. This leads to combustion with longer wavelength emissions that are characteristic of soot radiation—a consequence of the limited time for mixture preparation leading to an element of diffusion combustion. However, this radiation was attenuated over a short period of time, indicating complete soot oxidation. Many of the techniques developed for studying diesel sprays are also relevant to direct-injection gasoline engines [38].

5.7

CONCLUSIONS

Optical engines, with natural light photography and with laser-based combustion and flow diagnostics, are becoming an increasingly important means of elucidating spark-ignition engine performance. Until recently, the only readily available in-cylinder technique was pressure measurement, from which it was possible to estimate the burn rate. The only other measurements were of the inputs and the outputs, and engine designers had to hypothesize as to the in-cylinder processes. In parallel with the development of the in-cylinder diagnostics, there have been formidable developments in the computational predictions of flows, sprays, and combustion. The combination of advanced experimental techniques and computer codes will now help engines to be designed on a more rational basis.

REFERENCES

1. Kume, T., Iwamoto, Y., Iida, K., Murakami, N., Akishino, K., and Ando, H., "Combustion Control Technologies for Direct Injection SI Engines," *SAE Paper* 960600, 1996.
2. Ando, H., "Combustion Control Strategies for Gasoline Engines," IMechE Seminar Publication *Lean Burn Combustion Engines, Paper* S433/001/96, pp. 3–17, London, 1996.
3. Tomado, T., Sasaki, S., Sawada, D., Saito, A., and Sami, H., "Development of Direct Injection Gasoline Engine—Study of Stratified Mixture Formation," *SAE Paper* 970539, 1997.
4. Jackson, N.S., Stokes, J., and Lake, T.H., "Stratified and Homogeneous Charge Operation for the Direct Injection Gasoline Engine—High Power Output with Low Fuel Consumption and Emissions," *SAE Paper* 970543, 1997.
5. Edwards, N., Ellis, S.R., Frost, J.C., Golunski, S.E., Petch, M.I., and Reinkingh, J.G., "Hot-Spot Technology for On-Board Hydrogen," *European Fuel Cell News*, Vol. 3, no. 2, pp. 17–18, 1996.

6. Jamal, Y., and Wyszynski, M.L., "On-Board Generation of Hydrogen-Rich Gaseous Fuels—A Review," *Int. J. Hydrogen Energy*, Vol. 19, no. 7, pp. 557–572, 1994.
7. CSTI (Council of Science and Technology Institutes), "The Greenhouse Effect: Fact or Fiction?" Environmental Information Paper 1, (available from the UK Institute of Physics), 1992.
8. *Flyer*, March 1996, Vol. 8, no. 3, p. 82, 1996.
9. Li, Y., Shih, T.I.P., Schock, H.J., and Willis, E.A., "A Numerical Study on the Effects of Apex Seal Leakage on Wankel Engine Flow Fields," *SAE Paper* 910703, 1991.
10. Zhao, H., Collings, N., and Ma, T., "Two-Dimensional Temperature Distributions of Combustion Chamber Surfaces in a Firing Spark Ignition Engine," *Proc I Mech E* Part 208D, pp. 99–108, 1994.
11. Reeves, M., Garner, C.P., Dent, J.C., and Halliwell, N.A., "Particle Image Velocimetry of In-Cylinder Flow in a Multi-Valve Internal Combustion Engine," *Proc I Mech E* Part 210D, pp. 63–60, 1996.
12. Adrian, R.J., "Particle Imaging Techniques for Experimental Fluid Mechanics," *Annual Review of Fluid Mechanics*, Vol. 23, pp. 261–304, 1991.
13. Heywood, J.B., and Vilchis, F.R., "Comparison of Flame Development in a Spark-Ignition Engine Fueled with Propane and Hydrogen," *Combust. Sc. Technol.*, Vol. 38, pp. 313–324, 1984.
14. "Engine Videoscope 513D," *AVL Focus*, no. 15, December 1996.
15. Merzkirch, W., *Flow Visualization*, 2nd ed. New York: Academic Press, 1987.
16. Konig, G., and Sheppard, C.G.W., "End Gas Autoignition in a Spark Ignition Engine," *SAE Paper* 902135, 1990.
17. Hicks, R.A., Lawes, M., Sheppard, C.G.W., and Whitaker, B.J., "Multiple Laser Sheet Imaging Investigation of Turbulent Flame Structure in a Spark Ignition Engine," *SAE Paper* 941992, 1994.
18. Witze, P.O., "Techniques for Detecting Flame Arrival in Premixed-Charge Spark-Ignition Engines," in D.F.G Durao et al. (eds), *Combusting Flow Diagnostics*, pp. 481–93, Kluwer, 1992.
19. Philipp, H., Plimon, A., Fernitz, G., Hirsch, A., Fraidl, G., and Winklhofer, E., "A Tomographic Camera System for Combustion Diagnostics in SI Engines," *SAE Paper* 950681, 1995.
20. Chigier, N., ed, *Combustion Measurements*, Hemisphere Pub. Corp., 1991.
21. Taylor, A.M.K.P., ed., *Instrumentation for Flows with Combustion*. New York: Academic Press, 1993.
22. Eckbreth, A.C., "Laser Diagnostics for Combustion Temperature and Species," *Combustion Science and Technology Book Series*, 2nd ed. Gordon and Breach, 1996.
23. Felton, P.G., "Laser Diagnostics for Direct-Injection Gasoline Engines," *Lean Burn Combustion Engines*, IMechE Seminar, London, December 1996.
24. Berckmuller, M., "A Study of Mixture Formation in a Lean Burn Research Engine Using Laser Fluorescence Imaging," Ph.D. Thesis, Cranfield University, 1996.
25. Stone, R., *Introduction to Internal Combustion Engines*. New York: Macmillan, 1992.
26. Bachalo, W.D., Brena de la Rosa, A., and Sankar, S.V., "Diagnostics for Fuel Spray Characterization," in N. Chigier, ed., *Combustion Measurements*. Hemisphere Pub. Corp., 1991.
27. Arcoumanis, C., Imperial College, London, Private Communication, 1977.
28. Carabateas, N.E., Taylor, A.M.K.P., Whitelaw, J.H., Ishii, K., and Miyano, H., "Droplet Velocities Within the Pent-Roof of a Honda Vtec-E Cylinder Head," *Lean Burn Combustion Engines*, IMechE Seminar, London, December 1996.
29. Stone, R., Ball, J., Hatchman, A., McCord, R., Pashley, N., Raine, R., and Williams, R., "Combustion Analysis in the Rover K4 Optical Engine," AUTOTECH '97 Paper L16/C524-33.
30. Meyer, R., and Heywood, J.B., "Liquid Fuel Transport Mechanisms into the Cylinder of a Firing Port-Injected SI Engine During Warm-Up," *SAE Paper* 970865, 1997.
31. Witze, P.O., and Green, R.M., "LIF and Flame-Emission Imaging of Liquid Fuel Films and Pool Fires in an SI Engine During a Simulated Cold Start," *SAE Paper* 970866, 1997.
32. Green, R.M., and Cloutman, L.D., "Planar LIF Observations of Unburned Fuel Escaping the

Upper Ring-Land Crevice in a SI Engine," *SAE Paper* 970823. Also in SP-1244, *Analysis of Engine Combustion and Flow Diagnostics*, 1997.

33. Swindal, J.C., Furman, P.A., Loiodice, M.E., Stevens, R.W., Liu, P.C., and Acker, W.P., "Fuel Distillation Effects on the Outgassing from a Simulated Crevice in a SI Engine Measured by Planar Laser-Induced Fluorescence," *SAE Paper* 970825. Also in SP-244, *Analysis of Engine Combustion and Flow Diagnostics*, 1997.
34. Berckmuller, M., Tait, N.P., and Greenhalgh, D.A., "The Influence of Local Fuel Concentration on Cyclic Variability of a Lean Burn Stratified-Charge Engine," *SAE Paper* 970826. Also in SP-1244, *Analysis of Engine Combustion and Flow Diagnostics*, 1997.
35. Meyer, J., Graul, W., Kiefer, K., Thiemann, J., and Landry, M., "Study and Visualization of the Fuel Distribution in a Stratified Spark Ignition Engine with EGR Using Laser-Induced Fluorescence," *SAE Paper* 970868. Also in SP-1244, *Analysis of Engine Combustion and Flow Diagnostics*, 1997.
36. Knapp., M., Luczak, A., Beushausen, V., Hentschel, W., Manz, P., and Andresen, P., "Quantitative In-Cylinder NO LIF Measurements with a KrF Excimer Laser Applied to a Mass-Production SI Engine Fueled with Iso-octane and Regular Gasoline," *SAE Paper* 970824, Also in SP-1244, *Analysis of Engine Combustion and Flow Diagnostics*, 1997.
37. Tanaka, T., Fujimoto, M., and Tabata, M., "Planar Measurements of NO in an SI Engine Based on Laser Induced Fluorescence," *SAE Paper* 970877. Also in SP-1244, *Analysis of Engine Combustion and Flow Diagnostics*, 1997.
38. Kamimoto, T., and Kosaka, H., "Imaging Diagnostics for the Study of Diesel Combustion," IMechE Seminar Publication *Lean Burn Combustion Engines*, Paper S433/014/96, pp. 229–244, London, 1996.

CHAPTER 6

Combustion-Related Emissions in SI Engines

Simone Hochgreb
Department of Mechanical Engineering, MIT, Massachusetts, USA

ISBN: 0-12-639855-0
$25.00

6.1 INTRODUCTION

The design of spark-ignition (SI) engine systems has been increasingly constrained by the need to minimize vehicle pollutant emissions. Regulations requiring ever more stringent maximum emission levels have posed difficult challenges to the automotive industry. Significant engineering advances in combustion systems and emission control methods have led to major improvements in emissions per vehicle over the past 20 years, particularly in ambient carbon monoxide levels [1, 2]. Owing to the increase in vehicle miles traveled, however, these improvements have not always translated into improved urban air quality.

Emissions from properly operating modern vehicles are at least one order of magnitude lower than original precontrol levels. Further reductions are being mandated, calling for better fueling and operation control, new catalytic materials, and innovative engineering solutions.

In this chapter, the production mechanisms of combustion-related pollutants from four-stroke, spark-ignited engines are discussed, as well as techniques for minimizing sources upstream of the emissions aftertreatment process. This section offers an overview of the mechanisms leading to emissions in SI engines. Section 6.2 describes the process of nitric oxide formation, and Section 6.3 discusses the origins of carbon monoxide. Sources of hydrocarbon emissions are examined in Section 6.4.

6.1.1 Main Regulated Pollutants: NO_x, CO, HC, and Toxics

The main pollutants of concern in the case of spark-ignition engines are nitric oxides (NO and NO_2 in general, but primarily NO in spark-ignition engines), carbon monoxide (CO), hydrocarbons (HC), and organic toxics (currently regulated toxics in the United States are benzene, acetaldehyde, formaldehyde, and 1,3-butadiene). Nitric oxides, carbon monoxide, and toxic compounds have direct effects on human health. Nitric oxides and hydrocarbons contribute, in addition, to the formation of tropospheric ozone and smog, which lead to deleterious effects on visibility and human health. Particulate matter, a very important pollutant in the case of compression-ignition engines, is produced only in very small amounts in SI engines. However, concerns about health effects of small-size particulate matter ($<2\ \mu m$) have been increasing, and may lead to stricter standards.

Nitric oxides and carbon monoxide are formed during the combustion process, leaving the engine with other combustion products. Hydrocarbons and toxics emissions, however, may originate both from the tailpipe in the form of unburned or partially burned fuel, as well as in the form of evaporative emissions from the fuel tank, fuel lines, and losses during the refueling process. Evaporative losses of HC are estimated to be about the same order of magnitude as the contribution from the

Fig. 6.1.

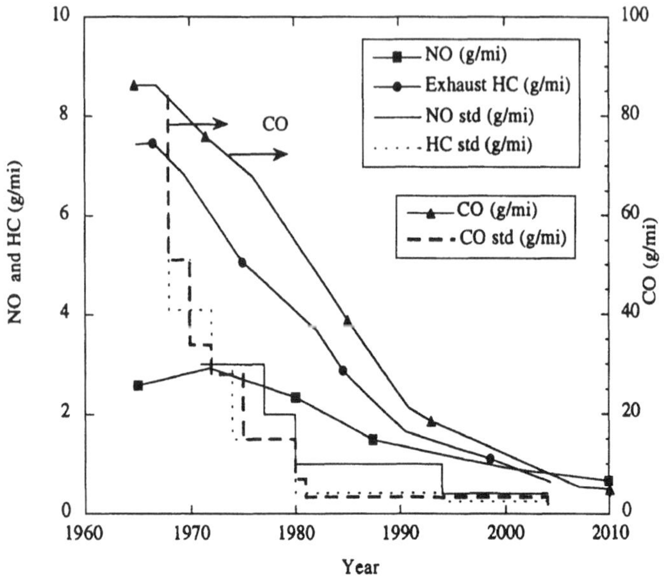

Evolution of aggregate light-duty vehicle fleet emissions and associated uncertainties, as estimated by fleet aggregate simulations (symbols and lines) and historical evolution of standards (lines) [2].

exhaust [2]. In this chapter, however, only combustion-related (nonevaporative) sources are considered.

Enormous progress has been made in reducing tailpipe emissions since the introduction of catalytic converters 25 years ago (Figure 6.1). Emissions from properly operating modern vehicles are at least one order of magnitude lower than precontrol levels.

6.1.2 Maximum Allowed Emission Levels

Emissions standards are permanently evolving throughout the world, reflecting demands for air quality as well as expectations on the minimum technically feasible emission levels. The state of California has led the way in lowering maximum allowable emission levels to address the severe air quality problems in southern California; however, many other countries have continuously moved to increasingly lower admissible levels.

Maximum allowed emissions levels in the largest consumer markets (the United States, Europe, and Japan) govern manufacturers' decisions in emission control implementation. Standards are defined relative to representative driving schedules in each region, so that absolute emission levels are not directly comparable. The different schedules for passenger vehicle tests, as well as the corresponding emission levels for spark-ignition engines are outlined in Figure 6.2 and Table 6.1. These driving schedules, designed to represent a "typical" driving pattern, include a cold-start period, followed by moderate accelerations and

Fig. 6.2.

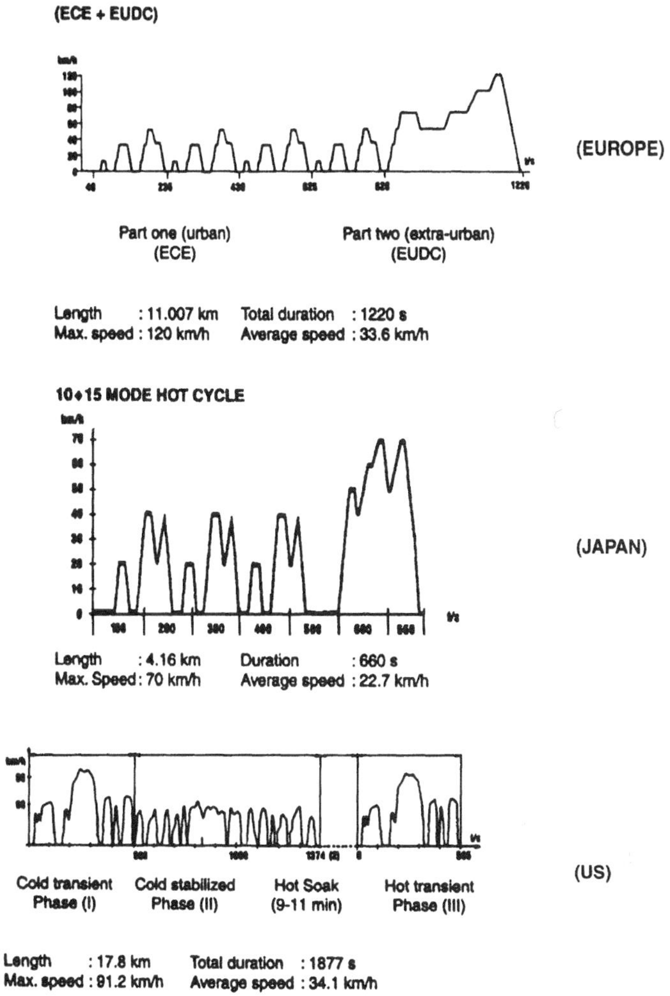

Driving schedules for emissions testing in Europe, Japan, and the United States.

decelerations. The cold-start period contributes a disproportionate fraction of the emissions (particularly hydrocarbons), as the catalytic converter requires tens of seconds to warm up and reach maximum removal efficiency. Therefore, much of the effort in designing effective emission controls has been targeted at the initial cold-start phase. This will become clear in the following discussions and in the description of emission control technology in Chapter 8.

6.1.3 Overview of Emission Sources

The origin of tailpipe (i.e., nonevaporative) emissions from spark-ignition engines is shown schematically in Figure 6.3. Nitric oxides and carbon monoxide are formed via oxidation of molecular nitrogen and fuel in the bulk gases, whereas

Table 6.1

Emissions standards for passenger vehicles

EU

Test	Emissions	Units	1993[a,b] Gasoline (=Diesel)	1996 Gasoline	1996 Diesel	2000[c] Gasoline	2000[c] Diesel
	$HC+NO_X$		0.97	0.5	0.7	0.2	0.5
EU	CO	g/km	2.72	2.2	1.0	1.7	0.5
	PM		0.14	—	0.08	—	0.04

[a]For passenger cars (<6 passsengers, 2.5 tons gross weight).

[b]Deterioration factors for 80,000 km: Gasoline: 1.2; Diesel: $HC+NO_X$:1.1, CO:1.1, PM:1.2.

[c]Proposed.

Japan

Vehicle[a]	Test	Emissions	Units	Mean[b]	Max[b]
Gasoline	10·5 mode	HC		0.25	0.39
		CO	g/km	2.1	2.7
		NO_X		0.25	0.48
Diesel	10·5 mode	HC		0.4	0.62
		CO	g/km	2.1	2.7
		NO_X		0.5	0.72
		PM		0.2	0.34

[a]Light-duty vehicles.

[b]Mean applies to imports, max to nonimports.

US

Test[a]	Emissions	Units	Durability (miles)	1991	1994	2003	ULEV[b]
US	HC	g/mile	50,000	0.41	0.41	0.125	—
FTP	NMHC		50,000		0.25		0.040
			100,000		0.31		0.055
	CO		50,000	3.4	3.4	1.7	1.7
			100,000		4.2		2.1
	NO_X		50,000	1.0	0.4		0.2
			100,000		0.6	0.2	0.3
	PM		50,000	0.2	0.08	0.08	—
			100,000		0.10		0.04

[a]For light duty vehicles and trucks (<12 passsengers).

[b]ULEV: ultralow vehicles emissions (California), mandated manufacturers' fleet average.

HC: hydrocarbons, NO_X: nitric oxides, PM: particulate matter, NMHC: nonmethane hydrocarbons.

Fig. 6.3.

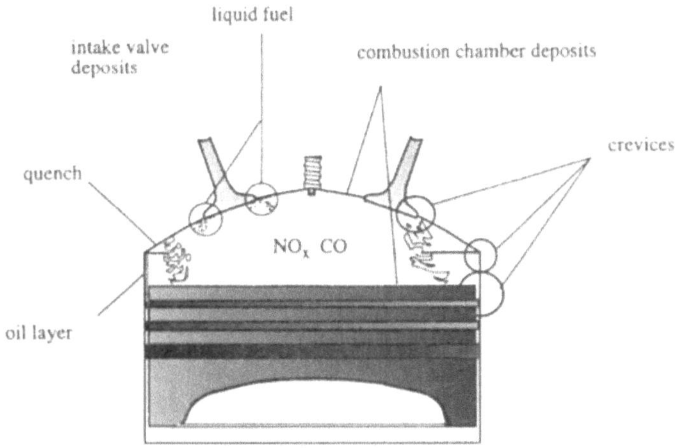

Sources of pollutant formation in spark-ignited engines.

unburned hydrocarbons are produced primarily around cold regions where the flame does not propagate.

Nitric oxide (NO) is formed during combustion as the high flame temperatures break down molecular oxygen and nitrogen from the inducted air, which then recombine into NO (although nitric oxides are usually referred to as NO_x, the level of NO_2 formed in spark-ignited engines is negligible). As will be shown, the production of NO depends primarily on the peak temperatures achieved during combustion.

Carbon monoxide (CO) results from the incomplete oxidation of the fuel to carbon dioxide. Carbon monoxide formation increases steeply with decreasing air-fuel ratio, as not enough oxygen is available to completely oxidize the mixture. Excursions into fuel-rich operation during cold start and transients are responsible for the bulk of carbon monoxide emissions in modern engines.

Unburned hydrocarbons (HC), unlike CO and NO, result from multiple processes in which fuel escapes the main combustion event during flame passage. As will be discussed in greater detail in the following sections, these processes include flame quenching near cold walls and within narrow gaps on the surfaces of the combustion chamber, absorption of fuel on layers such as lubricating oils and combustion chamber deposits, and the presence of liquid fuel, particularly during cold start. A large fraction of the hydrocarbons escaping combustion through one of these processes reenters the combustion chamber during the cycle, mixes with the burned gases, and is partially or completely oxidized before the exhaust gases leave the system. The so-called *engine-out* hydrocarbon emissions (i.e., emissions before catalytic converter treatment) are composed of the original fuel compounds, as well as partially oxidized hydrocarbons not originally present in the fuel.

Figure 6.4 shows the typical behavior of exhaust emissions with air-fuel ratio in SI engines. The exhaust gases collected in the exhaust manifold then flow

Fig. 6.4.

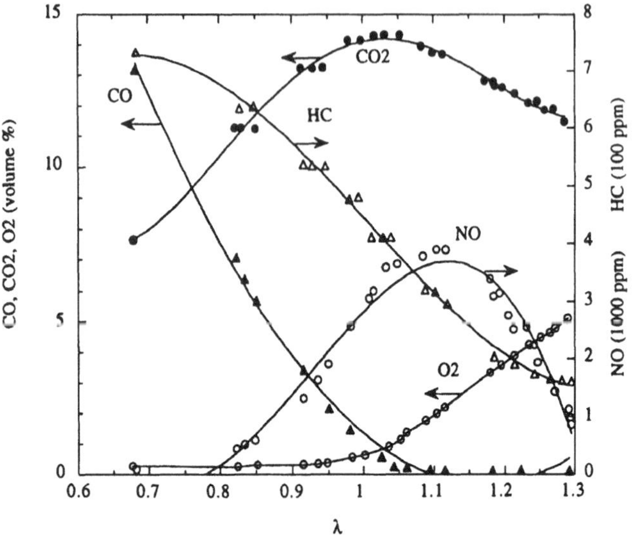

Variation of exhaust emissions with relative air-fuel ratio (Harrington and Shisu [3]).

through the catalytic converter, where most of the NO, CO, and HC is converted to molecular nitrogen, carbon dioxide, and water vapor.

Nitric oxide concentrations peak around the lean side of the stoichiometric air-fuel ratio,* since that is close to the point where the mixture temperature is highest. Carbon monoxide and HC levels increase at fuel-rich conditions due to the lack of oxygen and low temperatures. Clearly, one would like to operate under fuel-lean conditions from the point of view of minimizing emissions. However, the process of conversion of NO back to nitrogen and oxygen in the catalytic converter requires the presence of hydrocarbons as reducing agents for best efficiency. The maximum air-fuel ratio is also limited by the occurrence of misfires, which lead to poor operation and high hydrocarbon emissions. Most engines are therefore designed to operate under stoichiometric conditions. The use of lean burn engines is currently limited by their inability to meet the strictest NO emission levels, since there are at the moment no high-efficiency NO_x catalytic converters for fuel-lean operation.

6.2 NO_x FORMATION

6.2.1 NO_x Production Mechanism

The process of NO_x formation during combustion has been extensively studied since the initial work by Zeldovich [4]. Although one generally refers to nitric

*The stoichiometric air-fuel ratio $(A/F)_s$ is the theoretical proportion of air required for complete conversion of fuel to carbon dioxide and water. The equivalence ratio ϕ and relative air-fuel ratio λ are defined as $\lambda = 1/\phi = (A/F)/(A/F)_s$.

oxides, only NO is of relevance in spark-ignition engines operating under approximately stoichiometric conditions, while NO_2 is present in a substantial amounts only in compression-ignition engines. NO is formed in the hot burned gases during combustion. The formation rate is slow, relative to the overall combustion process, and the rates increase exponentially with burned gas temperatures. The major route to the formation of NO in a combustion system is the so-called *thermal route* or the *Zeldovich-Keck* mechanism [4, 5]:

$$N_2 + O \leftrightarrow NO + N \tag{R1}$$

$$N + O_2 \leftrightarrow NO + O \tag{R2}$$

$$N + OH \leftrightarrow NO + H \tag{R3}$$

The oxygen atoms present behind the flame in the burned gases initiate the decomposition of molecular nitrogen to form N atoms. The activation energy of reaction R1 is of the order of 75 kcal/mol (Table 6.1), which is typically the rate-limiting step in the formation of NO under most engine operating conditions. A smaller contribution arises from the formation of NO at the flame front through reaction of the CH radical with N_2.

$$CH + N_2 \leftrightarrow HCN + N \tag{R4}$$

This so-called *prompt* mechanism [6] has a lower activation energy than the thermal mechanism, and is particularly active within the flame front, where the short-lived CH radicals appear at peak concentrations of the order of a few parts per million. The contribution of the prompt mechanism in stoichiometric laminar flames has been estimated to be of the order of 5 percent to 10 percent [7]. Lavoie and Blumberg [8, 9] have made approximations to calculate the prompt NO contribution in engines, and concluded that it becomes important under conditions of high dilution (through exhaust gas recirculation, lean operation, or non-MBT[†] spark timing), under which thermal NO production is low. The comparison of model and experiments suggests that the prompt NO contribution cannot be excluded whenever NO emission levels are of the order of 100 ppm or lower [8].

An alternative route has been proposed for NO formation through N_2O under lean, high-pressure conditions via recombination of N_2 and O:

$$N_2 + O + M \leftrightarrow N_2O + M \tag{R5}$$

$$N_2O + O \leftrightarrow NO + NO \tag{R6}$$

Drake and Blint estimated that this particular route can contribute up to 10 percent to 15 percent in high-pressure, stoichiometric flames [7]. Similar routes involving the recombination of N_2 and H have been recently proposed [10]:

$$N_2 + H + M \leftrightarrow NNH + M \tag{R7}$$

$$NNH + O \leftrightarrow NO + NH \tag{R8}$$

[†]MBT: Maximum brake torque spark timing.

A quantitative assessment of the role of this path in NO formation in SI engines has not yet been made. However, one would expect that the contribution of the NNH path is important only under extremely dilute conditions, in which the thermal NO production is much reduced.

6.2.2 Modeling NO Formation

The process of NO formation in SI engines under premixed charge conditions is in general fairly well understood, even though details about the particular contribution of the thermal, prompt, and N_2O mechanisms may not be entirely resolved. Several investigators have successfully predicted the levels of NO emissions from spark-ignition engines [8, 11, 14] using the extended Zeldovich mechanism (R1–R3) and in some cases, making approximations to determine the prompt NO contribution through R4. Quantitative predictions of the expected levels of NO formation require the following information:

1. *Thermal state of the mixture.* The temperature of the burned gases is the most important factor determining the rate of formation of NO. Therefore, an appropriate characterization of the temperature and pressure of the burned gas after flame passage is essential for calculations of NO formation rates. Calculations can be based on actual experimental pressure traces—typically assuming that the post-flame mixture compression and expansion is adiabatic, while making assumptions about the overall rate of heat transfer—or from calculated pressure traces based on the operating and design characteristics.

Substantial temperature gradients can prevail within the burned gas elements after combustion as mass elements burn at different times after spark. Gases burning before peak pressure are compressed further to higher temperatures, whereas regions where the flame passes at later times will experience expansion. In the limit of very fast mixing between mass elements burned at different times, a single burned gas zone can be considered in the calculation. At the other extreme, a multiple zone model (corresponding to the minimum unmixed length) can be used, and the final NO production integrated over all burned gas elements.

Recent experiments by Schultz et al. showed that temperature gradients near the spark plug are much smaller than NO concentration gradients, which are highest near the origin of the flame [17]. This suggests that the extent of mixing is relatively small, and the controlling parameter is the temperature history of the mass element. The actual behavior is somewhere in between the fully mixed and stratified range, and in practice it has been shown that the use of five or ten different temperature elements is an adequate compromise [18].

2. *Chemical kinetic model.* The rate of NO production is given by the integrated contribution of the reactions listed in Table 6.2 throughout the history of the burned gases. At conditions of low dilution, MBT spark timing, and air-fuel ratios around stoichiometric, the thermal NO production dominates, and inclusion of the thermal

Table 6.2

Estimated reaction rate constants for NO formation (units: mol-cm^3-s-kcal)

	Reaction	Forward[a]			Reverse			k(2000 K)	ΔH°	F[b]	Ref.
		A	n	E	A	n	E				
R1	$N_2 + O = NO + N$	*1.82E+14*	*0.00*	*76.241*				8.47E+05	75.0	35%	[12] (review)
		1.49E+13	0.30	75.257	*3.27E+12*	*0.30*	*0.000*	8.67E+05			[13] (review)
		1.40E+14	*0.00*	*75.400*				8.06E+05			[14][c]
R2	$N + O_2 = NO + O$	*6.40E+09*	*1.00*	*6.280*				2.64E+12	−31.8		[13]
		1.72E+10	1.00	6.615	*3.80E+09*	*1.00*	*38.747*	6.50E+12		30%	[12]
R3	$N + OH = NO + H$	*3.80E+13*	*0.00*	*0.000*				3.80E+13	−48.6		[15]
		6.56E+13	0.00	1.002	*1.70E+14*	*0.00*	*48.801*	5.10E+13			[12]
R4	$N_2 + CH = HCN + N$	*3.00E+11*	*0.00*	*13.600*					3.3	50%	[12]
		4.20E+12	*0.00*	*20.400*							[16]
R5	$N_2 + O + M = N_2O + M$	2.23E+20	−2.50	24.925	*6.92E+23*	*−2.50*	*64.995*	5.87E+13	−39.9	40%	k_o [12][d]
	$N_2 + O + M = N_2O + M$	4.19E+17	0.00	19.540	*1.30E+11*	*0.00*	*59.610*	6.00E+09			k_∞
R6	$N_2O + O = NO + NO$	*1.00E+14*	*0.00*	*28.017*				2.43E+15	−36.0		[13]
		6.92E+13	*0.00*	*26.626*				2.27E+15		50%	[12]

[a] Italicized values are those given in the original reference.

[b] Estimated uncertainty, when available.

[c] Estimated from [13].

[d] k_o: low pressure limit, k_∞: high pressure limit.

mechanism (R1-R3) is sufficient for adequate predictions [8, 14]. In this case, the molar rate of NO formation per unit volume of the local mixture $\dot{\omega}_{NO}$ is given by adding the contribution of each reaction:

$$\dot{\omega}_{NO} = k_1[O][N_2] - k_{-1}[NO][N] + k_2[N][O_2] - k_{-2}[NO][O] + k_3[N][OH] - k_{-3}[NO][H] \tag{6.1}$$

where the expressions in brackets are the molar concentrations of each species, and k_i the reaction rate constants for each reaction i. Approximations have also been made to include an additional term to the rate of formation in equation (6.1), to account for the prompt NO contribution. This extra term, which can be represented as a function of pressure, temperature, and equivalence ratio, has been shown to be a key factor in order to capture the appropriate behavior at very dilute or lean conditions [7, 8].

3. *Steady-state and equilibrium approximations.* In order to use equation (6.1), the concentrations of N_2, O_2, and the radicals O, N, H, and OH must be estimated. The N_2 molar fraction can be considered approximately constant, since its depletion due to reactions is much smaller than the initial concentration. In addition, since the only relevant reactions involving N are R1–R3, with very small resulting concentrations, a steady-state approximation (i.e., the rate of change of N is much smaller than its production or destruction rate) is appropriate:

$$\dot{\omega}_N = k_1[O][N_2] - k_{-1}[NO][N] - k_2[N][O_2] + k_{-2}[NO][O] - k_3[N][OH] + k_{-3}[NO][H] \approx 0 \tag{6.2}$$

which can be solved for $[N]_{ss}$:

$$[N]_{ss} = \frac{k_1[O][N_2] + k_{-2}[NO][O] + k_{-3}[NO][H]}{k_{-1}[NO] + k_2[O_2] + k_3[OH]} \tag{6.3}$$

Finally, an approximation for the O, O_2, H, and OH concentrations must be made. Although equilibrium concentrations for the radicals are achieved at relatively short distances from the passing flame, the fast expansion of the burned gases after peak pressure leads to a deviation from expected equilibrium levels at the local pressures and temperatures. Third-body recombination reactions of the type $X + Y + M \rightarrow XY + M$, where M is any molecule in the system (say, $O + O + M \rightarrow O_2 + M$), are slow compared to the time available for equilibration. Superequilibrium levels of radicals H, O, and OH are thus found, with approximately frozen concentrations of these radicals at levels one to two orders of magnitude above equilibrium as the temperature drops below 1400 K. Nevertheless, since most reactions take place at the highest temperatures, good agreement with experimental results for stoichiometric operation at low dilution levels is obtained by using equilibrium levels for O and OH concentrations in the so-called *equilibrium Zeldovich* approximation [8, 11, 14], leading to a simple expression for the rate of NO production:

$$\dot{\omega}_{NO} = \frac{2\left[1-([NO]/[NO]_e)^2\right]}{(1+K[NO]/[NO]_e)}\dot{\omega}_1 \tag{6.4}$$

where

$$K = \frac{\dot{\omega}_1}{\dot{\omega}_2+\dot{\omega}_3} = \frac{k_1[N_2]_e[O]_e}{(k_2[O_2]_e+k_3[OH]_e)[N]_{ss}} \tag{6.5}$$

$\dot{\omega}_j$ is the reaction rate calculated using the equilibrium concentrations of the species involved in reaction j, and the subscripts e and ss indicate equilibrium or steady-state, respectively.

A more detailed treatment can also be used by introducing nonequilibrium calculations for the radical recombination process, as well as the previously mentioned approximations for prompt NO contributions as a function of flame temperature and pressure [8]. Figure 6.5 shows the evolution of the measured pressure, calculated temperatures, and NO as a function of crankangle degree. NO forms quickly during the highest temperature period, but freezes after the temperature has dropped to about 1400 K for stoichiometric mixtures (somewhat later for fuel-rich

Fig. 6.5.

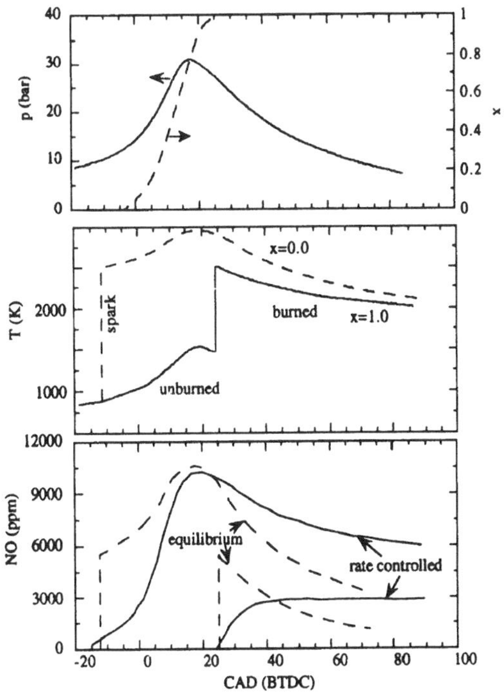

Measured pressure, calculated mass fraction x, temperatures of the unburned and burned gas, and NO concentrations as a function of crankangle degree [14].

Fig. 6.6.

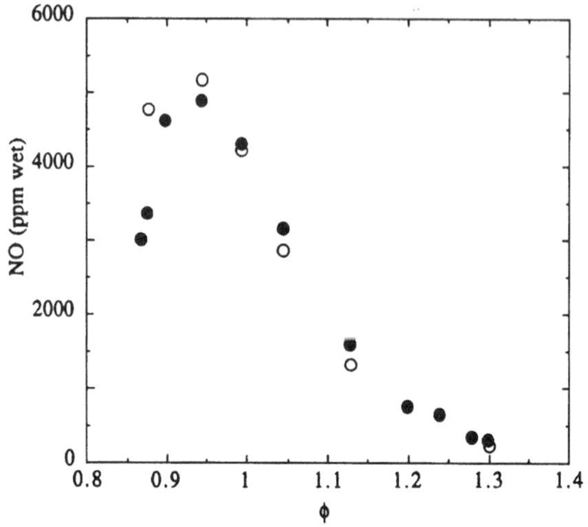

Comparison of model (open symbols) and experiment (filled symbols) as a function of diluent fraction [14].

mixtures). Therefore, NO formation in lean mixtures is more sensitive to peak temperatures, whereas in rich mixtures, it is more sensitive to temperatures after combustion [11, 14].

Figure 6.6 shows a comparison between data obtained in a single-cylinder engine and calculations performed using the equilibrium Zeldovich approximation described previously. There is perhaps surprisingly good agreement within these operating conditions, although some deviation (of about 30 percent) is observed under lean conditions. Lavoie and Blumberg [8] used a similar method (although with calculated rather than measured pressure profiles) (Figure 6.7) for more extreme cases of dilution. The baseline calculation clearly cannot account for the discrepancy at higher dilution levels (Figure 6.7(a)), whereas the calculation including the prompt NO approximation (Figure 6.7(b)) more closely approaches the measured concentrations.

6.2.3 Effect of Operating Conditions on NO Formation

Once the mechanism for NO formation is understood, it is a simple matter to verify that the primary factor governing NO formation is the temperature history of the burned gas, and in particular, the peak temperature. Therefore, any operating parameter leading to a change in the temperature history of the burned gas will have a pronounced effect on the NO yield. These factors are primarily dilution (via added air, or internal exhaust gas recirculation, provided by the trapping of residual gases, or externally imposed), spark timing (or more specifically, combustion phasing), and speed. At a fixed combustion temperature, oxygen concentration is the second most important variable affecting NO emission levels.

The engine design parameters compression ratio and valve overlap strongly affect emissions. Compression ratio affects the maximum temperature achieved by the burned gases, and the extent of valve overlap affects the amount of residual gases in the charge mixture, and thus the final gas temperature. Specific design characteristics (such as burn rate increase through swirl and the degree of stratification) also have an effect in SI engine NO emissions by changing the phasing of combustion. Stratified combustion in SI engines has been long suggested as one of the potential methods for controlling NO emissions by avoiding stoichiometric combustion at peak temperatures. Difficulties related to increased HC emissions have prevented its broader introduction. The effect of the most important operational variables is discussed next.

6.2.3.1 Equivalence Ratio

The air-fuel equivalence ratio λ (or its reciprocal $\phi = 1/\lambda$) affects both the gas temperature and the oxygen concentration in the burned gas. Peak temperatures occur around stoichiometric values, whereas NO emissions peak at slightly lean conditions ($\phi < 1$). The corresponding NO emissions as a function of equivalence ratio are shown in Figure 6.7. A change of equivalence ratio of the order of 10 percent around stoichiometric at mid-speed leads to a decrease of 30 percent to 40 percent in NO emissions, although the brake specific value varies with engine conditions.

Whereas lean operation would be desirable to reduce NO levels, a breakthrough is still required in NO catalytic converter technology to achieve high enough removal efficiencies under lean conditions, so as to achieve the most stringent emission levels.

Fig. 6.7.

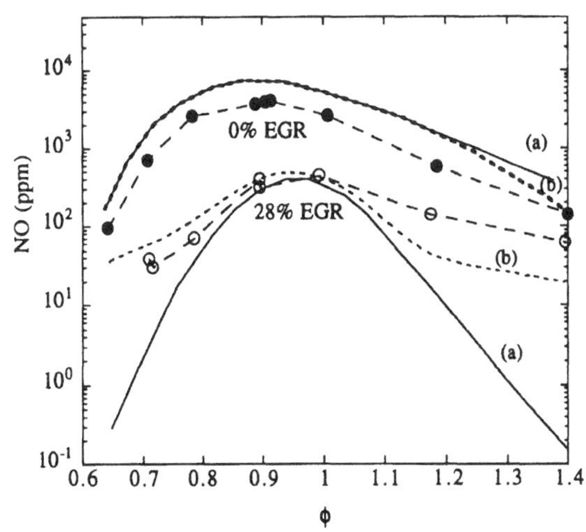

Comparison of experiments (symbols) and model (lines): (a) no prompt NO; (b) inclusion of prompt NO approximation [8].

6.2.3.2 Exhaust Gas Recirculation (EGR)

Exhaust gases can be recirculated into the fresh charge to act as a charge diluent. The high heat capacity of the gases (primarily due to the triatomic molecules CO_2 and H_2O) reduces the peak burned gas temperature, therefore producing a marked reduction in NO. Exhaust gases are retained in the charge both passively (i.e., through incomplete scavenging of the fresh charge) or through active recirculation via an EGR control valve, a strategy that is now routinely used for NO production control at selected operating conditions.

Quader systematically investigated the effect of dilution on NO emissions using different gases, and concluded that it could be entirely attributed to the diluent heat capacity of the gases used (Figure 6.8) [19]. Therefore, under excess oxygen conditions, an effective diluent gas-fuel ratio (including both EGR and air dilution) is sometimes used to collapse all effects into a single dilution effect, using the effective heat capacity of the mixture. Because of the high specific heat of water vapor, ambient moisture also decreases the final temperature of the mixture and therefore the NO production. A value of 10 percent EGR leads to about 30 percent to 50 percent decrease in NO emissions at a fixed spark timing. However, the effect is nonlinear, with benefits decreasing at higher levels. Penalties in combustion stability beyond 15 percent to 20 percent EGR limit the amount of exhaust gas that can be practically recirculated.

6.2.3.3 Spark Timing

The final temperature of the burned gases is a function of the time at which combustion occurs. Charge elements burning early in the cycle achieve higher

Fig. 6.8.

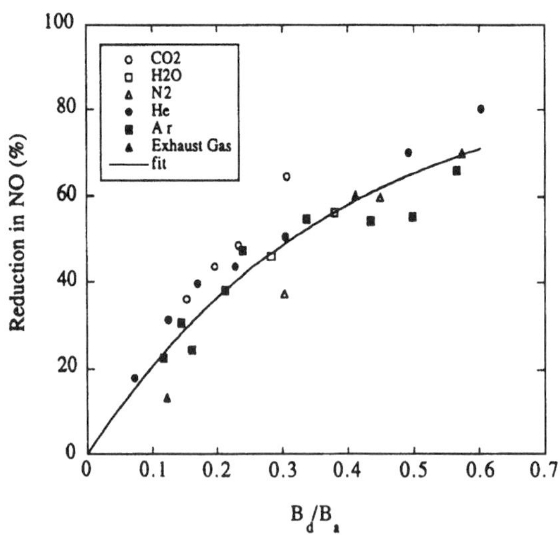

Effect of dilution on NO emissions. B_a and B_d represent the heat capacity (mc_p) of the charge mixture and diluent, respectively [19].

Fig. 6.9.

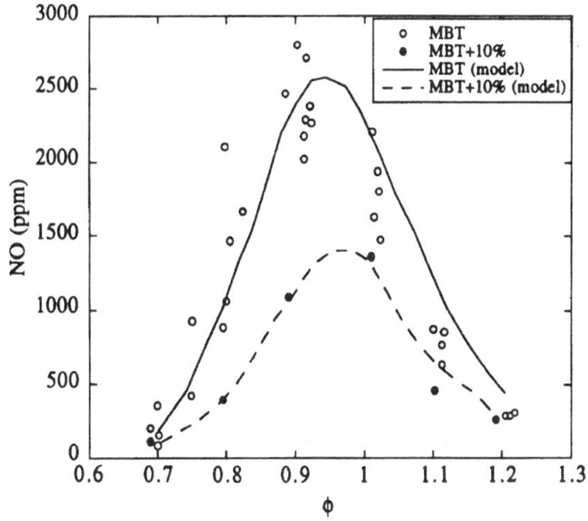

Effect of spark timing and equivalence ratio on NO emissions. MBT + 10%: spark timing change required for 10 percent increase in specific fuel consumption [9].

temperatures (Figure 6.5), as the burned gas is further compressed after combustion. Therefore, advancing the spark from MBT leads to higher peak temperatures and thus higher NO levels. Retarding the spark timing decreases the peak temperatures (and increases the exhaust temperatures), leading to lower NO, and lower HC, subject to a penalty in fuel efficiency. Lavoie and Blumberg showed that the effect of spark timing on NO emissions is much more pronounced in the case of lean operation, as shown in Figure 6.9 [9]. For a spark timing advance leading to a 10 percent penalty in fuel economy relative to MBT, a 50 percent reduction in NO is obtained at $\phi = 0.9$, whereas a smaller reduction is obtained in fuel-rich operation. This effect is a result of the much higher sensitivity to temperatures in the lean range, where thermal NO production is highest.

6.2.3.4 Speed

Engine operating speed has effects on several parameters that affect NO formation in opposite directions. Increases in engine speed lead to decreases in volumetric efficiency, therefore increasing the fraction of residual gases. Although heat transfer rates are increased with speed, the reduced time per cycle decreases the average heat transfer per cycle, leading to higher gas temperatures. Finally, the time available for NO formation also decreases. Therefore, the effect of speed on NO emissions depends on the specific operating point and the geometric details of the engine (valve overlap, compression ratio, and so on). Increasing speed, therefore, has a moderate effect on NO, which can be offset by the shorter reaction times, especially at lower temperatures. Predictions of the sensitivity of NO with speed have in general been adequate [8, 14], in spite of difficulties in determining heat transfer rates precisely (Figure 6.10).

Fig. 6.10.

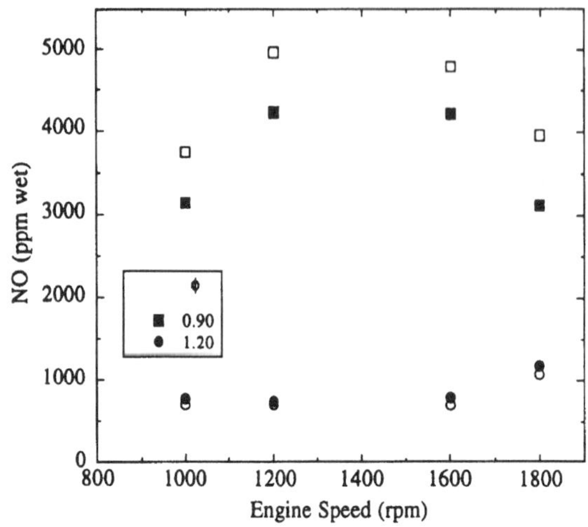

Effect of speed and equivalence ratio on NO emissions. Filled symbols: experiments, open symbols: model [14].

6.2.3.5 Load

The effect of increasing load on NO emissions comes about primarily through the decrease in the fraction of residual gases due to the higher pressures prevailing during the valve overlap period. Therefore, increasing load leads to higher NO emissions. Huls and Nickol measured a pronounced increase in NO with load at lean conditions, and a very moderate increase under stoichiometric conditions [20].

6.2.3.6 Geometric Design Factors

Engine geometry and valve timing have a role in NO production inasmuch as the final burned gas temperatures are affected. Higher compression ratios increase engine efficiency but also increase the final temperature of the burned gas, provided all other parameters are kept constant. Therefore, NO increases on a volumetric basis but decreases on a brake-specific basis due to the higher efficiency.

Valve overlap angles have a strong effect on dilution: Larger fractions of burned gases are trapped in the cylinder with increasing overlap angles and, therefore, decrease NO emissions. Benson and Stebar showed that, at a given power, air-fuel ratio and MBT spark timing, the effects of valve overlap, compression ratio, and external EGR can all be collapsed into a single curve when the dilution fraction is accounted for [21].

Changes in the inlet geometry, asymmetric EGR injection, and stratification can be used to increase tolerance to dilution and affect combustion phasing in order to decrease NO emissions.

6.2.4 Advanced Low-NO Engine Designs

The need to achieve ULEV levels of NO emissions (below 0.25 g/mile) in SI engines requires pushing the limits of charge dilution tolerance. Several concepts have emerged to enable low-NO operation while keeping engine stability. The addition of high-velocity tumble ports [22, 23, 24] can provide increased EGR tolerance at idle conditions. Lean operation offers advantages for fuel economy as well as low-NO operation, but a high-efficiency lean catalyst capable of achieving ULEV values has yet to appear [25, 26]. The use of stratified EGR concepts [27], coupled with sophisticated EGR and tumble motion control, has been shown to achieve ULEV NO levels in relatively conventional combustion systems.

6.3 CARBON MONOXIDE

6.3.1 Mechanism of CO Formation

The fuel conversion process can in general be described as a conversion of the primary fuel to smaller intermediate hydrocarbons, followed by oxidation to aldehydes or ketones, and finally to CO, which is then oxidized to CO_2. The conversion of the original fuel to intermediates is typically fast compared to the final CO to CO_2 conversion process. Carbon monoxide emission results from the incomplete conversion to carbon dioxide, as a result of the lack of oxygen, when the mixture is fuel-rich, as well as of the fast expansion of the burned gases, which freezes the final oxidation reaction.

In modern engines operating under closed-loop control, CO emissions are only prominent during cold start and hard accelerations, during which fuel enrichment is usually applied. The current emission target levels are easily achievable with closed-loop operation and an oxidation catalyst.

The introduction of oxygenated fuels in winter CO nonattainment regions in the United States was aimed at reducing carbon monoxide emissions under cold-start conditions. The enleanment of the fuel with the addition of oxygenated species to the fuel can be effective in reducing CO emissions, as long as engine controls are not adjusted to compensate for it.

6.3.2 CO Formation Models

Harrington and Shisu conducted an extensive survey of the effect of fuel type on CO emissions [3], showing that, for a range of fuel compositions and C/H

Fig. 6.11.

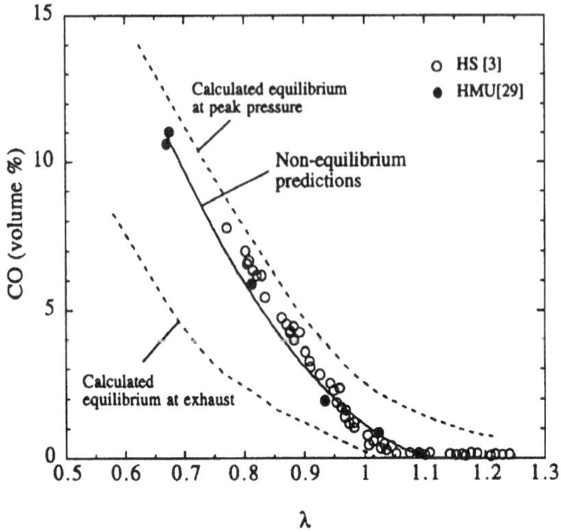

Measurements [3, 29] and predictions [28] of CO concentrations under steady conditions as a function of λ. Open symbols represent 11 fuels of different composition and volatility.

ratios, the amount of CO emitted is a function only of the relative air-fuel ratio (Figure 6.11).

Newhall established that the emitted CO at the end of expansion is higher than what would be expected at equilibrium conditions corresponding to the average temperature and pressure conditions [28]. Conversion of CO to CO_2 is governed primarily by reaction R9.

$$CO + OH \leftrightarrow CO_2 + H \tag{R9}$$

This reaction is fast in both directions at typical reactant concentrations. Newhall proposed that the CO-CO_2 ratio can be determined by assuming equilibrium for reaction R9, while using the nonequilibrium values of the concentrations of H and OH. He further determined that the main reactions controlling disequilibrium are the third-body recombination reactions of H, OH, and O.

The assumption clearly seems to hold throughout the prevailing operating conditions (Figure 6.11), as the predictions agree well with experiments. Further studies have been performed by Keck and Gillespie [30] using a rate-constrained equilibrium method. Measured CO values under lean conditions are much higher than the very low CO levels expected based on the partial equilibrium assumption for reaction R9. Under these conditions, the formation of CO becomes controlled by the kinetics of the forward and backward reactions, as concentrations of H atoms become very low. A relatively significant contribution to CO under lean conditions stems from the incomplete oxidation of unburned hydrocarbons. However, the very low final concentrations are of much less importance than at rich conditions.

6.4

HC EMISSIONS

The reduction in HC emissions from SI engines has been the focus of intense research and engineering over the past 30 years. Dramatic reductions in tailpipe hydrocarbon emissions were achieved by the introduction of the oxidation catalyst in the seventies. HC emission levels have continued to drop over the years, currently aiming toward the challenging ULEV levels of 0.040 g/mile. Whereas the catalytic converter removes over 90 percent of all hydrocarbons under warmed-up conditions, the challenge remains to further reduce engine-out HC levels (i.e., hydrocarbon emissions before the catalytic converter), particularly during cold start, when the catalytic converter is not yet operating at maximum efficiency.

Combustion of the fuel-air mixture in port-injected SI engines involves air and fuel induction into the cylinder, flame propagation through the charge after the spark, followed by discharge of the burned gases into the exhaust manifold. Whereas in an ideal combustion process all the induced charge would be consumed by the flame, several processes contribute to the escape of part of the induced fuel from the main combustion process, eventually leading to unburned hydrocarbons emerging into the exhaust.

Most of the processes leading to HC emissions—here called HC *sources*—result from contact of the fuel or fuel-air mixture with cold surfaces or layers, which prevents oxidation of the fuel during the main flame passage. After flame passage, and especially during expansion, unburned hydrocarbons emerge from the sources and are partially oxidized in the burned gas mixture. A fraction of the remaining HC is retained in the cylinder, while the remainder leaves the cylinder as engine-out emissions.

Figure 6.12 illustrates the main sources of hydrocarbon emissions in SI engines. The major contribution to HC emissions during steady-state operation arises from the presence of crevices, which are narrow passages around the combustion chamber where the flame cannot propagate. Absorption of fuel on lubricant oil layers and deposits, incomplete fuel vaporization, and other incidental sources such as valve leakage contribute various amounts to the total engine-out HC emissions. Under very dilute conditions, there is evidence that the flame may extinguish during expansion, leading to bulk quenching of the reacting mixture. The relative contribution of the different sources changes with operating conditions and engine design characteristics. Nevertheless, the vast number of experiments performed in different engines show that the processes through which HC emissions arise are similar across different engine designs.

The relative contributions of HC sources are quite different during cold start than under warmed-up conditions. Poor fuel vaporization is the main source of HC during cold start, whereas during warmed-up conditions, the effect of combustion chamber crevices dominates.

Fig. 6.12.

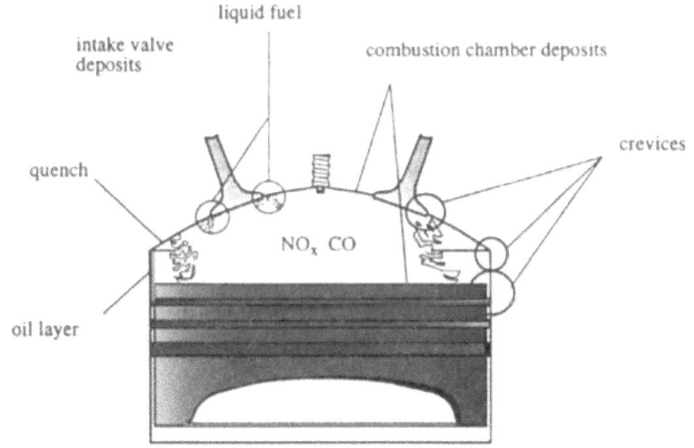

Main sources of HC emissions.

Recent surveys of the mechanisms leading to HC emissions have estimated contributions to HC emissions during warmed-up, mid-load, mid-speed conditions [31, 44]. An expanded diagram of the estimated contributions of the various mechanisms is shown in Figure 6.13. The estimates are based on a range of studies performed by various investigators. One of the most striking conclusions (as pointed out by Cheng et al. [31]) is the assessment that only about 90 percent of the charge is burned during the main combustion event. Whereas about three-quarters of the HC not consumed by the flame eventually burn in the cylinder and exhaust, the fact that the oxidation of the remaining HC occurs late in the cycle means that a good fraction of the energy is not used to power the shaft but simply to increase the enthalpy of the exhaust gases, contributing to a loss in the overall efficiency.

A useful structure to quantify the extent to which each HC source contributes to the final engine-out HC emissions is to consider the sequence of processes involved: (a) *storage*, (b) *in-cylinder postflame oxidation*, (c) *residual gas retention*, and (d) *exhaust oxidation*. Each source of hydrocarbons (e.g., crevices, liquid fuel) may store a maximum fraction ε_s of the original fuel. Of the amount stored ε_s, a fraction (ε_c) actually survives postflame oxidation in the cylinder. A fraction of the latter (ε_m) leaves the cylinder to the exhaust manifold, and a final fraction (ε_e) survives oxidation in the exhaust manifold to yield an overall fraction of the charge emitted, ε. This fraction is customarily called the specific HC emission index (frequently denoted by EIHC in units of grams per kilogram of fuel) $\varepsilon = \varepsilon_s \varepsilon_c \varepsilon_m \varepsilon_e$.

The first factor ε_s depends on the capacity for storage of the source, which is a function of geometric details, fuel type, and operating conditions. The extent of oxidation in the cylinder ε_c depends on the source release rate characteristics, fuel type, and operating conditions related to the burned gas temperature. The fraction of that that enters the manifold ε_m depends on the overall residual fraction, the location of the source relative to the exhaust valve gas volume, and the operating

Fig. 6.13.

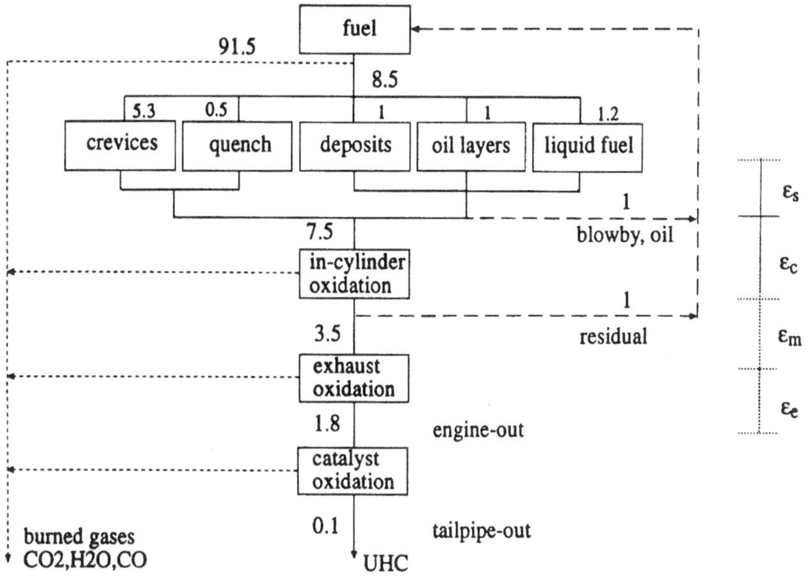

Schematic of pathways to formation of HC in SI engines: estimated magnitudes of contributions of HC emission sources during mid-speed, mid-load conditions [31].

conditions determining the residual gas fraction. Finally, the fraction surviving oxidation in the exhaust manifold ε_e is similar to ε_c, since it depends on the temperature and oxygen availability.

Since the specific fuel consumption in general changes with operating conditions, the emission index provides a more physical idea of the fate of the fuel and the efficiency of conversion than the absolute emission levels. One must keep in mind, however, that standards are established in grams *per unit distance traveled* during a standard test. Therefore, any achievable fuel economy at a fixed emission index translates into an immediate net emission level decrease.

6.4.1 HC Sources

6.4.1.1 Quench Layers

The first studies investigating the possible sources of HC emissions from spark-ignition engines identified *quench layers* as possible sources of unburned HC. Quench layers are regions very close to the cooled engine walls through which a premixed flame can no longer propagate. Although it has been shown since those early studies that quench layers contribute only a small fraction of the HC emissions, the discussion of flame quenching will be useful here in connection with the more important sources discussed in the next section, combustion chamber crevices.

Adamczyk et al. [41] and Lavoie et al. [32] investigated the contribution of quench layers relative to that of combustion chamber crevices. The minimum distance away from a wall into which a flame can propagate is determined by the location where energy release by the flame balances heat loss to the wall. Clearly, the definition of the quench distance may vary (e.g., the distance of maximum energy release, or the point of total HC disappearance), and the phenomenon is necessarily unsteady. Correlations for the minimum flame standoff distance have been developed by several investigators. In the simplest terms, the quench distance is the point where energy release is balanced by heat losses to the wall, so that:

$$q \cong \lambda \frac{(T_f - T_o)}{\delta_q} \cong \rho S_\ell h \tag{6}$$

where λ is the thermal conductivity, T_f and T_o the temperatures of the flame and wall, respectively, δ_q the quench distance, ρ the unburned mixture density, S_ℓ the local laminar flame speed and h the energy release per unit mass of charge burned. This is the same overall expression as that for free flame propagation, which is simply an expression of the fact that the quench distance should be proportional to the flame thickness, so that:

$$\delta_q = \kappa\alpha / S_\ell \tag{7}$$

where κ is a proportionality constant, α the average thermal conductivity, and S_ℓ the laminar flame speed. For normal fuels burning in an engine at average pressures of 10 bar, S_ℓ is of the order of about 20 cm/s and α around 10^{-5} m^2/s, thus leading to $\delta_q \sim$ 50–100 μm for κ of order unity. Since the thermal conductivity has inverse dependence on pressure, the quench distance also scales approximately with the inverse of pressure for flame speeds with only small dependences on pressure. Measurements of standoff distances for methane and acetylene made at 1 bar are of the order of 250–500 μm, which would correspond to 25–50 μm at pressures of 10 bar [32–36], in good agreement with the preceding estimated values.

The heat loss from the flame decreases with increasing wall temperature, thus leading to a smaller standoff distance. The maximum contribution of the quench layer to emissions during normal, stoichiometric engine operation will take place under low load (low-pressure conditions). For an average pressure of 10 bar, assuming $\delta_q = 50$ μm, and using the maximum surface-to-volume ratio at top dead center, the contribution ε_s of the quench layer is of the order of 0.1 percent to 0.2 percent of the inducted charge, or about 10 percent to 20 percent of the total HC sources (a good review of the different models assumed to calculate quench distances is given by Alkidas et al. [45]). The overall contribution of quench layers to the *final* unburned hydrocarbons, however, is considerably smaller, since the remaining HC is free to diffuse into the high-temperature burned gases as soon as the flame is quenched. Adamczyk et al. [37] showed experimentally, and Westbrook et al. [33] concurred with calculations showing that in the case of methane, most of the hydrocarbons are oxidized after 2–3 ms at the high pressures and temperatures encountered during flame passage. Kiehne et al. [38] showed that a similar fraction remains present in the case of propane.

The final contribution of quench layers after postflame oxidation is estimated to be of the order of 0.05 percent to 0.1 percent, a fraction of the order of 5 percent to 10 percent relative to the total. The presence of crevices along combustion chamber surfaces provides a much larger contribution. A possibly important fraction of the hydrocarbons, however, may result from *bulk quenching* or *misfires* during operation at very dilute conditions. This condition, although not widely investigated, can represent an important source of HC during idling, start-up conditions or at high-dilution conditions [9, 39].

6.4.1.2 Crevices

Crevices—narrow volumes present around the surface of the combustion chamber—are regions of high enough surface-to-volume ratio to prevent flame propagation. Crevices are present between the piston head and liner, along gasket joints between the engine head and block, along the seats of the intake and exhaust valves, and between the spark plug threads.

During compression and combustion, these volumes are filled with unburned charge. During expansion, part of the unburned HC-air mixture leaves the crevices and is oxidized in the hot burned gas mixture. The final contribution of each crevice to the HC emissions depends on its volume, location relative to the spark (crevices closer to the spark fill up with larger fractions of burned gases), and to the exhaust valve (gases trapped in crevices closer to the exhaust valve are less likely to be retained in the cylinder as residual and have less time to oxidize, thus contributing a higher proportion of the final unburned HC), and to the operating conditions governing the postflame temperature and oxygen availability.

Several investigators have sought to quantify the role of crevices on the overal HC emissions. Wentworth [40] performed one of the earliest systematic investigations of the role of piston head crevices, followed by a series of experiments by Adamczyk, Kaiser, and Lavoie [41–43], which sought to separate the contributions of wall and crevice quenching. More recently, a number of investigators have tried to determine how reductions in the different crevice volumes, especially piston head crevices, can lead to reduction in HC emissions [44–48].

A crevice will contribute to emissions if the flame does not propagate through its volume [49, 50]. One measure of the ability of the flame to penetrate a volume is the two-plate quench distance, which is the smallest width of a two-plate gap between which a flame can propagate. It is usually estimated as twice the quench distance, and depends primarily on the thermal conductivity, flame speed, and the temperature of the wall. Saika et al. [49] and Sterlepper et al. [50] showed that under normal warmed-up operating conditions, the flame does not penetrate the crevice if the radial clearance is smaller than the expected two-plate quench distance (of the order of 0.4 mm at typical operating conditions). In this case, a maximum fraction of the charge HC will be stored in that volume according to:

$$\varepsilon_s = \frac{m_c}{m_o} = \frac{V_c}{V_o}\frac{T_o}{T_c}\frac{p_{\max}}{p_o} \tag{8}$$

where m, V, T, and p represent mass, volume, temperature, and pressure, respectively, the subscripts c and o denote crevice and charge conditions at intake, and *max* refers to the maximum pressure. The maximum storage, of course, occurs at peak pressure. For a central spark plug, the crevices will be filled with fresh charge; however, depending on the location of the spark plug, the crevices may be filled with burned gases rather than fresh charge as the flame progresses. Assuming a total crevice volume of 1 cm^3, a total engine volume of 500 cm^3, intake temperature of 300 K and crevice temperature equal to coolant temperature around 350 K, and a ratio of maximum to intake pressure of 40:1 yields a storage factor of the order of 6 percent of the charge. Not all the material that flows into the crevices during the combustion process is fresh charge. Apart from the residual gas, which can be of the order of 10 percent to 15 percent of the mixture in the cylinder, part of the volume will be filled by burned gases as the flame propagates over the inlet of the crevice gap.

Clearly, a reduction in crevice volumes would in general be expected to lead to a reduction in HC emissions. Several studies confirm this expectation, although the benefit depends on the geometry and location of the crevice, as well as on operating conditions. Limits on reduction of crevice volumes are placed by clearances required by expansion of adjoining metal parts. Radial clearances between the piston and liner in typical production engines (Figure 6.14) are in general not smaller than 0.4 mm (cold). Top land piston crevice heights are of the order of 6 mm, height reduction being limited by the material stresses associated with the ring movement. Both top and second land crevices contribute to emissions, the latter contributing less than the top land. Typical crevice volumes

Fig. 6.14.

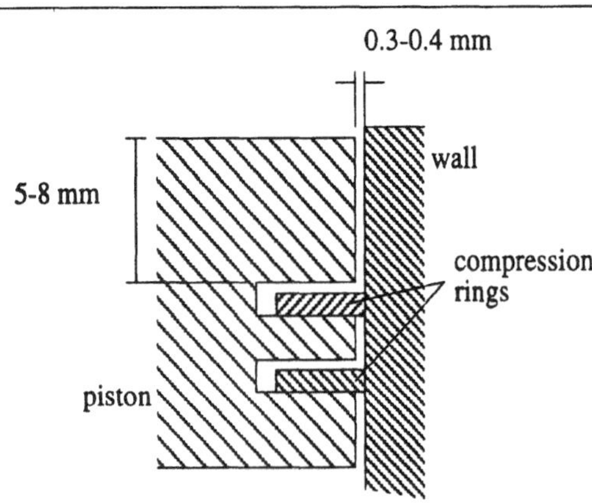

Sketch of typical dimensions of top land crevices.

(piston headland, gasket, and spark plugs) in production SI engines are of the order of 1 cm^3, but efforts are underway to reduce them. About 70 percent to 80 percent of this volume typically resides in the piston headland clearance.

Most of the studies performed in engines and combustion bombs indicate that the sensitivity of the engine-out emitted HC to a change in volume in the piston top land crevice is of the order of 10 percent to 30 percent, provided that the crevice is smaller than the two-plate quench distance. The actual reduction is significantly dependent on engine operating conditions, particularly EGR, and crevice aspect ratio (see Table 6.3).

The influence of crevice geometry, more specifically the radial clearance between the piston and liner, has been investigated by Sterlepper et al. [50], Alkidas et al. [45], Saika and Korematsu [49], and Thompson and Wallace [47]. Increases in radial clearance beyond the two-plate quenching distance (around 0.4 mm) lead to a *decrease* in HC emissions due to flame penetration, as measured by Sterlepper et al. [50] and Saika and Korematsu [49]. Nevertheless, for a low crevice volume piston headland (height of 3 mm), Alkidas et al. observed an *increase* in emissions for radial gaps increasing from 0.4 to 1.4 mm [45]. Although the reasons for the discrepancy in the studies are not yet clear, it is possible that differences in the residual gas fraction or combustion phasing may be responsible for part of the difference. Addition of chamfers to crevice top lands allows for flame penetration and lower emissions. However, the benefits depend strongly on the operating conditions. At very dilute conditions (high EGR or idle), flame quench distances become very large, so that large volume, large radial clearance chamfered crevices can lead to an increase in HC emissions, since the flame no longer penetrates the crevice (as it may have in fact quenched before arriving near the wall).

There is evidence of a certain nonlinearity in the extent to which HC emissions respond to crevice volume reductions. This was originally observed by Adamczyk et al. [42], who showed in a combustion bomb that for relative crevice volumes of the order of 0.1 percent or less of the combustion bomb, reductions were proportional to the fractional volume reduction with unity sensitivity, whereas a smaller change (sensitivity of 0.3) was obtained for larger relative volumes. Min et al. suggested that the nonlinear effects of oxidation (which increase with reactant concentration as the energy release contributes to oxidation) could be responsible for the change [51].

Whereas the effects of spark plug and valve seat crevices appear to be relatively small, head gaskets can contribute disproportionally high amounts of HC. Min et al. [46] showed that the sensitivity to head gasket crevice volumes is about 1.2 overall, with greater weighting toward clearances on the exhaust side (Figure 6.15). This has been attributed to the shorter residence times available for oxidation, as well as a lower fraction of the HC being retained in the residual gas.

Crevice volumes offer the largest contribution to the HC emissions during steady state operation (30 percent to 60 percent). Reduction of crevice volumes is in general advantageous, and parts manufacturers are continually working toward reduction in dead volumes without penalty in durability or performance.

Table 6.3

Measured effect of crevices on engine-out emissions[a]

Study	Description	Operating conditions	$S = -\frac{\Delta(HC)/(HC)_{(b)}}{\Delta V/V}$	Notes
Adamczyk et al. [41, 42, 43]	Combustion bomb	$\phi = 0.7 - 1.0$ propane	1.0 0.3	for $V_{cr}/V_{tot} \approx 0.1$ for $V_{cr}/V_{tot} > 0.1$
Boam et al. [44]	Reduction of piston headland crevice via PTFE seal four-cylinder engine	CR = 10 $n = 2000$ rpm $p_i = 0.47$ bar gasoline	0.3	Single experiment, one condition
Min et al. [46]	Modification of piston head volume and head gasket crevice volume single-cylinder engine	CR = 8.3 $n = 900, 1600$ rpm $P_i = 0.4, 0.7, 1.0$ bar $\phi = 1.0$	0.2 (piston head) 1.2 (head gasket)	Experiment with propane revealed that slope does not extrapolate to zero HC at zero crevice volume
Woods et al. [48]	Reduction of piston headland volume using special ring single-cylinder engine	CR = 10 $n = 1500, 2000$ rpm BMEP = 1, 2 bar $\phi = 1.0, 0.8$ gasoline	0.5–1.0	
Sterlepper et al. [50]	Increase in headland crevice radial spacing; chamfered headland gap four-cylinder engine	CR = 10.1 $n = 2000$ rpm IMEP = 2 bar $\phi = 1.0$ gasoline	−0.1	Increase in radial gap from 0.4 to 1.0 mm: decrease in emissions.
Alkidas et al. [45]	Change in radial spacing at two fixed headland heights; chamfered headland gap four-cylinder engine	CR = 9.5 $n = 1300, 2200$ rpm and idle BMEP = 3.3, 4.5, 6.5 bar $\phi = 1.0$ gasoline	0.1 −(0.1−0.3) −0.10 to + 0.50	Change in *head land gap height* Change in radial gap (gap height: 6 mm) Change in radial gap (gap height: 3 mm)

[a] CR: compression ratio, n: speed, ϕ: equivalence ratio, P_i: intake manifold pressure, IMEP: indicated mean effective pressure, BMEP: brake mean effective pressure.

Fig. 6.15.

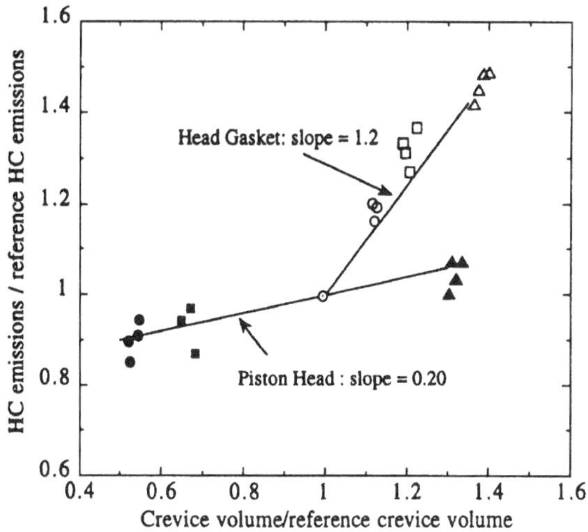

Effect of crevice volume on HC emissions [46].

6.4.1.3 Lubricant Oil Layer

The presence of lubricant oil on the liner walls creates the opportunity for fuel to be absorbed before flame passage. Although it is clear that the mechanism of fuel absorption on the oil layer is responsible for some fraction of the resulting HC emissions, the magnitude of this contribution continues to be debated.

Wentworth [52] replaced the lubricant oil in an engine with a graphite-water solution and found no change in emissions. However, Kaiser et al. [53], Haskell and Legate [54], and Ishizawa and Takagi [55] showed that addition of a small volume of oil to a combustion bomb or engine combustion chamber increases the remaining concentration of the specific fuel added, and that the contribution increases with fuel solubility in the oil. Adamczyk and Kach showed in a similar experiment that the increase in emissions is proportional to the volume of oil present and to the solubility of the fuel in the oil [56–57].

The potential contribution of the oil layer absorption to emissions in SI engines was examined in a number of papers, both experimentally [58–59, 65–66, 68] and analytically [60–64, 67]. Gatelier et al. [65] showed that removal of lubricant oil from the engine (by using graphite rings) decreases emissions by 30 percent. In addition, significant differences were found in the emission index from engines lubricated with mineral oil versus low-solubility polyglycol. The estimates for the contribution of the oil layer absorption mechanisms in SI engines in these studies vary from 5 percent to as much as 30 percent (for isooctane), and that the increase is in the same direction as the solubility of the fuel used. More recently, however, careful studies of the effects of block and head coolant temperatures on speciated emissions using specially blended fuels for different solubilities have suggested that the effect may be much smaller than was originally

Fig. 6.16.

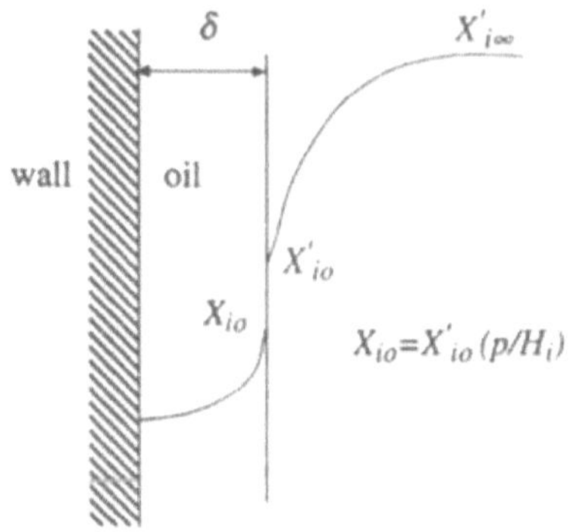

Diagram of liquid-gas diffusion.

estimated [66]. Several factors may contribute to the discrepancies, as discussed next.

Hydrocarbons are absorbed from the fuel into the lubricant oil in the vapor phase or liquid phase (by droplet impingement, for example). All models to date involve the assumption that most of the absorption takes place by equilibrium absorption of the vaporized fuel onto the oil film [60–64, 67, 79], with mass diffusion resistance in the liquid and gas phases (Figure 6.16).

The maximum amount of fuel component i per unit volume of oil that can be stored in this manner can be estimated as the equilibrium value. For a dilute mixture, the quantity can be approximated as:

$$n_i = n_o X_i p / H_i$$

where n represents the number of moles per unit volume, and the subscripts i and o the fuel components and lubricant oil, X_i is the molar fraction of fuel component in the gas phase adjacent to the oil, p is the instantaneous total pressure, and H_i is the so-called Henry's constant, which is a measure of the fugacity of the fuel species in the liquid phase, or the inverse of the solubility. Values of the solubility parameter H for a variety of fuel components and oil types (or in some cases, large hydrocarbons) have been measured by several investigators [69–71] and are shown in Figure 6.17.

The value of Henry's constant decreases exponentially with molecular mass and increases with temperature at a rate similar to the vapor pressure. This implies that a larger fraction of the heavier ends of the fuel are expected to be found in the lubricant oil than lighter components, especially at lower temperatures. This fact is in general confirmed by measurements of fuel concentrations in the lubricating oil [72, 73]. As the pressure increases, the partial pressure of the fuel in the mixture increases, pushing the fuel into the oil layer. After combustion, the partial pressure of the fuel in the burned gases is close to zero, and the fuel can diffuse out of the oil layer toward the burned gases, where it may be oxidized.

Fig. 6.17.

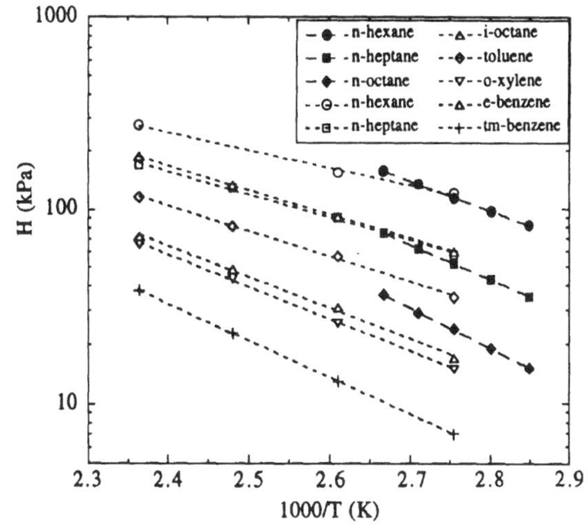

Henry's constant for typical fuels dissolved in lubricant oil. Solid symbols are by Sugiyama et al. for large paraffins (C32)[71]. Open symbols are by Schramm and Sorenson for 10W-30 lubricant oil [72].

At time scales prevailing in typical engine operating conditions, however, the process is also controlled by the diffusion rate of hydrocarbons through the liquid and gas layers, decreasing the net amount of fuel that can be stored in the oil. Simple analytical models [64] of a one-dimensional stationary oil layer show that under warmed-up conditions, the amount of fuel that can be stored within a cycle scales approximately with $(D/\delta^2\omega)^{-1/2}$ where D is the diffusion coefficient of the fuel in the oil layer, δ the oil layer thickness, and ω the angular speed of the engine. The ratio of diffusion in the gas and liquid layer (here represented by the Biot number Bi), further decreases the flux as the gas-phase resistance increases (say, at low temperatures or low turbulence) at a rate proportional to $Bi\ \omega^{-3/2}$.

The lubricant oil layer thickness is a strong function of temperature and engine speed (through the dependence of viscosity on both parameters), so that the dependence of the total amount of fuel absorbed is rather more complex than the preceding simple relationship implies. More sophisticated numerical models have been used [74, 75] to treat the absorption-desorption mechanism, which however do not alter the broad conclusions that the main factor expected to control the adsorption rate is the solubility of the particular fuel component in the lubricating oil.

By doping the lubricant oil with fuel, and operating the engine with hydrogen, Norris and Hochgreb showed that the total amount of fuel desorbed from the oil layer is governed primarily by the thickness of the oil layer rather than by diffusion times through the very thin lubricant layer (0.5–1.0 μm at warmed-up conditions) [76]. Diffusion coefficients of liquid fuels in oil are of the order of 10^{-10} m^2/s. Therefore, the time scale for diffusion is of the order of 5–10 ms, which is the same order of magnitude of the expansion cycle. The extent of oxidation of the emerging HC, however, depends strongly on the rate of desorption because of the rapid change in the burned gas temperature.

Using a combination of stationary oil layer modeling and experiment, several investigators have estimated the oil contribution as of the order of 10 percent to 30 percent [44, 61, 62]. Other investigators [58, 59, 61] determined that changes in oil solubility of factors of the order of 50 percent could be responsible for 0 to 10 percent reduction in HC emissions. However, recent experiments by Kaiser et al. [66] were unable to detect differences in the ratio of speciated emission indices from fuel components of vastly different solubility.

The discrepancy in the various estimates is a result of additional processes thus far not well integrated into models, and which are typically not very well controlled and quantified in usual experiments: The lubricant oil layer is not stationary but mixes dynamically with the lubricant oil in the oil control ring. Part of the fuel desorbs into the crankcase gases, and part returns to the liner. Recent results [73] have shown that the more soluble fuel compounds tend to accumulate in the lubricant oil layers, both in the liner and sump, which might explain in part why Kaiser et al. [66] were unable to detect significantly higher concentrations of more oil-soluble compounds in the exhaust gas.

Although it is undoubtedly true that higher-volatility fuels lead to higher emissions, and that oil layers do act as a source of HC, conclusive evidence of the magnitude of the contribution of the lubricant oil layer has yet to be produced, although a guess of under 10 percent of the total HC for modern engines is probably within the right range.

6.4.1.4 Deposits

Intake and combustion chamber deposits are formed on the surfaces of valves and deposits after several thousand miles of engine use. The nature of the deposits and the routes to their formation are still poorly understood. The deposits that are formed outside of the combustion chamber, on the intake valves and fuel injector, are most likely formed from polymerization and oxidation of the heavy fuel ends into a varnishlike coating. Combustion chamber deposits are more sooty, porous, and carbon-rich, and are formed by a combination of fuel condensation and oxidation during the repeated firings (see extensive review on deposit formation by Kalghatgi [77]). Each of these deposit types brings a different contribution to HC emissions.

Intake valve deposits (IVDs) contribute to driveability and performance problems, interfering with the fuel and air delivery during transients. Combustion chamber deposits (CCDs) have a well-documented effect on NO emissions and produce an octane requirement increase (ORI), both of which have the same origin: the insulating layer formed by the deposits, preventing heat transfer and increasing the charge temperatures. NO emission increases are of the order of 10 percent to 30 percent after 100 hours of operation [78, 79].

The NO emissions effect has been investigated extensively, both experimentally and analytically [78, 80]. Although the effect can be quantitatively accounted for, variations in measurements are large enough to preclude more precise predictions.

Contributions of CCDs to HC emissions have been suggested in the literature since the 1960s, particularly in connection with deposits from leaded gasolines [81]. Adamczyk and Kach showed in combustion bomb experiments that deposits can contribute to HC emissions in a manner similar to that of the oil layer contribution. Indeed, their measurements showed a direct correlation of emissions increase with the saturation pressure of the particular component, which is typically proportional to the solubility of the compound [82, 83].

Since that study, several systematic investigations have been made on the contribution of CCD deposits to emissions. In general, these studies appear to confirm that there is a measurable increase in HC emissions of the order of 10 percent to 25 percent in controlled engine tests, and of the order of 10 percent to 15 percent in actual vehicles as well as test engines [78, 79, 84]. However, there have been experiments where *no increase* in HC emissions was detected after 100 hours of operation [85]. Clearly, this effect is not as well characterized as some of those previously discussed. Nevertheless, new requirements for 100,000-mile vehicle emissions certification indicate that even small increases in emissions may not be easily ignored. Several detergent formulations have been successfully developed for maintaining intake valves clean. The problem of CCD build-up, however, has proved more challenging, and there are currently no means of preventing accumulation.

The mechanism of fuel storage in deposits is not yet well defined, but a few guesses based on the available experimental data allow a reasonable quantification of the expected effect. CCDs consist of a porous carbonaceous material that attaches itself to the piston, head, and valve surfaces after repeated operation. The composition is reminiscent of soot, although oxygen is present in much higher concentrations, thus rendering its rather polar characteristics (identified by solubility of deposits in polar compounds). Deposit build-up rates are very dependent on fuel type and operating conditions. Heavier, olefinic, and aromatic fuels lead to faster build-up than paraffinic compounds [86]. After 50 to 100 hours of operation under alternating load conditions, deposit thicknesses reach values of the order of 100 μm, with variable deposit thicknesses around the perimeter of the piston and head, depending on engine geometry and operation [84, 87].

Since deposits affect not only the storage of hydrocarbons but also the access of hydrocarbons to crevices (by partially filling the crevices), as well as the burned gas temperatures (and therefore the oxidation rates of the unburned hydrocarbons), it is perhaps not surprising that there is a wide discrepancy in the quantitative estimates of their contribution to HC emissions.

Continuous monitoring of the deposit thickness and HC emissions shows that the latter grow during the first 50–100 μm, then level off at a steady-state value (Figure 6.18) [84, 87]. This type of behavior seems to indicate that the HC

Fig. 6.18.

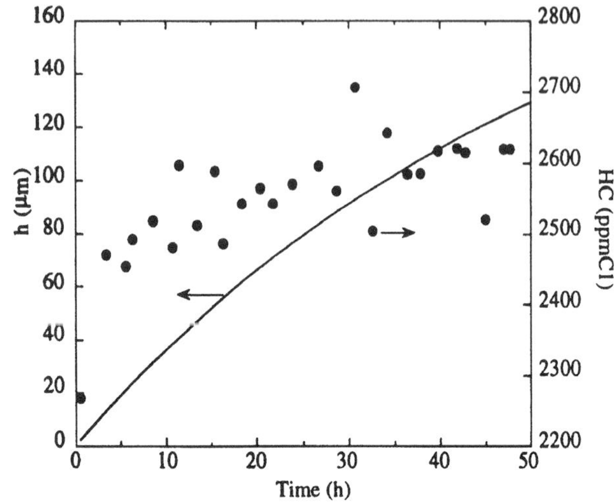

Deposit thickness growth and HC emission increase as a function of time [84].

emission mechanism is governed by some type of diffusion mechanism (as suggested originally by Adamzyk and Kach [82]). A simple model can be constructed to show that, for reasonable values of diffusion and adsorption rates through the porous deposit medium, adequate predictions of the HC build-up time and magnitude can be constructed. The solution is not unique, however, and models involving crevice storage with porous diffusion resistance can yield similar results.

6.4.1.5 Liquid Fuel and Mixture Preparation: Cold Start and Steady State

The largest contribution (>90 percent) to HC emissions from SI engines during a standard test occurs during the first minute of operation. Two important factors are responsible for this: (a) During this period, the catalytic converter is not yet warmed up, and (b) a substantially larger amount of fuel is injected than the stoichiometric proportion in order to guarantee prompt vaporization and starting.

The mixture preparation process in typical modern port-fuel injected engines relies on the vaporization of the fuel on warmed-up surfaces of the back of the valve, followed by mixing with the incoming airflow and hot gases flowing back into the manifold (Figure 6.19). During cold start, when surfaces are cold, additional fuel must be injected to ensure adequate vaporization for immediate start-up. Typically 8 to 15 times the stoichiometric amount of fuel is injected during the first few cycles. Much of the injected fuel remains in the cylinder and port for many cycles, vaporizing during and after the combustion event, thus contributing to HC emissions. Although this contribution decreases as the engine warms up, liquid

Fig. 6.19.

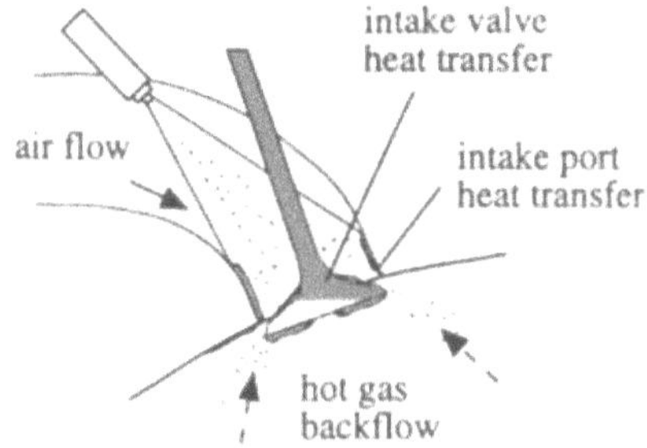

Fuel vaporization process in manifold.

fuel is believed to contribute significantly to hydrocarbon emissions even during warmed-up operation, as discussed next.

The concept of port-fuel injection (PFI) was introduced for more precise fuel metering independently of the airflow, better transient and start-up control, and better driveability as compared to carbureted engines. Nevertheless, the PFI strategy in general leads to higher HC emissions than premixed fuel preparation systems. A large part of recent research and development efforts has been devoted to determining how to minimize emissions while providing adequate driveability (i.e., prompt start-up). This involves understanding how the liquid fuel is vaporized and transported into the cylinder prior to and after the combustion event. Studies in this area are relatively recent but developing very quickly.

Quader originally investigated the effect of injection system design and fuel type on smoke and HC emissions, establishing a connection between injection targeting and HC emissions [88]. Yang et al. [89] and Alkidas [90, 91] investigated the effect of different injector types on HC emissions—premixed fueling (PMX), air assist injection (AAI), air forced injection (AFI) as compared to conventional port-fuel injection (PFI). The premixed systems typically involve thorough vaporization of the mixture. The AAI and AFI techniques produce small mass-averaged mean diameter droplets (40 μm for the AAI and about 15 μm for the AFI, which operates at a higher pressure, compared to about 300 μm for the conventional PFI studied). In all of these studies, which were performed under steady-state conditions, differences between premixed fueling (PMX) and conventional PFI were significant primarily when fuel was injected at open-valve injection (OVI)—a 50 percent increase for the PFI compared to the AFI [89].

At warmed-up conditions (90°C), there is essentially no difference between the injection strategies (PFI and AFI) tested in all studies with closed-valve injection (CVI), indicating that the liquid fuel contribution under these conditions is probably small. As the coolant temperature decreases, however, a dramatic increase in the contribution of liquid fuel is observed.

Quader originally determined a 57 percent decrease in HC for premixed fueling relative to PFI at 24°C [88]. Yang et al. determined about 15 percent increase in HC emissions produced by the PFI at closed-valve injection conditions compared to the AFI, at 30°C, whereas Alkidas still observed no significant differences between the AFI and PFI injectors at 40°C coolant temperatures. The more recent study by Alkidas and Drews [91] compared the increase in HC emissions of AAI versus PFI injectors on gasoline and propane, at coolant temperatures as low as −6°C. Based on the results, the authors concluded that the liquid fuel contribution increases from 0 percent at 90°C to 20 percent to 25 percent at 20°C and up to 46 percent to 54 percent at −6°C (Figure 6.20).

There is a clear benefit to improved vaporization under cold conditions, particularly when combined with an open-valve fueling strategy for quick starting. Given the penalty on fuel consumption for operation of an AFI (2 percent to 3 percent), however, the emissions benefit must be weighted against the additional expense.

Quader et al. [92] and Yang et al. [89] determined that the increase in HC emissions was accompanied by an increase in the fraction of heavy ends of the fuel used. This suggests that the mechanism leading to increased HC emissions involves incomplete vaporization of the fuel, perhaps, or increased absorption on lubricant layers, since solubility typically scales with boiling point.

Skippon et al. [93], Hardalupas et al. [94], Posylkin et al. [95] and Meyer et al. [96] measured droplet size distributions inside the cylinder during steady-state and warm-up conditions, and estimated the liquid fuel fractions in the cylinder. According to Skippon et al., a significant fraction of the fuel appears to still be present in liquid form just before spark during steady-state operation at 40°C, at levels increasing with the boiling point of the fuel. Incomplete vaporization of

Fig. 6.20.

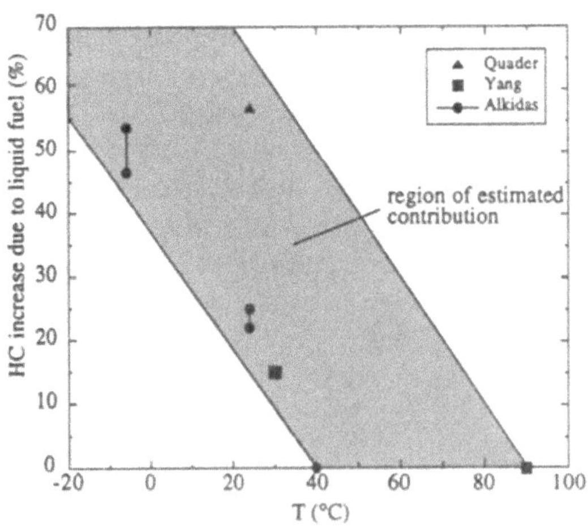

Estimated contribution of liquid fuel to HC emissions as a function of coolant temperatures [88–91].

the liquid fuel during or after combustion can lead to rich pockets, which may ultimately lead to hydrocarbon emissions.

Meyer and Heywood [96] used phase Doppler particle analysis to determine the amount and distribution of size of the droplets entering the cylinder during cold start. As previously suggested by Shin et al. [98], liquid enters the combustion chamber both during the injection period as well as through entrainment and by squeezing of the fuel film present around the valves. Figure 6.21 shows that, in the case of open-valve injection (OVI), the droplets are smaller than in closed-valve injection (CVI), and that droplets enter the cylinder during the injection time. In the case of CVI, three phases are observed. After intake valve opening, the airflow strips the liquid film off the port walls and valves, resulting in a first pulse of droplets. A second pulse appears at the point where the valve reverses direction, thus shaking off droplets, and a third spray appears when the valve closes, as a

Fig. 6.21.

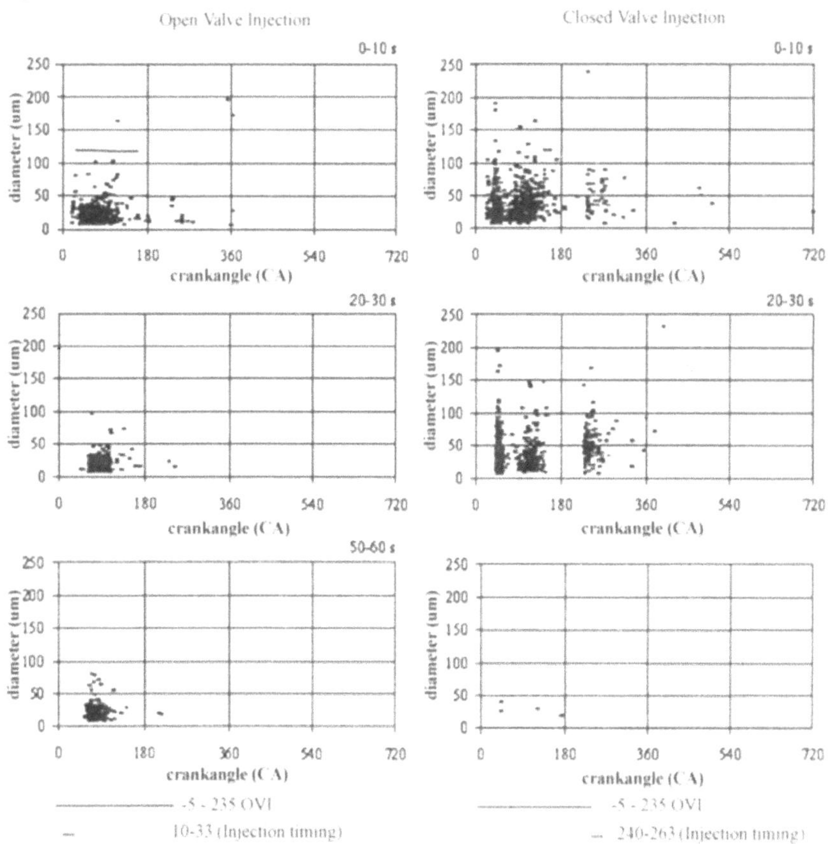

Liquid fuel droplet measurements during cold start and steady-state operation for open-valve injection (left) and closed-valve injection (right) [96].

result of fuel film squeezing. Notice that larger droplets are seen in the case of CVI than OVI: Droplets from film stripping near the valve are larger than those coming directly from the injector. The entrained liquid fuel is accumulated in the cylinder, evaporating as the flame front reaches the cold walls, and contributing to HC emissions. The initial amount of liquid measured as droplets is larger in the case of CVI than OVI. However, the integrated volume decreases rapidly as the engine warms up with CVI injection, whereas in the case of OVI, a steady amount of liquid is introduced in the cylinder at every cycle.

Whereas the contribution of unvaporized fuel to HC emissions in the case of warmed-up operation is small, the importance of injection strategy on HC emissions during cold start is well established. Saito et al. [97], Shin et al. [98, 99], Witze and Green [100, 101], and Swindal et al. [102] used different visualization techniques to determine the behavior of liquid fuel in the engine cylinder and intake port during cold start. In all cases, significant amounts of liquid fuel were observed in the cylinder and intake port for many cycles following start.

Two recent studies by Takeda et al. [103] and Schurov and Collings [104] attempted to quantify the amount of fuel that is retained in the cylinder and in the intake port during cold start in normal production engines using PFI injection strategies. In both studies, fast shut-off valves were used to isolate the cylinder/exhaust and the inlet port regions after a well-defined number of cycles, followed by vaporization of the remaining fuel by using high-temperature coolant and heated air. The total emerging concentration of hydrocarbons from each region was then integrated to determine the amount of fuel present. Takeda et al. used an engine starting calibration strategy using the minimum fuel for stable combustion, while Schurov and Collings used the standard production engine fueling schedule. Although the fueling strategies in the studies are different, the broad features in the fuel distribution are similar.

Typical fueling schedules in the first few cycles call for about ten times the stoichiometric amount of fuel. Of this, about 80 percent to 90 percent remains in the cylinder or intake port through the next cycle, or is moved to the crankcase via blowby or oil, a small percentage (10 percent to 15 percent) burns, and the remainder (about 5 percent to 10 percent) leaves unburned through the exhaust port.‡ During subsequent cycles, the liquid fuel carried over vaporizes and burns during the main combustion event and in the postflame burned gas, contributing to higher emissions.

Whereas there seems to be little benefit to be drawn from AFI injectors under warmed-up conditions, there is strong evidence that the fine atomization provided by these injectors can indeed decrease emissions during cold start (defined as the time between start to steady-state emissions) by about 50 percent [103, 105]. The finely atomized spray injected during the open-valve period enhances vaporization, leading to lower overfueling requirements for first-cycle starting, and a reduction

‡Schurov and Collings [104] indicate that about 50 percent of the injected fuel could not be accounted for. No such details are indicated by Takeda et al. [103].

in the amount of liquid induced into the combustion chamber. In order to be effective, however, close matching of the injection spray and the geometry of the port must be made, lest the benefits be minimized by wall impingement.

Song et al. [106], Shin et al. [98], Saito et al. [97], and Witze and Green [100, 101] provide dramatic visual evidence of the contribution of liquid fuel to HC emissions during cold start. A sooty flame zone is observed after flame passage near the liquid pools. This is evidence that the rich fuel pockets in the mixture lead to incomplete combustion and high HC emissions. Witze and Green used color video and laser-induced fluorescence of the gasoline fuel to show that liquid fuel is splashed into the cylinder during cold start, with puddles forming around the intake valve and on the piston head (Figure 6.22). With closed-valve injection, liquid fuel first appears around the rims of the intake valve at about the fifth cycle, then appears as a puddle on the piston surface and head. With open-valve injection, the spray momentum also carries liquid fuel to the wet exhaust side of the piston head surface.

Fig. 6.22.

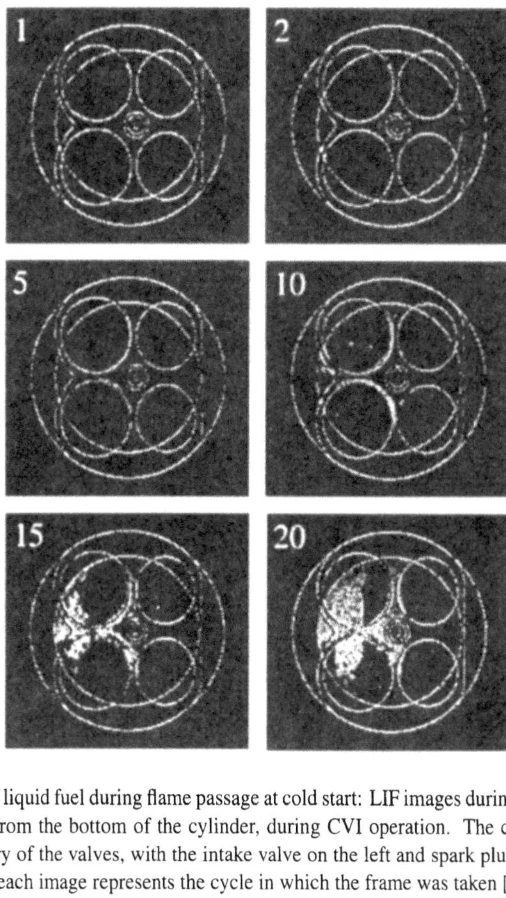

Behavior of liquid fuel during flame passage at cold start: LIF images during the first 20 cycles, as viewed from the bottom of the cylinder, during CVI operation. The clear lines represent the geometry of the valves, with the intake valve on the left and spark plug at the center. The number on each image represents the cycle in which the frame was taken [101].

The reduction in HC emissions during cold start thus depends heavily on understanding the dynamics of fuel transport and vaporization in the cylinder to ensure immediate starting at the minimum overfueling. Currently, the determination of the fueling schedule is done largely empirically through engine calibration techniques.

Nevertheless, models for the fuel behavior and vaporization during cold start are being developed to aid in the design and control of fuel delivery strategies during cold start. Aquino [107] and Hires and Overington [108] suggested a physically based first-order model aimed at predicting the behavior of the vaporizing fuel by determining the time scales for mass transfer to the cylinder (τ) and ratios of mass vaporized to mass injected (x) during start-up. These so-called x-τ models or variations thereof are frequently used in connection with engine control. Experiments are usually performed involving step changes in fueling to extract the appropriate time scales and storage parameters for use in specific engine configurations. Increasingly, physical models are being considered to take into account the heat transfer in the port [109, 110], the properties and characteristics of the fuel film on the wall [111, 112, 113] as well as fuel properties [114, 115].

Mixture preparation and fuel vaporization are important not only during cold start, but particularly during transients. Typical time scales for adjustment of airflow rates are much shorter than that for evaporation of the fuel. Therefore, sudden changes of air and fuel flow during acceleration typically lead to a temporary enleanment of the mixture, whereas decelerations typically produce enrichment, with time constants of the order of 100 ms. Careful attention to the fueling strategy is necessary to compensate for such excursions and achieve tight control of equivalence ratio. Control strategies that take into account the time delays of fuel delivery to the cylinder clearly offer lower HC emissions [116, 117, 118], particularly during hard accelerations, when mixture enrichment is usually applied to compensate for the time lag in fuel vaporization.

Mixture preparation during cold start is still the most important issue dominating control of engine-out HC emissions. Reducing the amount of overfueling during cold start and providing good atomization is key to optimizing the system. In spite of remarkable progress in understanding some of the practical issues, there is still ample room for improvement in models and the design of fueling strategies and control.

6.4.2 Effect of Operating Conditions on HC Emissions

In the most general terms, any factors that contribute to the increase in temperature or availability of oxygen to the postflame region lead to lower HC emissions. Therefore, air-fuel ratio, spark timing, EGR, speed, load, and coolant temperature all have some influence on HC emissions. Because of the large dependence of oxidation on temperature, these sensitivities are usually in the opposite direction to the influence on NO formation. Therefore, simultaneous reduction of NO and HC can rarely be achieved without significant compromise. The extent of the influence on HC emissions, however, usually depends on the particular design features of the engine. An overview of the operating parameters on HC emissions is given in this section.

Relative Air-Fuel Ratio (λ)

Hydrocarbon emission levels decrease with increasing air-fuel ratio (Figure 6.23), and are lower for lower-volatility, nonaromatic fuels. Although temperatures peak around stoichiometric or slightly rich values of air-fuel ratio, the excess oxygen contributes to a lower initial HC level in the charge, as well as additional postflame oxidation. For air-fuel ratios λ higher than about 1.3, however, misfires can become more frequent, leading to increased HC emissions. Lean operation is obviously desirable from the point of view of HC emissions; as mentioned in the previous sections, however, difficulties arise in achieving high NO removal efficiencies in the catalytic converter.

Exhaust Gas Recirculation (EGR)

Increases in EGR lead to a decrease in temperature of the burned gases without the additional oxygen. Whereas EGR promotes a substantial decrease in NO emissions, HC levels tend to increase with EGR, with increasing sensitivity as dilution increases (Figure 6.24) [9, 39, 119]. Beyond about 20 percent EGR, misfires due to dilution and decreased postflame oxidation lead to significant increases in HC emissions, while operating at MBT. Fast-burn engines are designed to compensate

Fig. 6.23.

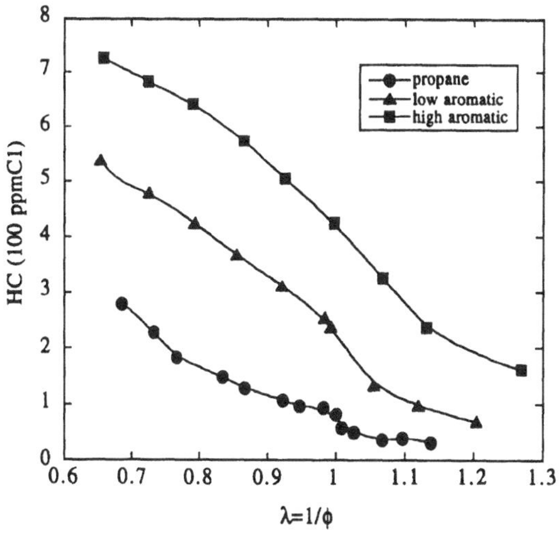

Effect of λ and fuel type on HC emissions [3].

Fig. 6.24.

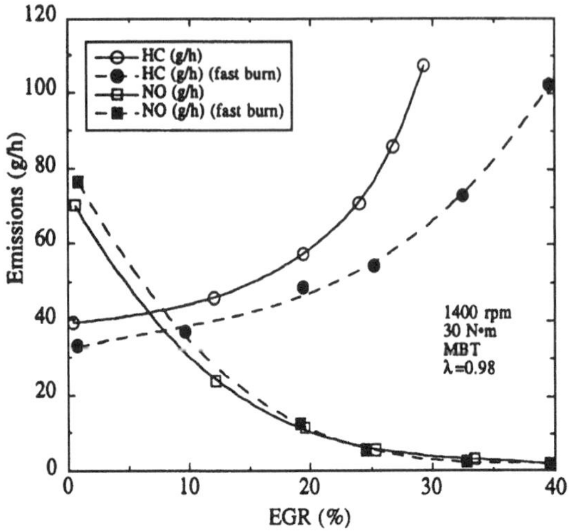

Effect of EGR on HC emissions [39].

for these effects by increasing the flame speed through more vigorous fluid motion.

Spark Timing

Spark timing affects timing of combustion and the maximum pressure, both of which influence storage effects. More important, retarding spark timing from MBT increases the burned gas temperatures during the exhaust phase (as mentioned previously in connection to NO, at the expense of fuel economy). Spark retard is often used to decrease HC emissions during certain operating conditions, such as cold starts. A retard of 10 degrees from MBT leads to a decrease in HC concentrations of the order of about 30 percent to 40 percent [8, 119, 120], at an efficiency penalty of the order of about 5 percent, which implies a correspondingly smaller reduction in the HC emission index at a fixed load.

Speed

The primary effect of increases in engine speed on HCs is to reduce total heat transfer from the burned gases, therefore increasing exhaust temperatures and the rate of oxidation of unburned hydrocarbons. The particular rate of decrease depends on the particular engine, coolant temperatures, and rates of heat transfer. Decreases in HC emission indices of the order of 20 percent to 50 percent are observed for increases in speed from 1000 to 2000 rpm [8, 119, 120], with decreasing sensitivity at higher speeds.

Load

The dependence of HC emissions on load is somewhat more complex, and no general rules for emission indices apply. Residual gas retention decreases for increasing load, so temperatures are higher, and so are oxidation rates. On the other hand, lower residual fractions lead to higher emitted fractions into the manifold (higher ε_m). Both small increases and decreases in HCs with load have been observed, depending on the particular operating conditions; there is typically a sharper increase at very light loads, followed by a plateau [8, 119, 120].

Wall Temperature

Coolant temperatures have significant effects on engine performance and HC emissions. Increasing liner engine temperatures decreases friction losses and lowers fuel consumption. Both cylinder liner and head temperatures in general affect HC emissions. Increased coolant temperatures lead to a lower HC emission index by about 0.4–1.0%/K [121, 122, 123], with only a small effect on NO. The effect appears to be a combined effect of lower storage of HC in crevices (by decreasing the density of the stored gas), earlier desorption or vaporization of HCs from lubricant oil layers (higher wall temperatures lead to decreased oil layer thickness and lower solubility), as well as higher temperatures available for postflame oxidation.

6.4.3 Postflame Oxidation

Only a fraction of the hydrocarbons that escape combustion during flame passage actually leave the cylinder, and an even smaller fraction leaves the exhaust section before reaching the catalytic converter. The overall fraction oxidized at mid-load, mid-speed conditions is of the order of 50 percent to 90 percent [31, 119, 120]. Many factors contribute to determine the final fraction surviving, the most important of which are the temperatures experienced by the emerging unburned hydrocarbons.

The postflame oxidation process is also responsible for the production of partial combustion products, which typically represent 40 percent to 60 percent of the exhaust HC. A large portion of these unburned hydrocarbons are olefinic species, aromatics, and oxygenated HC, which typically have higher tropospheric ozone production potentials than the original fuel [124, 125]. In addition, most of the regulated toxic species (1,3-butadiene, benzene, formaldehyde, and acetaldehyde) are produced during incomplete oxidation of the hydrocarbons.

In spite of the importance of the postflame oxidation processes, the details involved are poorly understood and have only recently been investigated. The presence of hydrocarbons near the cold walls has been investigated using fast sampling valves [126, 127, 128], fast-flame ionization detectors (FIDs) [129, 130] and, more recently, imaging techniques [131].

6.4.3.1 In-Cylinder Oxidation

Most of the fast sampling measurements were made in connection with investigations on the role of the quench layer on HC emissions by LoRusso et al. [132] and Weiss and Keck [126]. Fast sampling valves were placed on the engine liner above the piston top dead center, and samples were collected from a radius of influence of about 300 μm at different times during the cycle. Both sets of measurements reveal that the postflame oxidation takes place within 20–40 milliseconds under the conditions of the study. A sharp decrease in HC concentrations takes place during flame passage, followed by an increase as the piston moves up and pushes HCs from headland crevices into the sampling zone. Products of incomplete oxidation (e.g., ethene, propene, acetaldehyde) were measured, along with the original propane fuel.

Peckham and Collings measured the gradients in total HC concentrations in the cylinder near the exhaust valve [129]. By extending the sampling probe into the cylinder, they showed that HC concentrations approached burned gas concentrations at a distance of about 1 mm from the wall, corresponding to HCs from quench layers and crevices leaving the cylinder during the exhaust cycle.

Green and Cloutman used laser-induced fluorescence imaging to determine the spatial distribution of postflame HC emerging from crevices in an optically accessible engine using graphite piston rings [131]. The images, obtained using a mixture of propane and acetone (as a fluorescing agent), show that essentially no HC is visible prior to bottom dead center (BDC) after combustion in the crevice region away from the ring gap (Figure 6.25). After BDC, however, a plume of HCs becomes visible as it leaves the crevice and expands into the combustion chamber. Later in the cycle, the plume is engulfed by the vortex created by the piston moving upward. After 50° ABDC, substantial mixing has taken place, and there is no apparent decrease in the signal intensity, indicating that oxidation is very slow. A plume of HC is visible emerging from the ring crevice gap region about 90 degrees before bottom dead center, suggesting that the crevice gap contributes a larger fraction of HC than the remainder of the crevice volumes. It has been suggested that the clockwise-rotating plume that becomes visible after BDC is a result of a forced flow through the crevice, perhaps because of the reversal of the ring gaps.

The combined evidence suggests that the postflame oxidation process in the cylinder can be described as follows. After flame passage, hydrocarbons from oil, deposit, and quench layers diffuse into the burned gas mixture due to concentration gradients and, in the case of crevices, after peak pressure due to expansion. At very high burned gas temperatures (higher than 1300–1400 K), oxidation takes place in a layer very close to the wall (of the order of 1 mm). After the temperature drops to about 1200 K and lower, little oxidation takes place, and mixing is promoted by the motion of the piston upward, scraping a boundary layer vortex that eventually decays into smaller scales and promotes molecular mixing.

Fig. 6.25.

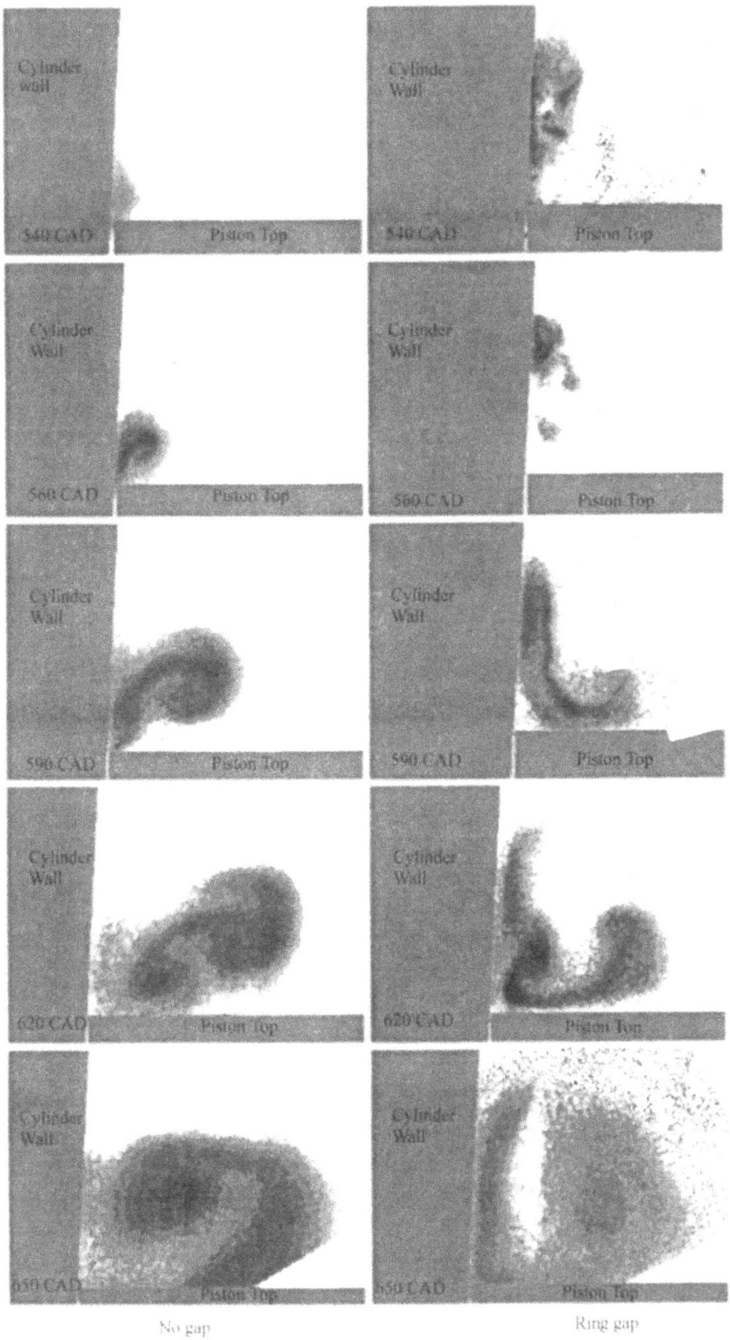

Laser-induced fluorescence images of HC leaving crevices. The image intensity is representative of the concentration of dopant. 540 CAD = BDC exhaust. Jet flow emerging from crevice away from (left) and at the ring gap (right), followed by the roll-up vortex scraped by the piston [131].

Clearly, the extent of oxidation of hydrocarbons from different sources and the production of intermediate species must depend on the burned gas temperatures, rate of emergence, extent of mixing, and the location of the source relative to the exhaust valve. Furthermore, the type of fuel clearly influences both the extent of oxidation and the type and amount of products formed in the exhaust.

Information about the extent of oxidation in the cylinder and exhaust port must in general be obtained indirectly by estimating the sources of unburned HC. Crevice sources can be isolated by running the engine on gaseous fuels (to eliminate the influence of liquid fuel, oil layer, and deposits), and estimating the volumes, densities, and fractions of unburned mixture in these regions. Min et al. [51] estimated that in the case of propane, about 60 percent to 80 percent of the crevice hydrocarbons are oxidized before leaving the engine at warmed-up, mid-load, mid-speed conditions (this includes oxidation in the cylinder and exhaust port). As mentioned in Section 6.4.2.2, crevices of the same volume can produce different final amounts of emissions, depending on crevice geometry and location, primarily through the extent of postflame oxidation and retention.

The extent of oxidation from liquid hydrocarbons desorbed from oil was investigated by Norris and Hochgreb by adding a single-component dopant fuel to the oil while running the engine on hydrogen. Values of 50 percent to 80 percent in-cylinder oxidation were determined, with the lower values corresponding to low-reactivity, high-solubility fuels (xylene), and the high end to lighter, high-reactivity fuels (Figure 6.26) [76].

Estimates of *in-cylinder* oxidation levels are not very reliable, given the difficulty in evaluating the magnitude of each original HC source and the exhaust oxidation. About two-thirds of the hydrocarbons are expected to be oxidized in-cylinder for normal gasoline-fueled engines under mid-load, mid-speed, steady-

Fig. 6.26.

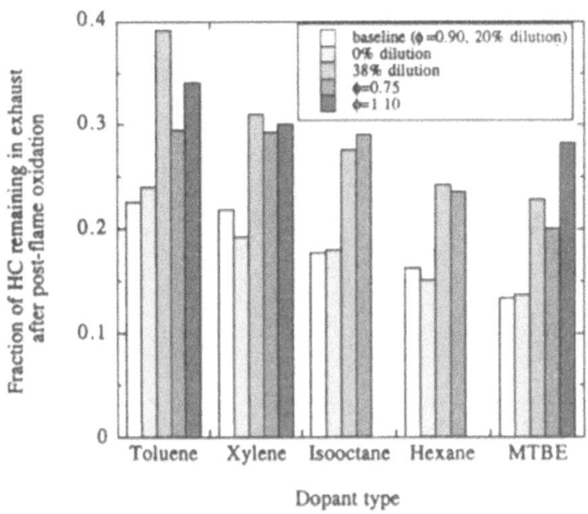

Fraction of HC remaining after oxidation in the cylinder and exhaust, after desorption from lubricant oil. Oxidation rate increases with fuel reactivity toward the right [76].

state conditions. Increases in speed and coolant temperature, retarding spark timing, and decreases in dilution all lead to increased postflame oxidation due to the higher burned gas temperature. Small increases in the air-fuel ratio lead to increases in oxidation (in spite of lower temperatures), and excursions into the rich side obviously decrease the extent of oxidation (due to the lack of oxygen). The fraction of nonfuel hydrocarbons also increases as the temperature increases, since the fuel quickly decomposes into intermediate hydrocarbons, but with insufficient time for complete oxidation. The level of oxidation and partial oxidation is very fuel dependent and has been one of the driving factors in defining the optimum parameters for reformulated fuel composition [124, 133, 134, 135].

6.4.3.2 Exhaust Port Oxidation

Whereas most of the oxidation of source HC takes place inside the cylinder, up to 45 percent of the HCs leaving the cylinder have been measured to oxidize in the exhaust port in a single-cylinder engine by Drobot et al. [136]. The experiments were made by quenching the exhaust at the cylinder exit using cold carbon dioxide and comparing the HC measurements to those without quenching. That is a surprisingly large fraction, considering the relatively low temperatures of exhaust gases. The values measured are somewhat higher than those suggested by Mendillo and Heywood [137] in a previous experiment using the same approach. However, the results in Drobot et al. [136] have also been corroborated by measurements of hydrocarbons along the exhaust port [130, 138].

6.4.3.3 Modeling Postflame Oxidation

Given the complexity of the oxidation process, it is not surprising that only simplified or semi-empirical models [9, 119, 139, 140] or very complex CFD models [141, 142] have appeared to explain the levels of oxidation in engines. However, the increasing evidence based on imaging and sampling experiments suggests that a possible compromise in the models may be achieved by assuming a one-dimensional model for the oxidation along the interface of the emerging unburned gas during the expansion stroke.

A more complex model is clearly required during the upstroke to account for mixing. However, it can be shown that very little oxidation takes place at temperatures below 1000 K in the available time scales, so that the mixing process may be treated without the complexities of the chemistry of oxidation. Such an approach has been explored by several investigators with encouraging results [51, 74, 143].

6.5 SUMMARY

The lightweight, low-cost characteristics of SI engines have made them the workhorse of light-duty vehicle transportation throughout the world. Environmental pressures to reduce tailpipe emissions, as well as to improve fuel consumption and minimize carbon dioxide emissions, will continue to challenge the imagination of engineers in designing solutions to satisfy the competing demands on efficiency, weight, and emission levels.

Substantial reductions in all pollutant emissions have been obtained over the last 25 years. The introduction of the catalytic converter is certainly responsible for a very large fraction of this reduction. Further reductions increasingly depend on careful control of the design details of SI engines.

The pollutants from SI engines of greatest concern in urban areas, with respect to the ability to achieve the necessary targets, are NO and HC. Nitric oxides are formed in the high-temperature flame environment and increase steeply with peak gas temperatures. Increasing exhaust gas dilution levels, while keeping early combustion phasing for good fuel economy, can reduce NO levels down to ULEV standards, with clever use of fluid mechanics, port flow management, and engine control technology.

The greatest lever in the HC control problem is the control of fuel delivery during cold start. Engine calibration is done largely empirically, but improved understanding of the dynamics of fuel vaporization characteristics will lead to advances in the design of more effective fuel management systems. Additional improvements, such as low crevice volumes, reduced lubricant oil volumes, and valve deposit control can also contribute to a cleaner engine.

Recent calls for increased fuel economy have led to the reintroduction of four-stroke direct-injection engines by different manufacturers [144, 145]. These engines offer up to 30 percent higher fuel economy over conventional SI engines. However, it remains unclear whether NO and particularly HC emission levels can be brought to strict levels likely to be enforced over the turn of the century. The very high level of activity in the emission control area and the knowledge developed over the past 25 years of emissions reduction suggest that the pace of invention and design will continue to bring out cleaner internal combustion engines.

REFERENCES

1. Calvert, J.G., Heywood, J.B., Sawyer, R.F., and Seinfeld, J.H., "Achieving Acceptable Air Quality: Some Reflections on Controlling Vehicle Emissions," *Science* **261**: 37–45 (1993).
2. Fox, J.F., Heywood, J.B., and McRae, G.M., "Aggregate Vehicle Emission Estimates for Evaluating Control Strategies," *SAE Paper* 940303 (1994).
3. Harrington, J.A., and Shishu, R.C., "A Single-Cylinder Engine Study of the Effects of Fuel Type, Fuel Stoichiometry and Hydrogen-to-Carbon Ratio on CO, NO and HC Exhaust Emissions," *SAE Paper* 730476 (1973).
4. Zeldovich, J., "The Oxidation of Nitrogen in Combustion and Explosions," *Acta Physiochem. URSS*, **21**:577 (1946).
5. Lavoie, G.A., Heywood, J.B., and Keck, J.C., "Experimental and Theoretical Investigation of Nitric Oxide Fromation in Internal Combustion Engines," *Combust. Sci. Technol.*, **1**:313–326 (1970).

6. Fenimore, C.P., "The Formation of Nitric Oxide in Premixed Hydrocarbon Flames," *Thirteenth Symposium (Int'l) on Combustion*, The Combustion Institute, Pittsburgh, 1971.
7. Drake, M.C., and Blint, R.J., "Calculations of NO_x Formation Pathways in Propagating Laminar, High Pressure Premixed CH4/Air Flames," *Combust. Sci. Tech.* **75**:261–285 (1991).
8. Lavoie, G.A., and Blumberg, P.N., "Measurements of NO Emissions from a Stratified Charge Engine: Comparison of Theory and Experiment," *Combust. Sci. Tech.* **8**:25–37 (1973).
9. Lavoie, G.A., and Blumberg, P.N., "A Fundamental Model for Predicting Fuel Consumption, NO_x and HC Emissions of the Conventional Spark-Ignition Engine," *Combust. Sci. Tech.* **21**:225–258 (1980).
10. Bozzelli, J.W., and Dean, A.M., "O+NNH—A Possible New Route for NO_x Formation in Flames," *Int. J. Chem. Kin*, **27**(11):1097 (1995).
11. Blumberg, P., and Kummer, J.T., "Prediction of NO Formation in Spark-Ignited Engines—An Analysis of the Methods of Control," *Comb. Sci. Tech.* **4**:73–95 (1971).
12. Hanson, R.K., and Salimian, S., *Survey of Rate Constants in the N/H/O System*, Ch. 6 in *Combustion Chemistry*, Gardiner, W.C., ed., Springer-Verlag, New York, 1984.
13. Baulch, D.L., Drysdale, D.D., and Hornge, D.G., *Evaluated Kinetic Data for High Temperautre Reactions*, Vol. 2, Butterworths, London (1973).
14. Komiyama, K., and Heywood, J.B., "Predicting NO_x Emissions and Effects of Exhaust Gas Recirculation in Spark-Ignition Engines," *SAE Paper* 730475 (1973).
15. Miller, J.A., and Bowman, C.T., "Mechanism and Modeling of Nitrogen Chemistry in Combustion," *Prog. Energy Combust. Sci.* **15**:287–338 (1989).
16. Dean, A.J., Davidson, D.F., Hanson, R.K., and Bowman, C.T., "Development and Application of CH Laser Absorption Diagnostic for Shock Tube Kinetic Studies." Western States Section/Combustion Institute Paper WSS/CI 88-91 (1988).
17. Schultz, C., Sick, V., Wolfrum, J., Drewes, V., Zahn, M., and Maly, R., "Quantitative 2D Single-Shot Imaging of NO Concentrations and Temperatures in a Transparent SI Engine," *Twenty-Sixth Symposium (Int'l) in Combustion*, The Combustion Institute, Pittsburgh, 1996, pp. 2597–2604.
18. Raine, R.R., Stone, C.R., and Gould, J., "Modeling of Nitric Oxide Formation in Spark Ignition Engine with a Multizone Burned Gas," *Combust. Flame* **102**: 241–255 (1995).
19. Quader, A.A., "Why Intake Charge Dilution Decreases Nitric Oxide Emission from Spark Ignition Engines," *SAE Paper* 710009 (1971).
20. Huls, T.A., and Nickol, H.A., "Influence of Engine Variables on Exhaust Oxides of Nitrogen Concentrations from a Multicylinder Engine." *SAE Paper* 670482 (1967).
21. Benson, J.D., and Stebar, R.F., "Effects of Charge Dilution on Nitric Oxide Emissions from a Single-Cylinder Engine," *SAE Paper* 710008 (1971).
22. Morris, G. "The Ford Zeta Engine Family, Paper 3—Engine Development," *I. Mech. E. Autotech* 1991, C427/243.
23. Larsson, T., Bergstrom, K., Hinderman, T., Hauptman, L.-G., Wennstrom, L., Denbratt, I., and Higgins, N., "The Volvo 3-L, 6-Cylinder Engine with 4-Valve Technology," *SAE Paper* 901715.
24. Stokes, J., Lake, T.H., Christie, M.J., and Denbratt, I., "Improving the NO_x/Fuel Economy Trade-Off for Gasoline Engines with the CCVS Combustion System," *SAE Paper* 940482 (1994).
25. Iwamoto, Y., Danno, Y., Hirako, O., Fukui, T., and Murakami, N., "The 1.5-Liter Vertical Vortex Engine," *SAE Paper* 920670 (1992).
26. Horie, K., Nishizawa, K., Ogawa, T., Akazaki, S., and Miura, K., "The Development of a High Fuel Economy and High Performance Four-Valve Lean Burn Engine," *SAE Paper* 920455 (1992).
27. Kuroda, H., Nakajima, Y., Sugihara, K., Takagi, Y., and Muranaka, S., "The Fast Burn with Heavy EGR, New Approach for Low NO_x and Fuel Economy," *SAE Paper* 780006 (1978).
28. Newhall, H.K., "Kinetics of Engine Generated Nitrogen Oxides and Carbon Monoxide," *Twelfth International Symposium on Combustion*, pp. 603–695 (1969).
29. Huls, T.A., Meyers, P.S., and Uyehara, O.A., SAE Technical Progress Series, Vehicle Emissions, part II, p. 71, 1967.
30. Keck, J.C., and Gillespie, D., "Rate-Controlled Partial-Equilibrium Method for Treating Reacting Gas Mixtures," *Combust. Flame* **17**:237–241 (1971).

31. Cheng, W.K., Hamrin, D., Heywood, J.B., Hochgreb, S., Min, K., and Norris, M.G., "An Overview of Hydrocarbon Emissions Mechanisms in Spark Ignition Engines," *SAE Paper* 932708 (1993).
32. Lavoie, G.A., "Correlations of Combustion Data for SI Engine Calculations—Laminar Flame Speed, Quench Distance and Global Reaction Rates," *SAE Paper Transactions Paper 780229*, Vol. 87, pp. 1015–1033 (1978).
33. Westbrook, C.K., Adamczyk, A.A., and Lavoie, G.A., "A Numerical Study of Laminar Flame Wall Quenching," *Combust. Flame* **40**:81–90 (1981).
34. Ferguson, C.R., and Keck, J.C., "Stand-Off Distances on a Flat Flame Burner," *Combust. Flame* **34**:85–98 (1979).
35. Cleary, D.J., and Farrell, P.V., "The Effects of Wall Temperature on Flame Structure During Flame Quenching," *SAE Paper* 940683 (1994).
36. Cleary, D.J., and Farrell, P.V., "Single-Surface Flame Quenching Distance Dependence on Wall Temperature, Quenching Geometry and Turbulence," *SAE Paper* 950162 (1995).
37. Adamczyk, A.A., Kaiser, E.W., Lavoie, G.A., and Isack, A.J., "Single-Pulse Sampling Valve Measurements of Wall Layer Hydrocarbons in a Combustion Bomb," *Combust. Flame* **32**:1–9 (1983).
38. Kiehne, T.M., Matthews, R.D., and Wilson, D.E., "The Significance of Intermediate Hydrocarbons During Wall Quench of Propane Flames," *Twenty-First (Int'l) Symposium on Combustion*, The Combustion Institute, Pittsburgh, 1996, pp. 481–489.
39. Kuroda, H., Nakajima, Y., Sugihara, K., Takagi, Y., and Muranaka, S., "The Fast Burn with Heavy EGR, New Approach for Low NO_X and Fuel Economy," *SAE Paper* 780006 (1978).
40. Wentworth, J.T., "Piston and Ring Variables Affect Exhaust Hydrocarbon Emission," *SAE Paper* 680109 (1968).
41. Adamczyk, A.A., "Hydrocarbon Emissions from an Annular Crevice: Effects of Spark/Insert Position, Equivalence Ratio and Pressure," *Combust. Sci. Tech.* **64**:263–276 (1989).
42. Adamczyk, A.A., Kaiser, E.W., Cavolowsky, J.A., and Lavoie, G.A., "An Experimental Study of Hydrocarbon Emissions from Closed Vessel Explosions," *Eighteenth Symposium (Int'l) on Combustion*, The Combustion Institute, Pittsburgh, 1981, pp. 1695–1702.
43. Adamczyk, A.A., Kaiser, E.W., and Lavoie, G.A., "A Combustion Bomb Study of the Hydrocarbon Emissions from Engine Crevices," *Combust. Sci. Tech.* **33**:261–277 (1983).
44. Boam, D.J., Finlay, I.C., Biddulph, T.W., Ma, T., Lee, R., Richardson, S.H., Bloomfield, J., Green, J.A., Wallace, S., Woods, W.A., and Brown, P., "The Sources of Unburnt Hydrocarbon Emissions from Spark Ignition Engines during Cold Starts and Warm-up," *I. Mech. E.* C448/064 (1992).
45. Alkidas, A.C., Drews, R.J., and Miller, W.F., "Effects of Piston Crevice Geometry on the Steady-State Engine-Out Hydrocarbon Emissions of a SI Engine," *SAE Paper* 952537 (1995).
46. Min, K., Cheng, W.K., and Heywood, J.B., "The Effects of Crevices on the Engine-Out Hydrocarbon Emissions in SI Engines," *SAE Paper* 940306 (1994).
47. Thompson, N.D., and Wallace, J.S., "Effect of Engine Operating Variables and Piston and Ring Parameters on Crevice Hydrocarbon Emissions," *SAE Paper* 940480 (1994).
48. Woods, W.A., Brown, P.G., and Crookes, R.J., "Measurements of Unburnt Hydrocarbons from a Four-Valve, Fuel Injection Spark Ignition Engine under Fully Warmed and Warm-up Conditions," *ASME Spring Technical Conference* 95-ICE-20 (1995).
49. Saika, T., and Korematsu, K., "Flame Propagation into the Ring Crevice of a Spark Ignition Engine," *SAE Paper* 861528 (1986).
50. Sterlepper, J., Neusser, H., and Ruhland, H., "HC-Emissions of SI Engines—Optical Investigation of Flame Propagations in Piston Top Land Crevice," *SAE Paper* 941994 (1994).
51. Min, K., and Cheng, W.K., "Oxidation of the Piston Crevice Hydrocarbons During the Expansion Process in a Spark Ignition Engine," *Combust. Sci. Tech.* **106**:307–326 (1995).
52. Wentworth, J.T, "More on Origins of Exhaust Hydrocarbons—Effect of Zero Oil Consumption, Deposit Location and Surface Roughness," *SAE Paper* 720939 (1972).
53. Kaiser, E.W., Adamczyk, A.A., and Lavoie, G.A., "The Effect of Oil Layers on the Hydrocarbon Emissions Generated during Closed Vessel Combustion," *Eighteenth Symposium (Int'l) on*

Combustion, The Combustion Institute, Pittsburgh, 1981, pp. 1881–1890.

54. Haskel, W.W., and Legate, C.E., "Exhaust Hydrocarbon Emissions from Gasoline Engines—Surface Phenomena," *SAE Paper* 720255 (1972).
55. Ishizawa, S., and Takagi, Y., "A Study of HC Emission from a Spark Ignition Engine," *JSME Intern. J.* **30**(260):310–317 (1987).
56. Adamczyk, A.A., and Kach, R.A., "A Combustion Study of Fuel-Oil Solubility and HC Emissions from Oil Layers," *Twentieth Symposium (Int'l) on Combustion*, The Combustion Institute, Pittsburgh, 1984, pp. 37–43.
57. Adamczyk, A.A., Rothschild, W.R., and Kaiser, E.W., "The Effect of Fuel and Oil Structure on Hydrocarbon Emissions from Oil Layers during Closed Vessel Combustion," *Combust. Sci. Tech.* **44**:113–124 (1985).
58. Schramm, J., and Sorenson, S.C., "Effects of Lubricating Oil on Hydrocarbon Emissions in an SI Engine," *SAE Paper* 890622 (1989).
59. Beckwith, P., and Cooper, J.H., "A Lubricant Formulation for Lower Unburnt Hydrocarbon Emissions," *SAE Paper* 950162 (1994).
60. Carrier, G., Fendell, F., and Feldman, P., "Cyclic Absorption/Desorption of Gas in a Liquid Wall Film," *Combust. Sci. Tech.* **25**:9–19 (1981).
61. Dent, J.C., and Lakshminarayanan, P., "A Model for Absorption and Desorption of Fuel Vapor by Cylinder Lubricating Oil Films and Its Contribution to Hydrocarbon Emissions," *SAE Paper* 830652 (1983).
62. Schramm, J., and Sorenson, S.C., "A Model for Hydrocarbon Emissions from SI Engines," *SAE Paper* 902169 (1990).
63. Korematsu, K., and Takahashi, S., "Effects of Fuel Absorbed in Oil Film on Unburnt Hydrocarbon Emissions from Spark Ignition Engines," *JSME Intern. J.* **34**(3), 362–368 (1991).
64. Linna, J.R., and Hochgreb, S., "Analytical Scaling Model for Hydrocarbon Emissions from Fuel Absorption in Oil Layers in Spark Ignition Engines," *Combust. Sci. Tech.* **109**:205–226 (1995).
65. Gatellier, B., Trapy, J., Herrier, D., Quelin, J.M., and Galliot, F., "Hydrocarbon Emissions of SI Engines as Influenced by Fuel Absorption-Desorption in Oil Films," *SAE Paper* 920095 (1992).
66. Kaiser, E.W., Siegl, W., and Russ, S.G., "Fuel Composition Effects on Hydrocarbon Emissions from a Spark-Ignited Engine: Is Fuel Absorption in Oil Significant?," *SAE Paper* 952542 (1995).
67. Korematsu, K., "Effects of Fuel Absorbed in Oil Film on Unburnt Hydrocarbon Emissions from Spark Ignition Engines," *JSME Intern. J.* Ser. II, Vol. 33, No. 3 (1990).
68. Trinker, F.H., Anderson, R.W., Henig, Y.I., Siegel, W.O., and Kaiser, E.W., "The Effect of Fuel-Oil Solubility on Exhaust HC Emissions," SAE Paper 912349 (1991).
69. Chappelow III, C.C., and Prausnitz, J.M., "Solubilities of Gases in High-boiling Hydrocarbon Solvents," *AIChE J.* **20**:1097–1104 (1974).
70. Sugiyama, T., Takeuchi, T., and Suzuki, Y., "Thermodynamic Properties of Solute Molecules at Infinite Dilution Determined by Gas-Liquid Chromatography." *J. Chromat.* **105**, 265–272 (1975).
71. Schramm, J., and Sorenson, S.C., "Solubility of Gasoline Components in Different Lubricants for Combustion Engines Determined by Gas-Liquid Partition Chromatography," *J. Chromat.* **538**, 241–248 (1991).
72. Murakami, Y., and Athara, H., "Analysis of Mechanism of Intermixing of Combustion Products in Engine Oil," *JSME Intern. J.*, Series II, Vol. 34, No. 4 (1991).
73. Frottier, V., Heywood, J.B., and Hochgreb, S., "Measurement of Gasoline Absorption into Engine Lubricating Oil," *SAE Paper* 961229 (1996).
74. Norris, M.G., Bauer, W., and Hochgreb, S., "Oxidation of Hydrocarbons from Lubricant Oil Layers in Spark-Ignition Engines," *Twenty-Sixth Symposium (Int'l) on Combustion*, The Combustion Institute, Pittsburgh, 1996, pp. 2645–2652.
75. Shih, L.K., and Assanis, D.N., "Modeling Unburned Hydrocarbon Formation Due to Absorption/Desorption Processes into the Wall Oil Film," *Symposium on the Mechanism and Chemistry of Pollutant Formation and Control from Internal Combustion Engines*, American Chemical Society, Div. Petroleum Chemistry, Washington, DC, August 23–28, 1992, pp. 1479–1497.
76. Norris, M.G., and Hochgreb, S., "Extent of Oxidation of Hydrocarbons Desorbing from the

Lubricant Oil Layer in Spark-Ignition Engines," *SAE Paper* 960069 (1996).
77. Kalghatgi, G.T., "Deposits in Gasoline Engines—A Literature Review," *SAE Paper* 902108 (1990).
78. Harpster, Jr., M.O., Matas, S.E., Fry, J.H., and Litzinger, T.A., "An Experimental Study of Fuel Composition and Combustion Chamber Deposit Effects on Emissions from a Spark Ignition Engine," *SAE Paper* 950740 (1995).
79. Bitting, W.H., Firmstone, G.P., and Keller, C.T., "Effects of Combustion Chamber Deposits on Tailpipe Emissions." *SAE Paper* 940305 (1994).
80. Studzinsky, W.M., Liiva, P.M., Choate, P.J., Acker, W.P., Smooke, M., Brezinsky, K., Litzinger, T., and Bower, S.A., "Computational and Experimental Study of Combustion Chamber Deposit Effects on NO_x Emissions," *SAE Paper* 932815 (1993).
81. Gagliardi, J.C., "The Effect of Fuel Anti-Knock Compounds and Deposits on Exhaust Emissions," *SAE Paper* 670128 (1967).
82. Adamczyk, A., and Kach, R.A., "The Effect of Engine Deposit Layers on Hydrocarbon Emissions from Closed Vessel Combustion," *Combust. Sci. Tech.* **47**:193–212 (1986).
83. Adamczyk, A., and Kach, R.A., "Fuel Storage in Combustion-Chamber Ceramics and HC Emissions: a Connection," *Combust. Sci. Tech.* **51**:251–263 (1987).
84. Haidar, H., Heywood, J., and Hochgreb, S., "Combustion Chamber Deposit Effects on Engine HC Emissions," *Symposium on the Mechanism of Combustion Chamber Deposits*, presented before the Division of Petroleum Chemistry, Inc., American Chemical Society, New Orleans, LA, March 24–29, 1996.
85. Bower, S.L. Jr., Litzinger, T.A., and Frottier, V.A., "The Effect of Fuel Composition and Engine Deposits on Emissions from a Spark Ignition Engine," *SAE Paper* 932707 (1993).
86. Price, R.J., Wilkinson, J.P.T., Jones, D.A.J., and Morley, C., "A Laboratory Simulation and Mechanism for the Fuel Dependence of SI Combustion Chamber Deposit Formation," *SAE Paper* 952445 (1995).
87. Kalghatgi, G.T., McDonald, C.R., and Hopwood, A.B., "An Experimental Study of Combustion Chamber Deposits and Their Effects in a Spark-Ignition Engine," *SAE Paper* 950680 (1995).
88. Quader, A.A., "How Injector, Engine and Fuel Variables Impact Smoke and Hydrocarbon Emissions with Port Fuel Injection," *SAE Paper Transactions Paper* 890623 (1989).
89. Yang, J., Kaiser, E.W., Anderson, R.W., and Siegl, W.O., "Effects of Port-Injection Timing and Fuel Droplet Size on Total and Speciated Exhaust HC Emissions," *SAE Paper* 930711 (1993).
90. Alkidas, A.K., "The Effects of Fuel Preparation on Hydrocarbon Emissions of a SI Engine Operating under Steady-State Conditions," *SAE Paper* 941959 (1994).
91. Alkidas, A.K., and Drews, R.J., "Effects of Mixture Preparation on HC Emissions of a SI Engine Operating under Steady-State Cold Conditions," *SAE Paper* 961956 (1996).
92. Quader, A.A., Sloan, T.M., Sinkevitch, R.M., and Olson, K.L., "Why Gasoline 90% Distillation Temperature Affects Emissions with Port Fuel Injection and Premixed Charge," *SAE Paper* 912430 (1991).
93. Skippon, S.M., Glover, A.R., Cooney, P.J., and Boyle, R.D., "Studies of Mixture Preparation in a Spark Ignition Engine Using Interferometric Laser Imaging for Droplet Sizing (ILIDS)," *SAE Paper* 950457 (1995).
94. Hardalupas, Y., Taylor, A.M.K.P., Whitelaw, J.H., Ishii, K., Miyano, H., and Urata, Y., "Influence of Injection Timing on In-Cylinder Fuel Distribution in a Honda VTEC-E Engine," *SAE Paper* 950507 (1995).
95. Posylkin, M., Taylor, A.M.K.P., Vannobel, F., and Whitelaw, J.H., "Fuel Droplets inside a Firing Spark-Ignition Engine," *SAE Paper* 941989 (1994).
96. Meyer, R., and Heywood, J.B., "Liquid Fuel Transport Mechanisms into the Cylinder of a Firing Port-Injected SI Engine During Start-Up," *SAE Paper* 970865 (1997).
97. Saito, K., Sekiguchi, K., Imatake, N., Takeda, K., and Yaegashi, T., "A New Method to Analyze Fuel Behavior in a Spark Ignition Engine," *SAE Paper* 950044 (1995).
98. Shin, Y., Cheng, W.K., and Heywood, J.B., "Liquid Gasoline Behavior in the Engine Cylinder

of a SI Engine," *SAE Paper* 941872 (1994).

99. Shin, Y., Min, K., and Cheng, W.K., "Visualization of Mixture Preparation in a Port-Fuel Injection Engine during Engine Warm-Up," *SAE Paper* 952481 (1995).
100. Witze, P.O., and Green, R.M., "LIF Visualization of Liquid Fuel in the Cylinder of a Spark Ignition Engine," *Proceedings of the Eighth International Symposium on Applications of Laser Techniques to Fluid Mechanics*, Lisbon, July 8–11 (1996).
101. Witze, P.O., and Green, R.M., "LIF and Flame-Emission Imaging of Liquid Fuel Films and Pool Fires in an SI Engine During a Cold Start," *SAE Paper* 970866 (1997).
102. Swindal, J.C., Dragonetti, D.P., Hahn, T.T., Furman, P.A., and Acker, W.P., "In-Cyliner Charge Homogeneity During Cold Start Studied with Fluorescent Tracers Simulating Different Fuel Distillation Temperatures," *SAE Paper* 950106 (1995).
103. Takeda, K., Yeagashi, T., Sekiguchi, K., Saito, K., and Imatake, N., "Mixture Preparation and HC Emissions of a 4-Valve Engine with Port Fuel Injection during Cold Starting and Warm-up," *SAE Paper* 950074 (1995).
104. Schurov, S.M., and Collings, N., "An Experimental Investigation of Fuel Transport in a Port Injected Engine," *SAE Paper* 952485 (1995).
105. Hilbert, H.S., Boggs, D.L., and Schechter, M.M., "The Effects of Small Fuel Droplets on Cold Engine Emissions Using Ford's Air-Forced Injection System," *SAE Paper* 952479 (1995).
106. Song, K., Clasen, E., Chang, C., Campbell, S., and Rhee, K.T., "Post-Flame Oxidation and Unburned Hydrocarbon in a SI Engine," *SAE Paper* 952543 (1995).
107. Aquino, C.F., "Transient A/F Control Characteristics of the 5.0 Liter Central Fuel Injection Engine," *SAE Paper* 810494 (1981).
108. Hires, S.D., and Overington, M.T., "Transient Mixture Strength Excursions—An Investigation of Their Causes and the Development of a Constant Mixture Strength Fuelling Strategy," *SAE Paper* 810494 (1981).
109. Shayler, P.J., Colechin, M.J.F., and Scarisbrick, A., "Fuel Film Evaporation and Heat Transfer in the Intake Port of an SI Engine," *SAE Paper* 961120 (1996).
110. Almkvist, G., and Eriksson, S., "An Analysis of Air to Fuel Ratio Response in a Multipoing Fuel Injected Engine under Transient Conditions," *SAE Paper* 932753 (1993).
111. Almkvist, G., and Eriksson, S., "A Study of Air to Fuel Transient Response and Compensation with Different Fuels," *SAE Paper* 941931 (1994).
112. Johnen, T., and Haug, M., "Spray Formation Observation and Fuel Film Development Measurements in the Intake of a Spark Ignition Engine," *SAE Paper* 950511 (1995).
113. Almkvist, G., Dembratt, I., Josefsson, G., and Magnusson, I., "Measurements of Fuel Film Thickness in the Inlet Port of an SI Engine by LIF," *SAE Paper* 952483 (1995).
114. Chen, K.C., Dewitte, K., and Cheng, W.K., "Fuel Effects and Enrichment Effects on Engine Starting and Warm-up Behavior," *SAE Paper* 950065 (1995).
115. Chen, K.C., DeWitte, K., and Cheng, W.K., "A Species-Based Multi-Component Volatility Model for Gasoline," *SAE Paper* 941877 (1994).
116. Matsumoto, M., Saitoh, M., Matsumura, R., and Yukawa, F., "Improvement of Lambda Control Based on an Exhaust Emission Simulation Model that Takes into Account Fuel Transportation in the Intake Manifold," *SAE Paper* 900612 (1990).
117. Horie, K., Takahasi, H., and Akazaki, S., "Emissions Reduction During Warm-Up Period by Incorporating a Wall-Wetting Fuel Model on the Fuel Injection Strategy During Engine Starting," *SAE Paper* 952478 (1995).
118. Nasu, M., Ohata, A., and Abe, S., "Model-Based Fuel Injection Control System for SI Engines," *SAE Paper* 961188 (1996).
119. Trinker, F.H., Cheng, J., and Davis, G.C., "A Feedgas HC Emission Model for SI Engines Including Partial Burn Effects," *SAE Paper* 932705 (1993).
120. Thompson, N.D., and Wallace, J.S., "Effect of Engine Operating Variables and Piston and Ring Parameters on Crevice Hydrocarbon Emissions," *SAE Paper* 940480 (1994).
121. Kobayashi, H., Yoshimura, K., and Hirayama, T., "A Study on Dual Circuit Cooling for Higher

Compression Ratio," *SAE Paper* 841294 (1984).
122. Russ, S., Kaiser, E.W., and Siegl, W., "Effect of Cylinder Head and Engine Block Temperature on HC Emissions from a Sincle Cylinder SI Engine," *SAE Paper* 952536 (1995).
123. Guillemot, P., Gatellier, B., and Rouverirolles, P., "The Influence of Coolant Temperature on Unburned Hydrocarbon Emissions from Spark Ignition Engines," *SAE Paper* 941962 (1994).
124. Lowi, A., Jr., and Carter, W.P.L., "A Method for Evaluating the Atmospheric Ozone Impact of Actual Vehicle Emissions," *SAE Paper* 900710 (1990).
125. Bowman, F.M., and Seinfeld, J.H., "Atmospheric Chemistry of Alternate Fuels and Reformulated Gasoline Components," *Prog. Ener. Comb. Sci.* **21**(5):387–417 (1996).
126. Weiss, P., and Keck, J.C., "Fast Sampling Valve Measurements of Hydrocarbons in the Cylinder of a CFR Engine," *SAE Paper* 810149 (1981).
127. Sellnau, M.C., Springer, G.S., and Keck, J.C., "Measurements of Hydrocarbon Concentrations in the Exhaust Products from a Spherical Combustion Bomb," *SAE Paper* 810148 (1981).
128. Adamczyk, A.A., Kaiser, E.W., Cavolowsky, J.A., and Lavoie, G.A., "An Experimental Study of Hydrocarbon Emissions from Closed Vessel Explosions," *Eighteenth Symposium (Int'l) on Combustion*, The Combustion Institute, Pittsburgh, 1981, pp. 1695–1702.
129. Peckham, M., and Collings, N., "Study of the Engine Wall Layer Hydrocarbons with a Fast Response FID," *SAE Paper* 932237 (1993).
130. Finlay, I.C., Boam, D.J., Bingham, J.F., and Clark, T.A., "Fast Response FID Measurement of Unburned Hydrocarbons in the Exhaust Port of a Firing Gasoline Engine," *SAE Paper* 902165 (1990).
131. Green, R.M., and Cloutman, L.D., "Planar LIF Observations of Unburned Fuel Escaping the Upper Ring-Land Crevice in an SI Engine," *SAE Paper* 970823 (1997).
132. LoRusso, J.A., Kaiser, E.W., and Lavoie, G.A., "In-Cylinder Measurements of Wall Layer Hydrocarbons in a Spark-Ignited Engine," *Comb. Sci. Tech.* **33**:75–112 (1983).
133. Kaiser, E.W., Siegl, W.O., Henig, Y.I., Anderson, R.W., and Trinker, F.H., "Effect of Fuel Structure on Emissions from a Spark-Ignited Engine," *Environ. Sci. Technol.* **25**:2005–2012 (1991).
134. Kaiser, E.W., Siegl, W.O., Cotton, D.F., and Anderson, R.W., "Effect of Fuel Structure on Emissions from a Spark-Ignited Engine. 2. Naphtene and Aromatic Fuels," *Environ. Sci. Technol.* **26**:1581–1447 (1992).
135. Kaiser, E.W., Siegl, W.O., and Anderson, R.W., "Fuel Structure and the Nature of Engine Out Emissions," *SAE Paper* 952542 (1995).
136. Kaiser, E.W., Siegl, W.O., Trinker, F.H., Cotton, D.F., Cheng, W.K., and Drobot, K., "Effect of Engine Operating Parameters on Hydrocarbon Oxidation in the Exhaust Port and Runner of a Spark-Ignited Engine," *SAE Paper* 950159 (1995).
137. Mendillo, J.V., and Heywood, J.B., "Hydrocarbon Oxidation in the Exhaust Port of a Spark Ignition Engine," *SAE Paper* 810019 (1981).
138. Kayes, D., and Hochgreb, S., "Development of a Time and Space Resolved Sampling Probe Diagnostic for Engine Exhaust Hydrocarbons," *SAE Paper* 961002 (1996).
139. Wu, K., Hochgreb, S., and Norris, M.G., "Chemical Kinetic Modeling of Exhaust Hydrocarbon Oxidation," *Combust. Flame* **100**:193–201 (1995).
140. Lee, G.R., and Morley, C., "Chemical Modelling of Hydrocarbon Exhaust Emissions," *SAE Paper* 941958 (1994).
141. Reitz, R.D., and Kuo, T.W., "Modeling of HC Emissions Due to Crevice Flows in Premixed Charge Engines," *SAE Paper* 892085 (1989).
142. Hellstrom, T., and Chomiak, J., "Oxidation of Hydrocarbons Released from Piston Crevices of SI Engines," *SAE Paper* 952539 (1995).
143. Wu, K., and Hochgreb, S., "The Roles of Chemistry and Diffusion on Hydrocarbon Post-Flame Oxidation", Comb. Sci. Tech, In Press (1998).
144. Kume, T., Iwamoto, Y., Iida, K., Murakami, M., Akishino, K., and Ando, H., "Combustion Control Technologies for Direct Injection SI Engine," *SAE Paper* 960600 (1996).
145. Toyota Motor Corporation Press Release, August 5, 1996.

CHAPTER 7

Pollution from Rotary Internal Combustion Engines

Mark Dulger
Department of Mechanical Engineering, Ben-Gurion University of the Negev, Beer-Sheva, Israel

7.1 INTRODUCTION

The operation of the rotary internal combustion engine is illustrated in Figure 7.1. The intake process goes through (1), (2), (3), and (4); it is followed by compression at (5), (6), and (7); next is the combustion and expansion stage (8), (9), (10); finally the exhaust stroke takes place at (11) and (12). Accordingly, the rotary engine is a four-stroke engine. One of the specific features of this engine is that as the rotor makes one complete rotation, the output shaft accomplishes three revolutions. In

ISBN: 0-12-639855-0
$25.00

Fig. 7.1.

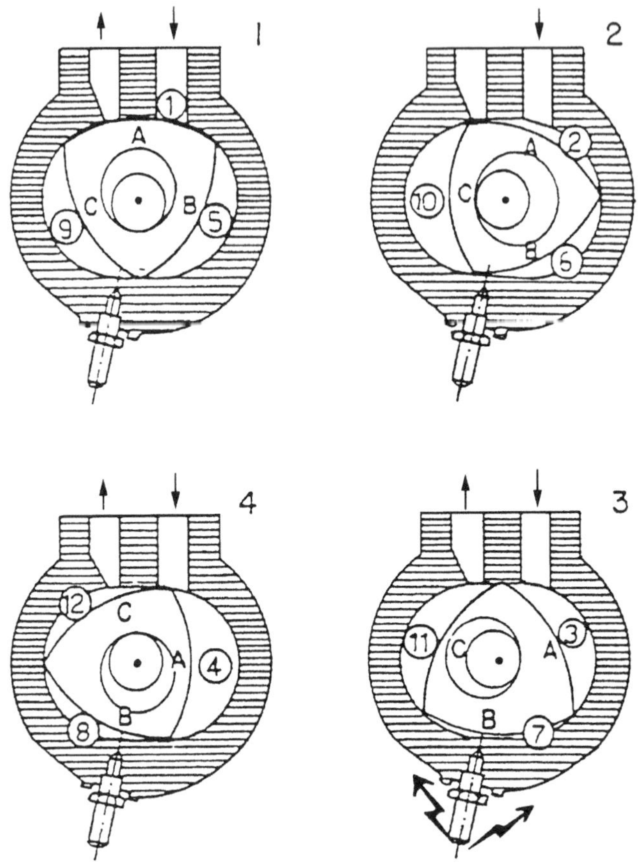

Operation principle of a rotary engine.

terms of the rotating angle of the output shaft, the time required to complete one stroke is 270 degrees, that is, 1.5 times more than for a four-stroke reciprocating engine.

From the design point of view, the rotary engine differs from the reciprocating piston engine in four primary ways:

- The rotor substitutes for the piston.
- The housing substitutes for the cylinder.
- The eccentric shaft substitutes for the crankshaft and connecting rods.
- Valves, camshaft, and timing belt are eliminated due to intake and exhaust ports in the housing; their opening and closing are accomplished directly by the motion of the rotor.

The rotary engines have several important advantages over piston engines [1]:

- Reduced size and weight, competitive with the gas turbine.
- Low vibration with as few as one or two rotors.

- Higher speed capability by virtue of complete balance.
- High volumetric efficiency through porting without the limitations of valve dynamics.
- Sizing flexibility.
- Mechanical simplicity.

In addition, due to the longer intake stroke duration in the rotary engine, high volumetric efficiency can be achieved even at high speeds, resulting in flat "torque versus engine speed" curve [2]. Also, the rotary engine is less prone to knocking and therefore more tolerable to the fuel quality. Another feature of the rotary engine is that its geometry is very suitable for charge stratification, as the rotor always moves the air charge past the stationary location of the spark plug and nozzles, and thus develops the necessary flow distribution for charge stratification within the chamber [1]. Finally, in recent years there is an increased interest in development of the rotary engine fueled by hydrogen. While the reciprocating four-stroke engine fueled by hydrogen is prone to preignition and backfiring through the intake ports, in the rotary engine the intake ports are separated from the combustion zone, so there are no hot spots in the induction phase.

The two main disadvantages of the rotary engines are the increased emission of unburned hydrocarbons and higher fuel consumption. Both disadvantages are related to the long and narrow shape of the combustion chamber, resulting in high surface-to-volume (S-V) ratio (large volume of the quenching layers), increased sealing perimeter (increased gas leakage from the combustion chamber), and long flame travel (lower thermodynamic efficiency).

At the same time, the content of NO_x in the exhaust gases of rotary engines is lower than in reciprocating engines. Regarding carbon monoxide, the level of its emissions is similar to that from piston engines. Figure 7.2 shows a

Fig. 7.2.

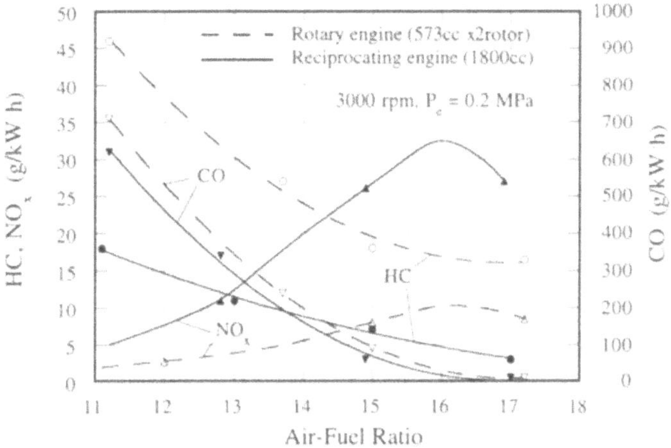

Comparison of exhaust emission characteristics [2].

comparison of the exhaust emission characteristics between rotary and reciprocating engines.

In the late 1970s, a thorough investigation of the sources of HC emissions was carried out at General Motors [3, 4]. The investigation was vital to the future of rotary engines, for unless the base engine hydrocarbon emissions (and fuel consumption) could be cut to a level near that of a reciprocating engine, it would have been extremely difficult for rotary engines to enter the engine market. The bulk of this section is based on the results of those studies.

Beforehand, however, a short note on the low NO_x emission rate in the rotary engines should be made. Plots shown in Figures 7.3 and 7.4 illustrate a comparison between nitrogen oxides emissions from carbureted rotary and piston engines. The tested engines were a four-cylinder, in-line 2300-cc displacement,8.44 compression ratio, water-cooled engine manufactured by Ford Motor Company and a two-rotor 1308-cc displacement, 9.4 compression ratio, water-cooled Model 13B engine manufactured by Mazda. The catalytic converter and exhaust gas recirculation apparatus were removed from both engines. The fuel used was a synthetic coal-derived gasoline. In the whole range of tested engine speeds and loads, the NO_x emissions from rotary engines were about two to three times lower. As the engine speed increased, the measured NO_x emissions from the rotary engine remained almost constant.

One of the reasons for the low NO_x emission rate underlined in the study is the high concentration of burned gas in the fresh mixture. This is due to the leakage across apex seals, which creates a built-in exhaust gas recirculation system in the rotary engine. Another important reason is the relatively high rate of heat transfer into the walls of the combustion chamber caused by its high S-V ratio.

Fig. 7.3.

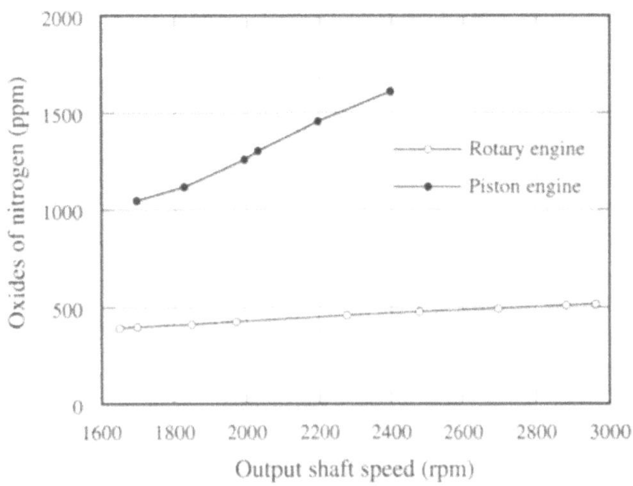

Engine speed effects on the oxides of nitrogen emissions [5]. (Reprinted with permission from SAE.)

Fig. 7.4.

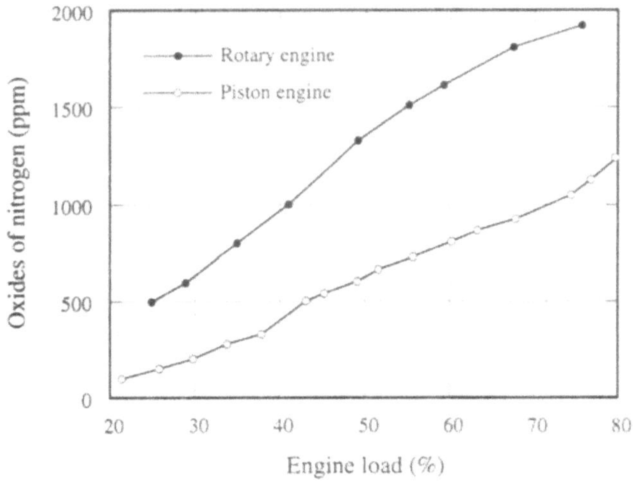

Variations of oxides of nitrogen emissions with engine load [5]. (Reprinted with permission from SAE.)

Both phenomena lead to low maximum combustion temperatures and, therefore, a low rate of NO_x generation.

7.2

SOURCES OF HYDROCARBON EMISSIONS

To assist the analysis of the reasons for high HC emissions and the main sources of hydrocarbons in a rotary engine exhaust, a cross-section of the General Motors Rotary Engine (GMRE) is shown in Figure 7.5.

This engine has two spark plugs per rotor, side inlet ports, and peripheral exhaust ports with 0° overlap between ports. It is worth noting that the spark plugs' position on the housing surface in the GMRE is different from that in the 12A or 13B engines of Mazda and it has a significant effect on hydrocarbon emissions. The reason for the latter is the different extent of flame propagation into the trailing part of the combustion chamber. Generally, the rotor motion creates an almost unidirectional flow of gas in the combustion chamber along the housing. Due to the varying distance between the rotor and housing surfaces, the flow velocity changes from cross-section to cross-section. At the stage of combustion, the velocity of the unburned gas in the trailing part of the chamber toward the flame front is significantly higher than the velocity of points on the rotor surface in the corresponding cross-sections. Thereby, a kind of squish is developed in the chamber trail. This squish pushes back the flame front as shown in Figure 7.6 and in the mid- and high-speed ranges prevents the flame from propagating to the combustion chamber trailing end before the exhaust port opening [6]. In that way,

Fig. 7.5.

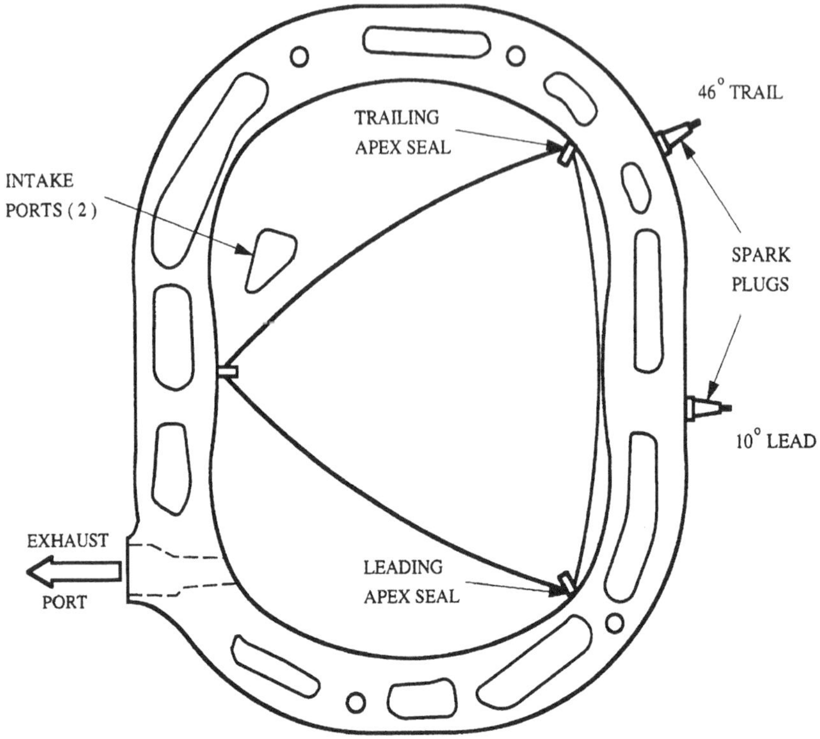

Cross-section of the General Motors Rotary engine with rotor face "A" at 0° rotor angle [3]. (Reprinted with permission from SAE.)

a certain portion of the fuel in the chamber trail remains unburned and is swept by the trailing apex seal into the exhaust manifold.

The unburned fuel in the trail of the chamber is considered to be one of the main sources of hydrocarbon emissions in rotary engines. Figure 7.7 illustrates this statement.

The plot represents time-resolved HC data taken from the GMRE exhaust port with the help of exhaust gas sampling during peak-pressure-triggered tests at an A-F ratio of 16.5:1. The curves obtained by the authors for other A-F ratios are of a similar shape. The plot shows HC concentration as wet hexane versus rotor angle position. The curve starts at the exhaust port opening and ends at the exhaust port closing. Because the apex seal never completely covers the exhaust port, there are periods of time for 6 degrees of rotor rotation at the opening and closing when the port is open to two adjacent chambers at once. At the port opening, concentrations are relatively high because the sample contains some gases from the latter part of the exhaust process from the preceding chamber. The low HC concentration in the sample from the center portion of the chamber suggests that these gases are relatively completely reacted and mixed with small amounts of quench layer gases. Close to the port closing, HC concentrations shoot rapidly upward. The remainder of the concentration curve stays at high HC concentrations.

Fig. 7.6.

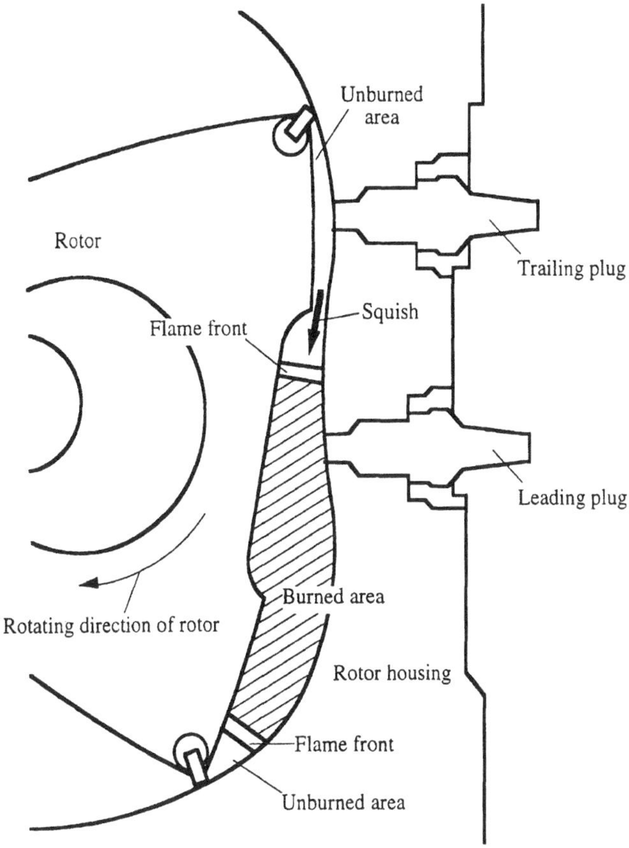

Pushing back of flame front by squish [6]. (Reprinted with permission from SAE.)

Besides the failure of the flame to propagate to the trailing apex seal, the authors of the study suggest three other mechanisms which could be responsible for high HC concentrations late in the exhaust process:

1. Unburned fresh mixture could leak across the apex seal from the trailing chamber when its compression pressure exceeds the pressure in the exhausting chamber.
2. The trailing apex seal could scrape the unburned quench layer from the housing wall and push it out of the port last.
3. Unburned HC could bleed from the clearance volume under the trailing apex seal, when the chamber pressure drops due to the exhaust port opening, and be swept out of the port just before the seal passes.

Another result of the study is that the peak pressure level has an insignificant effect on the time-resolved HC concentration curves.

Fig. 7.7.

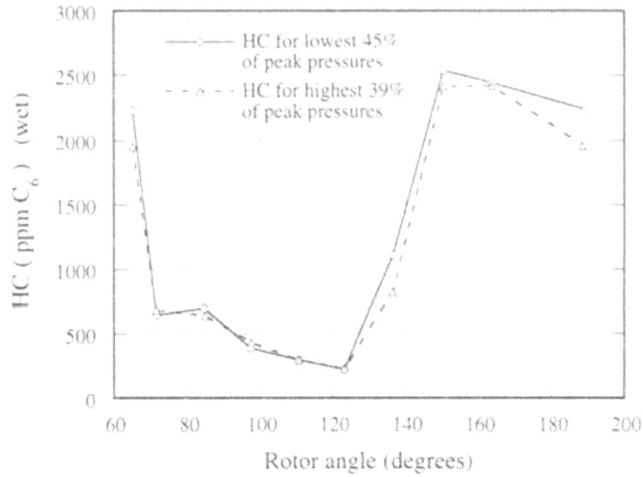

Time-resolved HC emissions at 16.5:1 air-fuel ratio [3]. (Reprinted with permission from SAE.)

The first of the preceding mechanisms—apex seal leakage due to its imperfect sealing to the trochoid surface—was also considered to be one of the main sources of HC emissions in the study [4]. Note that the leakage across the trailing apex seal is what was earlier called a built-in exhaust gas recirculation system in the rotary engine, contributing to an overall dilution of the charge by burned gases. Leakage from the hydrocarbon-rich narrow leading end of the combustion chamber past the leading seal directly contributes to the engine exhaust hydrocarbons. Various apex seal designs, shown in Figure 7.8, were examined with respect to exhaust hydrocarbons [4]. The split conformable seals of type GM-43 (left in the figure) showed the best performance. Their introduction reduced HC emissions by as much as 22 percent.

Another specific feature of the rotary engine is the presence of the so-called cold wall zone. Figure 7.9 compares the layouts of the carbureted reciprocating and rotary engines [4]. Both have two zones in common—an induction zone and a combustion zone. Both induction zones may, at times, have walls covered with liquid fuel. The rotary engine, however, has an additional cold wall zone, which could be wetted by fuel in a way similar to that of the induction zone. Then, however, the seals would carry the fuel to the next zone, where it is not properly burned because of its location in crevices next to the seals. It is then carried to the exhaust zone where it could contribute to the characteristic HC peak near the end of the exhaust stroke (see Figure 7.7). Note that in addition to the mechanisms mentioned previously, this mechanism is also very likely to cause high HC levels late in the exhaust process.

In the study [4], the contribution of the cold wall zone factor was assessed through comparison between HC emissions from the engines fueled by gasoline

Fig. 7.8.

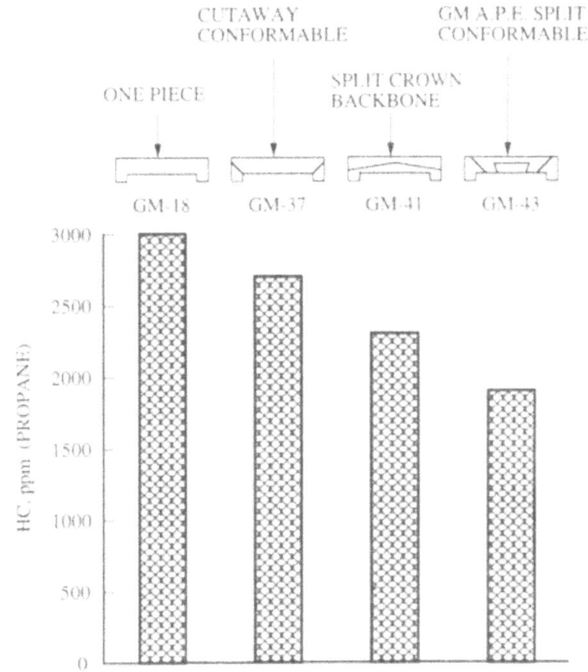

Performance of various apex seal designs with respect to exhaust hydrocarbons at 1500 rpm, road load [4]. (Reprinted with permission from SAE.)

and methane. Methane, being a gas, would not wet the walls. The results are shown in Figure 7.10. As was expected, the reciprocating engine, which has no cold wall zone, gave about the same HC emissions with both methane and gasoline. In the rotary engine, methane provided a remarkably lower HC output. The authors admit that part of this reduction may be due to improved air-fuel mixing, but suppose this factor to bring a minor gain.

One potential remedy is to raise the temperature of the cold walls by heating them with the exhaust gases. The effect from such an arrangement, together with the effect of coolant heating, is illustrated by Figures 7.11 and 7.12. The latter shows the results of steady-state chassis dynamometer tests on a two-rotor engine equipped vehicle. It is seen that heating both the cold wall zone and the coolant significantly reduces HC emissions. However, the authors point out that heating of the coolant is accompanied by a rise of NO_x emissions. (The extent of this rise is not reported.) In summary, they suggest that about 18 percent of the overall HC emissions could be attributed to the influence of the cold wall zone.

A well-known disadvantage of the rotary engine combustion chamber is its relatively high S-V ratio, causing increased surface quench area per unit displacement.

Fig. 7.9.

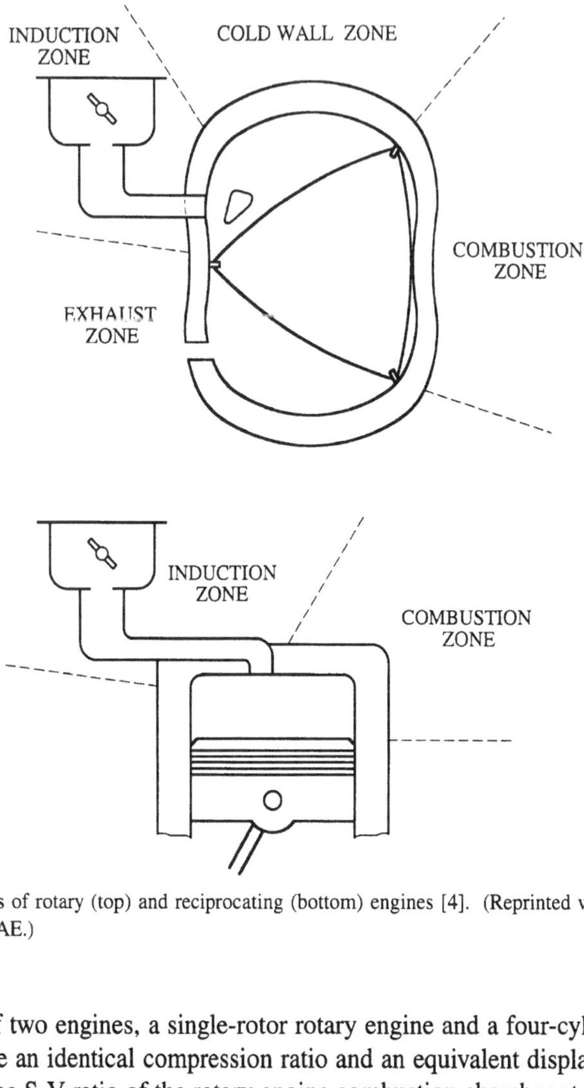

Layouts of rotary (top) and reciprocating (bottom) engines [4]. (Reprinted with permission from SAE.)

Typically, if two engines, a single-rotor rotary engine and a four-cylinder piston engine, have an identical compression ratio and an equivalent displacement volume, then the S-V ratio of the rotary engine combustion chamber at top center is about two times larger than that of the cylinder of a piston engine. To examine the effect of the S-V ratio on the overall HC emissions, several rotary engines with different S-V ratios were tested (Figure 7.13). As expected, the engine with the lowest S-V ratio had the lowest indicated specific hydrocarbons.

Furthermore, the authors considered the rotor surface and the housing and side walls as separate HC sources. At top center, the rotor surface accounts for about 40 percent of the combustion chamber surface. A standard rotor and a specially designed hot surface rotor with heat-insulating sealed air cavities were compared with respect to HC output (Figure 7.14). Along with the hydrocarbon reduction shown with the hot rotor was an increase in NO_x emissions of about 20 percent. At

Fig. 7.10.

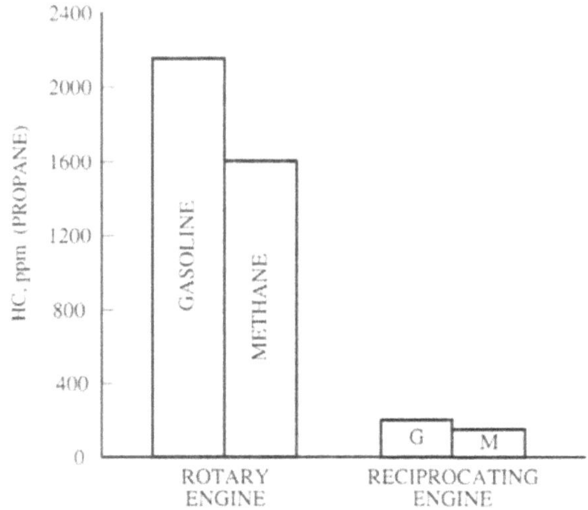

Gasoline and methane fuels compared in rotary and reciprocating engines at 1500 rpm, road load [4]. (Reprinted with permission from SAE.)

Fig. 7.11.

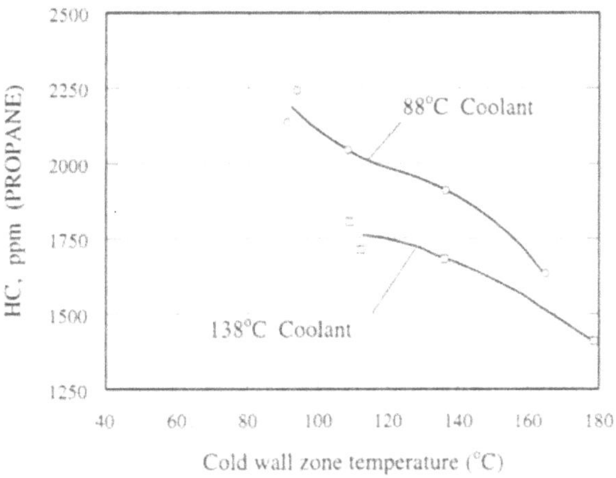

Effect of the cold wall zone temperature and the coolant temperature on base engine hydrocarbons [4]. (Reprinted with permission from SAE.)

Fig. 7.12.

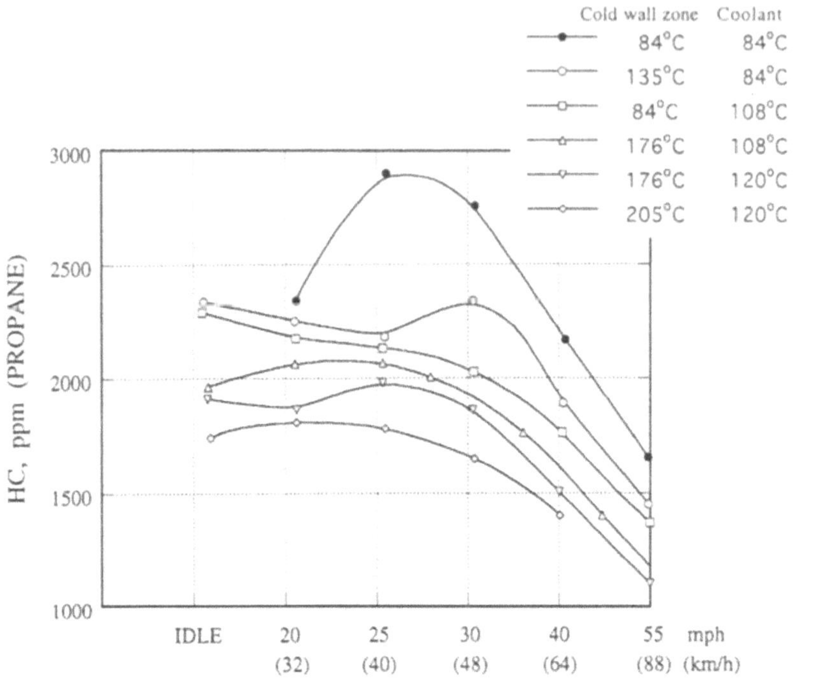

Effect of the exhaust-heated cold wall zone temperature and the coolant temperature on base engine HC emissions on a chassis dynamometer [4]. (Reprinted with permission from SAE.)

Fig. 7.13.

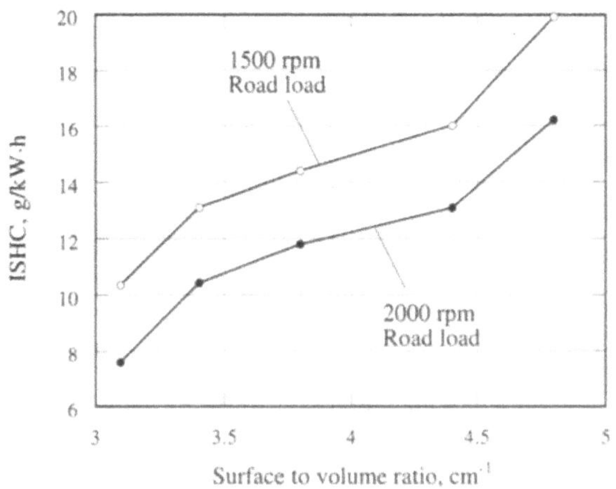

Effect of the S-V ratio on HC emissions in exhaust [4]. (Reprinted with permission from SAE.)

Fig. 7.14.

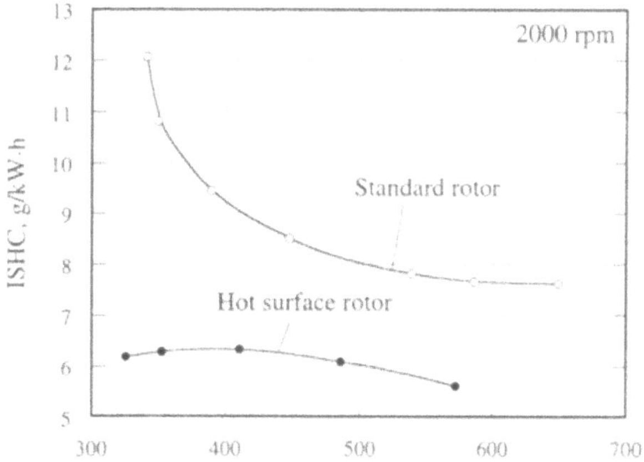

Indicated mean effective pressure, kPa

Reduction of indicated specific hydrocarbons through use of hot surface rotor [4]. (Reprinted with permission from SAE.)

heavy loads the rotor surface overheated. From the obtained data, the authors estimate the contribution of rotor surface quench to total HC output at about 10 percent.

Estimating the contribution of the housing and side walls to the HC emissions, the authors base their conclusion on the HC reduction achieved with increased coolant temperature. They tentatively attribute about 10 percent to this source at road load.

The next HC source remarked upon in the study [4] is a relatively high percent of residuals in the combustion chamber, which is believed to result from the specific rotary engine geometry, the gas exchange system, and the previously mentioned leakage past the trailing apex seal. A special test was run with the arrangement used to simulate residual change (Figure 7.15). Introducing various amounts of oxygen into the intake manifold provides replacement of the oxygen burned out of the residuals and thus simulates decreased residual concentration. The results of the test are shown in Figure 7.16, where V_{ept} stands for volumetric efficiency. The data led the authors to conclude that the contribution of the residuals to the total HC output is very high—about 26 percent.

One more source of hydrocarbons related to flame quenching is the crevice volume in the combustion chamber, which refers to zones too narrow to be penetrated by the flame. The crevice volume of a piston engine consists mainly of the volume above the top piston ring. In a rotary engine, the crevice volume includes the volume above the side seal and the volume below the apex seals. It was estimated [4] that the crevice volume in a rotary engine is responsible for about 6 percent of the hydrocarbon emissions.

Several other minor factors contributing to HC emissions in rotary engines are mentioned in the study. Among them, the lubricant coating over the combustion

Fig. 7.15.

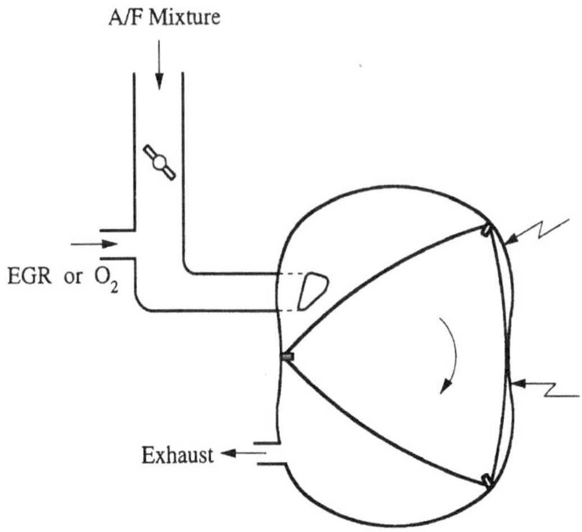

Arrangement for simulating residual change in the rotary engine [4]. (Reprinted with permission from SAE.)

Fig. 7.16.

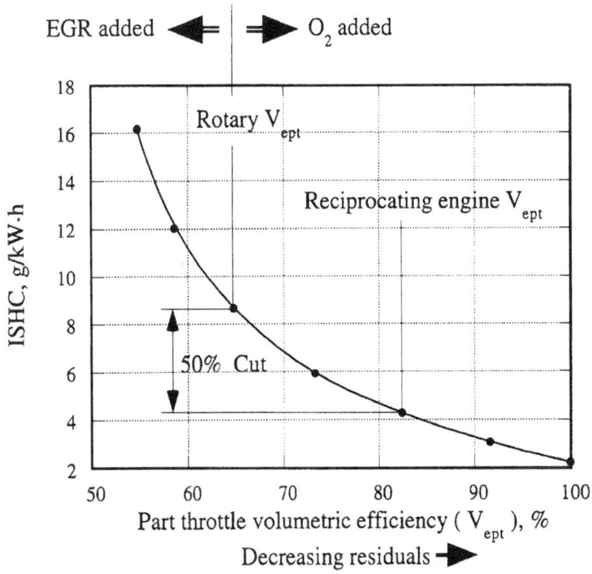

Change in hydrocarbons with changes in apparent residual level in the rotary engine [4]. (Reprinted with permission from SAE.)

chamber surface, which contributes through absorption of the fuel into the wall lubricant film during the compression stroke, combination of the spark plugs' location on the housing surface with the type of combustion pocket in the rotor, location of the spark plug gap with respect to surface on the housing, and others. The authors assigned a total contribution of 8 percent to the whole group.

A summary of the hydrocarbon sources in the carbureted GMRE [4] for 1500 rpm road load is shown in Figure 7.17.

Different methods to reduce HC emissions in carbureted rotary engines were suggested and implemented in different engine prototypes. Among them are catalytic converters and exhaust reactors, improved apex seals, exhaust-heated cold wall zone to overcome wall wetting effects, heated coolant to reduce combustion chamber surface quenching, raised side seals to reduce crevice volume, use of

Fig. 7.17.

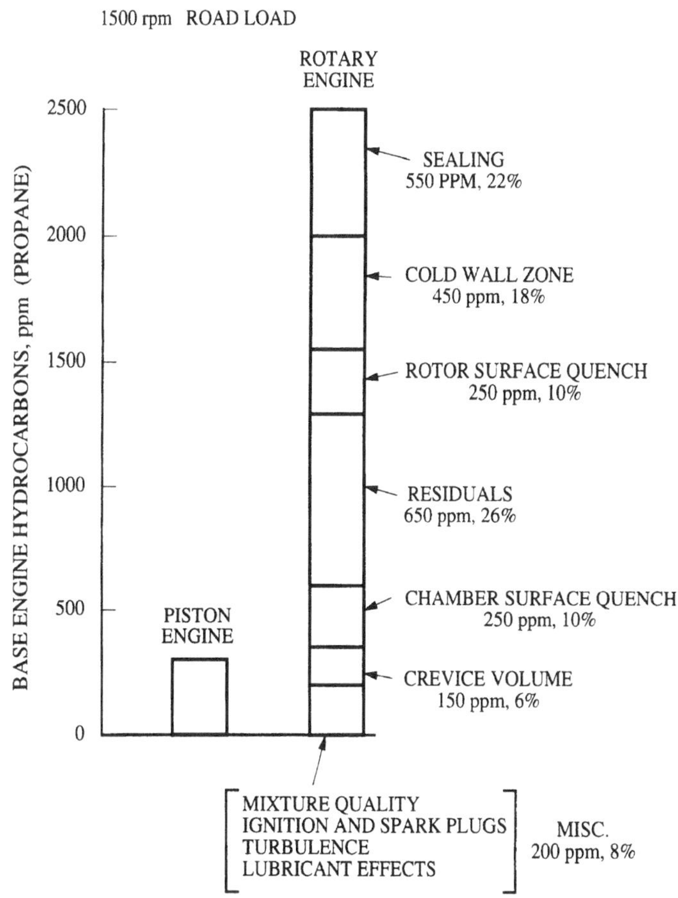

Summary of hydrocarbon sources identified in the rotary engine at 1500 rpm, road load [4]. (Reprinted with permission from SAE.)

hot surface rotors, and mounting of an additional far trailing spark plug. To date, however, none of the carbureted rotary engines have achieved the HC emission characteristics of reciprocating engines.

A somewhat different approach to reduce hydrocarbons in rotary engines is associated with direct fuel injection. In this way, some of the hydrocarbon sources could be eliminated, while others could be significantly suppressed. Thus, operating the engine unthrottled reduces the residuals problem. Injection into the combustion zone avoids the cold wall wetting problem, since the fuel never passes through the cold wall zone. It also reduces the crevice volume effect. Charge stratification in the combustion chamber allows the trailing part of the chamber to contain almost no fuel. Therefore, the negative effect of incomplete flame propagation into the chamber trail is reduced as well. Also, if the leading part of

Fig. 7.18.

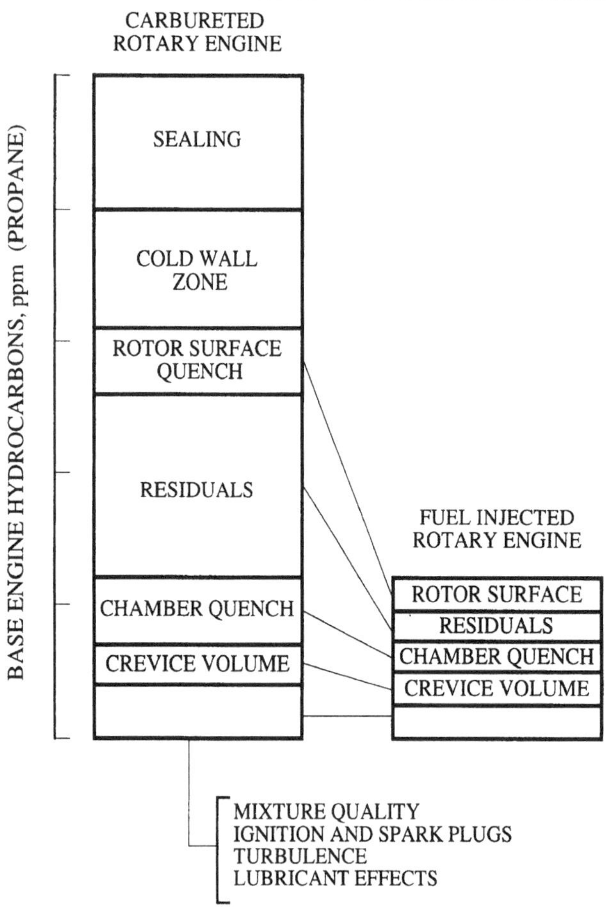

Estimate of the reduction of hydrocarbons possible in a rotary engine using fuel injection into the combustion chamber [4]. (Reprinted with permission from SAE.)

Fig. 7.19.

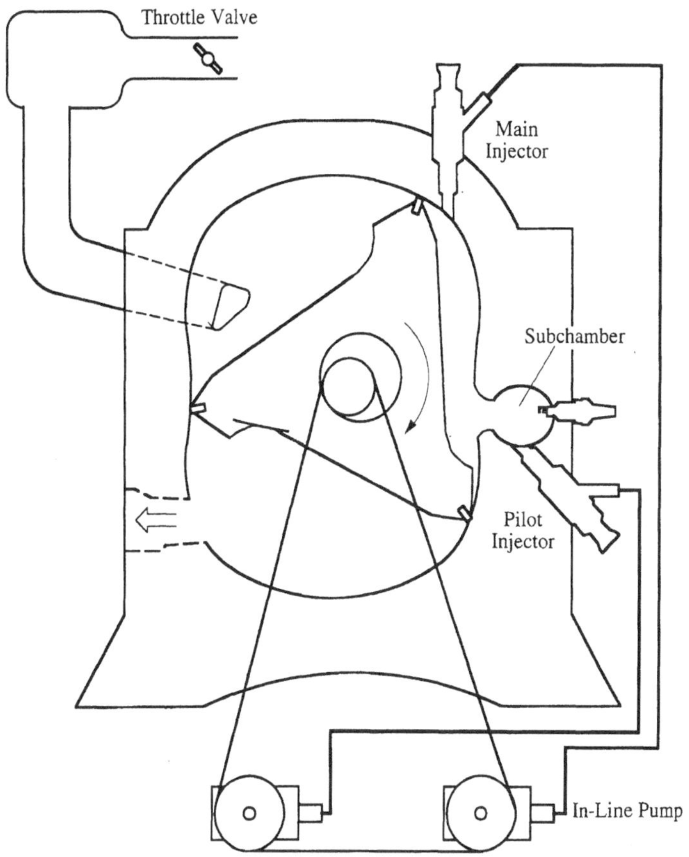

Schematic diagram of rotary engine with a direct injection stratified charge rotary engine [7]. (Reprinted with permission from SAE.)

the chamber is free of unburned hydrocarbons, leakage past the leading apex seal no longer contributes to HC emissions.

The authors of the study [4] estimated the effect of direct fuel injection on HC emissions from a rotary engine as shown in Figure 7.18.

To date, however, no reliable data on HC emissions from direct injection stratified charge rotary engines have been reported in the literature. Perhaps it is worth mentioning the study [7] carried out at Mazda Motor Corporation with the engine whose layout is shown in Figure 7.19. The authors reported on the changes relative to a conventional carbureted rotary engine of the performance characteristics including HC and NO_x emissions (Figure 7.20). It was emphasized in the study that the significant reduction of hydrocarbons achieved in the stratified charge process is contrary to the general tendency for HC emissions to increase with stratified combustion in piston engines. The authors consider it as a unique feature of the rotary engines and attribute the result mainly to their high S-V ratio.

Fig. 7.20.

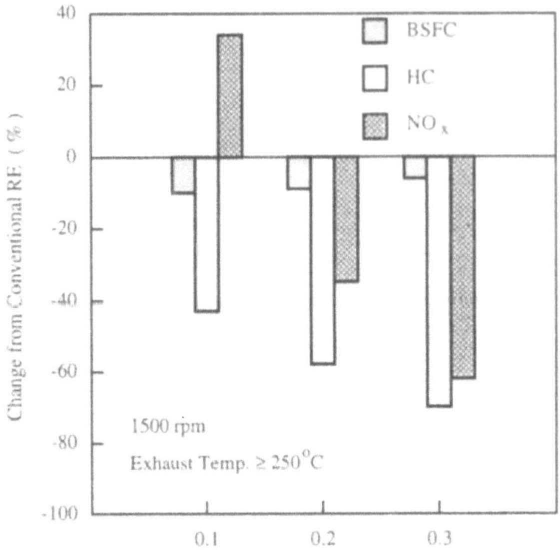

Break Mean Effective Pressure (MPa)

The fuel consumption and exhaust emissions performance of a direct injection stratified charge rotary engine compared to a conventional rotary engine [7]. (Reprinted with permission from SAE.)

REFERENCES

1. Jones, C., "Advanced Development of Rotary Stratified Charge 750 and 1500 HP Military Multi-Fuel Engines at Curtiss-Wright," *SAE Paper* 840460 (1984).
2. Yamamoto, K., *Rotary Engines*, Tokyo: Sankaido, 1981.
3. Bayer, R.J., S.F. DeNagel, and J.C. Steiner, "Rotary Combustion Engine Hydrocarbon Source Studies," *SAE Paper* 780965 (1978).
4. Burley, H.A., M.R. Meloeny, and T.L. Stark, "Sources of Hydrocarbon Emissions in Rotary Engines," *SAE Paper* 780419 (1978).
5. Kappos, C., and S. Rajan, "Relative Performance of Rotary and Piston Engines on Synthetic Coal Derived Gasoline," *SAE Paper* 890212 (1989).
6. Shimizi, R., T. Tadokoro, T. Nakanishi, and J. Funamoto, "Mazda 4-Rotor Rotary Engine for the Le Mans 24-Hour Endurance Race," *SAE Paper* 920309 (1992).
7. Kagawa, R., S. Okazaki, N. Somyo, and Y. Akagi, "A Study of a Direct-Injection Stratified-Charge Rotary Engine for Motor Vehicle Application," *SAE Paper* 930677 (1993).

CHAPTER 8

Control Technologies in Spark-Ignition Engines

Brian E. Milton
School of Mechanical and Manufacturing Engineering, The University of New South Wales, Sydney, Australia

ISBN: 0-12-639855-0
$25.00

8.1

GLOBAL AND LOCAL EMISSIONS: A BRIEF OVERVIEW OF THE PROBLEM

8.1.1 Background of the Problem

The exhaust emissions from internal combustion engines were perceived as a relatively minor problem until the 1960s. Prior to that period, the public was generally conscious of only two problematic exhaust emission factors from engines in general and gasoline engines in particular.

1. The exhaust of an engine contained substantial levels of carbon monoxide, (CO), which in a closed area was known to be a highly toxic substance in any but very low proportions. That is, it was general knowledge that engines should only be run in a well-ventilated space. The appropriate solution was, therefore, perceived as the individual taking appropriate care during operation of either vehicles in garages or engines used in the stationary mode in enclosed spaces.
2. Vehicles running on very rich fuel-air mixtures produced intolerable levels of black smoke in the immediate vicinity of the exhaust outlet. Also those where the piston rings, bore, and so forth were well worn produced blue/grey smoke in copious quantities. Neither of these was viewed as other than a minor irritant. The solution to both these problems lay in appropriate engine maintenance and tuning.

Air pollution, particularly in the form of smog, became a major problem in the 1960s due to increasing affluence and hence use of energy, industrialization, total population increases worldwide, and demographic redistribution to the cities. An early manifestation (Butler [1]) was the thick London fogs of around 1952 (the word *smog* does not seem to have been coined until a few years later) where an estimated 4000 people died of diseases associated with breathing. These were primarily due to the particulates from the domestic coal burners used at the time for space heating, and a combination of low wind strengths and near freezing conditions which caused a low-level, stagnant layer of highly polluted air. They were soon controlled by the compulsory change to smokeless fuel, the last of note occurring in 1962. Nevertheless, the primary cause, which is too many people in a confined region burning very large quantities of fuel inefficiently, was the same as that which affected Los Angeles with an entirely different climate a short time later. In this latter case, however, the secondary cause was different, being fundamentally a function of extensive automobile use. The solution was not as easy as in the original London case because of two factors. First, the details of the problem were more varied in that the emissions were made up predominantly of initially less visible products, unburned hydrocarbons and various nitrogen/oxygen compounds. Second, an easy switch to a different, less polluting energy source was not practicable for a number of reasons, these being that fuel type is restricted due to:

- The short time periods available in engines for fuel introduction and mixing.
- The combustion in engines providing a short residence time, a limitation on lean air-fuel ratios and necessarily high temperatures.
- The high fuel energy density necessary for adequate range in automobiles, aircraft, and other mobile applications of engines.

Thus, a major effort was required to control vehicle exhaust emissions by means which involved, at first, relatively minor changes to the engine and the fuel. As emission standards have tightened, the control strategies have changed and these have generally required more changes to engine design details, higher levels of control of the engine parameters, and increasingly sophisticated exhaust clean-up equipment.

Although there has been substantial progress in reducing the total emissions from individual engines, increasing traffic volume has meant that the Los Angeles type problem has not been solved and has in fact developed in many other cities, sometimes to an even greater degree. This is particularly noticeable in rapidly developing countries where population growth in the cities and the consequent traffic increase has been substantial over a very short period of time. Typical current (1996) examples of extreme air pollution from vehicle emissions are found in Bangkok, Seoul, and Manila, although no major city is immune. To varying extents, this has been exacerbated by other growth factors in the community, which include industry, other forms of transport (e.g., air and rail) and direct and indirect domestic fuel requirements. The contribution of vehicles can range from less than

Fig. 8.1.

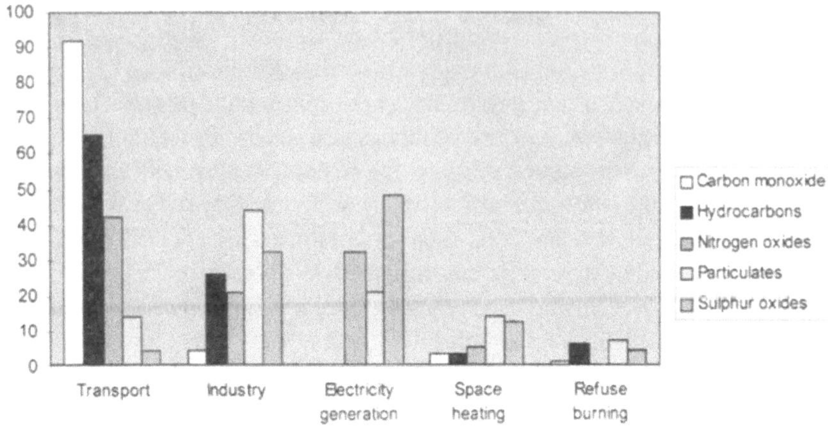

Relative emission production from various sources (data from Ledbetter [2]).

to substantially more than 90 percent, 60 percent and 40 percent for CO, HC, and NO_x respectively as shown on Figure 8.1 but is lower for particulates and sulphur oxides. Of the vehicle particulates, most are derived from diesel engines. Some of the nonvehicle emissions are potentially more controllable, either with a change of fuel type as in the London space heating example cited earlier, or by changing the locally used power to a different form of energy (e.g., combustion to electrical heating) where the production source is located well away from the city in less sensitive regions.

8.1.2 Global and Local Emissions

The pollution problems can be conveniently classified into two groups: *global* and *local*. The global variety receiving attention only since the 1980s are those which affect the atmosphere on a worldwide scale. The others, some of which have been referred to earlier, include carbon monoxide, unburned hydrocarbons, NO_x compounds, aldehydes, and particulates. The total amounts in a global sense are not significant but they create major problems due an intense accumulation in regions of high activity and population density. The compounds disperse away from these regions and have a minimal effect in an overall sense on the world's atmosphere. Hence, the products are sometimes termed *local emissions*. Each of these has an individual cause and methods of control related to the combustion process.

In summary then, global emissions have a universal effect, while local emissions have a regional one. A global problem associated with engines, although it is not directly an emissions problem, is the limited total fossil fuel resource that is available worldwide. This may force changes in engine technology to occur when alternative fuels are introduced whether they are of benefit in emission reduction or not. The main global emission problem is the dispersal into the atmosphere of relatively small quantities (compared to the total atmospheric mass) of exhaust gases such as carbon dioxide, CO_2 from fossil fuel combustion which remains there for extensive periods. These incrementally and progressively change the atmospheric composition as shown on Figure 8.2 with the likely result being a significant change in the overall climate indicated by both modeling and measurement. This phenomenon was first postulated in the 1970s but was not seriously considered until the 1980s. Since then, there has been considerable debate regarding whether the small temperature rises observed in many locations on the planet do in fact represent a global average rise and whether this greenhouse phenomenon is the major influence. Also, there are differences in opinion on what the ultimate effects will be. While the debate is likely to continue for some time, the general consensus seems to be that the high rates of fossil fuel use, the atmospheric levels of carbon dioxide, and the atmospheric temperature rises are indeed related and global with their effects unlikely to be generally benign.

8.1.3 Formation of Undesirable Emission Products

8.1.3.1 Local Emissions

The basic local emissions from vehicles with reciprocating engines (Ledbetter [2]) are shown in Table 8.1. For gasoline (i.e., SI engines) the emissions that occur in substantial quantities are carbon monoxide, hydrocarbons, and nitrogen oxides.

While the formation of the basic local emissions has been discussed elsewhere in this text, a brief discussion is included here. The principal products of concern are carbon monoxide, CO, unburned hydrocarbons (UHC), nitrogen oxides, called NO_x, and particulates. The unburned hydrocarbons are sometimes measured as NMHC (nonmethane hydrocarbons) because methane is essentially nonreactive under the circumstances prevailing for ozone and smog formation and it is, therefore, the nonmethane compounds that are significant. The first three

Table 8.1

Emissions from Uncontrolled Reciprocating Engines

Pollutant	SI engines kg/1000 liters fuel burned	CI engines kg/1000 liters fuel burned
Aldehydes	0.5	1.2
Carbon Monoxide	276	7.2
Hydrocarbons	24	16
Nitrogen Oxides	14	27
Particulates	1.4	13
Organic Acids	0.5	3.7
Sulphur Oxides	1.1	5

Fig. 8.2.

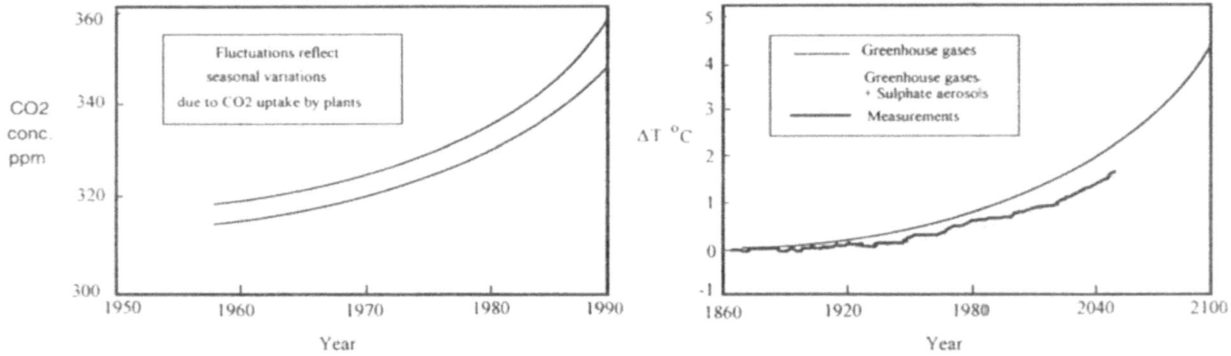

(a) Approximate change in CO_2 concentration

(b) Change in climatic temperature
Models use greenhouse gases only,
or greenhouse gases + sulphate aerosols

Typical estimates of increasing levels of atmospheric CO_2 and its possible effect (data from Carson [3]).

emission products are the most significant for SI engines (Heywood [4]) and will be discussed here. The particulates are more important in CI engines and gas turbine operation and are discussed in other chapters in this book.

Carbon Monoxide Carbon monoxide results from the combustion in insufficient oxygen of a fuel that contains carbon. The normal combustion route is that, when a hydrocarbon fuel and oxygen interact, carbon monoxide is preferentially formed, followed by water, with the carbon monoxide then being oxidized to carbon dioxide by any remaining oxygen. If the oxygen exists in less than stoichiometric proportions, this final process is incomplete with some of the molecules remaining as CO. A typical gasoline contains a large range of quite different hydrocarbons but has an average composition that may be typified by a formula such as $(CH_m)_n$ where m has a value of about 1.85 to 2.0 and n is approximately 7. To fully oxidize this substance (say $m = 1.85$, $n = 7$) to water and carbon dioxide, $1.4625n$ molecules of oxygen, O_2 or 6.96 molecules of air (air being taken as having approximately 3.76 moles of nitrogen to 1 mole of oxygen) are required. This is the stoichiometric ratio and may be expressed as an air-fuel ratio by volume (kmoles) of 6.96 or more usually, air-fuel ratio by mass where the value is about 14.50. If however, the air-fuel mass ratio for the foregoing fuel is lowered to about 6, the amount of oxygen available is so low that only water, hydrogen, and carbon monoxide are possible. That is, if only somewhere about $0.6n$ oxygen molecules would be available, only CO forms. Note that this is below the generally accepted rich flammability limit. From the rich flammability limit to stoichiometric, the combustion products must contain some CO, its proportion decreasing toward stoichiometric. That is, in order to contain CO to low levels, it is desirable to have a mixture with stoichiometric air as the minimum.

In practice, the situation is a little more complex. At the high temperatures and pressures that exist during the combustion process, significant quantities of CO form even when there is sufficient oxygen for complete combustion to occur. This is due to dissociation of the CO_2 molecules into CO and O_2 because of the high-temperature molecular vibration. In practice, the reaction at these temperatures is simply incomplete and ceases at the local chemical equilibrium composition. When the gas cools during expansion in the cylinder, the majority of the rate equations involved in CO and O_2 recombination to CO_2 are sufficiently fast so that equilibrium exists more or less continuously. That is, the recombination process occurs within the time scale of the engine processes. Thus, even as the products of combustion cool during expansion within the cylinder, the equilibrium curve is substantially followed and the CO is converted to CO_2. Once the exhaust valve opens, experiments show that the very rapid fall in pressure and temperature due to exhaust blowdown (i.e., the wave action that commences the gas exhaust) may cause some departure from equilibrium, particularly with rich mixtures. However, the known rates for the basic recombination reactions are fast enough so that in theory equilibrium should be maintained throughout. In general, departure from equilibrium is of only moderate significance; the reaction continues in the exhaust

Fig. 8.3.

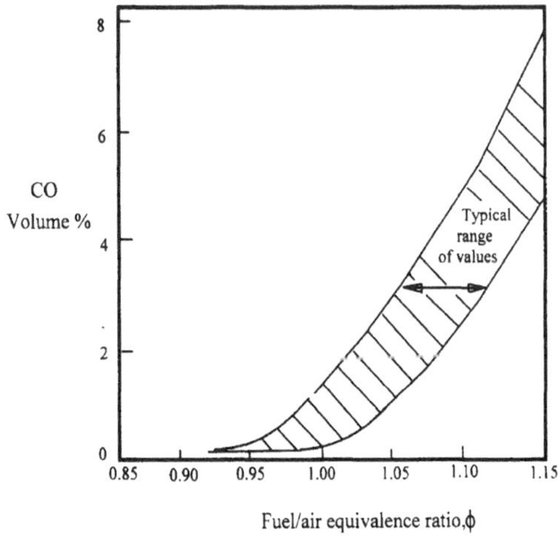

(a) Values for fuel of different H/C ratio

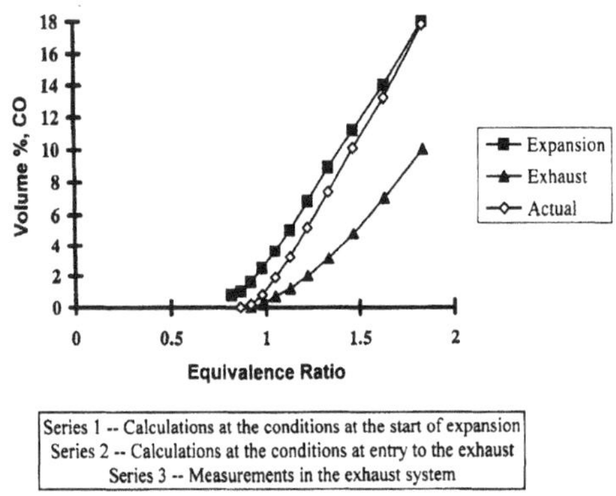

(b) Comparison of measured and equilibrium calculations of exhaust CO

Variation of exhaust CO with air-fuel ratio (based on data from Newhall [5]).

system and the final CO emission is predominantly though not entirely controlled by the equivalence ratio. A typical comparison from sources such as Newhall [5] and Heywood and Cheng [6] gives results which may be illustrated by the approximate range shown on Figure 8.3.

At slightly lean air-fuel ratios with fuel-air equivalence ratios of about $\phi = 0.9$ (note that $\phi = 1/\lambda$ where λ is air-fuel equivalence ratio), very little CO should exist in the exhaust gases. As the mixture becomes richer, the exhaust CO concentration increases at a progressively faster rate and rapidly becomes quite significant. Much of this is simply due to inadequate oxygen to complete the combustion. However, if the calculated curves for the equilibrium concentration at combustion and exhaust conditions such as those of Figure 8.3 are examined, it can be seen that the experimental measurements give values very close to calculated exhaust equilibrium values for lean mixtures but above them for rich mixtures. That is, the rate-controlled nonequilibrium behavior, which probably occurs during the blowdown process as described previously, is more evident as the mixture becomes richer, thus compounding the effect of insufficient oxygen.

Unburned Hydrocarbons (UHC) While much of the unburned hydrocarbons from engine sources originate as evaporation from the fuel tank and fueling system (which have separate controls such as venting to carbon canisters), they also exist in substantial quantities in the exhaust gas. While both are of interest, the former is essentially solved by nonthermodynamic methods and so will be considered later. Understanding the latter is more difficult and must be directly considered in relation to engine operation.

It is sometimes difficult to understand how fuels can remain unoxidized when there is sufficient air in a high-temperature, high-pressure environment suited to combustion. There are several reasons and the topic is complex. It is not simply a matter of control of the fueling process because, even with excellent control of the fueling and mixing, substantial quantities of unburned hydrocarbons appear in the exhaust. Much of this is due to what can be generally termed *quench zones*, which occur within the combustion space, and it is necessary to understand how these form.

In any confined combustion space such as that of an internal combustion engine, a severe temperature gradient always exists near the wall due to the large quantity of heat which is transferred from the burning gases to the walls, and finally to the engine's cooling system. Within the near-wall region, the temperature can reduce to less than that required to initiate combustion. When this occurs, no flame appears and it is essentially quenched; hence, the term *quench zone* is applied to it. With the more or less homogeneous mixture of an SI engine, it must contain fuel and some of the UHC may initially be contained here. Quench zones may exist in many places. A large area of quench zone exists completely around the confining walls of the combustion space but this may be quite thin in places. Hence, its total volume is likely to be less than smaller regions such as those termed *crevice zones*, which are much thicker and therefore more significant. These are spaces confined on several sides and the most significant is likely to be the space above the upper ring between the piston and wall (Wentworth [7]). Other significant crevice zones are behind the rings, between the upper rings, within the spark plug screw threads, in the locations where the head gasket has a slight mismatch to the block and head and where the different slopes of the valve and seat provide a small gap. These are noted by Heywood [8]. All quench zone regions contain a higher mass than volume proportion of the in-cylinder gas because, being at the same instantaneous pressure but at a lower temperature than the combustion chamber, they are at a higher density. Heywood estimates that about 8 percent by mass (which is about 3 percent of the cylinder clearance volume) of gas exists in the crevices but some of this, roughly 1 ± 0.5 percent, escapes past the piston to the crankcase. The gas returned to the combustion chamber unburned may be anywhere from about half of the remainder to all of it.

Typical values of wall quench zone thickness found from experiments for the conditions relevant to engine combustion are around 0.125 mm (Newhall [5]). While this may seem small, it needs to be remembered that for a typical 0.5 L cylinder swept volume in an SI engine with compression ratio of 9:1, the equivalent spherical combustion chamber diameter, which may be assumed to be an average of its actual dimensions, is about 50 mm. A surrounding quench zone of 0.125 mm then represents 1.5 percent of the chamber volume. The more flattened shape used in practice will increase these proportions. Thus, it can be seen that the quench zone is likely to be a significant proportion of the volume of the combustion space at the end of compression. As discussed earlier, the higher density of the gas increases their mass proportion compared to the volume proportion. The quench distance decreases with increasing fuel-air equivalence ratio from lean to a minimum at about the stoichiometric conditions, above which it is thereafter about constant thickness. It also decreases as pressure p and temperature T increase. While direct wall quench may provide some of the UHC under some operating conditions, much of it appears to be oxidized in the exhaust process as described later, and it is likely to be less significant than the crevice regions.

Another major source of unburned hydrocarbons is their absorption and desorption by the engine lubricating oil during a single engine cycle. Some of the HC in the mixture is absorbed into the lubricating oil film on the cylinder walls during intake. It does not take part in the combustion process and is then expelled during exhaust when the HC content of the gases is very low. This may be of similar magnitude to the total quench (wall/crevice) effect. It has been suggested (Ishizawa and Takagi [9]) that about 25 percent of UHC comes from this source under stoichiometric conditions, increasing to perhaps 50 percent with a lean fuel-air equivalence ratio of about $\phi = 0.75$.

Another possibility for UHC formation is carbon deposits on the cylinder wall, head, and piston crown. Some tests also show UHC increasing by up to 20 percent when extensive carbon deposits form on the walls, head, and piston crown (Wentworth [10]). Carbon is known to adsorb HC molecules from a fuel vapor onto its surface, and it is used in the carbon canisters to which fuel tanks are vented for this purpose. These are purged by an airstream to the inlet manifold. This process is likely to be followed within the cylinder during the engine cycle with adsorption onto the carbon during inlet with their release during exhaust. That is, the process is similar to that of the lubricating oil. It appears that a clean engine helps to prevent high UHC emissions.

It should be noted that, as the mixture becomes richer and/or more dense, the absolute amount of UHC in the quench/crevice zone regions increases even if the zone remains the same thickness. Unburned hydrocarbons from within quench regions in engines are expelled during the exhaust process. With slightly lean mixtures, although the quench zone is large, the excess oxygen reacts with much of the UHC when they are mixed later in the exhaust system. Also, with very lean mixtures, a bulk quench can occur with the flame partially or totally extinguishing during propagation. Combustion then generally becomes erratic and the amount

Fig. 8.4.

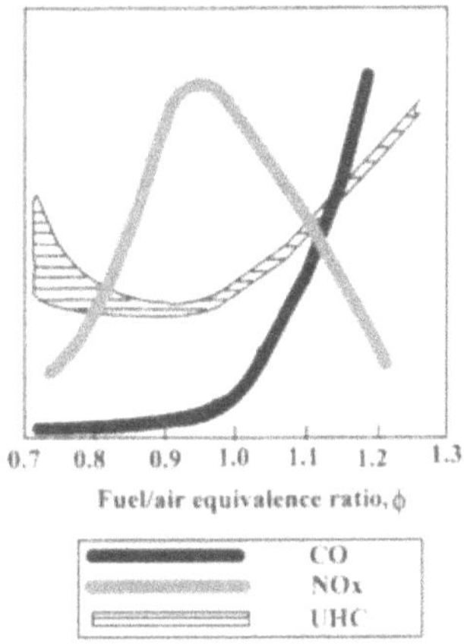

Representative trends (not scaled) for the variation of CO, NO_x, and UHC with fuel-air equivalence ratio (typical curves, for example, see Heywood [8]).

Fig. 8.5.

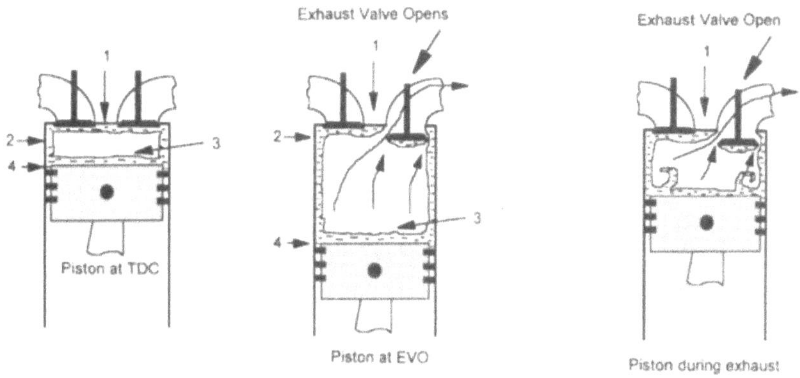

Schematic of flow of various UHC layers from an engine (adapted from Tabaczynski, Hoult, and Keck [11]).

of unoxidized UHC distributed more or less uniformly throughout the combustion space therefore increases. The combination of this and the thickening quench zone means that a minimum value of unburned hydrocarbons tends to occur at an equivalence ratio slightly less than 1 (i.e., slightly lean). This is shown in Figure 8.4.

It has been noted experimentally that in an SI engine the concentration of unburned hydrocarbons varies throughout any single exhaust stroke. This can be explained based on the work of Tabaczynski et al. [11] as follows. In general, four basic quench zone regions can be identified as on Figure 8.5. The gasdynamic flow to the exhaust from each of these could be expected to be different. The

Fig. 8.6.

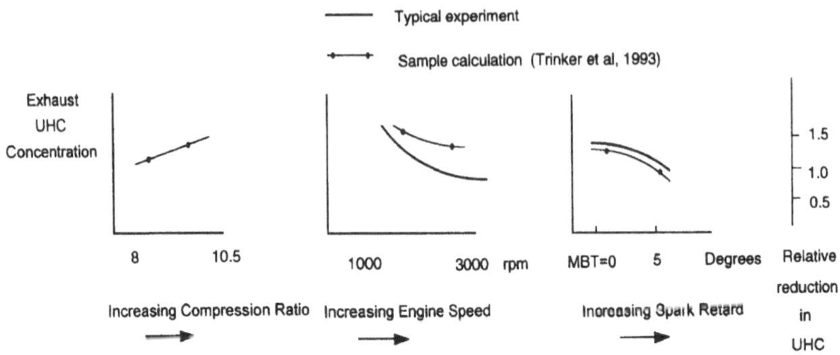

Typical trends on UHC formation of engine variables.

regions are: (1) the head, (2) the side wall of the cylinder, (3) the piston face, and (4) the quench zone around the piston above the top ring. When the exhaust stroke starts, region (1) and the top part of (2) are expelled early in the stroke. These are likely to be well mixed with the hot exhaust gases with substantial oxidation. During the piston movement toward top dead center, vortices form that collect the gas from regions (3), (4), and the lower part of (2) and this is expelled very near the end of exhaust with less chance of oxidation. This mechanism correlates with the observed concentrations of UHC, which are high near the end of exhaust. It explains the importance of crevice zones, in particular that of the gap above the upper ring, and why reducing it substantially reduces UHC in this phase.

The major engine variables affecting UHC in an SI engine are equivalence ratio, compression ratio, engine speed, and spark timing. Throttle position has little effect. The equivalence ratio contribution has already been discussed (Figure 8.4.). Typical trends associated with the other three are depicted in Figure 8.6. A high compression ratio increases unburned hydrocarbons because, with the smaller combustion space at TDC, the crevice regions, being of roughly constant volume for given engine conditions, make up a larger proportion of the total volume. Under some conditions, direct wall quench may have some significance, particularly noticeable in the squish region where the wall area to volume ratio is high, although the high turbulence there tends to mix quench zone and other gases, thereby promoting later oxidation. Unburned hydrocarbons decrease with engine speed because the greater turbulence promotes combustion as well as post-combustion mixing and oxidation. Also they decrease as the spark is retarded because more combustion occurs after TDC while the combustion space is enlarging with the crevice zones and desorption, then decreasing as a proportion of the combustion space. Also, exhaust gases are hotter with a retarded spark and more oxidation of the UHC is then likely in the exhaust system. It is noted that some small variation may occur between experiment and modeling, although the trends are generally similar (Trinker et al. [12]).

Oxides of Nitrogen, NO_x The oxides of nitrogen from combustion processes are a mixture of different products. From engines they are typically high in NO, some estimates being over 95 percent with some NO_2 and N_2O (Campau and Neerman [13]). However, under some conditions, N_2O may constitute over 10 percent of the total NO_x emissions. This may occur if there is flame quench and is not prevalent in normal operation. As such, the total are generally grouped and called NO_x. Under normal conditions, NO is not toxic but NO_2 adheres to the hemoglobin in the blood in much the same way as CO and is quite dangerous.

The formation of oxides of nitrogen is a highly temperature-dependent phenomenon. It occurs because the equilibrium concentrations of the various NO_x compounds formed when oxygen and nitrogen coexist at high temperatures of, say, 2000 to 3000 K are significant but are almost negligible at ambient temperatures of, say, 300 K. The forward reaction forming NO_x compounds is reasonably fast (due to the high temperatures) and the reaction moves substantially, though usually not completely toward equilibrium in the time available under most engine operating conditions. Newhall [5] shows that values reach about 90 percent of the equilibrium condition at engine combustion temperatures in about 15 ms. The backward (or reverse) reaction during which the NO_x breaks up into N_2 and O_2 is much slower (due to the gas cooling) and the NO_x is at least partially "frozen" (i.e., retained as NO_x) for a long period after it is exhausted from the engine. This gives it time to react with other substances such as the UHC compounds and ozone O_3, aided by the presence of sunlight, to form photochemical smog, which is very persistent.

The quantity of NO_x formed is coupled to many variables associated with the combustion temperature and air-fuel ratios in engines. Precombustion conditions, fuel-air ratio, and flame speed are all important. The in-cylinder turbulence is important due to the fact that turbulent combustion is faster and hotter, but it may have a direct effect other than the increase in flame speed. This is because at peak combustion temperatures small changes in temperature make significant differences to the amount of NO_x formed. Fluctuations above the mean increase the amount of NO_x more than those below decrease it. Total levels of NO_x in turbulent combustion are, therefore, likely to be higher than the use that a mean temperature would indicate and a full investigation requires the entire flowfield to be investigated. This is sometimes too complex and the detailed turbulent structure is often ignored, the mean time-dependent temperature being used in calculations.

The basic mechanism for NO formation is the extended Zeldovitch formulation (Heywood [8]). However, the supply of free radicals is sometimes controlling, and many mechanisms for rate studies of NO_x formation are available, which are usually adequate for calculations. However, extrapolation of measured values to the zero time point indicate that some NO_x appears much more quickly than these schemes show. This is called *prompt* NO_x (Fenimore [14]) and is a maximum at an equivalence ratio of about $\varphi = 1.4$ in a propane-air flame, which is reasonably representative. It has only been found in hydrocarbon flames. The most recognized theory for its formation suggests that additional routes of NO_x production occur via the development of intermediate substances during combustion, such as HCN. Also what is termed *superequilibrium* (i.e., above equilibrium) levels of atomic oxygen may form early in the combustion process due to its fast forward and slow backward reaction. This then could distort the production of NO_X early in the time period. This could be significant as most calculations of NO_X assume equilibrium atomic oxygen concentrations exist throughout. A further possible mechanism relates the prompt NO_X to additional fuel-bound nitrogen atoms. This may have a disproportionately greater effect with lean mixtures as its formation is not a temperature-dependent phenomenon.

While the most significant feature of NO_x formation is the temperature of the postflame gases, variation with equivalence ratio, as in Figure 8.4, is important.

It can be seen that NO_X concentrations peak at just lean of stoichiometric, approximately at $\phi = 0.95$. This is due to the fact that for lean operation the additional air acts as a buffer and reduces combustion temperatures, while for fuel-rich operation the latent heat requirements of the additional fuel also keep temperatures lower than stoichiometric. The peak value at less than the peak temperature is due to the additional O_2 available at lean mixtures. This helps promote the reaction, thereby offsetting the reducing temperature until the rate of temperature reduction becomes significant enough to become the controlling factor.

8.1.3.2 Global Emissions

The earth may be considered a closed system as shown in Figure 8.7 with the only energy transfers, both by radiation, being the heat received from the sun and that lost to space. The wavelength spectrum in which most of the energy enters is shown in Table 8.2. Radiation from the earth's surface is at a longer wavelength, also shown in this table (data from Zillman [15]). Radiated heat is strongly dependent on the temperature of the emitter, the relationship being proportional to T^4 where T is the absolute temperature in degrees Kelvin. For the incoming radiation, these temperatures, although fluctuating, have a constant average value and so the incoming heat is also constant when averaged over an extended time period. All

Fig. 8.7.

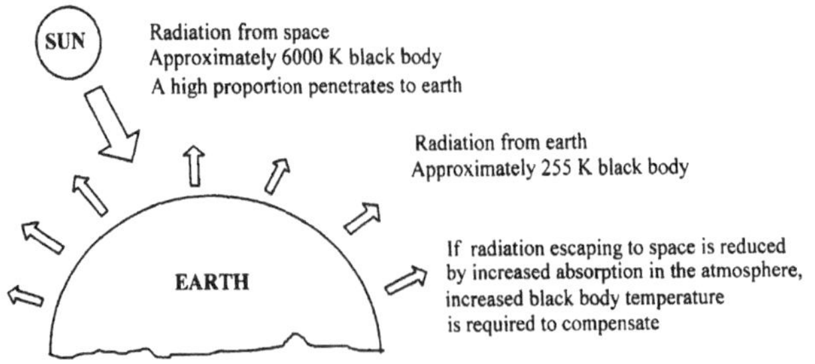

Schematic of energy transfer to and from the earth's surface.

Table 8.2

Radiation Spectra to and from the Earth and Gas Absorption Spectra

Radiation

Radiation to and from the earth	Range	Wavelength μm (approx.)
Incoming solar (6000 K black body)	peak	0.50–0.55
	>50% of peak	0.35–1.20
Reradiation from earth (255 K black body)	peak	12–16
	>50% of peak	8–28

Absorption

Substance	Major bands: Peak μm	Major bands: Approx. width μm
H_2O	1.1	0.1 (±0.05)
	1.9	0.2 (±0.1)
	2.6	0.8 (±0.4)
	6.0	0.1 (±1.0)
	20.0	>20
CO_2	2.4	0.05 (±0.025)
	4.4	0.8 (±0.4)
	15.0	6.0 (±3.0)

gases transmit and absorb radiant energy, the absorption being dependent on the incident wavelength. The gases from which the earth's atmosphere is composed absorb strongly in the ranges given in Table 8.2; that is, they absorb only moderate amounts of the incoming radiation. Much of the radiation from the earth escapes through the "atmospheric window" which is between 8 and 12 μm. However, it will be noted that the radiation above about 12 μm, that is, from the peak upward, falls substantially in the absorbing wavelengths. Hence, much of the energy is then retained. This is important as the effect has enabled the planet to remain warm, the temperature of the atmosphere reaching a value where it is high enough to radiate an energy quantity at the longer wavelengths equivalent to that being received at the shorter. That is, for a stable atmosphere, the incoming and outgoing heat flows are in equilibrium and the atmosphere acts like a *greenhouse* at all times. Details of the radiant energy flows may be found in Waters [16].

If a change takes place in atmospheric composition such that more absorption of the wavelengths radiated from earth occurs, the system is temporarily out of balance with less energy leaving. Energy then accumulates within the boundaries and the earth's temperature must rise, with the outgoing radiation increasing toward a new equilibrium with the incoming quantity. This continues until a new stable earth temperature is established at a higher level consistent with the new atmospheric conditions. This increase has been termed the *greenhouse effect.*

Prior to the widespread use of fossil fuels, during recent history on a geological time scale, the earth's atmosphere was fairly stable, although major natural catastrophes (e.g., volcanos, forest fires, etc.) could provide fluctuations after which a period for the old equilibrium to be re-established was required. Carbon dioxide (CO_2) is a major greenhouse gas and the very slow formation of fossil fuels over many millennia in fact slowly reduced its level by locking up the carbon. The extensive forests that previously existed over most of the earth also provided a temporary "bank" of carbon. This was always eventually returned to the atmosphere by burning or decaying of the timber but in a natural environment is continuously replaced by new growth. The use of fossil fuels in combination with extensive deforestation over recent years has released the carbon, which is locked up in either form, to the atmosphere but at an significantly faster rate than it was initially removed. That is, the amount of CO_2 added has increased from levels of reasonable stability at a high rate over a short time period as shown in Figure 8.2 and is still increasing. Table 8.3 (data from Zillman [15] and Haddad [17]) shows estimates of the effect of fossil fuel usage and deforestation. It will also be noted in the table that CO_2 is indeed absorbed back into the earth's surface predominantly in the seas, although at a noticeably lower rate than it is currently added to the atmosphere. That is, there is substantial accumulation.

In addition to CO_2, many gases including methane (CH_4), which also exists in reasonable quantities, have an absorption effect. Relative effects estimated on the basis of absorption and stability (and therefore the endurance of the species in the atmosphere) are shown in Table 8.4 (data from Walker [18]). That is, while

Table 8.3

Effects of Fossil Fuel Use and Deforestation

Source	Approximate quantities, GTC/year (worldwide estimate)
Fossil fuel burning	5
Deforestation	1–2
Absorption by ocean/soil	3–4
Net increase	3
CO_2 from fossil fuel burning	**Percent contribution, (Australian data)**
Transport	26%
Manufacturing industry	39%
Residential use	18%
Agriculture, mining	16%
Other	1%

GTC = Giga tonnes carbon.

Table 8.4

Relative Greenhouse Gas Effects Due to Absorption and Relative Endurance of the Species

Component

	Atmospheric absorption of additional gases	70–90%
	Deforestation	10–30%

Absorbtion contribution

Substance	Relative contribution/mole emitted	Overall effect (%)
CO_2	1	55
CH_4	6	20
NO	350	5
CFC	14,600–17,000	18
O_3	1	2

many have stronger basic absorption characteristics than CO_2, they remain stable in the atmosphere for shorter time spans and their relative detrimental effects are therefore reduced. Note that there is some variations on durability estimates from different sources. Also, it is obvious that the quantity of the absorbing gas in the atmosphere is significant. In addition, some exist in only small quantities. For example, it will be noted that the CFCs (chlorofluorocarbons) are high on the list but their total quantities are low. Their principal detrimental effect on the environment is not that of global warming but is due to chemical reactions causing ozone depletion in the stratosphere. This is independently of great environmental significance in that increased quantities of ultraviolet radiation can then penetrate to the earth's surface with effects such as increased skin cancer rates, cell mutations, and so forth. Also, it is worth noting that CFC emission is not directly a result of engine operation, although much of it is associated with vehicles due to their air conditioning.

8.2 GLOBAL EMISSIONS FROM SI ENGINES

8.2.1 The Contribution of SI Engines and Their Potential for Minimization

The major greenhouse gases are well known: water vapor, carbon dioxide, nitrous oxide, chlorofluorocarbons, and ozone, in descending order of importance. The predominant effect (apart from water vapor) is CO_2, not because it is the greatest absorber per mole of the radiation reflected from the earth's surface, but because of the high quantity of it and its persistence in the atmosphere. About 5 GTC (giga tonnes carbon) are emitted worldwide per year from fossil fuel burning and about 1 to 2 GTC are contributed by deforestation (Zillman [15]). As some is absorbed by oceans and soils, the net increase is about 3 GTC per year. Transport vehicles are responsible for about 20 percent to 25 percent of CO_2 (Haddad [17]). Note that methane absorbs at about 35 times the rate of CO_2, although it has a shorter life and therefore has been estimated to be only about six times as disadvantageous (Walker [18]). The combustion equation shows that 1 mole of CO_2 is produced for each mole of CH_4. Hence, there is a reduction in the greenhouse effect for every unit of CH_4 converted to CO_2 for energy use if it would have otherwise escaped to the atmosphere. That is, it is essential to control fugitive emissions of CH_4 wherever possible, particularly by converting them to useful power.

Global emissions come from many sources, the principal ones being industry, electric power generation, space heating, and sea, air, rail, and road transport. In larger power ranges, the steam power plant, the gas turbine, and the diesel (CI) engine predominate and it is only in the last of the group that the SI engine has the most significant role. Even here, a reasonable proportion is from CI engines, which predominate in truck and bus installations, and which have a small but increasing application in car power units. A breakdown of CO_2 production from different sources and from engine types will vary from country to country. Between 50 percent to 75 percent of transport vehicle CO_2 is likely to be from SI engines. That is, the SI engine is probably producing about 15 percent to 20 percent of the total CO_2 that is currently entering the atmosphere. While this is by no means the only controlling proportion, it is significant and attempts must be made to reduce it.

As the energy release in burning hydrocarbon fuels can only occur by oxidizing them to CO_2 and H_2O, CO_2 is a necessary end product of the process. The only way that it can be reduced is by improving vehicle efficiency or by changing to a fuel with lower levels of carbon for a given energy output. The former includes many factors, engine efficiency, improved use of the engine power by better transmission matching, improved drive train efficiency and lower vehicle mass, rolling resistance, and aerodynamic drag. As the tendency has been to improve many of these aspects simultaneously, it is not easy to determine their individual breakdown from an historical examination of recent trends. A typical trend is given in Figure 8.8. While the reduction has been very large on U.S. vehicles, it will be noted that some of this at least may be due to the modern preference for small van-type vehicles, which are not classified as cars but which replace the large station wagon which was. Nevertheless, it does represent a real trend, even if exaggerated. Figure 8.9 gives a better indication of changes in SI engine efficiency (although other factors such as drive train efficiency are included) over a similar period by considering vehicles of a given weight class. This, while significant, is noticeably less. It should also be noted that the modification of engines

Fig. 8.8.

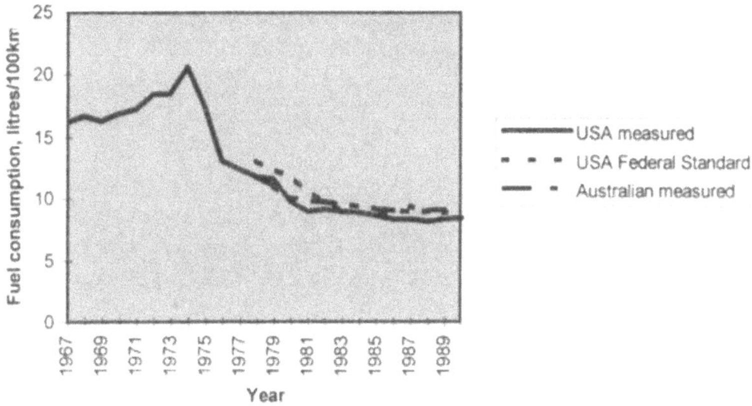

Trends in automotive fuel consumption (U.S. data from McTague [19] and Australian data from Wylie [20]).

Fig. 8.9.

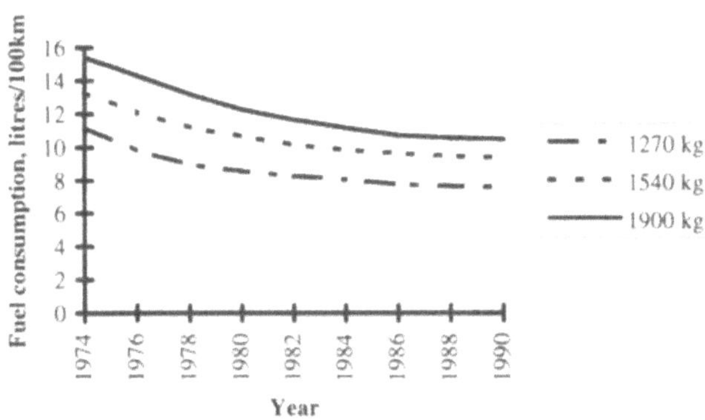

Trends in fuel consumption on vehicle weight class (U.S. data from McTague [19]).

for control of local emissions is generally negative from an efficiency point of view, although some modern approaches such as catalytic converters, lean burn techniques, and others have less impact than the methods used on older vehicles, therefore, showing a relative improvement when compared to them. Also, the changing refinery balance in producing the fuel needs to be noted as it is more energy intensive to produce a high-octane unleaded fuel than a leaded one, and a full scenario should include this in the assessment. That is, a full global picture of CO_2 production is only available by considering all engine types, the relative fuel use between them, and their changing requirements as local emission methods are adopted. This is an immense task and, to the author's knowledge, has not been attempted. Therefore, further comments here will focus on potential improvements to SI engine efficiency and ignore the other factors.

Table 8.5

Fuel Usage—Carbon Dioxide Per Unit of Energy

Fuel	Chemical composition	LHV*	HHV*
Hydrogen	H_2	0	0
Methane	CH_4	1.246	1.123
Ethane	C_2H_8	1.401	1.282
Propane	C_3H_8	1.468	1.351
Methyl Alcohol (gas)	CH_3OH	1.478	1.308
Butane	C_4H_{10}	1.505	1.390
Ethylene	C_2H_4	1.512	1.417
n-Pentane (gas)	C_5H_{12}	1.528	1.414
n-Hexane (gas)	C_6H_{14}	1.544	1.430
n-Heptane (gas)	C_7H_{16}	1.555	1.442
Propylene	C_3H_6	1.557	1.457
n-Octane (gas)	C_8H_{18}	1.564	1.451
Ethyl Alcohol (gas)	C_2H_5OH	1.566	1.419
Methyl Alcohol (liquid)	CH_3OH	1.566	1.376
n-Octane (liquid)	C_8H_{18}	1.576	1.462
Acetylene	C_2H_2	1.593	1.539
Ethyl Alcohol (liquid)	C_2H_5OH	1.620	1.463
Petrol (typical)	$(CH_{1.85})_{x\gg 7}$	1.639	1.537
Distillate (typical)	$(CH_{1.85})_{x\gg 13}$	1.684	1.577
Toluene (gas)	C_7H_8	1.856	1.773
Benzene (gas)	C_6H_6	1.893	1.817
Coal (typical black)	$(C_{1.4}H)_{x\gg 20}$	2.288	2.209
Carbon	C	2.541	2.541

*Values are $(\text{kmol } CO_2/\text{kJ}) * 10^6$.

Improvement examples: Propane for gasoline approximately 10 percent.
Methane for gasoline approximately 24 percent.
Methane for distillate approximately 26 percent.

It is also apparent that the fuel used will have an effect on the amount of CO_2 produced. For example, hydrogen burns without CO_2 in its products. In Table 8.5, a range of fuels has been classified on the basis of their higher and lower energies of combustion. Gasoline and diesel oil have been given an average classification as they consist of many hydrocarbons which may differ substantially from one batch to another. Apart from hydrogen, methane (CH_4) is the best fuel with about 76 percent of the CO_2 per unit energy (on LHV values) than that of gasoline. Methane is the major constituent of natural gas (NG) and a very significant proportion of biogas from various decay mechanisms. Propane and butane are also better than gasoline but to a lesser degree (an improvement of about 10.5 percent and 8 percent, respectively). Alcohols fare only moderately (about 4.5 percent better), although it can be argued that ethyl alcohol, which is produced from biomass, takes in its carbon from the atmosphere and so has no net effect. This, of course, depends on what fuels are used in its production, transport, and so on, which is quite labor intensive. Methyl alcohol can be produced from a number of sources such as wood, natural gas, and coal. As coal falls low on the list, it is necessary to question what happens to the remaining carbon in the processing and it should perhaps on that basis be placed lower in the table. In general, it can be seen from the table that the smaller molecules fare better, as do the paraffins, compared to olefins and other hydrocarbons. In both cases, this is because they have a lower C-H ratio.

8.3 ENGINE CONTROL FACTORS FOR LOCAL EMISSIONS

The settings of the engine parameters are, from the previous discussion, critical in determining the emission levels. It is important to understand what is required in the control of an engine and the emissions from it under different conditions. Two basic scenarios exist. These are steady-state operation and transient operation. The steady-state condition is self-evident and requires the engine operation to occur under constant speed and load conditions in addition to being fully warmed up, that is, with no changes to the temperature and pressure conditions for the gases used with it at any stage. Transient operation has a number of possible modes, as will be discussed in Section 8.4.

8.3.1 Steady-State Engine Operation

Under steady-state conditions, the engine parameters can be set for whatever conditions are regarded as optimum at the time, for example, maximum power, maximum efficiency, or minimum engine-out levels of any one of the foregoing emissions. However, it is not possible to optimize all the above emissions simultaneously. Emission control systems are, therefore, usually a mixture of engine design features, engine operational controls, and exhaust clean-up techniques.

8.3.2 Engine Transients

Two types of transient conditions are possible:

1. An acceleration or deceleration in engine speed or the change of load on the engine. The speed changes can occur at constant load and the load change at constant speed as long as the engine control system can respond quickly enough. Nevertheless, it is much more common for at least some coupling to occur between these two modes. It is evident that an engine is subject to these type of transients for much of its operation. This is particularly true of vehicle engines where traffic conditions, gradients, and driver psychology are predominant factors. Indeed, a vehicle engine is likely to spend more of its time in this type of transient mode than in the steady-state condition.

2. The warm-up period after a cold start or the cooling down (hot soak) period after operation. Even if speed and load are steady, all the engine components are then either increasing or decreasing in temperature, respectively, for the two cases and many emission-related phenomena follow due particularly to fuel evaporation or condensation effects on surfaces. This can also make up a considerable proportion of the total operation and is of major concern as both the cold start and hot soak periods are included in most emission testing schedules.

Although most vehicle emission testing is carried out on test cycles, which include a substantial component of transient operation of both the preceding types, much of the quantitative test data that are available have been obtained for steady-state conditions. This is because a fundamental understanding of the formation and perpetuation of the substance can be gleaned more readily from these type of data.

8.4 TRANSIENT OPERATION OF ENGINES AND THE EFFECT ON EMISSIONS

The control of emissions during steady-state operation can readily be achieved by an appropriate combination of techniques, which will be discussed in more detail in Section 8.5. However, it is the transient operation of engines that provides the most difficult problems. Engine transients, therefore, need to be understood.

8.4.1 Short-Term Transients

The transients that are imposed by the operator or control system to respond to changes in speed and/or load are short-term transients. They occur continuously during operation and may range from less than 1 to 10 to 20 seconds. Longer time transients are possible but are then likely to be indistinguishable from steady-state operation. Such control, while fundamentally simple, is more complex than at first appears and may be considered to be made up of four basic modes, which are as follows:

- Increase in engine speed.
- Decrease in engine speed.
- Increase in load (torque).
- Decrease in load (torque).

Generally, these will occur in some combination, that is, either one of the speed changes with either of the load changes, or vice versa; however, it is possible for engine speed change to occur at constant torque or a change in torque at constant speed. A typical example of the combination is a car accelerating at full throttle with the speed increasing. The torque then varies according to the engine characteristics, the torque first increasing with speed and then decreasing slightly toward the next gear change. An example of the latter is a vehicle maintaining a constant speed in the same gear when progressing from a flat to an uphill grade.

The question is: Why should the thermodynamic effects of a transient be different to those of steady-state operation? First, consider a change in engine speed when, for example, the throttle is suddenly opened and an increased pressure, as a wave front with a roughly constant pressure behind it, moves to the cylinder. Obviously, this wave will undergo some reflection and rereflections in the manifold system but these will adjust quickly, within milliseconds, due to the wave speed which is in excess of 340 m/s. For either steady-state or transient operation, as far as each individual engine cycle (either two- or four-stroke cycle) is concerned, the factors that are relevant are as follows:

- The initial pressure and temperature in the cylinder once the inlet valve closes.
- The fuel-air mixture ratio and its distribution within the cylinder.
- The in-cylinder turbulence, which is a function of the speed, valve lift, and engine geometry.
- The residual exhaust gas (plus any additional exhaust gas recycled for emission control purpose).
- The spark timing.

The actual change in piston speed from cycle to cycle is small, as can be seen in Figure 8.10, even for a fairly large acceleration of about 1000 rpm/s. Hence, the engine speed related factors such as turbulence and residuals (although not necessarily EGR) are largely unaffected by the direct mechanical effects of the acceleration. Even the initial cycle pressure and temperature will change only slightly from cycle to cycle after the initial adjustment period as long as the throttle valve remains at its new, constant position. Hence, as far as these factors

Fig. 8.10.

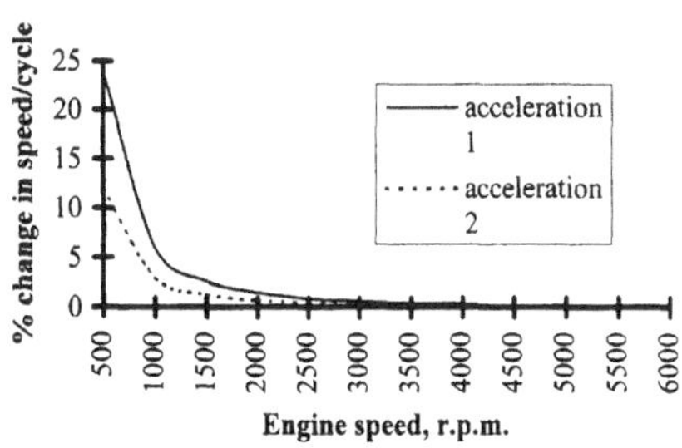

Percentage change in engine speed from cycle to cycle for constant engine acceleration rate—acceleration 1 = 1000 rpm/s; acceleration 2 = 500 rpm/s.

which dictate the general performance of the engine are concerned, a quasi-steady analysis (Milton et al. [21]) is possible with the engine being analyzed as a series of steady-state cycles. The important transient parameters for the engine performance are then the air-fuel ratio and its distribution within the cylinder, the EGR system, and the spark timing. These may vary from the best settings for either power or low emissions. If they have response times which are large compared to the duration of each cycle, they are therefore influential over a considerable time period. Outside the engine, the response of any exhaust clean-up system may also be of importance. These items are now considered.

8.4.1.1 Fuel-Air Ratio During Transients

The fuel for an automotive engine is almost always a volatile liquid—gasoline—although sometimes propane gas or occasionally natural gas is used. Liquid fuels offer the greatest advantage in energy density (i.e., the greatest energy per unit volume) and, hence, sufficient fuel can be carried to give substantial range for the vehicle. Control of a spark-ignition engine is via a throttling (pressure-reduction) procedure, which takes place prior to the air induction into the cylinders. An alteration in load requires a change in the fuel flow rate proportional to the change in airflow rate as the fuel-air ratio must be kept within narrow limits in order to support the flame front, which moves from the spark plug across the combustion space. That is, the response within the cylinders of the fueling system to a change should ideally be identical to that of the airflow to the throttling process. This in fact does not occur with liquid fuels.

8.4.1.2 Fuel Introduction Systems

Fuel introduction systems may be classified into several general types. These will have different effects on the air-fuel ratio distribution and hence on the emission levels. The common types are:

- Single-point (throttle body) systems, which are either the suction driven fixed or variable venturi carburetors or a positive pressure injection.
- Multipoint fuel injection at the inlet ports, which may be continuous, pulsed, conventional, or air assisted.
- Direct in-cylinder injection, which may be either single-fluid gasoline or air-assisted gasoline. These are currently still in the developmental phase and will be discussed in Section 8.7.

Single-Point Injection Systems (SPI) These are often designated throttle-body systems because of their location at that point. They represent the worst case for the response of the fueling system to a change and, therefore, provide the best illustration of the fueling excursions that can occur during a transient. Although throttle-body systems are now much less common than multipoint systems, it is worthwhile discussing them thoroughly here.

When a liquid fuel is sprayed into the incoming air, it is atomized in the jet spray to fine droplets. A droplet size distribution exists over quite a wide range of diameters. Measured values taken by Takeda et al. [22] give the Sauter mean diameters (SMD) as 320 μm for a conventional injector, 50 μm for an aspirated air-assist type (where the air is drawn by the port vacuum through the nozzle from atmosphere), and only 10 μm for a pressurised air-assist type. Shin et al. [23] give 130 μm for a conventional injector, which seems to be a more commonly accepted figure. The exact size depends on the viscosity and surface tension of the fluid, which does not vary much from one gasoline sample to another, and the injector characteristics, these being the nozzle design and the pressure difference across the nozzle, which is typically about 3 atmospheres in a fuel injection system but only about 0.5 atmosphere in a carburetor due to the venturi pressure drop. That is, a fuel injection system will have better fundamental atomizing characteristics than a carburetor fuel jet. However, the disadvantage of the higher pressure is that the jet velocity and, hence, droplet momentum are greater. Therefore, there is an increased probability of wall impact of the liquid fuel. Note that at part load the high airstream velocity past the throttle will reatomize the liquid vapor stream formed by impact on the throttle plate and the characteristics of throttle-body types, injection or carburetor, are substantially improved (Stone [24]).

Particles or droplets in an airflow do not exactly follow the airstream but gradually deposit on the walls. This has been observed for a very long time, an early report being by Mock [25]. The deposition can be due to the direction of the original spray, the turbulent nature of the airflow, or the change in direction of the airstream. With the first, it is noted that the jet distributes fuel over a range of angles, typically about 25° for a port injection system and 45° for a throttle-body type, but variable with the type of injector nozzle and injection pressure. Hence, some must be directed toward a surface. With the latter, either the random nature of the turbulence velocity vectors or those from the direction change of the airstream project some of the particles toward the wall where their inertia, being greater than that of the air, causes a proportion to deposit into a wall film. The flow in a manifold is usually turbulent. A typical Reynolds number range based on pipe diameter for the flow in a manifold is about 10^4 to 5×10^5, which means that a turbulent boundary layer starts to develop rapidly. Also such things as the air filter and the butterfly throttle valve in particular create a substantial wake region of a length equal to several duct widths directly downstream. For the butterfly valve, this consists of large eddies and reversed flow but the lower levels of turbulence to which it decays are transmitted further into the manifold. Typical measured turbulence values on the centerline from an experimental manifold simulation rig are 1.0 to 2.5 m/s (RMS) for mean flow velocities of 15 to 25 m/s under the worst conditions, which is with the butterfly valve setting at about 50° (Milton, Behnia, and Casey [26]). Thus, even in the relatively short

Fig. 8.11.

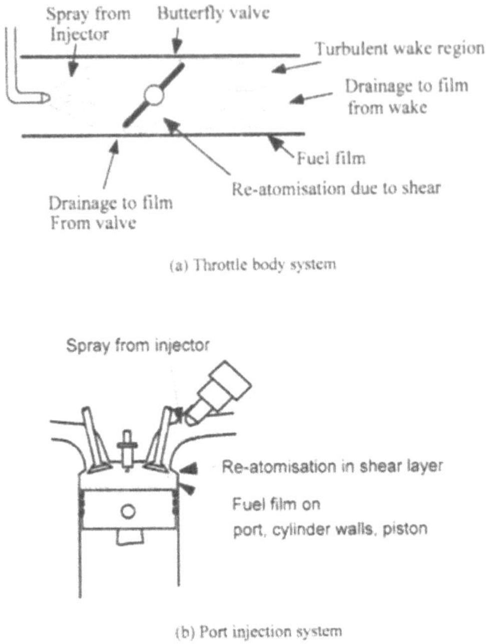

Spray and fuel deposition from a fuel injector.

length of a straight manifold, substantial fuel deposition may occur, particularly if the fuel is injected just upstream of the butterfly valve. A schematic of the film formation is shown in Figure 8.11. Collins [27] estimated from tests that, for a fully warmed carburetor engine, the film amounted to, as proportions of the fuel flow, around 3 percent to 8 percent for a lean fuel-air equivalence ratio of $\phi = 0.87$, 23 percent to 27 percent for a stoichiometric mixture and 20 percent to 40 percent for a rich mixture of $\phi > 1.1$ Kay [28] used visualizing windows in the manifold to observe fuel puddling. Tanaka and Durbin [29] using in-cylinder pressure measurements inferred that substantial film deposition takes place. While modern engine technology has improved this situation, it has not eliminated it. For a modern four-valve, port injected engine, Saito et al. [30] have shown that port wetting and direct fuel deposition on the cylinder walls and piston crown occurs, particularly during the start-up and warm-up periods. Other visualization experiments, (Shin et al. [31]) have also noted that a significant number of drops impact on the inlet valve and port surfaces and that the transport of the liquid fuel into the cylinder is via the wall film. Shearing during the flow past the inlet valve may aid reatomization.

The fuel is therefore transported from the injector to the engine as vapor, airborne liquid fuel droplets or as a liquid wall film. The vapor comes either from the evaporation of the drops or from the surface of the film. It needs to be noted

Fig. 8.12.

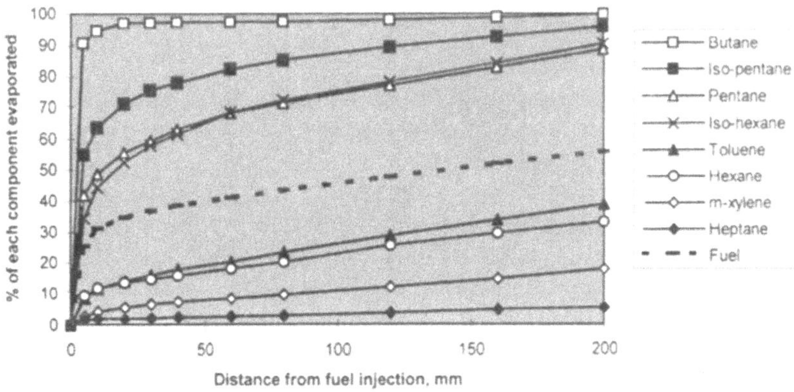

Calculation of evaporation of fuel components with distance from a fuel nozzle at atmospheric pressure and temperature, (initial airstream conditions, $T = 293$ K, $p = 98$ kPa, $v = 30$ m/s).

Fuel sample for Figure 8.12, analyzed by volume, consisted of:

- Components shown—butane, 11.4%; iso-pentane, 14.3%; pentane, 9.6 iso-hexane, 15.8%; hexane, 5.4%; m-xylene, 11.3%; heptane, 3.4%.
- Not shown due to minimal evaporation or low proportions—iso-butane, 3.6%; benzene, 8.0%; cyclo-hexane, 2.3%; octane, 0.9%; o-xylene, 2.9%; cumene, 2.6%; decane, 1.6%; other, 6.9%.
- Calculations by author.

that gasoline is not a simple substance and may contain perhaps 50 or 60 different compounds. Some 15 to 20 of these are in significant proportions (Boam and Finlay [32], Milton and Behnia [33]). Even if the fuel specifications are constant, these are merely a global specification and samples from different batches may contain some different compounds or compounds in different proportions. Unlike the viscosity and surface tension where an average value is appropriate, during evaporation each compound comes off separately and at substantially different rates and individual properties are required. A typical plot of evaporation with distance obtained from calculations is shown in Figure 8.12. Experimental measurements which distinguish the compounds are extremely difficult.

Once a wall film forms, several things may happen. First, evaporation may occur from the surface. Second, fluid may be dragged back into the flow as liquid droplets and this is probably related to disturbances on the surface, although the mechanism is not entirely clear and appropriate correlations are not available. Third, the fluid flows along the wall as a film of either increasing or decreasing thickness and extends peripherally around the manifold. It is noted that fluid deposited on the upper surfaces drains obliquely to the bottom. Typical measured velocities of a gasoline wall film dragged along a duct wall by the turbulent airstream are shown in Figure 8.13 (Behnia and Milton [34]). These are about two orders of magnitude smaller than the airstream velocity. The important features to understand about wall films is the rates of deposition into them, the evaporation and reentrainment from them, and their movement downstream.

Fig. 8.13.

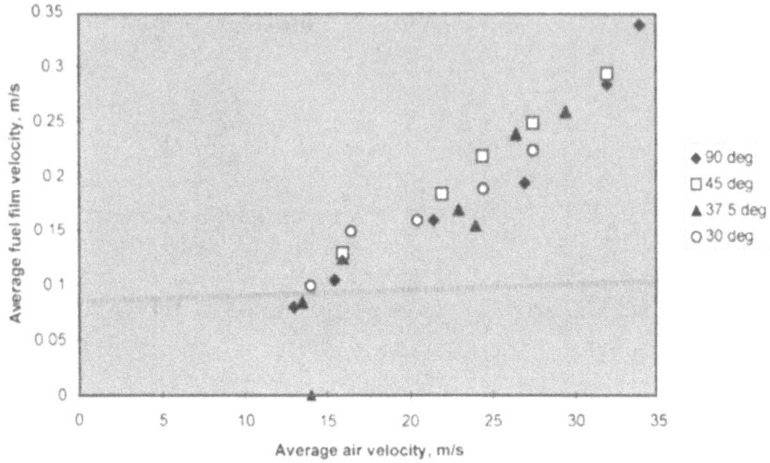

Fuel film velocity in a straight duct (from Behnia and Milton [34]).

Note: 90° = wide open throttle, throttle closed at 10°, square manifold 25 mm × 25 mm, fuel (unleaded gasoline) injected 30 mm upstream of throttle, airstream temperature, T, measured upstream of injection, pressure, p, and velocity, v, measured downstream of throttle.

Test conditions (ranges):

$\theta = 90^\circ$, $T = 298$ to 302 K, $p = 101.3$ to 99.6 kPa, $v = 12.9$ to 34.2 m/s
$\theta = 45^\circ$, $T = 297$ to 300 K, $p = 100.6$ to 92.5 kPa, $v = 13.2$ to 31.8 m/s
$\theta = 37.5^\circ$, $T = 296$ to 300 K, $p = 98.6$ to 87.8 kPa, $v = 13.2$ to 29.8 m/s
$\theta = 30^\circ$, $T = 295$ to 298 K, $p = 95.9$ to 77.6 kPa, $v = 13.8$ to 27.6 m/s

During steady-state operation, deposition into a wall film or evaporation and reentrainment from it is of no consequence to in-cylinder air-fuel ratios. Although fuel in a film moves more slowly toward the engine than airborne fuel, the deposition (or reentrainment) rate being constant means that fuel lost from the airstream into the film is exactly compensated for by film that has formed earlier, either when it flows into the engine or when it fully evaporates at some intermediate position. However, during a transient, it is of great significance.

Assume that a throttle valve is suddenly opened. Here the higher pressures result in more deposition in the film due to condensation and lower rates of film evaporation. Because the response of the film is slower than that of the fuel vapor or airborne droplets, the air-fuel ratio reaching the engine is temporarily lean if the fuel flow rate settings remain directly proportional to the air mass flow rate. This has been demonstrated by Tanaka and Durbin [29], Aquino [35], and Hires and Overington [36] and a typical situation is illustrated in Figure 8.14. The engine will then stumble. In order to maintain the correct airborne fuel-air ratios, it is necessary to enrich the mixture. However, this will cause even greater deposition into the film. Once the new steady-state condition has been reached, there will be

Fig. 8.14.

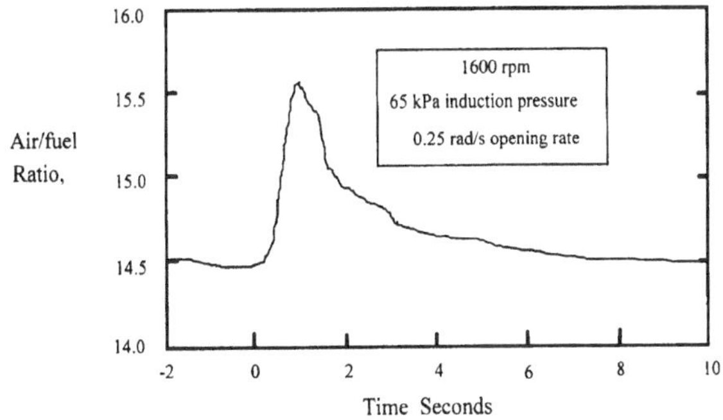

Carburettor engine - typical effect of rapid throttle opening - uncompensated air/fuel ratio

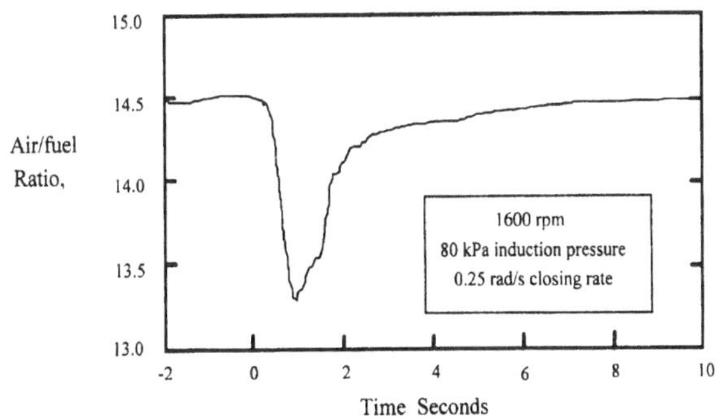

Carburettor engine - typical effect of rapid throttle closing - uncompensated air/fuel ratio

Air-fuel ratio excursions with time due to throttle movement in an uncompensated carbureted engine (data from Hires and Overington [36]).

an oversupply of fuel in the film reaching the engine and exhaust emissions will be high due to the rich air-fuel ratio. This usually does not cause engine malfunction as the combustion is more tolerant to rich excursions than lean ones. When the throttle is rapidly closed, the reverse process occurs and the air-fuel ratio becomes temporarily rich very rapidly. This is also illustrated in Figure 8.14. A possible effect of fuel-air ratio excursions is shown in Figure 8.15.

The problem can be minimized by using a heated manifold, which provides higher evaporation rates of the film. The less the extent and thickness of the film, the less important the problem is. Both water-heated (from the engine cooling system) and stove-heated (from the heat directly from the exhaust pipe) manifolds and stove-heated intake air have been used for this purpose. However, it is not a solution that is immediately useful in an initially cold engine. Also, heating the manifold or intake air reduces the air density and hence the air utilization by the engine. A higher intake temperature is also multiplied through the engine cycle giving higher peak temperatures during combustion with higher NO_x emissions. Thus, the heating needs to be kept to the essential minimum and to be well controlled.

Fig. 8.15.

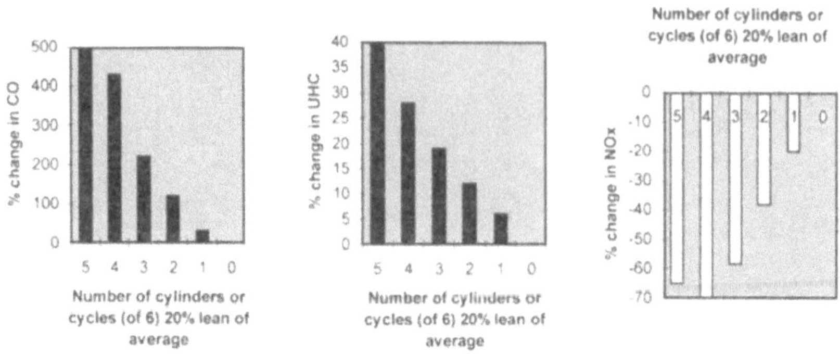

Effect of air-fuel ratio excursions on emissions.

Based on calculations using typical variation in emissions with air-fuel ratio. The base is stoichiometric when all cylinders are the same. When some cylinders or cycles become lean, the remainder become correspondingly rich. Calculations from Behnia and Milton [37].

The fuel introduction process is obviously of importance during these transients. The enrichment during acceleration needs to be exactly metered and a brief leaning at the end of the transient is helpful. During deceleration, some leaning of the mixture is required. This can be achieved by additional air to the manifold system, by preventing overly-rapid closure of the throttle, by partially or fully closing off the fuel supply, or by improved fuel scheduling for the injector using the engine management system. The last is the preferred option.

Multipoint Port Injection Systems (MPI) These are often designated as port injection systems. Obviously, the closer the injection point is to the cylinder intake valve, the shorter will be the transient effect. This is one advantage of multipoint port injection systems. Nevertheless, fuel deposition can still occur. This can be on manifold ports and on the inlet valve. Shin et al. [23] used a transparent, square section engine to demonstrate that a sequence of events is:

- Intake valve opens.
- Residual (exhaust) gas backflow takes place to the inlet manifold.
- Droplet stream and liquid film enter.
- Intake valve closes.
- Shearing of the film entering the cylinder and droplet splashing occurs in the cylinder.
- Liquid deposits on walls within the cylinder.
- Combustion occurs.
- Some film survival may occur until the next cycle.

Takeda et al. [22] and Saito et al. [30] have shown in a normal engine that direct fuel deposition on the cylinder walls does occur, particularly during the start-up and warm-up periods. That is, a liquid fuel flow takes place through the open inlet valve, which can impinge on the cylinder walls and piston crown. Takeda et al. [22] showed that in the first few cycles of a cold start more fuel is deposited in the cylinder than on the port, although this reverses after starting with the latter being about three times as great. They did not, however, consider acceleration-type transients, which would have modified these values. High engine-out HC emissions coincided with the greatest cylinder wall wetting. Nagaishi et al. [38] compared single- and multipoint injection systems and found that the wall wetting in the latter case is about 30 percent to 40 percent of the former. Other visualization experiments, (Shin et al. [31]) have also noted that a significant number of drops impact on the inlet valve and port surfaces and that the transport of the liquid fuel into the cylinder is via a wall film. It is also noted that fuel droplet flows reverse much less readily than the airstream in the pulsatile flow (Low et al. [39]) and their impingement on the intake valve is likely. Shayler, Teo and Scarisbrick [40] have shown that in-cylinder fuel-air ratio excursions similar to those noted earlier in carbureted types also exist in MPI engines with uncompensated excursions ranging from 20 percent to 80 percent maximum variation with total settling time of 4 to 12 seconds. These are quite considerable. Hence, similar phenomena to those in throttle-body systems can occur with port injection, particularly under cold conditions.

In summary, with multipoint systems, film formation in the port still occurs, and some fuel can be deposited on the upstream side of the inlet valve, it being noted that this valve is shut for about 70 percent of the time in a four-stroke cycle. When it is open, film flow can occur into the cylinder. While some benefit can be gained, particularly at part load, from timing the fuel injection so that flow coincides with the valve opening, it does not eliminate the problem. At full load, an intermittent injection approaches continuous operation. That is, it does not eliminate the problem. The not too inferior a performance from the continuous injector may be because the intermittent injector must provide a much higher fuel flow rate during the time that it is injecting. Either atomization may not be so good due to larger nozzle areas or the higher pressures required may cause greater immediate flow impingement on surfaces.

In-Cylinder Mixtures It is sometimes assumed that the mixture within the cylinder is uniformly homogeneous and fully vaporized at the point of ignition with this uniformity existing from intake through the compression stroke. This is not necessarily so and there is evidence that droplets exist within the cylinder up until the start of combustion. These may have a random distribution or undesired stratification throughout the cylinder at the time of ignition. The original experiments carried out on a single-cylinder, carbureted engine by Matthes and MeGill [41] showed deteriorating emissions and sometimes reduced power under nonhomogeneous conditions. More recent results on MPI engines with different port injector locations (Shayler et al. [42]) conclude that no single fueling installation is likely to minimise the emissions throughout the full operating range of the engine. This seems to be due to differences in mixture preparation, mixture inhomogeneity in the cylinder being one explanation of this influence.

Nonhomogeneity will be worse under cold engine conditions. While a controlled stratification of the fuel-air mixture during combustion is desirable (and is a feature to be designed into engines in the future—see Section 8.6), an incorrect distribution is likely to cause increased emission levels. That is, it is generally desirable to have as uniform a mixture as possible throughout the combustion space at the time of ignition unless the correct stratification can be guaranteed. Improved atomization from the injectors helps as small droplets (less than about 10-μm diameter) more readily follow the airstream. However, the reatomization of wall-deposited fuel by shearing from the inlet valve and ports is unlikely to result in that level of atomization.

As noted earlier, the basic atomization from the injector can give quite large droplets of well over 100-μm diameter (SMD). With the entering film flow, the droplet size range can extend to higher levels. Rink and Lefebvre [43] conclude that CO and UHC emissions are substantially reduced by improved atomization for equivalence ratios above about 0.8. This is due to the better vaporization and mixing of the smaller drops. Due to the reduced effect of evaporation rates on the combustion with leaner mixtures, there is little influence below this value. With NO_x emissions, the improvement occurs below approximately $\phi = 1.1$ with little effect above that value. Larger droplets are surrounded by flame envelopes, which burn locally in the diffusion mode at near stoichiometric conditions maximizing the NO_x formation. That is, in the normal operating range of an engine, $\phi = 0.8$ to 1.1, improved atomization within the cylinder is a benefit for all emissions.

8.4.1.3 Spark Timing

At different steady-state conditions, an engine requires altered ignition settings due to the different amount of residual gas remaining in the cylinder and the effect of the speed-dependent turbulence on flame propagation rates. In the older distributor ignition systems, the spark timing position was adjusted via centrifugal weights for speed and a vacuum advance from the inlet manifold for a more closed throttle position. The control operated via what were essentially ramp functions, which were crude approximations of the requirement to maintain the peak combustion pressure at the optimum position. While the response of these systems was moderate, of the order of a second, it was not instantaneous and could cause short periods with the incorrect timing. Incorrect spark timing can lower engine efficiency and, while it can sometimes have beneficial effects on the emission levels (see Section 8.5), a setting in the wrong direction can be detrimental. In modern engines, the control of the timing during acceleration-type transients is handled electronically and is more rapid and precise. Ignition system response is, therefore, not now a major problem affecting emission levels, although it is important that the timing system be optimized for emission control.

8.4.1.4 Emission Control Systems

The major emission control system items where the response times are important are:

- The EGR system.
- The fuel-air ratio excursions on three-way catalytic converters.

The latter only work efficiently in a narrow fuel-air ratio window around stoichiometric and can, therefore, be adversely affected by the transient problems discussed previously. In addition, during engine warm-up, overfueling can limit their effectiveness. With the EGR, its basic response due to the change in the mass flow rate of gas is more direct. This is because the quantity of exhaust gas in the cylinder directly modifies the NO_x produced during combustion. As the gas must be drawn through a valve from the exhaust, a short delay occurs until it readjusts to the new conditions. Typical flow velocities of perhaps 5 to 15 m/s give a total delay of the order of 0.1 to 0.2 seconds. The length of the pipe connecting the exhaust to the inlet manifold should be as short as possible and the valve should be rapid acting.

8.4.2 Long-Term Transients

The transients that occur when a major change of temperature takes place in the engine are long-term transients. It is therefore of particular concern during the warm-up period. This will obviously vary from engine to engine and with the ambient conditions but a typical value might be about 8 minutes (Pao [44]) from cold to a well warmed-up condition where a true steady-state operation is feasible. Much of the recent work discussed in Section 8.5.1.2 (Takeda et al. [22], Saito et al. [30], and Shin et al. [31]) is related to this aspect. In addition, Min et al. [45] have examined the effect of crevice volumes on cold and warmed-up engines and have concluded that, while the total UHC in the latter is much less, any variations stemming from crevice volume size are approximately the same for a warm as for a cold engine. In fact, the variation of UHC with crevice volume in both cold and warmed-up engines is surprisingly small.

When an engine is cold, only the most volatile fuel components such as butane evaporate readily (Figure 8.12) and are available in the gas phase for easy ignition and combustion. For a cold start, the immediately available air-fuel ratio is then very lean. To compensate for this, additional fuel has to be added from the fueling system so that sufficient evaporated components will bring the mixture to within the ignition limits. Hence, overall the mixture will be excessively rich.

Fox et al. [46] have found that abnormally high UHC concentrations exist in the cycle immediately prior to firing due to liquid fuel entering the cylinder. That is, the starting process is itself important. During the warm-up period, the problems associated with the short-term transients will also be greatly exacerbated because less evaporated components exist in the flow.

It would be expected that during a cold start the ambient conditions, particularly the temperature, are of great importance. There is evidence that this is the case with typical results showing that a drop from about 20°C to 0°C can more than double the UHC emissions (Pao [44]).

It is therefore critical that during a cold start (or even only partly warm), the fuel schedule is controlled so that the engine does not stumble and misfire but does not have more fuel than is necessary. Also, it is important that the engine design be such that the intake system warms up rapidly and that the engine cooling system comes up to temperature as fast as possible. At the same time, this cooling system must be such that it is sufficient to prevent overheating problems under severe, hot conditions. Details of such cooling systems will not be discussed here, but reference can be made to Stone [24] and Finlay et al. [47].

8.5 SOME DETAILS OF CONTROL SYSTEMS

8.5.1 Brief Historical Comments

Emission control from spark-ignition engines initially used relatively minor adjustments but has since escalated until it is now one of the major considerations in the design of engines. It may indeed put limits on the usage of combustion engines in inner-city regions in the future. For example, the legislation toward the use of a proportion of electrically powered vehicles within city areas, whether a rational decision or not, has been prompted by this factor.

The initial concerns in Los Angeles were that carbon monoxide and unburned hydrocarbons from the exhaust gases of automobiles were high. Thus, early control techniques and systems were aimed at minimization or oxidation of these emissions. Control commenced with the introduction of the positive crankcase ventilation (PCV) system in 1961 while the use of thermal reactors and modification of various engine design parameters followed shortly afterward. It was then several years more before the combined effects of the NO_x and UHC emissions on ozone levels and photochemical smog were identified. Their control immediately introduced significantly greater complexities. This is because reduction (as opposed to oxidation) of the NO_x to N_2 and O_2 is required and because the NO_x peaks, as discussed earlier, at the equivalence ratios near to those which are optimal for minimum CO and UHC, these being slightly lean of stoichiometric. However, in spite of the difficulties, fully enveloping techniques and controls have been developed and legislation requiring control of all types of emissions now exists in most countries. A range of emission control strategies and systems, both current and of historical interest, will now be discussed.

8.5.2 Emission Control Techniques

While some of the following are difficult to place in a single category, it is convenient to view the emission control techniques under a number of subheadings, such as engine design features, operational factors, engine treatment (i.e., add-on modifications to the engine or fueling systems), and exhaust treatment (clean-up).

8.5.3 Features of Engine Design

8.5.3.1 Compression Ratio

One of the easiest modifications that can be made to an engine design is to reduce the compression ratio and, in fact, this was an early method used for reducing the UHC. Experiments confirm that this method was effective. A typical result obtained by Saito et al. [48] can be examined here. These results were for a single-cylinder, hemispherical head, overhead camshaft engine of 73-mm bore by 77-mm stroke. The engine was tested without and with EGR, the results for the former being shown in Table 8.6. The values shown are estimations from the graphical data and are therefore approximate. Nevertheless, they indicate that with a lowering in compression ratio, UHC reduces substantially and NO_x reduces except at low speeds, while CO rises.

An initial reason for reducing the compression ratio was that much of the UHC formation is formed in the quench zones, which exist on the walls of the combustion chamber. As shown by Tabaczynski et al. [11], these are swept from the walls and mixed early in the exhaust process and so are probably not as significant as first thought. However, increasing the clearance volume relative to the swept volume (i.e., reducing the compression ratio) decreases the proportion of crevice zone to combustion zone volume. More recent results (Min et al. [45]) using a gaseous fuel (propane) indicate only modest sensitivity to crevice volume size and so this may not be the explanation. Another effect is that a lower expansion ratio maintains the exhaust gas at a higher temperature, which helps to promote the oxidation of both CO and UHC in the tailpipe or any chambers in the exhaust system. This is therefore a useful technique, particularly when combined with thermal rectors. A further advantage is that the lower end-of-compression temperatures reduce the peak temperatures reached during the subsequent combustion and some reduction in NO_x also ensues as indicated in Table 8.6. Obviously, lower compression ratios reduce engine efficiency and the consequence is greater fuel utilization and increased CO_2 output. One of the advantages of the use of catalytic converters for exhaust clean-up is that compression ratios can revert to those which are limited only by engine knock and higher thermal efficiencies can then be regained.

Table 8.6

(a) Effects of Compression Ratio on UHC Emissions

	Percentage reduction in UHC	
	Compression ratio 10:1 reduced to 8.5:1	Compression ratio 10:1 reduced to 7.0:1
1500 rpm	−7%	−14%
2000 rpm	−28%	−50%
2500 rpm	−50%	−70%

(b) Effects of Compression Ratio on NO_x Emissions

	Percentage reduction in NO_x	
	Compression ratio 10:1 reduced to 8.5:1	Compression ratio 10:1 reduced to 7.0:1
1500 rpm	+2%	+7%
2000 rpm	−10%	−17%
2500 rpm	+3%	−10%

(c) Effects of Compression Ratio on CO Emissions

	Percentage reduction in UHC	
	Compression ratio 10:1 reduced to 8.5:1	Compression ratio 10:1 reduced to 7.0:1
1500 rpm	+12%	25%
2000 rpm	+9%	+18%
2500 rpm	+8%	+16%

8.5.3.2 Cylinder Size and Engine Speed

The thickness of the quench zones is determined by the working substance pressures and temperatures and the wall temperatures. Hence, larger cylinders have a smaller surface area to volume ratio and their wall quench zones and crevice volumes will be proportionally less if other factors are kept constant. This will tend to reduce UHC emissions. That is, for the same overall engine capacity, with all other parameters identical, an engine designed with fewer, larger cylinders should provide a benefit in relation to UHC. However, it should be noted that other factors are influential and may be interdependent with cylinder size. For example, higher engine speed, to be discussed later, tends to reduce UHC emissions and this is often associated with smaller cylinder sizes for mechanical reasons. The larger the cylinder, the greater the proportion of burning that takes place in the high-temperature gases away from the walls and that will mean that more NO_x is formed. However, very little experimental data seem to be available on the scaling effect of engine size on emissions. A modeling study by Trinker et al. [12] discusses size effects.

Fig. 8.16.

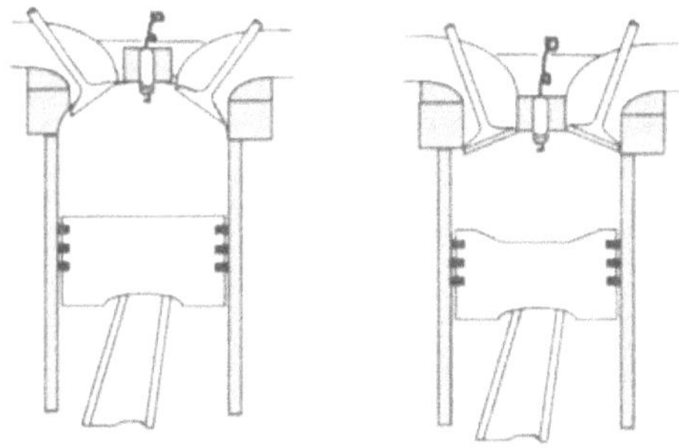

Compact combustion chambers hemispherical and pent-roof types.

8.5.3.3 Combustion Chamber Shape

Again, the surface area of the combustion chamber may be a factor in determining the quantity of UHC emissions. For a given volume, a spherical shape has the smallest surface area to volume ratio and hence is ideal from that point of view. Such a shape can be approximated by the use of a hemispherical head and recessed bowl in the piston. Hemispherical heads have often been used, mostly for high performance and minimum knock. However, from a practical point of view, a shallow angle pent-roof type (Figure 8.16) is a good approximation and allows in addition good valve positioning and size. As a general rule, it would be expected that compact chambers are preferable to extended chambers. It should be noted that these are also beneficial in reducing end-gas autoignition (knock), thereby allowing higher compression ratios and have a faster burn rate and lower heat transfer losses. That is, they improve engine efficiency. In addition, these heads provide higher volumetric efficiencies and tumble within the cylinder. This improves the burning rate and hence the thermal efficiency, and helps minimize quench zone thickness, thereby improving UHC emission levels. They are generally best designed with twin overhead camshafts, which add to both the original capital and the maintenance costs of the engine.

8.5.3.4 Engine Temperature Control

Engine temperatures both within the cylinders and in the induction system are important parameters in emission control. First, as noted previously, a very high proportion of the emissions are produced when the engine is either cold or only partially warmed up. Thus, it is important to bring an engine to the desired operating temperature as quickly as possible. This can be achieved by careful calibration of the thermostats on the cooling system and by improved use of the external airflow pattern across the engine and through the radiator from the cooling fan or fans. These are now often electrical rather than purely mechanical and can, therefore, be easily decoupled from the engine to eliminate their flow until it is required. With the induction system, the correct temperature is necessary as variations can cause density changes in the air. As most systems measure volume airflow rate to determine fuel mass flow rate, a density change can result in air-fuel ratio excursions. It is therefore desirable to bring the induction system to the temperature at which it will operate as quickly as possible, but then not to allow it to overheat past this point during hard operation. A further disadvantage of overheating is that NO_x emissions will rise. Water-heated intake manifolds or more generally exhaust-heated "stoves" or "hot boxes" using heat transfer from the exhaust manifold have been used. It is now common to use exhaust manifold heated air to mix with the cold air via a control valve. Port injection has minimized these problems. That is, inlet air and manifolds run at lower temperatures than those used previously in the early stages of emission control.

8.5.3.5 Crevice Regions

As discussed in Section 8.1.3, crevice regions provide a high proportion of the quench zones within an engine cylinder. Typical values are about 1 percent to 1.5 percent for the region above the upper piston ring, 0.25 percent for the spark plug screw thread, and 1 percent to 1.5 percent for other regions (percent of the clearance volume). These volumes are, therefore, quite significant and should be reduced to the minimum consistent with other design considerations such as mechanical strength in each case.

8.5.4 Fueling Systems

In 1996 the majority of automobiles used MPI (multipoint injection) systems. This is particularly true of the larger engine sizes, with only those below about 1.4-L capacity retaining an approximately equal share between SPI (single-point injection) and MPI. The suction-type carburetor is now little used. Fuel introduction systems via positively pressurized sprays have improved noticeably over the suction systems of carburetors during the decade or so from the early 1980s. The ability to control the air-fuel ratio at the fuel jet to the required specifications using feedback from exhaust measurements (oxygen sensors) is quite feasible. The problem is not so much in the fuel spray but, as noted previously, in the dynamic characteristics of the induction system itself.

8.5.4.1 Single-Point (Throttle-Body) Injection (SPI)

As discussed in Section 8.4.1.2, throttle-body injection systems are a simple replacement for the basic venturi-type carburetor. An injection system is readily adapted to feedback control from measurements of appropriate parameters in the exhaust gas whereas a carburetor is not. Under steady-state conditions, the replacement of a venturi-type carburetor with a throttle-body injector is an adequate solution from the emissions viewpoint but, while an improvement, it still retains the problems of maintaining the correct air-fuel ratio values under transient conditions. This can have direct effects on the production of the emissions in an engine. In addition, a tight control of air-fuel ratio is mandatory for a high-conversion efficiency in three-way catalytic converters.

8.5.4.2 Multipoint (Port) Injection (MPI)

This system has generally replaced the throttle-body type particularly on larger engines, as it minimizes the problems of single-point fuel introduction and, in particular, is the best current system to match the use of three-way catalytic converters. As pointed out in Section 8.4.1.2, the fuel distribution between cylinders and the timewise distribution to each cylinder are improved by this method. This makes three-way catalytic systems much more viable. Power levels of the engine are generally higher for a number of reasons. First, a fully vaporized fuel injected upstream (SPI system) displaces about 2 percent of the air coming into the engine. However, its cooling effect can more than offset this if the manifold is unheated. Generally, however, some manifold heating is required in these cases to avoid engine stumble during acceleration and the overall effect may lie in between, depending on the design and the conditions. That is, the net effect may be small. Of more importance is that the intake manifold can be better designed because it is for air alone and, hence, the volumetric efficiency is improved, less induction air heating is required, and fewer losses occur due to abnormal air-fuel ratios in the cylinder-to-cylinder fuel distribution.

8.5.4.3 Direct In-Cylinder Stratified Charle Injection (DISC)

With this system, the fuel is sprayed directly into the cylinder as in a diesel engine; the difference in a DISC engine is that the injection is under lower pressure and usually begins early in the compression process, allowing more time for evaporation and mixing of the fuel. Under these conditions, compression ignition would cause uncontrolled initiation of combustion and so a spark is still required for ignition. The concept, to be successful, requires the mixture at the time of ignition to be distributed (stratified) in an appropriate manner. Control of the fluid mechanics of both air and gas flows is critical. As far as the injection itself is concerned, problems of wall impact, droplet and film evaporation, and mixing are at least as important as in current systems. The cylinder warms up more quickly than other engine parts and the high temperatures there then enhance evaporation rates. Nevertheless, the DISC system may not completely eliminate lag-induced mixture variations during transients particularly under cold conditions if there is any direct impingement of fuel on the cylinder wall or piston crown.

A more complete discussion of DISC engines will follow in Section 8.6.3.

8.5.5 Engine Operational Controls

8.5.5.1 Lean Burn

For the greatest engine power, the equivalence ratio for an engine should be set at about 10 percent rich while, for best economy, it is about 5 percent to perhaps 15 percent lean depending on the load (Taylor [49]). If stable combustion can be maintained, higher engine efficiencies could be obtained from even leaner mixtures. Engines now predominantly operate near to stoichiometric largely because of the constraints imposed by the three-way catalytic converters used with them. One possible alternative is to use a lean-burn strategy, which is likely to produce an engine with low emissions as well as greater fuel economy. Two approaches are possible here. The first is that the mixture can be set to about 5 percent to 10 percent lean, which will mean that it is near the minimum in terms of UHC production and as well will be generating very low levels of CO. However, this will mean that it is near the peak NO_x emission levels and these will need to be separately controlled, usually by substantial use of exhaust gas recirculation (EGR). In the past, catalysts to reduce NO_x required a stoichiometric or rich mixture and a lean-burn approach was limited. It has basically been useful on smaller engines where the combination of short induction tracts, single-point injection, and lean burn provide an engine where low initial cost and excellent fuel economy make an attractive package. However, recent developments in catalysts allow NO_x reduction with lean mixtures giving greater versatility in the engine size which is reasonable for its adoption. The lean-burn approach is, therefore, receiving renewed attention.

To obtain a direct reduction in all emissions, even leaner operation is needed. For example, at 30 percent lean ($\phi = 0.7$), the NO_x emissions have fallen substantially (by approximately 85 percent from their peak at $\phi = 0.93$, see Blumberg and Kummer [50]) and the UHC emissions have risen only a small amount from their minimum at $\phi = 0.9$. This provides a possible mode of even leaner-burn operation and can be coupled with exhaust clean-up in the form of the new types of reducing catalysts for NO_x, such as the copper-exchanged zeolites, which will be discussed in Section 8.5.7.4, and an oxidizing catalyst to clean up the remaining UHC. Combustion irregularity (ignition failure, partial burn, cycle-by-cycle variability) is likely to become quite significant and many problems must be overcome before it is commonly used. The alternative is to move toward a stratified charge system, which will be dealt with in Section 8.6.3.

8.5.5.2 Spark Timing

The ignition timing is an important variable in obtaining the best performance from an engine. It is optimally set for efficiency to what is called MBT (a term used to mean either *minimum advance for best torque* or *maximum brake torque*, depending on the individual's preference). A small latitude in the actual setting exists as a few degrees either way from the best value makes a negligible difference, but the decrease in engine performance then starts to become marked quite rapidly with further changes of timing from the best value. The optimum position for efficiency is such that the peak pressure should occur some 10° to 15° ATDC. After ignition, the initial growth of the flame is slow and the pressure takes some time to maximize as the flame front moves across the combustion chamber. Hence, the actual firing of the spark is well before this, usually more than 20° of crankangle. Because the rate of pressure rise is dependent on the amount of residual gas in the cylinder, which in turn depends on the throttle setting (plus the EGR if any) and the turbulence, which is governed by the engine speed, the actual spark timing must vary quite considerably for each operational condition. However, it may be retarded from the optimum for efficiency to help minimize exhaust emissions. Due to the ease of this adjustment, it was one of the earliest modifications used with engines when emission control legislation was introduced. Note that it has always been normal to retard the spark to reduce engine knock. This is usually at low engine speeds and large throttle openings. The adjustment for emission control is in addition to this and may occur at different engine settings.

There can be several positive effects on emissions due to a retarded timing. The first is that the exhaust temperature is hotter and both CO and UHC oxidize more thoroughly in the exhaust system. This is more apparent when the engine is coupled to a thermal reactor where the high exhaust temperatures are essential. A second effect is that more of the burning now takes place once the piston has commenced its descent and, while the effective combustion chamber may not increase dramatically in size during this period, its increased volume helps reduce the proportion of the mixture in the quench zones. Finally, a retarded timing results in a lower peak engine pressure and, hence, peak temperature and the NO_x are reduced. Trends with ignition timing are illustrated in Figure 8.17. The problem with using

Fig. 8.17.

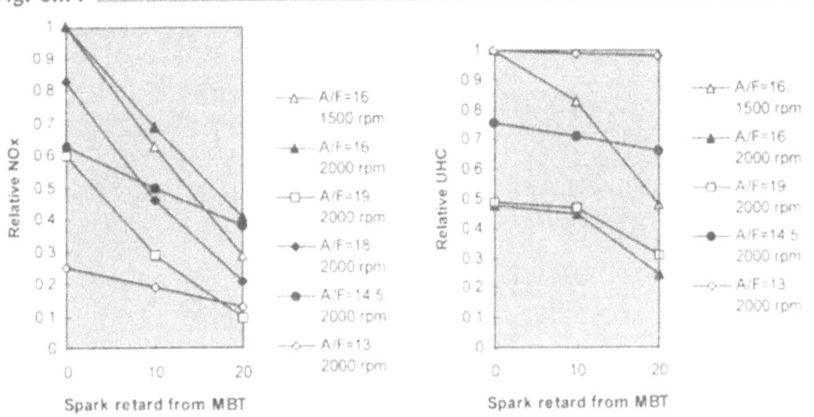

Effect of ignition timing on emissions relative to highest value of that emission in test (data from Saito et al. [48]).

it as a control mechanism is that the engine efficiency is reduced, often considerably as a result of substantial spark retard. It is one of the reasons for the increase in fuel consumption experienced with emission control prior to the introduction of catalytic converters, which allowed the timing to revert to optimum settings.

8.5.6 Treatment within the Engine

8.5.6.1 Crankcase Ventilation

This will be dealt with first as it is a simple, universal type of control applicable to all SI engines. It is also something which has been noted as important for reasons other than emission control purposes in all reciprocating engines since their early development.

In piston engines, a small quantity of the charge in the combustion space leaks past the piston rings to the crankcase due to the high pressure above the piston during the latter stages of compression and particularly during combustion. This is exacerbated as the engine wears and the piston ring gaps increase. To prevent build-up of pressure in the crankcase, it has always been essential that a crankcase ventilation system of some type be installed. Until emission controls were introduced, this generally consisted of a simple bent tube as shown in Figure 8.18. With this type of device, the gases which accumulated above the lubricating oil in the crankcase were expelled directly to the atmosphere.

In SI engines, the premixed nature of the charge means that any gas that leaks past the piston rings to the crankcase prior to ignition is high in unburned gases and hence in UHC. The leakage rate increases as the pressure rises. When combustion occurs, the flame front is initiated somewhere remote from the upper piston ring; hence, the gas in and around the piston ring crevice is likely to burn toward the

Fig. 8.18.

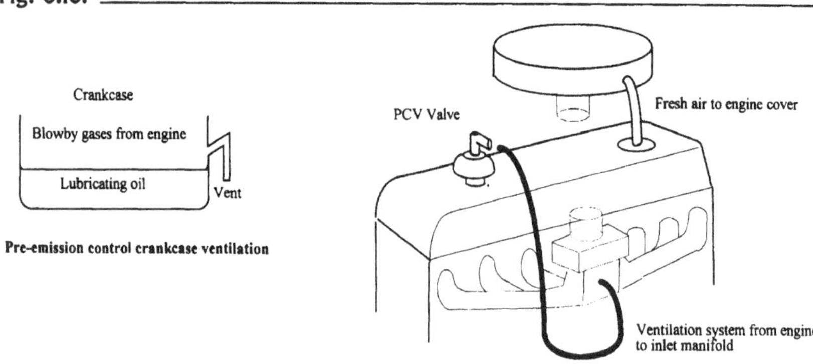

Pre-emission control crankcase ventilation

Modern Crankcase ventilation system

Types of crankcase ventilation systems.

end of the combustion process with some of that leaking past the rings, for much of the combustion process. In the later stages of combustion when the flame front reaches the zone above the piston rings, the gas contained between ring, cylinder wall, and piston may not burn at all, or may only be partially oxidized because the region within this crevice is a highly quenched zone. The gas forced through the ring gap, therefore, still contains high UHC levels with possibly some CO. The explosive nature of this gas, if it collects in the crankcase, is a major reason for venting the crankcase. However, from an emission viewpoint, it is undesirable for these exhaust products to be transmitted to the atmosphere, and the old style open ventilation system is not suitable.

Ventilation must, therefore, be achieved by removing the gases to where they can be oxidized to CO_2 and H_2O. This is accomplished by sucking off the gases using the lower than atmospheric pressure within the inlet manifold and returning them via the intake flow to the combustion chamber where they participate in the normal combustion process. The term *positive crankcase ventilation* (PCV) comes from this provision of a pressure drop to assist the flow.

The concept of providing positive suction from the inlet manifold to relieve the crankcase pressure was not new at the time emission regulation started. There are examples of it on vehicle engines in the 1950s and possibly earlier. Once the crankcase gases are drawn into the manifold, they mix with the incoming charge and are then burned in the cylinders. A one-way valve (the PCV valve) is required to ensure that flow can only take place from crankcase to manifold and that the flow is not excessive under high manifold vacuum conditions. In most modern systems, clean air is also ducted to the crankcase so that a continual purge exists. This is also shown in Figure 8.18. The emission advantages of a PCV system are that a significant source of UHC and CO is eliminated; the general advantages are in the slightly better ventilation and pressure reduction in the crankcase and the small saving of any wasted fuel. The disadvantage is in the slight extra cost involved, but this is small.

8.5.6.2 Diluents and Exhaust Gas Recirculation

NO_x emissions are essentially due to the O_2, N_2 combination that occurs at high temperature. As noted previously, they peak at slightly lean equivalence ratios around $\phi = 0.9$, which is the region where CO has become low and UHC is a minimum. Lean burn is therefore not yet a solution unless extremely lean values are resorted to at which point other problems such as low power output and high UHC appear. Other means of control of NO_x are therefore essential.

One common method is to provide a diluent to reduce the peak combustion temperature. The two major diluents that have been considered are:

- Water addition, either in the form of water injection or in an emulsified form with the fuel.
- Exhaust gas that is recirculated back into the cylinder.

Fig. 8.19.

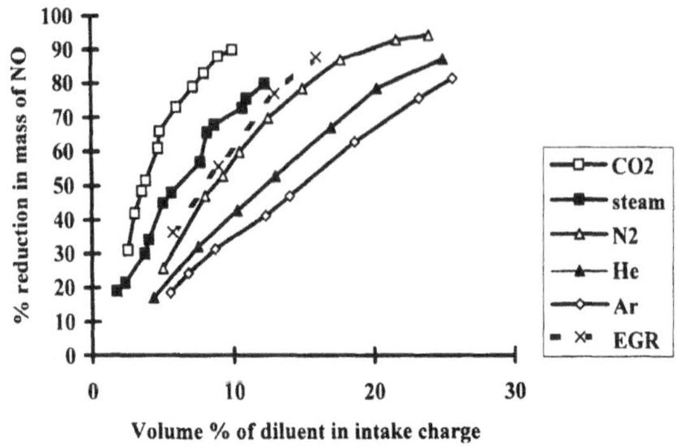

Trends in NO_x production with diluent addition (data from Quader [51]).

Probably the most extensive examination was by Quader [51] who used diluents such as CO_2, water (as steam), exhaust gas, nitrogen, helium, and argon. Some trends are illustrated in Figure 8.19. He found that the NO_x reduction from each indicated that the specific heat of the diluent was the dominant factor and that, therefore, they were effective in the order listed earlier, CO_2 being the best and argon the worst. However, that order was reversed for fuel consumption effects, with the CO_2 and H_2O increasing specific fuel consumption, exhaust gas and N_2 holding it roughly constant, while helium and argon reduced it. This is probably due to the cycle thermodynamics, as the value of the isentropic index is lower than air for CO_2 and H_2O, about the same for exhaust gas and nitrogen, and higher for helium and argon. Of these diluents, only water and exhaust gas are easily made available to an engine. The use of the former has some proponents, not only because of its ability to reduce NO_x emissions but for a variety of combustion-related reasons (e.g., microexplosions, Dryer [52]) and has been used in a number of situations in the past. The latter, because of its continuous availability in any engine, is by far the most commonly accepted diluent. For both, it is worth some further examination of their effects.

Water Addition A great many papers comment on the relative antiquity of water addition to engines and this is, indeed, justified. In an early work, it was examined by Hopkinson [53] as a means of providing cylinder wall cooling but, from the discussion, it appears to have been previously used even at that time. Later, it was a common addition to the high-performance, spark-ignition, aeroengines of World War II in order to help overcome the intense knock due to their large bores, supercharged (therefore, high pressure and temperature) conditions. In these cases, the water produces a lower-temperature combustion with a cooler end gas, thereby inhibiting end-gas autoignition (knock). In the early postwar years, the increase in compression ratios of automobile engines sometimes outstripped the availability of high octane number fuels and water or water/alcohol injectors were sold for automotive applications, again for knock control, mostly as a retrofit option. These operated intermittently as required, usually being controlled from the inlet manifold pressure, and the total amount of water consumed was small. A number of papers (Kettleborough and Milkins [54] and Weiss and Rudd [55]) considered the effect on power output with low-quality fuels and concluded that water-fuel ratios of up to the very high values of about 0.25 could provide substantial benefits if the compression ratio was raised from the relatively low values (about 7:1) used at that time. These researchers also reported that it reduced hard carbon deposits, a problem that has been substantially now overcome by the use of fuel-borne scavengers.

The effect of water addition on engine combustion and emissions received some attention in the 1970s (Dryer [52], Nicholls et al. [56], Quadar [51], and Peters and Steber [57]). The work is summarized by Milton and Carter [58] who also carried out further work related to NO_x reduction using either direct water injection into manifolds and ports or water-fuel emulsions. With the water injection, they showed for tests simulating a typical city driving cycle that NO_x emissions decreased dramatically, CO remained roughly constant, and UHC tended to rise. With the use of water-fuel emulsions, emulsion stability was a problem and the best results were found with volume proportions of about 2 percent. Emulsifiers tended to reduce octane number but had no effect on fuel viscosity. With these, NO_x reduced throughout the air-fuel ratio range but CO tended to increase with the water proportion. The general conclusion appears to be that, while benefits to NO_x emissions occur due to water injection or water-fuel emulsions, these may be offset by other factors, particularly UHC. When this is compounded by the fact that additional fluid tanks need to be carried (for injection) or that fuel stability problems may occur (with emulsions) and under some circumstances engine corrosion may be increased, a clear case for water addition is not established.

Exhaust Gas Recirculation (EGR) The use of a diluent is clearly effective in NO_x reduction. However, it needs to be one which is at least neutral in relation to the other engine emissions and, if possible, to the engine efficiency. It must also be a substance that is readily available. From the most likely possibilities examined by Quadar [51], only exhaust gas completely fits this requirement.

In engine cylinders, a certain amount of exhaust gas is retained at the end of the exhaust stroke in the clearance volume of the engine. This is termed *residual gas*. The residual gas fraction varies according to the intake and exhaust pressures, p_i and p_e, respectively, and the compression ratio r. If p_i equals p_e, an approximation for the mass fraction of residual in the new charge is the ratio of clearance volume to total cylinder volume. However, the temperature in the exhaust is higher than the inlet, and inertia effects in the flow, together with valve overlap, expel a little more exhaust than would otherwise occur. This then is usually somewhat lower than that value and is typically around 7 percent to 8 percent at WOT (see Heywood [8]). Lower inlet pressure, such as occurs when the engine is throttled, increases this proportion, while higher inlet pressures such as those in turbocharging reduce it. To provide adequate NO_x reduction, a total exhaust mass fraction is likely to be around 15 percent to 20 percent. (Figure 8.19). Thus, a continuous but variable supply of exhaust gas to the inlet is required. Also, the substitution of exhaust gas has other effects. First, it replaces air in the inlet system and reduces the amount of fuel which can be burned. Hence, the fuel flow rate to the engine will need to be varied from that which would be required without EGR. Second, the burning rate of the fuel-air mixture in the cylinder is lowered by the diluent, which may tend to reduce the thermodynamic efficiency of the engine and will certainly require a modification in the ignition timing. To obtain full engine power, it is therefore possible to temporarily inactivate the EGR. As most vehicle engines do not require full power at any point in a typical driving schedule for emission testing, this does not affect their emission rating, although it will mean that for brief periods in practice, higher NO_X levels will exist. It is also desirable to inactivate the EGR at idle conditions because the low manifold pressures mean that substantial residuals already exist here. With too much EGR, combustion irregularity becomes evident.

The EGR is usually taken directly from an exhaust tapping via a control valve to the inlet manifolds. It is usually controlled directly from inlet manifold pressure, although on modern systems, the engine management microprocessor may couple this with other variables such as engine speed.

8.5.7 Exhaust Treatment

8.5.7.1 Thermal Reactors

If the major exhaust emission products of concern were only CO and UHC, it is possible to oxidize them directly to CO_2 and H_2O in the exhaust system of the engine as long as the appropriate conditions can be obtained there. One method of achieving this is the use of a thermal reactor (or afterburner) in the exhaust system. For the appropriate oxidation reactions to take place, the reactor needs to be at a certain minimum temperature and there must be sufficient oxygen available to complete the oxidation process. These high temperatures are obtained from the already hot exhaust gases and the additional energy released by the exothermic reaction. Hence, for the first, a retarded spark is often required while for the latter, a rich mixture provides the additional energy.

Typical temperatures that are required are shown in Figure 8.20. It can be seen that a temperature of 500°C oxidizes about 40 percent of the UHC and 20 percent of the CO, the latter being too low to be really useful. To obtain 40 percent CO oxidation, this temperature must now reach about 580°C at which value nearly 80 percent of the HC is converted. To achieve a similarly high level of CO conversion (i.e., 80 percent), the temperature must be around 700° to 750°C. This is extremely high and, although the walls of the reactor will be at a lower temperature than this, it is well above red heat for steel and could cause problems in design, construction, durability, operation, and safety. In addition,

Fig. 8.20.

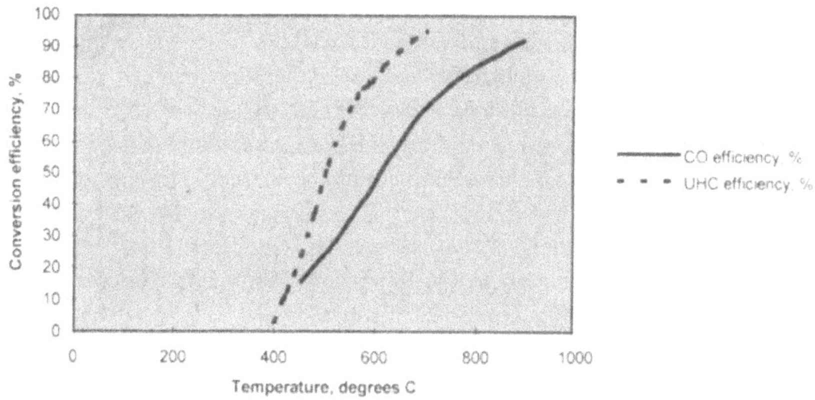

Typical conversion efficiencies at different temperatures for thermal oxidation of HC and CO.

the higher the temperature of the reactor, the lower the residence time required for the gas to achieve good conversion values and, hence, there is good reason to keep the reactor temperature near the maximum limits. For example, with CO an approximate 90 percent conversion rate is achieved after about 70 ms at 850°C but this level is not reached until well over 250 ms at 750°C. With UHC, the residence times are shorter, the approximately 90 percent conversion being achieved in less than 50 ms at 850°C, 100 ms at 750°C.

A typical mean temperature range for the combustion product gases leaving the exhaust port is about 350°C to 700°C but this may reach as high as 900°C under a full-power, high-speed operation due to the greater mass flow rate at the higher inlet manifold pressures and the shorter time available at higher piston speeds for heat transfer to take place to the cylinder walls. It should be noted that the exhaust temperatures vary throughout an engine cycle by about ±200°C and that the exhaust cools rapidly in a normal exhaust pipe as it moves downstream from the exhaust port, dropping by around 50°C to 100°C/meter. The reactor therefore must be placed close to the manifold outlet. The highest exhaust temperatures are achieved under stoichiometric conditions. A spark retarded from the normal (MBT) setting will increase exhaust temperatures due to the later burning and lower power extraction during expansion while an advanced spark, in spite of lower power, will decrease the exhaust temperature as more heat is then transferred to the cylinder walls. In order to obtain the required reactor temperatures, it is usually necessary to run the engine rich so that the extra fuel can be burned in the reactor. Under some circumstances, an additional spark retard is used to provide high temperatures in the gas entering the reactor. Obviously, if the engine is rich, the exhaust gas contains little or no oxygen and additional air is then needed. Depending on how rich the mixture is, this may amount to some 10 percent to 20 percent of the normal engine air supply. Thermal reactors therefore require an air pump and air nozzle to introduce the air into the reactor. In order to maintain a high enough temperature in the reactor, this air may need heating. Its flow rate needs to be controlled for different conditions. All this adds to the complexity, cost, and bulk. The maintenance of the necessary high temperatures under all conditions creates control problems for the reactor and high levels of oxidation cannot always be guaranteed. In addition, both the rich mixture and the spark retard reduce thermal efficiency and it is noted that the fuel consumption of automobiles in the United States (Figure 8.8) increased during the period to 1974 that thermal reactors were in common use. Given that this has an undesirable effect on both fuel resources and CO_2 emissions, thermal reactors are untenable as a long-term solution. Thus, while thermal reactors were a common early solution to the UHC-CO problem, they are now no longer generally used.

8.5.7.2 Catalytic Treatment of Exhaust Systems

Thermal reactors were only a transitional arrangement for two reasons. The first is that it was difficult to obtain high oxidation rates without running the engine rich and using retarded spark timing, both of which were inefficient; the second was that it was not useful to control NO_X. Hence, an alternative method of exhaust clean-up, the catalytic converter was introduced and this is now the most common method of exhaust emission control. This has the advantage that exhaust temperature can be kept moderately low and the engine parameters, compression ratio, and spark timing can be retained at levels which are optimized for the combustion factors alone and, hence, can provide the best efficiency for the air-fuel ratios used.

Catalytic converters use metallic catalysts to promote the desired reactions at lower temperatures than would otherwise be needed. Typical values are shown in Figure 8.21. The catalyst types could be a base metal (e.g., copper, Cu, or

Fig. 8.21.

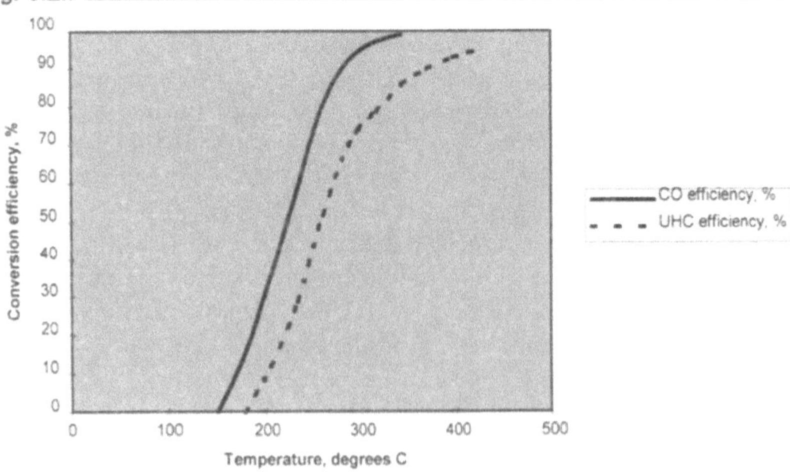

Typical conversion efficiencies at different temperatures for catalytic oxidation of HC and CO (data from Heywood [8]).

chromium, Cr) but are more usually noble metal (platinum, Pt, palladium, Pd, rhodium, Rh). The transition metal oxides of copper, cobalt, iron chromate, and vanadium are also useful as catalysts and have been considered. However, in spite of their expense, the noble metal types are commonly used as these have been found to be better catalysts. Reactors usually use some combination. The noble metals are preferred because:

- They are much less deactivated by sulphur in the fuel at temperatures <500°C.
- They have a higher specific activity for hydrogen oxidation.
- They can be made more thermally resistant to reduced low-temperature activity.
- They have a superior cold-start performance.
- They provide equivalent catalytic activation with smaller volumes than do base metals.

The amount of noble metal is kept small by plating on the surface of a pellet bed, ceramic honeycomb structure, or metal matrix. The pellet types originally used generally consisted of particles of approximately 3-mm diameter of a substance such as aluminium oxide (Al_2O_3). Because of their high-pressure drop, they have largely been replaced by the honeycomb or matrix types, which have an inside diameter of the sections of about 1 mm. Ceramic honeycombs now in use have wall thicknesses of about 0.15 mm with 62 cells/cm^2 giving a flow area of about 70 percent of the cross-section. This compares with about 90 percent for metal types (Heck and Ferranto [59]), which have a lower pressure drop. They use a catalyst wash coat over the honeycomb based on γ-Al_2O_3 but which contains the catalyst materials of about 0.1 percent to 0.15 percent, 20 percent cerium oxide (CeO_2), and stabilizers such as barium oxide (BaO). Rare earth and alkaline ions can improve stabilization (Church et al. [60]). The cerium oxide has the ability to store oxygen under lean conditions and return it to the flow under rich conditions and so extend the air-fuel ratio window at which the catalytic converter works (Padeste et al. [61]). The wash coat thickness may range from about 10 μm to 150 μm thick but is more commonly in the middle of that range at 20 μm to 60 μm (Heck and Farranto [59]). The largest metal mass is platinum, usually about five times that of the rhodium, although the ratio may vary considerably. Tests by Lox et al. [62] show that Pt-Rh catalysts and Rh-only catalysts have similar effectiveness, which is superior to Pt only. High-temperature performance may be enhanced by the appropriate combination of metals (Trimm [63]).

Flow through the converter may be either radial or axial but is usually the latter, as shown in Figure 8.22. The noble metal is at least partially recoverable at the end-of-life of the converter, a typical reclamation of about 90 percent being achievable using acidic solutions. This value varies little with the vehicle mileage and the degree of poisoning of the material that has taken place during its lifetime.

Fig. 8.22.

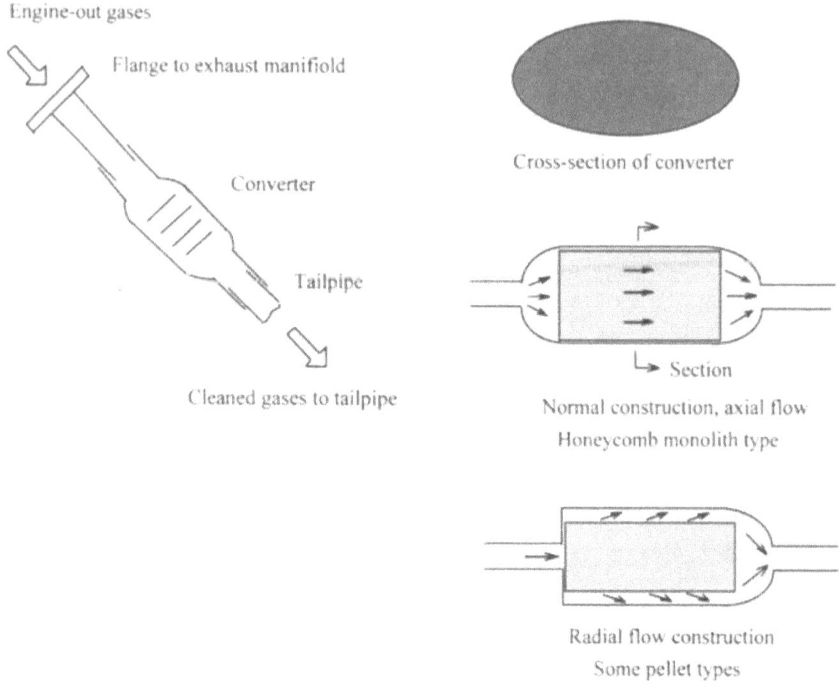

Catalytic converter construction.

The catalyst efficiencies using the noble metals for oxidation are high at moderate temperatures. For example, a typical 90 percent oxidation of CO occurs at about 300°C while for UHC the same level is reached at about 350°C to 370°C. The conversion efficiency falls to zero at about 150°C to 200°C, some 200°C lower than the thermal reactors. Best oxidation levels occur with lean mixtures (unless an air pump is used to provide additional air) as there is then sufficient oxygen available for oxidation in the normal exhaust gases. Catalyst efficiencies of 90 percent for UHC and 95 percent for CO are possible for air-fuel ratios that are either stoichiometric or lean.

The same catalyst materials also promote the reduction of NO_x to N_2 and O_2. For reduction, the exhaust CO and any H_2 (and UHC as these contribute to the CO) are used to remove the oxygen, perhaps directly or via a route which first produces N_2O. The final products in the exhaust gases are N_2, and CO_2 and H_2O. As CO is necessary for this process, it occurs much more efficiently when no oxygen exists in the exhaust gases, that is, when the mixture is rich. The noble metals of platinum and palladium are good catalysts for this reduction process, although again rhodium (Rh) is superior. This is because it produces less ammonia (NH_3) which can be reconverted to NO. Base metals, such as copper oxide (CuO) and nickel oxide (NiO), will accelerate the reduction process catalytically but are less efficient than the noble metals. Nevertheless, Ni compounds have been used widely because of their ability to convert H_2S to SO_2 but there is concern in some countries regarding their health risks and they are being phased out (Cooper [64]). Typically, a peak conversion efficiency of around 90 percent occurs at air-fuel ratios from zero to about 0.5 percent richer than stoichiometric (Heywood [4]). As the mixture enriches further, a small reduction in conversion efficiency occurs, while for leaner mixtures the fall-off is rapid. For a mixture about 1.5 percent lean of stoichiometric, the conversion efficiency is down to about 20 percent. Thus, the use of catalytic converters for NO_x reduction requires tight control of any lean-burn combustion.

8.5.7.3 Oxidizing Catalytic Systems

From the preceding discussion, one possible way to reduce tailpipe emissions is to use a lean burn system. This is coupled to a catalytic converter for the oxidation of CO and UHC, which simply provides a similar function to a thermal reactor except that it operates at noticeably lower temperatures for the same oxidation efficiencies and can be used with engines operating at or lean of stoichiometric rather than fuel rich. That is, the detrimental effect on fuel consumption and the difficult operational aspects are overcome. However, NO_x reduction must be achieved in other ways, usually by the use of exhaust gas recirculation (EGR), although some new catalysts (zeolites) offer promise of good NO_x reduction under lean conditions. It should be noted that while CO and UHC are a minimum some 10 percent lean of stoichiometric, which suits an oxidizing catalytic system quite well, the NO_x proportions reach a peak at that point. Hence, substantial EGR is likely to be required.

8.5.7.4 Reducing Catalytic Systems

A reducing system to control NO_x can be used in an engine as a separate converter bed from the system that controls the oxidation of CO and UHC. As the exhaust gas passing through it must be fuel rich and contain CO, it is essential that it is placed upstream of the oxidation system. The reaction will reduce the NO_x and in the process convert some of the CO to CO_2. However, in general, this does not remove all the CO. Also, all of the UHC remains. These can be oxidized by either a catalytic or thermal reactor downstream of the reduction bed but, as the system is fuel rich, additional air must be supplied via an air pump. A system with separate reduction-oxidation catalysts is complex and is therefore not commonly used. However, it was useful for the less precise air-fuel ratios that occurred prior to the introduction of λ (oxygen) sensors.

For reducing systems, a number of catalyst materials have been examined; some base metal such as inconel or monel (30 percent copper, 70 percent nickel) have been used, but the most successful have been from the platinum metal grouping. As mentioned earlier, recent developments in catalyst technology allow reduction of NO_x under lean conditions. Although a number of substances have been examined, the most likely are the copper ion-exchanged zeolite (Cu-ZSM-5) as the catalyst (Truex et al. [65]). Temperatures required are about 500°C for best conversion efficiency. Toyota has led the application of this technology introduction to vehicles and has been introducing vehicles with 1.6 and 1.8 liter lean-burn engines with catalytic reduction of NO_x to the market since 1994 (Toyota Annual Report [66]). However, there still seems to be some debate as to the durability of zeolite catalysts.

8.5.7.5 Three-Way Catalysts

It is apparent from the preceding discussion that there is substantial commonality in the catalyst materials which will reduce the NO_x and which will oxidize the CO and UHC. The noble metals of platinum, palladium, and rhodium provide a suitable combination, although a base metal combination is also possible but with less efficiency. The use of such an oxidizing-reduction system in the one bed is known as a three-way system. That is, it handles all three of the major emissions in the one bed. This is now the most commonly used approach to exhaust clean-up.

8.5.7.6 Fuel-Air Ratio Windows with Three-Way Systems

From the foregoing discussion on oxidizing and reducing systems, it is obvious that the engine should be stoichiometric to lean to provide high efficiencies for CO and UHC control, while it must be stoichiometric to rich for similar efficiencies in eliminating NO_x. If these are considered together, a small window in the air-fuel ratio range is possible where both will work simultaneously. That is, fuel-air equivalence ratios of about $\phi = 0.995$ (0.5 percent lean) to 1.008 (0.8 percent rich) will allow conversion efficiencies of greater than about 80 percent for all three emissions. Hence, the prime criterion for the use of a three-way system is tight air-fuel ratio control within the combustion chamber throughout the operating range. This is very demanding indeed. The converter works in what is termed an air-fuel ratio *window* as shown in Figure 8.23.

From the discussion of Section 8.4, it is clear that the major problem is transient operation, particularly short-term (acceleration) transients but also long-term (warm-up) transients. Fuel distribution from cylinder to cylinder is also a consideration. Good control using throttle-body injectors can only be achieved by manifold heating, which has deleterious effects on both volumetric efficiency and NO_x emissions. Also, a warmed-up engine is required for this to work well. With tightening emission control, much attention is being paid to cold starts where an abnormally high proportion of the UHC and to some extent CO emissions are

Fig. 8.23.

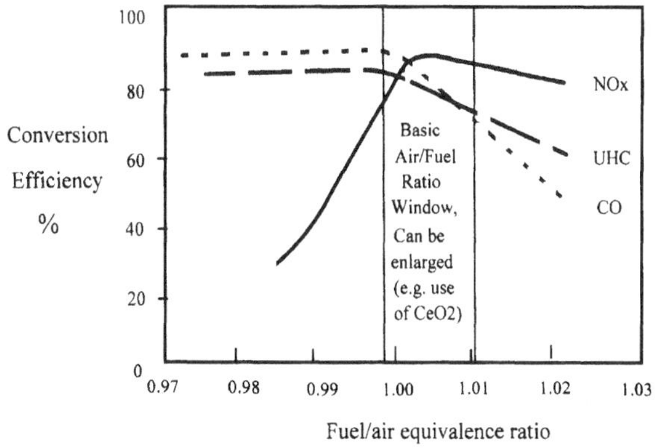

Window for three-way operation of a catalytic converter (TWC).

generated. Hence, the use of catalytic converters has moved fueling systems to multipoint port injection types with feedback control from exhaust sensors.

8.5.7.7 Light-Off Temperatures

From the preceding comments on the conversion efficiency of catalytic systems, it is apparent that a moderately high temperature of greater than about 300°C is required before reasonable conversion takes place. A usual definition of a light-off temperature is when the conversion efficiency reaches 50 percent. It is desirable to attain this within a short time after start to meet ULEV requirements, about 20 to 30 seconds. However, current systems are more likely to take about 2 minutes. Thus, the cold-start performance of catalytic converters is quite poor and this is in fact the time when they are most needed due to the likelihood of significant air-fuel ratio excursions and, hence, the high engine-out exhaust emission levels then occurring.

Two things are required. The first is that the catalytic reactor as a whole has low thermal inertia, that is, that it is constructed of materials which have a low specific heat but high thermal conductivity and, hence, will warm up to operating temperature very rapidly. This is one reason why washcoat materials such as aluminium oxide (Al_2O_3) in which the catalysts are incorporated have been chosen. Double, thin wall containing pipes with an insulating air gap and thin wall monoliths reduce the thermal inertia.

The second, and this is becoming more important as emission control legislation is being tightened, is that other means are found to promote pre-light-off emission reduction. The reactor can be warmed up prior to starting the engine or very rapidly once it has started. The most suitable means are electrical heating but this requires a large heating element and entails considerable additional cost as well as a major drain on the vehicle's battery at the time (e.g., subzero temperature start) when a large power drain may also be necessary for the starting itself. Battery

Fig. 8.24.

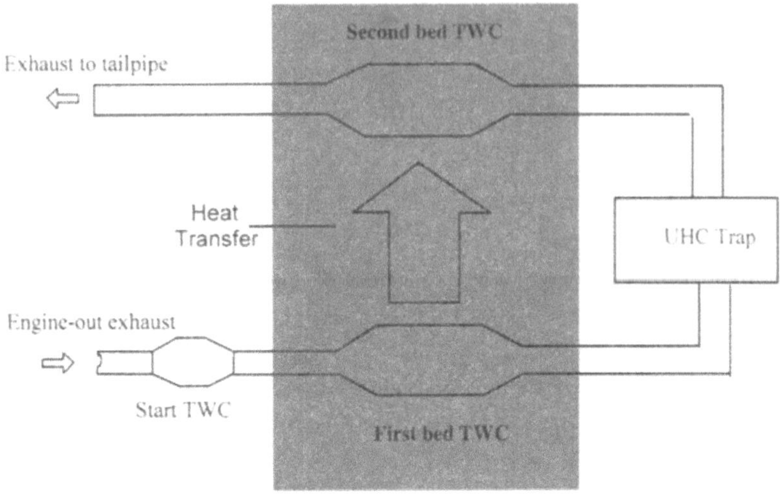

1. Cold start-exhaust unmodified by Start and first bed TWC
   - UHC adsorbed into HC trap
2. Engine warms- Start and first bed TWC reach light-off temperatures
3. Heat transfer occurs to second bed TWC
4. Clean gas from start and first bed TWC purge HC trap
5. Purged UHC is converted in second bed TWC
6. Fully warmed engine- start and first bed TWC provide most conversion

Exhaust system including a hydrocarbon trap.

systems are limited to about 1.5 kW for short periods only. An intelligent system has been proposed by Saab to meet ULEV legislation, which will heat the engine block, engine oil, and catalytic converter (Jost [67]). An alternative solution is to absorb the UHC during warm-up and then to release it to the catalyst when it reaches the required temperature (White [68]). The cold-start emissions have been reduced in tests by about 70 percent. However, a complex exhaust system such as that shown in Figure 8.24 is then necessary. One novel proposal (Takayama [69]) is to use engine rotation without fuel introduction for a number of cycles before starting actually commences. The heating by this method is limited and it also places demands on the battery but it may be a cheaper option. If mechanisms can be devised that will allow the compression pressure of air alone to be released as shock waves to the reactor, the higher temperatures due to the entropy increase across the shock could increase the heating rate by this method. However, this would require very rapid opening large valves from the high-pressure, high-temperature conditions and is mechanically complex. This has apparently been discussed with some Japanese automotive manufacturers.

8.5.7.8 Exhaust Sensors and Feedback Control

In order to maintain the air-fuel ratios within the optimum window for the three-way catalytic converters, a feedback system is required for their control. The options are a direct measure of the air mass flow rate, thereby requiring a real time calculation of the fuel flow rate that is needed or an appropriate exhaust parameter which will allow a direct assessment of the air-fuel ratio that is being provided. The second is the most convenient. If the mixture is rich, exhaust CO_2 correlates well with air-fuel ratio but it is a relatively insensitive measure around stoichiometric. Alternatively, for lean mixtures, exhaust O_2 is the appropriate parameter. For equivalence ratios of $\phi = \leq 1$, the residual exhaust oxygen changes linearly with air-fuel ratio and so is a useful measure. It is, of course not useful for values above $\phi = 1$. However, control at or just lean of stoichiometric is an ideal setting for operation as it gives good power output, good fuel economy, and reasonably low emission levels for CO and UHC directly from the engine. Exhaust oxygen sensors that are stabilized zirconium oxide detecting the oxygen partial pressure in the exhaust are, therefore, the most useful for providing direct feedback for the fuel injection process.

8.5.7.9 Catalyst Poisoning

One of the major problems with the use of exhaust catalysts is that their performance can be easily degraded. The principle ways that this occurs are due to:

- Overheating. This is often due to engine malfunction with substantial fuel quantities being passed through to the exhaust. For example, about 20 seconds of ignition failure on one cylinder at 4000 rpm or above may provide a sufficient temperature to destroy the catalyst.
- Substances in the fuel, in particular, sulphur, phosphorus, and lead. The first two can be reduced or eliminated in the refinery as they are not required for combustion. The lead is different because it has, in the past, been deliberately added as an antiknock measure.

Lead Lead in different values (e.g., 0.4 g/liter lead content in Australia up to the early 1980s) in the form predominantly of tetra ethyl lead ($Pb(C_2H_5)_4$) but containing other compounds (e.g., $Pb(CH_3)_4$) was used for many years as a convenient method of raising the octane number of the fuel. In some instances, leaded fuels are still available, although often now with lower levels. These provide the most rapid form of degradation of the catalyst as the surface becomes coated with lead which inhibits contact between the gases and the catalyst. A typical effect is shown in Figure 8.25. Here, 0.025 g/L of TEL (tetra ethyl lead) caused minor deterioration after 60 hours, 0.125 g/L results in a conversion efficiency drop from 90 percent to 70 percent in that time while 0.75 has halved the efficiency in only 5 hours. Thus, catalytic converters can only be used with lead-free fuels and it

Fig. 8.25.

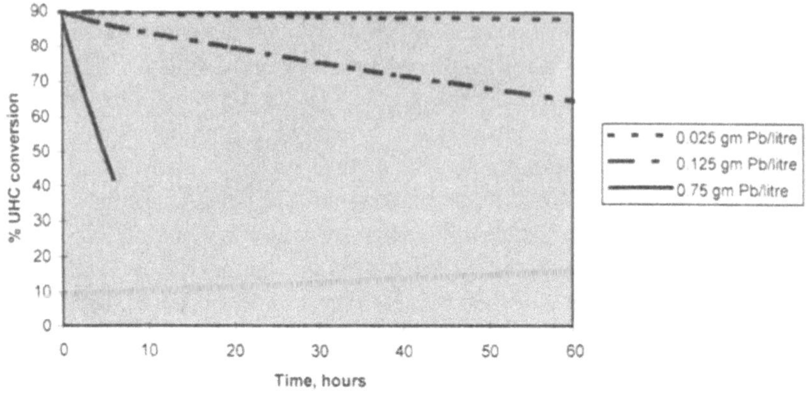

Typical rates of catalyst deterioration from lead in gasoline.

is this factor more than the direct health problems associated with lead ingestion that has promoted the introduction of lead-free fuels. Most legislation insists on deterioration factors at a fixed vehicle mileage (typically test degradation points around 50,000 miles or 80,000 kilometers are designated) where a minimum fall-off in performance is specified. Hence, all cars fitted with catalytic converters must run on unleaded fuel only. To inhibit the use of leaded fuels, unleaded gasoline bowsers in some countries are fitted with a smaller nozzle to match the smaller filling orifice of the vehicle tanks. The equivalent fuel when unleaded is of noticeably lower octane number than the original leaded fuel by (perhaps 8 or 10 points depending on the initial lead levels) but this may be partially compensated for by the reformulation of the fuel (usually at some cost). High octane number unleaded fuels with equivalent values to leaded are available in most countries but are considerably more expensive. Generally, unleaded fuels still have octane numbers some 5 points lower than leaded, although with further cost increases, the leaded values can be reached. Thus, compression ratios tend to be lower with engines fitted with catalytic converters with a consequent loss in efficiency. However, from U.S. data (McTague [19]), this seems to have been regained by the ability to use the optimum timing and air-fuel ratio settings when the thermal reactors are replaced by catalyst types. Another factor is that the lead that coats on the exhaust valve lubricates the valve seat providing it with some protection. Hardened valve seats or seat inserts are, therefore, required for unleaded fuel vehicles.

Sulphur Almost all crude oil contains some sulphur. Apart from the possible poisoning effect of sulphur on the catalyst, one on the effects of a catalytic converter is that it converts the sulphur to sulphur dioxide and hydrogen sulphide. This is unpleasant due to its odor, and it is an important contributor to acidic effects in the atmosphere. The removal of nickel oxide from catalysts for health considerations has the effect of making H_2S more likely as a tailpipe emission. Thus, the removal of sulphur in the fuel is even more important with unleaded fuel. As low a level as possible is necessary from the refinery.

8.5.8 Typical Installations for Emission Control on Current Vehicles

A large range of combinations is possible for different engines depending on the original design, the country of sale, the emission legislation, and the market forces. The following are intended as an indication of the essentials of each type and considerable variations may be found.

8.5.8.1 All Engines

In general, current vehicles are likely to be designed with:

- A compact combustion chamber to minimize wall effects on UHC emissions and to keep efficiency high.
- Electronic-type ignitions system.
- A closed-type positive crankcase ventilation system.
- An evaporative emission system to vent the fuel tank and supply.
- A throttle position sensor to allow precise control of idling speeds.
- A simple-type engine management system is all that is essential, although many now use multifunction microprocessor types.

In addition, the following are likely for the different types of engines.

8.5.8.2 SPI Engines

- A monolith catalyst system for oxidation of CO and UHC. This will be placed close to the exit of the exhaust manifold at the entrance to the exhaust pipe.
- Secondary air supply to the exhaust to provide additional air for CO and UHC oxidation.
- Centrifugal and dual vacuum advance mechanisms on the ignition system on earlier engines, although these are now generally replaced with electronic systems.
- A heating system for the inlet air. This is likely to allow either cold air only or a mixture of cold and hot air heated from the exhaust. The temperature will be regulated by a flap valve (usually thermostatically controlled) regulating the amount of hot air entering the system.
- An EGR system to limit the engine-out NO_x emissions. This will be controlled by a valve so that the regulated quantity of EGR is not excessive at low manifold pressures.

- A temperature control from the engine coolant to shut off the secondary air to the exhaust during warm-up. It will also limit the purge from the fuel tank vent carbon canisters and EGR under cold conditions.
- Fuel shut-off solenoid for deceleration conditions (carburetor types, now rare).

8.5.8.3 MPI Engines

- An engine management system, which is a microprocessor type with multi-functions.
- A three-way catalytic converter placed in the same position as the oxidizing reactor of the SPI engine.
- Ignition timing fully under the control of the engine management system with spark timing determined by the throttle position, engine speed, knock sensor, engine temperature, and manifold vacuum.
- A temperature control from the engine coolant of the purge from the fuel tank vent carbon canisters.

The MPI engine is unlikely to have a secondary air supply to the exhaust or a heated air inlet system. It may have EGR but this is often not required depending on the engine and the emission standards of the particular country.

8.6 DEVELOPMENTS FOR THE FUTURE

8.6.1 On-Board and External Monitoring Systems for Emissions

If emissions are to be fully controlled in an urban environment, it is imperative that the systems fitted to vehicles be maintained at a level consistent with their original design concept and the deterioration factors assessed for them. There is evidence, at least from Australian data, that the worst group of vehicles, perhaps 10 percent, is responsible for over 30 percent of the emissions (Paine [70]). In the United Kingdom, the Royal Auto Club has results, reported by White [68], which give figures of 12 percent responsible for 55 percent of emissions, which is not inconsistent with these. To regulate this, a program of monitoring the performance of individual vehicles must be maintained. Several systems are possible. These include:

- Regular simple checks of the vehicles on, for example, a yearly basis when their registration becomes due. At present, some countries do this through a simple exhaust check at a low and high idle engine setting. However, there is evidence (Paine [68]) that there is a high proportion of false-failures and false-passes with this type of test. This is inadequate and may indeed be misleading as many vehicles will be deemed to be satisfactory when their performance is below acceptable limits under load.

- Regular extensive checks of all vehicles on a chassis dynamometer test, which includes the full city and highway driving schedules specified for emissions. As this requires hot soak periods for evaporative and fugitive emissions, as well as collection and analysis of the exhaust samples, it is likely to be prohibitively expensive. A program targeted at the worst vehicles may be more cost-effective.
- Site monitoring in congested locations. That is, equipment can be located in key areas to check the emission levels of passing vehicles. However, it then does not indicate the vehicles performance under a range of speeds and loads, and may not discriminate the individual vehicle sufficiently. More development in this area is required. This will include instrumentation and monitoring techniques.
- On-board monitoring systems. These will require instrumentation on the vehicle to record point readings determined either at a specified mileage intervals or when certain levels of the CO, UHC, and NO_x emissions, either individually or grouped, are reached. While data could be collected intermittently rather than continuously and assessed numerically, sufficient points would be required to allow a reasonable estimation, within limits, of whether the emission control system of the vehicle is working adequately. This will require cheap instrumentation with a fairly high response time and a microprocessor with data storage capacity to be developed. The analysis could be either on-board or at a centralized location. Neither is a simple exercise. Also, the legal problems of how the data would be transmitted to the controlling authorities, and how they would then be used need careful consideration.

Whatever method is used, and it may vary from country to country, the maintenance of vehicle emission control systems at an individual level is important and needs careful consideration for the future. As ultralow emission legislation becomes widespread, this will be even more critical.

8.6.2 Potential for Improvement in the Efficiency of Current SI Engines

The possible engine developments for the future will be discussed in Section 8.6.3. However, some idea of the potential efficiency increase can be gleaned by examining the trends from Figure 8.9. It can be seen that overall vehicle efficiencies have improved substantially over the period. This is both because different vehicles have been substituted as well as improvements made to their power plants. Following this historical trend, it would seem that a lower plateau of at least 5 L/100 km is a likely value for moderate-sized cars and a target of around

3 L/100 km is possible (Watson and Watson [71]). However, historical trends are often discontinuous. That is, they may not necessarily decrease monotonically or they may be replaced by a step change due to some at present unanticipated factor. For example, a trend may occur back to heavier vehicles for perhaps safety reasons, or because of substitution of different energy sources is possible, with the consequent lowering of overall vehicle efficiency. Alternatively, new engine technology such as the DISC engines or those with total valve control (see Section 8.6.3.2) may provide substantial fuel economy improvements. A guess might be that average vehicle fleets of, say, the year 2010 will have a fuel economy some 10 percent to 20 percent better than at present but much will depend on legislation and crude oil economic pressures. This would contribute to a lowering of CO_2 emissions per vehicle of similar values. It could be easily outweighed by increases in vehicle kilometers traveled (VKT) due to:

- Greater individual need in a more complex society or greater population growth.
- The improving affluence in, at present, underdeveloped or partly developed countries.

It is, therefore, difficult to envisage a scenario, other than a substantial move away from individual vehicles to public transport, which will have a major beneficial effect on the global emission problem. Due to the popularity and convenience of individual transport, this is unlikely without a powerful driving factor, either economic or legislative. All that can be done in a technological sense is to minimize the impact of increasing population and affluence by achieving the best combination of engine/vehicle efficiencies possible.

8.6.3 Likely Changes to the Reciprocating SI Engine

There are two developments that are of particular interest here and these are ones which are most likely to have a significant effect worldwide on fuel use, greenhouse gas emissions, and local emissions. They are the direct-injection stratified charge (DISC) engines and engines with total inlet and exhaust valve control. The latter is best known as the existing Honda VTEC engine where there is partial timing and lift control on the inlet valves.

8.6.3.1 DISC Engines

The concept of stratified charge SI engine is not new, and an early mention of it dates from Otto's ideas in 1872. It was fully discussed by Ricardo [72]. Several experimental engines of the stratified charge type were developed during the 1960s, these being the Ford Proco (Simko et al. [73]) and the Texaco TCCS (Alperstein et al. [74]). The one successful introduction to the market until very recently was the Honda CVCC engine (Inoue et al. [75]) used in the United States in the late 1970s, but this relied on a divided combustion chamber. The ignition chamber had a separate inlet valve and was slightly fuel rich while the main chamber was lean. The combustion was transferred from the former to the latter via a torchlike flow through the throat between them. As with IDI diesel engines, this had significantly greater heat transfer and fluid flow losses at this point and was eventually discontinued. Recently, Mitsubishi has introduced a single-chamber, four-stroke, stratified charge 1.8-L engine to its commercial range (Thisdell [76]). The problems with a reducing NO_x catalyst for the overall lean conditions are highlighted. However, the introduction of stratified charge engines has taken longer than was thought in the early 1980s (see, for example, Heywood [77]). Other experimental engines include the two-fluid (air and fuel) injection system used with a two-stroke engine layout by the Orbital Engine Company in Australia (Houston et al. [78]). Two-stroke engines will be dealt with in Part IV of this book. Inevitably, a number of stratified charge engines will enter the market over the next decade. The preferred system, single-fluid, two-fluid, conventional or air-assist injection, four stroke or two stroke remains to be determined. It is not within the scope of the present work to evaluate these.

The importance of the stratified charge concept is that the air-fuel ratio can be varied within the combustion chamber in a desired pattern. The essential requirement is that the fuel is near stoichiometric or more likely slightly rich near the spark plug in order to provide good ignition characteristics. The rich mixture here tends to keep the NO_x at low levels. As the flame front progresses across the combustion space, it moves rapidly into a region that is considerably fuel lean. The high temperatures and pressures already generated within the combustion chamber allow the flame propagation to continue at an adequate speed. Thus, the benefits are:

- The overall air-fuel ratio can be quite lean allowing good economy.
- The burning is always in regimes where substantial NO_x is not produced.
- The CO and UHC produced early in the combustion process can be oxidized within the combustion chamber rather than externally, allowing a small but improved use of fuel.
- The gas in the crevice regions is very fuel lean and a lower source of UHC.
- The engine control can be obtained, at least partially, by varying the air-fuel ratios rather than by throttling, thereby minimizing pumping losses and improving engine efficiency.
- The end gas is very fuel lean and, therefore, less susceptible to end-gas autoignition. Hence, higher compression ratios are possible with a consequent improved engine efficiency.

The question is then: Why has stratified charge not been used earlier? The answer is that it is an extremely difficult concept to put into practice. This is because the stratification must be very tightly spatially controlled in the cylinder. The correct, slightly rich mixture is required near the spark; the immediate movement to a very lean mixture is essential but at no position can it become too lean so that flame quench occurs, as then significant levels of unoxidized or partially oxidized fuel are possible giving high CO and UHC emissions. It will be noted that the airflow patterns within the cylinder vary with speed and, if there is some throttling, with the load on the engine. The fuel flow patterns will always vary with speed and load. To obtain the correct stratification (and maintain it) requires an ability to control both the air and fuel flow patterns precisely under all operating conditions everywhere within the cylinder. This cannot be achieved with injection systems in the manifold or ports, as the incoming fuel is to a large extent randomly distributed in the jet flow through the inlet valve. Also, the distribution of fuel into vapor, droplet, and wall film components makes this harder to achieve. Hence, appropriate understanding of the engine fluid mechanics is essential for the engine design. In addition, the development of the appropriate injection systems that will provide the fuel distribution necessary for operation and not deteriorate under the severe conditions that exist within the cylinder is important. Their control is also critical. It is only now that these problems are being solved.

8.6.3.2 Total Valve Control

The valve systems in IC engines are one of the controlling factors on engine performance in general. It is obvious that a large valve area normal to the flow is essential to provide high volumetric efficiency, both in introducing the air and expelling the exhaust gases. Valve overlap exists in all engines whereby the valves open before and close after the respective inlet or exhaust stroke begins. A large valve overlap allows the inertia effects of the flow to continue the filling or scavenging process for a longer period, thereby increasing the volumetric efficiency at high engine speeds, but the flow inertia is less at low speeds and excessive backflow and contamination can take place. That is, the timing diagram should ideally vary with engine speed. Also, the turbulence in an engine cylinder is dominated by the inlet flow through the valve and is modified during the compression process to ignition. The correct level of turbulence is important for the appropriate combustion rate to occur. However, a valve lift which is suitable for the mass flow rate at high speeds and loads will not be suitable at low speeds and loads. Ideally, the valve lift should be varied throughout the speed/load range. Finally, it is noted that engines work most efficiently when under high load. It is possible to shut off the fuel flow to some cylinders while keeping others at full load in order to reduce the engine power output overall. This would have to be consistent with the engine balance. It may be possible with individual valve control to effectively close down cylinders as required so that their pumping losses are also minimized.

Engines with valve control are currently being developed. The original systems are two-step types where the operation switches from one cam profile to another at a specified speed. A typical system of this type is the Honda VTEC. Some systems currently under development are the Toyota VVT-i where a hydraulically controlled variable timing pulley drives the camshaft. Some earlier developments of this second type using mechanical systems such as that by Alfa Romeo in 1984 (see Stone [24]) have been in existence for some time. While some mechanical systems have been devised, it seems most likely that hydraulic or electrohydraulic systems will eventually be introduced that control both timing and lift continuously as the engine settings vary.

8.6.4 Other New SI Engines—How will they Compare?

New engine concepts of various types for the modification or replacement of the gasoline engine are continuously being put forward by inventors for evaluation. Most of these are different because of their mechanisms rather than their thermodynamics, although it is the latter which are the most limiting. Typical examples are rotary engines, orbital engines (rotary, with the central body always oriented in the one direction), and radial types, those with the conventional crank replaced by other mechanisms (e.g., cams or lobes) to allow variation in piston motion. While some of these may have merits in various respects, particularly for niche applications, it is noted that the basic arrangement and mechanisms of the SI engine have remained very similar for a very long period and are, therefore, hard to improve upon. One of the major problems with alternatives to the piston/cylinder types is that it is often difficult to obtain a well-shaped combustion chamber. The long, narrow chambers of rotary and orbital types have inherently high UHC emissions as well as potential knock and high heat transfer problems. These are discussed in Chapter 7 of this book.

8.6.5 Techniques for New Fuels—When and How?

8.6.5.1 The Use of Alternative Fuels in SI Engines

It is possible to envision a number of likely scenarios by which alternative fuels will be introduced, most probably due to the pressures from declining crude oil resources, although emissions may play some role in the change. One possible scenario is shown on Figure 8.26. The introduction of new fuels has been designated as a "fast" and a "slow" stream because, in the former, the technology already is widespread for both the engines and the fuel production, but for the latter this is not the case. It should be noted that, with all the fast-stream fuels, some reduction in CO_2 emissions results, as can be seen from Table 8.5. Some advantage is achieved with LPG, that with NG is substantially greater. However, the LPG has fewer disadvantages in terms of vehicle range as it is easily liquefied by pressure alone, propane having a critical temperature of 370 K. The liquefying pressures at atmospheric temperatures are moderate at around 10 atmospheres. Methane cannot be liquefied by pressure at atmospheric temperature as its critical point is only 191 K. At atmospheric pressure, it requires a temperature of -162°C. Hence,

Fig. 8.26.

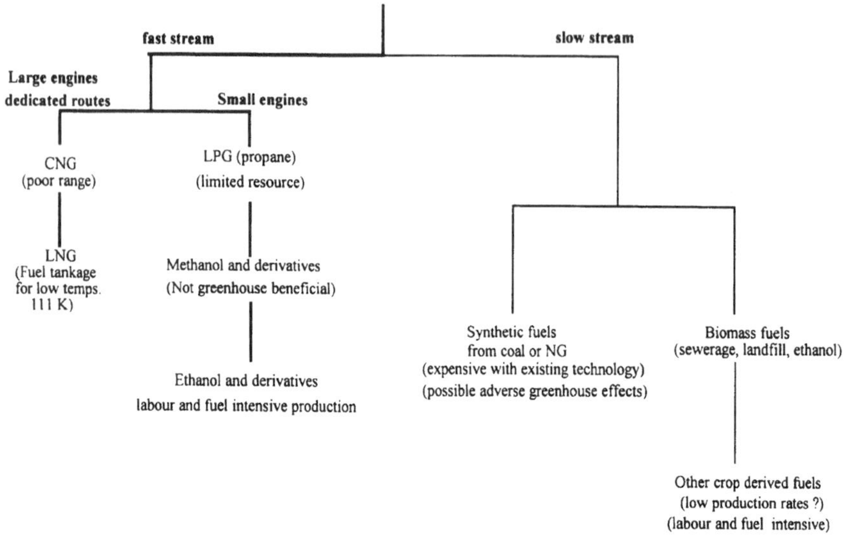

Possible scenario for introduction of alternative fuels.

its storage as either compressed natural gas (CNG) or liquefied natural gas (LNG) is a distinct disadvantage. Also, because of the inherent greenhouse tendency of methane, any leakage would need to be severely contained. With methanol, any greenhouse advantage would depend on the source of reactants for the process but it is unlikely to provide any greenhouse advantage. If obtained from coal, a detrimental effect when compared with crude oil is likely. Ethanol can be advantageous as it removes CO_2 during the growth phase of the biomass material from which it is produced but much depends on the agricultural procedures and energy use. With the slow-stream fuels, synthetics will also have a detrimental greenhouse effect during their production phase as they are most likely to be produced from coal or at best natural gas. Crop-derived oils are too far downstream to be seriously considered but may have an even higher agricultural energy intensity during their production than ethanol.

LPG is already widely used on SI engines and there are few problems as it has similar burning characteristics as gasoline. The gas is usually vaporized in a gas carburetor with some air capacity loss during induction due to the volume it displaces. This amounts to about 4 percent at stoichiometric conditions compared to 2 percent for fully vaporized gasoline and 1 percent for half-vaporized gasoline. Some differences in their cooling in the manifold due to vaporization may occur. In addition, when its lower fuel-air ratio is factored in together with the slightly higher heating value, a further loss of some 3 percent occurs. Therefore, maximum power is reduced by more than 5 percent. Deslandes and Chapman [79] found a 10 percent power loss on test at WOT. Systems with multipoint, port and liquid-phase LPG injection, which will minimize this problem, are under consideration but require high upstream pressures to prevent vaporization in the fuel lines. Ideally, the fuel-air ratio requires adjustment reducing the mass of fuel introduced by about 7 percent. Only minor ignition timing changes are desirable but are not essential. Typically, some additional advance is required (Deslandes and Chapman [79]). Due to its high octane number, compression ratios of one or two points higher than gasoline are possible, although this modification is usually not carried out because the engine is also required to revert to gasoline when necessary. With NG, all the foregoing are similar but more marked, its slow-burning characteristics and high-ignition energies requiring either different engine turbulence, substantial spark advance, and high energy spark. For a discussion of natural gas combustion in engines, the reader is referred to Bradley and Woolley [80]. It displaces an even greater air volume (about 9 percent at stoichiometric conditions), the loss in power can be over 15 percent, and a liquid-phase injection is extremely difficult, as substantial cooling of the fuel lines would be necessary. However, its very high octane number of about 130 RON and consequent high compression ratios (around 14:1 or even higher) and excellent lean-burn characteristics (which allow control via fuel-air ratio change to partially replace throttling) can provide much improved engine efficiencies. Watson et al. [81] have shown that thermal efficiencies for optimized NG SI engines can be substantially improved over the baseline gasoline engine. Overall, the mixing problems with gaseous fuels are minimized and emissions, both local and global, tend to be lower than with gasoline.

Alcohols may be used either as total alcohol or high alcohol content fuels (e.g., 90 percent methanol, 10 percent gasoline, designated M90) where the gasoline is there only to provide starting capabilities. More commonly, it is in low proportions of 10 percent to 20 percent as a crude oil extender or perhaps as a means of octane improver. Separation problems exist. The alcohols have a noticeably higher research octane number (e.g., RON $=$ 106 for ethanol, 110 for methanol) but the improvement in motor octane number MON is less (about 90 for both). Also, blend octane numbers cannot be directly proportioned from the fuel constituents (Bolt [82]) and are greater than would be expected from that calculation (Hagen [83]). The blend octane number depends on that of the base fuel with lower octane number fuels responding more noticeably. Its greater enthalpy of vaporization allows increased volumetric efficiency due to the cooler intake airflow. The mass air-fuel ratios for alcohols are lower (stoichiometric for methanol being about 6.44:1, ethanol 8.95:1, while gasoline is about 14.6:1). Alcohol addition, therefore, provides an effective leaning of the mixture if the fuel-supply injection settings are unchanged. Therefore, CO is reduced. The high latent heat and lower heating value of the fuel tend to produce a cooler flame with lower NO_X. UHC may become more of a problem due to the greater difficulty in vaporization. Overall, there is therefore some improvement in emissions, although aldehydes are likely to become a problem.

The introduction of such fuels with a lower greenhouse potential, at least in relation to CO_2 production, is possible as a control means with SI engines. However, the strategy by which they are introduced, either in an engine which reverts to gasoline when required or as dedicated engines for the fuel under consideration, is important as it is only with the latter that efficiencies can remain high or be improved.

8.6.5.2 The Use of Alternative Fuels in Alternative (Non-SI) Combustion Engines

It should be noted that the most likely short- to medium-term fuel substitutes are propane as a significant proportion of LPG, methane as the most important constituent of NG, or either methyl or ethyl alcohol. All these have good octane numbers and burn well in SI-type engines. It is noted that they all have low cctane numbers and, therefore, are not immediately suited to CI engines, although they can be used in engines of that type in the dual-fuel mode. Here, the gaseous fuel is introduced in the manifold but in a lean mixture so that autoignition does not occur prematurely. A small pilot distillate spray provides the ignition. In general, for truck-size engines, lean-burn SI engines have been preferred, although these may have difficulties in matching the fueling infrastructure. The preceding fuels are all well suited to the gas turbine or Stirling cycle type engines, which have continuous combustion and therefore no octane number or cetane number requirement. In summary, an SI engine replacement is not necessary for their widespread use although in the larger engine size range a replacement for the CI engine may well be required. The major inhibiting factor in the use of these fuels is their low energy density and, in some cases, their current cost.

8.6.6 Other Engine Types

It is convenient to divide engines which fill the operational envelope that is now occupied by the SI engine into two groups, those which are radical departures from the SI concept and those which are developments of it. The first group includes the Stirling cycle, gas turbine (usually using the regenerative cycle), and electric vehicle. This group is outside the scope of the present work and will not be considered other than to say that all have difficulties in spite of some immediately perceived advantages.

For example, the Stirling cycle has a high potential thermal efficiency and more easily controlled emissions from its continuous, external combustion process, but the size, mass, and complexity of obtaining the necessary high heat transfer and the sealing of the high-pressure, low molecular mass gas required as a working substance have, to date, prevented its general application. The regenerative gas turbine, with similar efficiency and steady, although generally not external, combustion advantages requires a regenerator of exceptional effectiveness to be viable. The continuous combustion associated with both these is not immune to emission problems. While lean operation is relatively easy under these conditions, high NO_x levels may result from continuously high temperatures. Also, particulates which are of relatively minor significance in SI engines may become a problem depending on the fuel and the method of injection into the high-temperature burning gases.

Electric vehicles, which are proposed as zero-emission vehicles (ZEV), have immense energy storage problems and, depending on the original source of electrical power, simply transfer their emissions problems to another area. For a nonnuclear power supply, their greenhouse gases performance may in fact be no better or even worse than a similar gasoline-powered vehicle. For example, the GM two-seater electric Impact (a developmental vehicle) has been calculated to result in 87 g (CO_2)/tonne-km at a steady 96 km/hr highway speed using an Australian power supply mix which is heavily coal based (Dobes [84]). However, the GM Ultralite, also a concept gasoline-powered vehicle, produces only 75 percent of the Impact's CO_2 emissions in comparable city driving conditions per seat-km while even current gasoline vehicles such as the Datsun Charade are 13 percent better than the Impact. It has been estimated (Gosden [85, 86]) that if natural gas is the source of electric power, that an EV would be 27 percent better than an internal combustion engine where the gas is burned in an IC engine. However, this is based on current IC engine technology and on nonoptimized NGV thermal efficiencies. It is likely that improvements in these areas will more than make up this difference in the near future. It may well be that other ZEV concepts such as the liquid N_2 powered engine (Knowlen et al. [87]) or the Ballard fuel cell engine using H_2 (Howard [88]) have more potential.

REFERENCES

1. Butler, J.D., *Air Pollution Chemistry*. London: Academic Press, 1979.
2. Ledbetter, J.O., *Air Pollution Part 1*. New York: Dekker, 1972.
3. Carson, D., "Global Warming: Basis for Concern," *RSA Journal*, pp. 52–61, June 1966.
4. Heywood, J.B., "Pollutant formation and control in spark-ignition engines," *Prog. in Energy and Comb. Sc.*, Vol. 1, pp. 135–164, 1976.
5. Newhall, H.K., "Combustion process fundamentals and combustion chamber design for low emissions," SAE Paper 751001, 1975.
6. Heywood, J.B., and Cheng, W.K., "Fundamentals of internal combustion engines: performance, efficiency and emissions," MIT Summer Session, Course Notes, June 1993.
7. Wentworth, J.T., "Piston and ring variables affect exhaust hydrocarbon emissions," SAE Paper 680109, 1968.
8. Heywood, J.B., *Internal Combustion Engine Fundamentals*. New York: McGraw-Hill Int., 1988.
9. Ishizawa and Takagi, *JSME International Journal*, Vol. 30, 260, pp. 310–317, 1987.
10. Wentworth, J.T., "More origins of exhaust hydrocarbons—effects of zero oil consumption, deposit, location and surface roughness," SAE Paper 720939, 1972.
11. Tabaczynski, R., Hoult, D., and Keck, J.C., "High Reynolds Number Flow in a Moving Corner," *Jnl. Fluid Mech.*, 42, pp. 249–255, 1970.
12. Trinker, F.H., Cheng, J., and Davis, G.C., "A feedgas HC emission model for SI engines including partial burn effects," SAE Paper 932705, 1993.
13. Campau, R.M., and Neerman, J.C. "Continuous Mass Spectrometric Determination of Nitric Oxide in Automotive Exhaust," SAE Paper 660116, 1966.
14. Fenimore, C.P., "Formation of nitric oxide in premixed hydrocarbon flames," 13th Int. Symp. on Comb, The Combustion Inst., pp. 373–379, 1970.
15. Zillman, J.W., "Transport and the Global Greenhouse," Aust. Academy of Tech. Sci. and Eng.

Symp., Melbourne, pp. 37–60, October, 1990.

16. Waters, M.H.L., "Road Vehicle Fuel Economy—State-of-the-art review 3, Appendix A—Global warming and road transport," Transport and Road Research Laboratory, Dept. Transport, HMSO Publications Centre, London, 1992.
17. Haddad, M., "Transport and the Environment—Balancing the Economics," Aust. Academy of Tech. Sci. and Eng. Symp., Melbourne, October, pp. 95–116, 1990.
18. Walker, I.J., "Potential to Reduce Emissions of Greenhouse Gases," DOPIE Rept., Comm. of Aust., May 1990.
19. McTague, J.P., "The customer, the environment and the gasoline powered vehicle," Aust. Academy of Tech. Sci. and Eng. Symp., Melbourne, pp. 129–145, October, 1990.
20. Wylie, D.K., "An automotive industry policy for an improved environment," Aust. Academy of Tech. Sci. and Eng. Symp., Melbourne, pp. 117–128, October, 1990.
21. Milton, B.E., Behnia, M., and Vidler, D.R., "An Examination of Transient Fuel/Air Flows in Manifolds." *Transport Phenomena in Thermal Engineering*, eds. Lee, Chung, and Kim. New York: Begell House, pp. 1247–1251, 1993.
22. Takeda, K., Yaegashi, T., Sekiguchi, K., Saito, K., and Imatake, N., "Mixture preparation and HC emissions of a 4-valve engine with port fuel injection during cold starting and warm-up," SAE Paper 950074, 1995.
23. Shin, Y., Cheng, W.K., and Heywood, J.B., "Liquid gasoline behavior in the engine cylinder of a SI engine," SAE Paper 941872, 1994.
24. Stone, R., *Internal Combustion Engines*, 2nd ed. London: MacMillan Press Ltd., 1992.
25. Mock, F.C., "Design of Intake Manifolds for Heavy Fuels," SAE Transactions, part 1, Society of Automotive Engineers, 1920.
26. Milton, B.E., Behnia, M., and Casey, R.T., "Deposition of liquid fuel droplets in turbulent SI engine induction airflows," *Turbulence, Heat and Mass Transfer 1*, ed. K. Hanjalic. New York: Begell House, pp. 401–406, 1995.
27. Collins, M.H., "A Technique to Characterize Quantitatively the Air/Fuel Mixture in the Inlet Manifold of a Gasoline Engine," SAE Paper 690515, Society of Automotive Engineers, 1969.
28. Kay, I.W., "Manifold Fuel Film Effects in an S.I. Engine," SAE Paper 780944, Society of Automotive Engineers, 1978.
29. Tanaka, M., and Durbin, E.J., "Transient Response of a Carburetor Engine," SAE Paper 770046, Society of Automotive Engineers, 1977.
30. Saito, K., Sekiguchi, K., Imatake, N., Takeda, K., and Yaegashi, T., "A new method to analyze fuel behaviour in a spark igntion engine," SAE Paper 950044, 1995.
31. Shin, Y., Min, K., and Cheng, W.K., "Visualisation of mixture preparation in a port-fuel injection engine during engine warm-up," SAE Paper 952481, 1995.
32. Boam, D.J., and Findlay, I.C., "A computer model of fuel evaporation in the intake system of a carburetted petrol engine." Proc. Inst. Mech. Eng. Conf., C89/79, pp. 25–37, 1979.
33. Milton, B.E., and Behnia, M., "A numerical study of the interchanging vapour, droplet and film flows in IC engine manifolds." *Heat and Mass Transfer in Gasoline and Diesel Engines*. eds. D.B. Spalding and N. Afgan. New York: Hemisphere Press, pp. 245–258, 1988.
34. Behnia, M., and Milton, B.E., "Experimental Determination of Fuel Film Flows in Manifolds," Proc. 11th Australasian Fluid Mechanics Conference, Hobart, December 14–18, Vol. 1, pp. 647–650, 1992.
35. Aquino, C.F., "Transient A/F Control Characteristics of the 5 Liter Central Fuel Injection Engine," SAE Paper 810494, Society of Automotive Engineers, 1981.
36. Hires, S.D., and Overington, M.T., "Transient Mixture Strength Excursions—An Investigation of Their Causes and the Development of a Constant Mixture Strength Fueling Strategy," SAE Paper 810495, Society of Automotive Engineers, 1981.
37. Behnia, M., and Milton, B.E., "The behaviour of fuel films on the walls of engine manifolds and ports," Proc. Mech 88 Conf., Thermodynamics in Australia, Inst. of Engineers, Aust., pp. 6–10, May 1988.

38. Nagaishi, H., Miwa, H., Kawamura, Y., and Saitoh, M., "An analysis of wall flow behaviour of fuel in induction systems of gasoline engines," SAE paper 890837, 1989.
39. Low, S.C., Baruah, P.C., and Winterbone, D.E., "Transportation of liquid fuel droplets in the pulsative air flow within the SI engine intake manifold," SAE Paper 810497, 1981.
40. Shayler, P.J., Teo, Y.C. and Scarisbrick, A., "Determining fuel transfer characteristics for mixture control during transient operation," *IMech E*, C469/008, pp. 81–87, 1993.
41. Matthes, W.R., and McGill, R.N., "Effects of the degree of fuel atomisation on single cylinder engine performance," SAE Paper 760117, 1976.
42. Shayler, P.J., Marshall, W.D.P., Turner, J., and Ford, K., "Exhaust emissions: the influence of fuel injection system details," *IMechE*, C394/015, pp. 169–179, 1990.
43. Rink, K.K., and Lefebvre, A.H., "Influence of fuel drop size and combustor operating conditions on pollutant emissions," SAE Paper 861541, 1986.
44. Pao, H.C., "The Measurement of Fuel Evaporation in the Induction System During Warm-up," SAE Paper 820409, Society of Automotive Engineers, 1982.
45. Min, K., Cheng, W.K., and Heywood, J.B., "The effects of crevices on the engine-out hydrocarbon emissions in SI engines," SAE Paper 940306, 1994.
46. Fox, J.W., Min, K.D., Cheng, W.K., and Heywood, J.B., "Mixture preparation in a SI engine with port fuel injection during starting and warm-up," SAE Paper 922170, 1992.
47. Finlay, I.C., Tugwell, W., Pirault, J.P., and Biddulph, T., "The influence of coolant temperature on the performance of a 1100 cc engine employing a dual circuit cooling system," *Heat and Mass Transfer in Gasoline and Diesel Engines*. eds. D.B. Spalding and N. Afgan, New York: Hemisphere Press, 1988.
48. Saito, A., Kawabata, M., Koga, K., Motomochi, M., and Kaneko, Y., "Relationships among engine variable, exhaust gas emissions and fuel economy in gasoline engines," Mitsubishi Heavy Industries, Technical Review, pp. 54–61, February 1978.
49. Taylor, C.F., "The internal combution engine in theory and practice," 2nd ed., Vols. I and II, Cambridge, MA: MIT Press, 1985.
50. Blumberg, P., and Kummer, J.T., "Prediction of NO formation in spark-ignition engines—an analysis of methods of control," *Combustion Sc. & Tech.*, 4, pp. 73–95, 1971.
51. Quadar, A.A., "Why intake charge diluation decreases nitric oxide emissions from spark ignition engines," SAE paper 710009, 1971.
52. Dryer, F.L., "Water addition to practical combustion systems—concepts and applications," *Power systems*, pp. 279–295, 1976.
53. Hopkinson, B., "A new method of cooling gas engines," *Proc. I. Mech Eng.*, pp. 679–715, 1913.
54. Kettleborough, C.F., and Milkins, E.E., "Water injection into internal combustion engines," *Commonwealth Engineer*, pp. 303–308, March 1955.
55. Weiss, K., and Rudd, J.W., "Engine performance improvement," *Automotive Engineer*, Vol. 49, no. 11, pp. 424–432, November 1959.
56. Nicholls, J.E., El-Messiri, A., and Newhall, H.K., "Inlet manifold water injection for control of nitrogen oxides—theory and experiment," SAE Paper 690018, 1969.
57. Peters, B.D., and Stebar, R.F., "Water-gasoline fuels, their effect on spark ignition engines," SAE paper 760547, 1976.
58. Milton, B.E., and Carter, E.A., "NO_X control using water injection and water-petrol emulsions in SI engines," Univ. of NSW report, FMT/1, 1987.
59. Heck, R., and Ferranto, R., "Automotive catalysts," *Automotive Engineering*, pp. 93–96, 1996.
60. Church, J.S., Cant, N.W., and Trimm, D.L., "Stabilisation of aluminas by rare earth and alkaline ions," *Appl. Catalysts*, no. 101, pp. 105–116, 1993.
61. Padeste, C., Cant, N.W., and Trimm, D.L., "Thermal decomposition of pure and rhodium impregnated cerium (III) carbonate hydrate in different atmospheres," *Cat. Letters*, no. 24, pp. 95–105, 1993.
62. Lox, E.S., Engler, E., and Koberstein, E., "The function of precious metals and washcoat oxides in modern three-way emission control catalysts," SAE Paper 910841, 1991.

63. Trimm, D.L., "Design of high temperature combustion catalysts," Proc. Int. Workshop on Catalytic Combustion, ed. H. Arai, pub. Catalysis Soc. of Japan, pp. 46–55, 1994.
64. Cooper, B.J., "Challenges in emission control catalysis for the next decade," *Platinum Metals Review*, 38, 1, pp. 2–10, 1994.
65. Truex, T.J., Searles, R.A., and Sun, D.C., "The opportunity for new technology to complement platinum group metal autocatalysts," *Platinum Metals Review*, 36, 1, pp. 2–11, 1992.
66. Toyota Annual Report, Toyota Motor Corporation, Toyota City, Japan, 1996.
67. Jost, K., "Reducing cold start emissions," *Automotive Engineer*, p. 22, December 1995.
68. White, D., "On the road to zero emissions," *The Chemical Engineer*, pp. 34–41, July 1996.
69. Takayama, K., Tohoku University, Institute of Fluid Science, Private Communication, 1995.
70. Paine, M., "Analysis of Australian in-service vehicle emission data," report for J. Haley, NRMA (National Roads and Motorist Association) Vehicles and Environment Dept., Sydney, Australia, April 1996.
71. Watson, H.C., and Watson, C.R., "Contribution of road transport to greenhouse gases," *SAE—Australasia Journal*, pp. 36–40, March/April 1990.
72. Ricardo, H.R., "Recent research work on the internal combustion engine," *SAE Trans.*, Vol 17, pp. 1–93, 1922.
73. Simko, Q., Choma, M., and Repko, L., "Exhaust emission control by the Ford programmed combustion process—PROCO," SAE Paper 720052, 1972.
74. Alperstein, M., Schafer, G.H., and Villforth, F.J., "Texaco's Stratified Charge Engine—Multifuel, Efficient, Clean and Practical," SAE Paper 740563, 1974.
75. Inoue, K., Ukawa, H., and Fujii, I., "Fuel economy and exhaust emissions of the Honda CVCC engine," *Comb. Sc. & Tech.*, Vol. 12, pp. 11–27, 1976.
76. Thisdell, D., "Nice engine, shame about...," *New Scientist*, pp. 21–23, January 1996.
77. Heywood, J.B., "Automotive engines and fuels: a review of future options," *Prog. in Energy and Comb. Sc.*, Vol. 7, pp. 155–184, 1981.
78. Houston, R., Archer, M., Moore, M., and Newmann, R., "Development of a durable emissions control system for an automotive two-stroke engine," SAE Paper 960361, 1996.
79. Deslandes, J.V., and Chapman, P.M., "Some aspects of LPG vehicle fuel economy," *Trans. Inst. Engineers*, Australia, pp. 32–38, 1983.
80. Bradley, D., and Woolley, R., "Methane as an engine fuel," Seminar S410/002/96, *IMechE.*, pp 9–15, 1996.
81. Watson, H.C., and Milkins, E.E., "Comparison and optimisation of emission, efficiency and power of five automotive fuels in one engine," *Int. J of Vehicle Design*, Vol. 3, no. 4, 1982.
82. Bolt, J.A., "A survey of alcohol as a motor fuel," *Alcohols and Hydrocarbons as Motor Fuels*, SAE Publication, SP-254, 1964.
83. Hagen, D.L., "Methanol as a fuel: a review with bibliography," SAE Paper 770792, Society of Automotive Engineers, 1977.
84. Dobes, L., *Alternative Fuels in Australian Transport*, Bureau of Transport and Communication Economics, Information Paper 39, Australian Government Publishing Service, Canberra, Australia, 1994.
85. Gosden, D.F., "Electric Vehicles and the Greenhouse Effect," Technical Note, Electric Vehicle Research Facility, School of Electrical Engineering, University of Sydney, 1990.
86. Gosden, D.F., "Generation of CO_2 resulting from the use of electric vehicles compared with gasoline fuelled vehicles in Australia," Trans of Inst. Eng., *Australia: Multi-disciplinary Engineering*, Vol. GE 15, no. 1, 99 83–90, 1991.
87. Knowlen C., Herzberg, A., and Mattick, A.T., "Cryogenic automotive propulsion," AIAA Paper 94-4224, 1994.
88. Howard, P., "Ballard Fuel Cell Powered Zero Emission Bus," Windsor Workshop on Alternative Fuels, Toronto, Canada, 1993.

PART IV

Compression-Ignition Engines

CHAPTER 9

Introduction

Franz F. Pischinger
FEV Motorentechnik GmbH and C. KG, Aachen, Germany

The diesel engine is known as the most fuel-efficient combustion engine. Its acceptance for use in passenger cars, however, varies geographically. Today the diesel car plays an important role in France, for instance, achieving a remarkable market share of about 42 percent, while in the United States its market penetration is negligible. Many questions are expressed concerning the future of diesel-powered cars.

ISBN: 0-12-639855-0
$25.00

Reprinted by permission of ASME from F. Pischinger, "The Diesel Engine for Cars – Is There a Future," Proceedings of the 1996 Fall Technical Conference of the ASME Internal Combustion Engine Division, Vol. 27-5, 1996.

The question affecting market acceptance is: Can the significantly better fuel efficiency of a diesel car outweigh perceived detrimental characteristics? Such unfavorable properties are thought to be low specific power, objectionable noise, higher exhaust emissions including smoke, and higher vehicle price. These features are closely influenced by the state of passenger-car diesel engine technology. This technology and its potential must be evaluated with respect to current and future demands, for instance, tighter exhaust emission regulations. In addition the commercial value and consumer acceptance of high fuel economy must be evaluated. It is clear that the ultimate result of weighing the pros and cons will depend not only on technological factors but also on political factors such as fuel taxation.

Regarding the state of technology, the diesel car is very promising. First, by employing a direct-injection combustion system, the fuel efficiency can be improved by about 15 percent over current swirl chamber engines. Furthermore, the specific power (hp/l) can be increased by efficient supercharging to achieve values of today's gasoline engines. By tuning the combustion system, using low-noise engine design features, and incorporating noise-reduction measures on the vehicle, the noise behavior of a spark-ignited vehicle can be reached. Exhaust emissions can currently be reduced to a level to satisfy today's European and U.S. Tier 1 emission limits.

However, significant development effort remains. More stringent emission levels (California U.S. Tier 2 ULEV, and Stage 3 in Europe) require further advancements in diesel combustion. The strong development potential of four-valve engines and new unique injection systems is evident. In addition there are promising developments with lean-NO_x catalysts and regenerative particulate filters. These technologies offer the potential to meet the very stringent future emission standards. It is anticipated that the sophisticated technology needed to meet these standards will make the future diesel car more expensive compared to a gasoline-fueled vehicle. This raises the issue of what price the consumer will pay for the higher fuel economy of a diesel car. In light of the worldwide rapid increase in passenger-car population and of the dwindling oil reserves and their global distribution, the fuel-efficient diesel engine will play an important role in the future of passenger cars.

9.1

THE DIESEL ENGINE FOR CARS—IS THERE A FUTURE?

The diesel engine is the most fuel-efficient combustion engine known. Its use in applications such as trucks, buses, agricultural equipment, ships, and power stations is firmly established. For passenger cars, however, its position differs geographically. As Figure 9.1 shows, the diesel car has a market share of about 20 percent in Europe, reaching in France, for instance, a record share of about

Fig. 9.1.

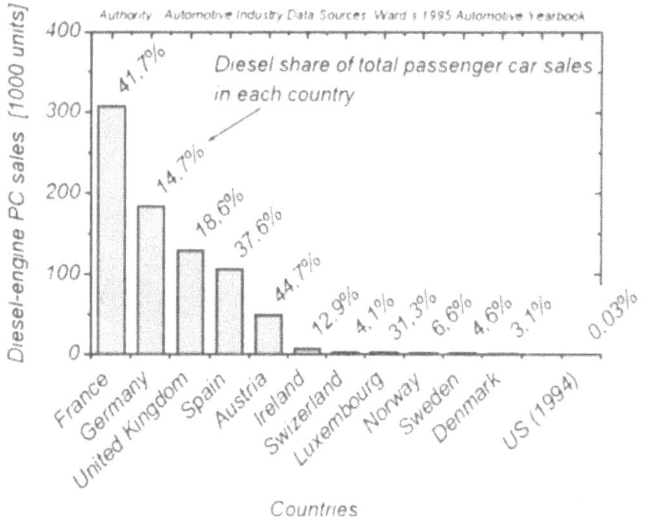

Diesel-engine passenger-car sales in Western Europe (January–April 1996) and the United States (1994).

42 percent (in 1996), while in the United States its penetration into the market is negligible.

A simple reason for this uneven distribution in market shares is different fuel prices and car tax rates. As an example, Figure 9.2 shows that in France and in Germany, due to lower taxes, diesel fuel is about 25 percent cheaper than the mostly used European premium gasoline, whereas in the United States fuel prices are much less loaded by taxes and there is no significant difference between prices of diesel and gasoline. As an example for differences in annual car tax rates, Figure 9.3 shows relative figures for comparable medium-sized cars, with the gasoline car being rated 100 percent. In Germany the diesel car is taxed nearly four times as high as the gasoline vehicle, which might explain at least partially the difference in diesel market share compared to France where there is no difference between diesel and gasoline car tax.

Beyond these differences in taxation, there are historical circumstances and technology-related pros and cons for diesel versus gasoline cars. For decades in most European countries, the diesel car has the image of an exceptionally fuel-efficient and durable vehicle. Over time an increasing number of European consumers have valued these properties more highly than historical drawbacks of the diesel car, such as lower power, higher noise, and higher price.

During the last decade in Europe, the introduction of the low-emission gasoline engine with three-way catalyst challenged the exhaust emission capability of the passenger-car diesel. Especially the particulate emission of the diesel has

Fig. 9.2.

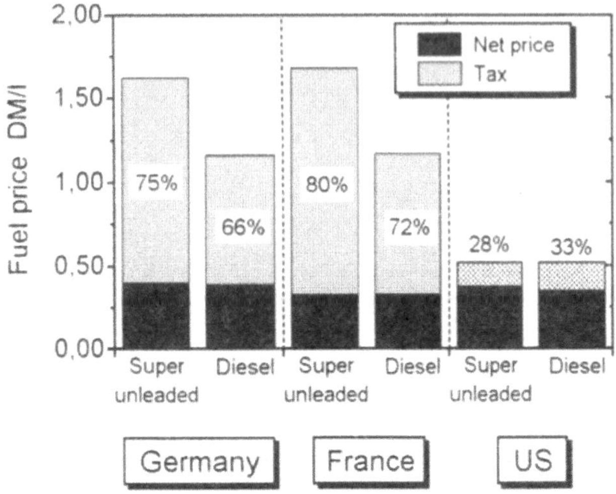

Comparison of fuel prices for diesel-engine and gasoline-engine passenger cars (tax in percentage of fuel price).

Fig. 9.3.

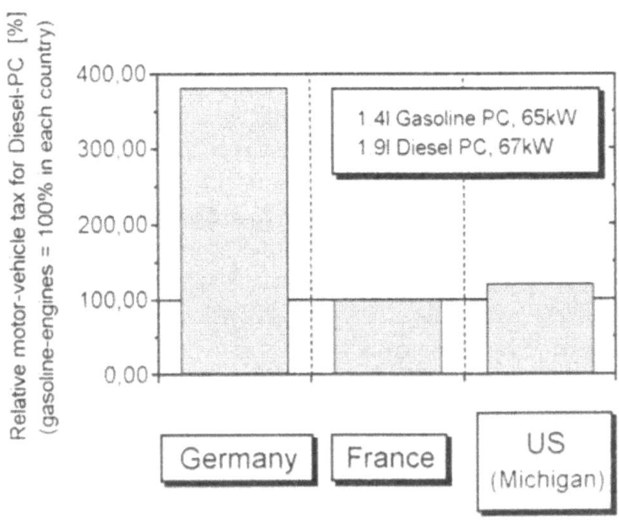

Comparison of motor vehicle taxes for diesel-engine and gasoline-engine passenger cars.

become the target of environmental concern. Even though no proof of serious health hazards caused by the particulate concentrations in urban areas could so far be found, such discussions contributed to a certain degree of uneasiness among some diesel-car buyers, especially in Germany, and gave rise to questions about the future of the diesel car.

This leads to the general questions: What will be the future of the diesel car? Will the trend in Europe toward increased diesel share continue, be stopped, or be reversed? Will the U.S. car market begin to accept diesel technology? Can diesel technology cope with future noise and exhaust emission regulation?

To predict the future role of the diesel car, a thorough evaluation of today's state of passenger-car diesel technology and its future potential must be compared to the existing and future requirements. Such a projection is presented here.

9.2 STATE OF TECHNOLOGY

9.2.1 Power Density

Among the historical drawbacks of the passenger-car diesel is the lower power density, even if this was accepted by diesel-car supporters. However, over the last few years a significant change took place in power density, as Figure 9.4 shows. The power output per liter increased up to 45 kW (60 hp) due to an improved combustion process, higher turbocharging, and use of improved materials. With direct-injected four-valve prototype diesel engines, even 50 kW/l (67 hp/l) can be reached, thus enabling the diesel engine to compete with many of today's gasoline engines, especially in the United States.

Fig. 9.4.

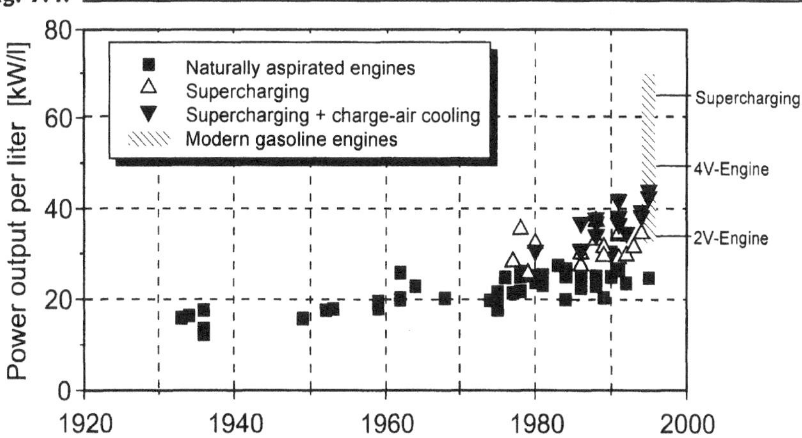

Power output per liter history of passenger-car diesel engines.

Fig. 9.5.

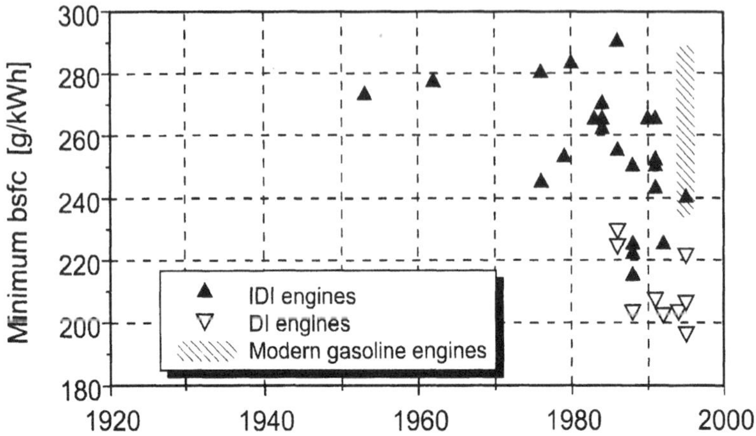

Minimum brake specific fuel consumption (bsfc) history of passenger-car diesel engines.

Fig. 9.6.

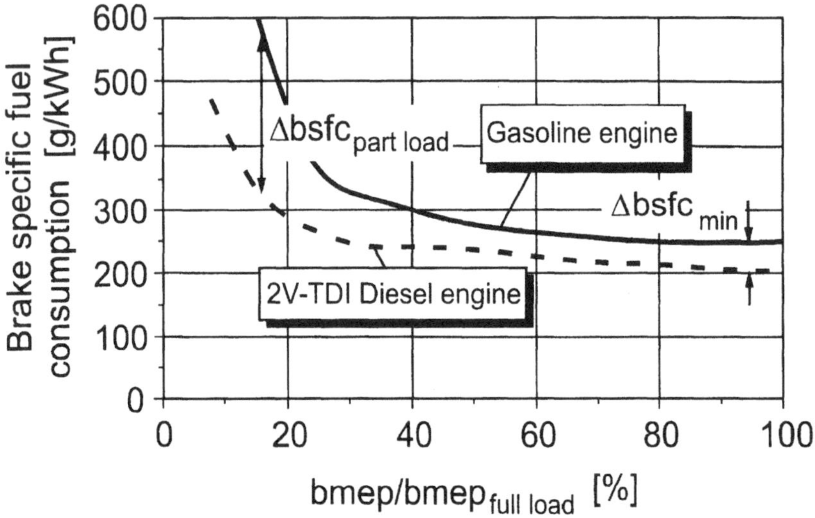

Brake specific fuel consumption of diesel engines compared to gasoline engines (part load, 2000 rpm).

9.2.2 Fuel Efficiency

The relative improvement goes even further with improved fuel efficiency, as shown in Figure 9.5, thereby improving still more the diesel engine's biggest asset. Especially the step from prechamber and swirl chamber (IDI) engines to direct-injected (DI) diesels improved the minimum effective specific fuel consumption by about 15 percent. Today's DI diesel engines reach values below 200 g/kWh (0.329 lb/hp). Still more impressive is the advantage of the DI diesel engine at part-load operation. Figure 9.6 shows a comparison of specific fuel consumption for a turbocharged, direct-injected (TDI) car diesel against that of a very good naturally aspirated gasoline engine versus load at 2000 rpm. For the full load range the diesel advantage is about 20 percent, whereas at lower part load, which is typical for city driving, it is in the range of 40 percent. It should be remarked here that due to the higher density of diesel fuel, which corresponds to a higher energy density, the volumetric fuel consumption of a diesel is already about 12 percent lower compared to a gasoline-fueled engine. A further bonus for the diesel comes from better fuel efficiency under cold starts compared to the SI engine. Altogether under average driving conditions the TDI diesel car has an advantage in volumetric fuel efficiency of between 30 percent and 40 percent compared to a similarly powered modern gasoline vehicle.

9.2.3 Exhaust Emissions

As is well known, there exists a trade-off between fuel efficiency and exhaust emissions. It is, therefore, remarkable that such fuel-efficiency improvements have been reached while meeting today's exhaust emission requirements. Compared to a decade ago, remarkable improvements in exhaust emissions have been reached as Figure 9.7 shows for particulate and NO_X emissions in the FTP 75 test. NO_X emissions could be reduced by about 30 percent while at the same time particulate emissions have reduced by more than 60 percent. However, it is evident that there is still significant improvement necessary, especially for NO_X emissions, to meet the stringent emission standards of the future such as California's ULEV standard. Concerning CO and HC emissions, diesel engines have an inherent advantage; however, these standards have been reduced over the last decade.

Fig. 9.7.

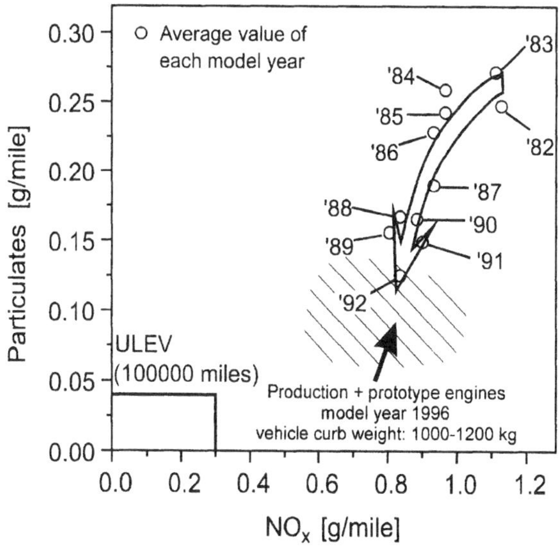

Pollutant emissions history of diesel-engine passenger cars (FTP 75 test).

Fig. 9.8.

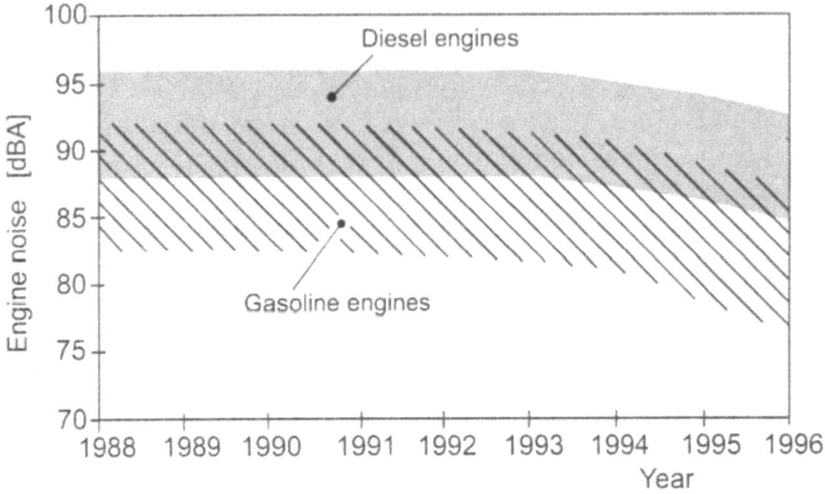

Engine noise emissions history of passenger-car diesel and gasoline engines 1988–1996 (full load, 3000 rpm, 1 m distance).

9.2.4 Noise Emission

During the last decade, noise emissions of passenger-car engines have improved considerably. This holds true for SI engines as well as for diesels, as can be seen in Figure 9.8. Mainly due to the lower combustion noise, SI engines show on average a lower level. However, the best diesel engines can compete with SI engines in their upper noise band. This could be achieved by tuning the combustion system and low-noise engine design features for DI passenger-car diesels. Together with careful noise-reduction measures on the vehicle, the noise behavior of a good gasoline car can be reached with a DI diesel under most operating conditions.

9.2.5 Technology

Today's technology to reach state-of-the-art figures in all preceding disciplines employs among other things two-valves per cylinder, direct injection, distributor-type fuel injection pump, turbocharging with charge air cooling, exhaust gas recirculation (EGR), oxidation catalyst, electronic control for fuel injection, EGR and charge air pressure, low-noise engine structure, and noise-silencing measures on the vehicle. It is clear that with these technologies it is difficult to keep the price of the diesel car close to the price of the gasoline version. However, it has been found in Europe that customers are willing to pay a higher price for the very fuel-efficient diesel car.

9.3 TECHNOLOGY FOR THE FUTURE

The goals that must be reached to assure a bright future for the car-diesel engine are numerous. The most important objectives are to meet future stringent emission standards, such as California's ULEV and European Stage 3, while achieving a further decrease in noise levels. While accomplishing these goals, the fuel economy should be further improved or maintained at today's low level. Further improvements in power density would, of course, be advantageous in combination with high torque at low engine speed. However, this should be achieved without much additional cost to the diesel car.

9.3.1 Chemical Thermodynamics

The crucial task for lowering diesel emissions is to improve particulate and NO_x emissions simultaneously—a difficult task, as there is usually a trade-off between these two emissions. To determine if this task is at all possible, basic fundamentals of chemical thermodynamics must be explored.

Figure 9.9 shows a typical mixture distribution in a diesel injection spray just prior to ignition, ranging from rich in the spray core to lean in the spray edge. The diagram on the left shows the temperature history of the zones with different relative air-fuel ratios (λ), proceeding from the curve of mixture temperature before combustion up to the curve of burned gases after combustion. The diagram

Fig. 9.9.

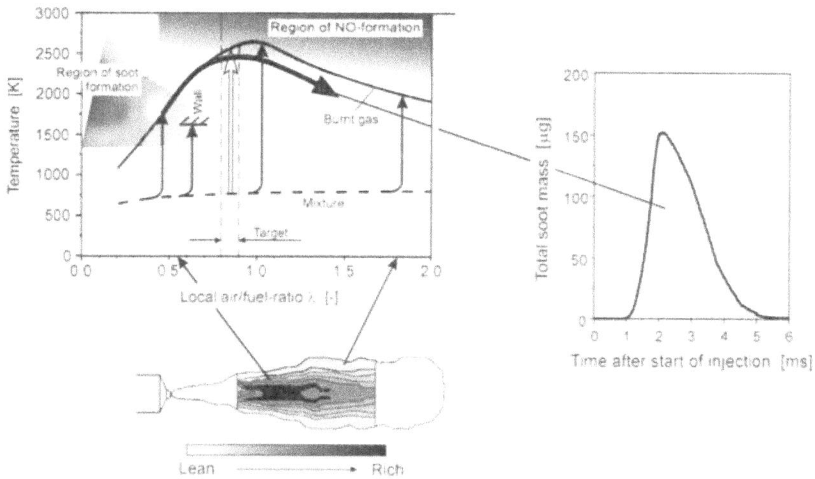

Soot formation and nitrogen oxide formation in a diesel fuel spray and its dependence on stoichiometric air-fuel ratio λ.

also shows the zones of soot and NO_x formation, the darker shadowed zones indicating higher formation rates. It can be seen that in an indicated target λ-range, combustion is possible without soot formation and only very low NO_x formation. Thus, in principle, smokeless, low-NO_x diesel combustion should be possible, if the mixture formation could be narrowed to the required mixture ratio. This may not be possible with the combustion of typical sprays. However, even if soot formation cannot be completely avoided, there is a possibility of burning the soot during the high-temperature combustion process, as is shown in the diagram on the right side of the figure. This, of course, requires intensive mixing at the beginning of the expansion process while the temperatures are still on a high level, although not too high because of NO_x formation. In this way the soot generated can be burned nearly completely during the process.

It can be concluded that a careful mixing strategy is the key to low-emission diesel combustion. Since mixing is controlled by fuel injection as well as by air motion in the combustion chamber, both processes must be developed concurrently. Future diesel combustion development approaches follow these guidelines. Typical design features addressing this fact are discussed next.

9.3.2 Four-Valve Technology

It is obvious that the mixing process during fuel injection into the combustion chamber is influenced by the position of the injector as well as by the symmetry of the airflow during injection. To achieve best mixing conditions for a typical multispray injection of a DI diesel engine requires equally good conditions for all sprays of the injector. Figure 9.10 shows the typical arrangement for today's two-valve passenger car diesel engines. The inclined injector is positioned off-center and sprays into the asymmetrical flow in an eccentric combustion chamber. By contrast, the four-valve design incorporates a centrally mounted vertical injector with completely symmetric sprays and a symmetric flow field in a concentric combustion chamber. In addition, the four valves instead of two valves improve the gas exchange process.

Figure 9.11 shows the influence of different four-valve arrangements on swirl number c_u/c_a and flow coefficient α_k as measured on a special steady-state flow rig. Swirl number c_u/c_a indicates the degree of air rotation around the cylinder axis produced by the intake process, which for efficient fuel-air mixture formation should be in the range of 1.5 and 1.9 for four-valve, high-speed, passenger-car diesel engines. The flow coefficient is influenced by the arrangement of the intake valves and the required swirl number. It can be seen in Figure 9.11 that the twisted valve arrangement produces the highest flow coefficient and thereby the best breathing capability of the engine at the required swirl intensity. The twin arrangement, which is known from larger diesel engines, and tandem arrangement are found to be inferior. The advantage of four valves over two valves also can be seen in Figure 9.11.

Fig. 9.10.

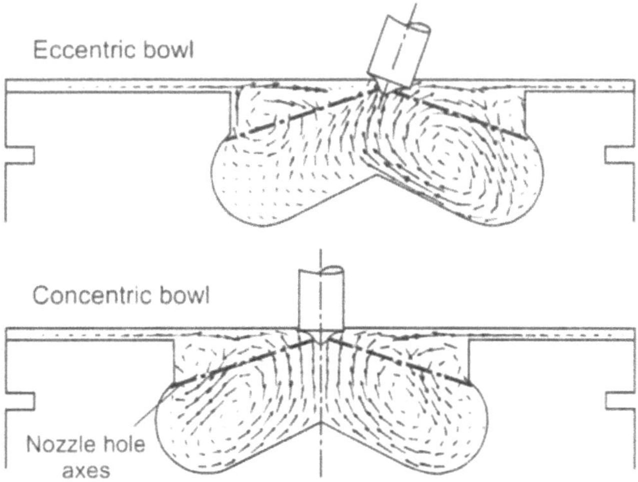

Comparison of two-valve and four-valve concepts with regard to flow in the combustion chamber (piston at TC).

Fig. 9.11.

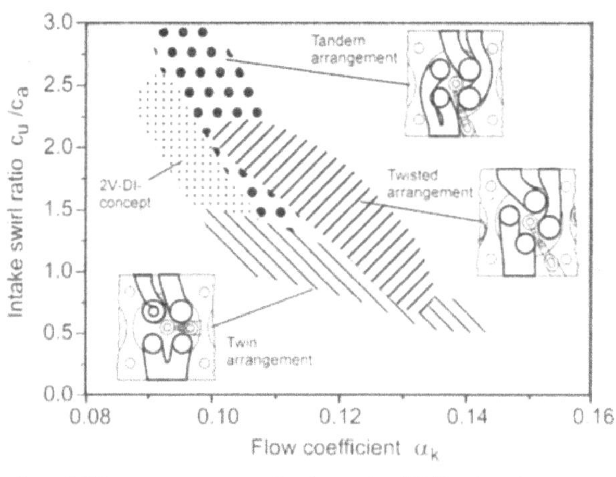

Flow quality of different port concepts for four-valve DI diesel engines.

Thorough investigations on a future diesel prototype engine conducted by FEV have proven that a twisted four-valve arrangement could be realized in a cost-effective and durable design. Views of the "Diesel Future" cylinder head are shown in Figure 9.12. It can be seen that even with a central vertical injector a single overhead camshaft design was possible.

Another advantage of the four-valve technology is the capability to produce variable swirl by port deactivation, as shown in Figure 9.13. By increasingly

Fig. 9.12.

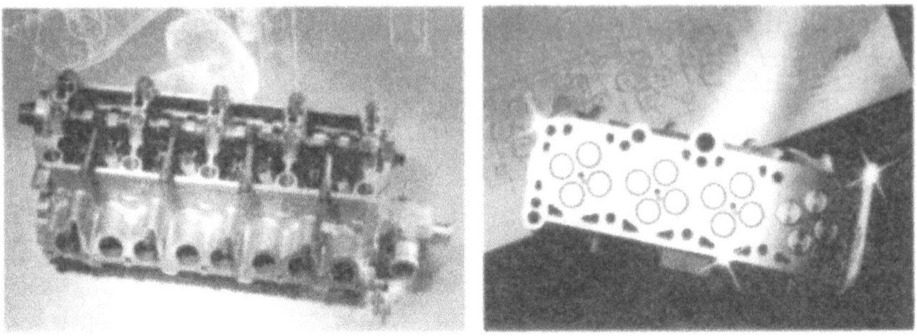

Cylinder head of FEV four-valve DI diesel engine for passenger cars.

Fig. 9.13.

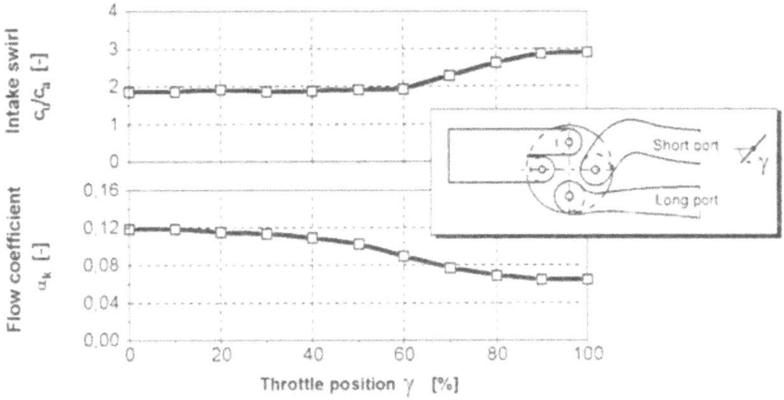

Flow characteristics of four-valve DI diesel engines for deactivation of short intake port.

closing the short intake port of the twisted arrangement, for example, by employing a throttle flap, the swirl number can be increased effectively. This is important in order to enhance fuel-air mixing and thereby to reach high torque at low engine speed. The associated decrease in flow coefficient is not detrimental for low-speed operating conditions.

By careful optimization of this four-valve arrangement, even in combination with current production distributor-type fuel injection pumps, an increase in engine power could be achieved. As full-load brake mean effective pressure versus engine speed curves show for such a turbocharged, four-cylinder engine (Figure 9.14), torque over a wide speed range and maximum power could be increased by about 9.15 percent over the two-valve engine. Of course, maximum cylinder pressure was limited to the same level for both versions.

Still more impressive are the associated improvements in exhaust emissions, as shown in Figure 9.14, which range between 20 percent and 30 percent. In

Fig. 9.14.

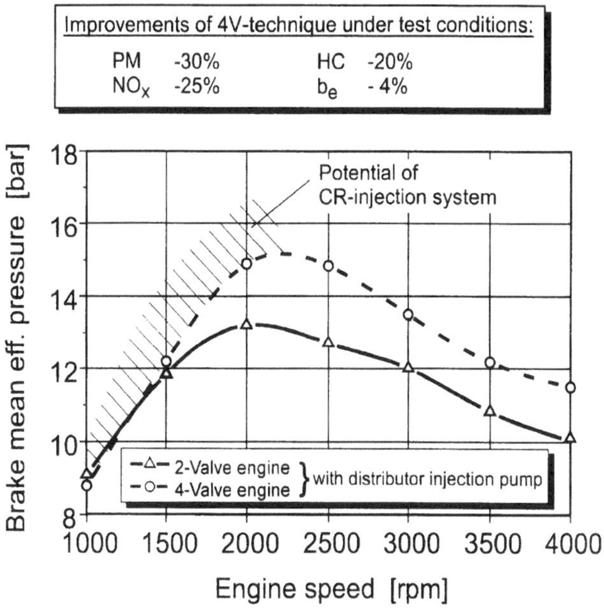

Improvements of four-valve DI diesel engines under test conditions.

addition the fuel consumption (specific fuel consumption b_e) could be improved by about 4 percent.

However, with a conventional cam-actuated injection pump, a drawback in combustion still remains at very low engine speeds, if the system is tuned for high power. This has a negative effect on low end torque, as Figure 9.14 shows. In this case advanced fuel injection concepts such as common rail (CR) systems can provide further improvement.

9.3.3 Fuel Injection Technology

For cam-actuated fuel injection systems there is always a strong dependence of injection pressure level on engine speed. Furthermore, the potential to shape the pattern of injection rate versus time is restricted by cam design. These limitations can be overcome by application of a common rail fuel injection system. Such systems can produce injection pressure levels largely independent from speed and load by feeding a fuel rail with electronically controlled permanent fuel flow from a high-pressure fuel pump. The fuel rail subsequently feeds the injector on each cylinder. Injection timing and injection quantity per cylinder and cycle is typically controlled by electromagnetic valves on each injector. Such systems for passenger-car diesel engines currently under development will soon be introduced in the market.

The application of common rail technology requires a special tuning of the combustion system to optimize mixture formation. By using the high-pressure

Fig. 9.15.

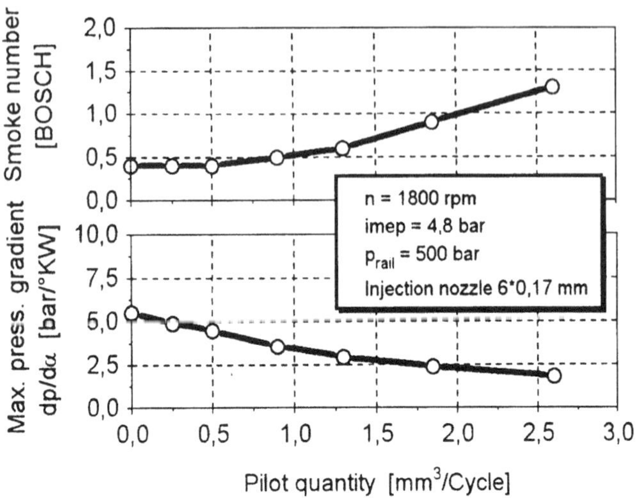

Common rail system: influence of pilot quantity on black smoke emission and pressure gradient.

capability of the injection system at low engine speed, low end torque can be improved considerably, as seen in Figure 9.14.

In addition the capability of CR systems for pilot injection can be used to reduce combustion noise. Here best results can be obtained by carefully tuning the pilot fuel quantity and separation time. As an example, the trade-off between exhaust smoke (Bosch smoke number) and maximum pressure rise as a function of pilot injection quantity is shown in Figure 9.15 for a typical city driving part-load operation point. To avoid an increase in soot formation while considerably reducing the noise-generating pressure rise, an injection quantity between 0.5 and 1.0 mm^3/cycle is desirable. This is a difficult task for an electromagnetic valve, which also tends to restrict rate shaping. Therefore, piezoelectric valve actuators for CR systems are under development and will offer additional advantages in the future.

Noise reduction achieved by application of an experimental piezoelectric CR system is shown in Figure 9.16. A reduction of about 10 to 15 dBA can be reached in the critical high-frequency range. Although a distinct disadvantage in combustion noise exists compared to low-noise, spark-ignited engines, the noise of the improved passenger-car diesel can be further compensated by low-noise design of the engine structure and cost-effective measures on the vehicle.

In addition to all these advantages, a common rail system offers the possibility of postinjection. Postinjection is an effective method to discretely supply the exhaust gas with a certain level of hydrocarbons as the reducing agent for a lean-NO_x catalyst. It is possible to synchronize the postinjected fuel for the lowest oxidation rate. Figure 9.17 shows the following quantities versus postinjection timing for a required HC level of 1500 ppm C_3H_8 at a part-load operating point: cylinder pressure, ratio of total supplied fuel mass to produced mass of HC, and

Fig. 9.16.

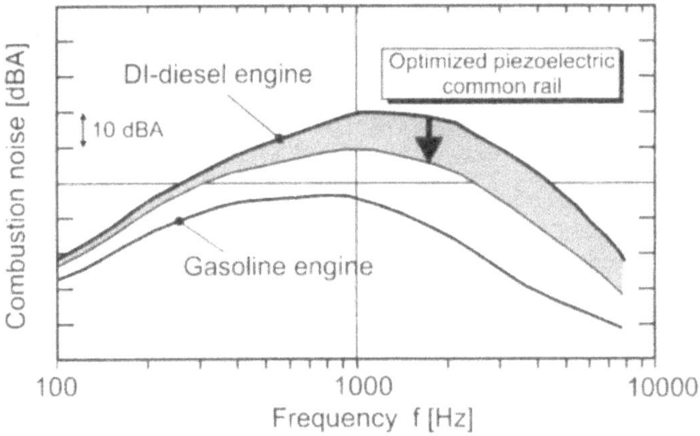

Combustion noise of modern passenger-car engines: potential of optimized piezoelectric common rail (part load, medium speed).

Fig. 9.17.

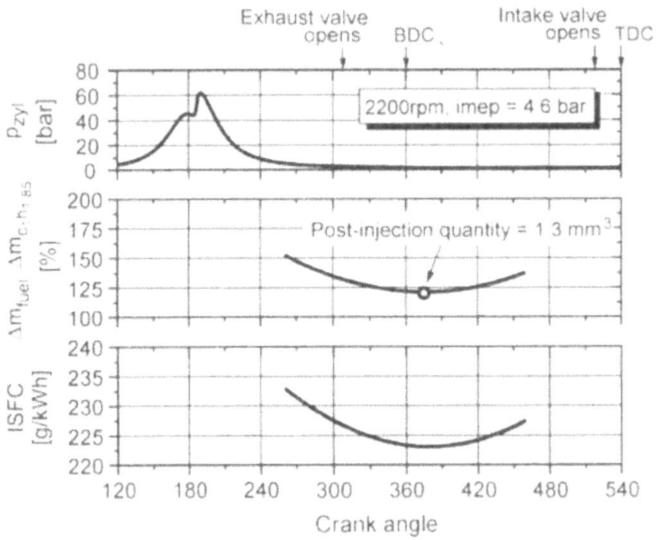

Common rail injection system: influence of postinjection timing (constant HC emission = 1500 ppm C_3H_8).

the corresponding indicated specific fuel consumption. The highest HC formation efficiency and correspondingly lowest fuel consumption is reached with a postinjection period near BDC.

9.3.4 Exhaust Aftertreatment

By application of the foregoing features in combination with state-of-the-art technology, not only are high specific output and improved fuel efficiency possible but also engine-out exhaust emissions will be improved considerably. However, these levels may not be sufficient to reach future emission goals even with the application of current production oxidation catalysts. This is especially true for larger passenger cars and light trucks. Therefore, additional measures for exhaust gas aftertreatment must be investigated. Two such techniques are particle storage reactors and lean-NO_x catalysts.

Particle storage reactors, also called particulate traps, collect particulates on a special surface, such as a porous ceramic, and burn them such that the unit is not overloaded. Overloading the particulate trap not only can cause plugging and severe loss in engine load and efficiency but also can destroy the unit by overheating in the case of quick oxidation. Today a rather promising method is under development that combines two methods for easy regeneration of the storage reactor, with the principle shown in Figure 9.18. A small amount of additive is added to the fuel, which after combustion makes the particles reactive for low-temperature oxidation in the trap. The additive can be carried as a lifetime supply on board the vehicle or it can be added to future diesel fuel in the refinery. Under most driving conditions

Fig. 9.18.

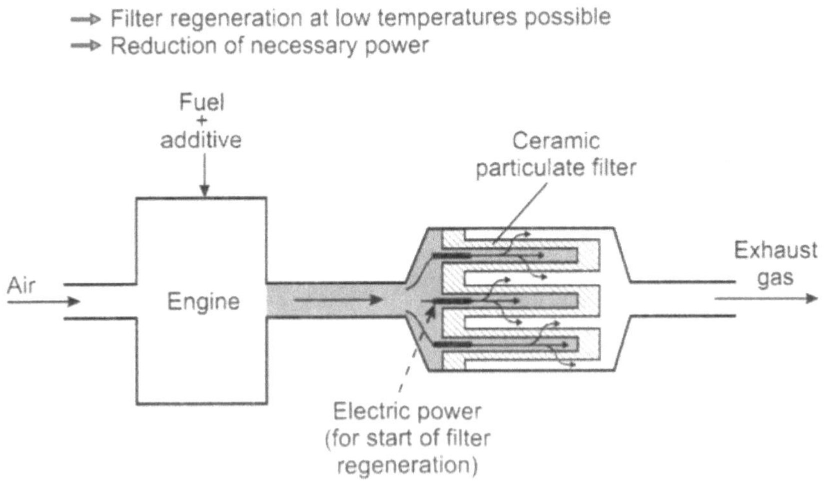

FEV particulate filter system (self-supported filter regeneration, SFR system) in combination with fuel additive.

Fig. 9.19.

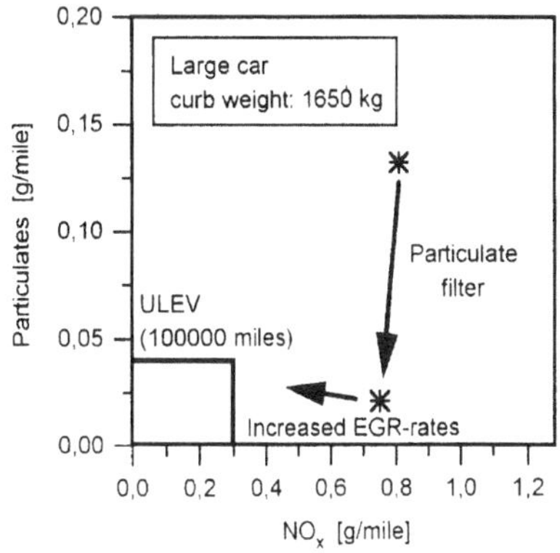

Influence of particulate filter and increased EGR rates on particulate and nitrogen oxide emissions (FTP 75 test).

exhaust temperatures are sufficiently high to burn all the particulates in due time. This has already been proven in ongoing vehicle tests. Under certain extreme driving patterns such as repeated severe winter cold starting followed exclusively by short-distance driving, this method may not function properly. Therefore, the patented SFR (*s*elf-supported *f*ilter *r*egeneration) system for burning the trapped particulate by electric flame initiation especially at low temperatures is combined with the additive method. The necessary energy for ignition is moderate.

With such a regenerative trap, particulate emission can be reduced by more than 80 percent. As indicated in Figure 9.19, for a large car this can be used to allow an increase in the degree of exhaust recirculation, thereby lowering NO_x emissions. As this may not be completely sufficient to reach very low emission limits (ULEV), catalytic NO_x converters for O_2-containing exhaust gases—so called lean-NO_x catalysts—are an important technology to be combined with future passenger-car diesels.

All lean-NO_x catalysts require a reducing agent to function. From a practical point of view, diesel fuel would be a desirable candidate, especially when it is supplied by a common rail system without the need for additional hardware. Under laboratory conditions conversion rates of up to 60 percent can be reached. Figure 9.20 shows, that for a diesel engine under steady-state operation, a maximum of 40 percent conversion rate is reached. Depending on the hydrocarbon-NO_x ratio and actual temperature levels, lower values are expected in practical applications. Presently conversion rates of 20 percent to 25 percent seem realistic for vehicle operation. Thus, a technological achievement is still necessary for future applications.

If urea is used as a reducing agent, such an achievement has already been reached. Urea is a nontoxic agent that can be dissolved in water. Unfortunately, an additional liquid on board a vehicle, which requires refilling in intervals (e.g., about 28 l/10,000 km) to function properly, cannot easily be accepted. Nevertheless, the

Fig. 9.20.

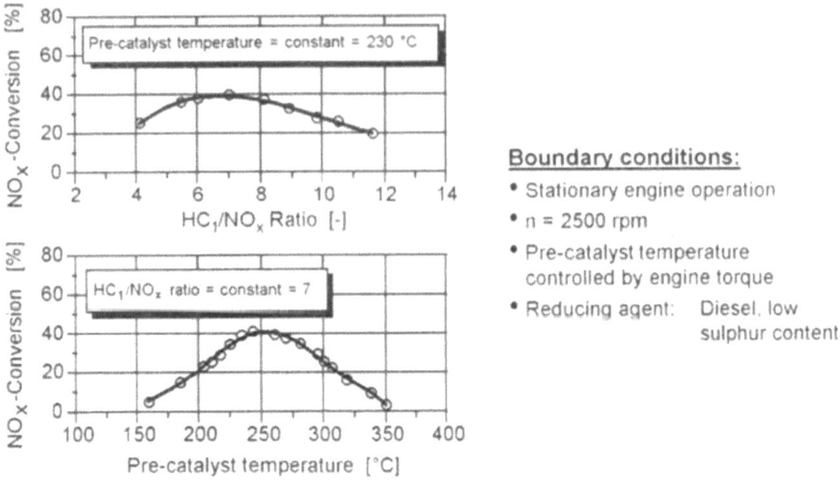

Characteristics of NO_x conversion with lean-NO_x/HC technology.

Fig. 9.21.

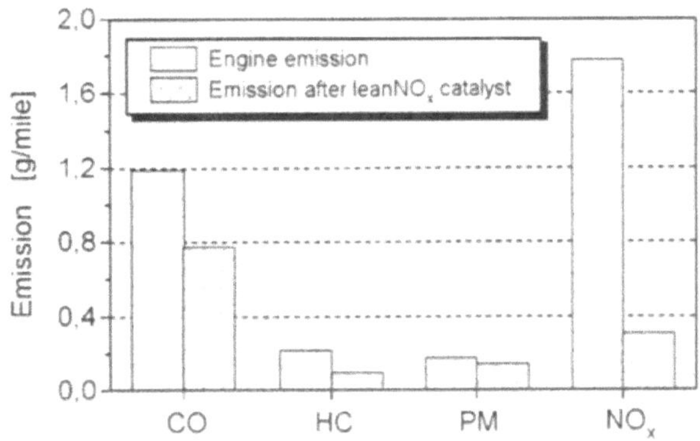

Reduction of emissions with lean-NO_x/urea technology (large car, FTP 75 test).

results with this technology for NO_x reduction are very encouraging. As shown in Figure 9.21 for a large car, an 80 percent NO_x reduction could be reached in the FTP 75 test. This can also be regarded as an encouraging result for further work on lean-NO_x catalytic systems in general.

9.4

SUMMARY AND CONCLUSIONS

The role of the diesel engine for passenger cars in the world market is characterized by historical and technological features as well as by other critical boundary conditions. In spite of the higher price of a diesel vehicle, its high fuel efficiency and durability combined with fuel price factors have led to an increasing market share in Europe.

Meanwhile car-diesel engine technology has further improved and mitigated many passenger-car historical problems with diesel engines. By introduction of a tailored, direct-injection, electronically controlled combustion system, the fuel efficiency is improved by about 15 percent over today's swirl chamber engines. Furthermore, the specific power (hp/ltr) could be increased by efficient supercharging to values of current gasoline engines. By tuning the combustion system, using low-noise engine design features, and incorporating careful noise-reduction measures on the vehicle, the noise behavior of a good gasoline car can be reached under most operating conditions. Exhaust emissions have been reduced to levels that satisfy current European and U.S. federal Tier 1 emission limits.

Meanwhile advancements to improve the properties of passenger-car diesel engines and comply with future, more stringent emission levels are ongoing. Great potential, especially with four-valve engines and new unique injection systems, is already evident. Other features not discussed within this paper, such as improved turbocharger technology, new EGR concepts, downsizing, and others will also contribute to the increased attraction of the diesel car. In addition there is promising work with lean-NO_x catalysts and regenerative particulate traps. These technologies offer good chances to meet the very stringent emission levels of the future.

Although there are promising technologies such as common rail injection to reduce the cost of a diesel engine, it can be expected that some highly sophisticated technology may be needed in the future to meet increasing consumer and exhaust emission standards. These technologies may continue to keep the future diesel car at a more expensive price compared to a gasoline-fueled vehicle. Whether or not this factor will be accepted by the consumer will strongly depend on other boundary conditions such as fuel price and taxes. In view of the worldwide rapid increase in passenger-car population and the limited world fuel resources, the boundary conditions of the future should favor the diesel car. Considering all these arguments, the future of the diesel car is expected to be very promising and challenging.

CHAPTER 10

Combustion-Related Emissions in CI Engines

J. Gary Hawley, Chris J. Brace, and Frank J. Wallace
Department of Mechanical Engineering, University of Bath, Bath, United Kingdom

Roy W. Horrocks
Diesel Engine Powertrain, Ford Motor Company Limited, United Kingdom

ISBN: 0-12-639855-0
$25.00

10.1 INTRODUCTION

For many years, particularly in the United States, diesel engines have had a reputation of giving poor performance and producing black smoke, an unpleasant odor, and considerable noise. However, most drivers would find it difficult to distinguish today's modern diesel car from its gasoline counterpart. Nevertheless, the challenges confronting the motor industry in terms of environmental protection are very substantial.

Catalysis of diesel exhaust remains a problem insofar as research has not yet been able to come up with an effective converter that eliminates both particulate matter (PM) and oxides of nitrogen (NO_x). Oxidation catalysts are well developed and used on current production engines. However, the realization of de-NO_x catalysts will be required for the more stringent emission standards being proposed beyond the year 2005.

Unlike the gasoline engine, which has a number of alternative fuel contenders, the diesel engine fuel substitutes are limited. Diesel fuel quality, in relation to the cetane number, which ultimately impinges on the sensitivity of the ignition delay period, is also a governing factor when considering emissions and performance.

To conserve natural resources and reduce emissions of the greenhouse gas carbon dioxide (CO_2), greater emphasis is being placed on fuel economy. The 3.0 liter/100 km car is now a well-publicized target for the industry to achieve. European Community (EC) Directive 93/115/EC mandates that CO_2 emissions be measured and fuel consumption calculated for all new vehicles from January 1, 1997.

For diesel engines the emphasis is to reduce emissions of NO_x and particulates, where these emissions are typically higher than those from equivalent port injected gasoline engines equipped with three-way catalysts. There is also a need for the passenger-car diesel to achieve gasoline engine levels of noise, vibration, and harshness (NVH) and provide equivalent acceleration response on demand.

The high-speed diesel engine is now well established in the European passenger-car market, currently accounting for around 22 percent of new cars [1]. The direct-injection (DI) version, first introduced for passenger cars by Fiat in 1988, is also becoming established with models offered by Fiat, Rover, VW-Audi, Mercedes, Volvo, and Opel/Vauxhall (see Table 10.1).

Table 10.1

High-Speed DI Diesel Engines

Year of introduction	Manufacturer	Layout	Capacity (liters)	Power (kW)	Torque (Nm)	Aspiration	Reference
1984	Ford	I-4	2.5	52	143	2-valve N/A	[9]
1985	Isuzu	I-4	2.5	50	152	2-valve N/A	[10]
1985	Isuzu	I-4	2.8	57	172	2-valve N/A	[10]
1985	Iveco	I-4	2.45	68	216	2-valve T/C	[11, 12]
1985	Iveco	I-4	2.5	76	225	2-valve T/C	[11, 12]
1986	Perkins	I-4	2.0	46	122	2-valve N/A	[29]
1988	Fiat	I-4	1.9	68	182	2-valve TCI	[13]
1989	Perkins	I-4	2.0	60	154	2-valve TCI	[15]
1989	Audi	I-5	2.5	88	265	2-valve TCI	[16, 17]
1989	Land Rover	I-4	2.5	82	260	2-valve TCI	[18]
1992	Ford	I-4	2.5	74	224	2-valve T/C	[20]
1992	VW-Audi	I-4	1.9	66	182	2-valve TCI	[21]
1995	Rover	I-4	2.0	77	210	2-valve TCI	[18]
1995	Mercedes	I-5	2.9	95	300	2-valve TCI	[24, 25]
1996	Iveco	I-4	2.8	76	240	2-valve T/C	[30]
1996	Iveco	I-4	2.8	90	285	2-valve TCI	[30]
1996	Opel/Vauxhall	I-4	2.0	60	185	4-valve TCI	[26]
1997	VW	I-5	2.5	55	160	2-valve N/A	[31]
1997	Audi	V-6	2.5	110	310	4-valve TCI	[27, 28]
1998	Mercedes	I-4	1.7	44	160	4-valve T/C	[32]
1998	Mercedes	I-4	1.7	66	180	4-valve TCI	[32]
1998	Opel/Vauxhall	I-4	2.2	86	260	4-valve TCI	[26, 33]
not known	Toyota	I-4	2.0	not known		4-valve T/C	[34]

Advances in engine technology have made it possible for the diesel engine to meet very stringent emissions levels. Further standards have generated interests in other technologies such as exhaust gas aftertreatment, fuel composition and quality, as well as additional engine improvements, and improved control strategies. Faced with the NO_X-PM trade-off challenge, engine manufacturers have several design choices while being concerned with increased production cost and complexity.

10.2 REVIEW OF CURRENT AND PROJECTED EMISSIONS CONCERNS—GENERAL CONSIDERATIONS

Social and political pressures on the automotive industry to reduce atmospheric pollution from internal combustion engines are being sustained, particularly to improve air quality in urban areas and to lower the emissions of greenhouse gases. This is forcing more stringent emissions legislation for motor vehicles. The impact of motor vehicles on the environment, particularly the use of natural resources and emissions of CO_2, is well documented. The 1992 Rio climate change treaty calls for nations to stabilize their greenhouse gas emissions at 1990 levels by the year 2000. The latest European Union (EU) report (May 1996) indicates that CO_2 emissions in the EU could increase up to 5 percent or more over the 1990 baseline. Because of these factors, there is an urgent demand to reduce fuel consumption from passenger cars to minimize emissions of CO_2 and the use of finite oil resources. Political pressures to meet the Rio CO_2 framework commitments led the EU authorities to insist on automotive-specific CO_2/fuel economy reduction commitments. The European Community is constantly in discussion with the European automotive manufacturers' organization (ACEA) with regard to measures for the reduction of fleet average fuel economy. At present there are no limits on fuel economy, but CO_2 is measured and fuel consumption calculated according to Directive 80/1268/EEC as amended by 93/116/EC, for EC-type approval.

As the number of vehicles on the road continues to rise globally, the U.K. increase is reckoned to be about 3.9 percent per annum [2]. In real terms there are 25 million vehicles registered for use on U.K. roads, which is 5 million more than ten years ago. The U.K. government green paper, "Transport—the way ahead" (published in May 1996), predicts a growth in traffic density of between 53 percent and 83 percent by 2025. An interesting statistic is shown in Figure 10.1, which compares the vehicles per mile of road in the developed world [3]. The continuing progress being made to develop less polluting internal combustion (IC) engines through legislative strategies must outpace the effect of transportation growth rates if a net reduction in emissions is to be achieved.

For air quality to reach an acceptable standard, incremental vehicle emission reduction standards, coupled with improved fuel quality, are essential. However,

Fig. 10.1.

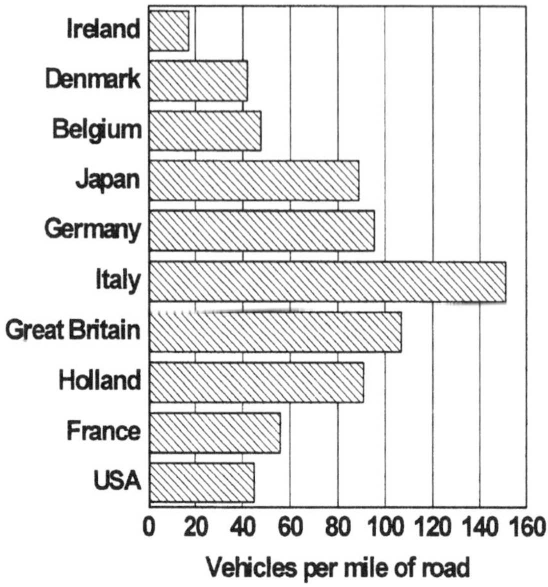

Vehicles per mile of road (taken from T. Fidell, "LP Gas as an Automotive Fuel," Automotive Fuels for the 21st Century, IMechE Seminar, January 21, 1997).

there is a contradiction between the air quality improvements required and the technology deliverables when setting vehicle emissions standards. What may be achievable with technology, might not result in the associated air quality expectations without drastic and unrealistic cost measures being enforced.

In the United States, cars, trucks, and off-highway vehicles are currently estimated to be responsible for about 40 percent to 50 percent of the hydrocarbons (HC) or volatile organic compounds (VOC), 50 percent of the NO_X, and 80 percent to 90 percent of the carbon monoxide (CO) emissions in urban areas [4]. The relative contributions in Europe and Japan are similar. In the United States it is estimated that half the total CO and HC emissions from on-highway sources come from the worst 10 percent of vehicles on the road.

In 1996 the Institution of Mechanical Engineers (United Kingdom) reported to the House of Lords' select committee studying zero emissions for road transport that little progress has been made toward an alternative to the internal combustion engine [5]. The committee was studying the state of technology and future prospects. Barriers to low-emissions technology largely rest on cost especially for smaller engines, and logistics and range for alternative fuel vehicles as with natural gas. The institution does not foresee more than a limited niche application for electric vehicles in the short term. It feels that the hybrid unit using an IC engine to provide battery charging and additional power is more likely. It does not believe, however, that the United Kingdom can reduce NO_X emissions without increasing fuel consumption.

It has to be stated that the use of alternative energy strategies in the automobile market has not maintained pace with the advanced technology associated with gasoline and diesel fueled cars. The associated high energy content of these fuels, coupled with safe and efficient storage methods and an established infrastructure, has yet to be bettered by any alternative. Also, there are few indications which point to any running cost savings. As with any new technology, car makers are probably nervous about funding new vehicle infrastructure over which they have little control.

Regardless of what alternatives are developed, it is important to consider fuel costs and the implications of the total energy chain. Electric vehicles provide the obvious answer to meeting zero emission targets from cars. However, the Californian Air Resources Board (CARB) has bowed to pressure from car companies to delay the introduction of legislation by five years to 2003. Then 10 percent of vehicles sold in California must produce zero emissions. Also, with electric vehicles, the energy provided has to be generated at some source. If a hydrocarbon fuel-burning power-generating station is that source, the emissions are merely moved from polluted urban areas to less sensitive regions. However, pollution knows no boundaries and unless nuclear power stations are the initial generators of electric power (which itself is another emotive issue) the problem is merely swept under the carpet. For instance, studies have been carried out which show that if electricity is produced in Germany, the direct-injection diesel-powered car creates less CO_2 than an electric one and in Norway the opposite is true [6].

10.3 HIGH-SPEED DI DIESEL DEVELOPMENTS

The following section presents an overview of high-speed, direct-injection (DI) diesel engine developments and is based on papers presented by Horrocks [7, 8]. The development of high-speed DI diesel engines is shown in Table 10.1 in chronological order of the year of initial introduction.

The Ford 2.5 DI naturally aspirated diesel engine was introduced in 1984 for the Ford Transit, as the world's first production high-speed DI diesel engine. The engine was a four-cylinder, in-line unit, with bore and stroke of 93.67 mm and 90.54 mm, respectively, giving a stroke-to-bore ratio of 0.97 and a capacity of 2.496 liters. Fuel was injected by Lucas DPS or Bosch VE rotary distributor fuel injection pumps through 17/21-mm Lucas or Bosch injectors. These were inclined at 23 degrees from the vertical and the nozzle was located 9.5 mm from the bore axis, giving a 10 percent offset. The bowl in piston was a conventional straight-sided toroidal design giving a 19:1 compression ratio. In-cylinder swirl, to support combustion, was generated by a helical intake port. The nominal power and torque was 52 kW at 4000 rpm and 145 Nm at 2700 rpm. In combination with revised drive ratios, this new engine gave improvements in vehicle fuel consumption of 20 percent to 24 percent for the Ford Transit [9]. This was a major milestone in the development of small high-speed diesels, which demonstrated that DI technology could be used for light-duty vehicles. It was the first of a new concept of fuel-efficient prime movers for cars and light trucks. The introduction of high-speed DI diesels from Isuzu, Iveco, Fiat, Perkins, Audi, Land Rover, VW, and Mercedes-Benz followed. Most recently an in-line, four-cylinder, 2.0-liter engine from Opel/Vauxhall and a new 2.5-liter, V-6 engine from Audi have been released with four valves per cylinder.

The Isuzu "J" series engines replaced the earlier "C" series IDI engines, with 2.5 and 2.8-liter capacities, in 1985. The 2.5-liter engine was very similar to the Ford unit, but had a square combustion chamber in the piston with a small squish lip. The 2.5 liter produced 50 kW at 3600 rpm and 152 Nm at 2000 rpm, the larger version giving 57 kW and 172 Nm [10].

The Iveco DI diesel engine 8140.21 was developed from the 8100 engine family and manufactured in volume production from 1985 for light-duty truck application. The 2.45-liter turbocharged engine produced 68 kW at 3800 rpm and 215 Nm at 2200 rpm. The stroke was also increased from 90 to 92 mm, which increased the capacity to 2.5 liters. This enabled an increase in power and torque to 76 kW and 225 Nm, respectively. This engine was subsequently used in passenger cars [11, 12].

Fiat is credited with producing the world's first DI diesel for passenger cars with the launch of the 1.9-liter TCI for the Croma in mid-1988. This engine delivered 68 kW at 4200 rpm and 182 Nm at 2500 rpm [13]. In 1991, this engine was upgraded by installing an Allied Signal variable geometry turbocharger (VNT), and the power was increased to 69 kW and torque to 200 Nm at the lower speed of 2000 rpm. Although constant-speed fuel consumption was unaffected, mixed driving was reported to be 7 percent to 8 percent lower [14].

The Perkins Prima 2.0-liter DI diesel, in naturally aspirated form, was available in the Maestro van in late-1986, but it was not introduced in the Maestro car until 1990. A turbocharged and intercooled version was installed in the Montego at the beginning of 1989. This engine produced 60 kW at 4500 rpm and set a new standard for fuel economy, for this size of car, of 5.8 liters/100 km on the European Urban Drive Cycle [15].

Audi introduced its new five-cylinder turbo diesel engine with second-generation direct injection at the 1989 Frankfurt Motor Show for the Audi 100. This engine featured pistons with a reentrant combustion bowl incorporating a Mexican Sombrero raised center portion, five-hole, two-spring injectors, electronically controlled Bosch VP-34 fuel injection pump, and an electronic engine management system, which provided improved driveability. Installation features for further refinement included electronically controlled engine mounts with variable damping and a fully enclosed engine compartment, with an automatically operated blind for the intercooler to reduce external noise at idle. Producing 88 kW at 4250 rpm and 265 Nm at 2250 rpm, the Audi 100 TDi established a high standard for DI passenger car diesels [16, 17]. Peak power was subsequently reduced to 85 kW for the next Audi 100. Land Rover introduced their 2.5-liter, in-line, four-cylinder,

82 kW DI TCI for the Discovery at the end of 1989 [18]. Following this, more recently in January 1995, Rover launched a 2.0-liter passenger car DI for the 620 models. This engine, known as the 'L' series, develops a maximum power of 77 kW at 4200 rpm and 210 Nm at 2000 rpm. It has a conventional aluminium cylinder head with two valves per cylinder [19].

Ford introduced a turbocharged version of the 2.5 DI diesel for the Transit in 1992 with the Lucas EPIC electronic fuel injection pump and engine management system. This was the first European light truck to use drive-by-wire controls [20].

In 1992 VW-Audi installed a 1.9-liter TDI Diesel in the Audi 80. The engine had a rating of 66 kW at 4000 rpm and a torque of 182 Nm at 2300 rpm. This engine used the Bosch VP34 electronic rotary distributor fuel injection pump, with the Bosch MSA6 electronic controller [21]. In the autumn of 1993, this engine was introduced in the VW Golf and Passat with revised turbocharger and fuel injection equipment to give an increased peak torque of 202 Nm and exhaust emissions to European Stage 2 standards (see Section 10.5.1).

In 1995 VW extended its DI diesel range by offering the TDI in the Golf convertible and introducing a naturally aspirated version of the 1.9 DI in the Golf, Golf Estate, and Vento. This engine produces 47 kW at 4200 rpm and 124 Nm at 2000 rpm and has a minimum full-load brake specific fuel consumption (BSFC) of 222 g/kWh. The company also reported plans to introduce an 81 kW version of the 1.9-liter engine to give GTI-type performance with good fuel economy [22]. This engine is now available for the Golf, Vento, and Passat. With a variable geometry turbocharger, it provides 235 Nm at 1900 rpm and 81 kW at 4150 rpm [23].

Mercedes-Benz was a recent entrant into DI passenger cars with the introduction of its 2.9-liter turbocharged intercooled DI diesel for the E-Class in 1995. This five-cylinder engine produces 95 kW at 4000 rpm and 300 Nm from 1900 to 2400 rpm. Like all the newer engines, it has a reentrant combustion bowl and uses the Bosch VP37 fuel injection pump with two-spring injectors. It is interesting to note that Mercedes-Benz has reverted back to two vertical valves per cylinder for the DI engine [24, 25].

Opel/Vauxhall has lead the industry by the introduction of the Ecotec 16-valve, 2.0-liter DI diesel engine in the autumn of 1996. This engine has been introduced in turbocharged form, producing 60 kW for the Vectra. The four vertical valves per cylinder are operated by a single camshaft, which is driven with a simplex chain from the injection pump. Bridge pieces actuate each pair of intake and exhaust valves. Bosch's new radial piston, high-pressure distributor injection pump, designated VP44, is used to inject fuel through two-spring, five-hole nozzles. A 74 kW TCI version and a 2.2-liter 88 kW TCI derivative are to follow [26].

Audi has recently announced the development of the first four-valve, V-6 TDI passenger-car diesel. This is a 90 degree V-6 with 30 degree offset pins and a single counterrotating balance shaft. The four vertical valves, two intake and two exhaust, are laid out obliquely to the engine axis around the central vertical injector. One tangential and a helical port are used to generate in-cylinder swirl. This engine uses a six-cylinder version of the Bosch VP44 radial piston distributor pump and an Allied Signal variable geometry turbocharger. The engine produces 110 kW at 4000 rpm and 310 Nm between 1500 to 3200 rpm [27, 28].

Putting aside fuel injection systems and lean-NO_x catalysts, the next key technological development for high-speed DI diesel design is undoubtedly the greater use of the four-valve and central vertical injector configuration to provide higher specific power combined with lower emissions and improved fuel economy.

10.4 OVERVIEW OF EMISSIONS FROM CI ENGINES

The environmental impact of passenger cars is a major problem that is being addressed with increasing urgency worldwide. In particular the contribution to atmospheric pollution from the motor car emphasises the need to reduce vehicle emissions. The second report of the Quality of Urban Air Review Group (QUARG) (1993) focuses on emissions from diesel vehicles in particular. Legislative developments envisaged up to and beyond the year 2000 have been assessed by Dunne [35]. The passenger car of the future must aim to have an exhaust virtually free of toxic substances. It must also have significantly improved overall fuel economy to limit the production of carbon dioxide. Carbon dioxide production is becoming increasingly important owing to its contribution to global warming. Vehicles that burn fossil fuels produce a net increase in atmospheric carbon dioxide.

Undoubtedly, improvements can still be made to engine design and exhaust gas after treatment. Horrocks sets out some of the challenges for automotive diesel engine designers seeking to comply with future emissions legislation [7, 8, 36, 37]. The requirements set out in the U.S. federal emissions legislation and the strong position of experimental DI diesel engines is highlighted by Burgler [38].

This section will consider the origin, measurement, and control strategies associated with diesel engine legislated emissions species. Although carbon monoxide (CO) emissions are regulated, they will not be considered here, as the diesel engine combustion process by definition inhibits the production of CO. As for CO_2 the only truly practical way is to reduce fuel consumption or to use fuels with a lower carbon content.

There are many classical texts, Heywood being just one example [39], which will explain the detailed processes that are taking place with regard to diesel engine combustion. These should be referred to for more detailed explanation of emission formation.

10.4.1 Oxides of Nitrogen

10.4.1.1 Production of NO_x

Nitrogen oxides, nitric oxide (NO) and nitrogen dioxide (NO_2), are formed during the combustion of oil by two mechanisms; high-temperature thermal fixation of molecular oxygen (O_2) and nitrogen (N_2) present in the combustion air and, second, reaction of atmospheric oxygen with nitrogen-containing compounds in the fuel. Both mechanisms result in NO because the residence time (the amount of time in the combustion cycle spent at high-temperature) in an internal combustion engine is too short for oxidation of NO to NO_2, even though NO_2 is thermodynamically favored at lower temperatures. However, NO does oxidize in the atmosphere to NO_2, which is a primary participant in the formation of photochemical smog. It is usual practice for NO and NO_2 to be grouped together and referred to as NO_x emissions.

The formation of NO_x is dependent on temperature, local oxygen concentration, and residence time. It is produced in a series of reactions, such as those defined by the Zeldovich mechanism [39]:

$$O + N_2 \leftrightarrow NO + N$$

$$N + O_2 \leftrightarrow NO + O$$

$$N + OH \leftrightarrow NO + H$$

The concentration of NO_x in engine exhausts is greater than those predicted by equilibrium thermodynamics. The elimination takes longer than the formation and the process is too rapid for equilibrium to be reached.

The formation of NO and other oxides of nitrogen increases rapidly with flame temperature. This may suggest that slightly rich mixtures with higher temperatures would produce the highest concentration of NO_x. This is not quite true, since NO_x formation is also influenced by flame speed. Lean mixtures have lower flame speeds, which gives more time for the NO_x to form.

The primary combustion mechanism in a diesel engine is the diffusion process, that is, that part which is controlled by the rate at which fuel and air are mixed. This is preceded by the so called premixed process in which a small number of fuel-air nuclei combust spontaneously at a rate controlled by reaction kinetics. A typical heat release diagram for a DI diesel engine is shown in Figure 10.2. In turbocharged engines the premixed phase is generally much less important than in naturally aspirated engines.

As fuel is injected into the charge air, it entrains air and diffuses, giving a widely varying equivalence ratio. At the core of the spray there is liquid fuel in droplet form; at the other extreme is a region with an equivalence ratio of less than unity. The flame is usually said to exist at an intermediate region where the equivalence ratio is unity. In practice the reaction will extend to either side of this zone into rich and lean areas. This means that NO_x is produced at varying rates according to the local equivalence ratio. There will be an averaging effect.

NO_x formation occurs predominantly in the lean flame region during the premixed combustion phase and particulate matter is formed mainly in the fuel-rich regions during the diffusion burning period. Actions to reduce NO_x formation, such as reducing the proportion of premixed combustion, will therefore have the

Fig. 10.2.

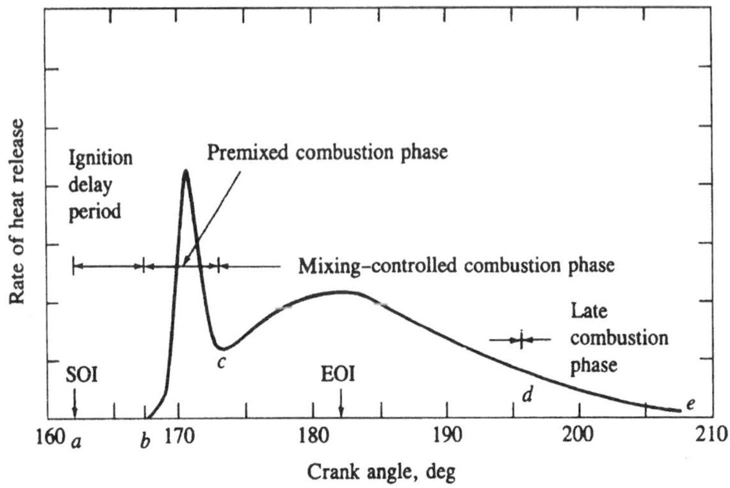

Typical heat release diagram for a DI diesel engine.

effect of increasing the diffusion burning phase and, hence, have the tendency to increase particulate formation and vice versa. This phenomenon is known as the NO_X-PM trade-off. However, combustion system and fuel injection development can by careful calibration and control improve the fundamental trade-off with minimum loss of fuel economy.

10.4.1.2 Points in Relation to NO_X Production

1. Lower engine speeds generally produce a higher concentration of NO_X for a given power due to the longer time available for the reactions.
2. NO_X reduction is generally attributed to less violent burning experienced with shorter ignition delay (ID) periods associated with high cetane number fuels.
3. Higher compression ratios reduce ID, which reduces the fraction of the fuel burned in the premixed region and allows more injection timing retardation to control NO_X, while maintaining a good NO_X-BSFC trade-off. Unfortunately, injection timing retardation is in conflict with good fuel economy and low particulate emissions.
4. NO_X production is also reduced by increasing the rate of injection. This reduces the duration of the diffusion controlled combustion by more rapid mixing.

Two of the most significant factors affecting NO_X emissions are injection timing and charge air temperature. The advantage of charge air cooling to increase power output for the same thermal loading of turbocharged engines is well known. However, intercooling offers significant reductions in NO_X emissions for the same

Fig. 10.3.

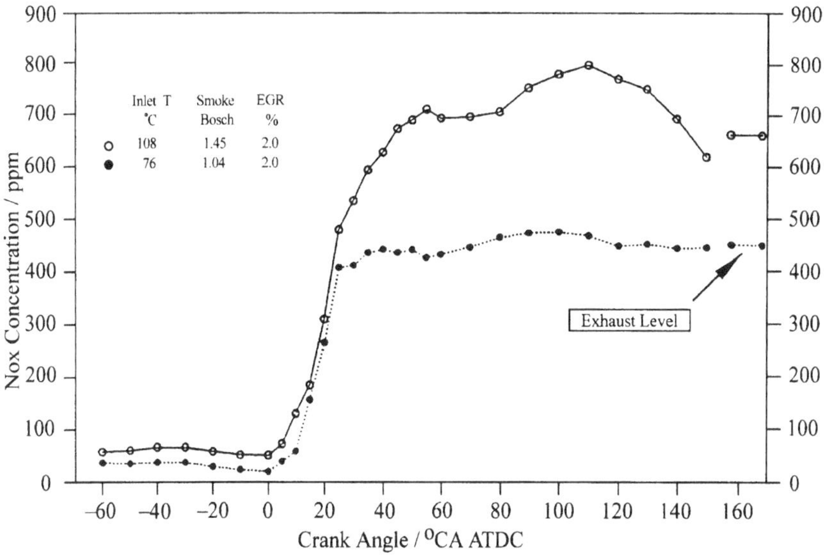

In-cylinder and exhaust NO_x concentrations for a Ford 2.5 DI TC diesel engine at 2000 rpm and 8 bar BMEP, with and without intercooling (taken from R.W. Horrocks, Figure 5, "Light duty diesels—an update on the emissions challenge," IMechE Seminar on Worldwide Engine Emissions Standards and How to Meet Them, London, 1992).

power output for turbocharged engines. This is highlighted in Figure 10.3, which shows in-cylinder NO_x concentrations and exhaust NO_x and smoke readings for a turbocharged 2.5 DI diesel engine with and without intercooling. Intercooling will also improve engine efficiency.

An effective method of reducing NO_x levels is to return some of the exhaust gas back into the inlet charge; this is called exhaust gas recirculation (EGR). This lowers both flame temperature and speed due to the dilution of the oxygen concentration, and the net effect is a reduction in NO_x. EGR also lowers the efficiency of the combustion process. Since the effective equivalence ratio is raised by EGR, it is not used at high loads, as this would result in an equivalence ratio of above the usual limit of 0.9. Beyond this point emissions of hydrocarbons and smoke will increase rapidly, as will fuel consumption. In any case the concentration of NO_x at high loads is generally less due to the faster flame propagation. EGR is covered in more detail in Section 10.6.2. A typical NO_x contour diagram for a Ford prototype 1.8-liter eight-valve DI engine, operating with EGR, is shown in Figure 10.4.

10.4.1.3 Measurement of NO_x

NO_x is measured using the chemiluminescent principle. If nitric oxide (NO) is mixed with ozone (O_3), a reaction takes place forming nitrogen dioxide (NO_2) and

Fig. 10.4.

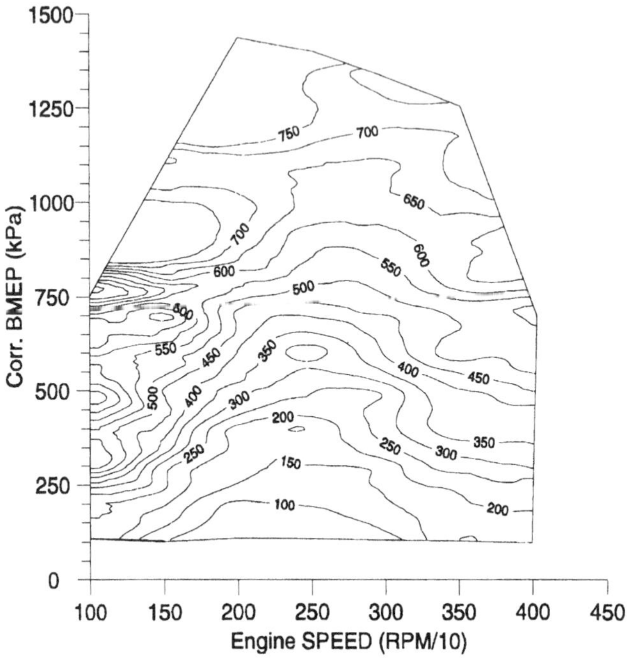

NO_x contours (ppm) taken from a Ford prototype 1.8-liter DI TC diesel engine.

oxygen (O_2). The NO_2 molecules produced are initially in an electrically excited state (NO_2^*). These revert immediately to the nonexcited state by emitting photons.

$$NO + O_3 \leftrightarrow NO_2^* + O_2$$

$$NO_2^* \leftrightarrow NO_2 + hv$$

where

h = Planck's constant
v = frequency, Hz

This light emission in the reaction chamber is directly proportional to the concentration of NO. The light is measured using a photodiode.

To measure NO_x the same method is used, although before entering the reaction chamber, the sample is routed through a converter where the NO_2 component is reduced to NO in the presence of a vitreous carbon catalyst at 400°C:

$$NO_2 \rightarrow 2NO + O_2$$

The analyzer output is thus a sum of the NO in the original sample plus the NO derived from the conversion of the NO_2. Any CO_2 present in the chamber will interfere with the reading by colliding with the (NO_2^*) molecules before they have given up their photons. This effect is typically less than 2 percent full scale for up to 16 percent CO_2.

10.4.2 Hydrocarbons

10.4.2.1 Production of Hydrocarbons (and other Organic Compounds)

This group of pollutants is perhaps the most difficult to characterize, as it contains so many different compounds whose abundance and environmental significance vary widely. Almost 400 organic compounds have been identified in vehicle exhaust [40, 41]. Hydrocarbons as a group of compounds range, in toxicological terms, from essentially nontoxic compounds such as methane to highly toxic compounds such as 4-hydroxybiphenyl. The role of hydrocarbons in atmospheric photochemistry is very important. The rate of formation of photochemical oxidants is closely related to the rate at which hydrocarbons are scavenged by hydroxyl radicals, since it is this scavenging that produces organic peroxy radicals, which subsequently produce ozone and other oxidants through the oxidation of NO to NO_2.

Hydrocarbon emissions arise on the one hand from unburned fuel in fuel-air mixtures that are too lean to combust, and on the other from fuel that overpenetrates and wets the cylinder wall during the ignition delay period. Both of these processes are heavily dependent on the length of the ignition delay period. In turn, the ignition quality of the diesel fuel used, defined by its cetane number, will have a significant effect on the ignition delay period and hence the HC emissions. These effects are amplified with the introduction of EGR. The HC emissions contours for a Ford prototype 1.8-liter DI turbocharged diesel engine, operating with EGR, are shown in Figure 10.5.

The five main sources of hydrocarbon emission from the diesel combustion process are generally identified as [39]:

1. Overmixing of the fuel and air beyond the lean limit of combustion.
2. Undermixing to an air-fuel ratio too rich for complete combustion.
3. Overpenetration of the fuel spray resulting in wall wetting.
4. Bulk quenching of the reaction due to the expansion of the mixture.
5. Fuel leaking from the injector sac and nozzle holes.

It has been established that overmixing plays a part in hydrocarbon emissions across a range of engine operating conditions and is most significant at idling and light loads. The mixture here is too lean to burn properly. Emissions from this source can be reduced by reducing the ignition delay. The delay reduces naturally as the load increases, resulting in cleaner combustion. Advancing injection timing raises temperature and pressure from an initially retarded condition, reducing the HC emissions.

The second distinct cause of excessive HC emissions is undermixing or overfueling. If there is insufficient air present, some of the fuel will not burn fully.

Fig. 10.5.

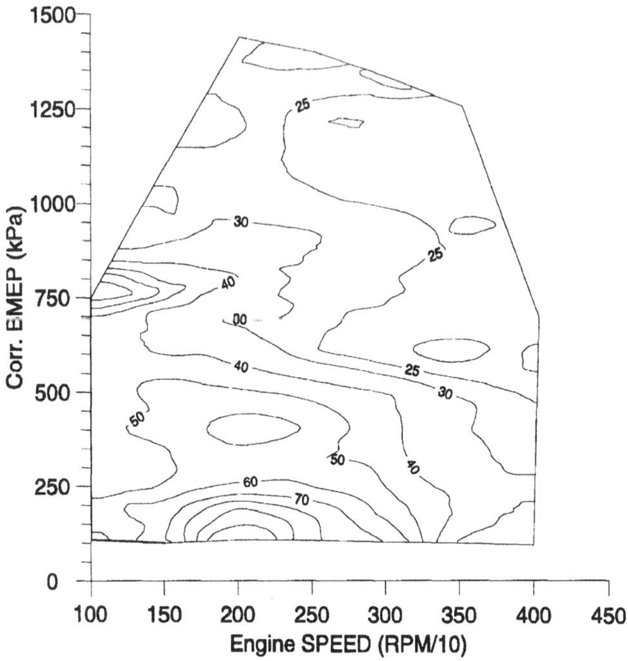

HC contours (ppm C^3) taken from a Ford prototype 1.8-liter DI TC diesel engine.

Black smoke is also produced. The transition between acceptable combustion and overfueling is quite pronounced. As the equivalence ratio rises above 0.9, the HC emissions rise rapidly. This is avoided by careful calibration of the fuel injection equipment. Overfueling can also occur if there is too much EGR, since the oxygen concentration will fall, raising the equivalence ratio. EGR is discussed in further detail in Section 10.6.2.

Third, in smaller-bore engines, overpenetration also has a significant influence on the hydrocarbon emissions.

Fourth, bulk quenching only contributes under very specific engine operating conditions.

Finally, fuel can be retained in the injector nozzle sac after injection. This fuel enters the combustion chamber later in the cycle and is not fully burned. Careful design of the fuel system minimizes the problem. The contribution from injector leakage is insignificant for modern injectors with valve-covered orifice (VCO) nozzles.

10.4.2.2 Measurement of Hydrocarbons

The most common method of measuring hydrocarbons is the flame ionization detector (FID). The analyzer makes use of the phenomenon that when a hydrocarbon is burned significant quantities of ions are formed. The number of ions produced is nearly proportional to the number of carbon atoms burned in HC form. This effect is related to a nonequilibrium chemi-ionization effect and not to the thermal (equilibrium) ionization. The latter results in negligible ionization at normal flame temperatures. For flames not involving hydrocarbons (such as hydrogen or carbon monoxide) this is the only mechanism and ionization levels are extremely low.

To measure this effect the sample gas is introduced into a hydrogen flame burning between two electrode plates with a direct current voltage applied between them. A small ionic current flows between the plates proportional to the hydrocarbon content of the flame. This current is measured and linearized to give a measure of the number of HC molecules burned. Care must be taken as some analyzers give a reading in ppm of Methane (CH_4), while others display ppm of Propane (C_3H_8). While the conversion is simple, the units need to be expressed to avoid confusion.

The sample gas conditioning and flow rate regulation in most commercial analyzers is very thorough, as increasing the flow rate of HCs will increase the FID reading. To achieve an accurate calibration the sample flow rate must be carefully controlled. Similarly, the analyzer must be maintained at a sufficiently high temperature to prevent hydrocarbons from condensing ("dropping out"). If they do, they will be released later, causing a smearing effect to be superimposed on and possibly swamping the output. Good filtration helps avoid contamination of the capillaries and chambers. All of these considerations enhance accuracy and reduce maintenance but, unfortunately, at the expense of transient response, which typically may be around 3 s delay time plus 2.5 s rise time. This can be problematic during transient tests.

There are other methods of measuring hydrocarbons including infrared absorption. This technique makes use of the fact that each species of hydrocarbon will absorb a slightly different wavelength of light. This allows the concentration of each major species of hydrocarbon to be determined continuously. However, the sum of the major components (separated by finding the Fourier transform of the signal) is always less than an FID reading due to the varying proportions of the minor species too numerous to measure transiently. For this reason commercial systems include a conventional FID to give a measure of total HCs.

10.4.3 Particulate Matter

10.4.3.1 Production of Particulates

Of all the emissions from the diesel engine, particulates are perhaps the most problematic. They have been identified as an irritant and possible carcinogen in a number of environmental health studies and are, therefore, subjected to very tight legislative limits. Currently, global emissions legislation considers the total mass of PM being emitted by a vehicle. Although a good starting point, this approach can give a false picture as two engines producing the same total mass of PM may, in fact, have completely different emission constituents and, hence, a markedly different reading on air quality.

Typical composition of diesel exhaust PM

Carbon, 31 percent
Unknown, 8 percent
Sulphate and water, 14 percent
Unburned fuel, 7 percent
Unburned oil, 40 percent

There are a number of concerns over possible adverse health effects of diesel particles. Potentially the most serious arises from published research in which mortality and morbidity have been shown to correlate with the concentrations of fine particles equal to or smaller than 2.5 microns in diameter (termed PM2.5). It is likely that these extremely small particulate sizes are the most detrimental to health as they can travel deep into the respiratory system, can become embedded more easily, and have a larger surface area-to-volume ratio so that they can carry more potentially carcinogenic chemicals.

In the United States the Environmental Protection Agency (EPA) sets National Ambient Air Quality Standards (NAAQS) for PM [42]. The EPA defines PM as either primary particles (direct particle emissions) or secondary particle precursors (e.g., NO_x, VOCs) in relation to all combustion sources, including motor vehicles and stationary applications. Single or multiple strategically positioned monitors collect samples of ambient air and these measurements are statistically adjusted to provide the quantity of PM (in g/m^3) collected over an annual period, and also a 24-hour period. The standards for annual primary PM2.5 and PM10s are set at 15 g/m^3 and 50 g/m^3 respectively [42]. The PM2.5 annual figure is intended to increase the emphasis for control of these small particulates and diesel vehicles could be especially targeted.

With regard to vehicle emissions testing, particulates are defined as any matter collected on a filter paper at a temperature of 325 K. The bulk of this matter is either unburned hydrocarbons or soot. The soot is also visible as smoke. Hence, measures aimed at reducing smoke will generally reduce particulates. A typical particulate emissions contour map measured from a Ford prototype 1.8-liter DI turbocharged diesel engine, operating with EGR, is shown in Figure 10.6.

Further particulate matter is added by the oxidation of sulphur in the fuel to form SO_2 (sulphur dioxide) and sulphates, which bind with water on the carbon nuclei. These particles can make up as much as 20 percent of the total PM produced. The use of low-sulphur fuel reduces this contribution and current European fuel standards specify a maximum sulphur content of 0.05 percent by weight.

Fig. 10.6.

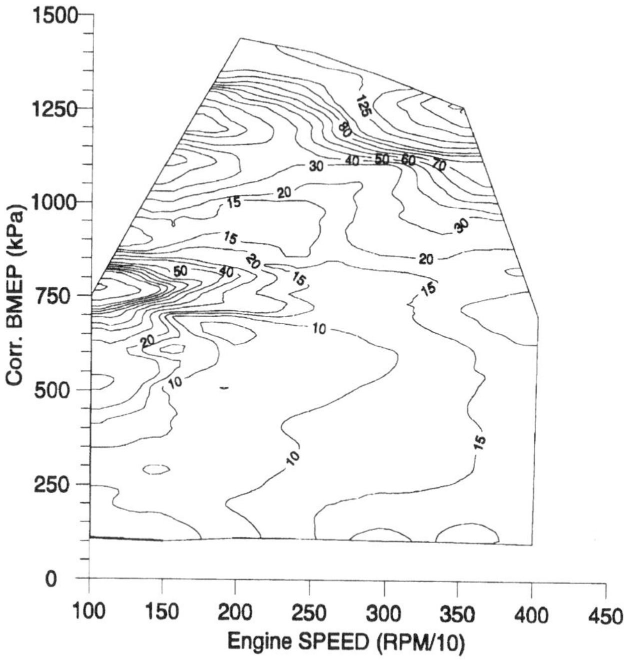

Particulate contours taken from a Ford prototype 1.8-liter DI TC diesel engine (g/kWh by Greeves and Wang).

Another major source of particulates is lubricating oil, which finds its way into the combustion chamber and is partially burned. The proportion of the PM attributable to this source is primarily dependent on the effectiveness of the piston ring pack and to a lesser degree on the valve stem sealing. Also important is the minimization of oil entrainment in the blow-by gases, which are recycled into the inlet, and the elimination of oil leaks from the turbocharger bearings. Particulates from the lubricating oil can make up as much as 40 percent of the total.

10.4.3.2 Measurement of Particulates

Particulates are measured by weighing a filter paper before and after a known volume of exhaust gas has been passed through. This is a slow process and not suitable for transient testing.

A more rapid technique is the Tapered Element Oscillating Micro-balance (TEOM) detector. Here a very small filter paper is fitted to the narrow end of a tapered tube in a 50°C constant-temperature oven. The tube is held at the wider end and free at the filter end. The flow of particulates is calculated by monitoring the change in mass of the assembly. To measure the mass the tube is excited to vibrate at its natural frequency as a cantilever. As the mass of the filter increases, the natural frequency decreases according to the relationship

$$f = \sqrt{\frac{k}{m}}$$

where

f = frequency (rad/sec)
k = spring rate (N/m)
m = mass (kg)

This process can give a figure for particulate mass flow updated at greater than 2 Hz.

10.4.3.3 Negative Particulate Mass Flows

Using the TEOM the mass of the filter paper is assessed continuously. When moving from a period of high load to one of low load, the rate of evaporation of volatiles may be higher than the rate of deposition of new particulate matter. This will result in the paper becoming lighter, shown as a negative mass flow rate. While this is clearly not an accurate reading, it is interesting since information about the relative proportions of the various components can be gathered. Work has been done using an FID before and after the filter paper to track the hydrocarbons in the flow. While not performed routinely, this may be a useful way of quantifying the effect. Once the volatile fraction has been driven off, the figure will become steady once again. It is debatable which method is nearer to the "real" measure of particulates, but since the legislation prescribes the conventional static filter method, it is this figure which must be optimized.

10.4.3.4 Greeves and Wang Correlation

Since the particulate mass consists mainly of unburned hydrocarbons and soot, it is possible to generate a correlation between HC concentrations, Bosch smoke, and particulate concentration. For small, high-speed, turbocharged diesel engines the following relationship has been developed (all figures in mg/m^3) [43]:

$$\text{PM} = 1.024\ \text{Soot} + 0.277\ \text{HC}$$

Concentration of soot (solid content) in the exhaust may be determined using a correlation against the Bosch smoke number and shown in Figure 10.7 [44].

10.4.4 Smoke

Although not a regulated emission for type approval, the measurement of smoke is a clear indication of the efficiency of the combustion process. A typical smoke contour man measured from a Ford prototype 1.8-liter DI turbocharged diesel engine, operating with EGR, is shown in Figure 10.8.

Fig. 10.7.

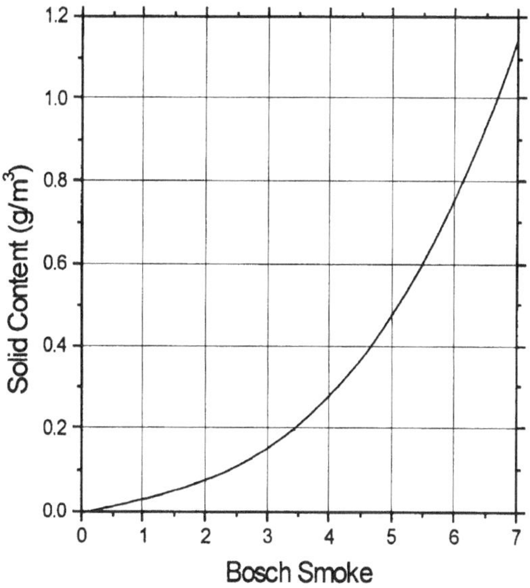

Fosberry and Gee correlation between Bosch number and solid content.

Fig. 10.8.

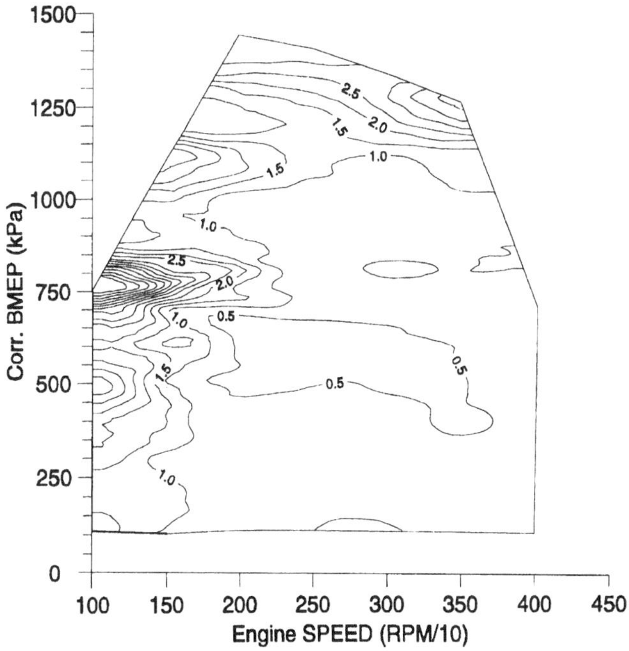

Bosch smoke taken from a Ford prototype 1.8-liter DI TC diesel engine.

10.4.4.1 Production of Smoke

The black smoke produced by a diesel is composed of carbon particles released by the thermal cracking of the large hydrocarbon fuel molecules. This process occurs on the rich side of the flame front during the diffusion combustion phase. The carbon agglomerates into particles visible as smoke in the exhaust. Some of the smoke particles are oxidized when they reach the lean side of the reaction. More are oxidized during the expansion stroke, after the diffusion flame has burned out. Smoke production is increased at elevated temperatures and higher equivalence ratios. Smoke production can be reduced by reducing the time taken by diffusion combustion. This can be accomplished by promoting more rapid mixing of fuel and air, that is, increasing swirl or increasing the injection rate and using a finer spray. Advancing injection allows more time during the expansion stroke for oxidation, reducing smoke.

Smoke production increases with load. This is due to the richer overall combustion and the increased diffusion combustion time, increased temperatures, and reduced oxidation during the expansion stroke caused by the lengthened diffusion phase and reduced oxygen concentration. This effect is generally the limiting factor on diesel engine brake mean effective pressure (BMEP). Improvements to engine fuel and air systems which reduce the diffusion combustion time can raise the allowable BMEP due to the reduced smoke potential.

Other factors that cause excessive smoke are common to those already mentioned which affect HC production due to an increased equivalence ratio. They include poor control of fueling during transients and excessive amounts of EGR. The design of the EGR map is generally limited by the trade-off between NO_x reductions and smoke and PM increases. Here again, technology that enhances mixture preparation can allow more EGR and, hence, less NO_x before smoke becomes a problem. This has resulted in a general move toward multivalve engines, higher injection pressures, and variable swirl mechanisms.

10.4.4.2 Measurement of Smoke

Smoke may be measured by opacity or reflectivity from a filter paper. The Celesco and Hartridge meters are optical instruments measuring opacity and the Bosch meter is based on the reflectivity of light.

The optical instruments determine opacity and are particularly suited to the assessment of smoke produced during transients. With the Bosch system a known volume of exhaust is drawn through a filter paper and its reflectivity assessed. A Bosch value of 0 is given to a perfectly clean paper; Bosch 10 indicates a paper which reflects no light at all. This method is not suitable for continuous monitoring since the paper needs to be indexed between readings, and the measurement itself takes a finite time.

A correlation can be developed linking opacity to Bosch number. An alternative technique is described by Kittelson and Collings [45] where the electrical charge associated with the particles is measured.

10.5 CURRENT AND PROJECTED GLOBAL EMISSIONS LEGISLATIVE REQUIREMENTS

The mandatory emission requirements for vehicles vary depending on the country (or state) of sale. The requirements are typically based around data from a chassis dynamometer test over a set schedule. The test schedules for Europe and the United States are shown diagramatically in Figure 10.9. The following information details the current and projected emissions standards in Europe, the United States, and also in California.

10.5.1 European Emission Regulations

10.5.1.1 Diesel Passenger Cars

The test procedure is the urban ECE-15 cycle followed by the high-speed Extra Urban Drive Cycle (EUDC), which has a maximum speed of 120 km/h. The cycle is designated as ECE-15 plus EUDC (Figure 10.9). The limit values are shown in

Fig. 10.9.

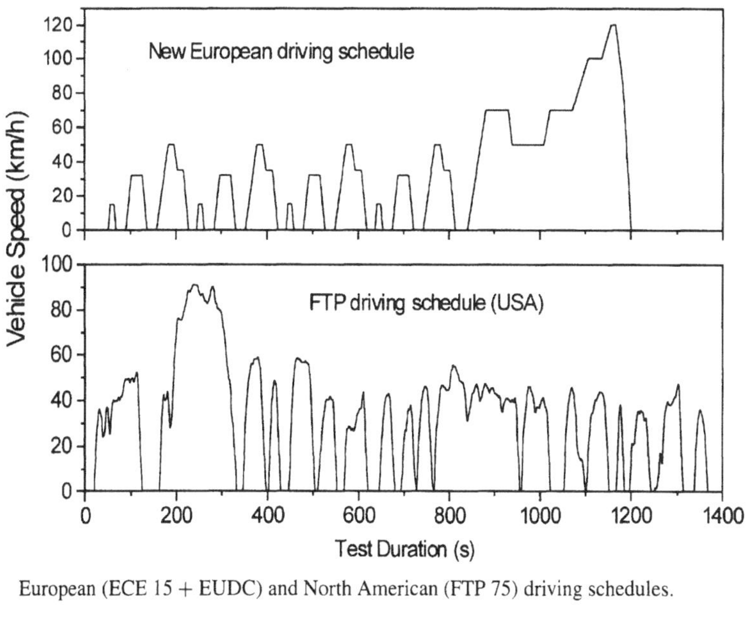

European (ECE 15 + EUDC) and North American (FTP 75) driving schedules.

Table 10.2

European Diesel Passenger-Car Emission Standards

	CO (g/km)	NO_x (g/km)	HC + NO_x (g/km)	PM (g/km)	Comments
Stage 1 (DI)	3.81		1.36	0.20	In force (1)
(IDI)	2.72		0.97	0.14	
Stage 2 (DI)	1.0		0.90	0.10	In force (2)
(IDI)	1.0		0.70	0.08	
Stage 3	0.64	0.5	0.56	0.05	(3)
Stage 4	0.5	0.25	0.30	0.025	(4)

(1) The Consolidated Directive 91/441/EEC.

Implemented on new models from July 1, 1992 and on new registrations from December 31, 1992. Included a derogation of 40 percent for all pollutants until 1994 and conformity of production allowance of 16 percent for gaseous emissions and 29 percent for particulates.

(2) In force.

Derogation on HC plus NO_x and particulates to continue for a further three years; however, separate limit values for conformity of production are deleted. This means that every vehicle produced must achieve the limit values for the "lifetime" of the vehicle. For the engine and vehicle manufacturer the lack of a production margin creates an extra task and will effectively lower the in-house development targets in order to achieve a significant margin for the production volume. In reality, these standards are measured against production limits and not certification standards. This has effectively meant a reduction of 20 percent on HC plus NO_x and 44 percent on particulates.

(3) Stage 3.

Draft proposals by the EU Commission have been published and will take effect from the year 2000. These standards are also accompanied by a change in the test cycle. As mentioned previously, the first 40 seconds of idle before emission sampling begins is to be deleted; the new cycle is intended to better reflect real-world operation. In consequence, this will also mean that the actual reduction in the numerical values from the current standards do not truly reflect the stringency of the limits. In real terms, the proposed limits for 2000 represent a severity of some 40 percent reduction over the present European Stage 2 standards [35]. Also, the concession on DI diesel engines will cease.

(4) Stage 4.

Emission legislation standards for 2005 are described as "indicative" with a rider to indicate that they will be reviewed by 1998 depending on the availability of the technology needed to achieve them.

There are also plans to specify CO_2 limits from diesel engines by specifying fuel consumption in relation to miles traveled. The figure of 5 liters/100 km has been mooted along with a 40 percent improvement in thermal efficiency by 2005, and 3 liters/100 km by 2010.

Table 10.2, together with draft limits for Stage 3 in 2000 and proposed for Stage 4 in 2005. Stages 3 and 4 standards will be accompanied by a change in the test cycle; the current first 40 seconds of idle before emission sampling commences will be deleted, so that sampling will start at cranking.

10.5.1.2 Light-Duty Trucks (LDT)

EU Council Directive 97/98EC for LDT classes is now in place with the limits shown in Table 10.3.

Table 10.3

European Light-Duty Truck Emissions Standards

	CO (g/km)	HC + NO_x (g/km)	PM (g/km)
LDT1 (up to 3.5t GVM up to 1250 kg ref mass)			
IDI Diesel	1.0	0.7	0.08
DI Diesel	1.0	0.9	0.10
LDT2 (up to 3.5t GVM over 1250 kg up to 1700 kg ref mass)			
IDI Diesel	1.25	1.0	0.10
DI Diesel	1.25	1.3	0.14
LDT3 (up to 3.5t GVM over 1700 kg ref mass)			
IDI Diesel	1.5	1.2	0.17
DI Diesel	1.5	1.6	0.20

Table 10.4

European Heavy-Duty Emissions Standards

	CO (g/kWh)	NO_x (g/kWh)	HC (g/kWh)	PM (g/kWh)
Euro 2 (current)	4.0	7.0	1.1	0.15*
Euro 3 (forecast for 2000)	2.5	5.0	0.7	0.10
Euro 4 (forecast for 2005)	1.0	3.0	0.5	0.10

*Particulates 0.25 g/kWh for engines up to 0.7 liter/cylinder and rated speed over 3000 rpm until October 1997 Homologations/October 98 Mandatory.

10.5.1.3 Heavy-Duty Commercial Vehicles

The current European heavy-duty emissions standards 96/1/EC, known as Euro 2, are shown in Table 10.4, together with forecast levels for Euro 3 and Euro 4, which are expected to be implemented for 2000 and 2005, respectively. Currently the ECE R49 13-mode test procedure is used, but the test cycle is due to be changed for 2000.

10.5.2 U.S. Emissions Regulations

A brief account of the current and projected emissions standards for U.S. passenger cars and light-duty commercial vehicles is in Tables 10.5, 10.6, 10.7, and, 10.8.

10.5.2.1 Diesel Passenger Cars

Table 10.5

Summary of Federal Emission Standards for Diesel Passenger Cars

THC (g/km)	NMHC (g/km)	CO (g/km)	NO_X (g/km)	Particulates (g/km)	Introduction
0.25	0.16 (0.19)	2.1 (2.6)	0.62 (0.78)	0.05 (0.06)	current
0.25	0.047 (0.056)	2.1 (2.6)	0.09 (0.13)	0.025	

Cycle: Federal FTP 75.

Durability: 80,000 km (160,000 km).

THC: Total hydrocarbons.

NMHC: Nonmethane hydrocarbons.

10.5.2.2 Light-Duty Commercial Vehicles

Table 10.6

Light-Duty Commercial Vehicles

Vehicle ref weight GVW (kg)	THC (g/km)	NMHC (g/km)	CO (g/km)	NO_X (g/km)	Particulates (g/km)	Status
0–1700	0.5	0.16 (0.19)	2.1 (2.6)	0.62 (0.78)	0.05 (0.06)	1995–2003*
0–1700	0.5	0.047 (0.056)	2.1 (2.6)	0.09 (0.13)	0.025	2004
1701–2608	0.5	0.20 (0.25)	2.7 (3.4)	(0.61)	0.05 (0.06)	1995–2003*
1701–2608	0.5	0.063 (0.071)	2.7 (3.4)	0.25 (0.31)	0.031	2004

*Phase-in schedule.

Cycle: Federal FTP 75.

Durability: 80,000 km (160,000 km).

THC: Total hydrocarbons.

NMHC: Nonmethane hydrocarbons.

10.5.3 California Emissions Standards

10.5.3.1 Diesel Passenger Cars

Table 10.7

Emission Standards for Passenger Cars

NMOG (g/km)	CO (g/km)	NO_x (g/km)	PM (g/km)	Formaldehyde (g/km)	Introduction
0.08(0.10)	2.11(2.61)	0.25(0.37)	(0.05)	0.009(0.011)	TLEV*
0.05(0.06)	2.11(2.61)	0.12(0.19)	(0.05)	0.09(0.011)	LEV*
0.025(0.034)	1.06(1.30)	0.12(0.19)	(0.25)	0.005(0.007)	ULEV*

*Phase-in schedule.

Cycle: Federal FTP 75.

Durability: 80,000 km (160,000 km).

NMOG: Nonmethane organic gas.

TLEV: Toward low-emission vehicle.

LEV: Low-emission vehicle.

ULEV: Ultralow-emission vehicle.

10.5.3.2 Light-Duty Commercial Vehicles

Table 10.8

Emission Standards for Light-Duty Commercial Vehicles

Vehicle GVW(kg)	NMOG (g/km)	CO (g/km)	NO_x (g/km)	PM (g/km)	Formaldehyde (g/km)	Introduction
0–1700	0.08 (0.10)	2.11 (2.61)	(0.375)	(0.05)	0.009 (0.011)	TLEV*
0–1700	0.05 (0.06)	2.11 (2.61)	(0.188)	(0.05)	0.009 (0.011)	LEV*
0–1700	0.02 (0.03)	1.06 (1.3)	(0.313)	(0.025)	0.005 (0.007)	ULEV*
1701–2608	0.10 (0.12)	2.73 (3.42)	(0.563)	(0.05)	0.011 (0.014)	TLEV*
1701–2608	0.06 (0.08)	2.73 (3.42)	(0.313)	(0.05)	0.011 (0.014)	LEV*
1701–2608	0.03 (0.04)	1.37 (1.74)	(0.313)	(0.025)	0.006 (0.008)	ULEV*

*Phase-in schedule.

Cycle: Federal FTP 75.

Durability: 80,000 km (160,000 km).

NMOG: Nonmethane organic gas.

10.6

ADVANCED EMISSION REDUCTION STRATEGIES FOR THE YEAR 2000 AND BEYOND

In order to respond to more strigent emission legislation throughout the world, vehicle manufacturers in conjunction with engine research organizations are exploring a wide range of approaches to develop advanced "clean technology" options. The focus, particularly in Europe, is toward zero levels of emissions [46]. This is proving to be extremely difficult, especially in the case of the light-duty, high-speed DI diesel engine, which will require highly developed and optimized combustion technologies coupled with exhaust gas aftertreatment. Subsequently, the cost of such an engine is estimated to be 35 percent to 40 percent more than an equivalent petrol engine [46].

10.6.1 Electronic Fuel Injection Systems

The move from IDI to DI technology in the light-duty diesel market, driven largely by the added fuel economy advantage of DI combustion systems, will probably result in even mid-range family vehicles being capable of 4.7 liters/100 km (60 miles per gallon). Increasingly stringent emission legislation means that the high-speed DI engine will have to employ advanced fueling technologies to meet these requirements principally by employing systems which use very high fuel injection pressures with electronic control of injection timing and fueling. These large pressure differences across the injector nozzle are required so that the injected liquid fuel enters the chamber at sufficiently high velocity to:

1. Atomize into small-sized droplets to enable rapid evaporation.
2. Traverse the combustion chamber in the time available and fully utilize the air charge.

Subsequently, by the year 2000 all high-speed DI diesel engines will use electronic fuel injection systems, replacing traditional mechanical systems, which are now approaching the end of their effective life. In what are termed *full authority fuel injection systems*, the benefits include emission reduction, improved driveability, smoother idling, integration with other vehicle systems, self-diagnostics, and performance checks.

The benefits of higher injection pressure on particulates and smoke are shown in Figure 10.10.

10.6.1.1 Electronically Controlled Distributor Pumps

Most fuel injection pumps fitted to current passenger vehicles are of the distributor type. The pump has one high-pressure pumping assembly, which is connected by

Fig. 10.10.

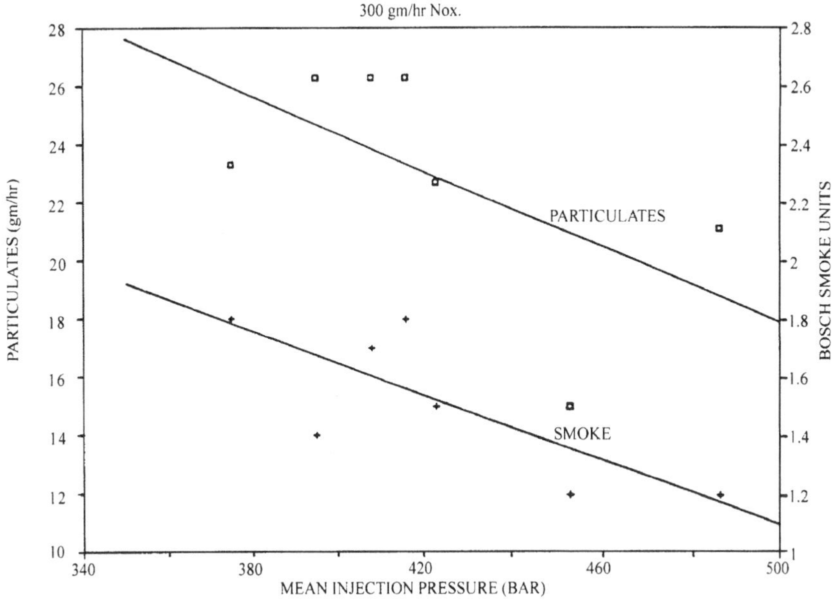

Benefits of higher mean injection pressure on particulates and smoke. Ford 2.5 DI N/A diesel engine at 3000 rpm and 7.2 bar BMEP (taken from R.W. Horrocks, Figure 7, "Light duty diesels—an update on the emissions challenge," IMechE Seminar on Worldwide Engine Emissions Standards and How to Meet Them, London, 1992).

drillings to each of the outlets in turn as the shaft rotates. There are two commonly used arrangements: an axial plunger and cam plate is commonly used in the Bosch fuel pump; radial plungers inside a cam ring are traditionally employed by both Lucas and Stanadyne Diesel Systems.

Mechanical fuel injection pumps are able to vary the fueling and injection timing using mechanical linkages and governors. Electronic pumps are controlled using electrohydraulic devices. Representative of this class of pump is the Lucas EPIC system [47, 48]. Here the cam ring is rotated to vary injection timing using a hydraulic actuator. The working fluid is diesel fuel, the pressure being regulated by a solenoid valve acting on a pulse width modulation (PWM) signal from the electronic control unit (ECU). Fueling is varied by moving the rotor mechanism axially using a second actuator.

This type of pump is now being superseded by the next generation that employs a solenoid-operated spill valve to control injection quantity. Typical of these pumps is the Bosch VP30 (axial plunger) and VP44 (radial plunger). Injection timing is still set by rotating the cam ring. As the cam starts to rise, pumping begins.

After the required stroke has been completed the spill valve opens, allowing the high pressure to decay and the injector needle to reseat, finishing the injection. When the cam has dropped back to its base circle, the spill valve is closed and ready for the next injection.

Future pumps will extend the use of the solenoid-operated spill valve to allow rate shaping and preinjection. Here the spill valve is not closed until the active portion of the cam is reached. This allows a specific section of the cam to be used. If the cam profile has a varying rate of rise, the injection rate can be varied by using the appropriate section of the cam. This technique is demanding, as there are two solenoid events per injection, each with associated errors. Errors in the timing of the valve operation will influence start of injection timing, injection rate, and injected quantity.

In addition to controlling the fuel pump, the ECU supervises a number of other engine systems. These are usually EGR control, inlet throttle control (if fitted), and turbocharger wastegate or variable geometry turbocharger (if fitted). The ECU will generally have other abilities including on-board diagnostics (OBD), cruise control, and networking with other controllers.

The high-pressure fuel lines and injectors are essentially similar to those found on mechanical pump systems, although detail changes make them suitable for the very high pressures common today.

One difficulty sometimes encountered with such systems is the high heat rejection rate to the diesel fuel returning to the tank. Heat is added by conduction from the engine and from the use of diesel fuel as the working fluid in the hydraulic control of the pump. Pump pressures of 10 bar are typical and on throttling back to return line pressure this energy is converted to heat. The same effect occurs when the high-pressure fuel lines are spilt back to low pressure. Although the flow is very small, the pressure is large.

The advantages of the distributor pump are that they are well established in the industry with a great deal of development invested in them; packaging problems have been solved by careful under-bonnet arrangements; and the same pump can be used with only minor hardware changes across a large number of engines and manufacturers. This can lead to economies of scale. Disadvantages include their bulk and weight and noisy operation when high pressures are required as well as their high cost. Newer technologies may be more cost-effective when production volumes rise. The advent of electronic control has opened up a variety of alternative approaches.

10.6.1.2 Electronic Unit Injector (EUI) System

The EUI is a radically different fueling system. Instead of having a central pump supplying individual injectors, both pump and injector are combined in a single unit for each cylinder (Figure 10.11). Each pumping plunger is driven directly by the engine's camshaft, and this enables very high injection pressures (in the region of 1500–2000 bar) to be obtained. The main benefits of these very high injection pressures are smaller droplet sizes, better fuel-air mixing, and reduced ignition delay. The timing and quantity of fuel can be adjusted to give optimum conditions over the complete engine operating profile. The system also has the capability to act as a

Fig. 10.11.

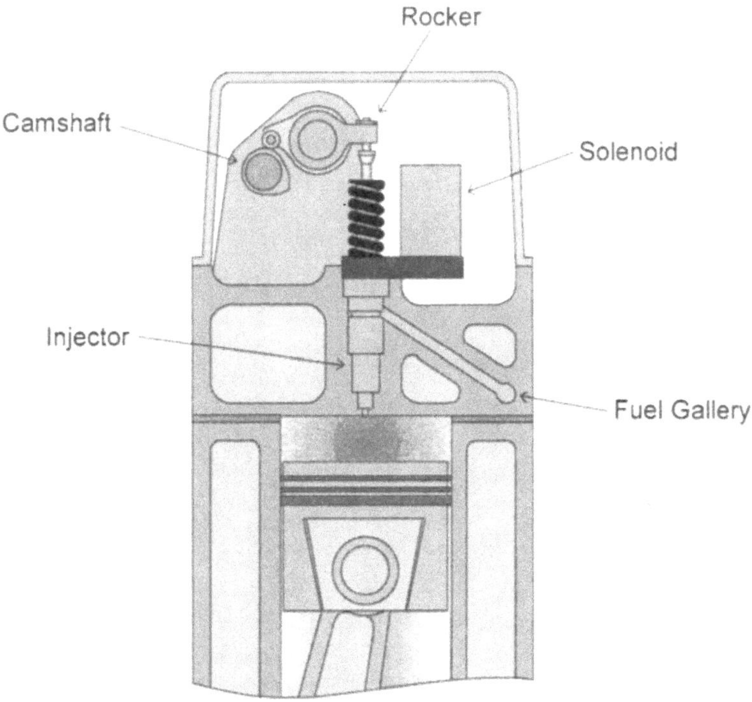

Schematic of EUI system.

pilot injection device to reduce NO_x and engine noise. Successful performance of diesel engines using unit injectors depends upon controlling heat transfer from engine to fuel such that there is a constant temperature of fuel reaching each fuel injector so that each one has the same mass and volume. Fuel feed and spill are through integral passages in the cylinder head. Sensors provide information to the electronic control unit (ECU) on the relevant functions of engine operation, which are:

1. Acceleration position
2. Engine speed
3. Camshaft position
4. Inlet manifold temperature and pressure
5. Coolant temperature

This information is continuously compared with the optimum values stored in the ECU memory. The result of that comparison is translated into signals instructing the unit injector solenoid-actuated spill valve system to deliver the fuel at the timing required by the engine.

The EUI range of electronic unit injectors was initially developed for truck applications and versions are now under development for the high-speed vehicle engine for both two-valve and four-valve cylinder heads. EUI technology is at the beginning of its life cycle and has significant further development potential. Developments for future-generation EUI systems will provide more complex patterns of preignition and postinjection and will provide higher pressures in the low- and middle-speed ranges.

10.6.1.3 Caterpillar HEUI System

Caterpillar Inc. has developed a high-pressure diesel fuel injection system, powered by hydraulic pressure and controlled electronically. This hydraulic electronic unit injector (HEUI) requires no mechanical actuation or mechanical control devices. Inherent features of the HEUI fuel system include injection pressure control independent of engine speed or load, totally flexible injection timing, and full control of injection parameters [49, 50].

High injection pressures, up to 1500 bar, are achieved by a 7:1 pressure intensification between an engine oil hydraulic circuit and the fuel in the injector (Figure 10.12). A high-pressure oil pump, driven by the engine, raises the system's oil pressure from typical engine operating levels to the actuation pressure level required by the injectors. High-pressure oil from the pump flows to a manifold or rail, which is connected to the injectors. A solenoid operates a control valve to start and end the injection process. Initial injection rate shaping is achieved with a

Fig. 10.12.

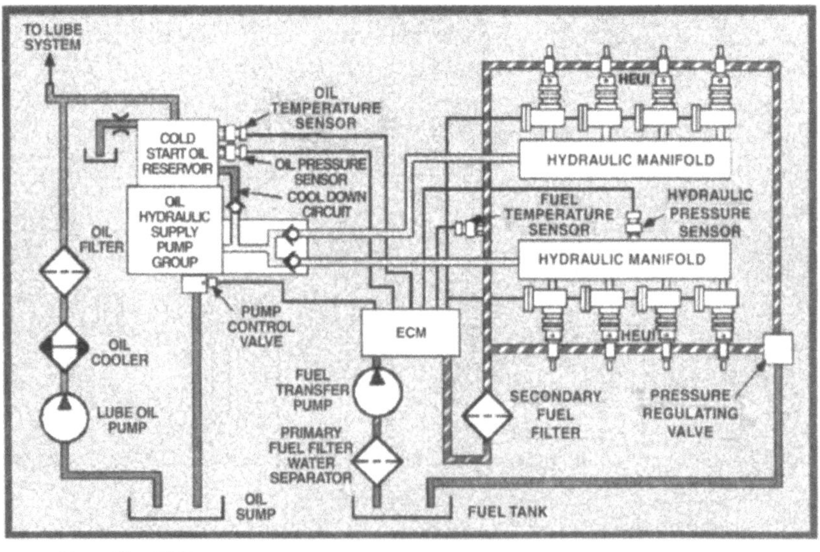

Caterpillar HEUI system.

ported spill control device in the plunger and barrel; this is called PRIME (preinjection metering). The injector uses a conventional nozzle. The HEUI system is time based, so injection characteristics are independent of engine speed [49, 50].

This system went into production in late 1993 for the Navistar T 444E 7.3-liter V-8 DI diesel used by Navistar and Ford, in the United States, for light trucks and buses [51].

Caterpillar has now designed and developed a smaller version of the original HEUI for use in automotive high-speed DI diesel engines. This injector has a 17-mm diameter lower body with a standard 7.2-mm diameter nozzle; the upper body increases to 25-mm diameter. The 1500-bar injection pressure capability is retained along with the PRIME initial rate-shaping device [52].

10.6.1.4 The Common Rail System

The competitor to the EUI is the common rail system in which high-pressure fuel is fed into a manifold (or rail), which then serves individual injector (Figure 10.13). Here the injector is simply a solenoid-actuated nozzle, which is opened under the control of the ECU when injection is required.

The common rail architecture has been widely recognized as a promising system for a long time; indeed, Rudolf Diesel tested injection directly out of a fuel line kept under constant pressure.

In 1989 a Consortium for Research and Technology, ELASIS, was formed in southern Italy. The Fiat company Magneti Marelli joined the consortium and

Fig. 10.13.

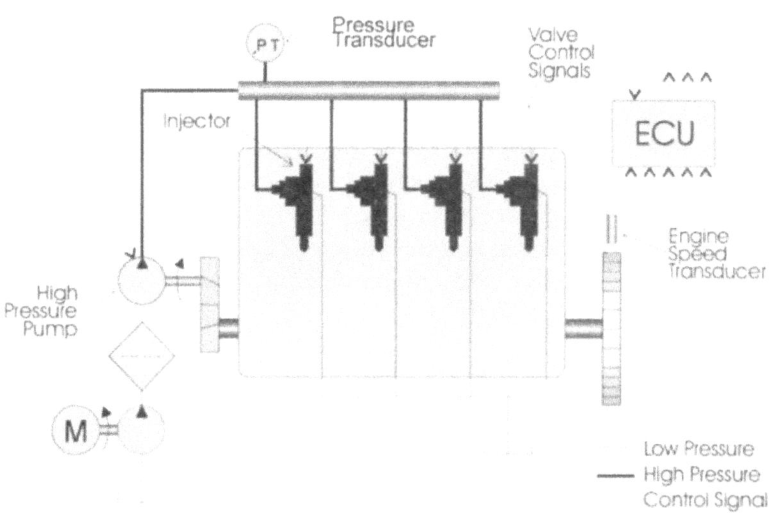

Schematic of a common rail system.

transferred its Unijet common rail fuel injection system technology. By the end of 1991 the second-generation Unijet system was demonstrating functional potentials. At the end of 1993 a preproduction version of Unijet was available. In the spring of 1994 the Fiat Group signed an agreement with Robert Bosch for further development and the volume production of the system [53].

Denso (formerly Nippondenso) has developed a common rail system for the 1990s' diesel requirements, called ECD-U2; this has been principally aimed at truck engines [54].

Ricardo and Renault have reported on a joint collaborative project to optimize a two-valve, medium-duty engine to achieve U.S. 1994 development goals without a catalyst, essentially by the application of a common rail fuel injection system featuring high mean effective injection pressure (MEIP), rapid injector spill, and flexible control of injection timing and MEIP according to load and speed [55].

The common rail fuel system consists of four component parts: the high-pressure pump, the high-pressure rail and pipes, the injectors, and the ECU. In principle the system is very simple, a low-drive torque mechanical pump pressurizes the distribution rail with diesel fuel at the desired injection pressure. Short pipes connect the rail to each injector. There is no pressure intensification within the injector. The injection process is controlled by a solenoid valve in each injector, the final fuel injected quantity being dependent on the opening period and system pressure. The solenoid is controlled by the ECU, as is the rail pressure via a pressure sensor. The ELASIS injector incorporates a two-way solenoid valve to control the injection process. Fuel pressure above the needle control rod is reduced when the solenoid valve opens to bleed fuel from the control volume, through an orifice, more quickly than it can be replaced from the rail via a smaller orifice. This causes the needle to lift and fuel is injected.

The Denso injector features a three-way, pressure-balanced, solenoid valve to switch fuel pressure on the back end of the injector needle control rod from rail to atmospheric pressure to control the injection process. By selection of the size of a one-way orifice, between the valve and the chamber above the control rod, the initial injection rate can be tailored [54].

The key advantage of the common rail fuel system, compared to a traditional pump-line-nozzle system, is that injection pressure is independent of engine speed and load. Injection timing, rate, and duration can be varied precisely over the operating range. The availability of very high injection pressure levels from low engine speed provides improvements in BMEP of 30 percent at 1000 rpm and about 20 percent in the range between 2000 and 3000 rpm, with better control of smoke [53]. At part load the higher injection pressure available gives a reduction of particulate emissions. The lower particulate emissions allow higher EGR rates, so the NO_x emissions can be reduced, leading to a superior NO_x-particulates trade-off [56]. Pilot injection for combustion noise control is feasible with common rail;

Fig. 10.14.

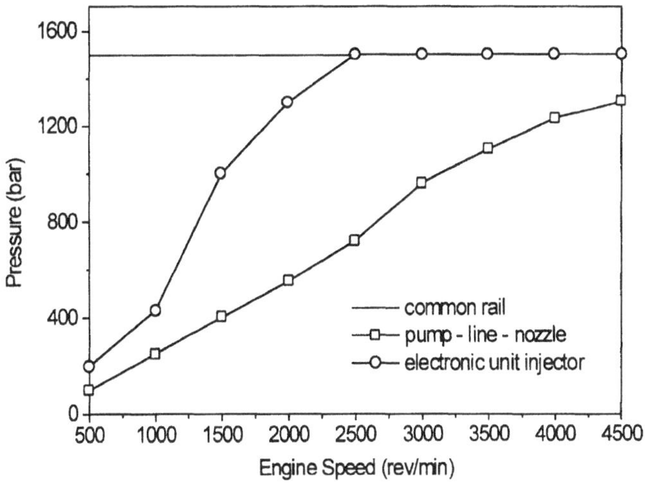

Comparison of different electronic fueling system capabilities.

small fuel quantities of 1 to 2 mm^3/stroke can be injected before the main injection [53, 54, 56]. Also the high-pressure pump acts continuously and so has a more uniform drive torque than the highly cyclic loads imposed by distributor and unit injector pumps.

A comparison between the capabilities of pump-line-nozzles systems, EUIs, and common rail systems is shown in Figure 10.14.

10.6.2 Exhaust Gas Recirculation

10.6.2.1 Background

Exhaust gas recirculation (EGR) is a well-proven technique for the effective suppression of NO_x emissions. The reductions in emissions due to the influence of EGR are shown in Figure 10.15 for a Ford prototype 16-valve, 1.8-liter TCI DI engine. For the automotive diesel engine to meet future European NO_x emissions standards, EGR will be the primary emission control technique. Most gasoline engines can recirculate up to a maximum of 20 percent of the exhaust gases without affecting combustion stability, while diesel engines are capable of recirculating up to 50 percent of the exhaust gases. Amounts greater than 50 percent EGR can be achieved by throttling the diesel intake. Since EGR reduces the oxygen content of the intake charge, it can only be used at part load, or when the fuel-air ratio is sufficiently high so as not to cause a dramatic increase in smoke, HCs, and PM. The EGR rate contours over the operating range for the eight-valve version of the engine considered in Figure 10.15 shown in Figure 10.16.

Fig. 10.15.

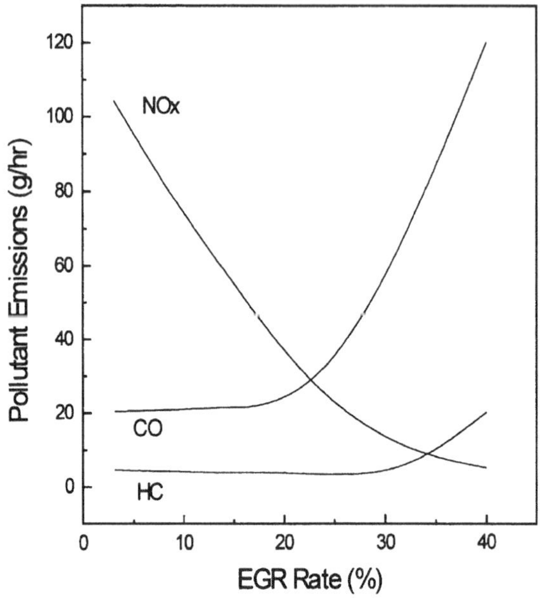

General trend on emissions due to the influence of EGR (taken from a Ford prototype 1.8-liter DI engine).

Fig. 10.16.

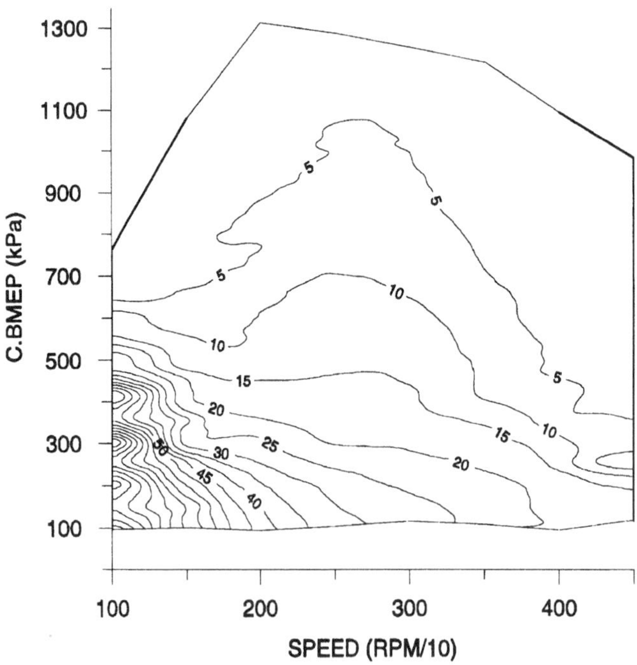

Contours of percentage EGR rate for a Ford prototype 1.8-liter DI diesel engine.

EGR has an effect on the flame structure, temperature, and species profiles following ignition. The reductions in NO_X emissions with EGR can be explained on the basis of a predicted decrease in peak flame temperature in a diffusion flame coupled with a reduced oxygen mass fraction. (The Zeldovich mechanism gives the NO_X formation rate rising exponentially with temperature.)

The reduction in NO_X due to EGR was for a long time thought to be a result of the increase in the heat capacity of the charge caused by the presence of recirculated CO_2. However, recent work has shown that the reduction of oxygen in the inlet charge (dilution effect) is the dominant factor [57]. The work investigated the way that EGR influences diesel engine combustion and emissions. Findings from the investigations included the effect of CO_2 dissociation (chemical effect) on exhaust emissions is small; the high heat absorbing capacity of CO_2 (thermal effect) had only a small effect on exhaust emissions including NO_X. The reduction in the inlet charge oxygen (dilution effect) is the dominant effect on emissions, resulting in very large reductions in exhaust NO_X at the expense of higher particulate and unburned hydrocarbon emissions and lower engine output and fuel economy.

10.6.2.2 EGR Cooling

The full potential of EGR in reducing NO_X will not be realized unless the recirculated exhaust gases are cooled. Cooling the EGR prior to mixing with the intake air lowers the combustion temperature by removing the "preheat" potential while at the same time increasing the inlet air charge and, hence, the oxygen-to-fuel ratio. The slight increase in available oxygen due to cooling may raise the flame temperature, but this is beneficial for soot control. As the bulk intake charge density is higher with cooled EGR, the engine volumetric efficiency is increased, which reduces pumping work leading to an improvement in specific fuel consumption.

Cooled EGR will also have a number of other complex effects, some of them undesirable. For example, the ignition delay can be increased and, subsequently, the amount of premixed fuel burned. This would tend to increase NO_X production depending on a number of factors including injection timing. With cooled EGR, the lower inlet charge and lower compressed gas temperature can also reduce fuel vaporization and mixing rate, which will lead to an increase in exhaust HC levels. However, in practice further significant improvements in NO_X particulates and improvements in fuel economy may be achieved with cooled as compared to hot EGR. The Ford 85PS turbocharged 2.5 DI diesel engine as fitted in the Transit vehicle was equipped with an EGR cooler in 1995. This enabled the vehicle to meet the European Stage 1 passenger car emission standards. The earlier model, without EGR cooling, achieved the same standards but only by taking advantage of the 40 percent DI derogation limits.

EGR cooling has also been applied to a Ford production 1.8-liter DI turbocharged and intercooled diesel engine installed in an Escort car. The benefit

Fig. 10.17.

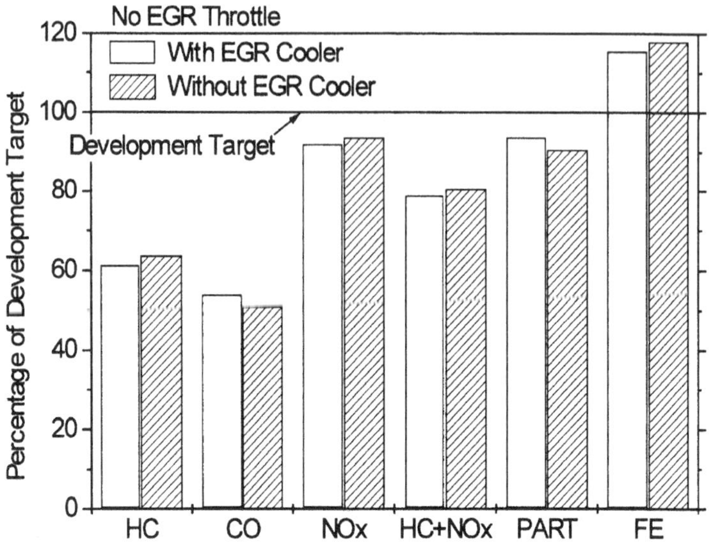

Effect of EGR cooling on vehicle emissions.

that the EGR cooler had on NO_x emissions and fuel economy without any deterioration in PM emissions is illustrated in Figure 10.17.

10.6.2.3 Additional EGR

Additional EGR is best applied to a turbocharged engine with a variable geometry turbocharger and a venturi in the intake system to provide a favorable pressure difference to aid the exhaust gas flow into the intake system. Such a system has been developed for a 12-liter heavy-duty diesel engine [58]. The objective was a 50 percent reduction in NO_x over the ECE R49 cycle without any increase in particulates. The system is shown in Figure 10.18, which consists of a short EGR route, tapping off exhaust gas before the turbine, cooling the gas and mixing it with the intake air by a venturi, after the compressor and intercooler. As the variable geometry flow area was reduced, by closing down the vanes, preturbine pressure increased. When this pressure exceeded that at the venturi throat, EGR followed. The result was a 15 percent additional EGR flow over most of the engine operating range.

10.6.2.4 Control of EGR

Current EGR systems consist of an electronically controlled, vacuum-operated poppet valve mounted on the exhaust or the intake manifold. A current-to-vacuum transducer (CVT) controls the amount of vacuum applied to the EGR valve diaphragm, which in turn lifts the poppet off its seat. The engine ECU contains

Fig. 10.18.

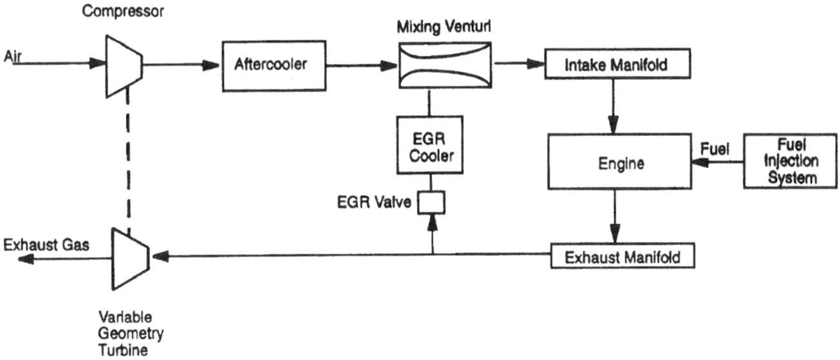

Use of a venturi system for increasing EGR levels (taken from Figure 2, R. Baert, D.E. Beckman, and A.W.M.J. Veen, "EGR technology for lowest emissions," IMechE Seminar on Application of Powertrain and Fuel Technologies to Meet Emissions Standards, Paper C517/034/96, p. 179, June 4–26, London 1996).

schedules (look-up tables), which relate engine speed, fueling rate, and mass air flow (MAF). The MAF sensor is located in the air intake system and provides the information to determine EGR valve position and, hence, EGR mass flow is calculated by implication.

Although current EGR control schemes using valve position have entered production, there are a number of disadvantages associated with this method, namely:

1. Volumetric characteristics of the engine can drift with time due to wear and tear and the build-up of deposits
2. The flow capacity of the EGR system can be affected by partial blocking through deposit build-up (i.e., soot, particulates, and so on and wear of the EGR valve)
3. Blockages in the inlet air filter can seriously affect EGR mass flow
4. Blockages in the exhaust can cause increased exhaust pressure

The goal is a real-time EGR mass flow measuring system that is not dependent upon the "health" of the engine. For the MAF systems the accuracy of the instrument at low airflows is a disadvantage in controlling EGR accurately under these conditions.

10.6.2.5 Transient Operation

Changes in engine conditions from low to high loads result in increased smoke during such a transient. The percentage of EGR is highest at low loads; an increase in load results in a nearly immediate response in fuel, while both EGR and turbocharger lag are experienced. An example of the transient EGR and fueling control strategy for a step increase in driver demand is shown in Figure 10.19a. This shows the effect of poor coordination of EGR and fueling. At the start of the transient (at light load) a high level of EGR is present in the engine. However, even if the EGR valve demand is reduced to zero immediately on detection of the transient, the exhaust gas in the intake manifold will take a finite time to pass through the engine. If the final steady-state fueling level is applied before the EGR has decayed, then a transient-rich AFR spike occurs resulting in transient smoke emissions. The effect of well-coordinated EGR and fuel control where the fuel is limited based on the estimates of air mass flow rate is shown in Figure 10.19b. This prevents the rich AFR spike and, therefore, reduces transient smoke production.

10.6.2.6 EGR Without External Pipes

Removing the need for external high-temperature pipework with the attendant expense, safety and leak-proneness risks by incorporating EGR arrangements within the cylinder head casting is an advanced technique being developed by a number of major engine manufactures. GM's new four-valve, direct injection diesel (the Ecotec) launched in 1996 incorporates such an EGR system [59]. This integral cylinder head arrangement also serves to warm the head quickly after a cold start, and cools the EGR gas by around a reported 30°C, which will further reduce emissions. Ford has also reported work on an integral cylinder head EGR system [60]. Internal galleries between the inlet and exhaust ports meet at a rotary valve for each cylinder, interconnected with those of other cylinders, so that simultaneous control of the complete engine EGR can be simply obtained. The rotor of the valve is designed so that it can be programmed to uncover an air port connecting with the exhaust manifold to achieve secondary air injection for burning exhaust gases during engine warm-up.

10.6.3 Exhaust Gas Turbocharging for Emissions Reduction

Turbocharging is a form of supercharging in which heat energy is extracted by a turbine in the engine's exhaust system. The turbine is directly coupled via a shaft to a compressor, which subsequently draws in a fresh charge of air, supplied to the cylinders under pressure (i.e., boost pressure). The air through flow, compared with naturally aspirated units, is therefore increased, which provides the capability to:

1. Increase the engine rating
2. Improve the limiting torque curve and torque back-up
3. Enhance the engine's dynamic response
4. Improve specific fuel consumption
5. Reduce exhaust emissions

Fig. 10.19.

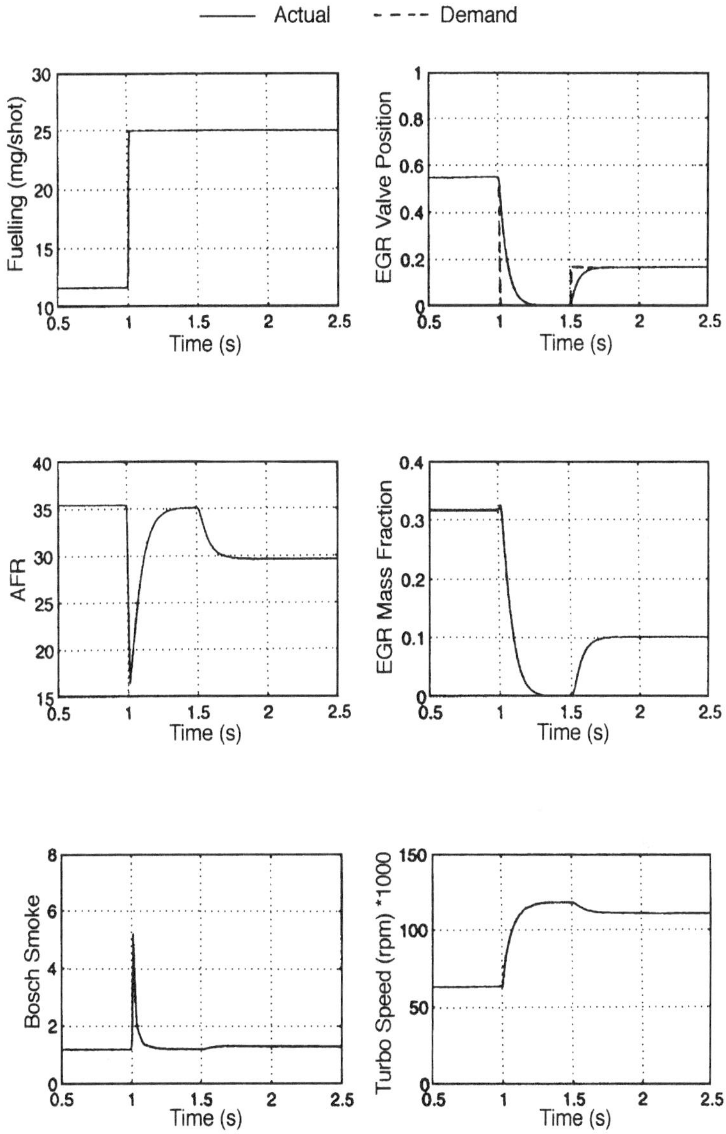

Transient EGR and fueling control strategy (taken from Figure 6a and 6b, Whelan et al., "Meeting the Passenger Car Challenge with the New Ricardo 16 Valve HSDI Diesel Engine," European Automobile Engineers Cooperation 5th International Congress, Paper SIA 9506A08, June 21–23, Strasbourg 1995). (Continues)

Fig. 10.19 (*continued*).

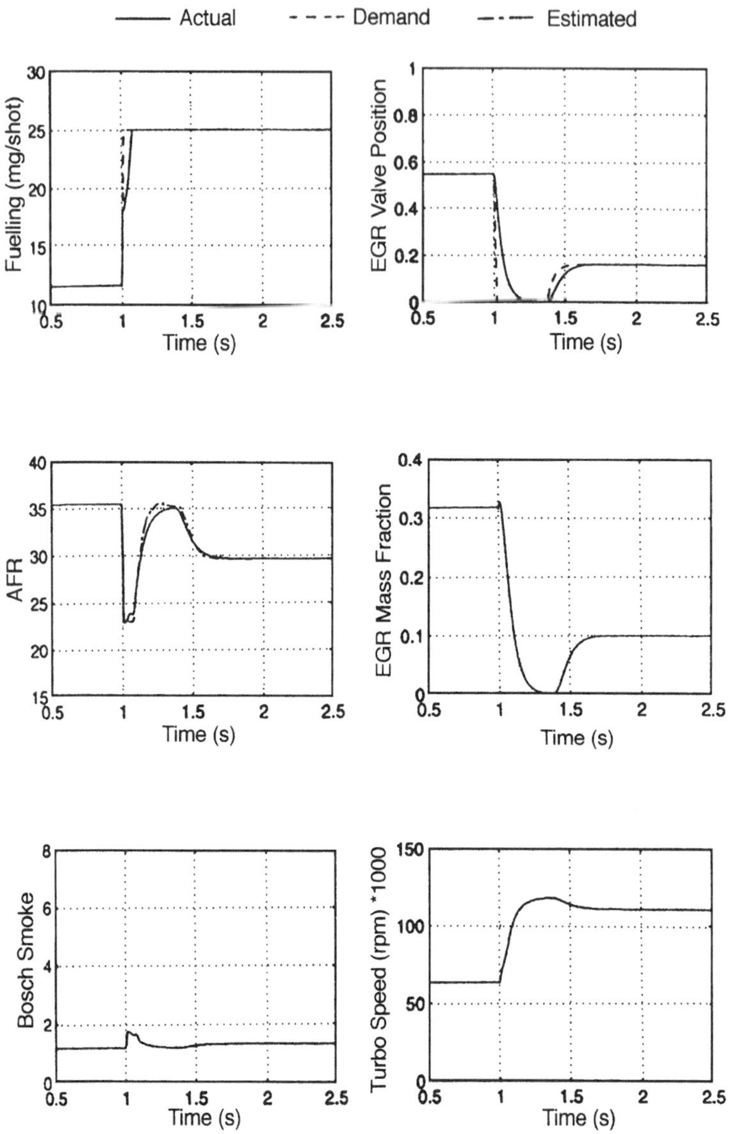

10.6.3.1 Conventional Turbocharging

The turbocharger's theoretical disadvantage is that at low engine speeds it can provide only low boost pressure due to the relative lack of available energy in the exhaust gases. The magnitude of boost pressure for a Ford prototype eight-valve, 1.8-liter DI engine is shown in Figure 10.20. Also due to turbocharger rotor inertia, there is invariably a delay in boost (turbo lag) when acceleration

Fig. 10.20.

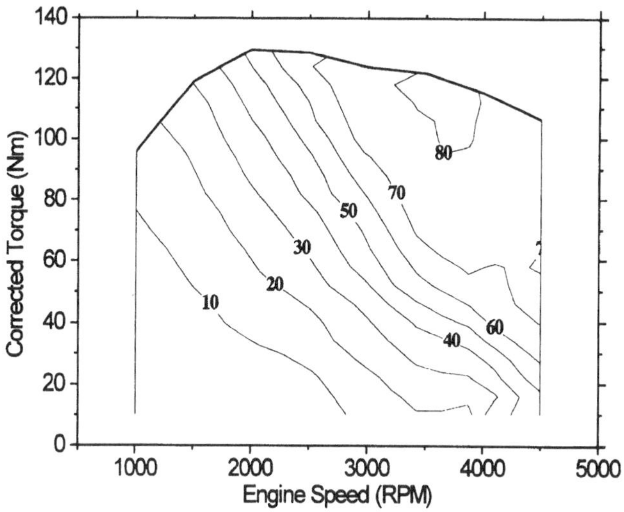

Typical contours of manifold pressure (kPa) taken from a Ford prototype 1.8-liter DI TC diesel engine.

is required. Comprehensive documentation of the science of turbocharging is provided by Watson and Janota [61].

Modern turbocharger technology has made it possible for turbocharger installations to be tuned for the production of relatively low-speed torque, while rotor size (and therefore mass) is coming down appreciably and reducing the magnitude of turbo lag. Also, the relative mechanical losses in the engine will be reduced and the specific fuel consumption will be further improved. However, at higher speeds the pressure ratio will be too great, and the turbocharger will require a by-pass valve (wastegate) and constant boost will thus be maintained.

Mechanical-driven (crankshaft) superchargers are capable of giving boost at low engine speeds, albeit with a significant performance penalty. At least 10 percent of rated power will be required to drive the unit with a resulting substantial increase in specific fuel consumption. Such units have been used and continue to be used especially in the two-stroke diesel marine market. However, the parasitic power requirements are not attractive to smaller-capacity engines and as such are not generally used for automotive applications.

On many light-duty, automotive turbocharged diesels there is little boost below 2000 rpm or so. Throttled gasoline engines have very low gas flows when the throttle is partially open and the corresponding speed of the turbocharger is very low. Before it can deliver useful boost, it has to react to the increase in gas flow that follows the throttle opening. A diesel engine has no throttle to close off the engine's air supply, and so there is always enough gas flow through the turbocharger to maintain its momentum. Conventional turbocharging systems for high-speed DI engines invariably include a wastegate, which by-passes an increasing fraction of the total exhaust gas flow from upstream of the turbocharger turbine direct to atmosphere. This allows the use of a smaller turbine, which improves low-speed torque as well as transient response and increases the effectiveness of EGR due to the higher turbine inlet pressures.

Fuel pumps on turbocharged engines usually have some means of limiting the fueling during periods of low boost pressure. These are encountered during transients from low to high speed/load since the turbocharger and manifold dynamics can be significantly slower than the fuel pump dynamics. Engines designed for automotive use must minimize this lag effect to retain good driveability characteristics. This is generally accomplished by using a small turbocharger with low inertia, which will accelerate quickly as the exhaust flow increases in temperature (as fueling rises) and volume (as the engine accelerates). Since a small turbocharger would give too much boost at higher engine speeds a wastegate is used. This opens when the boost pressure rises to allow some of the exhaust flow to by-pass the turbine. This has an additional benefit of reducing the back-pressure caused by the restriction of a small turbine at high flow rates.

10.6.3.2 Variable Geometry Turbocharging

To optimize the match of turbocharger to engine needs variable geometry turbochargers (VGT) are used. These are "flexible" devices which are capable of providing a range of pressure ratios for a given engine speed.

With conventional wastegated turbochargers or fixed geometry turbochargers (FGT), a turbine flow area cannot be adjusted during operation. However, a variable geometry turbine as the name implies, has moveable parts that can change the area in the turbine stage through which exhaust gases pass, or change the angle at which the exhaust gases enter or leave the turbine, or a combination of both. The result is a turbocharger that, depending upon its geometry, supplies varying amounts of boost to the engine. The effect is similar to a variable-pitch propeller.

Therefore, a VGT has an obvious advantage over a fixed geometry turbocharger in that it is possible to deliver optimum boost to the engine over a wide operating area. Furthermore, restricting the turbine flow area during transients increases upstream pressure and gas velocity at rotor entry providing the capability to rapidly accelerate the rotor group and thereby improve the dynamic response of the engine and diminish turbo lag.

The variable geometry multinozzle turbine, as shown in Figure 10.21, has the usual annular volute passage but the surrounding flow path to the turbine wheel is via a ring of pivoted nozzle vanes. In this system the nozzle area varies as a function of engine operating condition, thus controlling the available energy in the exhaust. A reduction in the turbine nozzle area alters the angle of airflow onto the turbine blades, which results in more energy transfer to the turbine and consequently higher turbocharger speed and boost pressure.

Until comparatively recently, the application of VGT was limited to a small number of large diesel engines for truck or off-highway applications in order to achieve improved torque back-up and transient response [62]. The use of

Fig. 10.21.

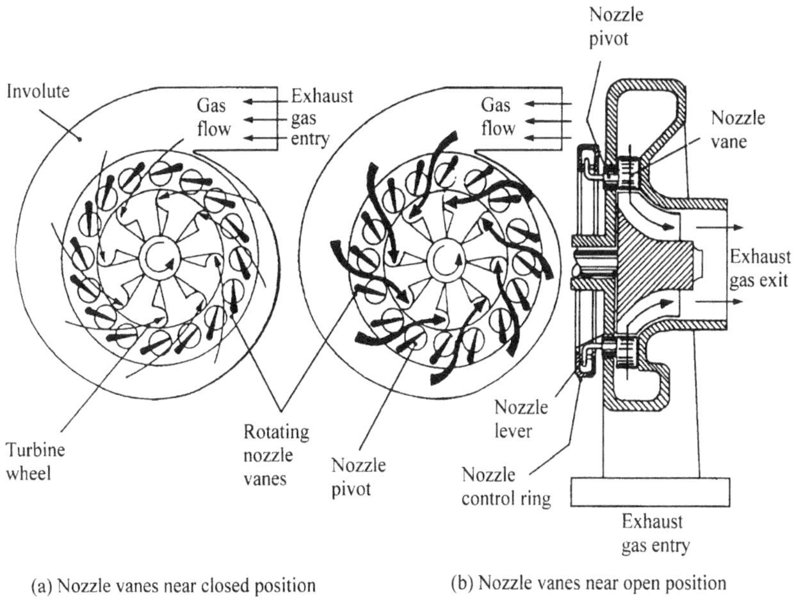

Variable nozzle geometry turbocharger (taken from Figure 6.68, p. 342, *Advanced engine technology* by Heinz Heisler, published by Edward Arnold, 338 Euston Road, London, NW1 3BH, ISBN 0 340 568224).

the VGT on production models of the high-speed, light-duty automotive diesel engine started to appear on the European market in 1997. Although the primary requirement is still aimed at improved lower-speed torque and transient response, the VGT offers the potential to reduce and control emissions in the critical part-load engine operating region. The following investigation highlights the gains to be made [63].

10.6.3.3 The VGT as an Emissions-Reduction Device

The following summarizes the test results obtained from investigations on a Ford prototype DI diesel engine capable of meeting European Stage 3 emissions regulations. The object being to evaluate the NO_X reduction potential at 11 discrete engine operating points, which are known to be significant with regard to emission production in the European emission test drive cycle. These modes are described in Table 10.9 in terms of percent rated speed and percent rated load.

Two series of tests were carried out. Initially, the FGT enabled a set of results to provide benchmark data. Second, the VGT was installed and a test pattern devised that enabled investigations to be undertaken at each of the modal points detailed previously. With the VGT, at each mode the vane position was varied over its full range, maximum opening (0 percent) to minimum opening (100 percent)

Table 10.9

Engine Operating Modes

Mode	Engine speed (percent rated speed)	Engine load (percent peak torque)
1	33	28
2	40	6
3	40	14
4	40	21
5	42	40
6	48	29
7	51	42
8	53	6
9	57	27
10	60	56
11	71	35

Table 10.10

Part-Load Comparisons of NO_X, Smoke, and Specific Fuel Consumption

	Specific NO_X (g/kWh)			Smoke (Bosch)			BSFC (g/kWh)		
Mode	FGT	VGT	Change (percent)	FGT	VGT	Change (percent)	FGT	VGT	Change (percent)
1	4.49	2.85	−36	0.7	0.7	0	259	259	0
2	4.36	3.90	−11	0.5	0.6	+20	547	488	−11
3	2.68	1.48	−45	0.5	0.5	0	332	326	−2
4	2.35	2.24	−5	0.5	0.5	0	285	282	−1
5	2.79	2.79	0	1.2	1.0	−27	242	241	−1
6	1.73	1.45	−16	1.1	1.1	0	264	268	+1
7	2.36	2.33	−1	0.9	1.2	+33	242	248	+2
8	1.93	1.68	−13	0.9	0.5	−44	463	459	−1
9	1.42	1.39	−2	0.9	0.9	0	280	282	+1
10	3.37	3.23	−4	0.4	0.6	+50	239	230	−4
11	3.36	2.87	−14	1.0	0.8	−20	271	260	−4

at a set of fixed EGR valve positions from fully closed (0 percent) to fully open (100 percent). The "optimum" vane and EGR valve positions were determined by a number of criteria based on achieving either the maximum reduction in NO_X or BSFC. This resulted in some 500 test points in total, which were all carried out with the same start of combustion (SOC) timing.

Modal Results The modal results are summarized in Table 10.10, showing the comparison of the FGT with the VGT results. For these particular test results the "optimum" criterion was the reduction of NO_X without severely compromising fuel consumption.

Table 10.11

Optimum Settings to Achieve NO_x Reduction Without Severely Compromising Fuel Consumption

Mode	VGT vane position (percent closed)	EGR valve position (percent open)
1	50	50
2	61	100
3	41	100
4	72	50
5	66	41
6	79	50
7	76	35
8	21	100
9	60	100
10	61	25
11	71	50

The results show that at all the test points considered, lower NO_x can be achieved (apart from mode 5 which did not show a deviation either way) with substantial reductions at modes 1, 2, 3, 6, 8, and 11 by the use of the VGT compared with the FGT. The tests were carried out such that fuel consumption over all the test points would not be compromised and this is evident from the results presented in Table 10.10. The smoke levels are virtually unaltered and the inclusion of the "percent change" can be misleading. The "optimum" VGT vane and EGR valve position for this particular set of results are detailed in Table 10.11.

The range of optimum VGT vane settings, as shown in Table 10.11, is 50 percent to 79 percent closed, giving a suitable margin for an improvement in transient response over the FGT. On acceleration from any of the foregoing modes, the EGR valve would be completely closed, and the VGT closed to obtain the maximum boost possible.

Mode 3 Results For the purposes of highlighting the influence of VGT vane and EGR valve position on emissions and fuel consumption, Mode 3 has been chosen for discussion. This mode is representative of the trends experienced at all the modes. The following observations are made by consideration of Figures 10.22 to 10.26.

Specific NO_x versus VGT Vane Position The trends in Figure 10.22 follow the expected course with NO_x reducing progressively as EGR flow increases by either larger EGR valve opening and/or greater manifold differential pressure, Δp. It also shows how a small EGR flow has a dramatic effect on the amount of NO_x produced. The benchmark figure from the FGT is included and it can be seen that a substantial part of the map lies under this level.

Fig. 10.22.

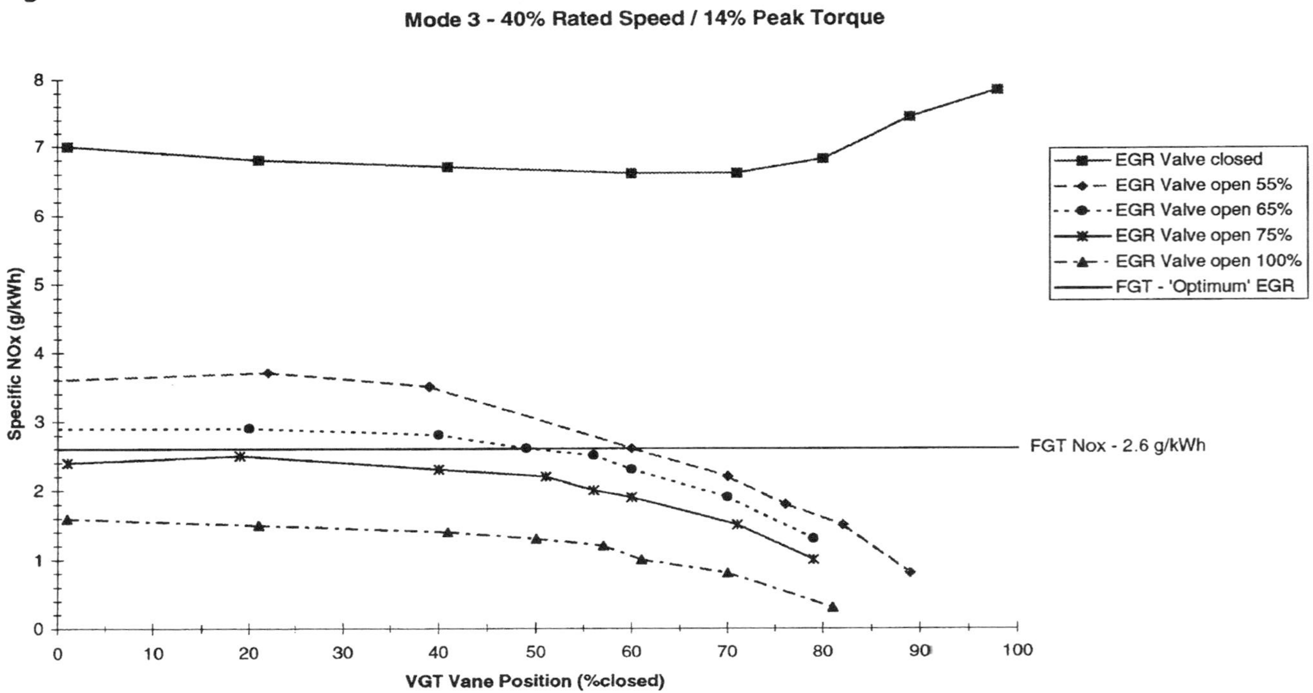

VGT vane position versus specific NO_x.

Fig. 10.23.

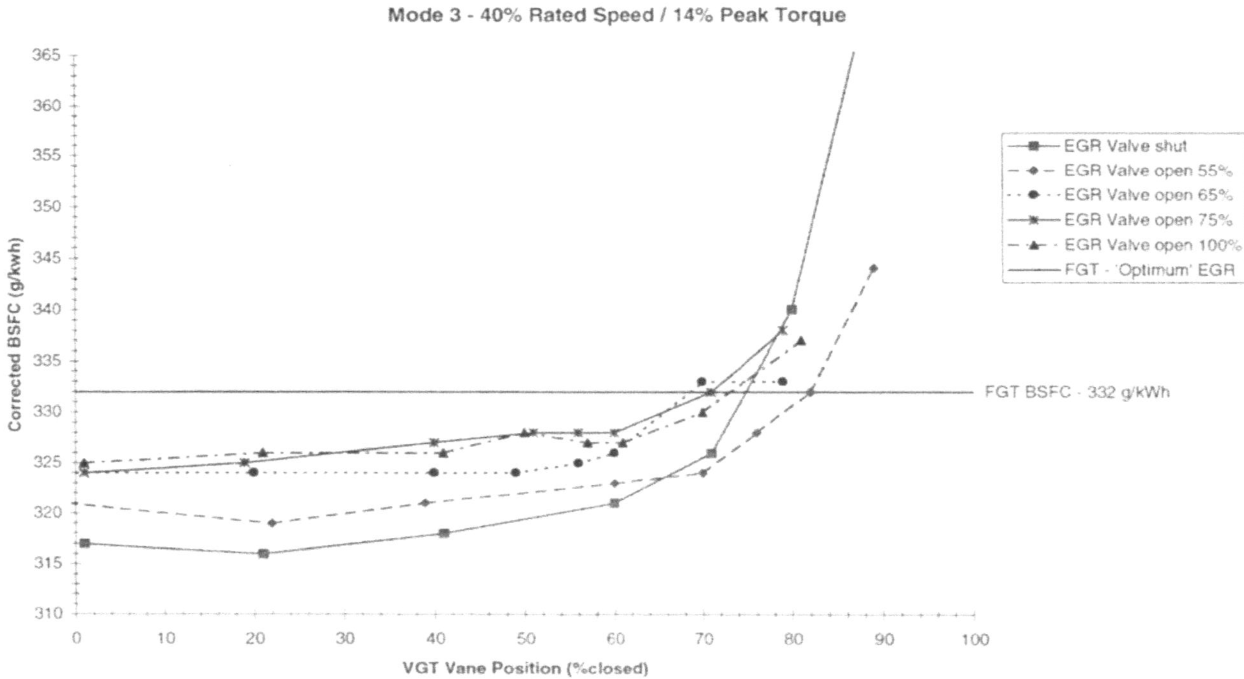

VGT vane position versus BSFC.

Fig. 10.24.

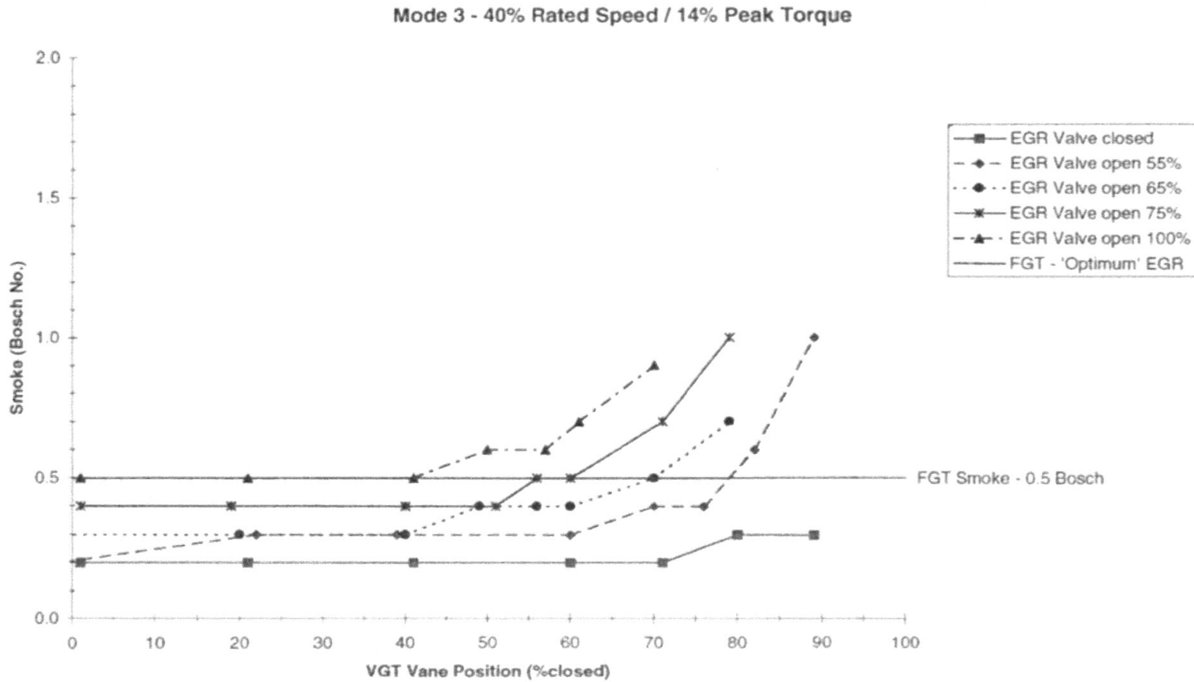

VGT vane position versus smoke.

Fig. 10.25.

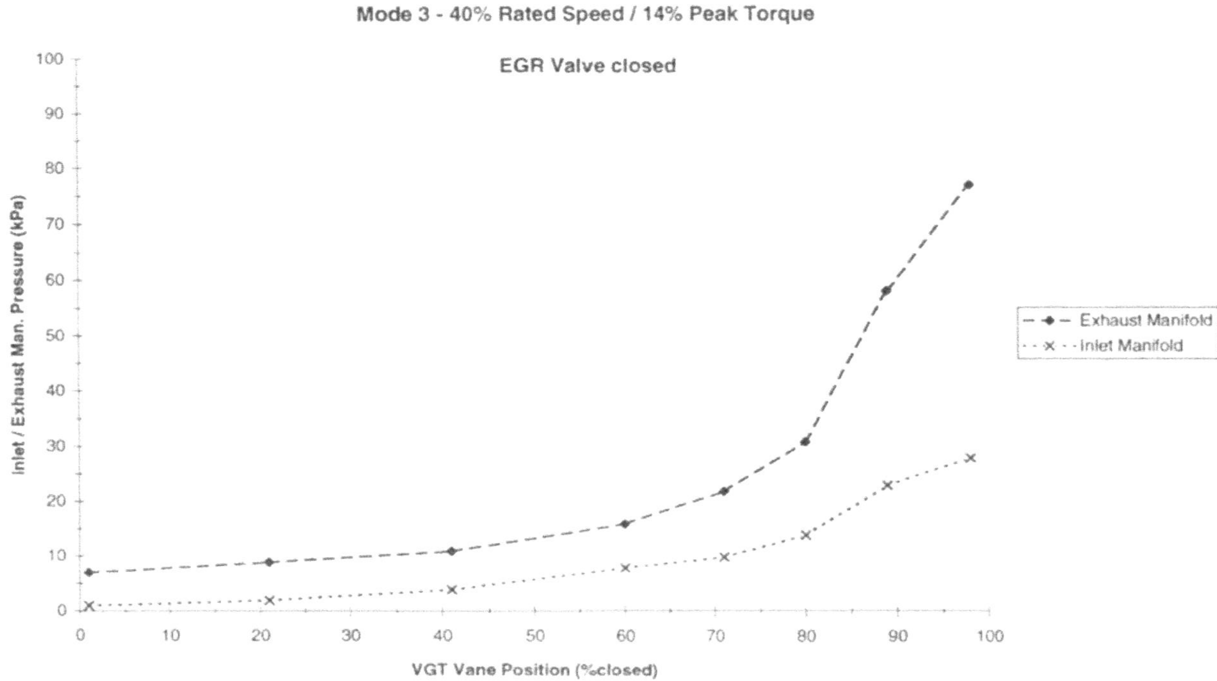

Inlet and exhaust manifold pressure versus VGT vane position.

Fig. 10.26.

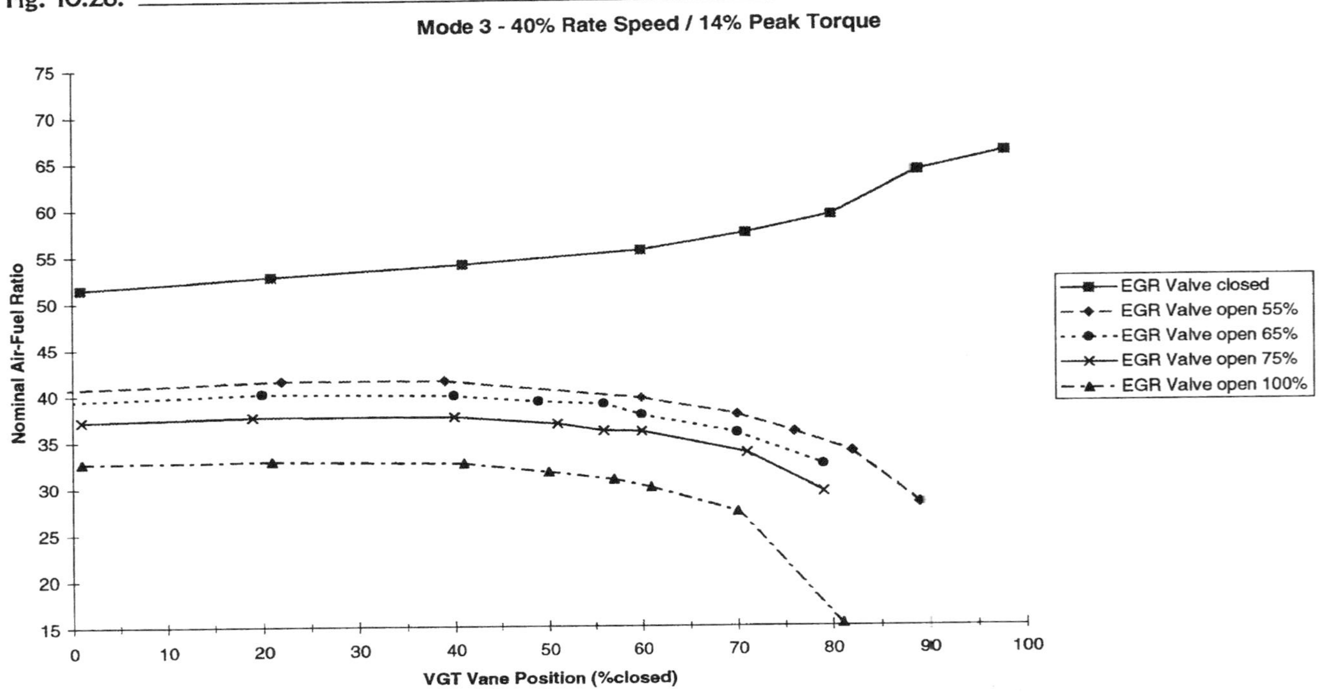

Norminal air-fuel ratio versus VGT vane position.

BSFC versus VGT Vane Position The large EGR flows associated with small vane openings give rise to a detrimental effect on the BSFC, as can be seen in Figure 10.23. Also, at these vane positions the pumping work is large and this further increases the BSFC. As before, a substantial part of the map lies under the target figure.

Smoke versus VGT Vane Position The effect on smoke is shown in Figure 10.24 where the trends have similarity to those for BSFC. As the AFR ratio reduces, drastically so at small vane openings, the smoke levels naturally rise. Once again there is a substantial part of the map under the target figure shown.

Inlet and Exhaust Manifold Pressure versus VGT Vane Position The difference between the inlet and exhaust manifold pressure over the full range of VGT vane positions is shown in Figure 10.25. This condition (with the EGR valve closed) is chosen as it most clearly shows the trend of pressure difference due to the absence of a gas leakage path. For other settings of EGR valve, the trend is less pronounced and reduces as the valve is progressively opened.

It can be seen that the pressure in both manifolds increases as the vanes are progressively closed, and that the pressure difference increases dramatically at the upper end of the closing scale. This results in a greater turbine speed and a higher level of inlet manifold boost; it also increases the pumping work, which leads to a increase in BSFC.

Nominal Air-Fuel Ratio versus VGT Vane Position In Figure 10.26, the line indicating the AFR with the EGR valve closed represents the true AFR. However, as soon as the EGR valve is opened, a nominal AFR ratio is recorded as it does not take into account the air in the recycling gas and, therefore, cannot be taken as true. This nominal value is still a useful measure of the engine operating conditions and is thus shown here.

As soon as the EGR valve is opened, the AFR reduces from the true level, but it remains relatively constant as Δp (the pressure difference across the manifolds) remains constant. As Δp increases, the EGR flow increases correspondingly, reducing the nominal AFR. This trend is exacerbated as the EGR valve is progressively opened further.

General Part-Load Trends Examination of the foregoing figures shows that there is a comparatively narrow range of EGR valve and VGT vane settings that correspond to an improvement on the FGT results. This is the trend for all modes. To obtain the best results, a small EGR valve opening is better in terms of BSFC and smoke, whereas a larger valve opening is preferable for greater NO_x reduction.

A further condition is that the smaller the VGT vane position selected, the greater the potential for improved transient response due to a greater boost differential available when the vane is closed down. From the preceding considerations the optimum settings for this mode are taken as 100 percent EGR valve and 41 percent VGT vane, giving an improvement in BSFC, NO_x, and smoke in this case. The use of a VGT in conjunction with a modulated EGR valve opening has shown that part-load reductions of NO_x can be achieved at discrete engine test points by as much as 11 percent to 45 percent without compromising either fuel consumption or smoke.

10.6.3.4 Charge Air Cooling

Usually referred to as *intercooling*, charge air cooling reduces the temperature of the intake air after it has been compressed (and hence heated), therefore increasing the charge air density and ultimately the volumetric efficiency. The effect of charge air cooling on NO_x reduction has been previously highlighted (Figure 10.3). The effect of full cooling (equivalent to isothermal compression) on charge air density is shown in Figure 10.27 along with the effect of compressor efficiency. Intercooling necessitates the use of heat exchangers, which for automotive applications are simply air-to-air devices.

A secondary effect of lowering the temperature of the intake charge is to lower the temperature of the exhaust side of the engine. Although this reduces the thermal loading on both the engine and the turbine, it may be advantageous to have a higher exhaust gas temperature under certain conditions.

Fig. 10.27.

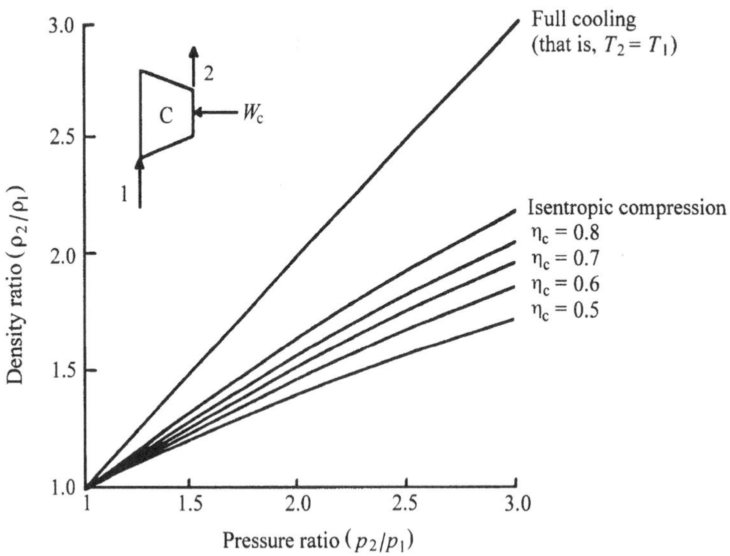

Effect of compressor efficiency on air density in the inlet manifold (taken from Watson and Janota, 1982).

10.6.4 Valves

Four valves per cylinder in gasoline engines have now become commonplace. The main point of using four valves is to achieve better airflow into the combustion chamber, particularly at high engine speeds. The benefits of having four valves in a DI diesel are different; a centralized combustion system offers reduced NO_X and particulate formation. This is by virtue of the lower swirl requirements which reduce the amount of fuel mixed with air during the delay period, coupled with the capability to vertically position the injector.

The benefits of four valves per cylinder are now summarized.

1. Research work performed on central combustion chambers with the injection nozzle being arranged as centrally and as vertically as possible optimizes combustion and reduces particulate formation. In a two-valve design the injector is always offset and inclined. However, the inclination of the injector leads to a particulate penalty, which can be seen in Figure 10.28 [64]. It can be seen that an injector angle of 10° gives a particulate penalty of some 5 percent compared with a truly vertical injector. The combustion effect is due to the inevitable asymmetry caused by an angled injector and the offset imposed by two valves [65].

2. There is more metal between four small valves than two big ones, reducing the large heat gradients that lead to thermal stress and head cracking problems.

Fig. 10.28.

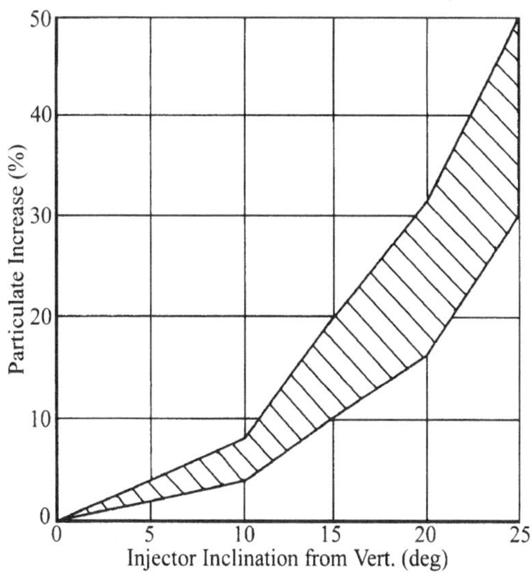

Effect of injector inclination on particulate production (taken from Figure 3, M.L. Monaghan, "High-speed direct injection diesels—present and future perspectives," IMechE Seminar on Lean Burn Combustion Engines, Paper S433/012/96, Dec 3–4, London 1996).

The same applies to the combustion bowl, formed in the piston crown: it can be absolutely central for even heat dissipation. As a result, two-valve engines tend to be limited to around 40 kW/liter [64].

3. The port cross-sections are not compromised as with two valves per cylinder, which reduce volumetric efficiency and inhibit scavenging. With four valves, the potential exists for increased volumetric efficiency in the lower speed range, enabling an increase in the maximum torque which will improve driveability.

4. The use of two inlet valves has also encouraged the more controlled use of swirl, with one helical inlet port and the other "straight-through" for greater mass flow at higher engine speeds and outputs, when the need for swirl is less. Other variations include the capability for controllable valve angle with one inlet valve on each cylinder fed angled tract making the air swirl vigorously in a whirlpool motion as it enters the cylinder. At low engine speeds, when vigorous swirl is needed to get good mixing of small amounts of fuel, all intake air comes through this valve. The other valve is fed by a straighter tract designed for maximum airflow: at low speeds it is shut off by a butterfly valve, which progressively opens as speed and load increase. The result, compared with a two-valve DI, is reported to be 50 percent more swirl at low speeds and much better cylinder filling at high engine speeds.

The drawbacks of four valves per cylinder, on the other hand, are obvious: the classic twin overhead camshaft layout is expensive, bulky, and complex; it also creates extra noise and friction.

10.6.4.1 The Camless Engine

The capability to control individual valve events without the need for cams and camshafts offers substantial improvements to many areas of engine operation. Such research is being carried out in the United States by the Ford Motor Company [66].

The avoidance of cams and springs allows the overall height of the engine to be reduced with the ability to vary the lift/timing events of both intake and exhaust valves, providing the following benefits:

1. Variable intake timing
2. Variable exhaust timing, which, facilitates optimization of the expansion ratio and also the provision of a certain amount of internal EGR
3. Variable valve lift. Air motion in a camless engine with two intake valves can be controlled by independently varying the lifts of the two valves as well as providing port deactivation
4. Variable valve velocity. Since the valve velocity is related to the actuation device and not engine speed, the valve can remain open for a greater proportion of the total valve event, resulting in improved volumetric efficiency

An electrohydraulic mechanism is provided for each valve and its utilization controls the sequence of valve events. Hydraulic forces open and close the valves, and the hydraulic system is operated under closed-loop control, using an electric pump and hydraulic accumulator. The hydraulic and electric losses of the valve train are quoted as equivalent to 50 and 100 mbar IMEP (indicated mean effective pressure), respectively, giving a total loss approximately equal to that of a conventional valve train.

10.6.5 Exhaust Gas Aftertreatment—A Brief Overview

10.6.5.1 Introduction

An overview of current and future European, U.S., and Californian legislative limits on emissions for diesel engine passenger cars, light-duty trucks, and heavy-duty vehicles is detailed in Section 10.5. The previous subsections of Section 10.6 discussed how current and near future emissions limits, especially for passenger cars, can probably be met by a combination of new electronically controlled fuel injection systems in conjunction with EGR and either fixed geometry turbochargers or variable geometry turbochargers.

However, it is clear that the even tighter requirements for the year 2005 and beyond will not be met by the these means alone and that some form of exhaust gas aftertreatment to reduce NO_x in addition to oxidizing CO, HC, and PM will be needed.

10.6.5.2 Oxidation Catalysts

In the case of a gasoline engine operating at, or close, to stoichiometric air-fuel ratios, the use of three-way catalysts for oxidation of CO and HCs, and simultaneous reduction of NO_x, is now well established. The situation is more difficult with the diesel engine, which always operates at air-fuel ratios greater than stoichiometric and, consequently, with free oxygen in the exhaust.

Oxidation catalysts have been used on diesel passenger cars since the early 1990s. Today they are fitted to all diesel passenger cars to meet European Stage 2 emission standards. In relation to the European drive cycle, typical conversion ranges are as follows:

HC:80–90%
CO:80–90%
PM:50–60%

These figures are based on a high platinum catalyst loading of 70 g/ft^3. The term *catalyst loading* refers to the amount of precious metal coated on the washcoat and the units of g/ft^3 are used throughout the industry.

10.6.5.3 NO_x Reduction Catalysts

Catalysts which are able to reduce NO_x in diesel engine exhausts (i.e., de-NO_x) are at an earlier stage of development, and the literature deals mainly with systems which have not, as yet, been fully proven in service. Furthermore, the subject of catalysis is a highly specialized one, involving complex chemical considerations. A detailed treatment is therefore considered to be outside the scope of this chapter. Instead, brief references are made to recently published research papers.

10.6.5.4 Catalyst Development

Brisley et al. of Johnson Matthey provide useful general background information on the special problems of catalysts for diesel engines [67]. With regard to HC control, which presents the most severe problems under start-up conditions, the main difficulty is that all catalysts become effective only at exhaust temperatures above 250–300°C, which generally occurs after some 20 to 30 seconds after cold start. The consequences of cold starting on emissions will be amplified by a change in the European test cycle for Stage 3 legislation in 2000. The first 40 seconds of idle before emission sampling begins is to be deleted; the new cycle is intended to better reflect real-world operation. In consequence, this will also mean that the actual reduction in the numerical values from the current standards do not truly reflect the stringency of the limits. Hence, some provision has to be made for enhancing catalysts performance during this period.

The first is to provide a secondary source of heat either by direct heating of the catalyst using electrical energy supplied by the battery or by burning a small quantity of secondary fuel upstream of the catalyst.

The second is to use catalyst bricks, which, due to their greater robustness, can be mounted very close to the turbocharger outlet; this will reduce warm-up times.

Catalyst materials are still being developed, with palladium-containing catalysts now favored in lieu of the more widely used platinum-rhodium combination. The control of NO_x by the appropriate catalysts under the strong oxidizing conditions of the diesel engine again demand special measures, although EGR will continue to make a substantial contribution to NO_x control. Again alternative approaches have been suggested. The first is direct decomposition of NO_x using copper-exchanged ZSM-5 zeolite, which unfortunately is very sensitive to the presence of sulphur dioxide and water in the exhaust; the two latter incidentally making a major contribution to particulate formation [68]. The second is the use of strong reducing agents such as ammonia (NH_3) or urea in conjunction with copper ZSM-5 catalyst, although clearly this introduces unwanted additional complications. Porter et al. report considerable success in meeting the proposed Euro Stage 3 legislation ($NO_x = 0.37$ g/km, PM $= 0.05$ g/km) in their CERES research vehicle using a modified four-valve Ford Mondeo engine in conjunction with VGT, EGR, and a lean-NO_x catalyst, which in the ECE-15 plus EUDC drive cycle (see

Section 10.5.1) gives figures of 0.312 g/km for NO_x and 0.035 g/km for PM that are well within the Stage 3 limits [69].

Finally, Shore et al. give the results of field tests carried out on buses under real city driving conditions both in London and Paris, using an AZ oxidation catalyst, and comparing the results obtained with those from chassis dynamometer measurements, simulating the European R49 13 drive cycle and the U.S. FTP-75 cycle. The former is a series of steady-state modes, the latter a true transient cycle. This being a pure oxidation catalyst, NO_x is not affected. The results obtained from these tests are highly variable, leading the authors to the conclusion that merely by changing the cycle, an entirely different view can be obtained of the potential benefits of oxidation catalysts [70].

Ford, with FEV, demonstrated this technology applied to a diesel transit [71]. An eutectic aqueous solution of urea was chosen for this exercise because it can achieve similar performance to ammonia, is nonpoisonous and does not present a fire hazard. The catalyst consisted of a vanadium-titanium substrate with a cell density of 200 cells per square inch (CPI) and the catalyst volume was optimized at 7.4-liters. Injection of the urea solution into the exhaust was controlled by a microprocessor. NO_x conversion efficiencies of 65 percent and 83 percent were achieved on the European and FTP-75 cycles, respectively. A reductant tank capacity of 40 liters would be required for the 2.5-liter diesel Transit to operate between the 15,000-km service interval. This tank would need replenishing at each service. Issues with this system concern operation at low ambient temperatures, the size of catalyst, and the practicability of an emission-related reductant carried additionally on board the vehicle [71].

10.7 STEADY-STATE AND TRANSIENT EMISSIONS

10.7.1 Comparison of Steady-State and Transient Emission Strategies

Conventional engine calibration techniques are centered around steady-state test work. Once the hardware specification is fixed, there are many control inputs which need to be mapped to allow the electronic fuel injection equipment to obtain the required performance from the engine.

The fundamental quantity, fueling, is set by the pedal position, which is combined with current engine speed in a map to determine the desired mass or volume of fuel to be injected. The shape of this map has a profound effect on the driveability of the vehicle and needs to be carefully designed. One of the main characteristics set by this map is the degree of torque back-up. If the engine speed drops when climbing a hill, for example, more fuel is injected. This increases engine torque output to help the vehicle feel more stable and easier to control. It is this effect which gives the diesel engine its characteristically strong driveability attributes.

The remaining control inputs are typically mapped against fueling and engine speed. The mapped variables are some or all of start of injection, rate of injection, degree of preinjection, EGR rate, turbocharger geometry or wastegate opening, inlet swirl control, and inlet throttle position.

There are many modifiers to the calibration, typically engine temperature, ambient conditions, and so on. The controller may also incorporate features allowing cruise control or communication with other vehicle systems such as the transmission.

The calibration procedure for steady state is heavily affected by emissions requirements, which are usually equally or more important than fuel economy and the limiting torque curve. The complete vehicle must pass a drive cycle test before it can be sold and the steady-state calibration will play a major part in achieving this. It is possible to represent a given drive cycle by a weighted average of a small number of carefully chosen steady-state points. Optimization of these points allows some confidence that the complete vehicle will perform as required. The rest of the speed/load range must also be considered to provide a smooth characteristic across the operating range.

The process of optimizing this small selection of points is generally an iterative process of locating the best trade-off between the various emissions by varying the principal engine control inputs. These are usually start of injection timing, EGR rate, and boost pressure. The best trade-off may not be immediately apparent. Reference must be made to the effect of the catalyst, if fitted, and to the legislative standards in force. Choosing the optimum trade-off between the various emissions and fuel consumption can be a matter for fine judgment.

10.7.2 Transient Calibration

Clearly the preceding technique is steady state based. The engine must work in a real situation which requires attention to the transient calibration.

At one extreme the transient calibration may be considered as a means of moving the engine from one steady-state point to another with the minimum deviation from design conditions. This is in itself a demanding task as the transient factors such as boost pressure will deviate grossly from their steady state values. The strategy must account for these effects.

To continue the preceding example, during an increase in power demand the strategy must limit fueling initially to the level which can be burned without excessive smoke production. Due to the low initial boost pressure this may be a lower fueling than may be expected from the steady-state calibration.

Today's control structures commonly use a combination of open- and closed-loop algorithms. Open-loop controllers simply schedule their output using a look-up table or simple algorithm taking as inputs the measured operating point and the demanded setpoint for the variable in question. The resulting control action is typically output as an analogue signal or pulse width modulation (PWM) signal.

Closed-loop controllers may use the same forward path structure as an open-loop controller but trim their output according to the error between the measured value (or feedback) and the setpoint. The most common algorithm used to do this trimming is referred to as PID control. Here the controller output is modified by an amount proportional to the error (P or proportional control), an amount proportional to the integral of the error over time (I or integral control), and by an amount proportional to the derivative of the error (D or derivative control). The gains of these modifiers may be individually set. Closed-loop control can only be implemented if the feedback is measurable; pressure, temperature, position, and speed control loops are relatively simple to close. The main problem is the harsh environment the transducers are required to work in and the cost of the extra components. Injection timing is more difficult to measure directly and so is often controlled in an open loop. An intermediate measure is to control the cam position in a closed loop. There is a good correlation between cam position and start of injection timing for a given engine operating point. This allows the required cam position for a given start of injection to be predicted empirically from a look-up table. Fueling can be handled in a similar manner. Some advanced control strategies use instrumented injectors and cylinder pressure transducers to improve the quality of these control loops.

Both open- and closed-loop controllers will use gains or constants to describe their operation. Setting these gains is an important part of the transient calibration task. In general the gains are scheduled according to engine operating point. Engine systems are likely to respond very differently at widely disparate operating points due to the highly nonlinear nature of the processes. There can be many hundreds of such parameters in a modern calibration. Arriving at satisfactory values for all of them is time consuming and expensive. This is done relatively rarely; usually a similar calibration is used as a starting point, requiring only a limited number of changes to be made.

Advanced control strategies aim to be more sophisticated by scheduling the change in engine control inputs in an organized manner. A good example of this is the EGR control system. Some codes can only move from one steady-state setpoint to the next directly. More recent codes allow a preset profile to be superimposed on the EGR demand during the transient. For instance, following a step-up in power demand from light to medium load it may be desirable to cut EGR entirely during the transient to raise the air-fuel ratio and help counter the effect of low initial boost pressure. Normal EGR rates could be resumed as the engine settles after the transient. Transient smoke would be reduced at the expense of a possible increase in NO_x.

These aspects of the strategy must be calibrated, preferably using a dynamic emissions test facility to evaluate the changes. Such facilities are scarce and expensive but allow the most benefit to be derived from the sophisticated algorithms now in use. More extensive use is also being made of advanced computer simulation techniques to allow an initial calibration before the hardware is even built. This technique relies on effective models of the plant, which again will require experimental data to validate them.

Some advanced control strategies incorporate models of the plant to improve the quality of their instructions. An example would be a model of the intake and exhaust systems, which allows the instantaneous air-fuel ratio to be estimated, allowing more precise control. The estimate may be used in place of a measured feedback signal in a control loop. This can lead to emissions improvements and enhanced driveability. Again these models need to be carefully validated against experimental data.

All control strategies have extensive driveability functions to perform. These risk conflicting with the emissions calibration. This effect is minimal in practice as the driveability functions usually only affect fueling levels. The emissions measures usually take engine speed and the postfilter fueling as their inputs and will thus not be greatly affected. There are conflicts, however. The limitation of transient smoke via fueling limits also has the effect of limiting transient torque. Usually this is not a problem as the rate of change of fueling is limited to avoid unpleasant powertrain oscillations, although there could be regions where the torque response is delayed sufficiently for subjective driveability to be affected.

10.7.3 Implications of Transient Emissions on Powertrain Control

Transient performance is very difficult to optimize using traditional techniques. It is, though, a crucial area of engine operation. For example, around half the total NO_x produced during a typical EEC-type approval test is attributable to the acceleration phases, even though they constitute only a quarter of the test duration. Any improvements here will have a significant impact on the overall results. Improving transient particulate production would have similar benefits. Around a third of the total PM mass is deposited during the acceleration phases. It may be argued that the ECE15 + EUDC cycle is comprised of predominantly steady-state running or slow transients, which may be approximated as steady state. This approach is invalid if the powertrain control is poor during transient maneuvers. Poor control may arise from a number of factors. A major area for improvement is in thorough calibration of electronic control strategies. This necessitates sophisticated experimental facilities equipped with emissions analyzers of sufficiently fast response to differentiate between rival calibrations.

Factors that affect the temperature and speed of combustion need to be considered when attempting to predict emissions formation. The major parameters are engine speed, injection timing, and fuel-air ratio; the latter in turn depends on fueling, boost pressure, EGR fraction, and charge air temperature. There are important heat transfer aspects which must be considered. These include heat transfer between the manifolds and the gas and between the combustion chamber walls and the reaction within. All of the factors discussed earlier will have a transient characteristic. It is the interaction of these characteristics that determines the transient emissions performance of the engine.

Overfueling can also occur if there is too much EGR, since the oxygen concentration will fall, raising the equivalence ratio. The EGR fraction may deviate significantly from the target during transients as the manifolds and pipes have volumes which take time to reach a new equilibrium. The control system itself may also introduce unwanted dynamic problems.

Many factors that cause excessive smoke are common to those already mentioned which affect HC production and result in an increased equivalence ratio. They include poor transient control of fueling during transients and excessive amounts of EGR. The design of the EGR map is generally limited by the trade-off between NO_x reductions and smoke increases. Of course, an engine which has excessive EGR during a transient will produce less NO_x. Similarly, too little EGR during transients will adversely affect NO_x production.

10.8 APPLICATION OF COMPUTATIONAL TOOLS TOWARD PREDICTING AND REDUCING EMISSIONS

10.8.1 Introduction

A vast number of engine models have been developed by researchers in many fields. The techniques employed are as diverse as the applications. Clearly, any engine model needs to be carefully designed and validated to suit both the particular characteristics of the engine in question and the intended application. The effort required at this stage is considerable. It is not generally possible to use another researcher's work verbatim, as the context will be different. As such, perhaps the most useful aspect of a review of previous work is to identify and evaluate the structure of the models considered. These structures may then be adapted to the particular case in question and calibrated as required.

Most engine models use a combination of analytical and empirical methods to represent the engine components but may be classified according to their majority content. Analytically based models tend to focus on the degree-by-degree variation of engine variables in some considerable detail and are hence generally slow running. Conversely, empirically based models tend to take a wider view, predicting the mean values or trends of the major engine variables with a subsequent loss of resolution in the limit. This usually allows much shorter run times.

10.8.2 Analytical Models

10.8.2.1 Filling and Emptying Models

A common approach to the modeling of IC engines is the filling and emptying method, Simulation Program for Internal Combustion Engines or SPICE [72] is an example of such a method. The engine is represented as a series of control volumes linked by nodes or valves. The principles of mass and energy conservation are applied to predict engine performance. Turbocharger performance is predicted by an empirical submodel. For emissions prediction a complementary multizone model is run alongside the single-zone model. This takes some inputs from the single-zone model. Rate of heat release can either be read from a data file or simulated using a homogeneous heat release function based on fuel burning rate. A suitable data file could be generated using a more detailed simulation or experimental data. This treatment achieves faster running than some more detailed simulations while retaining sufficient complexity to make accurate predictions. This type of model is useful as a teaching or engine design tool as it predicts the effects of varying engine parameters on the performance. A high level of analytical understanding of the engine operation is encapsulated within the code. This assists the user to appreciate the workings of the system.

10.8.2.2 Models with Pressure Wave Actions in Inlet and Exhaust System

The filling and emptying technique, by definition, treats the inlet and exhaust systems as reservoirs subject to the simple conservation laws of mass and energy. It cannot, therefore, distinguish between the strongly unsteady flow in the pipe systems upstream of the inlet manifold and downstream of the exhaust manifold, on the one hand, and the filling and emptying of the manifolds themselves.

In many cases the pressure wave action in these pipe systems greatly affects the performance of the engine and is indeed exploited to achieve substantial performance enhancements relative to "untuned" systems. Examples are:

1. Tuned individual inlet pipes in high-performance racing engines
2. Inlet resonance systems as used in certain truck engines [73, 74]
3. The pulse systems of turbocharging as widely used in light- and heavy-duty vehicular diesel engines [61]

In these systems pressure wave action initiated by the action of inlet and exhaust valves or ports results in pressure waves moving in both directions along the various pipes. This is subject to complex processes of partial reflection and partial transmissions at branches and reflection at the outer terminations of the pipe system, that is, the open ends in naturally aspirated engines, and in the case of turbocharged systems, the compressor or cooler on the inlet side and the turbine on the exhaust side.

The analytical techniques used in assessing these pressure wave effects are all designed to solve by numerical methods the conservation equations for mass, momentum, and energy for one-dimensional, unsteady, irreversible flow. The two main approaches are:

1. The method of characteristics, extensively described by Horlock and Winterbone [75].

2. The direct solution of the conservation equations by various numerical techniques as described by Chen et al. [76].

The relevant computer submodels based on these theoretical approaches are fully integrated into the complete engine performance prediction model, with the submodels for in-cylinder processes and for the turbocharging system, where applicable, being substantially similar to those used in filling and emptying methods.

10.8.2.3 Mean Value Models

To dramatically speed up a simulation it is possible to neglect the cyclic variation of engine parameters and to alternatively use a mean value model. Here the time scale under consideration is much longer than a single engine cycle and much shorter than, say, an engine warm-up time. The time scale is just adequate to describe accurately the changing mean value of the most rapidly changing engine variable. Hendricks discusses such a model of a spark-ignition engine [77]. It uses three differential equations and a number of instantaneous expressions to represent the main engine submodels, fuel dynamics, crankshaft dynamics, and manifold airflow. Model calibration constants are calculated from experimental data, although the structure retains a mainly analytical approach to the simulation process. In this manner the experimental data are expressed as relatively few constants, allowing the model to be compact and quick running while retaining considerable complexity and accuracy. The speed of the model is such that it is useful for control system design and possibly for a model-based controller. The complete model can achieve an accuracy for mechanical predictions of ± 2.0 percent in the steady state and similar in the transient.

Jensen applied the same technique to a small turbocharged diesel engine [78]. Components such as the turbocharger, manifolds, and restrictions are modeled in a similar fashion to that described by Hendricks [77]. Performance parameters are predicted to better than 10 percent during steady and transient operation. Emissions are not predicted but the technique could be extended for this purpose. The authors note that heat transfer in the manifolds can have a significant effect for certain transients. This effect is not modeled currently.

10.8.3 CFD Approach

Some of the most detailed approaches to engine modeling are those employing computational fluid dynamics (CFD). This technique is used to predict fluid motion and associated properties within the engine. The combustion chamber is split into many discrete volumes, typically as many as 40,000 elements are used. The partial differential equations of momentum, continuity, and energy with additional formulations for the creation and dissipation of turbulent energy are then solved by finite difference methods for each element. The results may be validated by comparing them to measured results using laser Doppler or hot wire anemometry.

This method has extremely large computing overheads, but is invaluable as a tool for greater understanding of the airflow patterns within the engine. Somerville used CFD techniques to investigate the air motion in an IDI diesel engine and compared the predictions with experimental measurements [79].

A technique described by Payri uses a similar elemental structure to model the complete combustion process [80]. The method uses many analytical and semi-empirical correlations to model the many processes which govern the evolution of the fuel from the start of injection until the end of combustion. The fuel is split into many "packets", one released at each degree of crank rotation. The evolution of each is predicted at each time step. The results can be used to predict engine performance and emissions through further correlations and analytical steps. The computational task is great due to the many calculations required at each time step.

Quio attempted to speed up the run times for such a model. Here the combustion chamber is represented by a two-zone model [81]. One zone represents the burning fuel spray; another models the unburned surroundings. Treating the fuel spray as a lumped volume makes the simulation run much more quickly while acceptable accuracy is claimed, except for smoke predictions. The simulation was implemented using a transputer to achieve concurrent processing. This enables the code to run still faster. The simulation still takes 24 minutes to simulate 10 seconds of transient running, as opposed to several hours on a conventional system.

10.8.4 Empirical Models

Models that use experimental data in a raw or processed form for most of the predictive process may be termed *empirical*. There are many forms of empirical models with different aims and attributes.

10.8.4.1 Interpolation from Steady-State Maps

Perhaps the most simple approach to empirical engine modeling is to interpolate from engine maps. Data are captured from an engine running at steady conditions, recording emission levels at a series of speeds and loads. These data can later be interpolated for speed and load to locate the instantaneous operating point of the engine during a transient test [82]. One map is required for each engine emission of interest and another showing fuel consumption. The major shortcoming of this approach is the transient accuracy. The emissions of a diesel engine are critically dependent on the equivalence ratio (Φ). For a turbocharged and intercooled engine, Φ is a function of turbocharger and intercooler performance. Both of these are very different during transients when compared to the steady-state or "design point" condition. Other important parameters are cylinder wall temperature, injection timing, EGR ratio, and exhaust back-pressure. These parameters also vary during engine operation. Clearly, an accurate transient model must have knowledge of these effects. This can be achieved using various extensions of the mapping method.

One such development of the steady-state mapping procedure is discussed by Watson [83]. The conventional engine map of emissions against torque and speed is extended to include speed and torque derivative axes. This necessitates a five-dimensional interpolation routine instead of the conventional three-dimensional structure. It also requires the collection of transient data, in this case from a transient chassis dynamometer. The technique shows a clear improvement in accuracy over the steady-state method when used to predict transient emissions. Errors occur because the four independent variables considered do not fully describe the system. More errors will be introduced by the wide scattering of experimental data points. The amount of testing and data processing required is considerable. This limits the feasible density of the maps to eight torques by eight speeds by three-torque derivative levels by three-speed derivative points. A higher density of data would improve the model accuracy but data collection, storage, and processing would become an increasing problem.

10.8.4.2 Polynomial Methods

Jiang describes a method combining analytical and empirical methods to model a diesel engine [84]. Models of combustion and the turbocharger assembly are used to predict an instantaneous equivalence ratio and engine speed. These values are then used to predict the particulate emission using a fourth-order polynomial curve fit of experimental data. Different polynomial coefficients are used for each of six engine speeds spaced over the operating range. The work showed a good correlation with experimental data (4.4 percent over the EPA transient cycle). The model is compact because of its use of analytical methods and the use of a polynomial to replace a look-up table. Possible limitations are that the model does not account for the effects of varying engine temperatures or injection timing. Both of these factors will also affect the particulate production. The engine considered did not use exhaust gas recirculation (EGR). EGR is an important method of reducing NO_X emissions, at the expense of increased particulates.

Stronach uses a polynomial fit method to predict engine torque from a turbocharged diesel [85]. Two pairs of polynomials are used. The first pair predicts the ratio of indicated torque to fuel rack position. These polynomials are in terms of the ratio of inlet pressure ratio (σ) to fuel rack setting. The second pair of polynomials predicts the ratio of net torque to fuel rack position. These are also in terms of σ. Engine delay is represented by a first-order lag. The value of σ is calculated using other engine parameters, such as exhaust pressure ratio and turbine performance, which are predicted by polynomial curve fits or other empirical models. The model is useful for control system design as it is very compact; all the experimental data are reduced to a few polynomial coefficients, making it very fast running. A major limitation is that no attempt is made at emissions prediction. The method could presumably be extended to address this shortcoming.

10.8.4.3 Modal Analysis

The legislative tests for vehicle emissions can themselves be used as the basis of a useful model. The drive cycle may be split into discrete segments classified as idle, acceleration, cruise, or deceleration. Cycles such as the ECE15 + EUDC are easy to divide in this way. Less regular cycles such as the FTP75 cycle are less well suited to this treatment. Once the cycle has been split as described earlier the emissions during each mode may be calculated. Measures aimed at improving one particular mode may then be simulated by scaling the experimental data by an appropriate factor. This is a relatively crude approach since each mode contains quite a range of operating conditions.

A refinement of the technique is described by Boam [86] and Ma [87]. They describe a model using modal emissions data from an ECE15 drive cycle test. Here the entire test is split into one-second segments. The contribution of each one-second segment together with the relevant engine performance parameters are presented in a tabular form. The data are used to predict the effect of modifications to the power plant. The modified system is tested in the steady state at a few representative operating points (four in this case). The original test data are then categorized to determine which calibration point best matches the operating point during each one-second segment. Any improvement observed during the calibration tests can then be applied to the experimental emissions data as a simple ratio of modified performance over original performance. This model regards the engine as being a quasi-steady-state device in that operating conditions are assumed to be constant during the segment. Errors will arise due to this assumption. Further errors are inevitable when a very small number of calibration points are used for the modified design. The model is nevertheless very powerful when used to predict drive cycle performance under various modified engine/catalyst conditions. The data manipulation is performed using a spreadsheet and is hence very simple and fast. Analysis may be undertaken for other drive cycles or different transmission strategies. The results from such work would need to be carefully validated, and larger errors would be expected. This is because different transient conditions will occur under the new operating cycle, which would alter the base data for a given engine operating point.

A more sophisticated variation on the same theme is described by Callahan [88]. Here linear regression analysis is used to predict real transient emissions using experimental data from a set of 120 discrete transient events. Each event is described by the length of the segment (in seconds), the average speed and load and the rate at which speed and load changed. Each emission is represented by an expression with an offset and polynomial terms for the important engine variables. First-order polynomials are used for the average power and the natural log of the time spent at that condition. Second-order polynomials are used for speed, torque, the first derivatives of torque and speed, and the first derivatives of torque and speed divided by time spent at that condition. The coefficients for each of these terms for each emission are calculated by the least squares linear regression method from the experimental data. This is effectively a lumped model, where the effects of temperature transients, boost pressure transients, and so on are accounted for in the data from the transient tests. The model is not transient in operation as it requires the driving pattern under investigation to be split into segments of known properties, which are then matched to the discrete transients. Each segment must therefore have finished before its emissions can be calculated. The simulation is used as a design tool to look at the total emissions from a drive cycle. The model is considerably more complex than the previous approach, but should be more general in its applicability. It may prove difficult to predict the effects of power plant modifications as the transient tests and subsequent calculations would need to be repeated. It would, however, be useful when studying the effects of different transmission strategies or other vehicle parameters.

10.8.5 Neural Networks

A useful technique for constructing nonlinear empirical models is the artificial neural network. A neural network is a structure consisting of a large number of very simple units, which combine to represent any given relationship of inputs and outputs. The name originates from the biological structures that inspired them.

The structure of a typical network is shown in Figure 10.29. It consists of three groups of nodes or neurons, the input layer, one or more hidden layers, and an output layer. Each layer is fully interconnected to the next via a series of connections, called synapses. The number of hidden layers and the number of

Fig. 10.29.

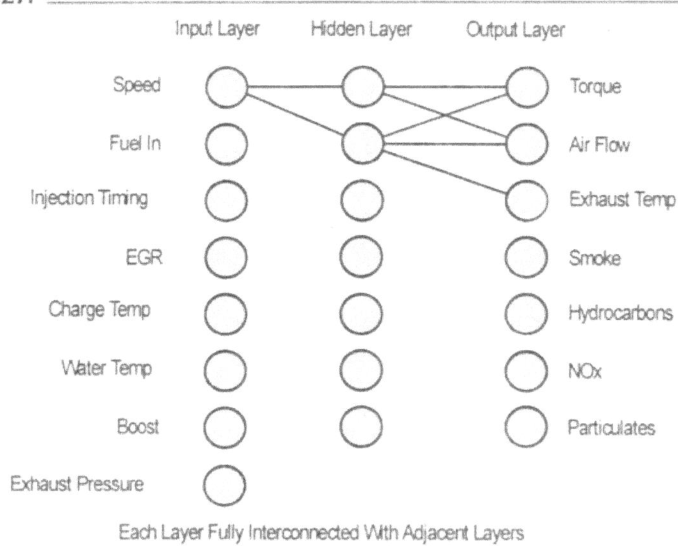

A typical neural network configuration.

neurons in each are optional. There is one input neurone for each input variable, which is presented to the neuron as a floating point number scaled from 0 to 1. These values pass via the synapses to the first hidden layer where the data from each input are weighted, summed, and passed through a nonlinear threshold function to give a single output value. This process is repeated at each subsequent hidden layer. The output neurons collect signals from the last hidden layer and present a value between 0 and 1.

To make a network represent a real system the weights and biases are optimized by an iterative training process. Once trained the network is treated as a black box. Input values are presented and the outputs are collected. A network can represent a huge volume of data in only a few lines of code and runs very quickly once trained. Networks have been used in many modeling and control applications, some of which are discussed here.

Scaife trained a neural network to successfully detect the onset of faults on a diesel engine by observing 17 engine parameters and comparing them with previous tests where known faults had been invoked [89]. Bacon investigated the potential for replacing elements of a spark-ignition engine controller with a neural network [90]. Large sections of current engine controllers are concerned with mapping setpoints such as ignition timing against operating conditions. The work aimed to replace such linear interpolation with a relationship learned by a neural network. Such an approach can make use of the inherent smoothing effect of neural networks to give a continuous transition between one area of the map and another. Shayler et al. applied neural networks to the fuel consumption prediction of an SI engine with varying operating temperatures [91]. The fuel consumption is strongly dependent on indicated mean effective pressure (IMEP), engine speed, and EGR rate. This type of multidimensional data manipulation is an ideal application of the technique and a correlation coefficient of 0.9985 was achieved when predictions were compared with experimental data. O'Reilly discusses a neural network based air-fuel ratio predictor for use in SI engine control [92]. The network models the complex behavior of the inlet manifold using a time history of previous inputs to predict the future AFR. Results were encouraging, with errors of less than 2 percent attainable for transient operation.

10.8.6 Hybrid Models Using Modular Structures

Each of the techniques described previously is suited to the representation of a particular group of processes. It is usual for a variety of these techniques to be combined within one simulation in order to apply the most appropriate tool to each of the system processes. A simulation environment that adopts a modular structure is useful in the development of these models. Most programming languages allow the use of subroutines or functions which, facilitate a modular structure. Many modern simulation tools use this modularity to present the simulation in a convenient format, which allows the user to develop each submodel in isolation. This can greatly simplify the design and validation stage of simulation work. A common approach is to define a global protocol for subroutine calls, which is then built into all submodels enabling easy assembly of individual blocks to form the overall simulation.

10.8.7 Remarks

A key weakness in most of the thermodynamically based available engine simulations is the way in which the combustion process is represented. The zero-dimensional or filling and emptying models rely on mathematical functions representing the overall engine heat release, implicitly embodying gross assumptions as far as engine combustion is concerned. Moreover, as a tool to investigate engine emissions, such models offer limited scope since they lack any detail of temperature and composition variability within the cylinder, a prerequisite to any successful emission predictions. However, a number of models employ multizone emission models which run in parallel to the zero-dimensional models [93, 94].

Diesel combustion is mainly of a diffusion nature and therefore is controlled by the processes that lead to fuel-air mixture formation. As such, the combustion chamber fluid mechanics including fuel spray formation and air motion characteristics significantly affect the combustion process. It is not surprising that the multidimensional CFD models supported by sophisticated laser and high-speed photography techniques have been used over the last decade to address the engine fluid flow and to a lesser extent combustion.

The CFD models aim at solving the mathematical equations describing fluid motion (primarily the Navier-Stokes equations) and use algorithms for solving the resultant algebraic equations. Although the detailed CFD models provide a formidable amount of predicted data regarding mainly the flow field, nevertheless, their accuracy is still limited by how accurately such phenomena as turbulence, heat transfer, and mass transfer are modeled and the accuracy of the boundary conditions presented to the system. Also, these models require a significant amount of computer power to predict even simpler fluid flow and are still under development for predicting spray behavior and combustion. As a result, the use of CFD models as part of a routine cycle simulation program is not feasible at this stage.

The alternative approach is to employ the multizone phenomenological concept to model diesel combustion and emissions. This approach has the advantage of providing in-cylinder temperature and concentration variations as compared to zero-dimensional models and needs significantly less computing power resources as compared to CFD models. Indeed, for routine engine cycle simulation work which involves appraisal of conceptual designs, performance and emissions analysis, control system development, component matching, and overall optimization studies, this type of model is the only practical choice at the present time.

Although multizone combustion models vary in complexity, the basic concept is to divide the fuel spray into various elements of fuel (referred to as zones or packets), which entrain air and ignite as the mixture is prepared for combustion. In these types of models, simplified quasi-steady equations are used to describe various processes of fuel injection, atomization, droplet formation, air entrainment, droplet evaporation, ignition, heat release, heat transfer, and so on. A synthesis of all these equations in an interactive manner together with equations for mass and energy conservation within each zone provide the detailed data needed for emission calculation.

10.9 ADVANCED ENGINEERING PROJECT

An advanced engineering project has been undertaken by the Ford Motor Company Limited to evaluate the latest high-speed DI diesel engine technology against the proposed changes in emission legislation while also assessing improvements in performance. The important findings from this project are now presented.

10.9.1 Engine Description

The engine used for these tests was a prototype 16-valve version of the 2.5-liter DI engine used in the Ford Transit. The engine has the following features [95, 96]:

- direct injection
- central vertical 17-mm Bosch injectors
- four-valves per cylinder, two intake and two exhaust
- separate intake ports
- intake port induced swirl level of around 2.0 with both ports operational
- Siamese exhaust ports
- intake port deactivation to provide variable swirl
- cross-flow configuration
- Stanadyne DS electronically controlled fuel injection pump
- electronically controlled EGR system
- EGR cooler
- turbocharged and intercooled

10.9.2 Performance Development

The performance objectives for this engine development were to pursue future emissions legislation and enhanced customer features such as improved specific power and torque, fuel consumption, and noise levels.

The NO_x-particulate trade-off is of prime importance in meeting current and future emissions legislation. As a result, features such as exhaust gas recirculation, EGR cooling, and air charge cooling were chosen to be incorporated in the development program and were thus included in the final engine specification.

Intake System The cylinder head was deliberately designed with two intake ports per cylinder to provide the capability of in-cylinder air swirl variation by port deactivation. Development trials were conducted into the effects of intake port configuration by combinations of single (by deactivation) and double ports, varying intake system pressure loss and swirl ratio, in combination with varying quantities of EGR at part load.

Fuel Injection System Electronic control of the fuel injection pump and EGR system allows accurate calibration of optimized requirements throughout the engine operating range. The use of second-generation electronic fuel injection pump technology has demonstrated, incorporated with the central injector location, significant improvements in performance and emissions, specifically in the low to mid-speed range.

The Standadyne DS rotary pump was used during the major part of the development program. This pump featured a single high-speed solenoid valve to control both fuel quantity and injection timing as well as variable injection rate. This combination allowed the choice of an effective region of pump cam profile.

Control of the pump was by an ECU, which together with pump control also included the capability of modulated control of the EGR valve and EGR throttle and, subsequently, control of intake port deactivation. The ECU took information from a complete spectrum of engine sensors including crankshaft position, coolant temperature, air charge temperature and pressure, EGR control position feedback, together with feedback from the pump itself. As part of the development program, a range of injector specifications were evaluated.

EGR System The EGR system consisted of an electronically controlled, vacuum-operated, poppet-type EGR valve mounted on the exhaust manifold. The EGR gases were then cooled by passing through a gas to water heat exchanger. Engine coolant was used for the water-side cooling media. To generate sufficient pressure drop to obtain the desired volume of EGR under certain part-load regions (where EGR optimum levels for best emissions necessitated higher flows than are naturally possible through the EGR valve), a throttle valve was incorporated in the induction system. This was operated when desired such that boost pressure and air charge flow were restricted to favor the flow of exhaust gases from the EGR cooler. A schematic diagram of the system used is shown in Figure 10.30.

10.9.3 Results

Development and optimization of calibrations for modern high-speed DI engines are now highly complex owing to the number of variables available to the

Fig. 10.30.

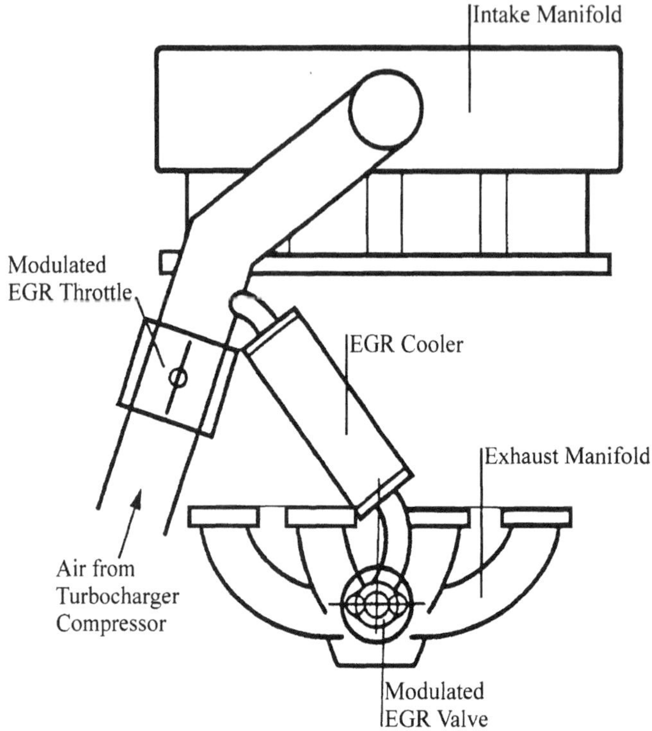

Schematic diagram of the EGR system used on a Ford prototype 16-valve, 2.5-liter DI diesel engine.

development engineer. However, electronics has enabled the precise control of a greater number of variables, with typical systems now covering:

- injection fuel quantity
- injection timing
- in-cylinder swirl level, by port deactivation
- boost pressure
- EGR quantity

Because of the number of variables being tested, a neural network computation technique was utilized for pattern recognition as an aid to results analysis. An example output from this analysis is shown in Figure 10.31, which shows NO_x plotted against particulates as a function of swirl level.

The most significant observations from these results were:

1. At low to mid speed, part-load investigations into the effect of increasing swirl without EGR showed a penalty on NO_x emissions.

Fig. 10.31.

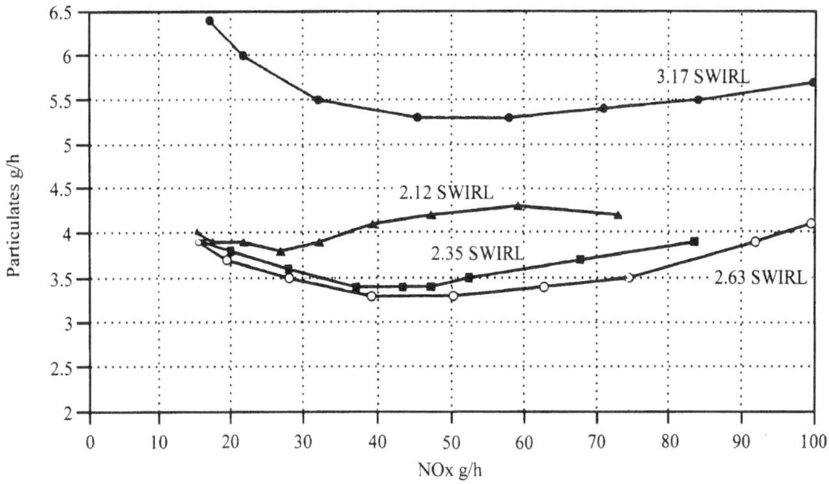

Example of the results from a neural network analysis for an engine operating condition of 2000 rpm and 2 bar BMEP.

2. Similar investigations using varying EGR rates showed a capability for an improved NO_x-particulates trade-off at increased swirl, but with more EGR.
3. Adopting port deactivation to increase swirl can increase pumping losses and, hence, worsen specific fuel consumption.
4. Adjustment of swirl without recourse to port deactivation could be more beneficial to fuel consumption.

REFERENCES

1. Love, M. et al., "Diesel Passenger Car and Light Commercial Vehicle Markets in Western Europe," Ricardo Consulting Engineers Ltd., February 1997.
2. Darkawa, K., and P.W. O'Callaghan, "Energy usage and environmental pollution in the United Kingdom: Which sector dominates?" *Proc. Institution of Mechanical Engineers, Part A, Journal of Power and Energy*, Vol. 210, pp. 3–14 (1996).
3. Fidell, T., "LP Gas as an Automotive Fuel," Automotive Fuels for the 21st Century, *IMechE* seminar, London (January 21, 1997).
4. Heywood, J.B., "Motor Vehicle Emissions Control: Past Achievements, Future Prospects," Combustion Engine Group Prestige Lecture, *IMechE*, London (May 21, 1996).
5. *Professional Engineering*, Vol. 9, no. 9, p. 5 (May 8, 1996).
6. Seiffert, U., "Environment: Continuous Challenge for Automobile R&D," IMechE International Seminar on Application of Powertrain and Fuel Technologies to meet Emission Standards, Key Note Address, London (June 24, 1996).
7. Horrocks, R.W., and P.S. Robertson, "Light-Duty Diesel Developments for Low Emissions and Low Fuel Consumption," 4th International Conference and Exhibition Automotive Industry and the Environment, Renaissance Hotel, Brussels, Belgium (September 17 and 18, 1996).
8. Horrocks, R.W., "Automobile and Engine Technology," 6th Aachen Colloquium, Eurogress Aachen, Germany (October 20–22, 1997).

9. Bird, G.L., "The Ford 2.5 Litre High Speed Direct Injection Diesel Engine—Its Performance and Future Possibilities," SAE Paper 850262, 1985.
10. Takeuchi, K., K. Kubota, M. Konagai, M. Watanabe, and R. Kihara, "The New Isuzu 2.5 Litre and 2.8 Litre 4-Cylinder Direct Injection Diesel Engine," SAE Paper 850261, 1985.
11. Knecht, W., T. Isik, and R. Tardi, "Development of Small High Speed Diesel Engines with Direct Injection for Low Emissions," SAE Paper 885152, XXII Fisita Congress "Automotive Systems—The Future," September 1988.
12. Knecht, W., G. Maul, and A. Schoni, "Development of a low emission small high-speed turbocharged diesel engine with direct injection," *IMechE* conference "Worldwide Engine Emissions Standards and How to Meet Them," September 1991.
13. "Fiat brings world's first DI car diesel to market in Italy," *Ward's Engine Update*, Vol. 14, no. 13, July 1, 1988.
14. "Variable Turbo Boosts Car Diesel Performance," *High Speed Diesels and Drives*, Vol. 10, no. 6, July–August 1991.
15. Stephenson, J.A., and B.A. Hood, "A high-speed direct injection diesel engine for passenger cars,"*Proc Inst. Mech Engrs, Part A*, 1988, 202 (A3), pp. 171–181.
16. Basshuysen, R.V., J. Steinwart, H. Stahle, and A. Bauder, "Audi Turbodiesel-motor mit Direkteinspritzung," *MTZ Motortechnische Zeitschrift* 50 (1989)12.
17. Stock, D., and R. Bauder, "The Audi 5-Cylinder Turbo Diesel Engine: The First Passenger Car Diesel Engine with Second Generation Direct Injection," SAE Paper 900648, February 1990.
18. "Rover RV market improves with 2.5 L diesel," *High Speed Diesels and Drives*, November–December 1989, VIII (6), 16.
19. "Rover's New 'L' Series the 'Best Practice' Diesel," Rover External Affairs, January 1995.
20. Bostock, P.G., and L. Cooper, "Turbocharging the Ford 2.5 HIGH SPEED DI diesel engine," I.Mech.E. Seminar on Diesel Fuel Injection Systems, April 14–15, 1992.
21. Rhode, W., S. Gokesme, J.R. Liang, and J.L. Schmitt, Der neue direkteinspritzende 1.9 Dieselmotor von Volkswagen, 3 Aachener Kolloquium, Fahrzeug- und Motorentechnik '91, 1991.
22. Jelden, H., and M. Willman, "Possibilities for Increasing the Power of the Volkswagen Four-Cylinder TDI Engine," 5th Aachen Symposium, Vehicle and Engine Technology, Aachen, October 1995.
23. Volkswagen Press Information, Volkswagen, Wolfsburg, March 1996.
24. Krause, R., and W. Saltzer, "A New Turbo Diesel Engine with Direct Injection for the Mercedes-Benz E-Class Part 1: Design and Mechanics," *MTZ Motortechnische Zeitschrift* 56 (1995) p. 6.
25. Peters, A., and W. Putz, "A New Turbo Diesel Engine with Direct Injection for the Mercedes-Benz E-Class Part 2: Combustion and Engine Management," *MTZ Motortechnische Zeitschrift* 56 (1995) 7/8.
26. "New Vauxhall direct injection Ecotec engines open a new chapter in the history of the diesel," Vauxhall News Release, July 1996.
27. Bauder R., W. Dorsch, H.-W. Polz, and L. Mikulic, "The New Audi V6 Turbo-Diesel Engine with Direct Injection and Four Valves per Cylinder, Part 1: General concept and thermodynamics," 18th International Vienna Engine Symposium 1997, Vol. 1, 48–69.
28. Bauder, R., H. Hoffmann, H. Stahle, and H.-W. Polz, "The New Audi V6 Turbo-Diesel Engine with Direct Injection and Four Valves per Cylinder, Part 2: Engine design and mechanical components," 18th International Vienna Engine Symposium 1997, Vol. 1, 70–87.
29. TurboDaily Iveco Information, 1985.
30. Iveco Daily Press Pack, July 1996.
31. Ricardo Information Services, *New Engine News*, January 1997.
32. Autocar, 26–31, December 1996.
33. Opel Press Information, Geneva Motor Show, March 1997.
34. "Toyota's New Direct Injection Diesel Engine," Toyota Press Information, March 1997.
35. Dunne, J.M., Status of Emissions Legislation, IMech E International Seminar, Application of Powertrain and Fuel Technologies to Meet Emission Standards, London (June 1996).

36. Horrocks, R.W., "Light Duty Diesels—the Emissions Challenge," *Proc. IMechE.*, Vol. 206 D01091, United Kingdom (1992).
37. Horrocks, R.W., "Light Duty Diesels—an Update on the Emissions Challenge," Worldwide Engine Emission Standards and How to Meet Them, *IMechE* seminar, London (May 1993).
38. Burgler, L., P. L. Herzog, and P. Zelenka, "Strategies to Meet US 1994/95 Diesel Engine Federal Emission Legislation for HIGH SPEED DI Diesel Engine Powered Vehicles," *Proc. IMechE*, Vol. 206 (1992).
39. Heywood, J.B., *Internal Combustion Engine Fundamentals*. New York: McGraw-Hill (1988).
40. Ball, D., J.P. Brimblecombe, and F.M. Nicholas, "Review of Air Quality Criteria for the Assessment of Near-Field Impacts of Road Transport," TRRL Report CR 240, Transport and Road Research Laboratory, Crowthorpe, United Kingdom (1991).
41. Graedal, T.E., D.T. Hawkins, and L.D. Claxon, *Atmospheric Chemical Compounds*. Orlando: Academic Press (1986).
42. U.S. National Ambient Air Quality Standards for Particulate Matter, *USA Federal Register*, Vol. 62, no. 138, pp. 38652–38762 (1977).
43. Greeves, G., and C.H.T. Wang, "Origins of Diesel Particulate Mass Emission," SAE Paper 810260 (1981).
44. Fosberry, R.A.C., and D.E. Gee, "Some Experiments on the Measurement of Exhaust Smoke Emissions from Diesel Engines," MIRA Report 1961/5, United Kingdom (1961).
45. Kittelson, D.B., and N. Collings, "Origin of the Response of Electrostatic Particle Probes," SAE Paper 870476 (1987).
46. Herzog, P., "Emerging Technologies to Meet EURO III and ULEV, and Their Implications for Engine Testing," AVL United Kingdom Ltd Seminar, Donnington Park (February 22, 1996).
47. Glikin, P. E., "An Electronic Fuel Injection System for Diesel Engines," SAE Paper 850453 (1985).
48. Martin, J., "Improvements in Transient Diesel Engine Performance by Electronic Control of Injection," IMechE Paper C382/140, London (1989).
49. Glassey, S.F., A.R. Stockner, and M.A. Flinn, "HEUI—A New Direction for Diesel Engine Fuel Systems," SAE Paper 930270, March 1993.
50. Stockner, A.R., M.A. Flinn, and F.A. Camplin, "Development of the HEUI Fuel System—Integration of Design, Simulation, Test, and Manufacturing," SAE Paper 930271, March 1993.
51. Hower, M.J., R.A. Meuller, D.A. Oehlerking, and M.R. Zielke, "The New Navistar T 444E Direct-Injection Turbocharged Diesel Engine," SAE Paper 930269, March 1993.
52. Glassey, S.F., R.H. Holtman, and R.A. Smith, "Development of HEUI Fuel System for 2-3L Automotive Diesels," *IMechE*, September 1995.
53. Rinolfi, R., R. Imarisio, and R. Buratti, "The potentials of a new common rail diesel fuel injection system for the next generation of DI diesel engines," 16th International Vienna Motor Symposium, May 4–5, 1995.
54. Miyaki, M. et al., "Development of New Electronically Controlled Fuel Injection System ECD-U2 for Diesel Engines," SAE Paper 910252, 1991.
55. Needham, J.R., and A. Bouthenet, "Competitive Fuel Economy and Low Emissions Achieved Through Flexible Injection Control," SAE Paper 931020, March 1993.
56. Stumpp, G., and M. Ricco, "Common Rail—An Attractive Fuel Injection System for Passenger Car DI Diesel Engines," SAE Paper 960870, February 1996.
57. Ladommatos, N., S.M. Abdelhalim, and H. Zhao, "The effects of carbon dioxide in EGR on diesel engine emissions," International Seminar on Application of Powertrain and Fuel Technologies to Meet Emissions Standards, Paper No. C517/028/96, pp. 157–174, June 24–26, London (1996).
58. Baert, R., D.E. Beckman, and A.W.M.J. Veen, "EGR Technology for Lowest Emissions," *IMechE* Seminar on Application of Powertrain and Fuel Technologies To meet Emissions Standards, Paper No. C517/034/96, London (1996).
59. Kerswill, J., "Raising the game," *Diesel Car & 4 × 4 Magazine*, pp. 50–51 (September 1996).
60. "EGR and SAI without external pipes," *Automotive Engineer*, p. 89 (June/July 1996).

61. Watson, N., and M.S. Janota, *Turbocharging the Internal Combustion Engine*. London: Macmillan (1982).
62. Wallace, F.J., D. Howard, and E.W. Roberts, "Variable geometry turbocharging optimisation and control under steady state conditions," *I.Mech.E.*, Int. Conf. On Turbocharging and Turbochargers, p. 215, London (1986).
63. Wallace, F.J., J.G. Hawley, A. Cox, A.C. Pease, R.W. Horrocks, and G.L. Bird, "Variable Geometry Turbocharging of a 1.8 DI Automotive Diesel Engine," 30th ISATA Conference, June 14–19, Paper No. 97EN054, Florence, Italy (1997).
64. Monaghan, M.L., "High-speed direct injection diesels—present and future perspectives," *IMechE* Seminar on Lean Burn Combustion Engines, S433/012/96, pp. 189–199, London (December 3–4, 1996).
65. Winterbone, D.E., D.A. Yates, R.A.O. Clough, P. Gomes, and J.-H. Sun, "Combustion in High Speed Direct Injection Engines—A Comprehensive Study," *Proc. IMechE*, Vol. 208, London (1994).
66. Schechter, M.M., and M.B. Levin, "The Camless Engine," SAE International Congress & Exposition, SAE Paper No. 960581, Detroit, Michigan (February 1996).
67. Brisley, R.J., D.E. Webster, and A.J.J. Wilkins, "Aftertreatment strategies to meet emission standards," International Seminar on Application of Powertrain and Fuel Technologies to meet Emissions Standards, Paper No. C517/047/96, pp. 95–114, June 24–26, London (1996).
68. Iwamoto, M., M. Yokoo, K. Sasaki, and S.J. Kagawa, *Journal Chemical Society, Farad. Trans*, 77, 1629 (1981).
69. Porter, B., T.J. Ross-Martin, and A.J. Truscott, "Control technology for future low emissions diesel passenger cars," International Seminar on Application of Powertrain and Fuel Technologies to meet Emissions Standards, Paper No. C517/035/96, pp. 125–146, June 24–26, London (1996).
70. Shore, P.R., and K.F. Brown, "Diesel Oxidation Catalysts for Heavy Duty Vehicles—A Critical Evaluation of Performance Under Real Driving Conditions and Over U.S. and European Legislation Cycles," 30th International Symposium on Automotive Technology and Automation, Paper No. 97EN003, June 16–19, Florence, Italy (1997).
71. Lueders, H., R. Backes, G. Huthwohl, D.A. Ketcher, R.W. Horrocks, R.G. Hurley, and R.H. Hammerie, "Urea lean NO_x catalyst system for light-duty diesel vehicles," SAE Paper No. 952493 (October 1995).
72. Charlton, S.J., "Simulation Program for Internal Combustion Engines," User Manual, University of Bath (1986).
73. Sato. A., K. Suenaga, M. Noda, and Y. Maida, "Advanced Boost-Up to Hino EP 100-11 Turbocharged and Charge Cooled Diesel Engine," SAE Paper No. 870298 (1987).
74. Cser, G., "Double pressurewave systems—a new way to improve the low speed operation of supercharged engines," 4th International Conference on Turbocharging and Turbochargers, Institution of Mechanical Engineers, Paper C 405/013, London (1990).
75. Horlock, J.H., and D.E. Winterbone, "The Thermodynamics and Gas Dynamics of Internal Combustion Engines," *Oxford Science Publications* (1986).
76. Chen, C., A. Veshagh, and F.J. Wallace, "A comparison between alternative methods for gas flow and performance predictions for internal combustion engines," SAE Paper No. 921734 (1992).
77. Hendricks, E., and S.C. Sorenson, "Mean Value modeling of Spark Ignition Engines," SAE Paper 900616 (1990).
78. Jensen, J. P., A. F. Kristensen, S. C. Sorenson, N. Houbak, and E. Hendricks, "Mean Value modeling of a Small Turbocharged Diesel Engine," SAE Paper 910070 (1991).
79. Sommerville, B., S. MacGregor, S. Charlton, and B. Nasseri, "A Study of Air Motion in an IDI Passenger Car Diesel Engine," IMechE Conference, Experimental and Predictive Methods in Engine Research and Development, London (November 17–18, 1993).
80. Payri, F., J. Benajes, and F.V. Tinaut, "A Phenomenological Combustion Model for Direct Injection, Compression Ignition Engines," *Appl. Math. Modelling*, Vol. 12 (June 1988).

81. Quio, J., J.C. Dent, and C.P. Garner, "Diesel Engine modelling Under Steady State and Transient Conditions Using a Transputer Based Concurrent Computer," SAE Paper 922226 (1992).
82. Waters, W.C., "General purpose automotive performance and economy simulator," SAE Paper No. 720043 (1972).
83. Watson, H.C., and B. Alimoradian, "A Transient Engine Mapping Model for Analysing and Predicting Fuel Consumption and Emissions," *IMechE* Paper C382/098, London (1989).
84. Jiang, Q., and J.H. Van Gerpen, "Prediction of Diesel Engine Particulate Emission During Transient Cycles," SAE Paper 920466 (1992).
85. Stronach, A.F., and R.J. Smith, "Development of a Simulation Model of Turbocharged Diesel Engine Prime Movers for Power System Studies," *Electrical Power and Energy Systems*, Vol. 10, no. 2 (April 1988).
86. Boam, D.J., I.C. Finlay, T.H. Ma, S. Wallace, J. H. Bloomfield, and R. Lee, "Modal Assessment of Hydrocarbons Reduction Technologies in a Vehicle During Engine Warm-Up," IMechE Paper C465/031 (1993).
87. Ma, T.H., and I.C. Finlay, "Modal Assessment of Exhaust Emissions Potential of a Lean Burn Petrol Engine Vehicle," *IMechE* Paper C465/030 (1993).
88. Callahan, T.J., T.W. Ryan, and S.F. Martin, "Comparison of Predicted and Measured Diesel Exhaust Emission levels During Transient Operation," SAE Paper 872140 (1987).
89. Scaife, M.W., "A Neural Network for Fault Recognition," SAE Int. Congress, Session 1, p. 23, Controls for Engines (1993).
90. Bacon, A., "Potential for Engine Control Using Neural Networks," IMechE Paper C448/057 (1992).
91. Shayler, P.J., N.J. Darnton, and T. Ma, "Predicting the Fuel Consumption of Vehicles for Drive Cycles Starting From Cold Ambient Conditions," SIA9506A27 EAEC 5th International Congress, Strasbourg, June 21–23, 1995.
92. O'Reilly, P., and S. Thompson, "Quantitive Robustness Testing of a Neural Network Based AFR Predictor," PD-Vol. 64-6, *Engineering Systems Design and Analysis*, Volume 6, ASME (1994).
93. Bazari, Z., "A DI diesel combustion and emission predictive capability for use in cycle simulation" SAE Paper 920462 (1992).
94. Bazari, Z., "Diesel exhaust emissions prediction under transient operating conditions," SAE Paper 940666 (1994).
95. Menne, R.J., P.J. Lawrence, R.W. Horrocks, and P.S. Robertson, "Ford 4-Valve Light-Duty DI Diesel Developments," SAE Paper 941926 (October 1994).
96. Horrocks, R.W., P.S. Robertson, A.M. Brohmer, and G. Heuser, Entwicklung schnellaufender 4-Ventil-Direkteinspritz-Dieselmotoren, Pkw-Dieselmotoren Haus der Technik, Essen, Germany (June 1995).

CHAPTER 11

Control Technologies for Compression-Ignition Engines

Stephen J. Charlton
Director, Advanced Diesel Engine Technology, Cummins Engine Company, Inc., Indiana, USA

ISBN: 0-12-639855-0
$25.00

11.1 INTRODUCTION

11.1.1 Control and Advanced Technologies

This chapter reviews the state-of-the-art in diesel engine control and advanced technologies as the twenty-first century approaches. Digital control of diesel engines can solve many of the traditional engine control problems: cold start, governing, torque curve definition, transient smoke emission, load acceptance and response, and basic engine protection, which were once controlled with mechanical or simple electrical elements. With the growth of digital control for on-highway automotive engines and increasingly for off-highway and industrial engines, the scope of control has broadened to include emission control systems, fuel systems, air-handling systems, on-board diagnostics, user-programmable features, adaptive control features, and management information retrieval systems. A key challenge facing diesel engine designers today and in the future is the development of technologies that simultaneously improve upon the traditional virtues of the diesel engine while meeting increasingly stringent emission standards, at manufacturing costs competitive with alternative prime movers.

Key technologies driving the dramatic progress in diesel engine performance, fuel economy, and emissions through the 1990s are:

1. Full-authority electronic controls and on-board diagnostics
2. Advanced electronic fuel systems
3. Exhaust gas recirculation
4. Variable geometry turbocharging
5. Advanced exhaust aftertreatment

This chapter will review each of these key technologies and relate them to the needs of both light- and heavy-duty automotive diesel engines.

11.1.2 Introduction to Electronic Control

The rapid growth of electronics and widespread application of microprocessor technology in everyday products has revolutionized the way we live our lives. This same revolution has allowed engine designers to improve dramatically the performance of diesel engines—initially through better control—and more recently through applications such as diagnostic and information retrieval systems.

For many years, prior to the widespread availability of microprocessors, diesel engines in generator-set applications used electronic governors to enhance control of electrical frequency—both in steady-state operation and during load applications. Initially these were analogue devices, which operated on the engine fuel system through an electromechanical actuator driving a mechanical linkage. This was a requirement at that time since fuel systems were generally under some form of mechanical control.

Other early examples of electronics applied to diesel engines included condition monitoring and remote operation of relatively large engine systems, notably power generation or marine applications. These applications allowed such engines to operate unattended and with greatly reduced risk of costly failure or mission-disabling breakdown. This approach also provided improved fuel consumption and reduced operating costs.

By the late 1970s a number of factors came together to encourage a more integrated approach to electronic control of diesel engines: first, an increasingly competitive environment, and the need to meet better the end-user requirements; second, increased awareness of environmental issues generally and the introduction of exhaust emission regulations for diesel engines; and third, the availability at low cost of increasingly sophisticated digital electronic hardware such as microprocessors, sensors, and actuators. Since that time the diesel engine industry has invested heavily in digital electronics and today almost all on-highway diesel engines—from passenger car to heavy-duty truck—have some form of digital electronic control.

This chapter will describe state-of-the-art diesel engine controls, with special emphasis on automotive applications, since these have tended to drive the technology development. The development of diesel fuel systems, which enable flexible control of variables such as fuel quantity, injection timing, fuel pressure, and rate shaping or pilot injection, will be discussed. Such systems provide the foundation for electronic control of diesel engines, just as the development of electronic injection systems for gasoline engines revolutionized the SI engine, and made the carburetor obsolete, in the 1980s. The diesel fuel systems that dominate as we move to the next century are the high-pressure rotary pump (HPRP), the electronic unit injector (EUI), and the emerging high-pressure common rail systems (HPCR). These fuel systems were developed for electronic control and take full advantage of integration to provide very high performance from the injection system.

As designers strive to achieve improved engine fuel economy, performance, and reliability, while simultaneously reducing exhaust emissions by 70 percent to 80 percent from unregulated levels, other elements of engine operation are coming under electronic control, for example, exhaust gas recirculation (EGR) systems, air-handling systems, active wastegate control, and variable geometry turbine control.

Electronic controls for diesel engines have undoubtedly brought benefits to the end user. Perhaps the most obvious of these benefits is the steady improvement in fuel economy, despite the fact that emission regulations in the United States have continued to tighten since the late 1970s. This has been possible through more accurate control of injection timing and fuel metering and the ability to optimize engine performance under all operating regimes, for example, warm-up, idling, cruise, hill climbing, and altitude. Accurate injection timing control from event-to-event, cylinder-to-cylinder, and engine-to-engine also improves the emissions capability of a diesel engine since both oxides of nitrogen (NOx) and particulates are highly responsive to this variable. Since production engines, by definition, meet emissions regulations, any lack of accuracy or flexibility in control will be reflected as a deterioration in other engine attributes, such as increased fuel consumption or first cost.

Other advantages of electronic control for heavy-duty vehicles have appeared in end-user programmable features, service bay diagnostics, information management systems, and duty-cycle adaptive "smart" controls. With the advent of standards for serial communication it is now possible for the makers of heavy-duty diesel engines to offer end-user programmable features, for example, the ability to select the maximum vehicle speed or choose between an automotive or variable-speed governor. This allows the end user to tailor engine parameters to match vehicle specifications, usage patterns, and environmental conditions.

Information management systems are becoming increasingly important for the operators of heavy-duty vehicles. Cummins Engine Company, Inc. has developed a number of support tools for its heavy-duty engines that allow drivers, service personnel, and fleet managers to gain access to valuable information about the operation of the engine. Drivers can access real-time trip information, such as average fuel consumption (MPG) and average trip speed. Service personnel can obtain diagnostic fault codes in the service bay that help with troubleshooting and allow faults to be located accurately and rapidly. Fleet managers can use information systems to understand better the operation of their vehicles, for example, average trip fuel consumption, percent idle versus cruise time, average vehicle speeds, and so on. Such data enable optimization of fleet usage patterns, maintenance schedules and new vehicle specifications—transmission ratios, engine power, and torque ratings.

11.1.3 Applications and End-User Duty Cycles

When developing an engine and its control system, it is of paramount importance to have a good understanding of the demands of the application—the total duty cycle to which the engine systems will be exposed and the end-user requirements.

only careful design of the combustion event and supporting fuel, air, and related emission controls, for example, EGR and aftertreatment, but is very demanding of accurate and repeatable control from event-to-event, cylinder-to-cylinder, engine-to-engine, and over the lifetime of the engine. Typically, the goal of the engine control system from an emission control viewpoint is to provide the demanded quantity of fuel, air, and EGR at the required time, in the required state (pressure and temperature), over the engine lifetime, compensating for wear-out and other forms of deterioration. Additionally, for vehicle applications, emission control systems should be supported by on-board diagnostic (OBD) systems capable of recording fault codes and lighting a malfunction indicator light (MIL), which is displayed until the fault is corrected.

The emissions compliance of an engine is measured by a test procedure, which defines a wide array of conditions under which the engine should be tested, and the methods by which the emissions should be measured and analyzed. For example in the United States, heavy-duty truck engines are tested on an engine dynamometer. The engine is subjected to a transient test cycle in which a schedule of engine speed and torque is followed, as shown in Figure 11.1. Light-duty vehicles are typically tested in-vehicle on a chassis dynamometer. In this type of test the vehicle has to follow a vehicle speed schedule. The European light-duty vehicle test cycle (ECE−15+EUDC) is shown in Figure 11.2. In both cases the

Fig. 11.1.

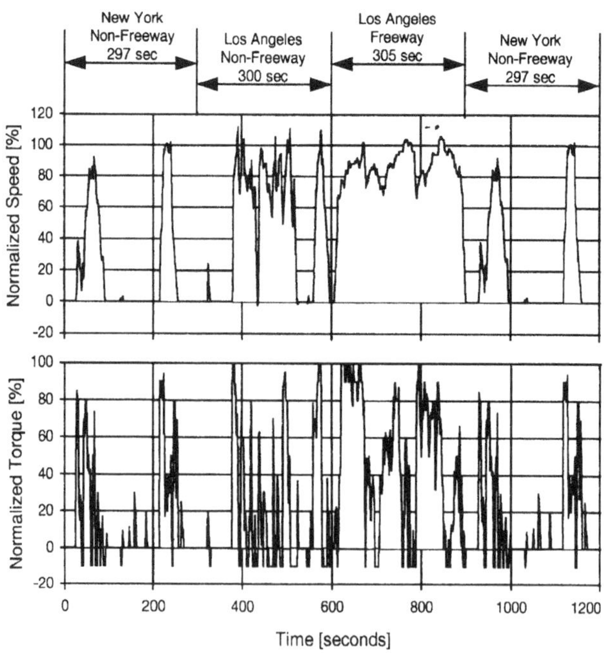

Engine dynamometer test cycle used to certify the emissions compliance of heavy-duty engines, US-FTP.

Fig. 11.2.

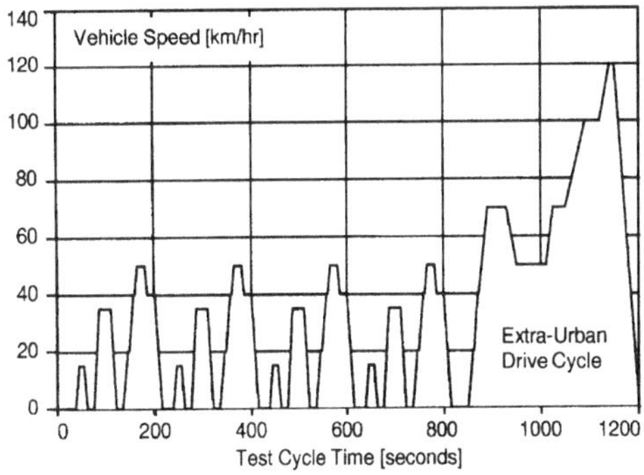

Cycle Analysis

Distance traveled		
ECE-15 cycle	1.013 km x 4 =	4.052 km
EUDC Cycle	=	6.095 km
Maximum speed		
ECE-15 cycle	50 km/h	
EUDC Cycle	120 km/h	
Average Speed		
ECE-15 cycle	19 km/h	
EUDC Cycle	63 km/h	
Acceleration	21.6 % (of cycle time)	
Deceleration	13.8 %	
Idle	35.4 %	
Steady speed	29.3 %	

Vehicle dynamometer test cycle used to certify the emissions compliance of light-duty vehicles, ECE−15+EUDC.

exhaust emissions produced during the cycle are measured for comparison with the standards required by the particular legislation.

The definition of the test cycle and data collection is quite different for these two cases. The challenge for the control system engineer is to provide optimum settings for the controlled variables, under steady-state and transient conditions, such that emissions compliance is achieved in volume production, both for new engines and over the lifetime of the engine, subject to an allowable deterioration factor. It will be obvious from the foregoing discussion that emission compliance should be achieved without undue sacrifice of fuel economy, performance, or reliability.

11.2

ELECTRONIC FUEL SYSTEMS FOR DIESEL ENGINES

11.2.1 Introduction

One of the keys to the success of the diesel engine has been the development of fuel injection equipment that can meet the many demanding requirements. In fact the development of a satisfactory fuel system was not achieved for almost 20 years after Rudolf Diesel first demonstrated the concepts of the diesel combustion system. Among the key requirements that differentiate diesel engine fuel systems from other types of fuel systems are:

1. Very high injection pressure (up to 2000 bar)—required to achieve a short injection period with acceptable fuel atomization and mixing.
2. High accuracy of injection timing—required to control engine fuel consumption, peak combustion pressure, and emissions performance.

3. High accuracy of fuel metering—required to control power output, limit smoke, and ensure cylinders develop equal power.
4. Ability to handle a range of diesel fuels, especially variations in density, viscosity, and lubricity.
5. Sharp end-of-injection to prevent nozzle fouling, smoke, and emissions problems.
6. Initial injection rate control for noise and emission control.
7. Ability to deliver excess fuel for cold start.

After many abortive attempts to develop a reliable and durable fuel system, the original technology that emerged was the jerk pump, in which cam-driven plungers are used to pump the fuel to the injectors through high-pressure lines. The start and end of injection were determined by the relationship of the top edge of the plunger and a lower helical slot to "fill" and "spill" ports machined into the barrel or cylinder in which the plunger reciprocated. The injection timing and fuel quantity information was contained in the often intricate shape of the plunger, in combination with other design features, such as the cam profile, injector nozzle flow, and the delivery valve characteristics. It can be appreciated that matching such fuel systems was a delicate balance of factors and consumed significant development effort.

Until electronic fuel systems became available, the fueling characteristics were determined for the entire engine operating range (load, speed, altitude, hot/cold, steady state, transient, and so on) by relatively simple hydromechanical elements. For example, the torque curve would be determined as a complete entity, which could be manipulated by changing such things as delivery valves, injector nozzles, and plunger geometry—often with outcomes that could not be predicted reliably. Similarly, injection timing schedules were determined by hydromechanical elements such as the plunger leading edge and helix profile, nozzle opening pressure, and the volume of the high-pressure line. As the demands on the engine became more stringent, for example, for smoke-free operation with turbochargers, or for better driveability, additional controllers were added above and beyond the basic features of the injection pump. Examples of these include boost pressure compensation, altitude compensation, and torque curve shaping.

Despite these obvious limitations, such systems have been made to give acceptable performance, and are used to the present day even in emission-sensitive, on-highway applications. However, the inability of such systems to meet the more stringent emission requirements of the late 1990s has driven the development of more sophisticated electronically controlled fuel systems.

Figure 11.3 is an attempt to classify diesel engine fuel systems according to the actuation and hydraulic design concepts. All of the cam-actuated fuel systems listed in Figure 11.3 have reached production either under mechanical control or some form of electronic control. The hydraulically actuated (common rail) fuel systems are, in the main, under development—with the exception of the HEUI

Fig. 11.3.

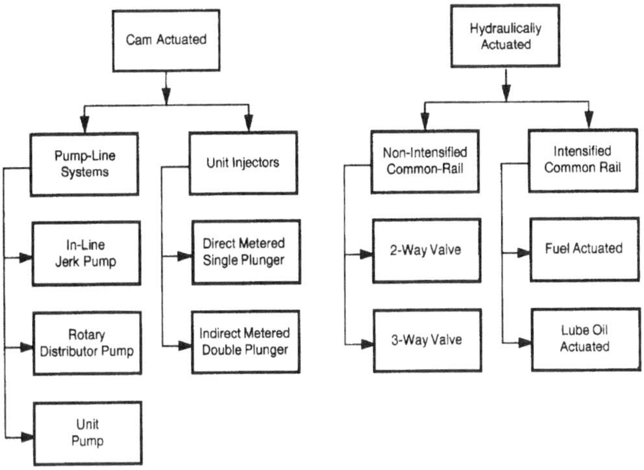

Classification of diesel engine fuel systems.

system [3, 4] produced by Caterpillar and the ECD-U2 system [5, 6] produced by Nippondenso. It is unclear which fuel systems will prevail into the first decades of the twenty-first century and, consequently, a very large development effort is under way in the search for fuel systems having improved injection performance, flexibility, and cost. The leading contenders are the high-pressure rotary pump, the unit injector, and the various common rail systems. These systems will be discussed in more detail in the following section.

11.2.2 Rotary Fuel Systems

This class of fuel system is typified by compact fuel pumps of four-, five-, six-, and eight-cylinder configuration—the main applications being turbocharged diesel engines used in passenger cars and light commercial vehicles. The rotary or distributor pump was popular for many years in cost-sensitive applications that did not require the highest levels of injection system performance, for example, in light marine, agricultural, and power generation applications. These pumps were generally mechanically controlled, fuel quantity was controlled either by metering the quantity of fuel entering the pumping element (inward radial plungers) or by sleeve control of effective stroke (axial plunger). Simple timing control, for example, cold-start advance or speed advance, was generally achieved by mechanical phasing of the fuel cam relative to the engine crankshaft.

The rotary pump is the fuel system of choice for the generation of passenger-car, high-speed, direct-injection (HSDI) engines developed from the mid-1980s to the present day for the European market. In order to meet the many needs of the passenger-car application, these mechanical pumps have become very complex as they attempt to perform complex functions with simple hydromechanical control

Fig. 11.4.

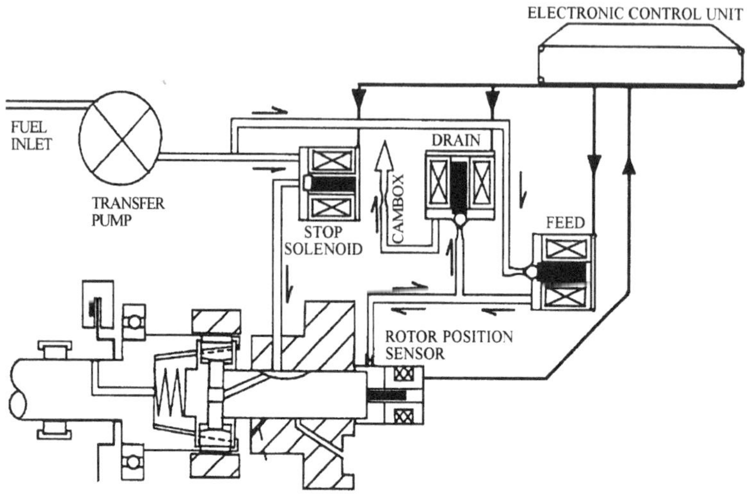

Lucas EPIC fuel metering control [7] *[Reprinted with permission of IMechE].*

elements. Typical mechanical control features include idle and high-speed governors and engine/ambient temperature compensation, including cold start excess fuel, transient air-fuel ratio (boost) control, and injection timing control.

With the introduction of emission standards for European light-duty vehicles, in the early 1990s, higher performance was demanded of the HSDI engine. Accordingly, fuel system manufacturers developed advanced versions of the rotary pump with electronic control features and improved hydraulic performance. Typically the injection event continued to be controlled by fuel ports and plungers known as "edge" control, but the actuation was no longer mechanical. In the Lucas EPIC rotary pump [7], solenoid valves are used to control fuel pressures in the pump body and these indirectly control fuel metering and fuel injection timing, as shown in Figure 11.4. The solenoid valves are under the control of an electronic control unit (ECU). The Lucas EPIC ECU monitors a number of sensors on the engine and in the fuel system and provides full-authority control of the engine, including the exhaust gas recirculation (EGR) system. For this generation of rotary fuel injection system, injection pressure at the pump was increased from typically 600 bar to around 900 bar. Through line amplification effects at high engine speeds, the pressure at the injection nozzle could attain in excess of 1000 bar.

Future rotary fuel systems, capable of meeting EURO-3 emission standards, will feature increased fuel system pressure and direct electronic control of the injection event. Fast-acting solenoid valves will operate directly on the high-pressure fuel to control both start and end of injection (Figure 11.5). This is expected to improve the rate of pressure decay at the end of injection and dramatically reduce response times for both fuel quantity and injection timing. Pump pressures in excess of 1000 bar are expected, allowing injector nozzle pressures as high as 1500 bar at high engine speeds [8]. This new generation of rotary fuel

Fig. 11.5.

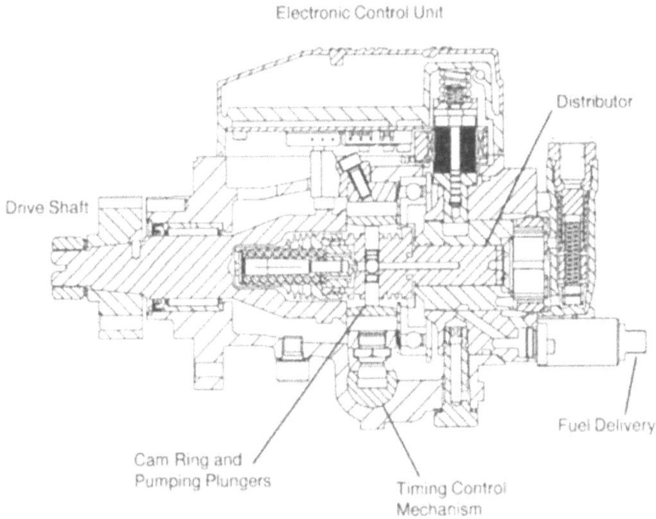

Cross section of an advanced rotary fuel pump [8] *[Reprinted with permission of IMechE]*.

injection pump will also feature integrated electronic control with serial data-link communication with the engine/vehicle controller.

11.2.3 Electronic Unit Injectors

Electronic unit injectors were first introduced in heavy-duty truck engines in the United States in the late 1980s. Although a number of implementations exist, having somewhat different characteristics, the underlying principle of operation and inherent advantages relative to pump-line-nozzle systems are the same.

The electronic unit injector (EUI) is a modular, integrated fuel injection system mounted directly in the cylinder head of the engine [9–13]. The pumping element and delivery nozzle are integrated into a single unit, the pumping plunger being driven by the engine camshaft, usually via a rocker lever. In the direct-metered unit injector, the space between the plunger and the nozzle is opened and closed by a fast-acting solenoid valve (Figure 11.6). The valve is closed to initiate injection and opened to terminate injection, thereby providing control of injected quantity and injection timing. Typically, injection pressure is related to injection timing since the cam profile does not generally allow constant injection rate over a wide range of timings. The key to the success of the unit injector is the very small dead volume between the pumping plunger and the injector nozzle. This allows very high-pressure operation with a relatively high efficiency. EUIs with pressure capability as high as 1600 bar are in service and pressures as high as 2400 bar have been demonstrated. The EUI is the preeminent fuel system for on-highway, heavy-duty trucks at this time, combining very high-pressure capability with low

Fig. 11.6.

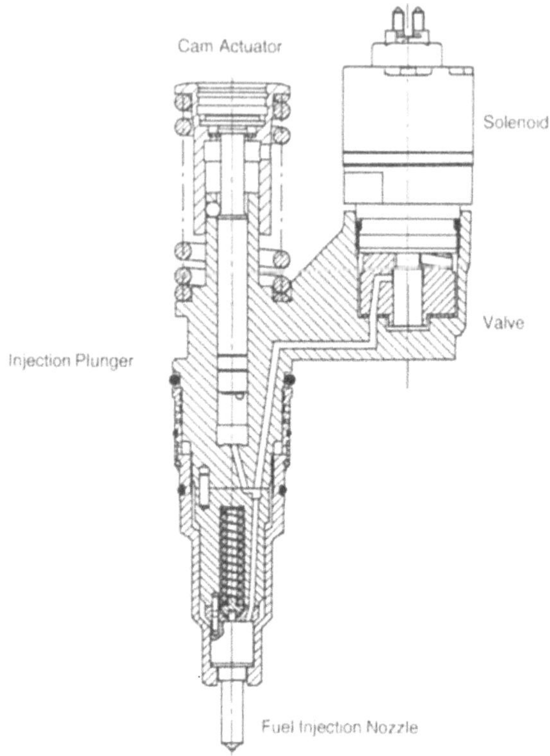

Example of a direct-metered EUI [10] *[Reprinted with permission of SAE]*.

power consumption and high reliability. The disadvantages of the EUI for smaller HSDIs are the difficulties of packaging, and the impact of the EUI on the total engine design and manufacturing process.

Cummins Engine Company, Inc. has developed an EUI for heavy-duty engines, known as CELECT, which employs two plungers within the unit injector, one controlling injection timing, the other controlling injection quantity [12, 13]. The high force needed to inject the fuel at high injection pressure is provided by an engine-driven cam. Both injection timing and injected quantity are electronically commanded and controlled by a solenoid-operated valve. A cross-sectional view of the Cummins CELECT EUI is shown in Figure 11.7. The working principle is illustrated in Figure 11.8, which shows the working sequence of the injector from the start of the filling stroke to the end of fuel delivery.

11.2.3.1 CELECT—Fuel Metering

Fuel metering is controlled by the upward movement of the timing plunger and the opening of the solenoid control valve. At the start of metering, the metering

Fig. 11.7.

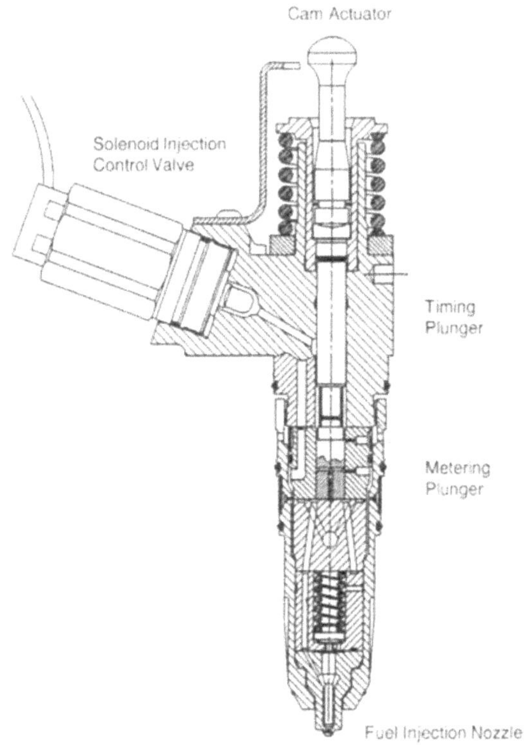

Double-plunger electronic unit injector [12] *[Courtesy of Cummins Engine Company, Inc.]*.

Fig. 11.8.

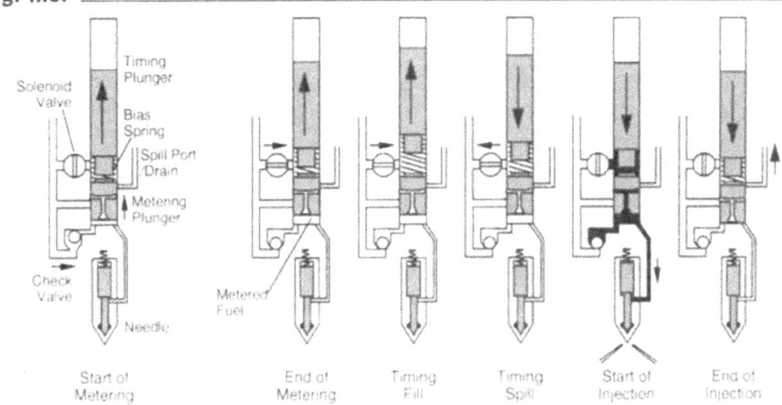

CELECT EUI working sequence [12] *[Courtesy of Cummins Engine Company, Inc.]*.

and timing plungers are bottomed in the injector and the solenoid valve is closed. As the cam rotates and relieves the force on the injector, the return spring forces the timing plunger upward following the cam movement. Fuel flows through the check valve into the metering chamber as long as the timing plunger is moving upward and the solenoid valve is closed. When the solenoid valve opens, metering ends. Fuel at supply pressure flows into the timing chamber, stopping the travel of the metering plunger. The bias spring ensures that the metering plunger remains stationary and does not drift up as the timing plunger continues to move up the injector bore. At the end of the filling process, the cam is fully retracted and the height of the metering plunger above the bottom of the stroke determines the quantity of fuel to be injected.

11.2.3.2 CELECT—Injection Timing

The start of injection is controlled during the downward movement of the timing plunger by the closing of the solenoid valve. As the timing plunger is forced downward by the cam, the solenoid valve remains open allowing fuel to flow from the timing chamber through the valve to the fuel supply passage. When the solenoid valve closes, fuel is trapped in the timing chamber, forming a hydraulic link between the timing plunger and the metering plunger. The metering plunger is forced to move downward as the timing plunger continues to move down, creating high pressure in the metering chamber. When this pressure reaches the preset opening pressure of the closed nozzle, the needle lifts, allowing fuel to be injected through the spray holes into the engine combustion chamber. Injection continues until the metering edge of the metering plunger passes the metering spill port. This allows a rapid decay of pressure in the metering chamber and allows the needle valve to close, resulting in a positive end of injection. The upper edge of the metering plunger passes the timing spill port immediately after the metering spill port is opened. This allows the fuel in the timing chamber to be spilled back to the fuel drain as the timing plunger completes its downward travel.

The characteristics of EUIs that are important for low-emission, heavy-duty diesel engines are high injection pressure capability, good injection quality, low power consumption, high reliability and durability, and the ability to control fuel metering and injection timing electronically. EUIs have proved capable of meeting these requirements even though emission standards have become more stringent through the late 1980s and early 1990s. Further developments of unit injectors are planned to enable heavy-duty engines to meet proposed emission standards beyond the year 2000.

11.2.4 Common Rail Fuel Systems

In recent years there has been a resurgence of interest in common rail fuel systems. Such fuel systems are characterized by separation of the pressure generation function and the fuel injection function. This separation allows the injection pressure to be controlled independently of engine speed and load. Common rail fuel systems date back to the early years of diesel engine development and some systems were produced as early as the 1930s. For various reasons they failed to compete with the ubiquitous in-line jerk pumps and distributor pumps. With the demands of meeting ever increasing customer expectations for vehicle refinement (low noise and vibration), fuel economy, and performance, as well as low exhaust emissions, common rail systems have come back into consideration. The development of electronic controls for diesel engines also makes control of the common rail easier to execute than in earlier times. Likewise, advances in high-speed solenoid technology (developed for EUIs) and manufacturing technologies further help to make common rail systems attractive.

The essential elements of a common rail fuel system are a high-pressure pump, which delivers fuel to an accumulator, or common rail, and injectors which can switch the rail pressure to the injection nozzle on demand. A high-pressure common rail (HPCR) fuel injection system is shown in Figure 11.9. In the Bosch HPCR system a three-plunger radial pump is used to raise the fuel pressure from the lift pump pressure to the rail pressure [9, 14]. A solenoid-controlled pressure control valve is used to regulate the rail pressure in a closed loop, using feedback from a pressure sensor placed in the rail.

Common rail fuel systems can be broken down into a number of subclasses depending on the detail design. The main subclasses are intensified or nonintensified injection. In the intensified systems the rail pressure is lower than the injection pressure and is multiplied, or intensified, by a stepped piston in the injector body. Typically, the intensification ratio is between 3:1 and 10:1. In the

Fig. 11.9.

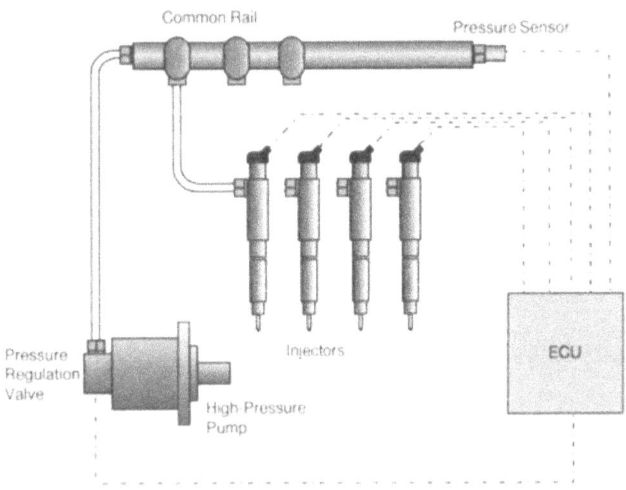

Layout of a high-pressure common rail fuel system.

Caterpillar system, released on the Navistar 7.3 liter engine, the rail pressure of 207 bar (3000 lbf/in^2) is boosted to 1450 bar (21000 lbf/in^2) by a 7:1 intensifier in the injector [3]. While there are some advantages in operating with a low rail pressure, the disadvantages include handling a large volume of actuation fluid, and the impact of the intensifier on the packaging (diameter) of the injector. In the nonintensified common rail system, the rail pressure is also the available injection pressure. Although the nonintensified common rail is not yet in series production, it has been written about extensively [5, 6, 9, 14] and its basic characteristics are well understood.

The nonintensified common rail may be further reduced to two subclasses depending on the detail design of the valve that controls the injection event. Both two-way and three-way valves have been developed and results from each have been presented in the literature [5, 6, 9, 14]. The ideal three-way valve allows the rail pressure to be switched to the nozzle for the injection event, and allows the nozzle to be switched to drain, to end the injection event. This would appear to be the most desirable option since it should prevent the rail pressure from leaking directly to drain throughout the injection event, thereby reducing power consumption. However, in actual configurations the three-way valve allows brief periods of short- circuiting between the rail and the drain as the valve opens and closes, such that the fuel used to operate the injector is considerably higher than the ideal case.

Figure 11.10 shows a cross section of the Bosch HPCR common rail fuel injector [14]. This injector has a two-way/two-position control valve (A). The valve connects the space above the control piston to drain. When the valve is opened, the pressure above the piston decreases rapidly, allowing the piston (P) to move upward commencing the injection event at the nozzle (D). Under this regime the flow continues through orifice (A) since it now essentially connects the rail to the drain through a series combination of orifices (Z) and (A). The detail design is such that the small valve orifice is partially obscured by the piston in its upper position such that drain flow is limited. In real designs it appears that the two-way valve is more efficient than the three-way valve, and in effect the success of the two-way valve has led to the keen interest in the simple nonintensified high-pressure common rail.

11.3 BASIC PRINCIPLES OF ELECTRONIC CONTROL FOR DIESEL ENGINES

11.3.1 Introduction

An electronic control system for diesel engines will consist, at a minimum, of a set of sensors, a microprocessor, and a set of actuators. The sensors enable physical variables in the engine system to be measured and made available to the controller as electrical signals. Examples include crankshaft speed, intake manifold pressure,

Fig. 11.10.

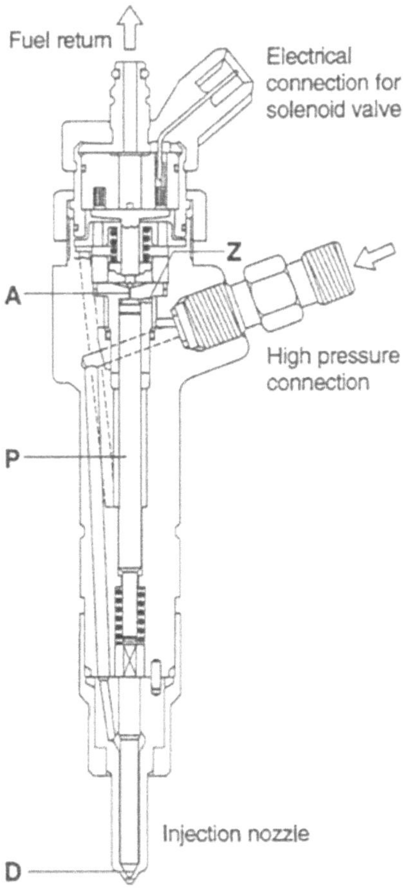

Cross-sectional view of the HPCR injector [14] *[Reprinted with permission of SAE]*.

and coolant temperature. Actuators convert command signals from the microprocessor into mechanical actions which essentially control the engine. A fuel system is no more than a very sophisticated actuator which controls fueling parameters, such as injection quantity and injection timing. Other examples of actuators used in diesel control include EGR valves and variable geometry (VG) turbochargers.

The microprocessor is typically supported by a number of electronic subsystems, which carry out functions such as initializing the system at start-up, storing programs and data, handling input and output conversions (I/O), and providing a stable power supply to the electronics. Together this collection of electronic hardware is called an electronic control unit or ECU, and it is common for it to be mounted on a single board and housed in a protective container. It is in fact more common to integrate many of the functions described previously onto a single chip known as a microcontroller. Placing special hardware for I/O and data storage on a single chip with the basic microprocessor functions has important advantages in high-speed control applications, such as engine control, the most significant being reduced cost and increased computation speed. A microcontroller architecture will be described in Section 11.4.1.

In operation, the ECU reads the sensor signals at regular intervals and executes a set of calculations or algorithms in order to determine suitable output values to control the engine system via the actuators. An example of a control algorithm is the calculation of maximum fueling during an engine transient acceleration. In a turbocharged engine, during a hard acceleration, it possible for the fuel rate required to meet the demanded acceleration to cause the fuel-air ratio in the cylinder to exceed that which can be burned cleanly and efficiently. In this case it is necessary for the engine controller to temporarily limit the fuel rate to the highest level that can be accepted. This can be accomplished by using the measured engine speed and intake manifold density (calculated from intake manifold pressure and temperature) to compute an estimate of the instantaneous airflow rate to the cylinders. In addition to the speed and intake density, the controller must have a prior knowledge of the variation of volumetric efficiency of the engine with speed. Such an example is illustrated in Figure 11.11. This example is one of many algorithms typically used in diesel engine control. Others will be discussed in the following sections, including fuel quantity control, injection timing control, boost pressure control, and EGR control.

11.3.2 Fuel Quantity Control

Fuel quantity is controlled by a governor or a series of governors, each of which handles a particular regime, for example, cold start, low-speed idle, or high-speed

Fig. 11.11.

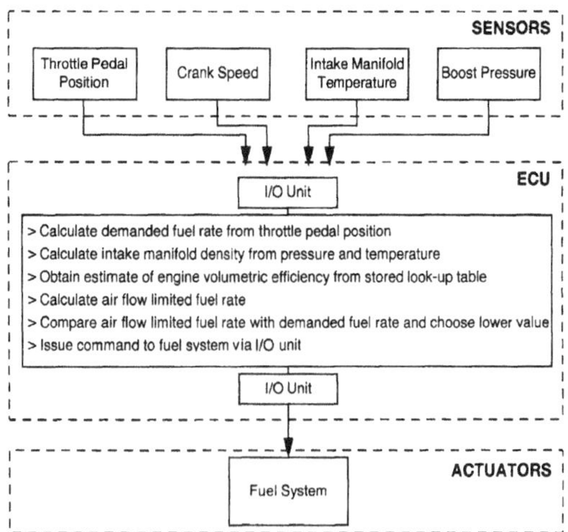

Example of diesel control—transient air-fuel ratio.

Fig. 11.12.

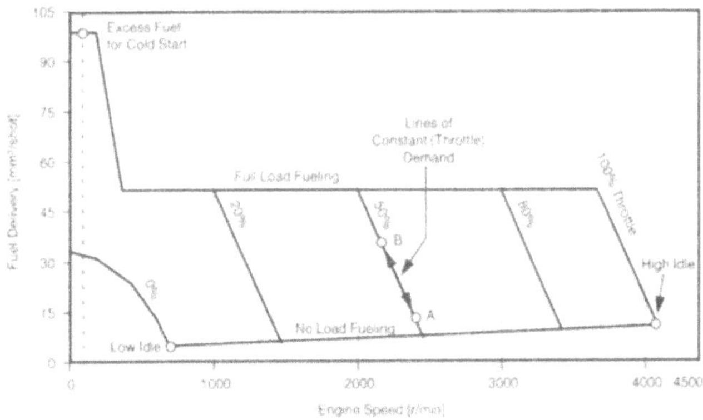

Characteristic curves for a variable-speed (all-speed) governor.

idle. There are two types of governors in regular use: variable speed (all-speed) governor, and the minimum-maximum speed (min-max) governor, which controls only the low-speed and high-speed idle regimes, such that at speeds between these extremes the delivered fuel is proportional to the input (throttle) demand.

The characteristic of a typical variable-speed (all-speed) governor is shown in Figure 11.12, for a naturally aspirated engine. The constant throttle curves are seen to slope (or droop) such that the speed is allowed to fall as load is applied at constant throttle demand. This is a simple form of proportional control in which the fueling is proportional to a speed "error." The steady-state speed "droop" with load is defined as the reciprocal of the gradient of the throttle curves.

For the simple example shown in Figure 11.12, the droop is $(2400 - 2000)/2000 = 20\%$. Values of 5 percent to 10 percent steady-state droop are common; however, steeper droop curves can be used for tighter control of speed with load. The variable-speed governor is used in applications where the load can change very rapidly, for example, agricultural tractor, earthmoving equipment, or applications where it is required to maintain an approximately constant speed, for example, power take-off or cruise control. The benefits of this type of governor for earth-moving equipment can be seen in Figure 11.12. Assume the throttle is at 50 percent with steady load state A (approximately 15 mm^3/shot of fuel), when a sudden load is applied requiring higher fueling. With variable speed governing the speed droop curve allows the fueling to be increased to state B (approximately 38 mm^3/shot of fuel), without any change of throttle position. This characteristic makes the engine much more responsive to rapid load changes and prevents stalling.

The min-max governor is used extensively in automotive applications, especially passenger-car diesel engines. The min-max governor provides automatic control of engine speed at idle and regulates the engine at high speeds to prevent damage. Between these conditions, fueling is proportional to the throttle position; that is, it allows the driver to control the torque output of the engine directly. The characteristics of a typical min-max governor are shown in Figure 11.13.

Fig. 11.13.

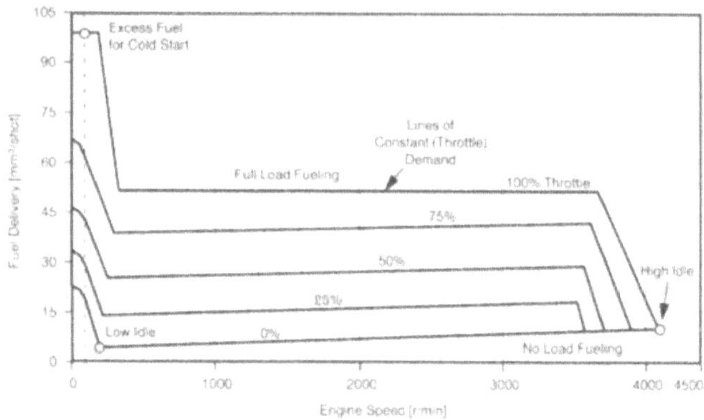

Characteristic curves of a minimum-maximum (min-max) speed governor.

Fig. 11.14.

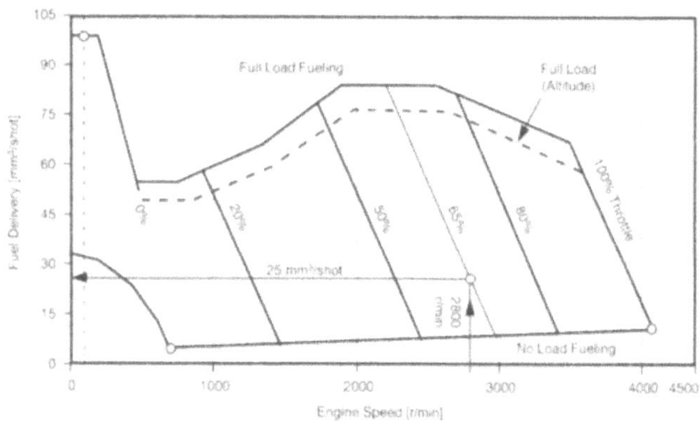

Example of governor characteristic determining fuel quantity (35 mm^3/shot) from instantaneous engine speed (2800 r/min) and throttle position (65 percent).

Figure 11.14 shows the governor characteristics more typical of a turbocharged engine. In this case the maximum fuel curve rises to give more fuel at speeds where the turbocharger is providing high boost pressure. Also shown is the derating of the maximum fueling curve required with altitude. This curve compensates for the reduced airflow that occurs with altitude and minimizes the risk of producing smoke under those conditions, or overspeeding of the turbocharger.

Figure 11.14 also shows how the governor characteristic is used to determine the fuel rate from the instantaneous engine speed and throttle position. The governor characteristic is normally stored as a three-dimensional fuel map versus engine speed and throttle position (Figure 11.15). Having calculated the fuel rate demanded, the engine controller will in general compute the fuel rate to satisfy other engine conditions, for example, the airflow limited fuel rate. The various computed fuel rates are compared and the lowest value is selected. In the example

Fig. 11.15.

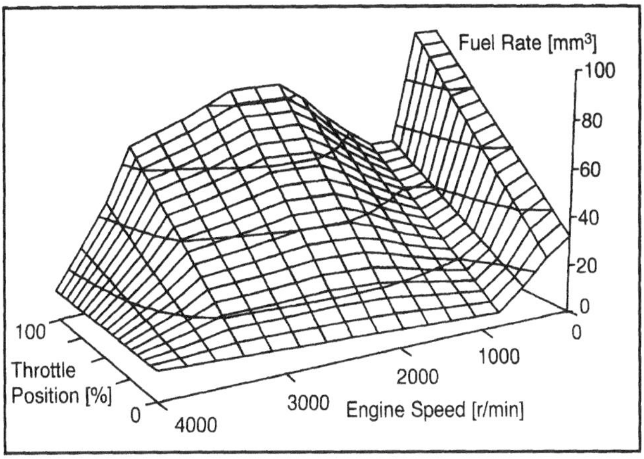

Typical fuel map—min-max governor.

shown, the engine speed is 2800 r/min, the throttle position is 65 percent, and when the governor characteristic is interrogated, a fuel rate of 25 mm^3/shot is returned.

With increasing demands for vehicle refinement, especially in terms of noise, vibration, and harshness (NVH), the flexibility of electronic fuel systems is being widely used to achieve adaptive control of fuel metering. Especially at low-speed idle, variations in cylinder-to-cylinder fuel metering accuracy may give rise to significant variations in engine speed, which may be perceived by the driver as unpleasant NVH. Full-authority electronic control systems can monitor the time between crank position signals to estimate variations in crank speed during an engine cycle, thereby identifying weak- or strong-firing cylinders caused by variations in fuel delivery. By adjusting the commanded fueling to certain cylinders (i.e., "trimming" the fuel delivery by cylinder), it is possible to greatly reduce fluctuations in crank speed and improve NVH. There are, however, limitations on what can be achieved in this regard, set by the controllability of the fuel system itself at very low fuel delivery. Typically, fuel systems are designed to have a linear response of fuel delivery to input command from minimum to maximum fueling: However, at minimum fueling almost all systems suffer nonlinearities that make control difficult. With solenoid-controlled systems the problem is often strongly related to the fact that the solenoid valve does not travel between its limits at very light fuel levels. This implies that the actual delivered fuel quantity is determined by the "ballistic" motion of the fuel control solenoid, which will naturally tend to be more variable than an event controlled by a positive limit stop.

11.3.3 Fuel Timing Control

Injection timing is a critical determinant of the combustion event and the resulting emission products. The connections between injection timing and the combustion event are many, and a list of the main connections follows.

1. Injection timing determines the relationship of the fuel spray to the combustion bowl in the piston as it travels toward and away from the top dead center position.
2. Injection timing will influence the interaction of fuel spray and air motion, since the level of swirl varies with time and reaches a peak close to top dead center.
3. Cam-driven fuel systems normally exhibit a relationship between injection timing and fuel injection rate, which can influence injection pressure and duration.
4. Start of injection timing influences the ignition delay period—very advanced or very retarded timings tend to prolong the ignition delay period and increase the fraction of fuel burned in the premixed mode.
5. Start of injection determines the start of combustion via the ignition delay period.
6. Start of combustion timing has a first-order influence on flame temperature.
7. End of combustion is strongly influenced by injection timing.

This is far from a complete list of the interactions but it does serve to emphasize the importance of injection timing control for low-emission diesel engines. All important engine performance and emissions parameters are a strong function of injection timing: specific fuel consumption, oxides of nitrogen, particulates and hydrocarbons, lube oil soot contamination, and peak cylinder pressure.

From the foregoing discussion, it will be clear that electronic control of fuel injection timing is a basic requirement for low-emission diesel engines. The freedom to command injection timing as a function of engine speed and load, ambient conditions, or emission cycle requirements offers the engine designer and development engineer major benefits. To appreciate these benefits, it is helpful to consider the limitations of mechanical fuel systems. Typically, mechanical fuel systems require hardware changes in order to modify injection timing schedules. Furthermore, the timing schedule of a mechanical fuel system has very few degrees of freedom, so that handling anomalies such as cold-start advance or low coolant temperature advance require added mechanical elements. Optimizing exhaust emissions over an emissions cycle that samples a wide range of engine speeds and loads is severely compromised by the inability to demand injection timings that optimize emissions and fuel consumption over local regions of the operating spectrum. This limits the extent to which an engine can meet customer requirements and deliver low emissions.

The ability to command injection timing in response to engine load, engine speed, and other factors is clearly essential for ultralow-emission diesel engines. Another factor which is almost equally important is the accuracy with which the commanded injection timing can be achieved. A small number of engines have used an injector needle lift sensor to provide feedback, for example, the Audi HSDI engine described in Bauder and Stock [15]. In the Audi engine only one cylinder is equipped with an instrumented injector and it is claimed that injection timing accuracy is within ±1 degree of crankshaft rotation. Advanced high-pressure rotary fuel pumps typically have a rotor position sensor built into the fuel pump itself. By comparing the crank position sensor signal with the fuel pump position sensor signal, it is possible for any angular errors between the pump and the engine, introduced during assembly, to be compensated for with the goal of improving the accuracy of control of injection timing.

11.3.4 Boost Pressure Control

Almost all automotive diesel engines are turbocharged, and many also utilize charge air cooling. These are effective means of boosting the power available from an engine of a given displacement. With the advent of the wastegated turbocharger (WGT) and variable geometry turbocharger (VGT), it has been possible to control the intake manifold (boost) pressure and, therefore, the mass flow rate of air to the engine cylinders. Wastegates allow engines to be fitted with small turbines that produce boost at low engine speeds and enable good low-speed torque to be achieved. At high speeds the wastegate by-passes some of the exhaust gas around the turbocharger turbine to prevent excessive boost pressure and airflow. In many implementations the wastegate is controlled by feedback of boost pressure to a pneumatic diaphragm-type actuator, which operates the wastegate valve through a linkage. This form of control is effective for regulating boost pressure to some predetermined upper limit. It is a low-cost solution that has limitations as engines are designed for more stringent emission standards. Although the feedback of boost pressure would suggest closed-loop control, in reality these systems are only as accurate at regulating maximum boost pressure as the set-up procedure during assembly allows. Any change in the characteristics of the pneumatic actuator or wastegate flow performance would not generally be compensated for and errors in maximum boost pressure could result.

In recent years there has been a trend to control wastegates electronically such that boost can be commanded as part of an integrated engine control strategy. Typically, the wastegate actuator is pneumatic, as in the simple boost feedback system described previously, but the air pressure fed to the actuator diaphragm is modulated by a pressure control valve. Such schemes allow flexible closed-loop control of boost pressure, or airflow, which allows compensation for the effects of manufacturing variability, environmental conditions (including altitude), and changes during engine lifetime. An intake manifold pressure (MAP) sensor signal is normally used to close the loop. Other advantages of such an approach include the ability to regulate airflow to suit better the prevailing demands on the engine.

Fig. 11.16.

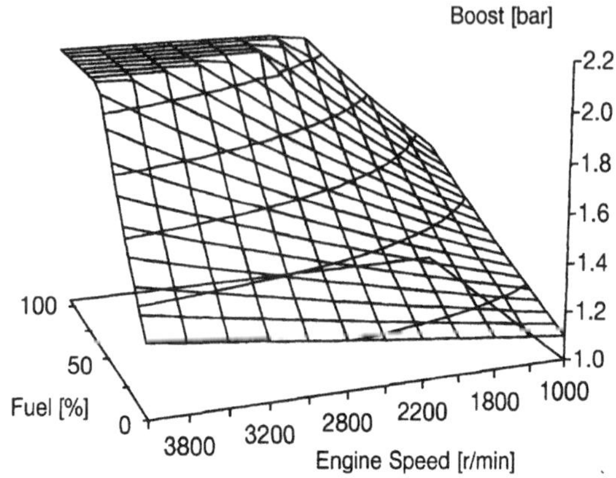

Typical boost pressure map (at sea level).

For example, during transient acceleration of the vehicle, the wastegate would typically be closed to provide a rapid build-up of boost pressure. With electronic control of the wastegate, the point at which the wastegate begins to open and the rate of opening may be determined by many factors besides boost pressure to improve the vehicle performance. It may be possible to allow boost pressure to overshoot the steady-state value, defined by the map, in order to improve acceleration—provided other constraints, such as turbocharger speed and peak cylinder pressure, are not exceeded.' Figure 11.16 shows a typical boost pressure map that would be used for determining the steady-state boost pressure schedule for an engine at sea level. The boost pressure is limited to a maximum value to prevent possible damage to the engine or excessive airflow at rated power with an accompanying increase in fuel consumption.

Variable geometry turbochargers also allow the boost pressure to be scheduled as a function of engine operation and ambient conditions. The control of VG turbochargers will be discussed in more detail in Section 11.4.3.2.

11.3.5 EGR Control

Exhaust gas recirculation (EGR) is an effective means of controlling the emission of oxides of nitrogen (NO_x) [16]. The role of EGR in the combustion event is to act as an inert diluent to the combustion air, reducing the oxygen concentration and resulting flame temperatures. The oxygen concentration controls flame temperature since, in the presence of a diluent, complete combustion of the fuel will involve a greater mass of charge. Relatively small reductions in flame temperature can have a large effect on NO_x, since it is exponentially related to the flame temperature. EGR allows NO_x emissions to be controlled without excessively retarding the injection timing, which would have a negative impact on fuel consumption and contamination of the lube oil with soot.

Fig. 11.17.

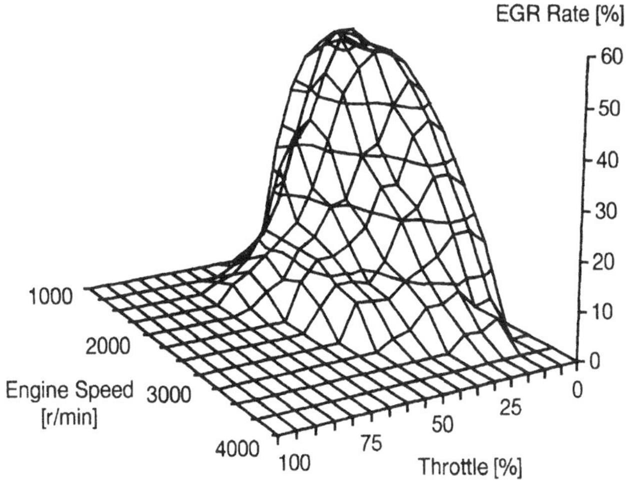

Typical light-duty EGR rate map using engine speed and throttle demand inputs.

The tightening of emission standards has led to increased application of EGR for automotive diesel engines. European passenger car diesel engines use EGR to meet ECE-15 Stage 2 emissions, whether IDI or DI combustion is used. A typical EGR control map for a passenger-car diesel engine is shown in Figure 11.17. In light-duty applications, the engine is used mainly in the lower load and speed range, where high EGR rates (30 percent to 60 percent) are easily achieved. At higher loads the delivery of EGR is not so readily achieved while maintaining adequate air-fuel ratio; hence, it is more usual to rely on alternative strategies such as retarded injection timing under those conditions. Since flame temperature is also a function of intake manifold charge temperature, further gains in NO_x control may be achieved by cooling the EGR prior to mixing with the charge air. An increasing number of passenger-car diesel engines are using cooled EGR as a means of improving the NO_x/fuel consumption trade-off at low emission levels.

In the United States, Cummins Engine Company, Inc. produces a version of its 5.9-litre B-series, DI diesel engine for the Dodge Ram light truck with uncooled EGR. The engine was developed to meet the stringent California Transitional Low Emission Vehicle (TLEV) standards for heavy-duty engines. In this application the engine is certified on the U.S. heavy-duty transient emission cycle shown in Figure 11.1. EGR is controlled electronically in response to throttle demand and ambient conditions. The EGR system is subject to on-board diagnostic requirements for emission controls as described in Section 11.3.7. The Cummins TLEV automotive engine uses a form of open-loop control of EGR as shown in Figure 11.18. EGR is drawn from the exhaust manifold upstream of the turbocharger turbine and transferred to the intake manifold via a flow control orifice and a vacuum-actuated EGR valve. The EGR valve is used to close the EGR path during start-up and during periods when the engine coolant is below a prescribed value. The venturi is

Fig. 11.18.

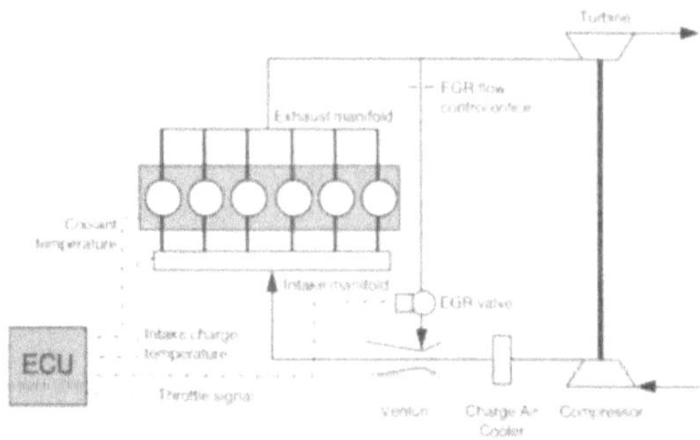

Open-loop EGR scheme used on Cummins TLEV diesel engine [*Courtesy of Cummins Engine Company, Inc.*].

used to increase the available pressure differential between the exhaust and intake manifolds, which can be limited on efficiently turbocharged engines. The EGR rate schedule with engine speed and load is determined by the natural characteristics of the EGR and air-handling systems.

A further elaboration of open-loop EGR control involves modulating the EGR valve according to a valve position schedule stored in the controller as a look-up table. The valve will normally be fitted with a position sensor to enable closed-loop control of valve position. Although the valve can be positioned accurately and repeatably, the actual EGR mass flow rate may vary with factors such as intake and exhaust restriction or EGR valve fouling. In order to compensate for system and environmental changes that affect EGR rate for a given valve effective area it is required to measure either the EGR flow or the airflow rate to provide closed-loop control.

Closed-loop control of the EGR rate is desirable as emission standards become even more stringent. The most effective variable to measure in order to close the loop is the EGR mass flow rate itself. In fact, EGR flow rate is measured in some gasoline engine applications using a critical flow orifice. Direct measurement of EGR is difficult on the diesel engine because the available pressure differential between intake and exhaust manifolds is normally very low. Any flow losses associated with an intrusive flow measurement technique would require an additional increase in exhaust manifold pressure relative to intake pressure in order to drive the EGR flow.

An alternative to measuring the EGR flow rate directly is to infer the EGR rate indirectly by measuring the airflow into the engine and the intake manifold pressure and temperature. The mass flow rate of EGR may be inferred by comparing the actual airflow rate with that which would be achieved under the prevailing boost conditions without EGR flowing. Since the density of the air charge is known, the volume occupied by the EGR may be estimated from knowledge of the volumetric efficiency of the engine. The expected density of the EGR at the point of entry to the intake manifold can then be used to compute the actual EGR mass flow rate.

If a mass airflow (MAF) sensor is used in the engine control scheme, it can also be used periodically to update the engine volumetric efficiency data stored in the controller as a look-up table, allowing compensation for lifetime changes.

11.3.6 High-Speed Serial Communication Networks

The first generation of electronic controllers was generally limited to control of the engine, in which an ECU was wired to a set of sensors and actuators through a complex wiring harness. Such an approach has limitations, primarily because it does not readily accommodate communication between the engine ECU and other ECUs that may be added to the vehicle system. The emergence of electronic transmission controllers, electronic traction control, antilock braking systems, and other subsystem controllers has necessitated the definition of standardized communication networks. Such networks allow manufacturers to develop their components to a single electronics interface standard. With the emergence of on-board diagnostics, computer-based service tools, and electronic management tools, the need for a more flexible, standardized serial network approach is clear. The serial data link overcomes the need to "hard-wire" ECUs together in order to exchange information, which simplifies wire harnesses and reduces the number of pins required on ECUs. The serial data link also has the advantage that if one ECU in the system should fail it will not affect operation of the other electronic devices in the system.

Efforts to define a serial communications network (also known as a data link) began in the mid-1980s with the development in the United States of the SAE J1709, J1587, and J1922 standards, and in Europe with the Robert Bosch Control Area Network (CAN) standard. The goals of the J1922 development were primarily general information sharing, diagnostics, and proprietary communications with limited capabilities for powertrain control. Before this standard was fully implemented a new initiative (J1939) was started, with the goal of providing high-speed communication capable of supporting control applications as well as the applications of the J1922 data link [17, 18].

The serial data link consists of a simple twisted pair of wires—one at ground (0V), the other being used to carry the binary signal (0 or +5V) used to transmit digital information. A number of electronic devices (typically ECUs) will be connected to the serial data link (or bus), as in Figure 11.19. At any given time only one device is permitted to transmit data onto the bus; all other devices are in receiver mode and ready to accept data over the bus. A device will construct the data it needs to transmit in the form of a message having a strictly defined protocol.

The CAN protocol uses content-based addressing, in which each message is assigned a unique identifier based on the type of message it is, for example, "engine torque." The other nodes on the network will process or receive the message only if that identifier appears in the list of acceptable messages stored locally in a format known as an acceptance list. The individual nodes, therefore, do not require unique addresses, and destination addresses per se are not required in the message. This has a number of advantages: First, a given message will be available to all nodes

Fig. 11.19.

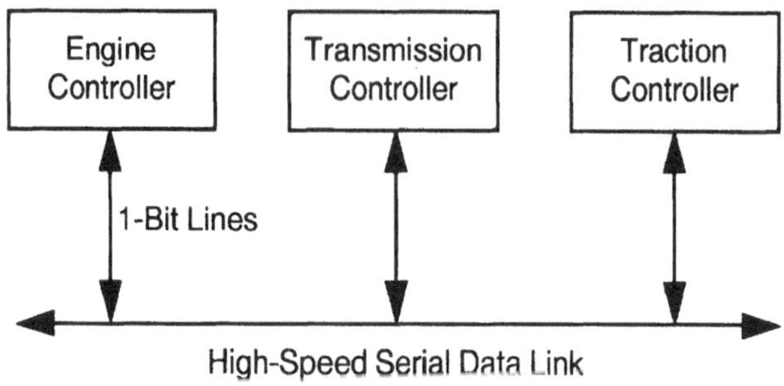

Serial data link layout for network communication between ECUs.

requiring the message without repeated sends or multiple addresses preceding the message; second, the network will accommodate variations in the components that are attached.

A device can gain access to the data link when it is not being used by another device. This event is detected by completion of a message and an intermessage pause. A device will attempt to access the data link by transmitting the message identifier prior to the complete message. The message identifier also carries a priority rating, to be used to resolve conflicts, where more than one node requires to transmit over the data link simultaneously. If more than one device is trying to access the data link, then the message with the highest priority is allowed to proceed. This occurs without delay to the transmission. All devices not transmitting are switched to receiver mode, awaiting the next message.

As end-user expectations for fuel economy and performance increase, and emission standards become more demanding, greater integration of engine and vehicle functions will be required. Furthermore, as engine control becomes more sophisticated it is conceivable that subsystems such as fuel injection, EGR and air handling, lube, and cooling will be endowed with local intelligence as in Figure 11.20. This will allow vehicle and engine control to be executed by a number of "smart" subsystems cooperating and communicating with each other over a simple two-wire, high-speed serial network. It is also likely that the smallest subsystem might be a smart sensor, a sensing element with local intelligence and communications.

11.3.7 On-Board Diagnostics

On-board diagnostics (OBD) refers to regulations appearing in the United States and Europe to define the in-service monitoring of emission control systems for gasoline, diesel, and alternate fuel engines for passenger cars, light-duty trucks, and medium-duty vehicles. The medium-duty vehicle (MDV) is a class of vehicle defined by the California Air Resources Board (CARB) with gross vehicle weights between 2720–6350 kg (6000–14,000 lb), in which the emissions can be tested on either a chassis or engine dynamometer. For other U.S. states, the

Fig. 11.20.

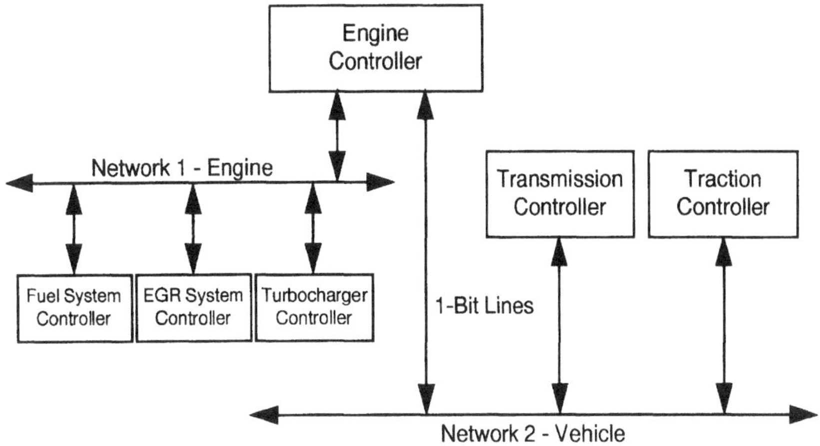

Example of hierarchical serial data link.

OBD regulations apply only to vehicles below 3850 kg (8500 lb). At present, on-board diagnostics regulations do not apply to heavy-duty vehicles above these classifications.

The primary goal of the OBD regulations is to enable malfunctions in the emission control system to be detected rapidly, in order to reduce the time over which increased emissions are produced prior to repair. The California regulations, known as OBD-2, are defined in *California Code of Regulations* [19]. The EPA (49 State) U.S. regulations, known as OBD, are defined in *Code of Federal Regulations* [20]. In Europe, the European Union (EU) is developing a set of regulations for on-board diagnostics for emissions controls that will be similar to those defined for the United States. It is expected that common standards will be used to define the EU OBD requirements such that engines developed for the United States and Europe will comply with OBD regulations in both regions.

On-board diagnostics monitoring requirements are listed in simplified form in Table 11.1, with comments regarding their applicability to diesel engines. Table 11.1 shows ten monitors, which are intended to verify that each listed subsystem is performing its prescribed function for the useful life of the vehicle. If a malfunction is detected that would cause emissions to exceed the regulated standard by more than 50 percent, a malfunction indicator light (MIL) must be lit on the dash-panel to warn the operator that repair is required. The "comprehensive component" monitor in Table 11.1, refers to any component that directly or indirectly affects emissions, for example, coolant or intake manifold temperature sensors. If such components affect emissions, they must be monitored. In the case of sensors, monitoring for open or short circuit is readily achieved by checking for an out-of-range signal.

Figure 11.21 shows a simplified flowchart representation of an OBD-2 system for a diesel engine. The flowchart shows the general operating principles. Not all monitors are active on a continuous basis; for example, the EGR system may be

Table 11.1

OBD-2 Emission System Monitoring Requirements (those which apply to Diesel Engines are Shaded).

Emission system monitors	Applicability to diesel engines
1. Catalyst	Applicable but no practical means of detecting malfunction is known
2. Heated Catalyst	Applicable but no practical means of detecting malfunction is known
3. Misfire	Malfunctions can be detected
4. Evaporative System	Not applicable
5. Secondary Air System	Not applicable
6. Air-Conditioning System	Not applicable if non-CFC refrigerant is used
7. Fuel System	Malfunctions can be detected
8. Oxygen Sensor	Not applicable
9. EGR System	Malfunctions can be detected
10. "Comprehensive Component"	Malfunctions can be detected

Fig. 11.21.

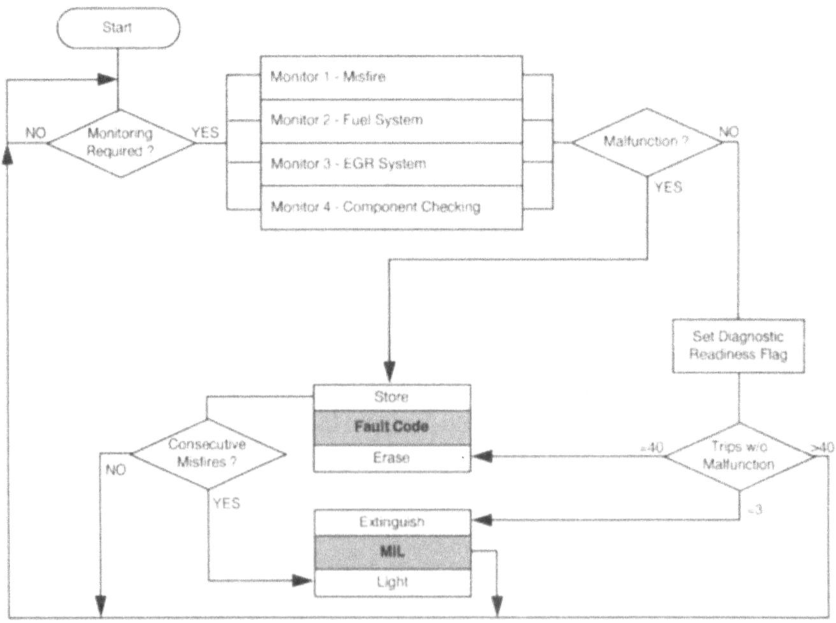

Simplified flowchart showing operation of OBD-2 system.

monitored only once per trip. Assuming the conditions are met for a subsystem to be monitored, the controller will initiate monitoring. If the conditions are met for a malfunction to be declared, the fault is stored in the ECU for later retrieval during servicing. The MIL light will be illuminated after two consecutive occurrences of the malfunction. If a malfunction does not recur in the next three trips, the MIL may be extinguished. If a malfunction does not recur for 40 trips, the fault code may be erased from the ECU.

Some diesel engines that qualify for OBD certification are fitted with oxidizing catalysts, the function of which is to reduce fuel- or lubricant-derived soluble particulates. At present a practical means to monitor the function of a lean catalyst does not exist, although research work is under way to find a commercially feasible solution.

Misfire is a term used to describe a wide range of possible malfunctions of the combustion system, but it is a term that applies more readily to gasoline engines. For example, if the mixture is out of limits or the spark-ignition system is faulty, then a complete or partial misfire can occur; under these conditions unburned fuel will be emitted from the exhaust. The diesel combustion process is inherently more robust than the spark-ignition combustion process, since it is designed to operate over a very wide range of air-fuel ratios and does not require spark ignition. In the diesel engine, misfire could conceivably occur under extremely cold conditions if injection timing is retarded or compression temperature is lower than intended. Misfire detection systems are typically based on analysis of the crank position signal to detect wide variations in instantaneous crankshaft speed during the cycle.

It is required to monitor the function of the fuel system for malfunctions that increase emissions. The detailed approach will depend on fuel system type and the sensors used in its control. In general, as a starting point, it is possible to check all sensors for a rational and in-range signal. It may also be possible to check the response of the fuel system to input commands; for example, commanding increased common rail pressure can be checked against the rail pressure sensor output. Some failure modes will not be amenable to detection using these approaches and will require additional sensors; an example might be injector needle lift. An injector could be fitted with a needle lift sensor for the purpose of diagnosing injection timing errors. This sensor could then be used to provide closed-loop control of injection timing, bringing additional system benefits.

OBD-2 requires that the EGR system is monitored for malfunctions that might cause an increase of emissions. The strategy for the EGR system could be similar to the strategy for the fuel system. Sensors can be checked continuously for rational and in-range signals. The overall function of the EGR system could be checked by opening the valve under steady engine operating conditions and looking for changes in sensor responses, for example, intake manifold temperature or MAF sensor responses. EGR valve position feedback alone is insufficient to detect all EGR malfunctions but can assist diagnosis of faults.

The successful implementation of OBD relies extensively on the definition of high-speed serial communications network standards as discussed in Section 11.3.6. Standards are required to define many features of an OBD system, for example, diagnostic trouble codes, data link message format, service tool interface, connector types, and terms and definitions. The standards adopted by EPA, CARB, and EU are either entirely common or compatible subsets [21].

Heavy-duty engines used in line-haul trucks do not currently come under direct regulation for on-board diagnostics for emission control. However, this class of engine currently makes extensive use of electronics for control and diagnostics, such that many of the OBD requirements are implicitly met [21].

11.4 ELECTRONIC HARDWARE FOR DIESEL ENGINE CONTROL

11.4.1 Electronic Control Unit

The electronic control unit (ECU) consists of a group of electronic components, usually assembled onto a single printed circuit board, that together provide all the control functions for a given control application, excluding sensors and actuators. Typically the ECU is housed in a protective casing with multipin connections to the rest of the system via a wiring harness. ECUs are used for many applications including engine control, antilock braking control, traction control, and transmission control. The particular control requirements will determine the make-up of the ECU in terms of processor characteristics, memory requirements, input/output (I/O) requirements, and peripherals, such as analogue-to-digital converters (ADCs).

Many ECUs make use of powerful integrated circuits known as microcontrollers [22]. A microcontroller is typically a complete, single-chip microcomputer. It will integrate almost all of the electronic components needed to perform the functions associated with engine control: processor (CPU), volatile memory (RAM), nonvolatile memory (ROM/EPROM), electrically erasable read/write memory, input/output control, and peripherals. In some applications additional memory may be provided outside of the microcontroller. By contrast, a microprocessor generally consists of a processor, registers, and interrupt handlers; other parts of the microcomputer, such as RAM, is provided outside the microprocessor. A simplified block diagram of a typical microcontroller architecture is shown in Figure 11.22.

Microcontrollers are used in control applications that demand fast response to real-time events and high-speed execution of algorithms. Most microcontrollers are considerably more complex than shown in Figure 11.22; however, the common elements are a CPU with an arithmetic logic unit (ALU), a variety of memory types, and a variety of input/output peripherals. The block diagram in Figure 11.22 shows

Fig. 11.22.

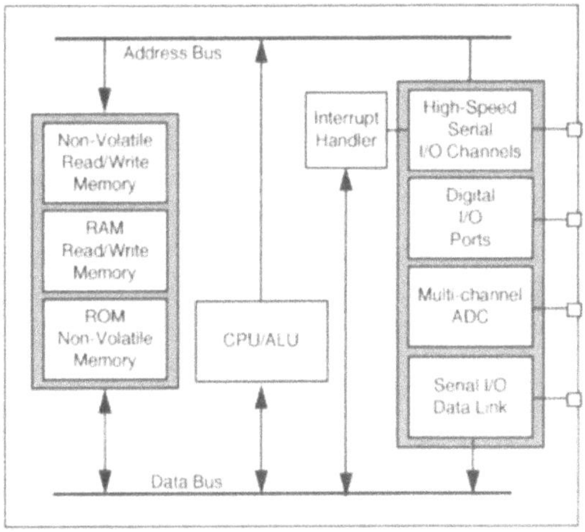

Block diagram of a simplified microcontroller architecture.

a data bus and an address bus. These buses allow data to be transferred between memory and I/O devices and the CPU. The address and data bus each consist of multiple wires that allow the transmission of multiple bits (binary digits) of information in parallel. The width of the data bus is typically 8 bits, 16 bits, or 32 bits. The greater the data bus width, the higher the rate at which data can be transferred to and from the CPU. The address bus is sized to allow an appropriate number of unique memory addresses to be handled. A 20-bit address bus will allow up to 1,048,576 memory locations to be directly addressed. Other considerations that determine the capabilities of a microcontroller are the cycle time, often called the clock speed, and the available set of primitive instructions, known as the instruction set. Clock speeds for microcontrollers are currently in the 12–32 MHz range, although this figure is rising in response to increasing demands from high-speed control applications.

The instruction set is a unique set of low-level commands, which the CPU can recognize and use to control the operations it performs, such as loading data into registers and performing simple arithmetic. Since most instructions require a number of clock cycles to complete, the combination of the clock speed and number of clock cycles per instruction is an important indicator of microcontroller execution speed. User programs are usually written in a language other than the instruction set. Assembly language, while also at a low level, is commonly used to program microcontrollers. The advantage of assembly language is that it allows the programmer to control each step in the execution of a program—generally providing opportunities for high processing speed. The disadvantages of assembly language are that it tends to be processor dependent, and is relatively slow to write compared with higher-level languages. Increasingly, engine control programs are being written in higher-level languages such as C. These languages are defined by international standards and are widely supported by chip manufacturers, such that any programs written in them are highly portable to other processors. A high-level language uses a computer program (compiler) to translate the high-level commands or statements into the low-level instructions required by the processor, widely known as machine or object code.

Engine control requires that the microcontroller be capable of responding rapidly to events that occur that are not preplanned or otherwise routinely expected, for example, throttle pedal being depressed or a sudden change in a sensor signal. Such events are handled by a software concept known as an interrupt. An interrupt allows higher-priority tasks to demand immediate attention from the processor. When an interrupt is received, the part of the program being executed will be suspended and the program will jump to the section of program that the interrupt requires executed. When the interrupt service has been provided, the program will continue at the point where it was suspended.

The microcontroller shown in Figure 11.22 uses three kinds of memory-storage device. If memory is *volatile*, it means that stored information will be lost when power is removed. Conversely, *nonvolatile memory* retains its stored data even when the ECU is switched off. Nonvolatile read/write memory is used to store important data that may require later retrieval from the ECU, such as OBD fault codes, service codes, or service diagnostic snap-shot data. The volatile read/write (RAM) memory is used to store the intermediate results of calculations, run-time data, and engine control variables, such as boost pressure or engine speed. The nonvolatile, read-only memory (ROM) is used to store the engine control program and any permanent data needed for control, such as look-up tables for injection timing or boost pressure control.

The input/output devices are very important in determining the performance of the microcontroller. Data are being transferred to and from the sensors and actuators and potentially between the engine ECU and other ECUs in the system many hundreds of times per second and at a variety of rates and in a variety of formats. The analogue-to-digital converter (ADC) allows the microcontroller to read and operate on analogue signals originating in sensors. An ADC will usually be multiplexed to enable it to handle multiple signals or channels. The analogue signal is translated into a binary number usually consisting of 8 or 10 bits. The number of bits used to represent the analogue value defines the resolution with which the actual value can be represented. An 8-bit converter will have a resolution of 1 in 256, which usually meets the requirements of engine control. A 10-bit conversion will resolve the analogue signal to within 1 in 1024. The high-speed I/O channels are critical for engine control. They allow fast transmission of critical events in the engine, such as timing wheel (pulse) information to enable crank angle to be tracked. In general, the high-speed serial I/O channels are controlled by interrupt services. The digital I/O ports allow communications with other devices, such as digital sensors and actuators, and other ECUs. Digital I/O ports are used for slower communications and are read and written to by the control program directly rather than an interrupt service. The high-speed serial I/O data link channels are used to support network protocols such as SAE J1939 and CAN 2.0b, which allow data to be exchanged between ECUs or for service diagnostic data to be input or output to a service bay computer.

11.4.2 Sensors

This section will present and discuss the key sensor technologies used in the control of advanced diesel engines: mass airflow sensors (MAF), crank position sensors, and temperature and pressure sensors.

11.4.2.1 Crank Position and Crank Speed Sensor

In order to control fuel injection timing, it is necessary to know the instantaneous position of the engine crankshaft. For a four-stroke engine it is also necessary to know which part of the cycle a particular cylinder is on—the pumping revolution or the compression-combustion revolution. The latter problem may be addressed by using the camshaft as an indicator of crank position, since the two shafts are driven at a 2:1 speed ratio. Since the camshaft rotates only once during a complete engine cycle, a unique relationship exists between the firing top dead center (tdc) of each cylinder and an angular location on the camshaft. The primary disadvantage of using the camshaft to provide a fuel injection timing signal is the greater error, compared with the crankshaft, due to backlash in the camshaft drive and potentially greater torsional oscillation. Typically, the rotating wheel will have as few as four targets (impulse generators) or as many as 60 targets, and will often include an additional target to allow the controller to reference the cyclic pulse train to the piston positions.

Crank position sensors in regular use are based on either magnetic or optical principles to detect the passing of targets placed on a rotating wheel. Crankshaft speed is measured with the same sensor by counting the number of pulses over a given time period [23].

Variable Reluctance Sensor (VRS) The variable reluctance sensor uses a magnetic circuit to detect the passing of ferrous targets placed around the periphery of a timing wheel attached to the crankshaft or camshaft. A permanent magnet is arranged as shown in Figure 11.23. The magnetic circuit is essentially made complete when the ferrous target is centered in the gap. As the target enters the gap the magnetic permeability is increased and the magnetic flux increases. With the target centered in the gap the magnetic permeability is at its highest and the magnetic flux at its highest. As the target moves out of the gap the magnetic permeability decreases and the magnetic flux is reduced back to zero. The coil has a voltage induced by the rate of change of magnetic flux as shown in Figure 11.23. The output voltage waveform produced by the coil is relatively easily converted

Fig. 11.23.

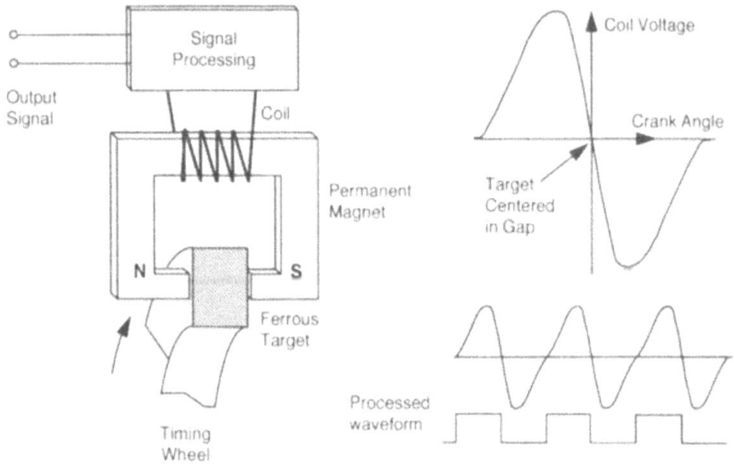

Variable reluctance crank position sensor and output voltage waveform.

to a logic signal or square wave, the leading edge of which can be used to indicate tdc if the targets are arranged to coincide with piston tdc events.

A disadvantage of the variable reluctance sensor for engine crank position measurement is that it requires the crank to be rotating in order to produce output, since the output voltage is proportional to the rate of change of magnetic flux density. In effect the output from the sensor is proportional to both crank position and crank speed. This prevents engine timing from being set statically during engine build, and produces a weaker signal at cranking and low idle speeds. The dependence of the output voltage on time rate of change of magnetic flux requires that the engine be rotating to produce a useable signal.

Hall Effect Sensor The Hall effect sensor is also used with a rotating wheel carrying ferrous targets. The Hall effect is the production of a voltage in a thin slice of semiconductor material when it is simultaneously exposed to an electrical current and a magnetic flux in perpendicular directions to each other. The voltage is induced in the third orthogonal direction, as illustrated in Figure 11.24. In the Hall effect sensor the strength of the applied magnetic field is varied by the absence or presence of a ferrous target within a magnetic circuit (Figure 11.25). This reduces the reluctance of the circuit and hence the magnetic flux is increased, increasing the output of the Hall element. The Hall effect is relatively small in comparison with the VRS principle, and so the output voltage requires amplification.

Hall effect sensors in production typically incorporate supporting circuitry in an integrated package. To enhance the overall accuracy of the sensor, a constant current supply is built into the sensor. The Hall effect sensor produces an output signal when the engine is at rest, which allows the engine timing to be set statically during engine build.

Fig. 11.24.

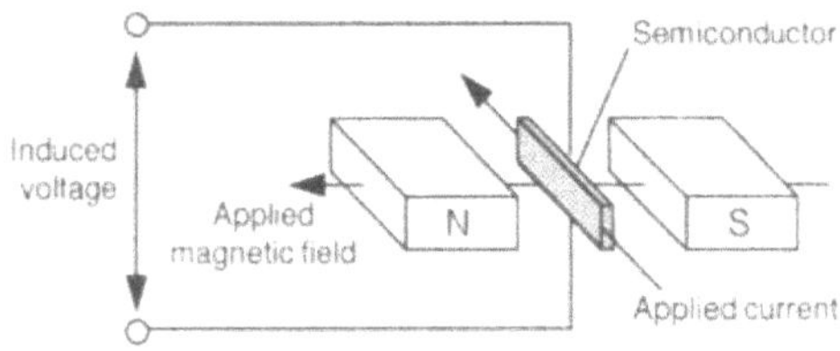

Hall effect is a small voltage induced by applied current and magnetic flux.

Fig. 11.25.

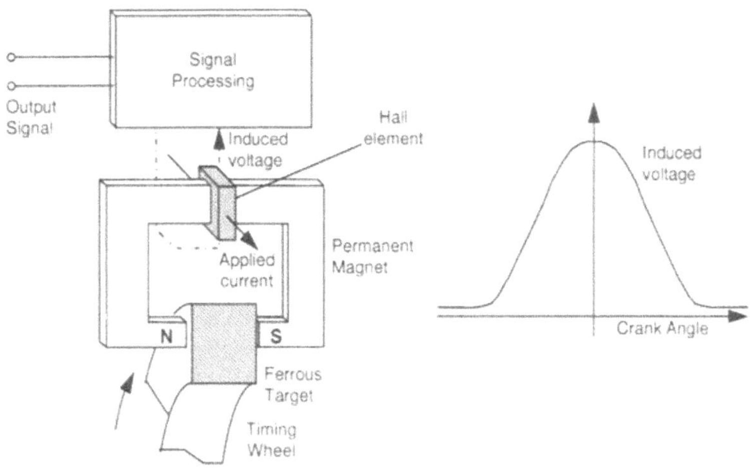

Hall effect sensor as used to measure crank position.

11.4.2.2 Temperature Sensors

Effective control of a modern turbocharged diesel engine requires a number of temperatures to be measured. At a minimum, the temperatures required include engine coolant, intake manifold air, and lubricating oil. The coolant and lubricating oil temperatures generally change relatively slowly over a period of minutes rather than seconds. Potentially, the intake manifold air temperature (IMT) can change much faster, but for engines with charge air coolers, the IMT changes over a relatively narrow range and changes occur at rates of the order of several seconds. Consequently, the response characteristic required is in general not a paramount concern in the design of automotive diesel engine temperature sensors. A possible exception might be the non-intercooled turbocharged diesel engine in which the IMT can change between ambient and 150°C in a matter of a few seconds. Since the IMT sensor will generally be used to calculate the instantaneous airflow, from which the smoke-limited fueling is calculated, sensor response becomes a consideration.

Fig. 11.26.

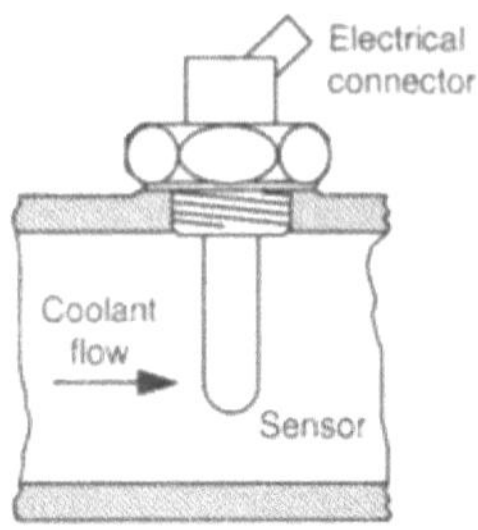

Engine coolant temperature sensor.

Fig. 11.27.

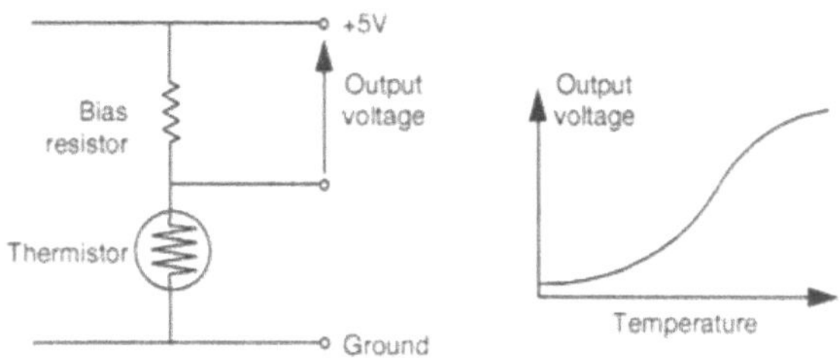

Thermistor circuit with bias resistor.

The majority of temperature sensors use a shielded thermistor [24] positioned in the fluid stream through a threaded access hole, as shown in Figure 11.26.

The thermistor is a semiconductor element having a negative temperature coefficient, which means that the electrical resistance falls with increasing temperature. Typical resistance values are [25]:

Temperature	*Resistance*
−40°C	100 kΩ
130°C	70 kΩ

The characteristics of the thermistor that make it attractive for automotive diesel engine applications are its high reliability, relatively low cost, and high sensitivity to temperature. A typical electrical circuit in which the sensor element is located includes a bias resistor, as shown in Figure 11.27.

11.4.2.3 Pressure Sensors

The effective control of turbocharged and intercooled diesel engines requires the measurement of intake manifold absolute pressure (MAP) and ambient pressure. Intake manifold pressure is required for the calculation of instantaneous airflow rate. This measurement is combined with intake manifold temperature to determine intake manifold air density. The ambient pressure is required to allow the

Fig. 11.28.

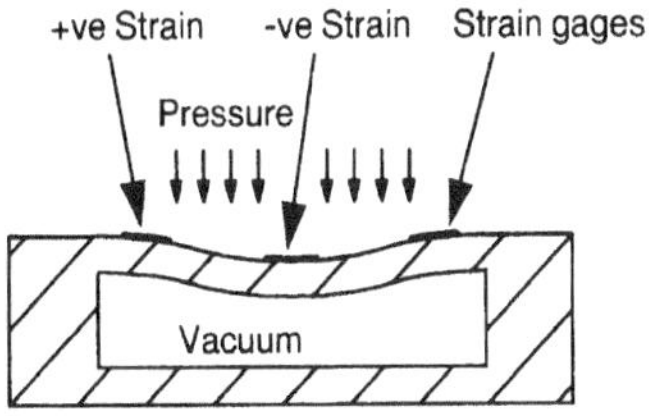

Piezoresistive pressure sensor showing elastic diaphragm and semiconductor strain gauges.

engine controller to compensate for high-altitude operation. A particular concern at high altitude (2000–3000 m) is protection of the turbocharger, which will run at increased speed and boost pressure.

There are two types of pressure sensors in regular use in automotive control applications—the semiconductor piezoresistive sensor and the capacitive sensor. Both pressure sensor concepts use the pressure to cause deflection of an elastic diaphragm, the deflection and strain of which are used to produce electrical changes. Figure 11.28 shows schematically the layout of a piezoresistive pressure sensor. It may be desirable to locate the strain gauges within the sealed vacuum space to prevent attack from moisture or dirt-laden air.

The piezoresistive sensor uses the change of electrical resistance of small silicon strain gauges caused by the mechanical strain in the diaphragm. Typically the gauge factor (K = electrical strain/mechanical strain) is as high as 100. This high sensitivity allows the diaphragm design to be stiff, giving low stress levels and good durability. The strain gauges are arranged in a Wheatstone bridge such that gauges on opposite sides of the bridge strain in the same sense (tension or compression) and gauges on adjacent sides strain in opposite senses. This arrangement reduces the sensitivity of the sensor to temperature changes. As pressure is applied, the bridge becomes unbalanced producing a net voltage output, which can be detected by a differential amplifier as shown in Figure 11.29.

The capacitive sensor also uses the deflection of an elastic diaphragm under pressure loading, in this case, to change the gap of a capacitor and, hence, its capacitance. Typically the sensor has integrated signal processing electronics, which convert the changes of capacitance to an analogue voltage of 0–5 V, which is proportional to applied pressure.

11.4.2.4 Mass Airflow Sensor

In order to meet customer expectations for performance while meeting the exacting emissions standards, it is essential to measure the instantaneous airflow into the engine cylinders. Currently this is achieved by two different approaches. The vast majority of heavy-duty engines typically do not measure airflow rate directly, but rely on the direct measurement of physical variables from which the airflow rate can be calculated. This approach, known as the speed-density method, requires that the air temperature and pressure in the intake manifold and engine crankshaft

Fig. 11.29

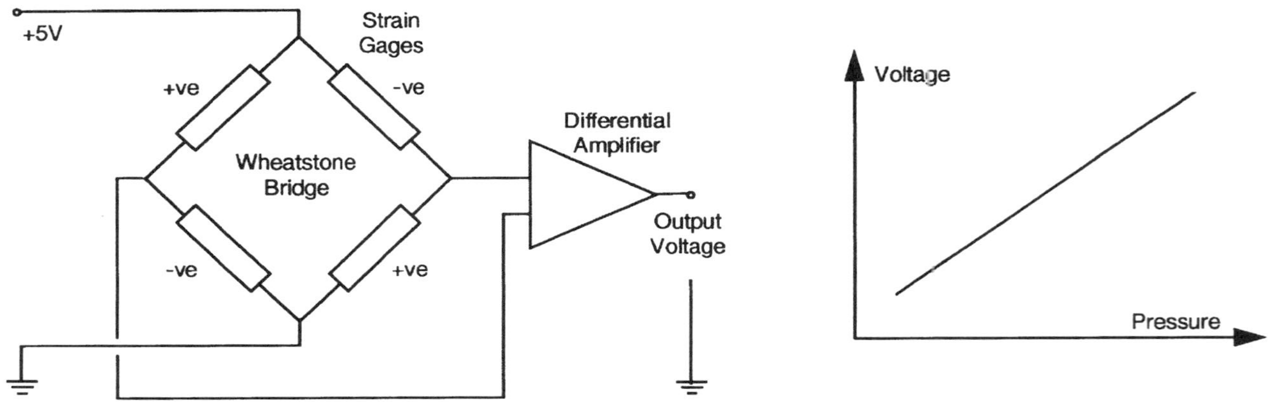

Piezoresistive pressure sensor electronic circuit, showing strain gauges loaded in tension (+ve) and compression (−ve).

Table 11.2

Range of Airflow Sensors for Engine Application [22, 25, 26]

Working principle	Description
Plate and Funnel	Airflow is accelerated past a pivoted bluff plate such that the position of the plate can be calibrated to indicate airflow rate—used in Bosch KE-Jetronic gasoline electronic control system.
Rotating Flap	Airflow deflects a rotating flap, positioned in the stream, against a spring. A potentiometer converts the angle of the flap to a voltage. Air temperature is sensed locally to allow compensation for ambient air temperature—used in the Bosch L-Jetronic gasoline electronic control system.
Hot Wire, Hot Film	Both hot wire and hot film MAF sensors use the relationship between heat transfer and air velocity as their working principle. These are described later.
Karman Vortex	This MAF sensor uses the Karman vortex principle to measure airflow rate. Vortices are shed from the trailing edge of a cylinder placed perpendicular to the flow. The frequency of the vortices is measured by an ultrasonic sender and receiver. Signal processing is required to convert this information into a signal proportional to airflow.

speed are measured. From this data an estimate of the airflow can be made—subject to assumptions about the engine volumetric efficiency. This makes the approach unable to compensate for changes in engine breathing resulting from variations in the manufacturing process or from deterioration in use. A further disadvantage of the speed-density method is its inability to differentiate between airflow and EGR flow other than by a crude calculation based on air and EGR temperatures.

To overcome these disadvantages it is desirable to measure the airflow directly using a mass airflow sensor (MAF). MAF sensors were first developed for accurate control of gasoline engines to meet emission standards in the mid-1980s. A number of different concepts have been developed and these are listed in Table 11.2.

Hot Wire/Hot Film Working Principle Hot wire and hot film MAF sensors are robust, high-volume, low-cost versions of the well-known hot wire/film anemometers used in research and development for measuring fluid velocity and turbulence. The working principle is based on the relationship between the rate of heat transfer and the fluid velocity in convective heat transfer [27]. The hot wire has electrical resistance and when a current is passed through it heats up. The wire temperature increases until the heat produced is exactly balanced by the convective heat transfer from the wire surface, which depends on the fluid velocity and on the temperature difference between the wire and the fluid. In the hot wire MAF sensor the temperature difference between the intake air and the wire is maintained constantly

Fig. 11.30.

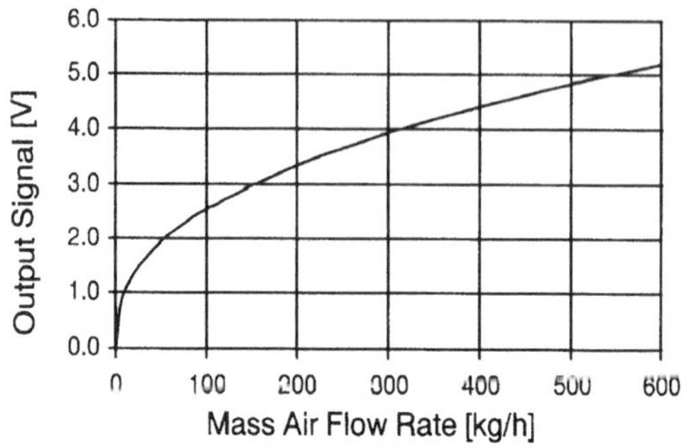

Mass airflow sensor operating characteristic.

Fig. 11.31.

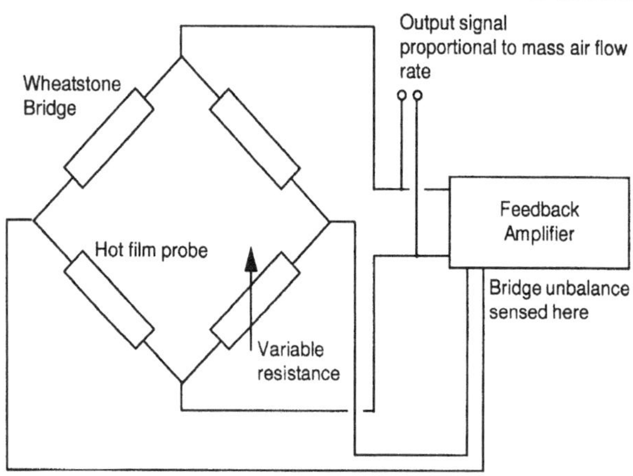

Typical hot film MAF sensor circuit.

at about 200°C. This is achieved by measuring the air temperature at the MAF sensor.

Figure 11.30 shows a typical calibration for a hot film mass airflow sensor. The hot wire forms one side of a Wheatstone bridge, which is powered by an amplifier. Any change in air velocity tends to change the temperature of the wire and hence its resistance, tending to unbalance the bridge. The unbalanced voltage is compensated by the amplifier so as to maintain a constant wire temperature such that the amplifier output voltage is proportional to mass airflow rate. A Wheatstone bridge circuit for airflow measurement is shown in Figure 11.31.

In the latest designs the bridge circuitry is deposited as a thin film on a ceramic substrate. The traditional problem of deposits forming on the film, which change the heat transfer characteristics and hence the calibration of the sensor, are largely overcome by moving the film away from the leading edge of the sensor. New developments allow MAF sensors to retain accuracy even when the flow direction reverses, as it can in four-cylinder naturally aspirated engines. The transient response of a hot film MAF sensor to a step change in mass flow rate is typically better than 75 ms to 95 percent of the final steady-state signal. The pressure loss incurred by an MAF sensor is typically less than 15 mbar at maximum rated airflow.

11.4.3 Actuators

11.4.3.1 EGR Valve

As diesel engine emission standards for oxides of nitrogen (NO_x) become more stringent, an increasing number of engines are being equipped with exhaust gas recirculation (EGR) systems. These systems return a fraction of the engine exhaust gas to the intake manifold via corrosion-resistant tubing and a control valve, known as an EGR valve. EGR control was discussed in an earlier section; this section will discuss the valve itself and the different approaches adopted.

To the present day, the only diesel engines fitted with EGR systems have been for light-duty applications, notably diesel engines for passenger cars in Europe, and for light-duty trucks and vans in Europe and the United States. In Europe virtually all diesel passenger cars have EGR systems, some also using EGR cooling to gain additional NO_x control. In the United States, the Dodge Ram pick-up truck powered by the Cummins 5.9L DI diesel engine has EGR to meet the California TLEV standards, and the GM 6.5L light-duty diesel engine used in sport-utility and pick-up truck applications also uses EGR for emission control [28]. In developing an emission control strategy that uses EGR, the goal is to find the optimum trade-off between fuel consumption, NO_x control, and particulate control. In general, adding EGR allows injection timing to be advanced for improved fuel consumption, but at some cost in particulate control. When all of the trade-offs have been made, the emission control strategy will usually require EGR to be scheduled as a function of engine speed and load (or throttle position), as shown in Figure 11.17 for passenger-car application. In addition, EGR may be scheduled during transient operations, although EGR is not normally used during hard accelerations, during start-up or periods of motoring or overrun.

The role of the EGR valve is to control the flow of EGR between no flow and maximum flow according to the EGR rate demanded by the engine controller. This requires that the valve is accurate and rapid in achieving the positions commanded by the controller. It also demands a high level of repeatability and usually fine resolution so that EGR rates can be controlled accurately. The accuracy required depends on operating conditions and the emission stringency but a figure of $+/-3$ percent is not uncommon. Unwanted forms of valve behavior, such as hysteresis, temperature drift, and dead bands, should also be minimized.

The most common configuration used for diesel engine EGR valves is the vacuum-actuated poppet valve, either inward or outward opening, as shown in Figure 11.32. In this design the poppet valve is actuated by vacuum to open, and

Fig. 11.32.

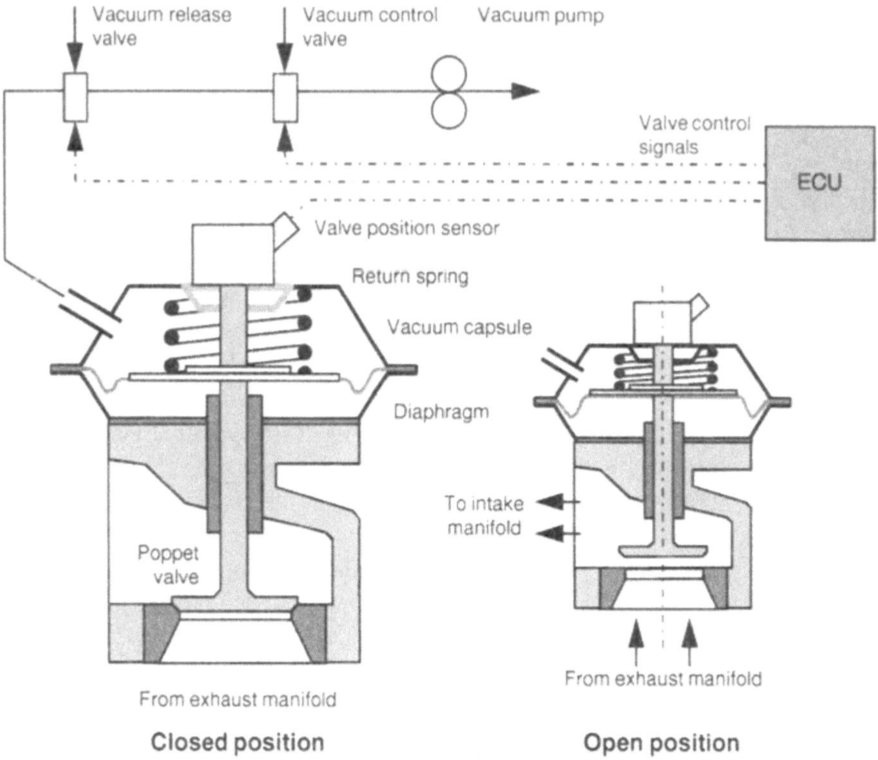

Vacuum-operated EGR valve with inward opening poppet valve.

is closed by a return spring as the vacuum is removed. The other significant force acting on the valve is the net gas loading force. This is generally greatest when the valve is closed and depends on the net valve area exposed to the gases on either side, and on the difference between exhaust and intake manifold pressures. The vacuum is normally provided by a fixed-displacement vacuum pump driven by the engine or by an electric motor. In order to limit the size of the EGR valve and provide positive actuation, the design vacuum is often as low as 0.3 bar-abs. A typical vacuum circuit is also shown in Figure 11.32. For fast opening of the valve, it is necessary to optimize the displacement of the pump and minimize the volume of the vacuum tubing and diaphragm chamber. The addition of a vacuum reservoir may also help response. For rapid closing it is usual to add a vent valve to admit air to the vacuum line when closing is required. This collapses the vacuum very quickly and enables the return spring to close the valve. Typical response times are 100–200 ms for opening and 50–100 ms for closing.

Although some applications do not require modulation of the valve between fully open and closed, by far the majority of applications are modulated. This allows a more flexible approach when calibrating the engine controller and enables EGR rates to be compensated against system changes, such as increased intake or exhaust restriction.

The valve shown in Figure 11.32 is modulated by controlling the vacuum above the diaphragm to adjust the position of the valve. Vacuum is modulated by the vacuum control valve, which usually consists of a solenoid valve operated by a pulse width modulated signal, such that air is admitted to the vacuum line at a controlled rate to modulate the vacuum. In this example, valve position is under closed-loop control. A valve position sensor provides feedback to the controller so that valve position can be controlled accurately. The position sensor is typically a linear potentiometer or an LVDT.

Other types of EGR valve are being developed in an attempt to overcome the limitations of the vacuum valve. The main limitations of vacuum technology are speed of response, cost and complexity. The complexity derives from the need to provide a vacuum system, and the distribution of vacuum around the engine compartment. The most desirable systems from control and installation points of view are actuated directly by electrical means. The candidate actuators include stepper motors, proportional solenoid valves, and DC motors.

11.4.3.2 Variable Geometry Turbocharger

Almost all automotive diesel engines are turbocharged. The primary benefits are greatly increased power density from the engine and lower system cost per horsepower. Turbochargers use the expansion of the exhaust gases through a turbine stage to provide power to drive a compressor placed in the intake stream [29, 30]. The performance of the turbocharger depends on the isentropic efficiencies of the turbine and compressor. Modern turbochargers have high overall efficiencies (50 percent to 65 percent) such that at high engine load and speed the turbocharger not only provides adequate airflow but also contributes significantly to the thermal efficiency of the engine system by providing a positive pumping loop.

The flow versus pressure ratio characteristic of a typical turbocharger turbine is nonlinear, such that at low gas flow rates the pressure ratio is small. As the gas flow rate is increased, the pressure ratio rises as a power law, until eventually the unit will choke and mass flow cannot be increased by pressure ratio alone. Since turbine power is closely related to expansion ratio and gas flow rate, turbochargers are generally unable to provide adequate airflow at low engine speeds. The precise relationship of gas flow rate and pressure ratio is controlled by several factors, of which the most significant is the effective flow area of the nozzle upstream of the turbine wheel. Smaller nozzle areas will provide higher upstream pressures and, therefore, turbine expansion ratios.

The inability to deliver adequate airflow to the engine at low speeds results in reduced torque and reduced driveability, and possible smoke emission. Traditional ways to solve this problem are to match the turbine at low speed (i.e., use a small turbine casing size) and either accept penalties of high boost pressure and airflow at high engine speeds, or use a turbine by-pass (wastegate) valve at high engine speeds to reduce turbine power. These approaches generally involve compromises between low-speed and high-speed operation, which may be significant to the end user depending on the intended application of the engine.

A number of variable geometry turbine concepts have been developed [31–34] in an attempt to overcome the inherent limitations of the fixed geometry turbine, described earlier. The most common approach is to use a vaned nozzle in which the nozzles are able to rotate about an axis parallel with the turbine rotor [31]. This allows the effective flow area of the nozzle to be increased or decreased to suit prevailing engine demands, thus varying the power developed by the turbocharger turbine. It will be observed that in this approach, the gas inlet angle to the turbine wheel is also a function of nozzle blade position and this may affect isentropic efficiency in a nonoptimal way. An alternative to the rotating nozzle vane approach is to employ a moving side wall in the throat section of the nozzle. This can be used to control the nozzle effective area and hence turbine power. Various designs have been reported in the literature, both with and without nozzle vanes. Holset has developed a moving side-wall design that also features fixed nozzle guide vanes [34].

The Holset variable geometry (VG) turbocharger has been developed to overcome the limitations of the fixed geometry turbocharger [34]. Figure 11.33 shows the variable turbine nozzle ring of the Holset HX40V variable geometry turbocharger. This design uses radial turbine and compressor stages. The variable geometry is provided by a moveable nozzle ring, which translates axially to control the effective area of the nozzles upstream of the turbine. The nozzle ring has fixed blades which slide into a close-fitting shroud plate, which is fixed in the turbine housing.

Also shown in Figure 11.33 is a schematic representation of the control system used to control the VG turbocharger. The figure shows the axially moving nozzle ring, which is actuated in proportional control mode by a pneumatic actuator through a lever mechanism. The pneumatic actuator is controlled by an electropneumatic control valve, which regulates a high-pressure air source to provide the pressure required to position the nozzle ring against a return spring. The VG turbocharger essentially allows closed-loop control of the engine boost pressure. This is handled by the engine ECU, which monitors all engine sensors and computes the optimum required boost pressure for the prevailing conditions. A control loop within the ECU uses the demanded boost pressure signal and modulates the VG turbine to eliminate errors between the actual boost pressure and the demanded pressure. At rest the turbine is fully open. Increasing the control air pressure moves the nozzle ring toward the shroud plate, reducing the nozzle width and increasing boost. In an automotive application, the turbine would run fully open at rated power and be progressively closed at low engine speeds, during load-on transients, during high-altitude operation, and during engine braking operations.

Figure 11.33 also shows control air pressure and turbocharger speed sensors. These are provided for protection and diagnostic purposes. VG turbocharger speed

Fig. 11.33.

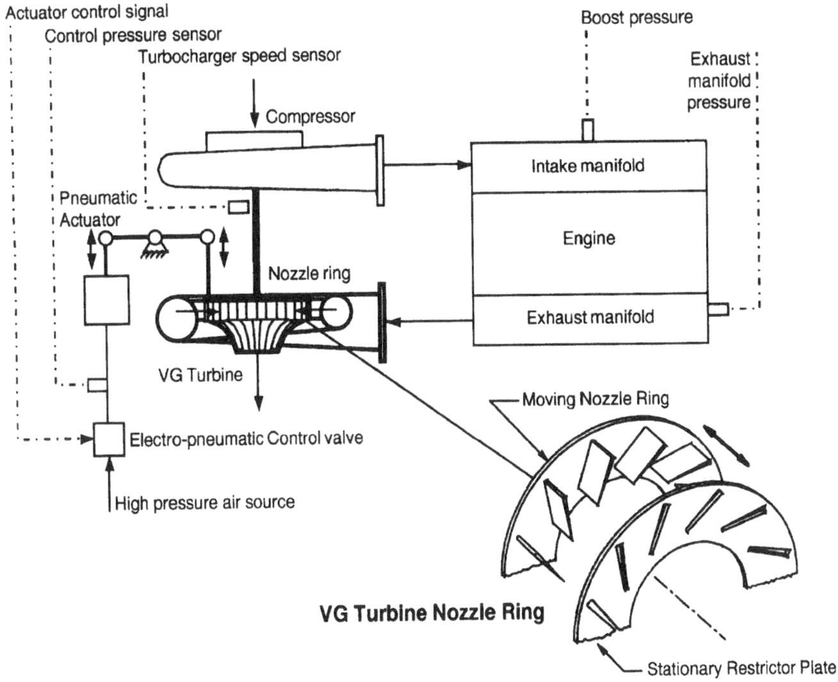

VG turbine nozzle ring arrangement and control system for a variable geometry turbocharger [34, 37] *[Courtesy of Holset Engineering Company Limited]*.

is used as part of a control scheme to protect the turbocharger in the unlikely event of the nozzle ring failing in the closed position and overspeeding the turbocharger. If such a failure is detected, the engine controller can take appropriate action, such as reducing the engine fueling to limit peak cylinder pressure until the problem is corrected.

The VG turbocharger may be used to enhance a number of engine attributes [29–37]. Low-speed torque may be improved dramatically, as mentioned earlier, without the compromises in high-speed fuel consumption or high airflow associated with fixed geometry turbochargers. If the engine is matched correctly to the transmission, the additional low-speed torque will improve driveability and reduce the number of gear shifts required. The VG turbocharger can also be used to improve the transient performance of a turbocharged engine by providing more boost at light loads and engine speeds. During a hard acceleration transient, the VG turbocharger should be controlled to provide maximum airflow to the engine throughout [34, 37]. This will normally require modulation of the nozzle ring from almost closed to fully open depending on the end states of the engine. The exhaust manifold pressure sensor, shown in Figure 11.33, is optionally provided to enhance control of the VG turbocharger during hard accelerations in automotive applications. VG turbocharging can also reduce the engine power derate at high altitude, and improve the effectiveness of engine compression braking systems by providing higher boost pressures during braking operation.

A number of approaches have been proposed by various authors for the control of engines having variable geometry turbochargers [32, 34, 35, 37]. Caddy [37] discusses the advantages of sensing the exhaust manifold pressure, upstream of the VG turbine, in addition to boost pressure, for control purposes. The main advantages are in achieving maximum airflow during a transient as outlined earlier, and in maximizing braking torque. During a transient acceleration, it is important to maximize the torque delivered to the engine flywheel throughout. Using a simplistic approach to VG control during a transient, the VG mechanism would be moved to the closed position as soon as a transient event is detected by the ECU. In this strategy the exhaust pressure rises very rapidly compared with the intake pressure, giving rise to a large negative pressure difference across the engine cylinders. The flaw in this strategy is that it overlooks the relationship between the negative pressure differential and volumetric efficiency, and the negative work incurred through the pumping loop. As the exhaust pressure rises, the engine's volumetric efficiency can fall to surprisingly low levels (< 50 percent). If the pressure difference across the engine cylinders is calculated by the ECU from exhaust and intake pressure sensor signals, it can be used to determine the optimum setting for the VG turbine in order to deliver torque to the flywheel.

If the VG turbocharger is used to enhance engine braking, it is very desirable to have knowledge of the exhaust and intake pressures, so that braking torque can be produced repeatably despite manufacturing variations in the engine or turbocharger, variations in ambient conditions, and lifetime changes in engine and turbocharger.

11.5 EXHAUST AFTERTREATMENT

Exhaust aftertreatment encompasses any form of exhaust gas processing aimed at reducing the emission of one or more exhaust components. Aftertreatment devices for diesel engines include diesel catalysts and soot filters—sometimes known as particulate traps. Diesel catalysts may be further subdivided into oxidation catalysts and lean-NO_x catalysts, although development of catalysts capable of both oxidation and reduction is reported [38].

The aftertreatment device is normally installed in the exhaust system under the floor-pan of the vehicle. In some diesel catalyst applications the process is divided into two parts: a primary catalyst installed close to the outlet from the turbocharger turbine, and a secondary catalyst located in the tailpipe.

The goal of any diesel engine development program is to achieve the emission standards set for the product without recourse to exhaust aftertreatment. However, in some cases it is either beyond the current technology to meet the emission standards using in-cylinder emission controls, or the trade-offs between in-cylinder emission control and exhaust aftertreatment favor inclusion of aftertreatment. A key trade-off involves product cost. If the cost of the in-cylinder technology required to meet the standards exceeds the cost of the aftertreatment technology, then provided fuel economy, packaging, and other requirements are met, aftertreatment can be a highly effective solution. If aftertreatment is required, then it should be viewed as an important engine subsystem requiring integration with the engine control strategy—no less important than any other emission control element.

11.5.1 Diesel Exhaust Composition

The exhaust from a diesel engine is a complex mixture of organic and inorganic compounds in solid, liquid, and gaseous phases. (See Table 11.3.) The organic compounds originate with the fuel and lube oil, some of which reach the tailpipe without modification, some of which are modified by combustion or by reactions in the catalyst. They are typically hydrocarbons ranging from lightweight fractions with high hydrogen:carbon (H:C) ratios, through heavier fractions with very low H:C ratios. The hydrocarbons usually contain aldehydes, akanes, alkenes, and aromatics [39].

The inorganic compounds are largely derived from the fuel and lube oil, although other sources include the intake air and wear products from the engine hardware. Depending on the fuel formulation, sulfur in the fuel can be a significant source of exhaust particulate [40]. This can take the form of gaseous sulfur dioxide (SO_2), sulfates (SO_3, SO_4), and sulfuric acid (H_2SO_4). Fuel and lube oil provide a source of carbon which can form soot particles in the combustion process through gas- or liquid-phase reactions. Although the soot particles formed initially are less than 0.1 μm in diameter, they can rapidly (<10 ms) combine with other soot particles by a process of agglomeration to form larger particles, typically up to 1 μm in diameter [39]. Most of the soot is oxidized to carbon dioxide during the combustion process; however, complete oxidation is difficult to achieve because combustion is limited by the rate of mixing toward the end of combustion, and temperatures are falling due to expansion. A very small amount of ash can also

Table 11.3

Diesel Exhaust Composition

Solids	Liquids	Gases
Soot	Hydrocarbons (SOF)	Nitric Oxide (NO)
• nuclei	• fuel derived	Nitrogen Dioxide (NO_2)
• agglomerated particles	• lube oil derived	Hydrocarbons
Sulfates	Sulfuric Acid	Carbon Monoxide
Ash	Water	Carbon Dioxide
• oil additives		Water
• engine wear products		Oxygen
		Nitrogen

be found in diesel exhaust, which is derived from engine wear products or from additives in the lube oil.

For the purpose of engine development, *particulate matter* is defined by the relevant test standard used to measure it. *Total particulate matter* (TPM) consists of any exhaust component, other than water, filtered from a sample drawn from a dilution tunnel, operated at the conditions specified in the relevant test procedure. Dilution tunnel conditions simulate real conditions at the tailpipe of a road vehicle where exhaust products mix with the surrounding air.

Diesel particulate matter typically consists of small carbon spheres (0.010–0.080 μm) with adsorbed hydrocarbons, agglomerated carbon particles (0.05–1.0 μm) with adsorbed hydrocarbons, liquid condensed hydrocarbons, sulfates, and ash. The precise analysis of diesel particulate will depend on many factors, including fuel formulation, fuel injection quality, lube oil control in the engine, the nature of the combustion event (e.g., premixed versus diffusion burn fractions), air-fuel ratio, air-fuel mixing rate, and flame temperature history. The flame temperature history is important because the oxidation of soot occurs at a finite rate that depends on flame temperature. The residence time of soot at high temperature is inversely proportional to engine speed, making soot oxidation difficult as engine speed is increased. It is vitally important to consider the composition of diesel particulate matter when developing catalysts or filters for emission control. It is usual to divide particulates into three parts: soot, soluble organic fraction (SOF), and sulfates. The SOF consists of the hydrocarbons adsorbed on to the soot particles, and liquid hydrocarbons. SOF is so called because it may be measured by extraction using an organic solvent.

Oxides of nitrogen form during the diesel combustion process in the high-temperature flame as nitric oxide (NO) and nitrogen dioxide (NO_2). Collectively these two gases are known as NO_x. NO is the predominant form since the reactions from which it is produced proceed at higher rates than the reactions for NO_2 at the conditions prevailing in the diesel combustion process. Theoretical considerations based on chemical kinetics would suggest NO_2/NO ratios less than 5 percent. In reality NO_2 typically accounts for 10 percent to 20 percent of the NO_x.

Emission standards regulate the sum of all oxides of nitrogen as NO_x; some standards regulate the sum of NO_x and unburned gaseous hydrocarbons (HC). Diesel engines tend to emit very small quantities of carbon monoxide (CO) compared with gasoline engines, and meeting the regulated CO standards is not normally difficult.

11.5.2 Exhaust Aftertreatment Design Requirements

The design requirements for aftertreatment systems depend not only on the exhaust composition, but also on the duty cycle that the emissions test and the end users impose on the engine. The aftertreatment device should function correctly when new and continue to meet emission control requirements over the design life of the engine. The emission cycles discussed earlier in this section were for heavy-duty engines and light-duty vehicles. The implications for the design of the aftertreatment system are somewhat different for these applications. The heavy-duty cycle, shown in Figure 11.1, imposes a high load factor on the engine such that the average power can be as high as 50 percent of the engine rated power. This produces relatively high engine-out exhaust temperatures, typically 400–700°C. The light-duty cycle, shown in Figure 11.2, applies to passenger cars and light-duty trucks and vans having vehicle test weights up to approximately 3850 kg (8500 lb). Such vehicles typically demand engine power in the range of 10–25 kW (13–33 BHP), averaged over the emission cycle. For most light-duty engines, that represents relatively light load operation with relatively low engine-out exhaust temperatures at 200–400°C. The engine-out exhaust gas temperature is normally higher than the gas temperature at the catalyst because of expansion through the turbocharger turbine and heat loss from the exhaust system. The transient nature of most emission cycles, including cold start, exacerbates the problem. Clerc [41] reports heavy-duty engine gas temperatures at the catalyst between 150°C and 400°C during the emission cycle. Over the European light-duty cycle, Smedler et al. [38] suggest catalyst temperatures as low as 150°C to 200°C for the urban (ECE-15) drive cycle and 240°C to 350°C for the high-speed extra-urban drive cycle. The low temperature of the feedgas poses a major challenge in diesel catalyst design, which partially explains the development of the close-coupled diesel oxidation catalyst (DOC) for passenger-car diesel engines. While exhaust temperature may be one notable difference between the light- and heavy-duty cycles, other differences are also relevant such as soot content, fuel and lube oil content, durability, and fuel economy requirements.

In all diesel applications the overall air-fuel ratio is lean of stoichiometric, varying from 1.25 to 5.0:1 excess air ratio. This has limited the application of highly effective three-way catalysts (HC, CO, NO_X) developed for gasoline engines. The three-way catalyst requires engine operation very close to stoichiometric in order to drive oxidizing and reducing reactions simultaneously.

11.5.3 Diesel Catalysts

Reaction rates between exhaust components are typically governed by Arrhenius equations in which the rate at which a given reaction proceeds is a function of the Arrhenius parameters and the temperature [43]. The Arrhenius parameters are constant for a given reaction, unless a catalyst is present. The Arrhenius parameters include a property known as the *activation energy*, which is the minimum energy that reactants must have in order to form products. The reaction rate is related to temperature exponentially such that, at normal catalyst temperatures, the oxidation of HC and CO would proceed very slowly. Catalysts are used essentially to lower the activation energy of the overall reaction such that the reaction proceeds at a high rate even at low temperatures. A catalyst is a material that accelerates the

Fig. 11.34.

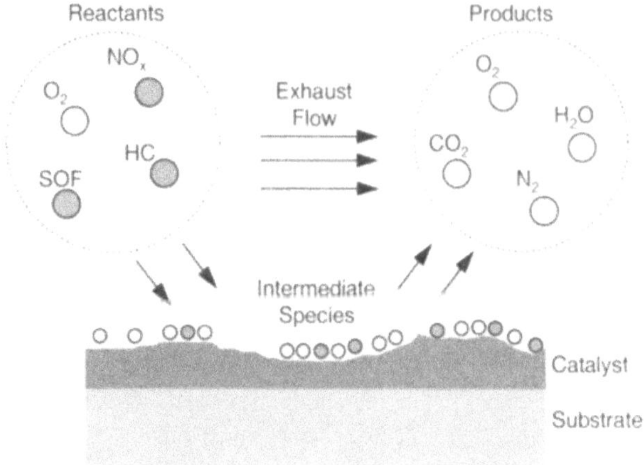

Simplified model of catalytic action in a diesel oxidation catalyst.

rate of a given reaction without itself being consumed by the reaction. Catalysts increase the rate of a given reaction by providing alternative paths with lower activation energies, as shown in simplified form in Figure 11.34.

In exhaust aftertreatment the catalyst and reactants are in different phases—the catalyst being in the solid phase, while the reactants are gaseous and liquid. The presence of the catalyst can dramatically increase the rate at which the reaction proceeds. The action of the catalyst is to adsorb one or more of the reactants onto the surface of the catalyst at so-called active sites. Usually this will involve some kind of chemical breakdown of the molecule at the catalyst surface, which is known as chemical adsorption. For example, an oxygen molecule might be adsorbed onto the surface of a catalyst and in the process dissociate into two separate atoms of oxygen. The actual behavior at the surface of a diesel oxidation catalyst is complex and is believed to support several different adsorption processes.

Diesel catalysts in volume production for automotive applications primarily control emission of the soluble organic fraction of particulate (SOF), gaseous hydrocarbons (HC), and carbon monoxide (CO). Some catalysts also reduce NO_x emissions by relatively small amounts [38].

A DOC is a passive device, installed in the engine exhaust system, through which all of the exhaust gas flows (Figure 11.35). It consists of a metal outer casing that contains a core or substrate, usually of ceramic. The substrate forms a matrix of flow passages parallel to the direction of flow, having as many as 60 cells/cm^2 (400 cells/in^2). The object of the matrix design is primarily to provide a large surface area exposed to the exhaust gas. The substrate is coated with a washcoat consisting of base metal oxides, for example, silica and alumina, and small quantities of precious metals—typically a combination of platinum, palladium, and rhodium—which, together with the base metal oxide, act as the catalyst.

Fig. 11.35.

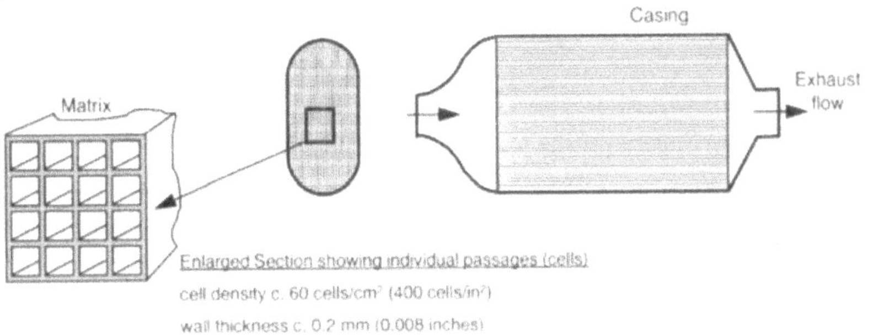

Typical diesel oxidation catalyst.

The challenge in any catalyst application is to maximize the efficiency of the desired reactions while minimizing the efficiency of any undesired reactions [38]. These desired and undesired reactions include:

Desired Reactions		
$C_nH_m + (n + m/4)O_2$	$> nCO_2 + m/2\ H_2O$	HC oxidation
$CO + 1/2\ O_2$	$> CO_2$	CO oxidation
NO_x reduction		
Undesired Reactions		
$SO_2 + 1/2\ O_2$	$> SO_3$	sulfate formation
$SO_3 + MO$	$> MSO_4$	sulfate storage

The problem of formulating catalysts that promote desired reactions and inhibit undesired reactions, known as selectivity, is one of the most challenging areas of catalyst development. A key challenge for both light- and heavy-duty catalyst applications is obtaining a balance between high activity for HC/CO/SOF at all temperatures and resistance to sulfate formation at high temperatures.

In addition to the selection of catalyst materials, the geometry of the catalyst plays a very important role in the overall performance. The most important geometric factors are:

1. catalyst volume
2. cell density and active surface area
3. net flow cross-sectional area
4. inlet diffuser design

Factors 1 and 2 determine the performance potential of the catalyst since they essentially define the residence time and space velocity, and the probability of a given molecule coming into contact with the catalyst. Space velocity is defined as volume rate of gas flow divided by the volume of the catalyst. In the Cummins

B-series application, described by Clerc [42], the space velocity exceeded 150,000 1/hr giving residence times of the order of 24 ms. Good performance with a high space velocity enables the catalyst volume to be reduced, which is a key design requirement in all vehicle applications where space is at a premium.

For both heavy-duty and light-duty applications, it is important that the catalyst does not unduly restrict the gas flow, since this will apply back-pressure to the engine and increase fuel consumption. This requires a large net cross-sectional area and efficient expansion and contraction of the flow at inlet and exit. It will also be appreciated that uniform flow through the catalyst core is essential for good performance and durability. This is achieved by careful attention to the inlet diffuser design.

Based on the foregoing comments, it will be readily appreciated that the design of a catalyst is a compromise between many factors including the chosen geometry and the activity of the catalyst coating.

11.5.3.1 Design of Catalysts for Heavy-Duty Vehicles

Clerc [42] and Voss [44] describe the development of diesel oxidation catalysts for heavy-duty engines certified by the U.S. EPA heavy-duty engine emission cycle as shown in Figure 11.1. A number of washcoat and precious metal catalysts formulations were screened with the primary goal of finding formulations that have high activity for SOF particulate while inhibiting oxidation of sulfur dioxide to sulfate particulate. It was also required to retain activity for gaseous HC and CO emissions, although for 1994 U.S. heavy-duty (HD) emissions, these were of secondary importance. This work led to the development of a new family of catalysts based on base metal oxide washcoat formulations, with very small quantities of precious metals. The screening process considered catalyst performance over an aging period of 1000 hours, that is, the resistance to poisoning by oil additives such as zinc, potassium, and calcium, and storage of sulfates in the washcoat. The formulations that showed the best overall performance consisted of a base metal oxide washcoat (containing alumina) with small quantities of platinum or palladium. The performance of these catalysts after 1000 hours is shown in Table 11.4.

Table 11.4

Performance of Catalysts after 1000 hours of Aging, Expressed as Percent Emission Reduction over the U.S. HD Emission Cycle using 0.05 Percent Sulfur Fuel [44]

Catalyst formulation	Gaseous HC	CO	SOF	Sulfate (SO_4)	Total particulate
Base Metal Oxides and Platinum 0.014 g/m^3 (0.5 g/ft^3)	29%	8.7%	51.8%	23%	32.8%
Base Metal Oxides and Palladium 1.4 g/m^3(50 g/ft^3)	49%	8.7%	57.0%	25%	32.3%

The results were obtained by testing a Cummins 5.9L HD diesel engine over the U.S. HD emission cycle, using fuel containing 0.05 percent sulfur by weight. The results show very high activity for SOF and a net decrease in sulfates, together providing a 32 percent to 33 percent decrease in total particulates.

Pataky et al. [45] describe the effects of an oxidation catalyst on both regulated and unregulated emissions from a heavy-duty diesel engine. They used a precious metal catalyst having a relatively high platinum loading of 0.85 g/m^3 (30 g/ft^3). It displayed high activity for HC (60 percent to 70 percent) and SOF (53 percent to 71 percent), and provided significant control of total particulates (27 percent to 54 percent). The sulfate production was low except for the highest exhaust temperature conditions, close to rated power. The emission of polynuclear aromatic hydrocarbons (PAHs), which are not explicitly regulated, was generally reduced by the catalyst. The particle size distribution of the particulate was also largely unaffected by the catalyst.

11.5.3.2 Design of Catalysts for Light-Duty Vehicles

Smedler et al. [38] describe the development of a catalyst for European light-duty diesel applications, certified to the ECE−15 + EUDC cycle shown in Figure 11.2. They adopted novel approaches that allowed them to improve upon the performance of traditional platinum/alumina catalysts commonly used in passenger-car applications. They report the results summarized in Table 11.5, which show very high activity for gaseous HC and CO, coupled with a 15 percent conversion efficiency for NO_X and 35 percent conversion efficiency for total particulates. The NO_X performance was achieved at an average molar HC/NO_X ratio of 0.3, that is, without addition of hydrocarbons into the exhaust stream. The catalyst formulation was not revealed but it was suggested that platinum was a key precious metal for low-temperature HC and CO oxidation, and NO_X reduction. Support materials referred to included modified alumina and silicon carbide. The key technologies were:

1. Combined metal oxide promoters, which selectively promote precious metal activity for desired reactions including NO_X reduction while inhibiting SO_2 oxidation and sulfate storage.

Table 11.5

Summary of Emission Conversion Efficiencies over the European Light-Duty Emission Cycle for a Diesel Passenger Car [38]

HC	CO	NO_X	Total particulate
90%	95%	15%	35%

2. A method for introducing precious metal by which a high average dispersion and a narrow particle size distribution was achieved.
3. Stabilized support system, which is less prone to thermal sintering and poison uptake.

This advanced catalyst technology will probably allow European passenger cars to achieve projected Stage 3 emissions standards, when combined with advanced engine technologies such as high-pressure fuel injection and cooled EGR.

11.5.3.3 Lean-NO_x Catalysts

Unlike HC and CO, which require an oxidizing atmosphere for conversion to water and carbon dioxide, NO_x conversion requires a reducing atmosphere. Since diesel combustion occurs with an overall lean mixture, the exhaust products contain excess oxygen, which makes the reduction of NO_x very difficult. Gasoline engines can be controlled to operate at the stoichiometric air-fuel ratio such that the exhaust will not contain excess oxygen. This facilitates the coexistence of oxidizing and reducing reactions which act on HC, CO, and NO_x—the so-called three-way catalyst.

The selective catalytic reduction (SCR) of NO_x is a well-developed process that requires the continuous addition of a reducing agent upstream of the catalyst, typically ammonia or urea. With this process the conversion efficiency can exceed 80 percent under steady conditions and as much as 70 percent to 75 percent over the U.S. HD emission cycle. The disadvantages that essentially prevent the widespread application of SCR for automotive applications are prohibitive cost, possible emission of ammonia during transient engine operations, and the logistics of supplying and storing the reducing agent. SCR is in use for stationary power applications where the investment can be justified by the fuel savings, compared with alternative NO_x controls, such as injection timing retard.

The main thrust of research on lean-NO_x catalysis for automotive diesel engines is aimed at using diesel fuel derived hydrocarbons as the reducing agent [46]. Typically, the hydrocarbons are required at between 2 and 5:1 molar HC/NO_x ratio to produce any significant NO_x control. At the present time the demonstrated conversion efficiency is less than 50 percent and this occurs over a very narrow temperature range. Although schemes exist where more than one formulation is used, in order to broaden the effective temperature range, the demonstrated conversion efficiencies and relatively high cost are not currently attractive for automotive applications. The fuel consumed at the required HC/NO_x ratios is sufficiently high that the recognized in-cylinder NO_x controls, for example, timing retard and EGR, remain the more attractive options.

In the catalyst development study reported by Smedler et al. [38], the catalyst formulation developed for European passenger cars, which exhibited passive control of NO_x emissions, was also tested as a lean-NO_x catalyst. In the lean-NO_x

Fig. 11.36.

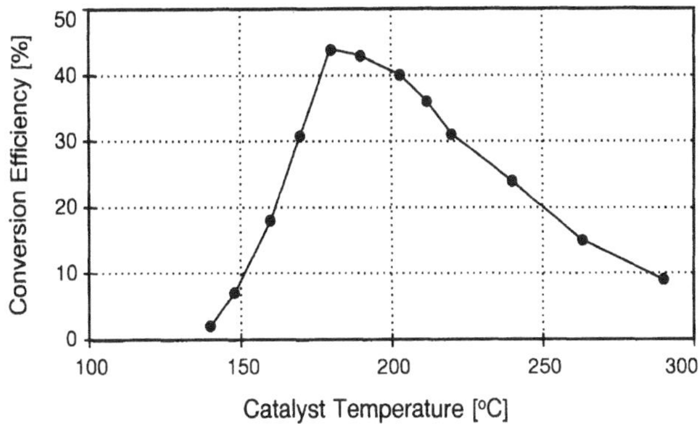

NO_x conversion efficiency for an advanced diesel catalyst with 2:1 molar HC/NO_x ratio by fuel addition ahead of the catalyst [38] *[Reprinted with permission of SAE].*

demonstration the volume of the catalyst was increased to reduce the space velocity to 30,000 1/hr and hydrocarbons were added to the exhaust flow upstream of the catalyst to increase the NO_x reduction. Hydrocarbons were added in the form of diesel fuel to give a HC/NO_x ratio of 2:1 (Figure 11.36). The NO_x conversion efficiency reached a maximum of 45 percent at a temperature of 190°C. The temperature range at which the efficiency exceeded 25 percent was approximately 175°C to 235°C.

11.5.4 Soot Filters

A soot filter is a device designed to remove soot particles, formed during the combustion process from the exhaust stream. Typically a soot filter is located in the vehicle tailpipe. There have been many different approaches to soot filter design attempted over the last 10 to 15 years. The challenges that have been difficult to overcome are chiefly the cost of the unit and the ability to regenerate the filter reliably and without applying excessive back-pressure to the engine. The majority of soot filters work by filtering the soot particles (i.e., storing the soot in the filter) until a prescribed level of flow resistance is detected, at which point a regeneration process is initiated. *Regeneration* is the term used to describe the process of removing soot by oxidizing it to carbon dioxide at a suitably high temperature.

More recent work has focused on passive soot filters, which use catalysts to reduce the oxidation temperature to levels low enough to fall within the normal exhaust gas temperature range. In this scenario the heat contained in the exhaust gas itself is used to oxidize the soot, which removes the need for active control of a regenerative heating system as in earlier designs. The catalyst can be part of the filter construction or could be delivered with the exhaust gas, for example, as a fuel additive [47, 48].

Fig. 11.37.

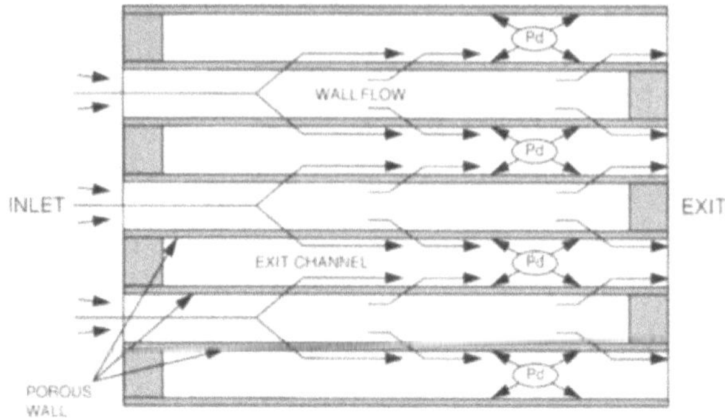

Wall flow particulate filter with catalyzed exit channels [47] *[Reprinted with permission of SAE]*.

Rao [47] presents an assessment of a palladium catalyzed, wall-flow soot filter for light-duty passenger diesel engines for European and U.S. applications. The design they evaluated is shown in Figure 11.37.

As the name suggests, a wall flow filter has porous walls through which all of the exhaust gas flows. This is in contrast to the through-flow catalysts discussed earlier in which the exhaust gas flows through passages and reacts with catalysts coated onto the walls. Unlike this device, through-flow diesel catalysts oxidize only the SOF content of the exhaust particulate, having little effect on the soot. The wall flow filter described here oxidizes both the soot and the SOF content of diesel particulate.

The operation of this device relies on the addition of a very small quantity of catalyst material to the diesel fuel, in this case copper octoate at 0.13 g/liter of fuel. During operation, the fuel-borne catalyst and the soot particulate are deposited on the inlet channel walls, being too large to pass through the porous walls. Gaseous HC and CO pass through the porous walls and are oxidized by the palladium catalyst with which the walls are impregnated during manufacture.

The fuel-borne catalyst, which is deposited with the soot, essentially reduces the ignition temperature of the soot to 290°C to 315°C such that when the exhaust gas temperature exceeds this temperature range the soot will oxidize automatically. For light-duty vehicles much time can be spent with exhaust temperatures below the autoignition temperature; however, during normal accelerations or during highway driving the temperature will be high enough to cause oxidation. The reaction is relatively fast, requiring as little as 2 to 6 seconds to occur once the temperature is sufficiently high. Since the oxidation of soot is an exothermic reaction, the heat released during regeneration can significantly increase trap temperatures, although special heat shielding from the vehicle was not required. The results presented by Rao et al. [47] showed a substantial reduction of exhaust particulate to below the projected CARB ULEV standard of 0.04 g/mile. This allowed EGR rates to be increased somewhat to produce a modest reduction of NO_x. Both CO and HC were reduced with CO being below the ULEV standard and HC remaining slightly above the standard. The impact on vehicle fuel consumption was minor, returning over 40 mpg during the federal test procedure. As an additional benefit exhaust noise was attenuated compared with a conventional muffler.

REFERENCES

1. Greenbaum, J.J., "Light Truck Powertrain Optimization Techniques and Procedures," SAE Paper 872232, 1987.
2. Durnholz, M., E. Helmut, and P. Frisse, "Pre-Injection—A Measure to Optimize the Emission Behavior of DI Diesel Engine," SAE Paper 940674, 1994.
3. Glassey, S.F., A.R. Stockner, and M.A. Flinn, "HEUI—A New Direction for Diesel Engine Fuel Systems," SAE Paper 930270, 1993.
4. Hower, M.J., R.A. Mueller, D.A. Oehlerking, and M.R. Zielke, "The New Navistar T 444E Direct-Injection Turbocharged Diesel Engine," SAE Paper 930269, 1993.
5. Miyaki, M., H. Fujisawa, A. Masuda, and Y. Yamamoto, "Development of New Electronically Controlled Fuel Injection System ECD-U2 for Diesel Engines," SAE Paper 910252, 1991.
6. "Injection Rate Shaping Technology with Common Rail Fuel System (ECD-U2)," *IMechE Seminar* 1995-3, 1995.
7. Lewis, G.R., "The Development of the Lucas EPIC System for HSDI Diesel Engines," *IMechE Seminar* On Diesel Fuel Injection Systems 1992-4.
8. Fenton, G.N., "High Pressure Rotary Spill Pump with Electronic Control," *IMechE Seminar* 1995-3, September 28–29, 1995.
9. Krill, W., "New Bosch Solenoid Valve Controlled Fuel Injection Systems," *IMechE Seminar* 1995-3, September 28–29, 1995.
10. Greeves, G., and S. Tullis, "Contribution of EUI-200 and Quiescent Combustion System Towards US94 Emissions," SAE Paper 930274, 1993.
11. Tullis, S., and G. Greeves, "Improving NO_x Versus BSFC with EUI 200 Using EGR and Pilot Injection for Heavy-Duty Diesel Engines," SAE Paper 960843, 1996.
12. Lannan, R.B., A.E. Sisson, and W.G. Wolber, Cummins Electronic Controls for Heavy-Duty Diesel Engines, *IEEE Proceedings*, International Congress on Transportation Electronics, Dearborn Michigan, October 17–18, 1988.
13. Amlung, R.J., J.P. Seger, T.A. Warlick, and D.D. Willhelm, "Cummins Quantum CELECT Fuel System Description for the QSK19 Diesel Engine," SAE Paper 961748, 1996.
14. Stumpp, G., and M. Ricco, "Common Rail—An Attractive Fuel Injection System for Passenger Car DI Diesel Engines," SAE Paper 960870, 1996.
15. Bauder, R., and D. Stock, "The New AUDI 5-Cylinder Turbo-Diesel Engine: The First Passenger Car Diesel Engine with 2nd Generation Direct Injection," SAE Paper 900648, 1990.
16. Ropke, S., G.W. Schweimer, and T.S. Strauss, "NO_x Formation in Diesel Engines for Various Fuels and Intake Gases," SAE Paper 950213, 1995.
17. Moore, A., H. Kaess, and G. Ruhnau, "SAE J1922 Data Link Control of Electronic Engine Torque Applied to Automatic Traction Control," SAE Paper 902212, 1990.
18. Stepper, M.R., "J1939 High Speed Serial Communications for Next Generation Network for Heavy Duty Trucks," SAE Paper 931809, 1993.
19. *California Code of Regulations* (CCR [CARB]), Title 13, Section 1968.1—Malfunction and Diagnostic Requirements—1994 and Subsequent Model Year Passenger Cars, Light Duty Trucks, and Medium Duty Vehicles and Engines (OBD-2).

20. *Code of Federal Regulations* (CFR [EPA]), Title 40, Part 86, Section 86.094-17—Control of Air Pollution from New Motor Vehicles and New Motor Vehicle Engines; Regulations Requiring On-Board Diagnostic Systems on 1994 and Later Model Year Light Duty Vehicles and Light Duty Trucks (OBD).
21. Stepper, M.R., S.R. Butler, and G.G. Zhu, "On-Board Diagnostics, A Heavy Duty Perspective," SAE Paper 951947, 1995.
22. Jurgen, R., *Automotive Electronics Handbook*. New York: McGraw—Hill, 1995.
23. Wolber, W.G., "Sensors for Heavy-Duty Engines and Vehicles," SAE Paper 870815, 1987.
24. Andrews, E.B., and D.V. Nishar, "An Improved Temperature Sensing System for Diesel Engine Applications," SAE Paper 940635, 1994.
25. Ribbens, W.B., *Understanding Automotive Electronics*, 4th ed. Place: Publisher, 1992.
26. *Automotive Handbook*, 3rd ed., Robert Bosch GmbH, 1993.
27. Lomas, C.G., *Fundamentals of Hot Wire Anemometry*. Cambridge, England: Cambridge University Press, 1986.
28. Richeson, D.A., and R.W. Amann, "The Electronically Controlled 6.5L Diesel Engine," SAE Paper 932983, 1993.
29. Watson, N., and M.S. Janota, *Turbocharging the Internal Combustion Engine*. New York: Macmillan Press, 1983.
30. Horlock, J.H., and D.E. Winterbone, eds., *The Thermodynamics and Gas Dynamics of IC Engines, Volume II*. Oxford, England: Oxford University Press, 1986.
31. McCutcheon, A.R.S., and M.W.G. Brown, "Evaluation of a Variable Geometry Turbocharger Turbine on a Commercial Diesel Engine," Paper C104/86, IMechE Conference—Turbocharging and Turbochargers, London, May 6–8, 1986.
32. Wallace, F.J., A. Baghery, and M.R. Ziarati, "Variable Geometry Turbocharging for Transport Engines," Paper C38/82, IMechE Conference—Turbocharging and Turbochargers, London, April 26–28, 1982.
33. Flaxington, D., and D. Szczupak, "Variable Area Radial-Inflow Turbine, Paper C36/82," IMechE Conference—Turbocharging and Turbochargers, London, April 26–28, 1982.
34. Weber, H.G., C.J. Marston, and D.A. Ruthmansdorfer, "Variable Geometry Turbocharger and Control System for Reduced Emissions on a Heavy Duty Automotive Diesel Engine," *Proc. Diesel Engine Emission Reduction Workshop*, University of California—San Diego, July 19–22, 1993.
35. Winterbone, D.E., and S. Jai-In, "The Application of Modern Control Theory to a Turbocharged Diesel Engine Powerplant," *Proc.IMechE. Part 1: Journal of Systems and Control Engineering*, Vol. 205, 1991.
36. Weaving, J.H., ed., *Internal Combustion Engineering: Science and Technology*. "Chapter 14: The Turbocharger," by B.E. Walsham and D.E. Winterbone, Elsevier Applied Sciences, 1990.
37. Caddy, S.J., Variable Geometry Turbocharger Control, Patent Application, 1996.
38. Smedler, G., G. Ahlstrom, S. Fredholm, J. Frost, P. Loof, P. Marsh, A. Walker, and D. Winterbone, "High Performance Diesel Catalysts for Europe Beyond 1996," SAE Paper 950750, 1995.
39. Johnson, J.H., S.T. Bagley, L.D. Gratz, and D.G. Leddy, "A Review of Diesel Particulate Control Technology and Emissions Effects," SAE Paper 940233, 1994.
40. Wall, J.C., and S.K. Hoekman, "Fuel Composition Effects on Heavy-Duty Diesel Particulate Emissions," SAE Paper 841364, 1984.
41. Clerc, J.C., "Catalytic Diesel Exhaust Aftertreatment," *Applied Catalysis B: Environmental* **10** (1996) pp. 99–115, Elsevier, 1996.
42. Clerc, J.C., R.K. Miller, A.C. McDonald, and H. Schlamadinger, "A Diesel Engine/Catalyst System for Pick-up and Medium-Duty Trucks," SAE Paper 932982, 1993.
43. Atkins, P.W., *Physical Chemistry*, 4th ed., Oxford, England: Oxford University Press, 1990.
44. Voss, K., B. Yavuz, C. Hirt, and R. Farrauto, "Performance Characteristics of a Novel Diesel Oxidation Catalyst," SAE Paper 940239, 1994.

45. Pataky, G., K.J. Baumgard, L. Gratz, S.T. Bagley, D.G. Leddy, and J.H. Johnson, "Effects of an Oxidation Catalytic Converter on Regulated and Unregulated Diesel Emissions," SAE Paper 940243, 1994.
46. Kawanami, M., M. Horiuchi, J. Leyrer, E. Lox, and D. Psaras, "Advanced Catalyst Studies of Diesel NO_x Reduction for On-Highway Trucks," SAE Paper 950154, 1995.
47. Rao, V.D.N., H.A. Cikanek, and R.W. Horrocks, "Diesel Particulate Control for Ford 1.8L Sierra Turbo-Diesel to Meet 1997–2003 Particulate Standards," SAE Paper 940458, 1994.
48. Lemaire, J., and M. Khair, "Effect of Cerium Fuel Additive on the Emissions Characteristics of a Heavy-Duty Diesel Engine," SAE Paper 942067, 1994.

PART V

Two-Stroke Engines

CHAPTER 12

Introductory Chapter: From a Simple Engine to an Electronically Controlled Gasdynamic System

Cornel C. Stan

FTZ Research and Technology Association Zwickau, Westsaxon Institute of Zwickau, Zwickau, Germany

ISBN: 0-12-639855-0
$25.00

12.1

INTRODUCTION

The future trend in developing internal combustion engines will be predominantly determined by emission regulations, which are becoming more and more stringent as a reaction to the environmentally unfriendly functioning of the engines in the past. Consequently, due to the apparent differences between the pollution level of the two-stroke and four-stroke engines, it seems (at first sight) that the two-stroke engine is nearing its end.

When banning two-stroke engines as driving systems for small two-wheel vehicles and marine and garden equipment, we may solve one problem but face another one; among the various kinds and types of power units, the power-to-weight ratio and the power-to-bulk volume ratio of the two-stroke engine are definitely the highest.

Since both the two-stroke and four-stroke engines are technological products aimed at accomplishing thermodynamic cycles, neither of them should be ignored when research and development and improvement programs are considered.

In the year 1997 the region of Zschopau in Saxony, Germany, known as the cradle of the motorcycle, celebrated its seventy-fifth anniversary since motorcycles and engines were first manufactured there. From DKW to MZ, the two-stroke engine has been playing a dominant role in various interpretations—from support of revolutionary technical ideas to the simple and cheap engine for everyone. In these two-stroke engines, the cylinder swept volumes spanned between 30 cm^3 and 300 cm^3, and either cross and loop scavenging systems have been employed. A representative DKW engine is shown in Figure 12.1. The two-stroker engine for everyone, as presented in Figure 12.2, has been widely used for decades—from MZ and Simson to Wartburg or to the legend of Trabant. This kind of two-stroke engine was not only cheap but simple to maintain in a simple workshop's infrastructure. In view of the present situation of energy and ecology, the specific fuel consumption and the pollutant emission, especially the HC level, of such two-stroke engines require quite a different and updated approach.

The two-stroke engine has been practically eliminated from car use. A similar exclusion procedure, however, cannot be applied for small two-wheelers, marine engines or garden equipment without serious implications on engine weight and volume. The question could be reversed: Can we imagine that the popular elegant scooters (see, for example, Figure 12.3) will be prohibited? And to what extent will small two-wheelers be demanded in the future?

When comparing cars and powered two-wheelers, it is interesting to note that during the same journey the two-wheelers consume between 55 percent and 81 percent less fuel than cars (despite the fact that scooters are equipped with two-

Fig. 12.1.

DKW two-stroke, two-cylinder engine with scavenge pump.

Fig. 12.2.

MZ 250 two-stroke, one-cylinder engine (courtesy of the Motorcycle Museum Augustusburg, Saxony).

Fig. 12.3.

Modern scooter with two-stroke engine (courtesy of Peugeot Motocycles).

stroke engines)! They are up to 46 percent quicker and have a considerably shorter idle time. Nevertheless, the present two-stroke engines cannot comply with the two-step EU regulations of limiting exhaust gas pollution: By the year 1999, CO emission will be limited to 6 g/km, and (HC + NO_x) emission to 3 g/km. By the year 2002, CO emission will be limited to 1 g/km, and (HC + NO_x) emission to 1.2 g/km.

In this context the two-wheeler, the marine engine, and the garden equipment industry are fighting to save the two-stroke principle, being the first in the front of searching for new concepts. Being first does not mean being alone, since current results very much encourage research of similar solutions in classic sectors of four-stroke engines. Meeting the challenge of the future, two-stroke engines will have to achieve an improved gasdynamic behavior during the cycle along with keeping the mechanical components as simple as possible.

12.2

POLLUTION FORMATION

For a long time, the definition of four- and two-stroke engines has been a kind of comparison of different engine designs. But the extension of this definition to two- and four-stroke principles is doubtful. In four- and two-stroke engines the major basic processes are essentially the same: intake/compression/heat induction/expansion/heat exhaust/gas exhaust; six elementary components of a cycle which occur in one crank revolution (two strokes) or in two revolutions (four strokes). The advantage of the two-stroke concept is that these six successive processes occur in half the cycle period as compared with the four-stroke concept. This results in some advantages regarding the power-to- weight ratio, but this is on the expense allowing a shorter time for each process. Alternatively, overlapping processes is unavoidable. It is also possible that some of the processes will be performed out of the working cylinder. For example, a crankcase pump or a screw blower may replace the expansion stroke of the working piston to induce fresh charge, and reciprocally a tuned exhaust pipe may replace the pumping stroke to displace the exhaust gases. It is not necessary that the engine be installed with intake and exhaust ports for such processes; this is also possible by an appropriate overlapping of exhaust and intake valves opening time. A Toyota engine with four valves for each cylinder and with all other classic features of a modern four-stroke engine design is a good example of this principle. But in extending this example, this principle can also be applied for other stages, not only for scavenging. Present developments of mixture formation systems show that the very short time allowed for this process in a two-stroke engine can be similarly compensated by a *partial mixture* formation before the mixture enters the working cylinder, as shown in Figure 12.4. In classic two-stroke engines the overlapping of scavenging phases

Fig. 12.4.

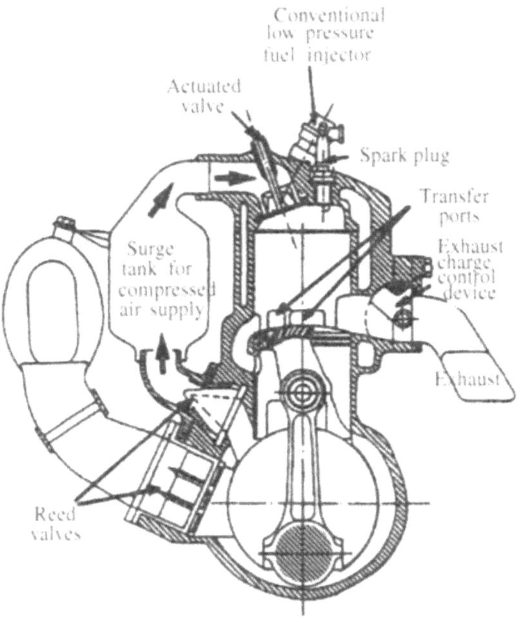

Formation of a partial mixture of fuel and air before the mixture enters the working cylinder, IAPAC concept.

is applied but the mixture formation is performed completely before the mixture enters the working cylinder.

Would it be sufficient for the future two-stroke engines to improve the scavenging process in order to achieve the pollutant limitation level?

The overlapped exhaust and charging processes during scavenging in typical two-stroke engine have some important advantages but also some disadvantages. In general terms, the overlapping results in a lower maximum bmep of about 70 percent to 80 percent of the values of four-stroke engines. The reasons are not only associated with the shorter time allowed for the scavenging process, but also because of related gasdynamic problems.

This is not a disadvantage regarding the power because, in contrast to the four-stroke cycle, every revolution in the two-stroke engine is a working stroke; therefore, power-to-volume ratio appears to be some 1.4–1.6 higher than in four-stroke engines, while the weight-to-power ratio is about 0.44–0.5 lower for the two-stroke engines. A higher bmep for two-stroke engines is certainly achievable by different methods of scavenge tuning, but this usually results in higher temperatures, which consequently result in higher NO_x emission.

With a proper engine design, it is possible to significantly reduce the short-circuiting losses in a small operation range of engine speeds and loads, although the average scavenging losses over the entire range of operating conditions may still remain five to six times higher than for typical four-stroke engines. Different methods of variable exhaust tuning with engine speed may extend the range of low losses but not decisively. The technical complexity of such systems is another issue to be considered.

Table 12.1 demonstrates the improvement in pollution emission and bsfc of a two-stroke engine when the scavenging system is improved. The improvement is significant but far from the values of typical four-stroke engines having 15–30 g/kWh HC and a bsfc of 300–400 g/kWh.

The comparison with the four-stroke engine is less dramatic when considering the pollutant limitations for the next few years. The California Air Resources Board (CARB) stipulates legal limitations beginning in 1999, as given in

Table 12.1

Comparison of bsfc and Pollutant Emission of Two-Stroke Engines with Classical and with Improved Scavenging Systems

Two-stroke engine	HC [g/kWh]	NO_x [g/kWh]	CO [g/kWh]	bsfc [g/kWh]
Classic scavenging	100–250	1–3	250–600	500–700
Improved scavenging	60–140	2–10	20–120	400–470

Table 12.2

Pollutant Limitations of CARB 99

<table>
<tr><th>CARB'99</th><th>HC
[g/kWh]</th><th>NO_x
[g/kWh]</th><th>CO
[g/kWh]</th><th>Particulates
[g/kWh]</th></tr>
<tr><td>Hand held</td><td>67</td><td>5.4</td><td>174.3</td><td>0.34</td></tr>
<tr><td>Non hand held</td><td colspan="2">4.3</td><td>134.1</td><td>0.34</td></tr>
</table>

Table 12.2. It shows that even with an improved scavenging system, the future of two-stroke engines is apparently in doubt.

12.3

METHODS OF MIXTURE FORMATION

For a realistic evaluation of the various concepts of mixture formation within the working cylinder of a two-stroke engine, two extreme models are representative.

12.3.1 Mixture Formation at the Intake Manifold

Mixture formation at the intake manifold by carburetor or low-pressure injection in the intake duct has the advantage of sufficient fuel atomization and a good distribution of the fresh air, leading to a homogeneous mixture. However, the inclusion of fuel in scavenging losses is a disadvantage.

12.3.2 Mixture Formation after Scavenging

An advantage with mixture formation after scavenging by direct fuel injection into the working cylinder is that fuel is not included in scavenging losses (for an appropriate injection angle). However, because only a very short time is allowed for mixture formation, gasdynamic problems result causing a tendency toward incomplete mixture or insufficient mixture quality, with implications for the combustion, and exhaust gas composition.

It can clearly be seen why the direct injection techniques for two-stroke engines are polarized around two concepts, as follows:

- Formation of a partial mixture out of the working cylinder with the desired amount of fuel but with a significantly reduced part of air, and mixture induction to the cylinder after scavenging. In this arrangement, the time allowed for mixture formation is prolonged in a supplementary space, where the thermodynamic conditions permit a good mixture.

- Mixture formation in the working cylinder, after scavenging, by direct fuel injection. This method requires such injection systems that can assure an extremely short injection time in all speed ranges and a sufficient fuel atomization. Such requests are practically achievable if the injection law is independent of the engine speed.

Methods of charge stratification and liquid-fuel injection are described later.

12.3.3 Partial Mixture Formation

In this method, a very rich mixture is prepared out of the working cylinder, while the scavenging process is performed with the main part of fresh air. This part is introduced to the cylinder first. A good fuel atomization in the range of 4 to 12 μm SMD (Sauter mean diameter) can be assured by this method. The premixture can be transferred into the working cylinder after scavenging through a duct where the opening time can be controlled mechanically or electronically. Such a concept has been successfully applied in the fifties by Puch/Germany. The simplest concrete solution is to fit a carburetor for rich mixture, whereas the mixture is formed in a small additional cylinder and then pumped into the working cylinder by a piston-controlled duct, as shown in Figure 12.4. Despite its simplicity, this technique leads to interesting results, as demonstrated in the figure.

With this arrangement, the air-fuel ratio is between 0.48 and 1.18, and the premixture, which is to be transferred into the working cylinder after scavenging, has a pressure of 0.3–0.6 MPa. The volume ratio is typically 1:3, and the reduction of the bsfc and HC emission is around 30 percent.

Despite the promising results at high engine speed and torque, Figure 12.5 shows a different trend in the low engine speed and torque regime. The reason is associated with the fact that the two components of the premixture (liquid

Fig. 12.5.

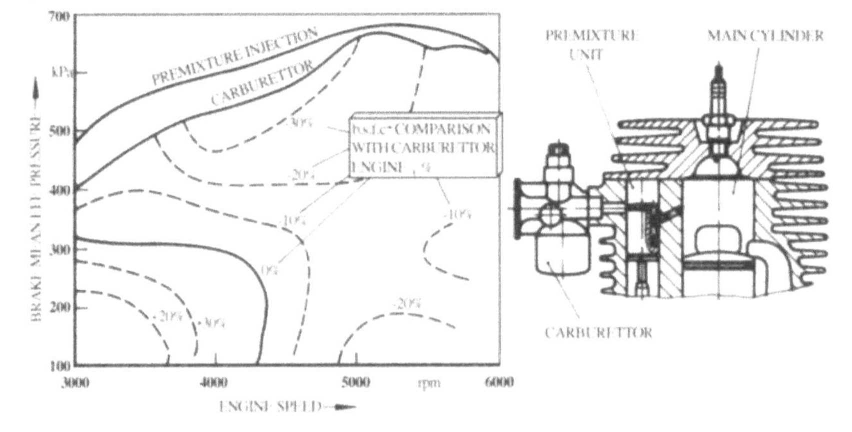

MZ engine with injection of premixture made by the University of Zwickau.

and gas) have different fluidity behavior during the admission into the working cylinder.

12.3.4 Direct Injection of Liquid Fuel

This concept could appear to be simpler and more promising than the formation of a premixture, as usually applied for diesel engines. The problem is that conventional injection systems like those in the diesel engines cannot be applied in their present form to fuel injection systems in SI two-stroke engines having a wide speed range because of the strong dependence of the injection law on the engine speed. In Figure 12.6, the time-related and the angle-related injection rates are illustrated.

In diesel engines the angle related injection rate is the usual way to determine the behavior of injection pumps. In such interpretation, the injection rate decreases and the injection time increases with the engine speed as shown in the figure. For high-speed, two-stroke applications, the time-related diagram shows that the injection rate is higher for high speed, which means that the fuel velocity at low engine speed is very low. Consequently, the atomization of fuel will be poor just in the speed range where the fresh air energy is also reduced. Also, the strong variation of spray velocity with the engine speed means different penetration length of the spray into the combustion chamber, which is a problem for SI engines with their fixed spark plug position. The strong variation of penetration length with the engine speed is the reason why high-pressure pumps that can assure good fuel atomization at low speed are also difficult to adapt to SI engines. Recent tests with adapted plunger pumps for two-stroke SI engines have shown values of bsfc between 400 and 500 g/kWh and HC emission of 68 to 135 g/kWh in a speed range of 3000–7500 rpm, which still do not satisfy future requirements.

It seems to be quite a logical consequence that for unchanged spray length and fuel atomization for the entire engine speed range the pressure in the injection system must be constant at a sufficiently high level. A constant fuel pressure in the range of 6 to 7 MPa, which results in fuel droplet size of 5–25 μm SMD, can be assured with different common techniques. The requested timing of the injector, which is also independent of the engine speed but with optimized injection beginning in every torque/speed point, is possible when using mechanical or magnetical devices. The last one is more advantageous by permitting an accurate electronic control.

The problem of such systems—similar to the modern common rail in the diesel engine—is the relatively high power input for the injection system itself, assuring that the high-pressure level must also be held in the time between injections. This means a low energetic efficiency, which is unacceptable for small two-stroke engines. Considering, for example, a speed of 3000 rpm and a usual injection duration of 0.3 ms, a constant pressure of 6 to 7 MPa will be used only for 1.5 percent of the time in a cycle! Consequently, for a constant atomization and spray length in the entire range of engine speed, the maximum fuel pressure, independent of the engine speed, must be produced only for a period covering

Fig. 12.6

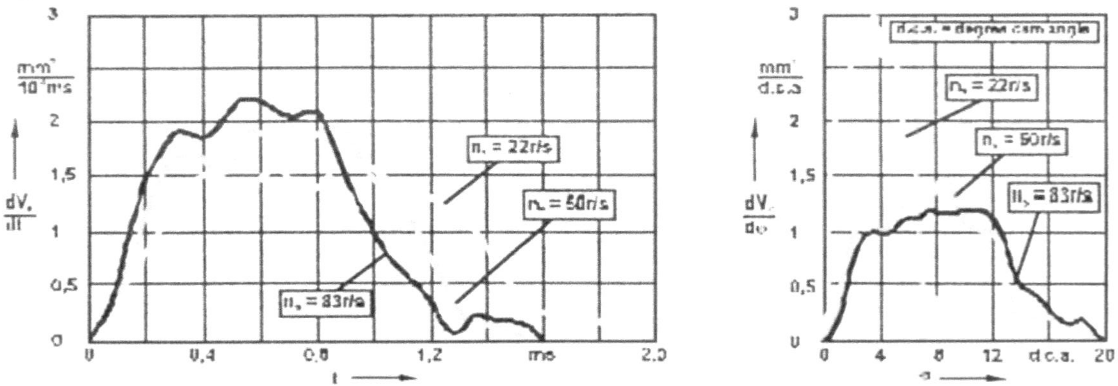

Time-related and angle-related injection rate of a mechanical injection pump with cam-driven plunger.

Table 12.3

Pollutant Emission and bsfc of SI Two-Stroke Engines with Improved Scavenging and Direct Air-Fuel Mixture Injection

HC [g/kWh]	NO_X [g/kWh]	CO [g/kWh]	bsfc [g/kWh]
5–20	8–17	10–20	260–300

more or less the injection time in order to maintain high energetic efficiency. This means a pressure wave modulation, which can be carried out, for example, on the basis of the water hammer effect.

Such a solution could appear much more complex than the simple and cheap carburetor. The two-stroke engine has to survive in relatively simple machines such as scooters or boats. Is it justified to develop concepts, theories, and finally systems of such complexity in this framework? Why should we not try to improve the scavenging system? Table 12.3 presents exhaust gas emission and fuel consumption of two-stroke engines installed with an improved scavenging system, and a device to form mixture after scavenging.

When comparing the values in Tables 12.1 and 12.3, the reason for the current efforts regarding mixture formation becomes clear. In this context, there is a hope for the survival of the two-stroke engine.

12.4 TECHNIQUES TO REDUCE POLLUTION

This introduction is not aimed at giving a detailed overview of the current successfully applied techniques for improved scavenging and mixture formation. This will be done in the following chapters. However, the methods of mixture formation described previously will be presented as some concrete examples, illustrating that the concept is realistic for the immediate future. Moreover, these examples show possibilities of an interactive influence between the types of mixture formation as mentioned.

The following direct partial mixture injection systems have been developed during the last few decades:

AVL	System of the Gesellschaft für Verbrennungskraftmaschinen und Messtechnik m.b.H. Prof. Dr. H.C. List (AVL), Austria
DCC	Delayed Charging System, developed by the University of Paris, France
FAST	Fully Atomized Stratified Turbulence, developed by Piaggio, Italy

GE	Gemisch-Einspritzung, developed by the University of Zwickau, Germany
IAPAC	Injection Assisté Par Air Comprimé, developed by the Institut Français du Pétrole (IFP), France
ORBITAL	System developed by the Orbital Engine Corporation (OMC), Australia
QUB	System developed in the Queen's University of Belfast, Northern Ireland
SEFIS	Version based on the ORBITAL System of OMC, Australia

The technical features of these systems can be classified as follows:

12.4.1 Control of Injection

- By installing a valve in the cylinder head, as shown in Figure 12.4, in order to minimize the scavenging losses (AVL, FAST, IAPAC, ORBITAL, QUB, and SEFIS). In some of the systems (IAPAC, ORBITAL, QUB, and SEFIS), the valve is controlled electromagnetically by an electronic unit, or mechanically by a camshaft (AVL, FAST), which is a less expensive solution.
- By controlling the opening of the mixture transfer duct using the working-cylinder-piston (DCC, GE), as shown in Figure 12.5.

12.4.2 Fuel Supply

- By low-pressure fuel pump, 0.25–0.6 MPa (AVL, IAPAC, ORBITAL, QUB and SEFIS).
- By a carburetor fitted for rich mixtures (DCC, FAST, and GE).

12.4.3 Air Supply for Partial Mixture

- By mechanical compressor with a pressure range of 0.3–0.7 MPa (FAST, GE, ORBITAL, and QUB).
- By gasdynamic resonance leading to a pressure wave (DCC and IAPAC).
- By capturing air from the working cylinder during compression stroke in a premixture chamber (AVL and SEFIS).

In fact, any of these features are previously known solutions from other engineering applications. This affirmation will not diminish a sincerely great appreciation for the effort of engineers and technicians who have been involved in the development of these promising systems for so many years. Consequently, technical risks for series application of such systems are not excessive. The problem to be solved is the optimization of system technology, costs, energy input, precision, and reliability as well as adaptation to the engine.

The preceding classifications could show after a detailed analysis of each system the system or the combination of elements of different systems that could be better adapted to a special engine application. Regarding the injection of liquid fuel, two main advantages appear in comparison with the air-assisted systems: the elimination of pressurized air supply and mixture chamber, and the avoiding of mixture formation problems caused by the relatively low injection energy.

The problem of liquid injection is to reach a droplet atomization in the same range of the air-assisted systems, which means a high fuel pressure but with minimum energy consumption. Therefore, the adaptation of common rail systems characterized by constant pressure at maximum value is very limited in this field. A maximum fuel pressure independent of the engine speed only for a period corresponding to the injection time can be achieved by a pressure wave modulation. The most representative examples producing such pressure waves are the following systems:

Ficht PDS—Pumpe-Düse System, developed at the University of Zwickau and adapted for fuel direct injection in series applications in the Ficht Company, Germany.

Zwickau ram-tuned system—Druckstoßeinspritzsystem, developed at the University of Zwickau, Germany for direct injection in spark-ignited and diesel engines.

Both systems use the effect of the impact between fuel and a solid body. The relative velocity between the fuel jet and the solid body can be generated either by the accelerated fluid against a fixed valve (Zwickau Ram Tuned), or by an accelerated piston against a stationary fluid (Ficht PDS). The two systems are characterized by the main common elements as follows:

Solenoid valve—to stop the accelerated fluid (Zwickau ram-tuned system), and accelerating a piston (Ficht PDS).

Fuel injector—mechanical injector actuated by a spring, connected to the pressure pipe

Low pressure fuel pump—pressure of 0.3 to 0.5 MPa leading to fluid acceleration (Zwickau ram-tuned system), and for the fuel transfer from the tank to the injection system (Ficht PDS).

The most important difference between these systems is the kind of energy transformation in order to obtain the pressure wave. While in the Ficht system the pressure wave is provided practically by the transformation of the electrical energy into kinetic energy in the solenoid, the Zwickau ram-tuned system uses both the electrical energy in the motor of the low-pressure pump, and the electrical

Fig. 12.7.

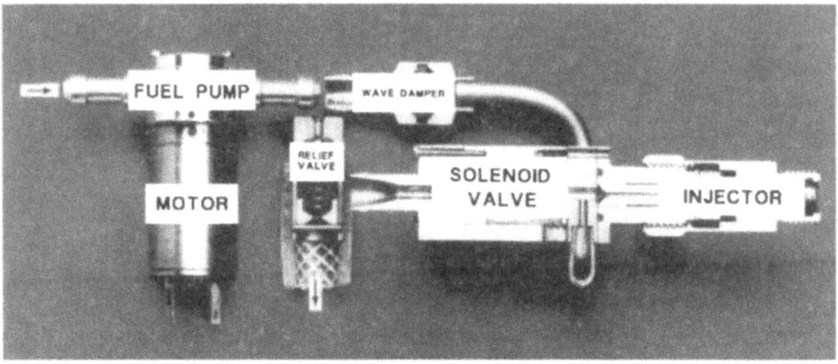

Zwickau ram-tuned injection system for a single-cylinder engine application.

energy in the solenoid to produce the high-pressure wave. This last method permits energetic transformation under the saturation level of each element, which leads to a good energetic efficiency. On the other hand, the fluid circulation through the pipes of the ram-tuned system in each injection cycle is a method of avoiding damp formation for high temperature of the injector, depending on the engine application. In Figure 12.7 the main elements that are common to both systems are exemplified in the case of a Zwickau ram-tuned system for single-cylinder engine application.

Similar to the air-assisted systems, there are elements in both configurations that are previously known from other technical solutions such as pumps, solenoid valves, and injectors. The fact that additional air pressure supply and a mixture chamber are not necessary is significant. But most important is the advantage of the velocity difference between fuel and air in the combustion chamber caused by the high-pressure fuel injection, leading to promising mixture stratification concepts.

12.5

THE FUTURE OF THE TWO-STROKE ENGINE

The scavenging improvements as well as the various methods of mixture formation concepts are decisive arguments for the survival of two-stroke engines. But despite the development of direct-injection systems and improved scavenging systems in two-stroke engines, a major problem still remains to be solved in the near future: the correlation between injected fuel flow and gas flow within the cylinder. Experimentally obtained results for direct-injection engines are encouraging but the potential is even greater.

The computer-aided process simulation offers more and more powerful tools to efficiently adapt the injection system to the gasdynamic flow in the cylinder. Current analysis and optimization methods using CFD codes such as KIVA III or

Fig. 12.8

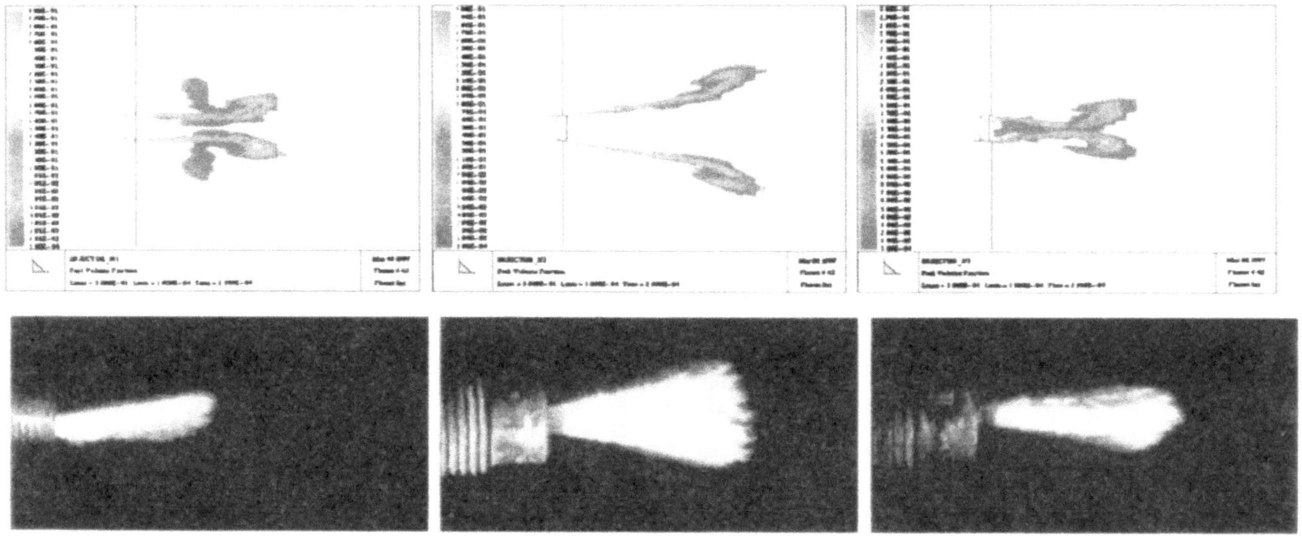

Fuel spray propagation during direct injection by Zwickau ram-tuned injection system for different injection laws and injector geometries, a comparison between computational and experimental results.

Fig. 12.9.

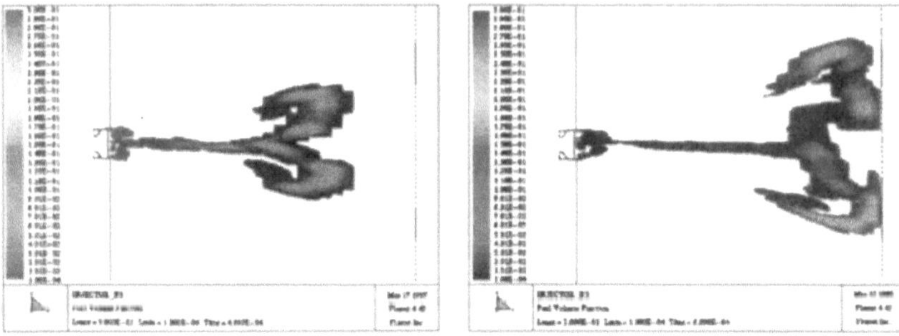

Wall wetting by impact of the injected fuel on the chamber surface.

FLUENT are very convenient for engine application. One of the problems when applying direct injection, especially in engines with small swept volume, is the optimization of the spray form to the chamber and its time-related development in order to achieve an appropriate mixture formation. This implies an adaptation between injection law and injector form, which is difficult to accomplish experimentally. Examples follow that illustrate the advantages of computer-aided processes.

Figure 12.8 presents results of a simulation of fuel mass distribution for different configurations of injector law and injector form, obtained by the CFD code FLUENT and confirmed experimentally as shown, in different time phases of injection.

Not only the optimization of the jet expansion zone in the combustion chamber but also the avoidance of possible impact with a chamber wall are decisive for a good combustion process. Figure 12.9 shows such a predicted impact during the spray penetration.

Both the form of the jet in the combustion chamber and the possible wall impact can be additionally strongly influenced by the gas flow in the chamber. This depends on the scavenging method and on the combustion process as well. Figure 12.10 illustrates how the gas flow within the cylinder changes the jet form of the first configuration in Figure 12.8 for different time sequences of the mixture formation.

The fact that the optimization of the mixture formation and combustion is the decisive factor for the reduction of the two-stroke pollutant emission shouldn't minimize the importance of the catalytic aftertreatment as a final security measure. Figure 12.11 shows a catalyst for two-stroke engine applications, which demonstrates not only an important contribution to emission reduction but also good reliability.

The combined methods of internal mixture formation by direct injection and the catalytic aftertreatment give the two-stroke engine a real chance to survive.

As a final observation it is worthwhile to return to the basic advantages of the two-stroke cycle (i.e., the power-to-weight and the power-to-volume ratios). These advantages recommend the use of the advanced two-stroke engines also

Fig. 12.10.

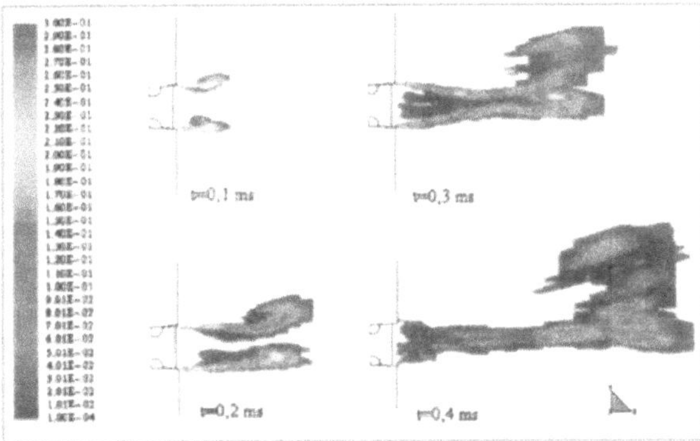

Influence of the gas flow in the combustion chamber on the fuel spray form.

Fig. 12.11.

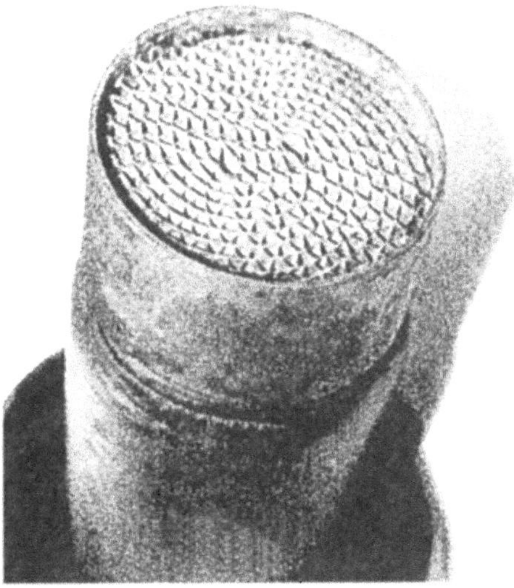

Catalyst for two-stroke engines (courtesy of Johnson-Matthey).

in the new field of hybrid drive systems, where the storage of sufficient electric energy in batteries is still a difficult problem. The two-stroker used as generator for electric energy is smaller, lighter, and less complex than any other drive, such as the four-stroke engine or gas turbine. And there is another interesting possibility to improve its behavior: *the function of two-strokers in operating windows of torque and speed.* For example, the first window is at idle and the second one is at maximum torque and nominal speed with a narrow field for parameter control and correction. In this manner the mixture formation by direct injection and the catalytic aftertreatment can be controlled with much higher accuracy.

There is no easy route from a simple two-stroke engine to an electronically controlled two-stroke gasdynamic process. Not each solution or supply in itself but the synthesis and adaptation to each other seems to be the problem. By stimulating creativity in a very pragmatic sense the two-stroke drive can be changed *from a noisy power to a clean art to mobilize.*

REFERENCES

1. Stan, C., Perspectives of two-stroke engine development based on the optimization of scavenging and mixture formation, DSc Thesis, Technische Hochschule Zwickau, Germany, 1989.
2. Blair, G. P., *Design and Simulation of Two-Stroke Engines*, SAE 1996.
3. Stan, C., "Concepts for the Development of Two-Stroke Engines," SAE Paper 931477, 1993.
4. Carson, C.E., R.J. Kee, R.G. Kenny, C. Stan, K. Lehmann, and S. Zwahr, "Ram-Tuned and Air Assisted Direct Fuel Injection Systems Applied to a SI Two-Stroke Engine," SAE Paper 950269, 1995.
5. Martorano, L., A. Franco, C. Stan, and H. Eichert, Simulation Concept for the Optimization of a Direct Injection Configuration, ISATA Congress, June 1997, Florence, Italy.

CHAPTER 13

Air Pollution from Small Two-Stroke Engines and Technologies to Control It

Yuji Ikeda and Tsuyoshi Nakajima
Department of Mechanical Engineering, Kobe University, Rokkodai, Nada, Kobe, Japan

Eran Sher
Department of Mechanical Engineering, The Pearlstone Center for Aeronautical Engineering Studies, Ben-Gurion University, Beer-Sheva, Israel

ISBN: 0-12-639855-0
$25.00

13.1 POLLUTANT FORMATION

Essentially, there is no difference between the mechanism of pollutant formation in two-stroke-cycle engines and that in its counterpart, the four-stroke-cycle engine. There are, however, some important differences in the exhaust gas composition between the two. Owing to the imperfect gas exchange process in SI two-stroke-cycle engines, a higher fraction of residuals remains inside the cylinder and the maximum temperature in the cycle is lower than in a comparable four-stroke-cycle engine. As a result, lower NO_X emission is expected. In small two-stroke-cycle engines, the lubricant agent is traditionally provided through the fuel system and higher HC emission is inevitable. In the present section, a short outline of the general mechanisms of pollutant formation in SI two-stroke-cycle engines is given. Following is a brief review of up-to-date technologies of pollutant control in two-stroke-cycle engines.

The spark-ignition engine's exhaust gases contain three major types of pollutants in the following relative amounts: oxides of nitrogen, NO_X (mainly NO and small amounts of NO_2), 500 to 1000 ppm; carbon monoxide, CO, 1 percent to 3 percent; and unburned or partially burned organic compounds, HC, 500 to 5000 ppm (as C1). In addition, there are three classes of spark-ignition engine particulate emissions: lead, organic particulate (including soot), and sulfates. For traditional two-stroke-cycle engines, where the lubricant is mixed with the fuel, higher HC levels up to 8000 ppm are expected. The relative amounts depend on the engine operating cycle, engine design, and operating conditions. One of the more important variables in determining emissions' levels in spark-ignition engines is the fuel-air equivalence ratio, $\phi (= 1/\lambda$, where λ is termed the relative air-fuel ratio). In general terms, leaner mixtures yield lower emissions until the combustion quality becomes poor and the HC emission rises sharply. Richer mixtures yield low NO emission but steadily rising HC and CO emissions. Understanding the formation mechanism of these components may provide a grounded explanation of this behavior. In the present section, a short outline of the general mechanisms of pollutant formation will be given. For more details, the reader is referred to Heywood [1].

A typical example of an exhaust gas analysis for a crankcase-scavenged, two-stroke-cycle engine is shown in Figure 13.1. A 175-cc engine with a rated output of 4.9 kW at 5000 rpm was tested under WOT conditions and a fixed air-fuel ratio of 1:14. Two additional series of tests were carried out with two modified versions of the engine. In the first, the exhaust port was controlled by an auxiliary

Fig. 13.1.

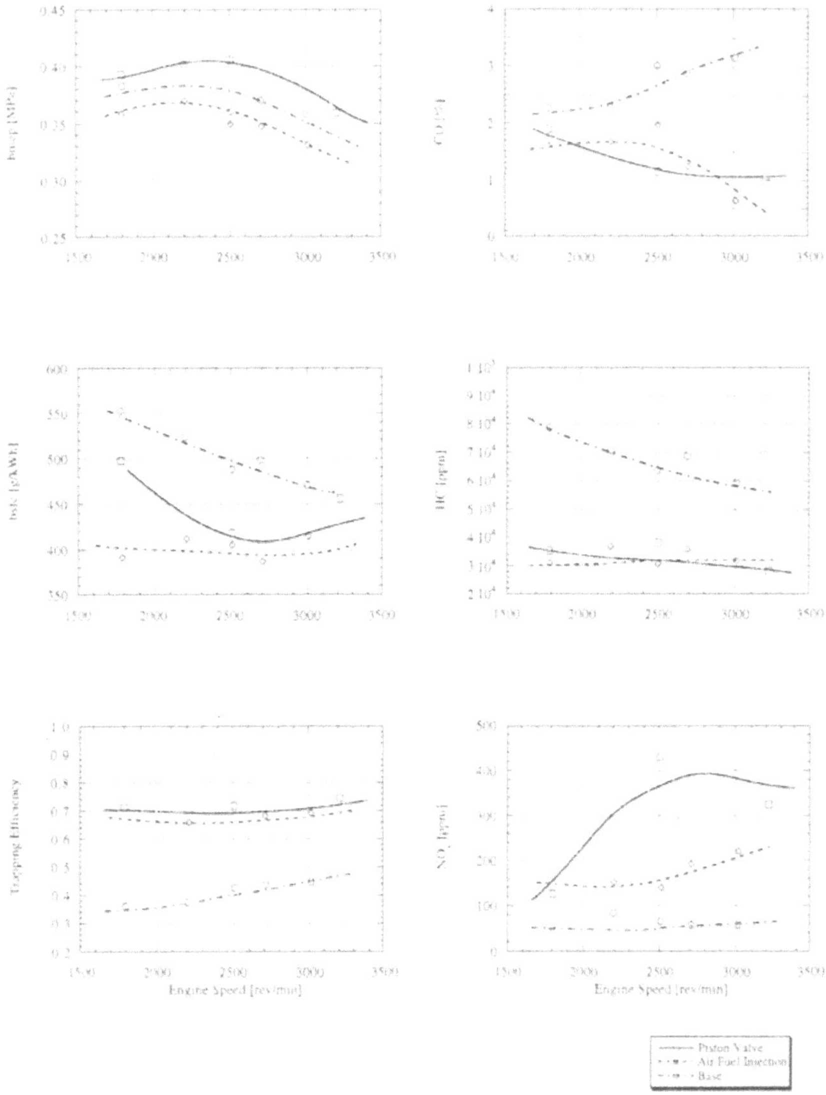

Example of a typical exhaust gas analysis for a crankcase-scavenged, two-stroke-cycle engine [2]. Also shown are results obtained with an exhaust port control system, and results obtained with an air-assisted fuel injection system. Data for HC are presented as ppm $C_{8.26}$; to convert to C_1 ppm divide by 8.26 (reprinted with permission from SAE).

piston and the timing was optimized for highest bmep at each point. In the second, the basic engine was modified to include an air-assisted fuel injection system. It seems that the HC emission is dramatically suppressed with either the exhaust port controller or the fuel injection system, while the trapping efficiency increases considerably. This may suggest that short circuiting of the fresh charge through the exhaust port is a major drawback of the basic engine. High CO emission levels at high engine speeds may suggest that poor mixing results in highly diluted pockets (with residuals), thus creating regions of nonflammable mixture. The bsfc seems to be improved when the auxiliary piston or the fuel injection system is employed. However, due to friction losses with the auxiliary piston, fuel injection is preferable.

13.1.1 NO_x Emissions

NO_x emissions are mainly composed of NO molecules and a small amount of NO_2, in a typical ratio of 1 to 100. NO is one of the intermediate products of chemical reactions involving nitrogen and oxygen atoms and molecules, which occur at the high-temperature burned gases behind the flame. Rapid cooling (below 1800 K), due to the expansion stroke, does not allow the products to attain chemical equilibrium and some intermediate products (including NO) "freeze" and leave NO concentrations far in excess of levels corresponding to equilibrium at exhaust conditions. The high residual fraction, which is typical to two-stroke-cycle engine, results in lower maximum temperatures and therefore lower NO production.

The strong effect of the temperature is evident; a reduction of the maximum temperature from 2500 K to 2300 K results in a cutback of 90 percent in the initial rate of NO production. Increasing the residuals mass fraction is an effective and practical method of reducing the maximum temperature and thus the NO emission level. Due to the peculiar gas exchange process in two-stroke-cycle engines the recycled exhaust gas is an inherent property, and the typical low NO emission level of a two-stroke-cycle engine is anticipated. Owing to the exceptionally high value of water-latent heat, introducing water spray into the induction manifold during the cylinder (or crankcase) charging process is another practical method to depress the adiabatic flame temperature and, hence, the NO emission level.

13.1.2 CO Emissions

Carbon monoxide is also one of the products of the combustion process. It forms mainly under rich-mixture conditions but may be detected in small quantities also under lean conditions. With rich fuel mixtures, there is insufficient oxygen to burn fully all the carbon in the fuel to carbon dioxide. With lean fuel mixtures, the high temperatures prevailing inside the combustion chamber result in a partial dissociation (although in small quantities) of carbon dioxide to carbon monoxide. The measured exhaust CO concentrations for fuel-rich mixtures are close to equilibrium concentrations in the burned gases during the expansion process. For close to stoichiometric and fuel-lean mixtures, measured CO emissions are substantially higher than the equilibrium concentrations. Here again, rapid cooling of the gases due to the expansion stroke does not allow the products to attain chemical equilibrium and the CO concentration level is left far in excess of levels corresponding to equilibrium at exhaust conditions. Another possible explanation [1] is that only

Table 13.1

Comparison Between Gas Chromatographic Analysis Results in a Traditional Crankcase-Scavenged, Two-Stroke-Cycle Engine, and a Similar Capacity Four-Stroke-Cycle Engine [4]

HC component % weight	Two-stroke	Four-stroke
Methane	1.00	4.21
Acetylene	0.99	7.81
Isopentane	8.42	5.38
2 Methyl pentane	4.13	2.98
3 Methyl pentane	2.67	2.06
n-Hexane	2.85	2.38
3 Methyl hexane	1.71	1.31

partial oxidation to CO may occur of some of the unburned hydrocarbons emerging during expansion and exhaust from crevices in the combustion chamber and from any oil layers (and droplets in fuel-oil mixtures in two-stroke-cycle engines) or deposits on the chamber walls. Exhaust CO emission level is practically determined by the fuel-air equivalence ratio. For slightly fuel-rich mixtures, $\lambda < 1$, as commonly used in spark-ignition engines to ensure smooth and reliable operation, CO concentrations in the exhaust increase steadily with increasing amounts of excess fuel. For this reason, it is a common practice to tune up small spark-ignition engines by using an on-line device for the measurement of CO concentration. The following global chemical reaction may clearly demonstrate this point. If we consider a stoichiometric oxidation of a general hydrocarbon molecule, having C carbon atoms and m hydrogen atoms, we obtain:

$$C_nH_m + \left(n + \frac{m}{4}\right)O_2 + 3.76\left(n + \frac{m}{4}\right)N_2 \rightarrow nCO_2 + \frac{m}{2}H_2O + 3.76\left(n + \frac{m}{4}\right)N_2 \tag{13.1}$$

for a fuel-rich mixture, $\lambda < 1$, and eq. (13.1) takes the form of:

$$C_nH_m + \lambda\left(n + \frac{m}{4}\right)O_2 + 3.76\lambda\left(n + \frac{m}{4}\right)N_2 \rightarrow \left[n - 2\left(n + \frac{m}{4}\right)(1-\lambda)\right]CO_2$$

$$+2\left(n + \frac{m}{4}\right)(1-\lambda)CO + \frac{m}{2}H_2O + 3.76\lambda\left(n + \frac{m}{4}\right)N_2 \tag{13.2}$$

The CO concentration seems to be linearly dependent on the fuel-air equivalence ratio, λ.

13.1.3 HC Emissions

The unburned hydrocarbon emissions are usually measured with a flame ionization detector (FID), which is functionally a carbon atoms counter. One butane (C_4H_{10})

Fig. 13.2.

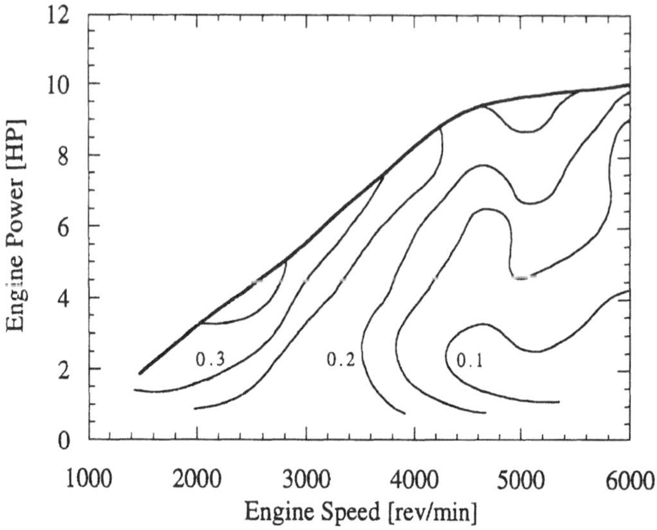

Map of short-circuiting ratios for a traditional crankcase-scavenged, two-stroke-cycle engine [4] (reprinted with permission from SAE).

molecule signals four times the signal of a methane molecule (CH_4). Various data in the literature are provided on different scales, and in order to facilitate quantitative comparison, it is practically convenient to convert all data to ppm C_1 by simply multiplying by the number of atoms in the counted molecule. As an example, if a ppm propane (C_3H_8) counter is used, multiply by 3 to convert to ppm C_1. Unlike the NO or CO mechanism of formation, HC, which is a term used to quantify a wide variety of hydrocarbon compounds, is formed through several different mechanisms.

The effect of the spark timing on the exhaust HC emissions has been found to be similar to that of injection timing [3]. For a later spark timing, or later injection timing, the total amount of HC increases. The gas includes more aromatic species, such as toluene, which mean that combustion has not been completed before the exhaust port is exposed. The residence time of the fuel molecule at high temperatures in the combustion chamber seems to determine to a large extent the amount of high-HC molecules in the exhaust gas. In a traditional crankcase-scavenged, two-stroke-cycle engine, short circuiting of the fuel directly to the exhaust port accounts for the heavier molecules that were detected [4]. Such are various derivatives of pentane and hexane (Table 13.1). Figure 13.2 demonstrates how important the short circuiting is and the extent to which it may reach. It shows that under low engine speeds and high loads, the short circuiting may account for fresh charge losses of up to 30 percent. In-cylinder fuel injectors are therefore good solutions to minimize HC emissions and fuel losses through short circuiting.

Another important HC emission source, which is typical of two-stroke-cycle engines operating particularly under low-load conditions, is associated with incomplete combustion and misfired cycles. Incomplete combustion may occur due to bulk quenching of the flame in that fraction of the engine cycle where combustion is very slow. Such conditions are most likely to occur also during transient engine operation when the air-fuel ratio, spark timing, and the fraction of residual gases may not be properly matched. Drake et al. [5] have employed a laser-induced, fluorescent imaging of gasoline combined with simultaneous exhaust HC sampling using a flame ionization detector to investigate the role of cyclic variability on HC emission level. It was shown that misfires and partial burns associated with cyclic variation in the local equivalence ratio at ignition contribute significantly to HC emissions, but only about 25 percent of the injected fuel appears in the exhaust immediately after a misfire because of the reduced exhaust mass flow combined with the two-stroke-cycle engine's high residual fraction. The results provided strong evidence that the dominant HC sources in a stratified in-cylinder injection, two-stroke-cycle engine are incomplete combustion of the injected fuel cloud and late heat release of fuel trapped in the injector nozzle-exit crevice. It is anticipated that this type of HC source is accountable for a significant portion of the total HC emission level under these conditions. However, because of technical problems associated with precise measurements in a real engine, there are no quantitative measurements yet of their relative importance.

Other important mechanisms for HC formation, which are relevant to the same extent in four-stroke-cycle engines, have been extensively investigated particularly in the latter. A detailed review may be found elsewhere [1]. A short outline is given here.

During compression and combustion, some of the fresh mixture inside the cylinder is compressed by the increasing cylinder pressure into small cavities connected to the combustion chamber. These include crevices, narrow volumes between the piston rings and small volumes captured in the cylinder head. Most of this gas is unburned fuel-air mixture, which is left unburned because the entrance of these crevices is too narrow for the flame to enter. The mixture that leaves these crevices later in the gas exchange process is one significant source of HC emissions.

Another source for the HC emission is the quench layer on the combustion chamber walls. This thin layer (< 0.1 mm), which is the distance at which the approaching flame front extinguishes, contains unburned and partially burned fuel-air mixture that doesn't experience an appropriate combustion. Porous deposits and oil left in a thin film on the cylinder wall, piston, and on the cylinder head can absorb fresh mixture during compression and desorb fuel hydrocarbon components after combustion, thus contributing to the HC emission level.

The unburned and partially burned fuel-air mixture that does not experience an appropriate combustion mixes with the products of combustion and leaves the cylinder during blowdown and scavenging periods. Substantial oxidation of these hydrocarbon molecules can occur during expansion and the gas exchange process.

The amount of oxidation depends on the temperature and oxygen concentration time histories of these HC as they mix with the bulk gases.

13.1.4 Particulate Emissions

Particulate emissions from spark-ignition engines may be classified into three general classes: sulfates, lead, and organic particulate (including soot).

Unleaded gasoline contains 150 to 600 ppm by weight sulfur (S). Sulfur is a yellow, nonmetallic element known to be a very reactive element that combines with most other elements, including oxygen (O_2). Sulfur is therefore oxidized within the engine cylinder to sulfur dioxide, SO_2, and in the presence of an appropriate catalyst surface (some vehicle catalyst types are among these), SO_2 can be oxidized to SO_3 under normal operating conditions of spark-ignition engines. At ambient temperatures SO_3 combines with water to form a sulfuric acid (H_2SO_4) aerosol, which is a strong dibasic acid. Levels of sulfate emissions depend on the fuel sulfur content, the operating conditions of the engine, and the details of the catalyst system used. Typical average automobile sulfate emission rates are 20 mg/km or less [1].

For spark-ignition engines operated with leaded gasoline, the particulate is dominated by lead compounds. The particle size is typically on the order of 1 to 2 μm in diameter. The particles are formed and grow in the exhaust system due to vapor-phase condensation enhanced by coagulation and, therefore, the exhaust temperature has a significant effect on particulate emission levels. Cold engines emit higher levels of particulate. Unleaded gasoline reduces particulate emissions.

Organic particulate (soot emissions) can result from the combustion of excessively rich mixtures or from the combustion of lubrication oil. In mixture-controlled engines in which the lubrication oil is not supplied through the fuel metering system, the exhaust gases contain negligibly small amounts of organic particulate.

13.2 POLLUTANT CONTROL

The composition of the exhaust gas of an SI two-stroke-cycle engine differs from that of a four-stroke-cycle engine by two major components: HC and NO_x. The unperfected gas exchange process in two-stroke-cycle engines results in a higher HC emission due to an inevitable partial short circuiting of fresh charge through the exhaust port. In small two-stroke-cycle engines, the lubricant agent is traditionally provided through the fuel system and provides a significant contribution to the HC emission level. Owing to the high residual fraction that remains inside the cylinder prior to the compression stroke, the maximum temperature in the cycle is lower than in a comparable four-stroke-cycle engine and a lower NO_x emission is expected. In large CI engines, high stroke-to-bore ratios can be employed. More complete combustion is possible and lower CO and HC emissions, as compared with small-size engines, are achieved. In the present section, the effect of some important parameters on the exhaust gas composition and up-to-date technologies of pollutant control in two-stroke-cycle engines are briefly reviewed.

The means of controlling pollutants from internal combustion engines may roughly be sorted into three main classes: designated design of the cylinder and port assembly, an appropriate management of the operating condition, and afterburning exhaust gas treatment.

13.2.1 Engine Design Factors

The compression ratio is a decisive factor for the thermodynamic efficiency of the cycle. However, it affects the exhaust gas composition by two means: A high compression ratio increases the maximum temperature in the combustion chamber prior to combustion and thus enhances the NO_x formation during combustion; however, preignition of some portions of the cylinder charge may result. Preignition is associated with extremely high temperatures, which further increase the NO_x formation. Alternatively, demands for a higher-octane-number fuel may introduce other pollutant components to the fuel and thus to the exhaust gases.

The design of the combustion chamber shape also determines crevice volumes and, therefore, has an important influence on the emission level of HC. For this reason, compact combustion chambers with high volume to surface area ratios are preferable. Combustion chamber and scavenge port-entrance assembly are also important in the determination of the turbulence intensity prior and during combustion. High turbulence intensity ensures good fuel-air and residuals mixing and high flame speed, which are significant to minimizing cycle-by-cycle variations and, hence, HC emission.

The position of the spark plug in the combustion chamber affects the flame travel distance and, hence, the combustion duration and the formation of various species. A longer combustion duration is associated with lower maximum pressure and temperature and, thus, lower NO_x emission. At the same time it results in higher HC emission due to relatively uncompleted combustion. For this reason, dual spark plugs contribute to a lower HC emission and a higher NO_x emission.

In traditional crankcase-scavenged engines, the lubricating oil is introduced to the cylinder through the fuel system. In this case the oil undergoes a partial burning process and may contribute a significant portion to the HC emission. A separate lubrication system is by no means an important modification to decrease HC emission level.

The port's timing, scavenge ducts geometry, exhaust tuning, and the crankcase to cylinder volume ratio have crucial effects on the HC and NO_x emissions by eliminating short circuiting of fresh charge (and in some cases fuel) through the exhaust port and controlling the amount of residuals retained. In-cylinder fuel injection (high pressure, low pressure, or air-assisted injection) is one of the most important and remarkable solutions to minimizing fuel losses through short circuiting.

13.2.2 Operating Conditions

For the same intake pressure, lower engine speed results in a lower turbulence intensity and, therefore, a lower flame speed during combustion. In terms of crank-angle degrees, however, a lower engine speed results in a shorter combustion duration. Higher pressures and temperatures are involved and higher NO_x emission is anticipated. Mixing of the fuel, air, and residuals is less effective at low engine speeds and HC emission is, therefore, higher. The engine load has a significant effect on the HC and NO_x emission levels as well as on the fuel consumption of the engine (Figure 13.3a). Short circuiting of the fresh charge to the exhaust port is another significant mechanism of fuel losses through the exhaust, thus increasing HC emission and apparent fuel consumption. Figure 13.3 compares the performance of carbureted and in-cylinder, fuel injected scooter, crankcase-scavenged, two-stroke-cycle engines. At partial loads the residuals mass fraction in the cylinder at EC is high (may exceed 50 percent), and large-volume pockets of fuel-rich mixtures (lack of oxygen), which are incapable of being ignited, are trapped in various regions. These fuel-rich pockets are exhausted during the gas exchange process, thus constituting high HC emission and fuel consumption. Under these conditions, NO_x emission is extremely low.

Under acceleration conditions, emission values are considerably higher than under steady-state operating conditions. In manifold injection or carbureted engines, a quickly opened throttle results in an acceleration of the fuel-air mixture, and due to the high difference in their densities, some of the fuel lags behind the air. High enrichment of the mixture is, therefore, required in order to ensure that the cylinder-entering mixture is ignitable. The excess fuel remains in the intake system and is consumed in the next cycles. High HC and CO emissions are therefore expected. In this regard, in-cylinder injection systems have a remarkable advantage over other fuel charging methods.

The fuel-air equivalence ratio, $\phi (= 1/\lambda$, where λ is termed the relative air-fuel ratio), is one of the more important variables in determining emissions' levels in spark-ignition engines. In general terms, it shows that fuel-lean mixtures yield low emissions until the combustion quality becomes poor and the HC emission rises sharply. Rich mixtures yield low NO_x emission but with steadily rising HC and CO emissions. Minimal HC and specific fuel consumption are achieved near stoichiometric conditions. An earlier ignition timing results in an increase in the maximum pressure and temperature and, therefore, a higher NO_x emission level is expected. Also, higher conversion efficiency of the fuel is expected, thereby lowering CO emission. Under these conditions, the cycle approach closer to the ideal Otto cycle and higher thermal efficiencies are anticipated, thus lowering bsfc.

Since the exhaust gases are essentially inert gases, circulation of a measured portion back to the combustion chamber can significantly reduce the peak combus-

Fig. 13.3.

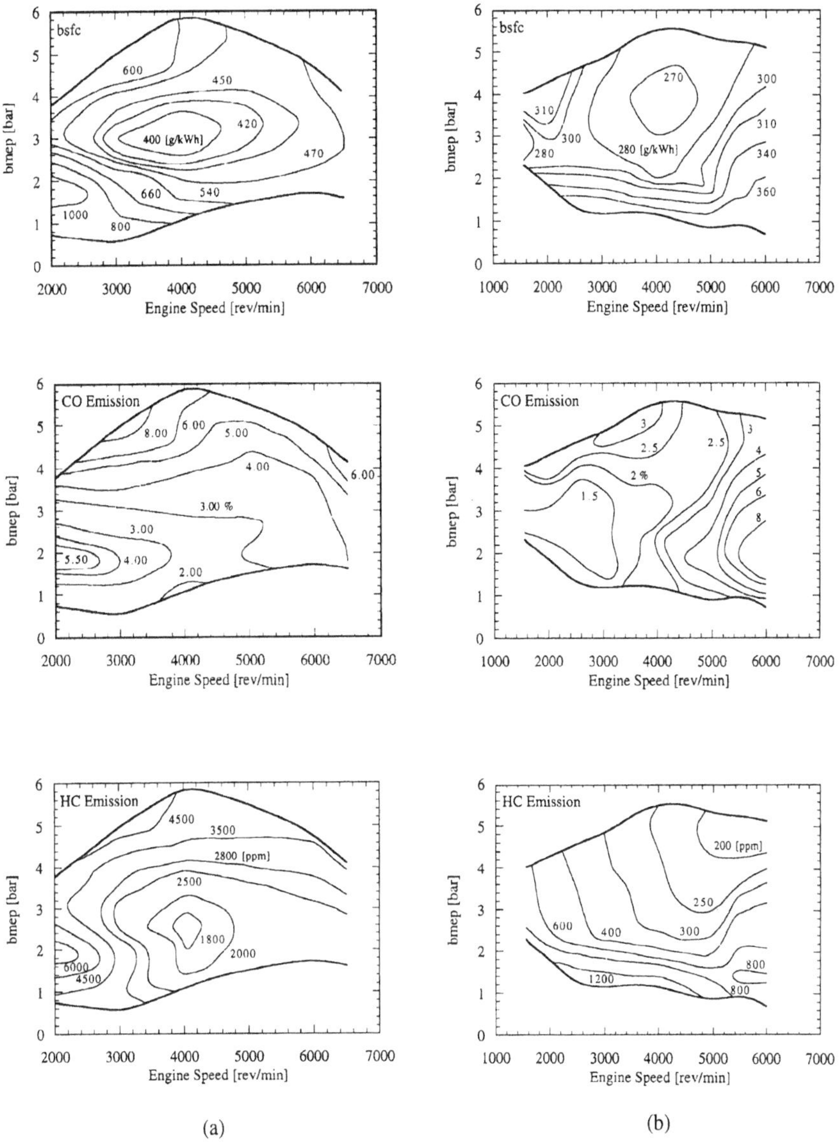

Performance maps of the engine shown in Figure 13.8 (a) carbureted engine, and (b) in-cylinder fuel injected engine [6]. Fuel consumption in [g/kWh], CO emission in [vol. %], and HC emission in [vol. ppm] (reprinted with permission from SAE).

Fig. 13.4.

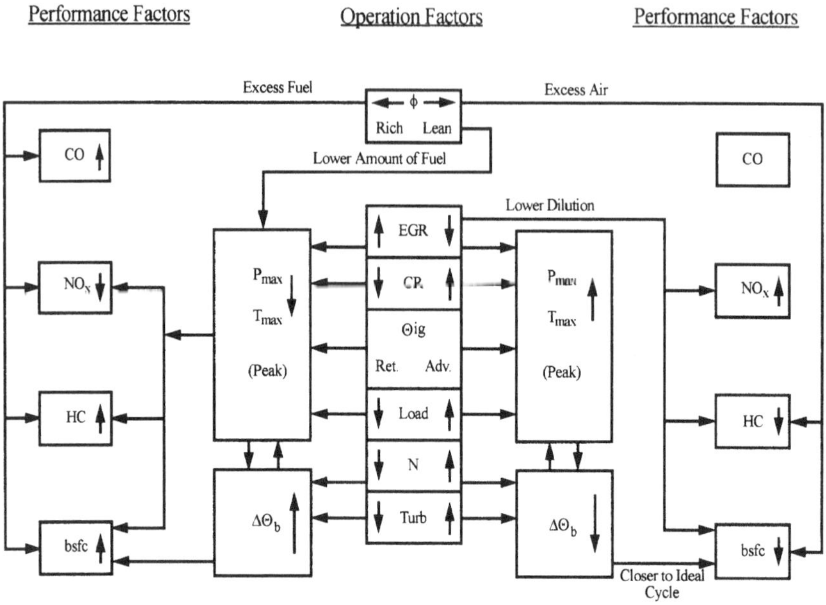

Schematic diagram showing the more important effects of the operating conditions on the exhaust pollutants and bsfc of an SI engine.

tion temperature and, hence, the formation of the NO_x component. Exhaust gas recirculation (EGR) is a very effective method commonly used in many four-stroke-cycle engines to reduce NO_x emission level. A decrease of the peak combustion temperature is associated also with an increase in the bsfc and HC emission to an extent. Exhaust gas recirculation can be achieved by either an external or internal means. Because the typical scavenging efficiency in two-stroke-cycle engines is much lower than unity, an internal EGR is an inherent property of two-stroke-cycle engines; thus, their typical NO_x emission is practically insignificant.

Exhaust throttling is another effective method to minimize fuel losses through the exhaust port, thus reducing HC emission and apparent fuel consumption.

Figure 13.4 illustrates the more important effects of the operating conditions on the exhaust pollutants and bsfc of an SI engine.

13.2.3 Afterburning Exhaust Gas Treatment

A significant reduction in emissions can be obtained by removing pollutants from the exhaust gases by using catalytic converters (this group includes oxidizing catalysts for HC and CO, reducing catalysts for NO_x, and three-way catalysts for all three pollutants), thermal reactors for HC and CO, and filters (or traps) for particulate.

A catalyst is a material that increases the rate of a chemical reaction while itself not undergoing any permanent change. For a molecule A to be catalytically converted, the following steps must occur: Molecule A diffuses through a stagnant thin layer of gas in close contact with the catalyzed carrier toward the active sites; once it arrives it is chemisorbed onto it and converted to product B. The product then is desorbed from the site and diffuses through the stagnant layer into the bulk gas.

A catalytic converter is a device through which the exhaust gases are directed to flow while converting some prespecified species into others. The catalytic converter is commonly installed along the exhaust system between the engine and the exhaust silencer. It consists of metal housing, a substrate (ceramic or metallic honeycomb), carrier (porous oxide particles such as Al_2O_3, SiO_2, and TiO_2), and an active catalytic material (noble metal catalysts such as Pt, Rh, Pd, and V_2O_5). The specially designed metal casing directs the exhaust gas flow through the catalyst bed. Since the number of molecules to be converted per unit time is directly related to the number of catalytic sites available, maximizing the surface area of the active catalyst material is an important design parameter. To accomplish this, the active catalytic components are often dispersed on high surface area (the carrier). The carrier is impregnated with the catalytic components. A typical pore size of a carrier is between 20 to 50 Å and a surface area of about 50 to 200 m^2/g. When a carrier impregnated with the catalytic components is bonded to a monolithic honeycomb support, it is called a washcoat.

Present available substrates can be divided into three different types: ceramic pellets, ceramic monolith honeycombs, and metallic monolith honeycombs. The ceramic pellet type has layers of spherical pellets resting on top of each other, with the pellets being made of ceramic such as high-temperature-resistant magnesium-aluminum silicate. The spaces surrounding the spherical contacting points provide a large surface area exposure to the exhaust gas. The pellet pack is packed in a perforated container. One half of the perforated container is exposed to the incoming exhaust gases, while the other is the exit leading to the exhaust silencer. The pellet type of substrate is steadily losing ground in favor of the ceramic and metal monoliths. A typical straight-channel monolithic honeycomb support is made of extruded ceramic material with about 24.4 cells per square cm (400 cells per square inch). The monolith has a large open frontal area to minimize flow resistance. The channel's shape can be triangular, hexagonal, squared, or any other convenient shape. The monolith is situated in a soft substrate material to absorb mechanical stresses, and the package is mounted in a metal housing. Metallic monoliths are much more durable and expensive. They are sometimes employed where higher cell densities with low-pressure drops are important, or as supplements to the main catalytic converter in order to provide rapid catalytic conversion in cold starts.

The oxidizing catalysts transform hydrocarbons and carbon monoxides into water vapor and carbon dioxides through the following reaction steps [7].

$$C_nH_m + \left(n + \frac{m}{4}\right)O_2 \longrightarrow nCO_2 + \frac{m}{2}H_2O \tag{13.3}$$

$$CO + \frac{1}{2}O_2 \longrightarrow CO_2 \tag{13.4}$$

and

$$CO + H_2O \rightarrow CO_2 + H_2 \tag{13.5}$$

For this goal they require an environment of excess air. The required oxygen is either provided by a lean mixture setting, or by secondary air supplied upstream of the converter. Self-inducting air valves and a centrifugal pump driven by the engine are commonly employed. In oxidizing catalysts, oxides of nitrogen (NO_x) remain practically unaffected.

Reducing catalysts for NO_x are sometimes used in dual-bed catalytic converters. In this arrangement, the converter consists of two catalysts connected in series with each other (Figure 13.5). The exhaust gases first flow through a reduction catalyst, where the oxides of nitrogen are converted into nitrogen, carbon dioxide, and water through the following reaction steps:

$$NO + CO \longrightarrow \frac{1}{2}N_2 + CO_2 \tag{13.6}$$

$$NO + H_2 \longrightarrow \frac{1}{2}N_2 + H_2O \tag{13.7}$$

and

$$\left(2n + \frac{m}{2}\right)NO + C_nH_m \rightarrow \left(n + \frac{m}{4}\right)N_2 + nCO_2 + \frac{m}{2}H_2O \tag{13.8}$$

With this arrangement, the engine must be operated with a rich mixture (Figure 13.5). Air is provided in between the two beds, and the hydrocarbons and carbon monoxide are converted in the second bed through reactions (13.3) to (13.5).

If an engine is operated at all times with an air-fuel ratio at or close to stoichiometric, then both NO reduction and CO and HC oxidation can be done in a single catalyst bed. Such a catalyst is termed a *three-way catalyst*. The three-way catalysts require that the air-fuel mixture supplied is at the stoichiometric ratio, and when combined with a lambda closed-loop control, it provides the most effective pollutant reduction system presently available.

The conversion efficiency of a catalyst is the ratio of the rate of mass removal in the catalyst of the particular constituent of interest to the mass flow rate of that constituent into the catalyst [1]. A conversion efficiency of 98 percent to 99 percent for the CO and about 95 percent for the HC can easily be achieved with a new commercial catalyst when operated at high enough temperatures. At low temperatures the catalyst become ineffective. The temperature at which the catalyst becomes more than 50 percent effective is referred as the light-off temperature.

Fig. 13.5.

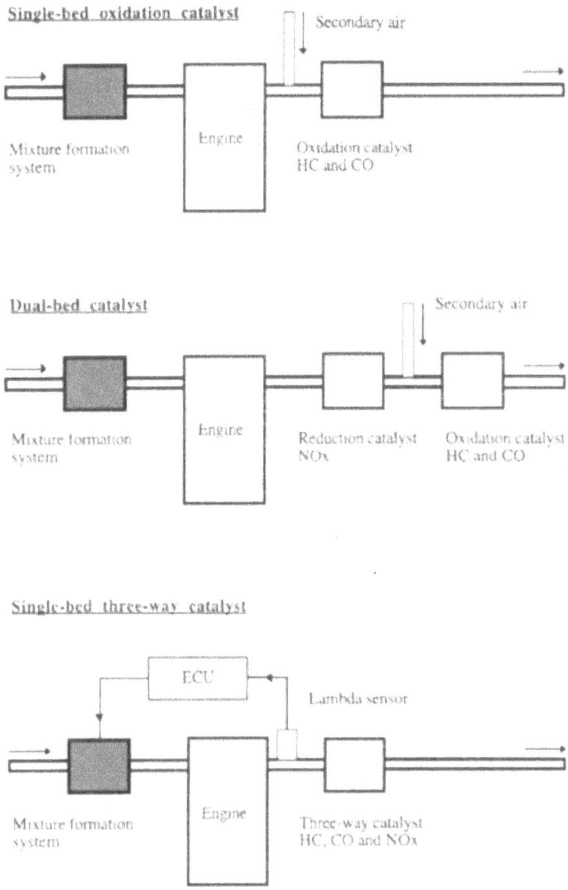

Installation of a catalytic converter system.

The ideal operating conditions for high conversion efficiencies and long life are between 400°C and 800°C. Above this range, the thermal aging process accelerates by sintering of the noble metals and the carrier, thus shrinking down the active surface area. Present NO catalysts operate effectively in the temperature range between 350°C to 600°C.

The pollution emission characteristics of HC, CO, and NO_x after a catalyst are shown in Figure 13.6. Low emissions of HC and CO are noticeable in lean fuel-air mixture regions. Conversion efficiencies between 92 percent and 97 percent have been achieved in the entire range of operating conditions [8]. The specific NO_x emission, however, seems to be still high in the regions where the bmep is high and the air-fuel ratio is near stoichiometric (between 14 and 15).

Fig. 13.6.

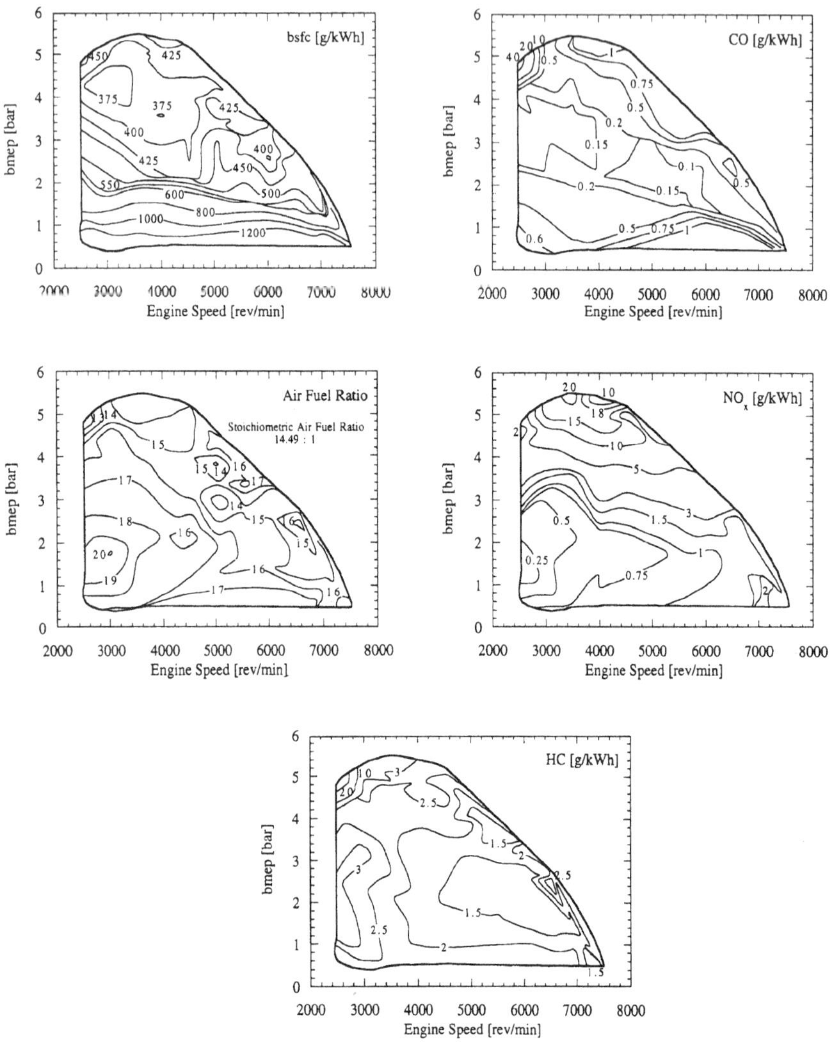

HC, CO, and NO_x emissions of the 2.7-hp moped engine [8] (reprinted with permission from SAE).

13.3

FLOW AND EMISSION DIAGNOSTICS (EXPERIMENTAL RESULTS)

The laser Doppler velocimeter [9–13] (LDV), phase Doppler technique [14–17], and laser-induced fluorescence [18] (LIF) are examples of scientific tools generally utilized in research and development of engines in order to understand flow and emission characteristics. This section discusses the experimental results obtained by using some of these tools on a small two-stroke engine. Specifically, data for the flow characteristics and CO, CO_2, and HC emission concentrations and their variations are presented in this section. The results were obtained from a crankcase-compression, 99-cc, two-stroke motorcycle engine [19–20].

When the scavenging port opens, the intake flow is forced into the engine cylinder due to the pressure difference between the crankcase and the cylinder. At the same time, the exhaust port is also open. This means that in two-stroke engines, both pressure wave propagation and scavenging flow characteristics should be taken into account when describing the flow. In four-stroke engines, however, only the in-cylinder flow is considered. When fresh fuel-air mixture enters the cylinder, there is a substantial amount of residual gas still trapped within the cylinder, and the performance of the engine and the emissions from the engine are, therefore, influenced by the flow dynamics between strokes. Based on this, two important design issues need to be considered to improve the emission characteristics of a two-stroke engine: the short circuiting of fresh fuel-air mixture into the exhaust manifold, and unstable combustion conditions at low loads, which yield troublesome cycle-to-cycle variations in engine operation. Monitoring pressure reveals useful information about the flow dynamics and combustion characteristics of an engine. However, pressure data alone are insufficient to explain the important characteristics of emission formation and pressure-driven flow characteristics. For instance, pressure increase *only* indicates that combustion has occurred. The relationship between pressure and the pressure-driven scavenging flow is important and is discussed here.

For studies of cyclic variation, time series data with high temporal resolution are required. Flow and emission characteristics are not well quantified in the control of combustion because it is such a complex phenomenon. Hence, current research and device development must rely on empirical knowledge. In practice, combustion control by adjusting the exhaust control valve only influences the scavenging flow timing [21]. Such a control scheme is determined by trial and error. However, in order to reduce CO and UHC emissions significantly it is necessary to demonstrate the flow characteristics and their relation to combustion.

The following sections present some experimental results demonstrating scavenging flow characteristics, misfiring detection, cyclic variations, and air-assisted fuel spray characteristics.

13.3.1 Scavenging Flow Characteristics

The intake flow from the crankcase into the cylinder and the exhaust flow from the cylinder to the exhaust pipe are not separated and are, therefore, greatly influenced by the pressure difference between the crankcase and the exhaust pipe. An illustration of the scavenging flow in a crankcase-scavenging, two-stroke engine is shown in Figure 13.7. When the exhaust port opens, the high-pressure burned gases blow out to the exhaust pipe. The pressure at the exhaust pipe varies during

Fig. 13.7.

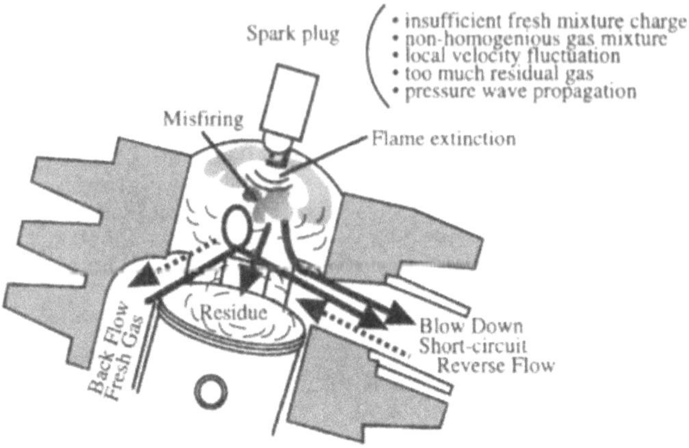

Illustration of scavenging flow in a crankcase-compression, two-stroke engine.

Fig. 13.8.

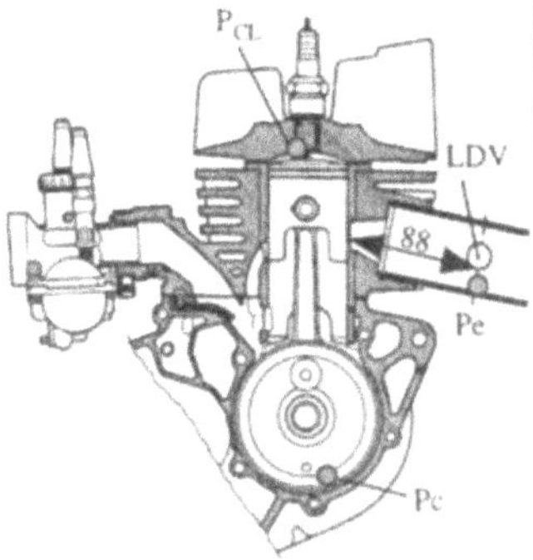

Type of Engine	Two-stroke Spark Ignition Crankcase Compression with Reed Valve Schnürle Scavenging	
Displacement	98.2 (cm^3)	
Bore × Stroke	50.0 (mm) × 50.0 (mm)	
Exhaust-Port Timing	Open	97.0 (C.A.)
	Close	261.7 (C.A.)
Scavenging-Port Timing	Open	122.7 (C.A.)
	Close	240.0 (C.A.)
Ignition Timing	BTDC 22.0 (C.A.) (4000 r/min)	
Compression Ratio	6.59	

Schematic layout of the engine examined and its specification (reprinted with permission from SAE).

the opening of the exhaust port and negative pressure plays a role in preventing fresh gas from short circuiting. The combustion pressure propagates to the end of the exhaust pipe, where it is reflected. The reflected pressure wave together with the pressure difference between the crankcase and the exhaust pipes create flow reversal into the cylinder from the exhaust pipe.

Although the flow through the scavenging port into the cylinder is charged, when the scavenging port is open, substantial backflow from the cylinder to the crankcase is observed.

These findings were obtained from the experimental results with the crankcase-compression, two-stroke engine [22–26] shown in Figure 13.8. The fresh gas mixture was charged into the cylinder through the scavenging port. The scavenging

Fig. 13.9.

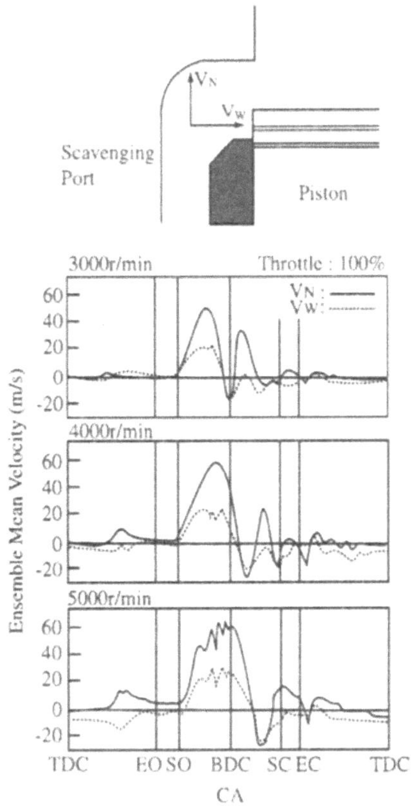

Velocity variation in the scavenging port at different engine speeds (reprinted with permission from SAE).

flow velocities, measured by LDV [12] at different engine speeds, are shown in Figure 13.9. It is clear that charging of the fresh mixture starts just after SO and reaches its maximum velocity before BC. In all cases negative velocity peaks were observed in both directions, which indicate the pressure of backflow from the cylinder into the crankcase. This backflow causes scavenging flow stagnation and charging with insufficient fresh mixture, which result in unstable combustion. The pressure difference between the crankcase and the exhaust pipe was found to be the dominant factor in the occurrence of this backflow, since backflow occurred near BC, at which time both the scavenging and exhaust ports are open and the combustion pressure has little direct influence on the scavenging flow.

Figure 13.10 shows [22] the blowdown flow characteristics in the exhaust pipe, which demonstrate the scavenging flow across the cylinder. Ten velocities were measured by LDV to demonstrate exhaust pipe flow distribution using ensemble averaged data. A strong blowdown flow was observed when the exhaust

Fig. 13.10.

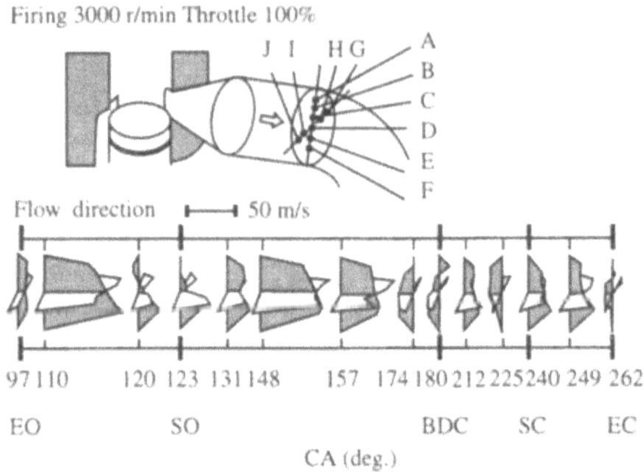

Velocity distribution in the exhaust pipe (reprinted with permission from SAE).

port opened, the maximum velocity of which was almost 100 m/s. However, there was then sudden flow stagnation in the exhaust pipe due to the reflected pressure wave, which caused flow reversal from the exhaust pipe to the cylinder near BC.

To better understand the characteristics of blowdown and reversed flow in the exhaust pipe, the pressure difference ($\Delta P = Pc - Pe$) between the crankcase and the exhaust pipe was used, as shown in Figure 13.11. The first valley in ΔP corresponds to the blowdown velocity peak. The maximum peak of ΔP is at the same crank angle as the reverse-flow velocity peaks. The second valley of ΔP shows the timing of the reverse-flow velocity peaks. The blowdown flow is stagnated due to pressure reflection from the exhaust pipe. This means that the ΔP peaks correspond to the timing of typical exhaust-pipe flow behavior, and this method can, therefore, be used to predict blowdown and reverse-flow characteristics. This correspondence has been examined at different engine speeds at both the scavenging port and at the exhaust pipe to demonstrate its usefulness, and the results show evidence of the advantageous relationship between ΔP and scavenging flow characteristics. However, the results were ensemble averaged so as to include large variations and cyclic variations. Time series analysis is required for cyclic variations and unstable combustion analysis.

13.3.2 Misfiring

One of the most difficult problems in two-stroke engines is misfiring and the unstable combustion induced by the former at low-load conditions. The misfiring mechanism should be considered a complex phenomenon, influenced by flow fields, pressure, fuel-air mixture, and residual gas and their interrelationship. This interrelationship is explained in the next section, where the characteristics of the scavenging and blowdown flows in misfiring are considered.

Fig. 13.11.

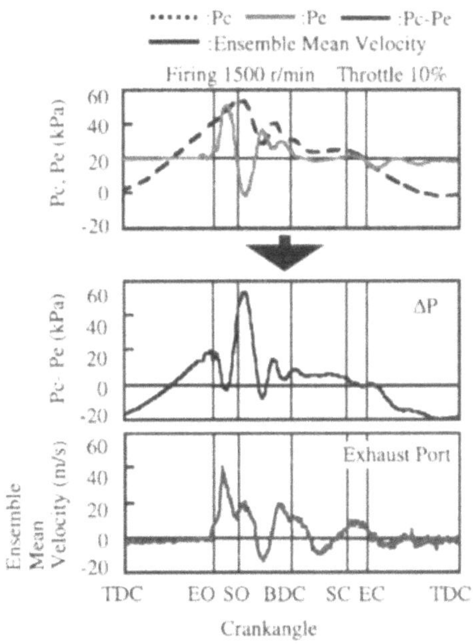

Pressure difference between the crankcase and the exhaust pipe, and ensemble averaged exhaust pipe flow velocity.

The same engine as that in Section 13.3.1 was used, and one condition was chosen to demonstrate misfiring phenomena. The in-cylinder pressure traces are shown in Figure 13.12. At this condition, various kinds of combustion states exist. Each pressure trace represents a combustion event that induces strong or weak blowdown flow. For a better understanding, time series analysis with high temporal resolution is recommended. Let's see the flow characteristics in the exhaust pipe. First of all, the ensemble averaged blowdown flow variation at different IMEP are shown in Figure 13.13. At IMEP < 0 there was no blowdown flow and a reverse flow was formed around SO and after BC, due to the pressure difference between the crankcase and the exhaust pipe. At high IMEP strong blowdown flow occurred after EO, and there was also a strong reverse flow before BC, due to combustion pressure wave reflection from the exhaust pipe.

The relationship of the pressure difference between the crankcase and the exhaust pipe, and the flow field in the scavenging port and in the exhaust pipe, is shown [23] in Figure 13.14. The pattern of the occurrence of misfiring was investigated and classified. The misfiring ratio at the condition was [23] 36.3 percent. Velocity measurements were carried out separately, but pressure and velocity measurements were also made simultaneously in both the exhaust pipe and in the scavenging port. The misfiring detection scheme was the same in order to show the ensemble averaged flow variation with pressure data.

Fig. 13.12.

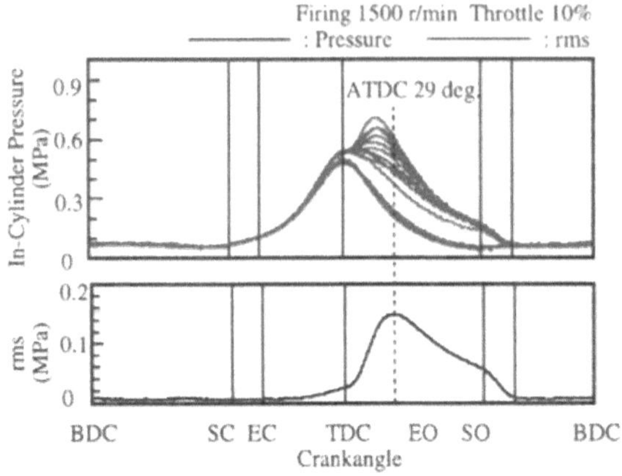

Combustion pressure variation and its rms (reprinted with permission from SAE).

Fig. 13.13.

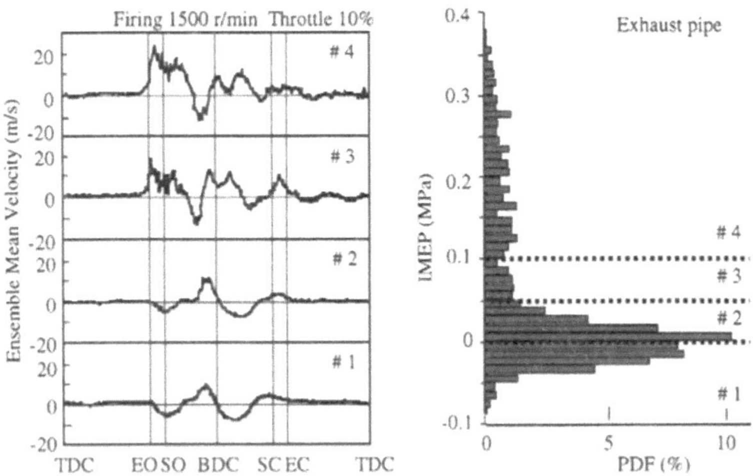

Blowdown flow velocities in the exhaust pipe and classified IMEP.

In the firing cycle, strong blowdown flow can be observed after EO, the timing of which can be predicted by the negative ΔP peak. At the crank angle of positive ΔP peak, the scavenging port flow and the blowdown flow show maximum acceleration and have their maximum velocity peaks. Just before BC, at which time ΔP shows a negative peak, backflow from the cylinder to the crankcase and reverse flow from the exhaust pipe to the cylinder take place. The timing of these flow field variations can be evaluated with the help of ΔP variations versus the crank angle.

Fig. 13.14.

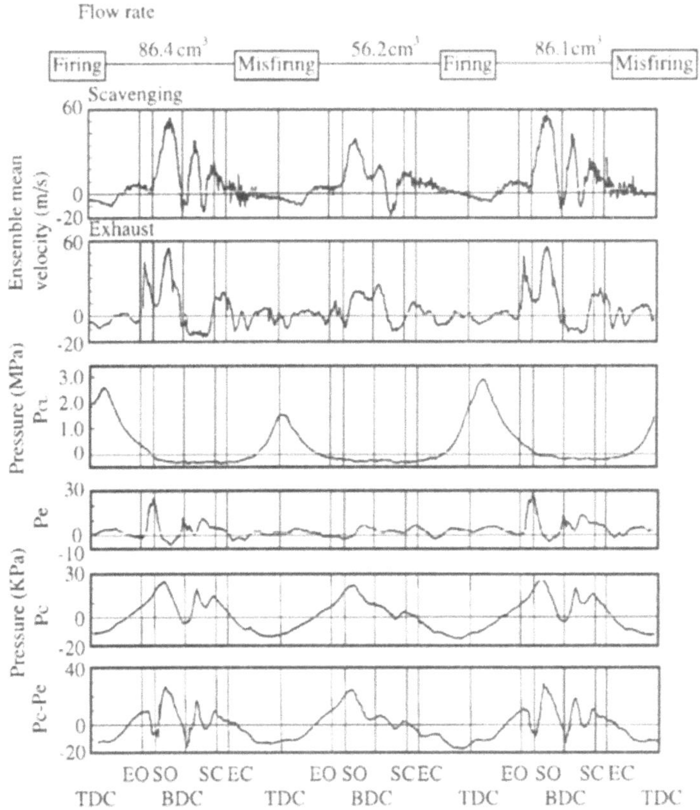

Velocity and pressure variation at misfiring cycle (reprinted with permission from SAE).

In a misfiring cycle, both flows show lower velocity peaks after EO and SO. Negative velocity peaks can also be observed after BC, at the time of negative ΔP peaks.

In this way, the scavenging port and exhaust pipe flow field behaviors, and their variation over time, can be followed and evaluated using the pressure difference between the crankcase and the exhaust pipe. These provide useful information for practical engine development and testing.

As shown in the preceding section, several velocities in the exhaust pipe were measured by LDV and their velocity distribution is shown in Figure 13.15. Very large discrepancies can be observed in the two velocity distribution variations. At the crank angle of 103°, combustion produced a large blowdown flow, while the reverse flow was formed due to the pressure difference between the crankcase and the exhaust pipe, and not because of combustion. During firing cycles, the timing of the reverse flow for firing and misfiring is very different.

Fig. 13.15.

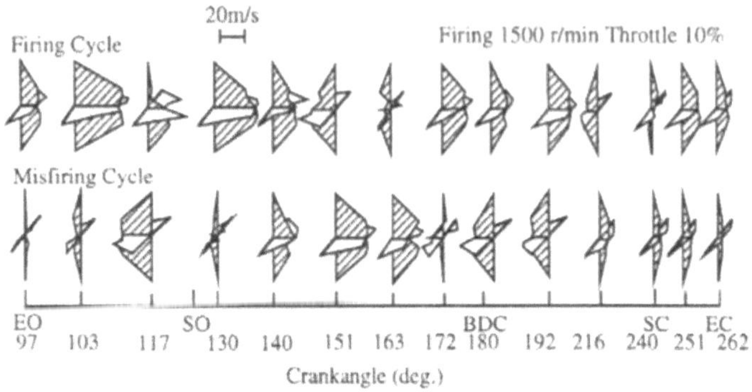

Blowdown flow velocity distributions both in combustion and in misfiring in the exhaust pipe (reprinted with permission from SAE).

Fig. 13.16.

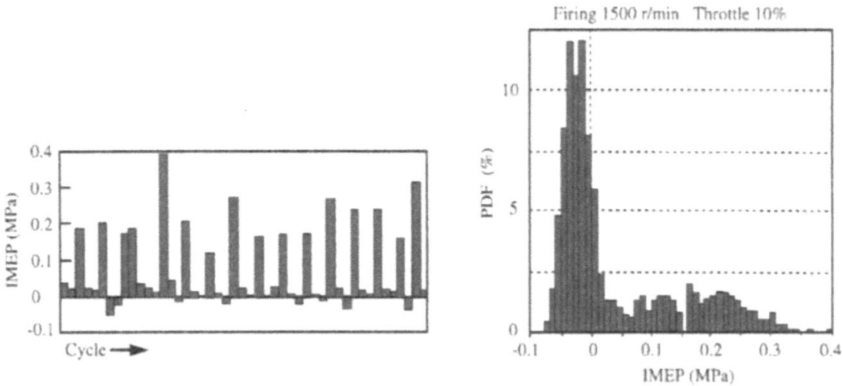

Cyclic variation in IMEP.

13.3.3 Cyclic Variation

Combustion status is different in each cycle and varies from cycle to cycle, which is termed *cyclic variation*. In order to understand cyclic variation, separate and simultaneous time series measurements of the flow characteristics, pressure, CO, and HC are required.

The IMEP variation of the same engine (Section 13.3.1) and the operational conditions are shown in Figure 13.16. This cyclic variation is caused by several factors: pressure wave reflection, insufficient fresh mixture charging, too much radical gas, flame extinction, and so on [25]. Pollutants such as CO, CO_2, and HC also vary from cycle to cycle.

The CO and CO_2 variation in the exhaust pipe is shown in Figure 13.17. At this operating condition, a remarkable pattern of cyclic variation was observed.

Fig. 13.17.

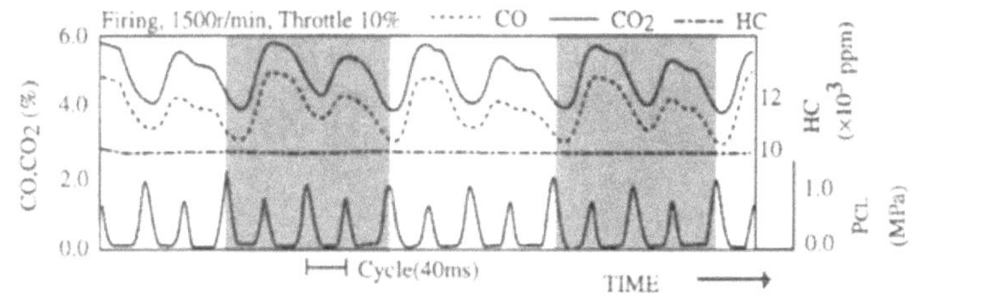

Cyclic variations of CO, CO_2, HC, and combustion pressure (reprinted with permission from SAE).

Fig. 13.18.

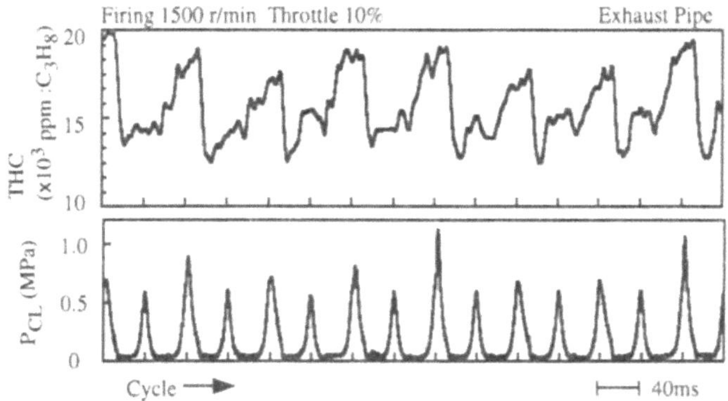

Cyclic variation of HC concentration in the exhaust pipe and in-cylinder pressure.

The cyclic variation patterns of CO and CO_2 were almost the same, having a period of four cycles. At the same condition, HC in the exhaust pipe was also varying as shown in Figure 13.18. These measurements of CO, CO_2, and HC can only be carried out by instruments capable of a very fast response [27], such as those used in the current experiments. However, some uncertainties remain in the data, due to the instruments' response time, pressure variation, and temperature variations. This experimental condition was chosen to demonstrate typical misfiring and its cyclic variation at a low-load condition. Because of several factors and their concurrent occurrence, strongly periodic variations of CO, CO_2, and HC can be observed. The key feature to control unstable combustion is to determine the sources of these variations. Are these variations due only to the combustion process, or are they induced by other factors? To determine the sources of misfiring and cyclic variation, HC measurements were performed near the spark plug, using the same experimental set-up.

Variation in HC concentration near the spark plug is shown in Figure 13.19, together with the IMEP. The temporal resolution of this measurement was [28–31] 4 ms. The charging of fresh mixture varies, inducing unstable combustion. Combustion is caused by the presence of a sufficient HC concentration near the

Fig. 13.19.

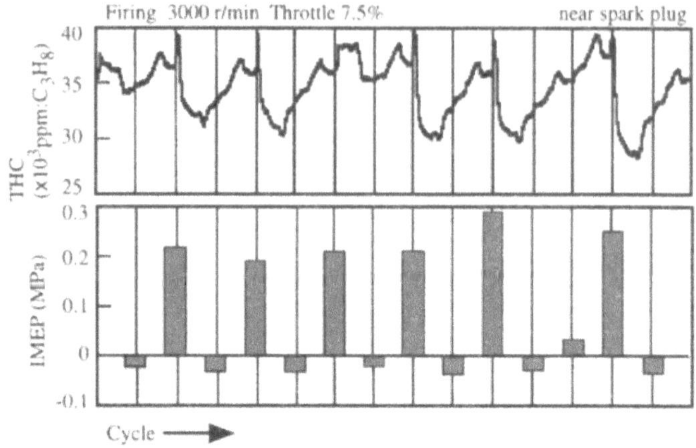

Cyclic variation of HC concentration near the spark plug and IMEP.

spark plug, but when the HC content is consumed by the combustion process resulting in high IMEP, then in the next cycle the fresh mixture is insufficiently charged to cause combustion, thus increasing HC concentration in the exhaust.

The reasons for HC variation near the spark plug were examined by simultaneous measurements of velocity in the exhaust pipe, pressure at several locations, and HC in the cylinder. Time series data acquisition was carried out in order to show the cyclic variation patterns over 1024 cycles.

Six consecutive cycles are demonstrated in Figure 13.20. In this figure, the relationship between the IMEP, the blowdown flow, the reverse flow, the crankcase pressure, and the HC concentration near the spark plug is demonstrated, both in the case of combustion and in the case of misfiring. These data were measured simultaneously, in a time series with high temporal resolution.

In the combustion cycle: ① High IMEP shows strong combustion, which consumes HC near the spark plug. ② Strong blowdown flow is produced by this combustion and reverse flow also occurs. ③ The strong blowdown flow in the exhaust pipe yields fresh mixture charging from the crankcase into the cylinder, so that the crankcase pressure decreases rapidly, ④ but when flow reversal in the exhaust pipe occurs, the crankcase pressure becomes positive, which causes insufficient fresh mixture to charge into the cylinder, ⑤ and this leads to misfiring in the next cycle.

In misfiring: ①′ Negative IMEP shows misfiring. ②′ The exhaust pipe flow has no blowdown because of the lack of sufficient combustion pressure. The flow is sucked into the cylinder and at the same time there is a negative peak at BC. The flow toward the exhaust pipe is short circuited before SC and EC. ③′ The crankcase pressure cannot be properly decreased because there is no sucking force in the cylinder and, therefore, no negative pressure in the crankcase occurs. ④′ Then the fresh mixture charging yields, ⑤′ and the HC concentration near the

Fig. 13.20.

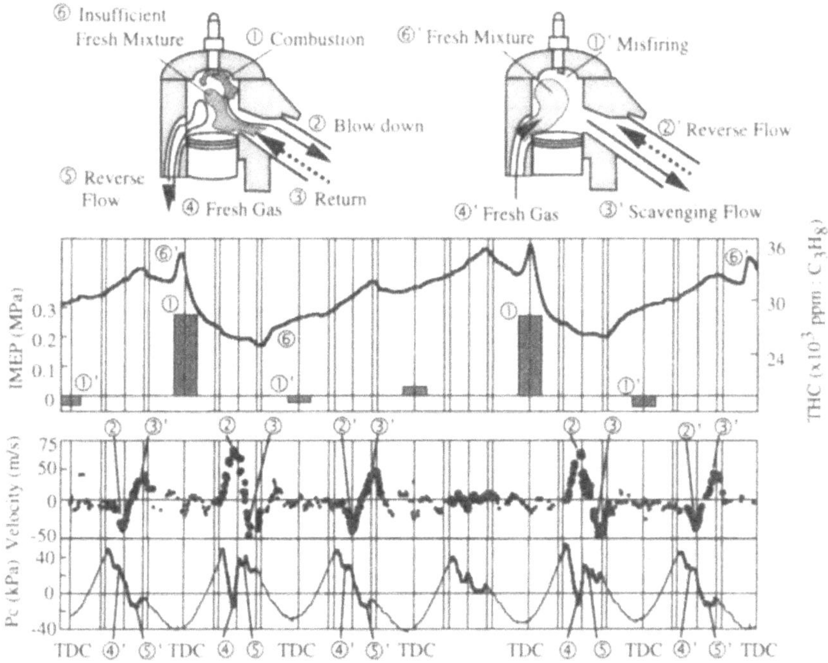

spark plug becomes sufficiently high as to allow combustion ⑥′ in the next cycle. There is another cyclic pattern after misfiring, which then causes misfiring again.

In this way, cyclic variations occur due to several overlapping factors. In practice, it is desirable to control combustion variation in various ways, utilizing the pressure data.

Current technology controls this unstable combustion by the use of an exhaust control valve [32] and radically induced [33, 34] combustion should take place as the next step.

13.3.4 Air-Assisted Fuel Injection

Fuel injection systems are indispensable to the improvement of two-stroke engines to enhance their advantages in automotive engine applications. There are numerous reports of injector developments [35–42] but very few contain sufficient information related to the detailed spray droplet characterization. Spray and injection systems have been thoroughly researched, especially in diesel engines. The two-stroke engine involves complex processes, such as the scavenging process, cyclic variation, and misfiring, which are closely related to pressure wave propagation and reflection. Although the scavenging process has been the key feature in the development of two-stroke engines [20, 22–24, 43–46], very little experimental evidence is available to explain the relationships between gasoline spray vaporization, mixture formation, and the scavenging process [47–54].

For small two-stroke engines, direct fuel injection has been considered as a way to solve the problems of incomplete combustion and excessive HC concentration in exhaust gases. In particular, air-assisted fuel injection was developed as a powerful tool in the creation of a more combustible fuel-air mixture at lean combustion conditions. Air-assisted injection uses compressed air to atomize fuel in the injector and enhance the penetration of small droplets. Many different kinds of injector mechanisms have evolved around the world. The spray formation of an air-assisted injector is dominated by the assisting airflow, so the droplet dispersion process and its atomization should be understood, together with droplet dynamics.

Laser diagnostic tools like laser sheet [55], exciplex [56], and LDV [14] can provide information regarding spray angle, spray shape, penetration, vapor region, and so on, but detailed spray information, such as the droplet diameter distribution and its velocity in a two-dimensional plane, has not yet been obtained. A visualization technique can supply sufficient spatial, but very poor temporal, information of spray characteristics. A phase Doppler anemometer (PDA) can measure the diameter of a droplet and its velocity with very high spatial and temporal resolution, but this is a single-point measurement method. An alternative technique is required to determine a two-dimensional spray image with detailed droplet characteristics.

In this section, the usefulness of Sauter mean diameter (SMD) [57, 58] in a periodic injector is proven and droplet size classes are implemented in order to better understand the momentum transfer between liquid and gas phases.

The air-assisted injector used in this experiment was a commercial injector for a two-stroke marine engine with over 22 kW (30 ps) per cylinder, as shown in Figure 13.21. Fuel is first injected into a cavity and the air injector is operated

Fig. 13.21.

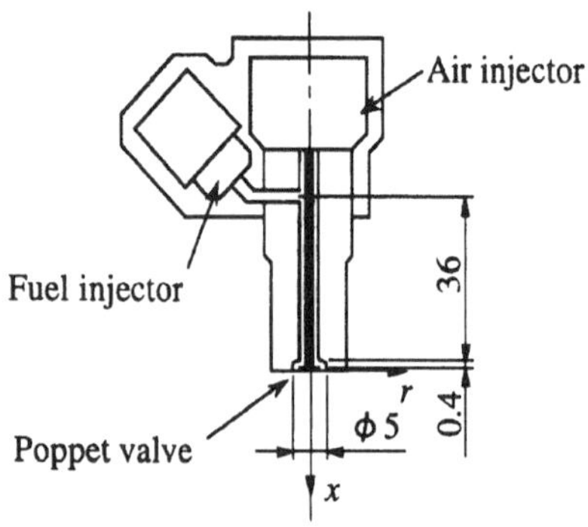

Air-assisted injector (reprinted with permission from SAE).

by opening a poppet valve. The air-fuel ratio can be controlled by changing the opening period of the valve when the pressure difference between the air and the fuel is set at a certain level. The injector has a 36-mm-long, straight pipe before the valve, where preatomization is carried out. Air-assisted fuel is injected from the 5-mm diameter poppet valve.

A dry solvent with a refractive index of 1.427 was used as fuel instead of gasoline. The specific density of the dry solvent is 0.77 g/cm^3, which is very similar to that of gasoline (0.7–0.8 g/cm^3). The scattering angle of 68° was determined by the first-order refraction angle [59]. For vector measurements, the one-component LDV was used by changing the incident beam angles of $\pm 45°$.

Direct photographs of the injected spray are shown [60] in Figure 13.22. It is clear that a mushroom vortex is induced by shear stress at the spray shell. The spray tip velocity calculated from these pictures is about 64 m/s. A YAG laser sheet was used to take a two-dimensional spray image, as shown in the same figure. These shots are direct pictures at a certain cycle. It is well known that there is cycle-to-cycle variation in this kind of air-assisted injector. The figure also shows two pictures at different cycles at the same timing. These photographs indicate the importance and need to analyze the spray by a two-dimensional image having high temporal resolution, as laser sheet visualization cannot provide time variation information and diameter information. Single-point measurements do not reveal the cycle-to-cycle variations and spatial structure variations. However, using single-point measurement with ensemble averaged data, the two-dimensional image of the spray can be demonstrated with its spatial structure as shown [61] in Figure 13.23. The Sauter mean diameter (SMD) and the relevant velocity vectors are also shown.

Spatial droplet dispersion is best explained using planar information sources such as a photograph or a laser sheet image. The PDA technique provides single-point information, but a phase-locked ensemble averaged method can demonstrate a two-dimensional image, as shown in Figure 13.23. The spray axisymmetry was checked by measuring at opposite locations up to $r = -3$ mm. This figure shows the SMD variation and its spatial structure versus time. The vector length was calculated as droplet trajectory length within 0.25 ms, and the color represents the SMD. The maximum SMD was set to be 130 micron.

At 1.6 ms after the injection signal, which was used as the assisted air drive signal, the first droplet was observed on the axis. After 0.25 ms, the spray tip velocity reached approximately 65 m/s and droplet dispersion in a radial direction was observed. The spray tip velocity of 65 m/s was almost the same as that calculated from the direct spray picture. The SMD was about 25 micron at the spray tip. On the center axis, droplet direction was parallel to the axis, while droplet direction at the spray shell region was over 45 degrees in a radial direction.

At 2.3 ms, the spray tip velocity on the axis increased and the next droplet from the nozzle formed a larger-sized droplet group. The region penetrated by the droplets resembled an umbrella. The small and fast droplets existed until 2.8 ms. After 2.8 ms, the spray tip velocity decreased and the SMD enlarged near the center

Fig. 13.22.

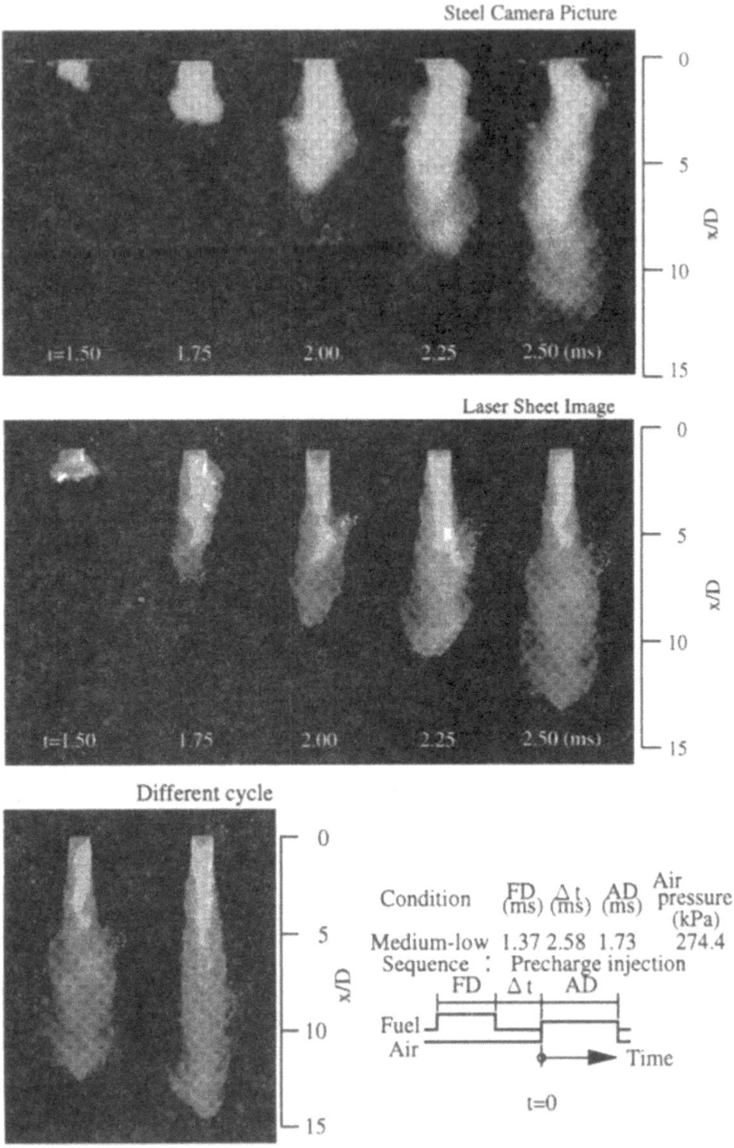

Images of injected spray structure (reprinted with permission from SAE).

Fig. 13.23.

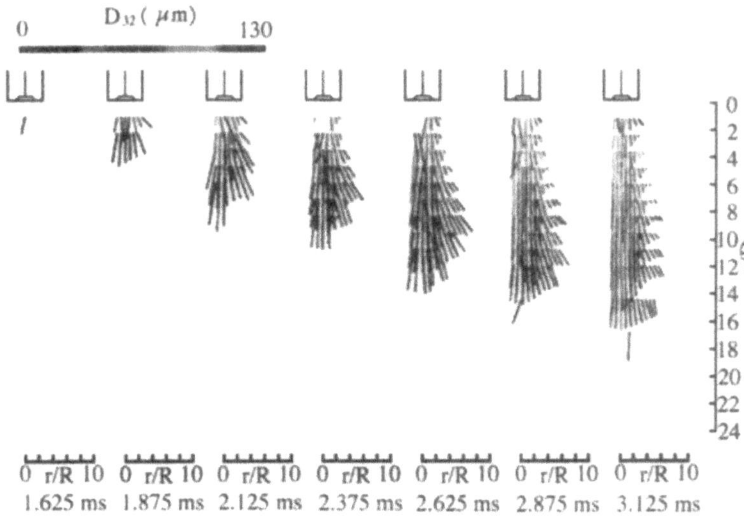

Droplet velocity vectors and SMD (reprinted with permission from SAE).

Fig. 13.24.

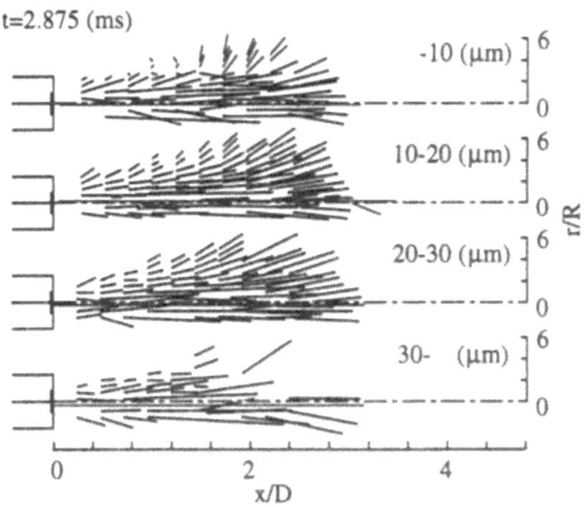

Size-classified droplet dynamics at 2.875 ms (reprinted with permission from SAE).

axis. The larger droplets caught up and collided with the smaller droplets and, hence, the diameter began to increase. The droplets of spray at the outer region had lower velocity due to the strong shear flows, and the direction of the droplets then showed a wavy spray structure. The very large droplet near the nozzle, which is red in color, was formed at 2.875 ms, when the spray tip droplet size was 30 micron.

Furthermore, spray droplets influenced by turbulent air tended to follow air motion but large droplets with a high momentum penetrated the highly turbulent

Fig. 13.25.

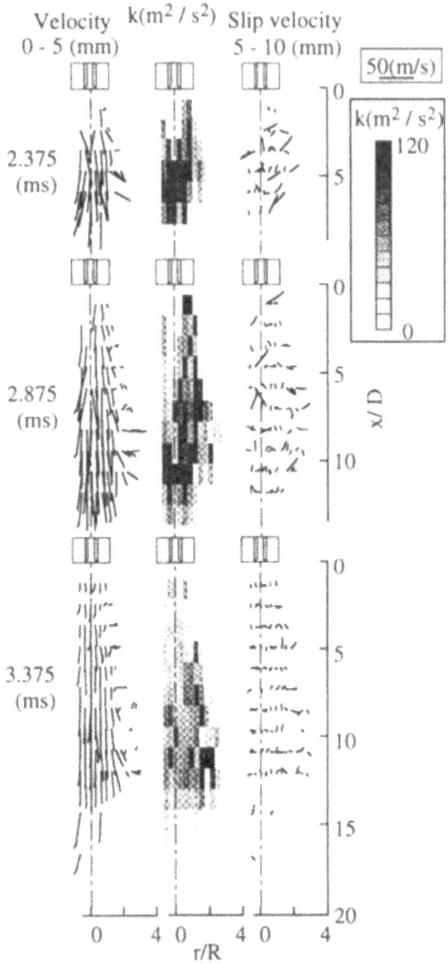

Airflow motion, turbulent kinetic energy, and slip velocity of small droplet (reprinted with permission from SAE).

flow regions, such as the recirculation flow areas. These droplet dynamics could not then be demonstrated by Sauter mean diameter alone but require other advanced methods such as size-classified analysis.

Four size-classified droplet velocity vectors are shown frozen at 2.875 ms in Figure 13.24. It is clear that there is a shear flow-induced mushroom vortex formation in the small droplet regions. At the spray tip, small droplets show larger velocity gradients than larger droplets. The larger droplet vectors have straighter and narrower injection angles. There are no droplets larger than 30 μm in the spray shell region.

The spray angle of each size class and the momentum decay should be quantified in order to understand evaporation processes and mixture formation.

The air motion and turbulent energy intensity profiles are shown in Figure 13.25. The large turbulent energy area revealed by the dark area in the figure indicates the presence of a strong shear flow region. At an early injection period, the larger spot is located at the center of the axis. The dark area appears at the spray shell region in the next stage. The slip velocity vector shows the large vector angle at the strong shear region.

The spray characteristics of an air-assisted gasoline injector were investigated by phase Doppler measurements. A summary of the foregoing results follows.

A two-dimensional planar image of size-classified droplets was used to demonstrate the spatial structure of spray formation. It was found that the Sauter mean diameter is not the best representation value at the acceleration region, and that the size-classification technique is very useful for understanding detailed spray characteristics. The slip velocity and relative Reynolds number were implemented to show the momentum transfer region due to strong drag force. A mushroom vortex was formed by the strong shear flow at the spray shell and consisted of small droplets of 10 to 20 μm. A double spray tip structure was found near the nozzle, which diminished rapidly with distance. Droplets larger than 30 μm penetrated almost straight through to downstream. It was found that this spray animation can be a most powerful tool in understanding momentum transfer processes.

REFERENCES

1. Heywood, J.B., *Internal Combustion Engine Fundamentals.* New York: McGraw-Hill, 1988.
2. Murayama, T., Y. Sekiya, B. Sugiarto, and T. Chikahisa, "Study on exhaust control valves and direct air-fuel injection for improving scavenging process in two-stroke gasoline engines," SAE Paper No. 960367, 1996.
3. Petit, A., J. Lavy, G. Monnier, and X. Montagne, "Speciated hydrocarbon analysis: a helpful tool for two-stroke engine development," Proceedings of the international seminar on "New generation of two-stroke engines for the future?" P. Duret, ed., Rueil-Malmaison, France, November 1993.
4. Nuti, M., and L. Martorano, "Short-circuit ratio evaluation in the scavenging of two-stroke S.I. engines," SAE Paper No. 850177, 1985.
5. Drake, M.C., D.T. French, and T.D. Fansler, "Advanced diagnostics for minimizing hydrocarbon emissions from a direct-injection gasoline engine," 26th International Symposium on Combustion, The Combustion Institute, 1996.
6. Nuti, M., "Direct fuel injection: Piaggio approach to small 2T SI engines," SAE Paper No. 880172, 1988.
7. Heck, R.M., and R.J. Arrauto, Catalytic Air Pollution Control. New York: Van Nostrand Reinhold, 1995.
8. Laimbock, F.J., and C.J. Landerl, "50 cc two-stroke engines for mopeds, chainsaws and motorcycles with catalysts," SAE Paper No. 901598, 1990.
9. Durst, F., A. Melling, and J.H. Whitelaw, *Principles and Practice of Laser Doppler Anemometry.* New York: Academic Press, 1981.
10. Bopp, S., et al., "In-Cylinder Velocity Measurements with Mobile Fiber Optic LDV System," SAE Paper 900055, 1990.

11. Obokata, T., N. Matumoyo, and Y. Hirao, "LDV Measurement of Pipe Flow in a Small Two-Cycle Spark-Ignition Engine," SAE Paper 840425, 1984.
12. Ikeda, Y., et al., "A Compact Fiber LDV with a Perforated Beam Expander, Measurement Science and Technology" (formally, *Journal of Physics E: Scientific Instrumentation*), Vol. 1, no. 3, 260–264, 1990.
13. Lorenz, N., and K. Prescher, "Cycle Resolved LDV Measurements on a Fired S.I-Engine at High Data Rate Using a Conventional Modular LDV-System," SAE Paper 900054, 1990.
14. Obokata, T., H. Tanaka, and T. Koyama, "PDA Analysis of Transient Spray Flows Initiated from Air-Assisted Injector," Seventh International Symposium on Application of Laser Techniques to Fluid Mechanics, 36.5, 1994.
15. McDonell, V.G., et al., "A Comparison of Spatially-Resolved Drop Size and Drop Velocity Measurements in an Isothermal Chamber and a Swirl-Stabilized Combustor," 21st Symposium on Combustion, 685–694, 1986.
16. Presser, C., et al., "Interpretation of Size-classified Droplet Velocity Data in Swirling Spray Flames," AIAA Paper 95-0283, 1995.
17. Seay, J., et al., "Atomization and Dispersion from a Radial Airblast Injector in a Subsonic Cross-flow," AIAA Paper 95-3001, 1995.
18. Fujikawa, T., et al., "Visualization Study of Misfire Phenomena under Light Load Condition in a 2-stroke Engine," *Transactions of JSAE*, Vol. 61-590, 1995.
19. Ikeda, Y., M. Hikosaka, and T. Nakajima, "Scavenging Flow Measurements in a Motored Two-Stroke Engine by Fiber LDV," *SAE Transaction*, Vol. 100, Section 3, SAE Paper 990-998, 910669, 1991.
20. Ikeda, Y., M. Hikosaka, T. Nakajima, and T. Ohira, "Scavenging Flow Measurement in a Fired Two-Stroke Engine by Fiber LDV," *SAE Transaction*, Vol. 100, Section 3, 981–989, SAE Paper 910670, 1991.
21. Asai, Masahiro, et al., "Analysis on Fuel Economy Improvement and Exhaust Emission Reduction in a 2-Stroke Engine by Using an Exhaust Valve," SAE Paper 951764, 1995.
22. Ikeda, Y., T. Ohira, T. Takahasi, and T. Nakajima, "Flow Vector Measurements at the Scavenging Ports in a Fired Two-Stroke Engine," *SAE Transaction*, Vol. 101, Section 3, 635–645, SAE Paper 920420, 1992.
23. Ikeda, Y., T. Ohira, T. Takahashi, and T. Nakajima, "Misfiring Effects on Scavenging Flow at Scavenging Port and Exhaust Pipe in Small Two-Stroke Engine," *SAE Transaction*, Vol. 102, Section 4, 648–661, SAE Paper 930498, 1993.
24. Ohira, T., Y. Ikeda, T. Takahashi, and T. Nakajima, "Exhaust Gas Flow Behavior in a Two-Stroke Engine," *SAE Transaction*, Vol. 102, Section 3, 672–680, SAE Paper 930502, 1993.
25. Ikeda, Y., T. Ito, and T. Nakajima, "Cyclic variation of CO and CO_2 Emissions and Scavenging Flow in a Two-Stroke Engine," *SAE Transaction*, SAE Paper 940392, 1994.
26. Ikeda, Y., K. Kakemizu, and T. Nakajima, "In-Cylinder Flow Measurement and Its Application for Cyclic Variation Analysis in Two-Stroke Engine," *SAE Transaction*, SAE Paper 950224, 1995.
27. Takeda, K., and J. Aoki, "Exhaust Gas Measurement in Transient Mode by Fast Response Exhaust Gas Analyzer," (in Japanese) *Trans. of Internal Combustion Engine*, Vol. 30, 380, 1991.
28. Henein, N.A., et al., "In-situ Phase-shift Measurement of the Time-Resolved UBHC Emissions," SAE Paper 950161, 1995.
29. Cheng, C.O., et al., "Intake Port Phenomena in a Spark-Ignition Engine at Part Load," *SAE Transaction*, SAE Paper 912401, 1991.
30. Peckham, M., and N. Collings, "In-cylinder HC Measurements with a Piston-Mounted FID," SAE Paper 932643, 1993.
31. Rose, D., et al., "In-Cylinder Mixture Excursions in a Port-Injected Engine During Fast Throttle Opening," SAE Paper 940382, 1994.
32. Ishibashi, Y., et al., "A Trial for Stabilizing Combustion in Two-Stroke Engines at Part Throttle Operation, A New Generation of Two-Stroke for the Future?" edited by Pierre Duret p. 113, 1993.

33. Onishi, Shigeru, et al., "Active Thermo-Atmosphere Combustion (ATAC), A New Combustion Process for Internal Combustion Engines," SAE Paper 790501, 1979.
34. Ishibashi, Y., and M. Asai, "Improving the Exhaust Emissions of Two-Stroke Engine by Applying the Activated Radical Combustion," SAE Paper 960742, 1996.
35. Gentili, R., et al., "Development of Pumpless Air Assisted Injection System for Two-Cycle S.I. Engines," SAE Paper 940397, 1994.
36. Morikawa, K., et al., "Comparison of Fuel Injection Systems and a New Combustion Method for a Direct Injection System and a New Combustion Method for a Direct Injection Two-Stroke-Cycle Automobile Engine," ISAE 9536304, 1995.
37. John Beck, N., W.P. Johnson, R.L. Barlhimer, and S.H. Patterson, "Electric Fuel Injection for Two-Stroke Cycle Gasoline Engines," SAE Paper 861242, 1986.
38. Sato, T., and M. Nakayama, "Gasoline Direct Injection for a Loop- Scavenged Two-Stroke Cycle Engine," SAE Paper 871690, 1987.
39. Nuti, M., "A Variable Timing Electronically Controlled High Pressure Injection System for 2T S.I. Engines," SAE Paper 900799, 1990.
40. Schechter, M.M., E.H. Jary, and M.B. Levin, "High Speed Fuel Injection System for 2-Stroke D.I. Gasoline Engine," *SAE Transaction*, SAE Paper 910666, 1991.
41. Nuti, M., "Direct Fuel Injection: An Opportunity for Two-Stroke SI Engines in Road Vehicle Use," SAE Paper 860170, 1986.
42. Duret, P., A. Ecomard, and M. Audinet, "A New Two-Stroke Engine with Compressed-Air Assisted Fuel Injection for High Efficiency Low Emissions Applications," SAE Paper 880176, 1988.
43. Kenny, R.G., R.J. Kee, C.E. Carson, and G.P. Blair, "Application of Direct Air-Assisted Fuel Injection to a SI Cross-Scavenged Two-Stroke Engine," SAE Paper 932396, 1993.
44. Blair, G.P., "The Basic Design of Two-Stroke Engines," SAE Publications R-104, 1990.
45. Wyzalek, F.A., "Two-Stroke Engine Technology in the 1990's," SAE Paper 910663, 1991.
46. Yamashita, H., et al., "Driveability Evaluation in Two-Stroke Engine at Part Throttle Operation by Torque-Map," SAE Paper 950226, 1995.
47. Kuo, T.W., and R.D. Reitz, "Three-Dimensional Computations of Combustion in Premixed-Charge and Direct-Injected Two-Stroke Engines," *SAE Transaction*, SAE Paper 920425, 1992.
48. Huang, H.H., Y.Y. Peng, M.H. Jeng, and J.H. Wang, "Study of a Small Two-Stroke Engine with Low-Pressure Air-Assisted Direct-Injection System," SAE Paper 912350, 1991.
49. Duret, P., and J.F. Moreau, "Reduction of Pollutant Emissions of the IAPAC Two-Stroke Engine with Compressed Air Assisted Fuel Injection," SAE Paper 900801, 1990.
50. Monnier, G., and P. Duret, "IAPAC Compressed Air Assisted Fuel Injection for High Efficiency Low Emissions Marine OutBoard Two-Stroke Engines," SAE Paper 911849, 1991.
51. Abats, D., and K. Wellenkotter, "Characterization of Ignition and Parametric Study of a Two-Stroke-Cycle Direct-Injected Gasoline Engine," SAE Paper 920423, 1992.
52. Diwakar, R., T.D. Fansler, D.T. French, J.B. Ghandhi, C.J. Dasch, and D.M. Heffelfinger, "Liquid and Vapor Fuel Distributions from an Air-Assisted Injector—An Experimental and Computational Study," SAE Paper 920422, 1992.
53. Ghandhi, J.B., P.G. Felton, B.F. Gajdeczko, and F.V. Bracco, "Investigation of the Fuel Distribution in a Two-Stroke Engine with an Air-Assisted Injector," SAE Paper 940394, 1994.
54. Gantili, G., S. Frigo, and L. Tognotti, "Development of a Pumpless Air Assisted Injection System for Two-Cycle, S.I. Engine," SAE Paper 940397, 1994.
55. Kuwahara, K., T. Kawai, and H. Ando, "Influence of Flow Field Structure after the Distortion of Tumble on Lean-Burn Flame Structure," International Symposium COMODIA 94, 84–94, 1994.
56. Felton, P.G., "Fluorescence Imaging of Engine Sprays," Res. Paper, ASI on Unsteady Combustion, L38, 1993.
57. Lefebvre, H., *Atomization and Sprays*, Place: Hemisphere Publishing Corporation, 1989.

58. Taylor, A.M.K.P., *Combusting Flow Diagnostics*, Place: Kluwer Academic Publishers, 233, 1982.
59. Pitcher, G., G. Wigley, and M. Saffmem, "Sensitivity of Dropsize Measurements by Phase Doppler Anemometry to Refractive Index-Changes in Combusting Fuel Sprays," Fifth International Symposium on Applications of Laser Techniques to Fluid Mechanics, 227–247, 1990.
60. Ikeda, Y., et al., "Spray Formation of Air-Assist Injection for Two-Stroke Engine," SAE Paper 950271, 1995.
61. Ikeda, Y., et al., "Size-Classified Droplet Dynamics and Its Slip Velocity Variation of Air-Assist Injector Spray," SAE Paper 970632, 1997.

CHAPTER 14

Air Pollution from Large Two-Stroke Diesel Engines and Technologies to Control It

Svend Henningsen
MAN B&W Diesel A/S, R&D Department, Copenhagen, Denmark

ISBN: 0-12-639855-0
$25.00

14.1 INTRODUCTION

This chapter concerns air pollution and control technologies for large two-stroke diesel engines, as well as for marine and stationary power plant use. The development of low-emission engines historically has been driven by existing and proposed regulation of exhaust emissions. However, today also the general awareness of the environment forces the engine builder to offer low-emission engines in a competitive market.

Large two-stroke direct-injection diesel engines are discussed in this chapter. The engines are all turbocharged and charge air cooled (intercooler) and usually directly coupled to either a propeller or a generator. Because of the oil crisis during the 1970s, all engines are also highly fuel optimized with efficiencies above 50 percent. The cylinder size ranges from 26 cm to nearly 1 m in the bore diameter and the engine revolutions from 250 r/min for the small bore size to roughly 75 r/min for the large bore size, depending on the layout point for the engine. The power output for one engine ranges from 1.5 to nearly 70 MW, depending on the number of cylinders used (from 4 to 12 varying with engine bore size).

The first section will outline the present international and national regulations in order to support the scope for the past development. Mainly, the interest in nitrogen oxides (NO_X) emission has been in focus, whereas particulate matter (PM), carbon monoxide (CO), or unburned hydrocarbons (HC) up to now have been of interest only to the few power plant applications. This will change in the future, however, when the NO_X emissions are generally under control and the concern for the environment becomes more acceptable for the marine industry.

The second section presents a status of the different emissions. The section is based on specific research on the large engines over the past ten to fifteen years, where the discussion for the coming international regulation has taken place.

Due to the relative small market for these engines, compared to the automotive field, and the difficulty and expense of testing these engines, research and development are still lagging behind the automotive engines. On the other hand, fuel economy and reliability by far exceed the level for the automotive area.

Finally, the third section outlines the different technologies that exist for emission reduction, mainly for NO_x emission and with emphasis on the methods that are well proven. The reader should keep in mind that due to the specific applications for these large engines, the reduction methods used are very conservative in order not to jeopardize the reliability of the engine.

14.2 REGULATED EMISSIONS

As in all aspects of development, unless regulations are set for the secondary effects also, like the small amounts of unwanted components in the exhaust gases, these may not be regarded as interesting enough to be investigated. Today, however, the unwanted emissions represent a challenge to both engine designers and to makers of exhaust gas treatment equipment.

Several different sets of regulations are enforced for the large two-stroke engines because of the different uses of these engines and the strong international market where they are sold.

It is not the intention of this book to present the numerous legislations or agreements but, by showing some of the differences in the important regulations, to highlight the difficulty for the engine builders when they have to optimize engines for the different applications.

14.2.1 Marine Engines

Different subcommittees of the International Maritime Organization (IMO)'s Marine Environment Protection Committee (MEPC) have been charged to propose regulations for pollution from ships. The outcome of this effort is to be a proposal for a new Annex VI on air pollution from ships to the International Convention for the Prevention of Pollution from Ships (MARPOL 73/78).

Since the 1979 convention, discussions on declarations have progressed on the transboundary air pollution of sulfur dioxide (SO_2) and nitrogen oxides (NO_x) emissions from marine engines. See, for example, Resolution A.719(17) adopted on November 6, 1991 and by MEPC 30 common goals were adopted for the individual countries on reduction of NO_x and SO_x emissions with 30 percent and 50 percent, respectively, before the year 2000.

A regulation on NO_x is almost finished, as included in a draft technical code for consideration and approval at MEPC 39 [1]. The NO_x limit, a function of the rated engine speed, is based on a stipulated 30 percent reduction from the existing fleet in 1990. The dependency of speed reflects the high efficiency of the large engines by allowing a higher emission of NO_x at low engine speed ratings.

The proposed NO_x limit is shown in Figure 14.1. At rated speeds lower than 130 r/min, the limit is set at 17 g/kWh, and above a rated speed of 2000 r/min,

Fig. 14.1.

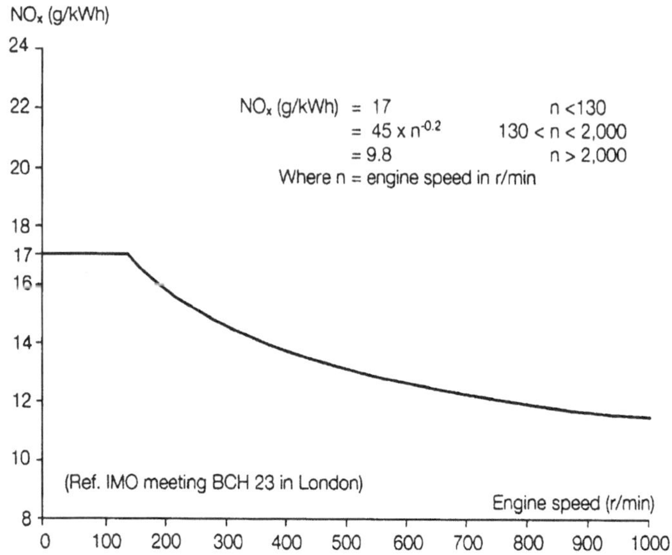

Proposed IMO NO_x emission limit [1].

the limit again is fixed now at 9.8 g/kWh. Between these two speeds, the limit is given with expression (14.1):

$$\begin{aligned} NO_x \text{ limit} &= 17(\text{g/kWh}) \text{ for } n <= 130 \\ NO_x \text{ limit} &= 45^*\text{n}^{**}(-0.2)(\text{g/kWh}) \\ NO_x \text{ limit} &= 9.8(\text{g/kWh}) \text{ for } n >= 2000 \end{aligned} \tag{14.1}$$

with n rated speed in (r/min).

Contrary to the NO_x regulation, the regulation on SO_2 emission still seems to divide the member states, depending on their involvement in the oil industry. The regulation was thought to act as a cap on the sulfur content in the fuel and be the easiest and cheapest way to limit the SO_2 emission from the ships. A cap from 3 percent to 5 percent has been suggested, but this will not prevent acid rain as a consequence of the marine traffic in certain coastal areas [2]. A definition of certain special areas that are especially fragile for acid rain from the SO_x content in the engine exhaust is included today in the proposal. Such area is, for example, the Baltic Sea. But the North Sea states have also proposed to declare the North Sea and adjacent waters a special area with regard to MARPOL 73/78 [3]. A sulfur cap from 0.5 to 1.5 has been discussed in this area.

Still individual states may introduce special, tougher regulations for engines that are operated entirely in their waters. This is the case with the Bodensee (Lake Constance), where the surrounding countries have agreed on a special regulation that has been in force since 1993. Also, the U.S. Environmental Protection Agency

Table 14.1

Proposed Regulation by the U.S. Environmental Protection Agency (EPA) for Marine Engines >560 kW, from Year 2000

Nitrogen Oxides (NO_x)	9.2 g/kWh
Hydrocarbons (HC)	1.3 g/kWh
Carbon Monoxide (CO)	11.4 g/kWh
Particulate matter (PM)	0.54 g/kWh
Smoke	20/50% opacity

EPA uses the ISO 8178 emission measurement standard, but a different cycle than IMO for marine engines. The standard may include corporate averaging.

(EPA) [4, 5] has announced a special regulation for marine engines on inland or territorial waters in line with the regulation for U.S. off-road diesel engines. Table 14.1 shows the proposed emission limits by the EPA for new marine diesel engines not covered or regulated by the IMO. A desire for harmonization between the EPA and the IMO is expressed by both sides, but how exactly it will be carried out is not yet clear.

The previous EPA/CARB proposal [6] gave certain incentives in the way of reduced harbor fees, if tough NO_x reductions were introduced on the engines. A similar incentive plan has been adopted by the Swedish authorities [7], where an agreement between owners and port authorities will restrict NO_x to 2 g/kWh and sulfur content in the fuel to 0.05 percent for all inland ferry traffic. In the future, we may see more and more of this type of collaboration in order to restrict emissions on a partly voluntary basis.

14.2.2 Stationary Engines

Contrary to the marine sector, power plants and special industries have been regulated for many years. But the increasing awareness of the environment has tightened most of the existing regulation. As the reduction technologies evolve, the economical penalties become smaller allowing the governments or funding banks to require the regulations to be met.

The World Bank (WB) guidelines [8] are frequently set as the standard for power plants in developing countries. The requirements depend on the state of technology in the country in question, and on an assessment of the amount of pollution for a sustainable environment. The WB requirements are historically closely in line with the U.S. EPA regulation [5]. A new proposal [9], recently published by the WB for comments, follows this line by adopting the EPA regulation set for small engines without regarding the quoted harmonization plans with IMO for large engines operated on heavy fuel oils.

14.3 EXHAUST EMISSIONS

The exhaust emissions from an internal combustion engine depend on the combustion process, the fuel used, the intake and engine conditions, and possible control technologies that are used for the engine. The major components are nitrogen (N_2) and oxygen (O_2) from the atmosphere and carbon dioxide (CO_2) and water (H_2O) from the combustion products when burning a hydrocarbon-based fuel. The incomplete combustion products constitute only a minor fraction of the exhaust gas. A typical composition of the major pollutants and input media is shown in Figure 14.2.

In the past, common usage regarded the incomplete combustion products as pollution only. But, since the discussion on global warming has heated up, also CO_2 is looked upon as an important component in the exhaust. For the common hydrocarbon-type fuels, the CO_2 content is related mainly to the efficiency of the combustion process [10, 11]. Whereas, for different fuel types, the carbon content will have an important effect on the CO_2 emission [12]. Figure 14.3 shows a four-stroke diesel comparison of the exhaust CO_2 amount from different types of fuels compared to a gas turbine [10, 12]. It should be kept in mind that the CO_2 from processing of the "clean" fuels should be added to the greenhouse effect [13] (the cradle-to-grave effect).

Fig. 14.2.

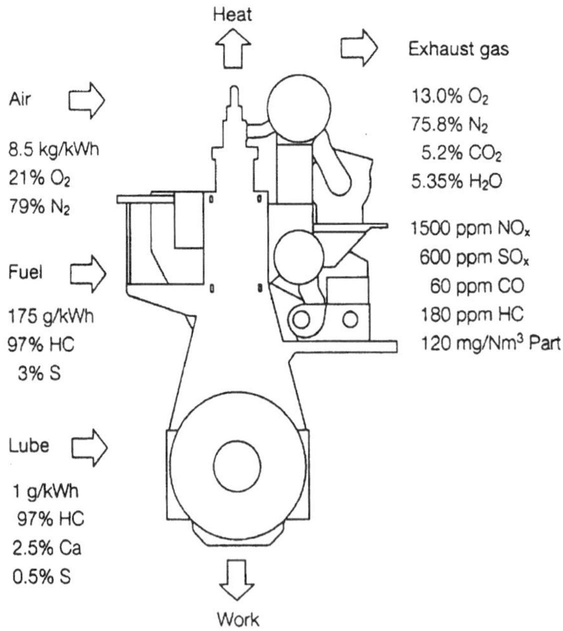

Typical emissions from an MC-type low-speed diesel engine [18].

Fig. 14.3.

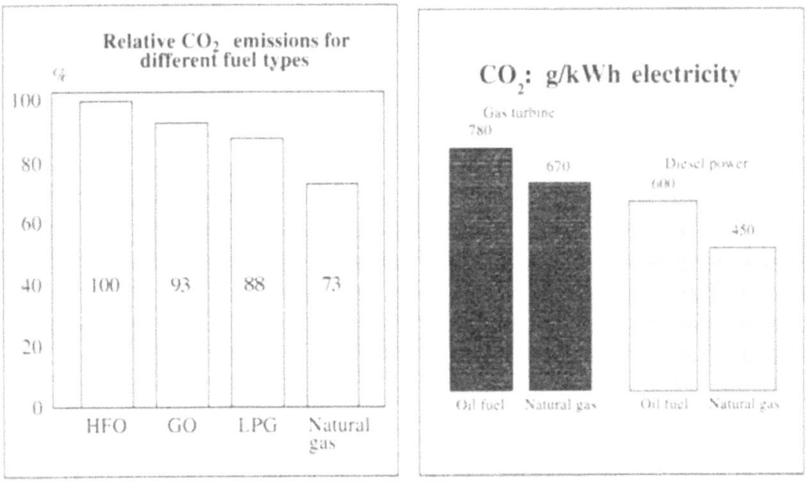

Comparison of CO_2 emissions levels from different fuels in a four-stroke engine [12] and between a four-stroke diesel and a gas turbine [10].

Due to the differences in the cylinder scavenging process, a certain reference is introduced for either O_2 or CO_2 content in the exhaust gas. In Europe, the most common reference is 15 percent O_2, related closely to an average O_2 concentration in the engine exhaust [14]. In Japan 13 percent O_2 is seen as a reference often, whereas for power plants, values related to coal-fired plants still take precedence in, for example, the German regulation (TA Luft [15]) with 5 percent O_2. Equation (14.2) presents the correction from measured to reference O_2 condition but also allows for conversion between the different references.

$$E\text{ref} = E\text{meas}^*(20.95 - \text{O}_2\text{ref})/(20.95 - \text{O}_2\text{meas}) \qquad (14.2)$$

where E is the selected emission component

The earlier Chapters of this handbook have described the basic emission formation mechanisms from combustion- or design-related sources (or see, for example, referenced textbooks [16, 17]). This chapter quantifies some of the emissions from large two-stroke engines and some of the important sources that may be candidates for the later reduction techniques.

14.3.1 NO_x Emission

The majority of the NO_x emission originates from the high-temperature reaction of atmospheric nitrogen with oxygen present during the combustion process. A secondary contribution arises from the fuel-bound nitrogen content. A rough assumption estimates a 100 percent conversion of the fuel-bound nitrogen, although a full conversion is not certain.

The combustion process produces almost entirely NO, which during the expansion process and in the exhaust system oxidizes further to NO_2. Depending on the retention time in the exhaust system (affected by the volume of the exhaust system) and the temperature of the exhaust gas, NO typically is converted to NO_2 in the order of 5 percent to 7 percent of the NO amount. The emitted NO continues to oxidize to NO_2 in the atmosphere, giving the exhaust plume a characteristic yellow-brownish color.

Depending on spray formation and mixing, individual combustion temperature distributions will be obtained in the combustion chamber for each engine, and because of the strong temperature influence on the NO_x formation, different NO_x characteristics will result. Furthermore, as far as the used fuels cause different combustion temperatures, also the fuel will be responsible for variation in the NO_x emission. In this connection, the family of different heavy-fuel oils or diesel fuels usually do not provide sufficient differences (except for the bound-N effect, see later). But, in case of gaseous or alcohol-type fuels, not only the adiabatic flame temperature but also the mixing process may be very different.

Most measurements on large engines are not performed using specified "calibration" fuel due to the extreme expense and the lack of availability of these fuels. For stationary engines (or marine engines in service), the use of site fuel is usually required in the engine or plant contract.

In order to correct the on-board measurements of NO_x emission from fuel influences, two approaches were proposed during the discussions for the IMO correspondence groups. A full adjustment based on the 100 percent conversion of the fuel-bound nitrogen by Equation (14.3):

$$\text{Bound NO}_x = (\text{Nfuel\%} - \text{Nref\%})/100\%^{*}\text{sfoc}^{*}46/14 \tag{14.3}$$

where: Nref% is percent bound N from test-bed measurements
sfoc is the specific fuel-oil consumption (g/kWh)
46/14 is mol weight ratio between NO_2 and N

or an assumption of a fixed contribution for each percent nitrogen in the fuel, without the knowledge of the bound nitrogen in the original test-bed measurements (14.4):

$$\text{Bound NO}_x = \text{Nfuel\%}^{*}\ 0.35\ (\text{g/kWh}) \tag{14.4}$$

The IMO technical code proposal [1] grants a 10 percent NO_x allowance for on-board measurements due to the possible influence of the heavy-fuel oil independent of the bound-nitrogen content.

Figure 14.4 shows typical NO_x versus engine-load dependencies for several different engine types. As indicated on the upper plot, when the engine is operated on a propeller curve, a maximum will be found in the 75 percent to 90 percent load area depending on the layout for the engine. The NO_x value will increase again at low-load operation due to the correction to 15 percent O_2 content in the exhaust.

Fig. 14.4.

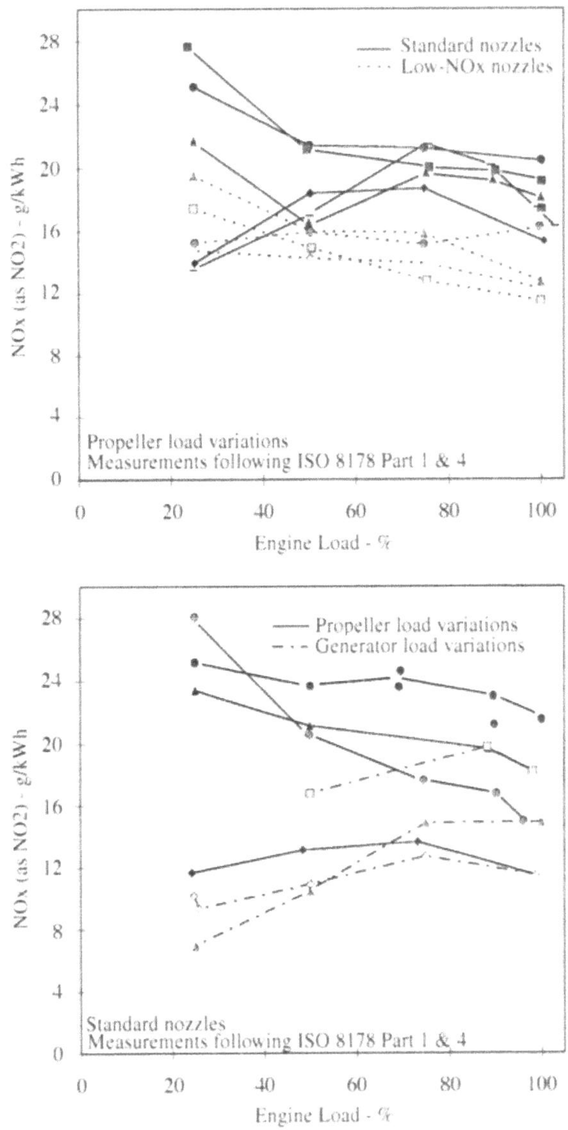

Typical NO_x emissions versus engine load behavior for different engines.

Fig. 14.5.

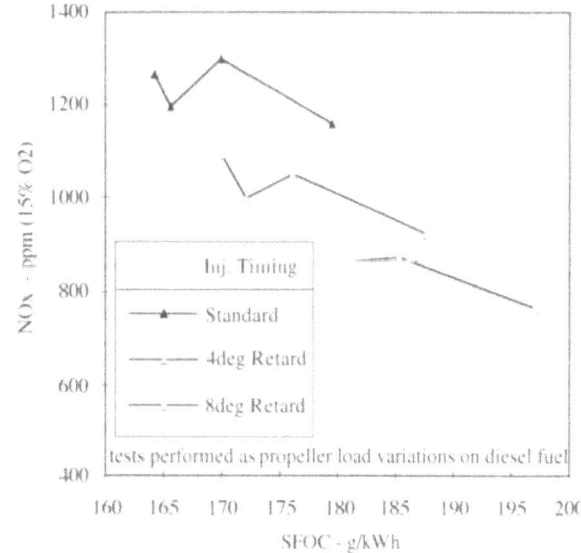

Influence of injection timing on NO_x and SFOC [56].

For an engine operated on the generator curve (constant speed), only the 100 percent load point will be the same as for the propeller curve, whereas the low-load behavior is different. A generator NO_X curve, as seen on the lower plot, decreases faster toward low loads and continues to decrease at low load.

Figure 14.5 shows typical NO_X versus injection timing dependencies [18, 19]. As indicated, late injection is an easy method to lower the NO_X emission from a standard engine. But as will be seen later, the trade-off between NO_X and specific fuel consumption (sfoc) and between NO_X and several different emission parameters (HC, PM, and CO) shows an increase in these emissions in exchange for the low NO_X emission.

14.3.2 HC Emission

As described for the small four-stroke DI engines, the hydrocarbon (HC) emission has several sources. The original two-stroke problem of scavenging unburned fuel through the cylinder was a typical premixed-type engine problem, which is nonexistent in today's direct-injection engines.

Unburned lube oil from cylinder lubrication (for large two-stroke engines direct lubrication of the liner is performed through special oil-feed holes) or from valve stem leakage is another major contributor to HC or particulate emission. In cross-head type engines, the piston-rod stuffing box may cause system oil to leak

Fig. 14.6.

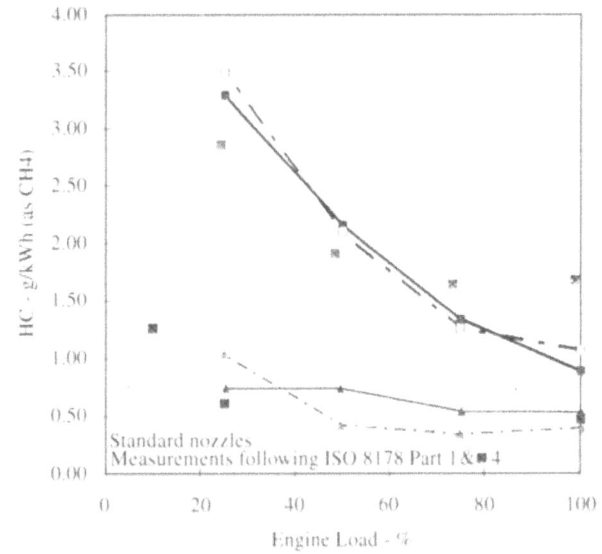

Typical HC emission versus engine load behavior for different engines.

into the scavenging box and further on into the cylinder during the scavenging process.

The nozzle sac has been known as a major source of HC emission from small engines [16]. This is also the case in large engines. Still, only a few power plant engines are operated with special mini-sac fuel nozzles to reduce this source, mainly because of the difficult production of the large mini-sac nozzles for the large engines. A later section (14.5.1) will report on the benefits of using these nozzles.

When comparing HC emissions from standard diesel oil (DO) with heavy-fuel oil (HFO), the HFO-operated engines usually show less HC emission [20, 21]. However, it should be kept in mind that the measuring method (the heated flame ionization analyzer and sampling hose) with the prescribed temperature of 180°C is developed for standard diesel fuel and not for HFO. The high boiling end fractions of the HFO, therefore, may condense on the particulates. Consequently, HC should be viewed together with the particulates in case of operation on HFO.

Figure 14.6 shows standard HC emission performance as a function of load for several engines. HC decreases with increasing load, similar to the behavior for small DI engines.

14.3.3 CO Emission

The formation of carbon monoxide (CO) is mainly a function of the excess-air ratio [16]. The formation is strongly affected by the local conditions in the combustion

Fig. 14.7.

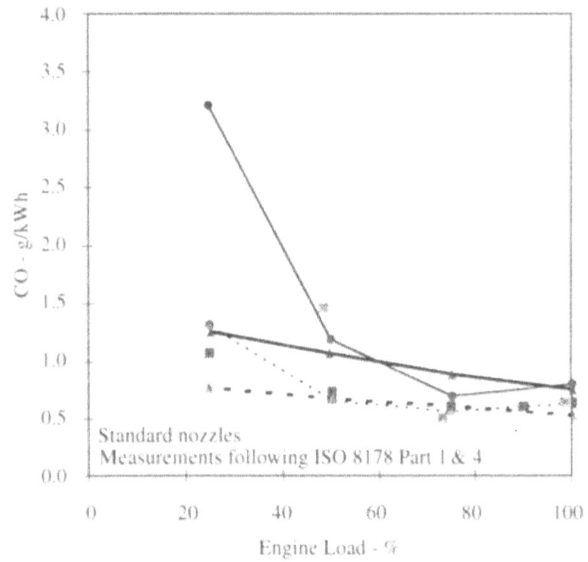

Typical CO emission versus engine load behavior for different engines.

chamber. Therefore, a good air-fuel mixing process and excess air, which usually is available in the large turbocharged two-stroke engines, is the basis for minimal CO emission.

Figure 14.7 shows typical CO emission performance as function of load for several engines. The CO emission decreases rapidly with increased engine load and above approximately 50 percent load becomes almost insensitive with load. Except for nozzle layout cases, where the individual sprays interfere either with each other or with the wall at higher loads, CO remains at a very low level.

14.3.4 SO_x Emission

Because of the organic origin of the standard hydrocarbon fuel oils, various amounts of sulfur are present in the fuel and injected into the combustion chamber. During the combustion process, the sulfur is oxidized to different sulfur oxides (SO_x), mainly SO_2 and SO_3. Therefore, the SO_x emission from the engine is a function of the fuel-sulfur content. Furthermore, as a result of the sulfur content in the lube oil, a minor contribution to the exhaust SO_x is caused by this sulfur. However, a part of the gaseous SO_x will condense on the particles in the exhaust and further attach a certain amount of water (see particulates later).

SO_2 and SO_3 will condense as sulfuric acid not only inside the combustion chamber on the cold cylinder walls [22, 23, 24] but also throughout the exhaust system, wherever the dewpoint is passed, because of the cold environments.

Therefore, to protect the engine from corrosion, alkalic additives are added to the cylinder lube oil in an attempt to neutralize the acid [25].

SO_x can be controlled either by removing the sulfur from the fuel or by removing the SO_x from the exhaust gas by different cleaning methods as will be discussed later.

14.3.5 Particulate Matter

The particulate matter in the exhaust gas consists of several different components. Not only the solid-phase material but also liquid material condensed on the originating nuclei-soot material make up the particulates. During the combustion process, soot is formed through a hydrocarbon cracking process and later consumed by oxidation of these products [16, 17, 26]. Attached to the soot conglomerates is the ash content consisting of several metal oxides. During the expansion process and later in the exhaust system (or the atmosphere), different hydrocarbons and metal oxides continue to condense on the particles, forming the final particulates.

As a result of different measurement procedures to collect the particulate matter and the conditions for these measurements, different amounts and also different materials are collected. Table 14.2 defines some of the commonly used methods and presents a short explanation of conditions and collected material.

In an attempt to obtain the actual particles that affect the environment (including humans), the ISO 8178 procedure [27, 28] has adopted a dilution method for measuring the particulate mass (PM) [28], which simulates the actual dilution process in the atmosphere. This method dilutes the exhaust and collects the material on a filter kept at a maximum temperature of 51°C. Therefore, all the different species defined as *particulates* are included by the measurement procedure. The major fractions are a solid carbon structure with different adsorbed or absorbed hydrocarbons, called SOF (soluble organic fraction), ash components from fuel and lube oil, sulfates, and attached water [29–34]. Because of the high sulfur content in the heavy-fuel oil (up to 5 percent), a considerable fraction of particulates will be sulfates and attached water.

Due to the uncertainty of the effects of the different dilution tunnel designs, and the complications of the sulfur content for the method, ISO recommends the use of one of the "hot-filter" methods for sulfur contents larger than 0.8 percent.

For power plants, historically, a hot-filter method [35–37] has been used. But, as discussed earlier, since the standards for this method specify different filter-sample temperatures, also different amounts and types of material (the condensed SOF) may be collected on the filters. Also, during the subsequent filter handling, different material may be kept on the filters and, therefore, regarded as particulate matter. The particles from the hot-filter method are usually referred to as *dust*.

Several methods to measure *soot* rely on a visual comparison of the degree of blackening of a white filter paper, where a certain volume has been drawn through [38–39]. The methods are either based on subjective judgment or on an assumption of a similar behavior of the fuel that was used. Needless to say, the

Table 14.2

Particulate Measurement Methods

Definition	Standard	Material	Conditions
Particulates	ISO 8178 [27]	carbon, SOF ash, SO_4, H_2O	multihole probe dilution 10–50(?) times, filter temperature maximum 51°C
Dust*	EPA method 5 EPA method 5B EPA method 17 [35] VDI 2066 [36] JIS Z8808 [37]	carbon, SOF ash SO_4 (partial) as EPA method 17 as EPA method 17	filter 120°C filter 160°C in-stack filter
Soot	Bosch [38] Bacharach [39]	not relevant but influences the reading	blackening of white filter paper
Opacity**	Hartridge [39] ISO 8178 [40] SAE J1667	not relevant but influences the reading	in-stack full flow or partial flow (outside stack) on-highway method
Visual	Ringelmann [41]	not relevant	sky background (subjective method)

*All dust methods use isokinetic sampling, adopted from early boiler-type power plants, in certain defined cross-sectional in-stack positions. For diesel engines isokinetic sampling is an illusion. Due to the different temperatures various amounts of SO_4 and SOF are collected.

**Necessary for transient measurements.

result may be quite erroneous as a measure for particulate matter unless a standard diesel fuel is used. The main advantage of the method is the ease of use.

Nevertheless, numerous correlations have been performed in order to compare one method with another [39]. But unless all the foregoing complications are considered, the quantitative result may be meaningless.

Finally, an optical method that measures the *opacity* exists. The opacity is the fraction of light (in percent) that is prevented from reaching the observer (or instrument receiver) when a source is transmitting light through a certain path length of exhaust gas. Different methods exist, either measured directly in the exhaust gas stream or in exhaust gas pumped through a certain calibrated device [40]. The method may be sensitive to certain gaseous molecules depending on the wavelength that is used or corrections may be introduced depending on the amount of these components in the exhaust.

The advantage of the opacity method is the fast and direct response to the gas condition in the exhaust pipe. The method, therefore, is the only usable method for the transient smoke measurements for which a proposal is being prepared within ISO to replace the earlier EU and EPA transient smoke measurements [42].

As described earlier, the total PM is a highly diverse material, and this material originates from different sources in the engine. Therefore, in order to be able to control (or reduce) PM, a detailed analysis of this material is necessary. Specifically, it is important to separate between fuel- and lube-related sources to address possible improvements most efficiently. Based on the small-engine experience, using dilution tunnels and engines operated on DO, several methods have been refined to make a breakdown of this matter [43–48]. Unfortunately, because of the differences between DO and HFO, some of these methods are not easily available on HFO.

One example is the vacuum-oven sublimation method that was proposed to simplify the standard Soxhlet analysis of soluble-organic material (SOF) [46]. The result, usually referred to as the volatile-organic fraction (VOF), is very similar to the soluble fraction from the Soxhlet method. But it is necessary to analyze for sulfates also after the process, since part of the sulfates will evaporate and therefore be counted as VOF fraction. The problem is accentuated by the usual high sulfur content in HFO. Another example is the GC method to separate between fuel and lube-oil fractions of the SOF [43–44, 49]. Since the distillation curves for HFO are very similar to the curves for lube oil, a calibration blend is not possible. Special tracer methods are therefore necessary [45, 48].

Today, total PM from several large engines operated on HFO as well as on DO have been obtained. An example of the fuel influence on the PM-analysis breakdown is shown in Figure 14.8, where one cylinder of a four-cylinder test engine (a 4T50MX [18]) was operated on four different fuels. Fuel DO is a

Fig. 14.8.

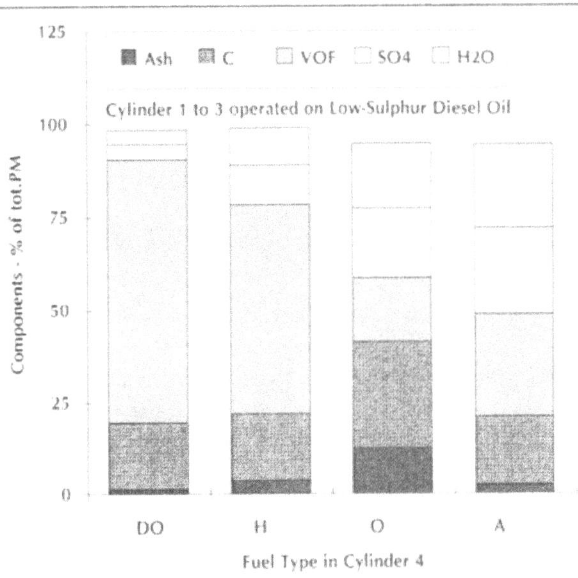

Particulate-component fractions on four different fuels at 100 percent engine load.

Fig. 14.9.

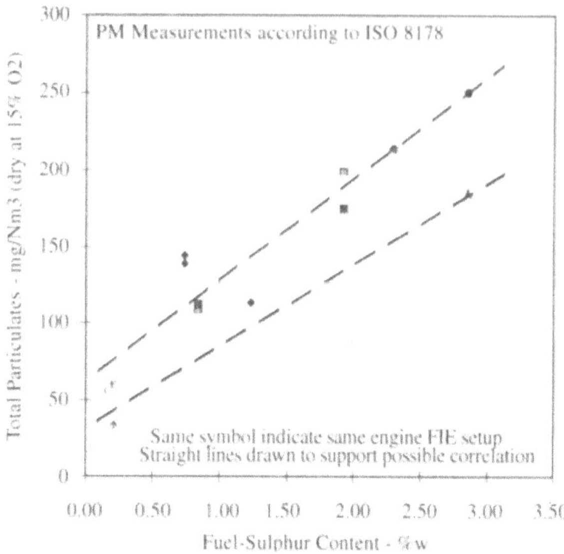

Influence of fuel-sulfur content on total PM emission.

0.05%w low-sulfur diesel fuel, which also was used as reference fuel on the three remaining cylinders.

As Figure 14.8 shows, the sulfates (and the attached water, usually approximately 1 g H_2O per 1 g H2SO4 [33]) represents a very large fraction of the PM for engines operated on HFO. To illustrate this further, Figure 14.9 shows a plot of PM as function of the fuel-sulfur content based on the experience from the several different engines and fuels. Total PM increases almost linearly with increasing sulfur content, as has been reported for several small engines. Of course, the PM levels are different for different engines due to other PM sources. An estimate of the PM amount from these sources for the individual engine can be obtained by extrapolating to zero-sulfur content.

The fuel-sulfur conversion rate (the measured PM sulfur divided by the fuel-sulfur content), found to approximately 5 percent [33], actually may increase with increasing engine load. A new study [34] indicated an apparent conversion rate between 1 percent to 8 percent, not accounting for the possibly increasing SOF fraction due to the sulfate content [50]. This compares to the 1 percent to 3 percent conversion rate reported for several small engines, but not all investigations have found a load-varying effect.

Part of the SOF fraction is polycyclic aromatic hydrocarbons (or PAH), of which certain compounds are related to mutagenic activity and therefore of specific interest. Several studies have been performed at different universities on small engines and using DO trying to address the effect of different fuel specifications and lube oils on the individual PAH compounds, see, for example, [49, 51–52]. The major part of the PAH is believed to originate from the fuel, even though,

Table 14.3

Organic Micropollutants

Engine		S5 [53]	1L42MC
PAH	$\mu g/Nm^3$	13–26*	50–100
nitro-PAH	$\mu g/Nm^3$	0.74–3.56	<0.04

*With phenanthrene, fluoranthene, pyrene, and chrysene being the major constituents.

Table 14.4

Fuel and Lube Oil Effects on Exhaust Emission

Effects on: of:	NO_x	SO_x	HC	CO	CO_2	PM	Smoke	SFOC
Fuel	***	***	**	*	**	***	**	*
Lube oil	*	*	**	*	*	(*)*	(*)*	*
Engine	***	*	**	*	(*)*	(*)*	***	***

***Indicates strong effect; **some effect; *little effect.

for small four-stroke engines, this may be added to the particulates from the oil sump [49, 52]. But, since both the measurement technique and the analysis of the compounds are very demanding (the amounts are usually measured in $\mu g/Nm^3$), only a few programs on large two-stroke engines and HFO have been conducted [53]. A comparison for two engines is shown in Table 14.3.

It is difficult to estimate the amount of particulates that will be formed during the combustion process. The volatile hydrocarbons (VOF) that are attached to the particles most likely relate to the HC emission and the ash content correlates to the amount of additives or ash in fuel and lube oil. Furthermore, it is likely that the amount of carbon can be expected to be a function of the fuel quality, even though a good correlation has not been found. The carbon-forming compounds are normally assumed to be expressed by the Conradson carbon residue (CCR) analysis or related to the asphaltene content. Most likely, several factors interfere with the assumption. Finally, the influence of the sulfur content on the particulates has already been discussed previously. Still the literature presents several correlations that may be useful as a starting point [30–32, 54–55].

14.3.6 Fuel and Lube-Oil Effects

From the foregoing discussion, it is clear that the fuel used in many cases has a dramatic effect on several of the emission components. Table 14.4 summarizes the effects on the different emission components and on SFOC [56–57].

The influence of fuel differences on emission values for an actual large two-stroke engine are illustrated in Table 14.5 from the study [20–21], given fuel specifications and measured emissions. The NO_x emission varies between 1460 and 1765 ppm (dry at 15 percent O_2) for the used HFOs. Neither ignition delay nor fuel-bound nitrogen seems to correlate with the NO_x emission. A clear difference in HC and dust is seen between the values for HFOs and DO (fuel #1 in Table 14.5). DO has twice the HC emission but only between 10 percent and 50 percent of the dust depending on the selected HFO. The Conradson carbon (CC) value as the asphalt content has an influence as well on the dust amount as assumed.

"Clean" fuels, like liquid natural gas (LNG), compressed natural gas (CNG), alcohols (methanol or ethanol), or dimethyl ether (DME) can of course be used in the large two-stroke diesel engines in multifuel or gas versions (see CNG in Section 14.4.1.3). But due to the expense of these fuels compared to the heavy-fuel oils, the actual interest is very low except under special circumstances.

14.4

EXHAUST EMISSION CONTROL TECHNOLOGIES—NO_x REDUCTION TECHNIQUES

The technologies or methods that are used to reduce the exhaust emissions are either direct engine-related methods, called primary methods, since they affect the emissions at the source of formation, or aftertreatment methods, or secondary methods, that reduce the emissions (clean the exhaust gas) after they leave the engine.

From the discussions on air pollution in the previous section, all parameters that influence the emission components can be used in the search for control technologies. Table 14.6 summarizes the parameters that affect the different emission components. The table includes parameters that primarily were meant for small four-stroke engines, since many of these for several years have been used on large two-stroke engines due to the effect they have on efficiency and power. Table 14.6 also shows the effect of the individual parameters on fuel consumption, since, in most cases, the emissions are in strong competition with the fuel consumption when optimizing the engine.

All emission components are included in the table, but only some of the NO_x reduction techniques will be discussed in this section, whereas the other emissions are discussed in the next section. Because of the stronger emphasis that has been placed on the NO_x emission up to now, the methods for NO_x reduction are more elaborate and the discussion in this section will include not only engine-adjustment methods and fuel-caused emission reduction but also some aftertreatment methods.

All parameters that affect NO_x emission are in the future IMO regulation required documented in the Technical File on board [1]. The Technical File shall include an analysis of the sensitivity of each parameter to allow a decision on emission limit compliance.

Table 14.5

Fuel Specification and Emissions for 1L42MC Engine [20, 21]

Fuel No.	1	2	3	4	5	6	7	8	9	10	11	12	13	Guiding specs.	Units
Viscosity	2.27	3.8	84	85	141	198	255	470	520	560	690	710	240	700	cSt/50C
Density	843	968	995	970	993	938	977	985	983	1010	1008	1030	988	991	kg/m^3 at 15C
Flash P.	65	98	84	80	103	100	106	90	95	90	79	84.90	60	deg.C	
LCV	42.6	40.7	39.8	40.7	40.3	41.4	40.8	40.4	40.4	39.7	39.5	38.9	39.7	–	MJ/kg
CC	0.01	0.3	17.2	12.1	13.3	9.4	14.5	16.8	14.8	17.3	22.1	24.7	14.9	22	% weight
Asphalt	0.00	0.78	15.1	8.9	9.2	3.7	10.0	11.3	12.8	14.6	19.3	29.0	10.4	14	% weight
Sulfur	0.22	0.10	2.72	1.16	0.91	0.83	0.87	0.90	1.18	2.22	3.52	3.30	3.17	5.0	% weight
Nitrogen	0.015	0.20	0.30	0.30	0.30	0.31	0.47	0.30	0.42	0.67	0.40	0.54	–	–	% weight
Water	0.00	0.01	0.01	0.01	0.00	0.01	0.02	0.02	0.01	0.00	0.00	0.00	0.00	1.0	% weight
Ash	0.00	0.00	0.065	0.025	0.03	0.03	0.025	0.03	0.035	0.04	0.07	0.09	0.055	0.2	% weight
Aluminum	–	–	–	–	–	–	–	–	–	–	–	–	30		mg/kg
Vanadium	0	0	220	20	23	12	17	24	45	122	300	370	148	600	mg/kg
Sodium	0	0	27	23	24	25	40	35	22	22	24	50	28	30% of V	mg/kg
CCAI	805	912	874	849	866	807	843	844	841	868	864	885	853		–
Ign.delay	1.7	1.3	1.4	1.6	1.4	1.4	1.3	1.4	1.4	1.8	1.4	1.5	1.6		deg.CA
pres.rise	3.3	3.7	2.6	2.8	3.0	2.5	3.3	3.0	2.9	2.6	3.0	2.5	3.3		bar/deg.CA
NO_x MCR	1660	1690	1700	1710	1660	1460	1765	1615	1655	1590	1710	1490	1500		ppm dry
CO MCR	16	18	21	28	28	20	34	26	26	40	26	27	26		ppm dry
HC MCR	94	98	42	51	47	45	34	22	44	52	52	47	41		ppm wet
Dust MCR	7	10	45	17	21	20	24	27	31	46	67	72	46		mg/Nm^3
Dust 25%	7	–	–	–	–	18	–	–	21	85	–	64		–	mg/Nm^3

All emissions corrected to 15% O_2.

Table 14.6

Emission Reduction Techniques

	Influenced parameter				
Technique	NO_X	HC	PM	Smoke	SFOC
Primary methods:					
Turbocharging	**		***	***	**
Intercooling	***				**
Compression ratio		**		*	**
Combustion-chamber design		**	**	**	
Swirl ratio			**	**	*
Nozzle layout	***	**	**	**	**
Injection pressure	*	**	**	**	
Injection timing	***	**	**	**	***
Electronic controls	**	**	**	**	*
Lube-oil control		**	**	**	
Reformulated fuels	*	*	***	**	
Water emulsion	***	*	**	*	**
Water or ammonia injection	***				
Exhaust gas recirculation	***	*	**	*	
Secondary methods:					
Aftertreatment	***	***	***	***	*

***Indicates strong effect; **some effect; *little effect.

14.4.1 Engine Adjustment Methods

14.4.1.1 Injection Timing

The injection timing, mentioned earlier, has been an integral part of engine optimization from very early engine history. Figure 14.10 shows, from the injection-timing investigation [19], changes in NO_X, Bosch number, and specific fuel-oil consumption (SFOC) as function of injection-timing delay in degrees crank angle (deg. CA). At both load conditions the NO_X emission is reduced, when delaying the injection timing from the original fuel-optimized injection timing, similar to the common behavior of small engines. But importantly, at the same time, both fuel consumption and Bosch number increase.

The injection timing has two effects on NO_X formation. When the combustion process is delayed, the rate of heat release is changed slightly, but more importantly, the maximum pressure and thereby the cylinder bulk temperature is changed during the combustion process. The delayed "end of combustion" may affect different emission components, whereas NO_X is related more closely to the early combustion phase, around TDC, where the temperatures are high.

For all large two-stroke engines the injection timing is used to adjust the maximum cylinder pressure to the designed maximum pressure in order to obtain optimum fuel consumption. Therefore, the NO_X emission for a specific engine

Fig. 14.10.

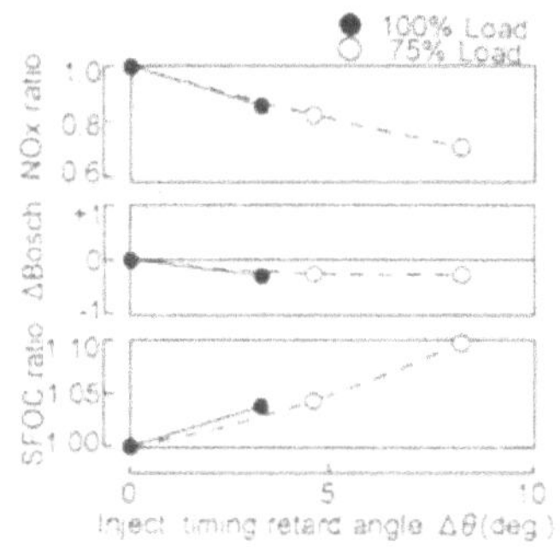

Results of injection timing retard test [19]: relative changes in SFOC, Bosch number, and NO_x with timing retard (deg. CA).

Fig. 14.11.

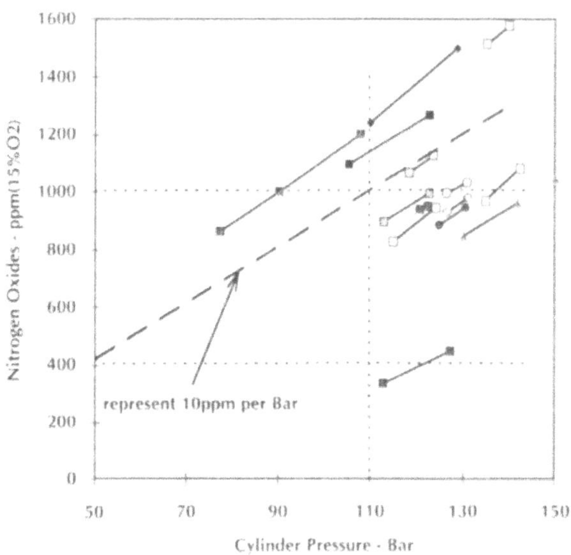

Influence of maximum cylinder pressure on NO_x emissions.

type will be closely related to the maximum pressure. The result from such a pressure adjustment on several different engines is shown in a correlation between the maximum cylinder pressure (or temperature) and the NO_x emission in Figure 14.11. As indicated, a rule of thumb can be deducted from the data, stating a 10 percent NO_x decrease for each 10 bar reduced maximum pressure. It can also be deduced from the large spread between the individual engine data that the cylinder pressure is not the "real" independent parameter for NO_x formation. But still, the pressure is a parameter that similarly to the NO_x is affected by the combustion temperature.

A constant pressure process may be an advantage for low NO_x formation, but the concept can only be accomplished by changing different parameters simultaneously.

14.4.1.2 Scavenging Temperature and Pressure, Compression Ratio

Similar to the temperature effect of the injection timing, also the simultaneous temperature effect of changing the scavenging air temperature or pressure, or the compression ratio will affect the NO_x formation. Intercooling is generally used on all large two-stroke engines, and because of the beneficial effect on the fuel consumption (about 1 percent decrease per 15°C scavenging air temperature decrease), the scavenging air should be cooled as much as possible. But the practical potential is limited by the available sea water temperature.

Different four-stroke developments indicate a substantial NO_x reduction potential, from 50 percent to 60 percent [10, 58], usually by involving several different simultaneous modifications of the engine design. A trade-off between higher compression ratios and faster injection rates has been ongoing for several years [59].

14.4.1.3 Fuel Spray and Air Swirl

The combination of the injected spray and the air motion in the combustion chamber has been the focus of small DI combustion research for many years. However, the situation is different for the large two-stroke engines, where several injection valves may be used in the periphery of the cylinder cover to allow for a single central exhaust valve. (A strong axial symmetry is always preferred in highly thermal-loaded engines to avoid heat-introduced stresses.)

To illustrate this, Figure 14.12 shows a comparison between the temperature distribution in one plane of the cylinder cross section caused by different air-excess ratios from a CFD calculation for an engine with either two or three fuel-injection valves.

The layout of the spray pattern of the injector nozzles is an easy method to affect the NO_x formation [18–19, 56–57, 60]. Use of a special "low-NO_x" nozzle design offers from 10 percent to 25 percent NO_x reduction depending on the engine type, but the method is accompanied by a certain fuel penalty.

Furthermore, it is necessary to verify heat release as well as heat load on the combustion chamber components to ensure a minimal fuel-consumption penalty and component temperature increases.

Table 14.7 shows a comparison of several different engine types of the ISO-E3 averaged NO_x values for a standard economy-optimized engine and a low-NO_x nozzle-optimized engine. The study was performed in order to investigate the possibility of meeting the coming IMO regulation without other NO_x reduction means. An actual optimization may still improve the performance results.

Fig. 14.12.

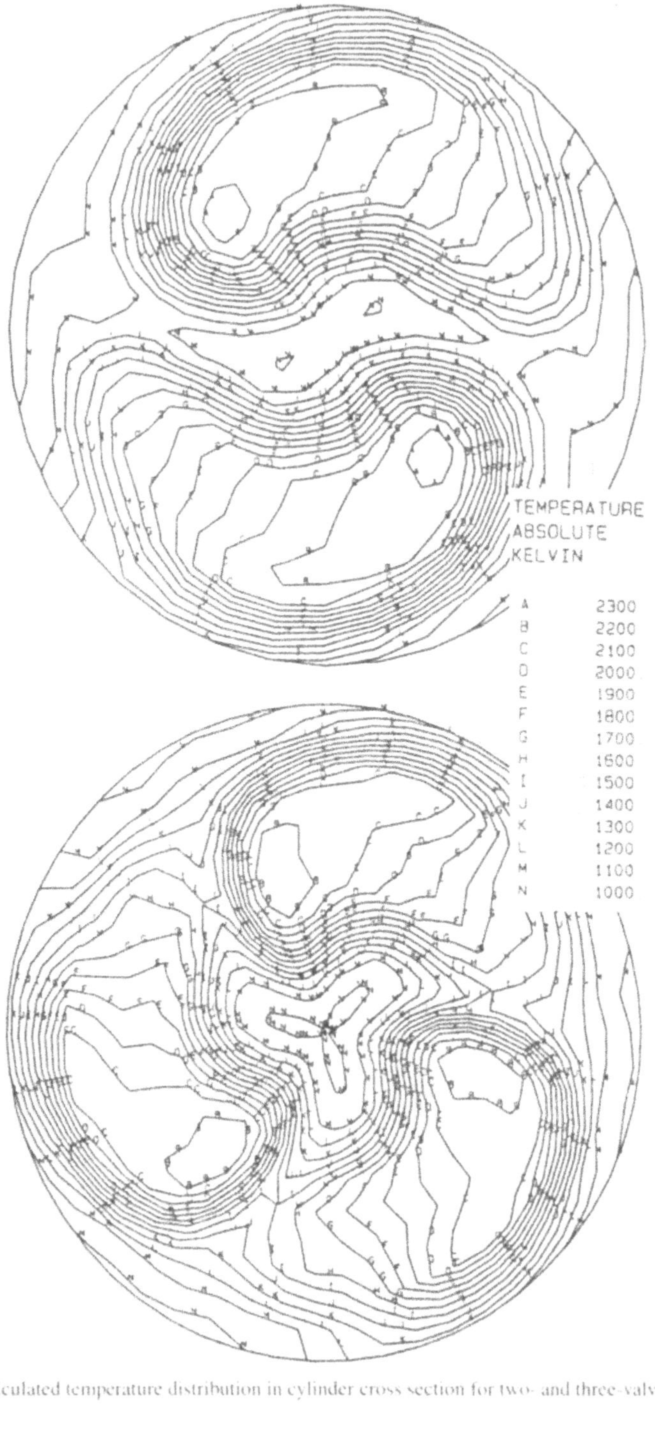

Calculated temperature distribution in cylinder cross section for two- and three-valve layout.

Table 14.7

Measured NO_x Emissions Compared to the IMO Limit

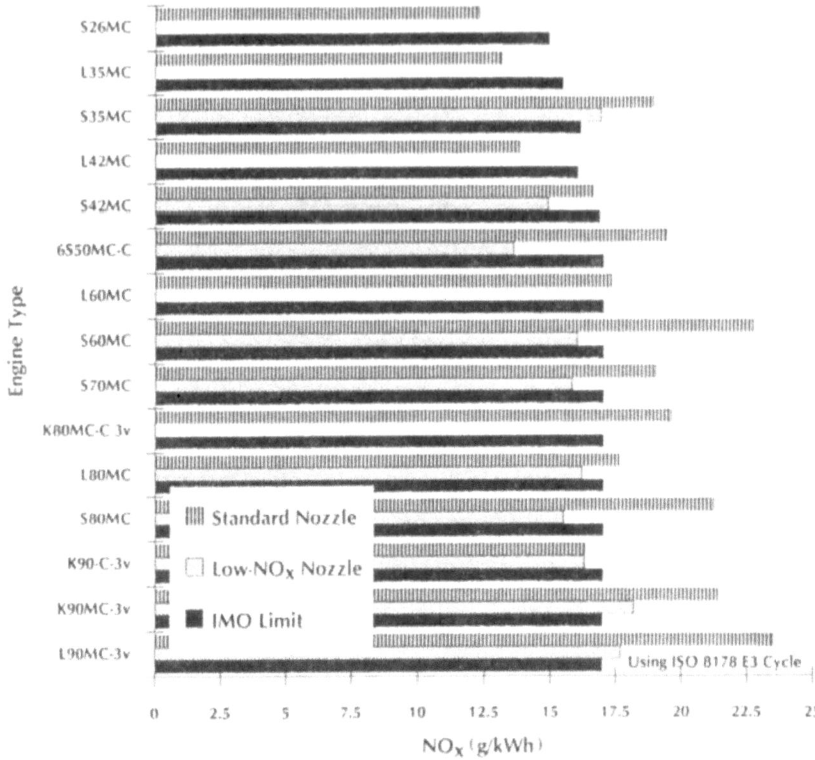

All two-stroke, uniflow, scavenged engines have a certain degree of swirl, though low, by small DI standards. The amount of swirl has been optimized for best fuel consumption, which also ensures the proper scavenging of the used exhaust gas. An overly high swirl will inevitably cause overly high heat transfer and a less efficient scavenging process. The injection process is fully capable (in itself) of creating the proper mix of fuel and air.

14.4.1.4 Injection Pressure and Rate Shaping

The fuel-injection equipment (FIE) has for many years been the key to improvements in diesel engine performance. Increasing injection pressures (up to 2000 bar) and electronic control of the opening and closing of the injector to optimize the speed-load requirements have revolutionized the obtainable emission requirements. The injection pressure for the large engines, though, is far from reaching the high levels necessary for the small engines [59, 61]. Still, an optimal distribution of fuel and air is always a goal when reducing NO_x emission. The spray penetration and the spray angles need careful attention.

Different advanced fuel valves, some combined with electronic control, are used on small DI engines to further improve on emission and noise performance. Injection-rate shaping or even split injection may also be in the future for large engines [18, 57]. A four-cylinder research engine (a 4T50MX) with a computer-controlled electronic fuel-injection system is used to investigate the possibilities for the large two-stroke engine. Figure 14.13 shows fuel-valve injection pressure and valve lift from a comparison between three different injection profiles at 75 percent load: a standard (slow-rising) injection profile, a "basic" electronic profile (with a very fast initial injection-rate rise followed by an almost square profile), and a pilot-injection pattern. The figure also shows the corresponding calculated heat release and the NO_x and fuel-consumption performance. The pilot-injection profile reduces NO_x by some 20 percent at a SFOC penalty of around 0.6 percent. Because of the difficulty of improving the SFOC much further for large engines, the flexibility of the electronic systems may instead be used to improve on the low-load emission performance. Low-load smoke is a potential problem, as mentioned earlier, that could be improved without sacrificing the good high-load performance because of the easy change of the injection characteristic of the electronic system.

Improvements in NO_x by changes in the heat release pattern, as seen in Figure 14.14 from the four-stroke development [59], is not obvious for the large two-stroke engines, where the premixed combustion phase always has mixed almost undetected into the diffusion-combustion phase.

14.4.1.5 Exhaust Gas Recirculation

Exhaust gas recirculation (EGR) is a very effective method for reducing NO_x in small engines. The exhaust gas reduces the oxygen concentration (the partial pressure) and, thereby, slows the combustion process down and decreases the maximum temperatures and the NO_x formation. EGR also has a great potential for large two-stroke engines [18, 19, 56–57]. An initial investigation was carried out several years ago with a moderate amount of EGR, but still a substantial NO_x reduction around 20 percent to 25 percent was obtained for only 10 percent EGR and with a fuel-consumption penalty of around 2 percent. Figure 14.15 shows the effect of EGR on a four-cylinder test engine operated at three different loads [18]. Again, this figure shows EGR as a very powerful method for reducing NO_x; 15 percent recirculated gas reduces NO_x up to 50 percent. The fuel consumption in this case was also increased about 2 percent. The moderate increase in SFOC is probably connected to the almost unchanged heat-release curves for these tests.

Different methods to obtain the EGR exist for turbocharged two-stroke engines. Figure 14.16 shows two cases of the EGR layout system [18]. In both cases (high- or low-pressure recirculation), the exhaust gas needs to be cooled and cleaned for both sulfur and particulates (see later section on aftertreatment methods) before it enters the scavenging air. The potential danger of forming sulfuric

Fig. 14.13.

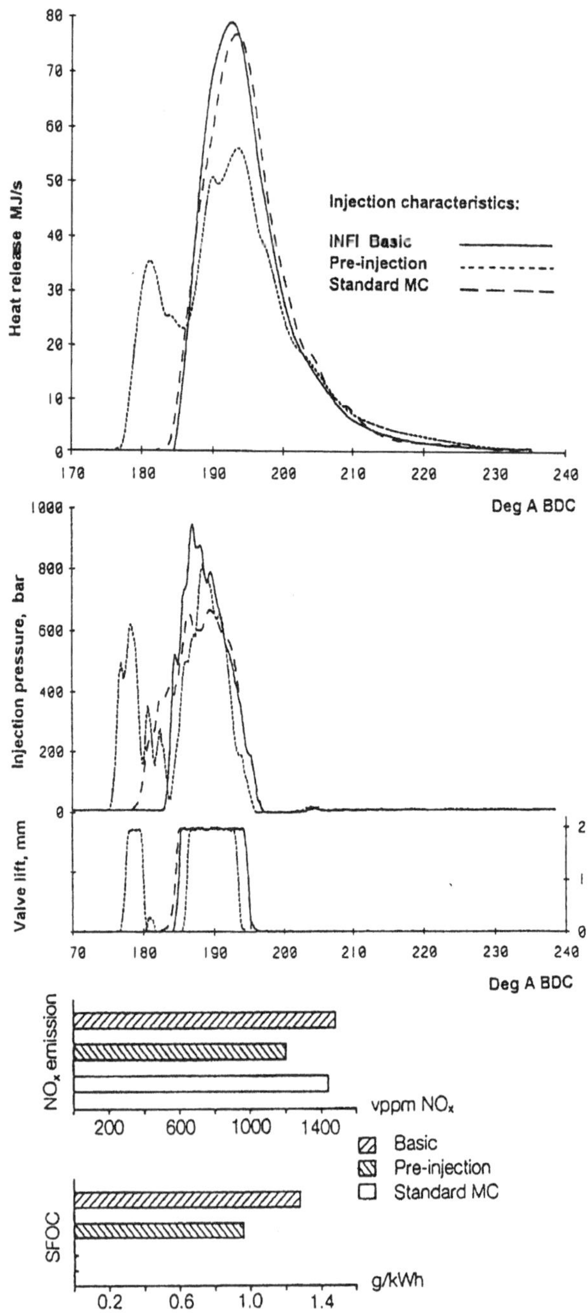

Heat release, injection pressure, and valve lift for three different injections on the 4T50MX test engine [18, 57] at 75 percent load, including effects on NO_x and relative SFOC.

Fig. 14.14.

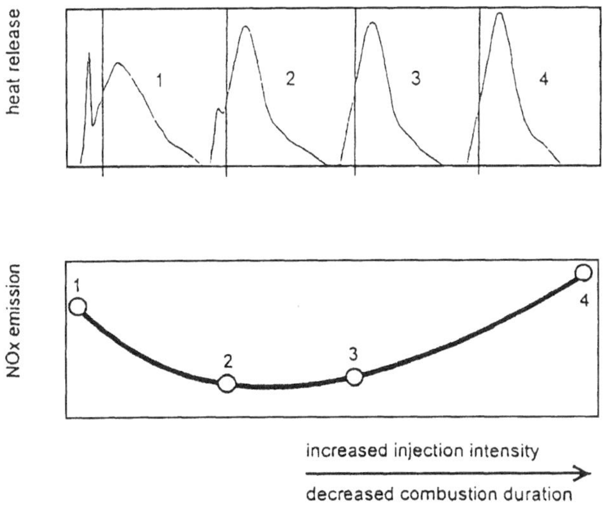

Typical relationship between heat release and NO_x emission for four-stroke engines [59].

Fig. 14.15.

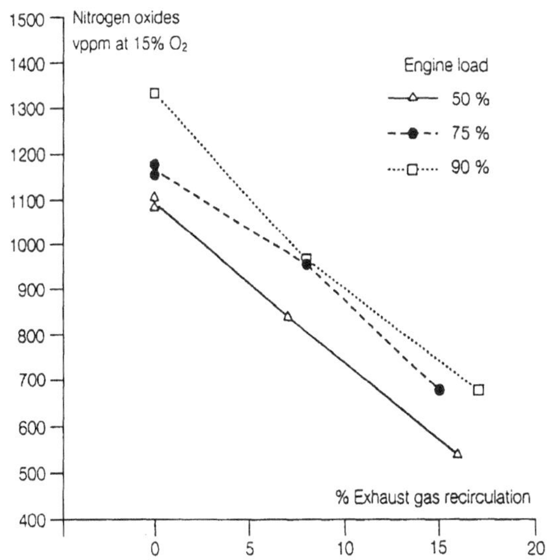

Influence of exhaust gas recirculation (EGR) on the 4T50MX engine at three different engine loads [18].

Fig. 14.16.

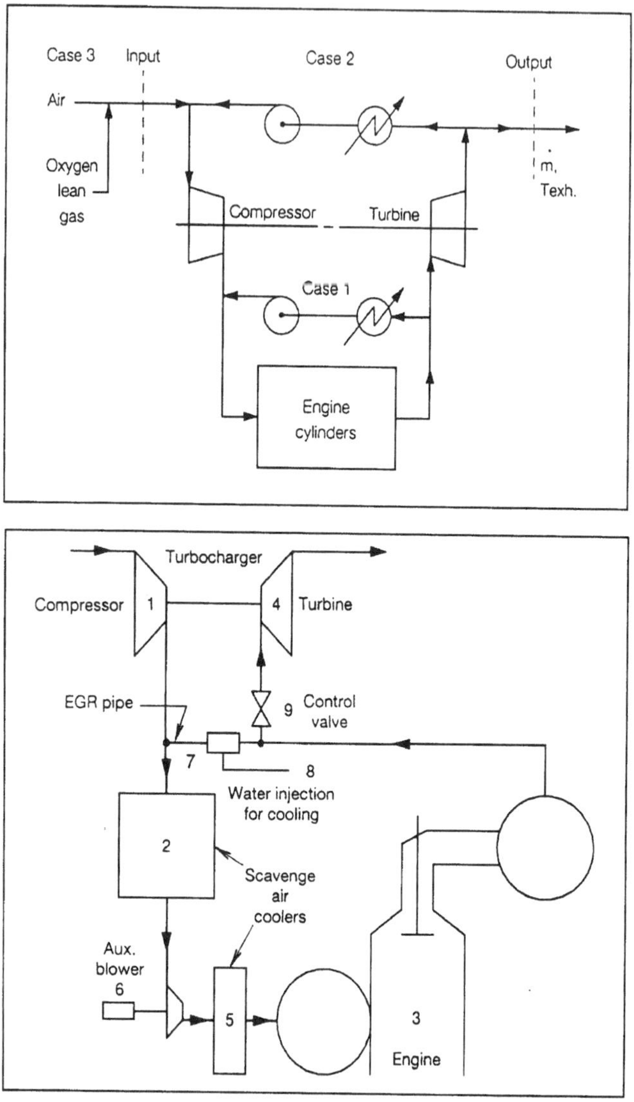

Different EGR system layouts (top) and EGR layout for the 4T50MX engine (bottom) [18].

acid in air coolers and on the cylinder liners has up to now prevented the availability of this method. The sticky particulates usually formed in a diesel engine operated on heavy fuel need more research before the method becomes available. Of course, if the engine is operated on a "clean" fuel or gas, the same restrictions do not exist, and, for example, in a process industry, where an inert gas is available, this gas may be used instead of gas from the exhaust.

The consideration of different emission components is another concern, when assuming EGR as an option for NO_x reduction. The trade-off between

NO_x and particulates (or soot) is well known, even though the increase in PM in the preceding tests was less than anticipated [56]. The investigation needs to be repeated with a properly cleaned exhaust gas.

14.4.1.6 Water and Ammonia Injections

Water can be used as an inert component, if injected directly into the combustion chamber. Actually, three methods using water have been investigated during the recent years [10, 18, 33, 56–57, 62–64]:

1. direct water injection
2. water-in-fuel emulsion
3. fumigation

Method 1 will be discussed in this section, method 2 in the next section. Method 3 is usually not considered attractive for two-stroke engines because of the potential risk of water condensation on the cylinder liners. The water causes possible destruction of the lube-oil film and, consequently, causes increased wear. Actually, water is always taken out after the charge air coolers with water mist catchers. However, humidification of the intake air may provide some reduction of NO_x at a small fuel penalty [62–63].

With the electronically governed injection systems that are becoming available for large engines, timed water injection becomes practical. Water can be injected by a separate valve [56], or by an integrated valve with both water and fuel injection to improve on the water-fuel mixing [10, 63], or finally by an injection where water is sequentially mixed into the fuel in an integrated valve [64].

These systems offer the flexibility of varying injection pressure and injection duration independently and, of course, independent of the fuel injection. The advantage is the elimination of an external homogenization unit and handling of the higher-viscosity water-fuel mixture (see later), but the disadvantages are the complicated or additional injector and for some systems the less effective mixing. To optimize the water injection for low NO_X, the water needs to be in the burning zone during the high-temperature time window. Excess water in the entire charge (as in method 3) cools the charge but does not even out the high-temperature peaks. The optimal time frame for the injection (at the highest local temperatures) is around top dead center (TDC) or shortly after, to position the water in the spray during the initial diffusion-combustion phase. If injected during the ignition delay period, an increased delay period may be avoided, especially for the faster four-stroke engines.

The direct-injected water may be timed better than in the emulsion case, but it is difficult to obtain the optimal position (mixing into the fuel spray) of the water that is obtained with a water-in-fuel emulsion.

Four-stroke as well as two-stroke experiments report on considerable NO_x reductions with the DI H_2O method. A 50 percent water addition causes a NO_x reduction of between 40 percent to 60 percent depending on engine conditions, which is similar to the NO_x reduction obtained using water-in-fuel emulsions (see later). But the effect on different emissions and SFOC varies between the studies. Some authors report reductions in soot, HC, and CO formation for some methods, while others find increases in the same or only some of the components. The penalty in SFOC varies between neutral to 5 percent, again depending on engine conditions.

Water injection was tested on a L35MC-GI engine with the gas valve replaced with a conventional-type injector for water. The tests were an extension of tests with direct-ammonia injection [56].

The reduction of NO_x by reaction with ammonia in an SCR reactor is an established and commercially available technique (see next section). By injecting ammonia directly, the purpose was to avoid the large catalyst and the transient problem in the SCR method. However, the ammonia has to be injected and distributed evenly in the combustion chamber within a rather narrow time window. If the injection is too early, when the temperature is too high (should not be higher than approximately 1100°C), the ammonia will burn instead of reacting with the NO_x. If the injection is too late, the temperature is too low (should be higher than approximately 900°C) and the reaction with the NO_x is too slow, causing a large slip of unused ammonia in the exhaust.

The tests were carried out on one cylinder of a six-cylinder engine. The test cylinder was equipped with a special cylinder cover developed for dual-fuel operation, a GI type (see the later section), where the high-pressure gas injectors were changed with injectors for an aqueous ammonia or urea solution. To separate the effects of water from the aqueous ammonia solution, injection of pure water was also tested.

The results showed (see Figure 14.17) that water alone had no effect on the NO_x emission, when injected at the same late timing as the ammonium or urea solutions. However, when water was injected during the fuel-oil combustion period, it was possible to reduce NO_x by up to 40 percent. The tests were not extended far enough to provide the optimal timing for the water injection.

Figure 14.17 shows the results of test for a fixed fuel-injection timing while varying the timing for the ammonia injection. Also, two different amounts of ammonia were tested, corresponding to a dosage of 2.5 and 4 times the necessary amount in case of assuming a stoichiometric reaction with the measured NO_x. There is clearly an optimum timing for ammonia injection, just as there were critical time and temperature factors for the reactions described earlier. It was possible to reduce NO_x by up to 50 percent, but on account of a very high ammonia consumption of four times the stoichiometric amount. Some of the excess ammonia will burn, and some of it will end up as an excessive NH_3 slip. The experiments further showed that there were only insignificant differences in NO_x reduction when using ammonia or urea.

Fig. 14.17.

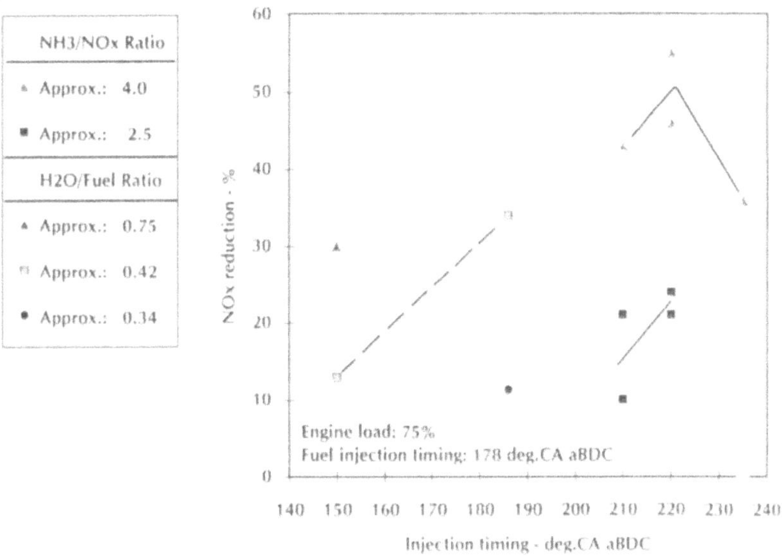

Timing influence of direct injection with ammonia/urea and water on NO_x reduction for different NH_3/NO_x and H_2O/fuel ratios [56].

Of course, the mixing and the injection timing of ammonia in the combustion area may not have been optimal. However, significant improvements will be necessary to reduce the ammonia consumption in order to make the method economically attractive.

14.4.2 Fuel Effects

This section is not intended to examine individual fuel component effects on the exhaust emissions. A much more detailed study is needed for this. However, some of the important aspects of the fuel components are discussed along specific topics in the different sections. The section concerns fuels not usually used for diesel engines, but which due to different circumstances are becoming interesting for use in diesel engines either due to availability and/or because of a NO_x reduction potential.

14.4.2.1 Water-in-Fuel Emulsion

Water emulsification has been used in diesel engines for several years. Verification of the method started more than 20 years ago, and two power plants have now been operated on water-in-fuel emulsion for more than 15 years together [18, 33, 56–57].

A standard engine allows the addition of some 20 percent water at full load because of the usual excessive volumetric capacity of the injection pumps.

However, the limit for water addition is much higher; 50 percent water added presents no problems for the engine and up to 70 percent (maybe more) is feasible, but at this amount, an increased ignition delay needs to be compensated. Since the NO_x emission typically increases at lower loads, increasing amounts of water at part loads could be beneficial. The required amounts may not be restricted by the fuel-pump capacity, but rather by the increasing ignition-delay consequences of the large water amounts.

Large amounts of water (more than 20 percent) usually will require a reevaluation of the engine design regarding forces in the pump-drive train (e.g., pump, camshaft structure, shaft drive). For new engines this will cause only a minor added expense, whereas, as retrofit, a restriction on the water amount may be necessary unless larger expenses are foreseen. Usually the fuel-pump plunger and the nozzle layout need a special optimization to allow for the best fuel consumption [33]. The drawback is that when optimized for water emulsion, the engine cannot operate efficiently on "pure" fuel. But, if the pump-nozzle change is not introduced, the fuel consumption is penalized by the increased injection duration.

A schematic of a fuel system with an added homogenization unit is shown in Figure 14.18. Depending on the required water amount, the pressure and heating capabilities of the fuel system may need adjustment. The viscosity of a water-in-fuel emulsion increases with increased water content (see Figure 14.19), which, as for standard heavy fuel, is compensated by increasing the mixture temperature. As a consequence, the fuel-line pressure needs to be increased to prevent vapor

Fig. 14.18.

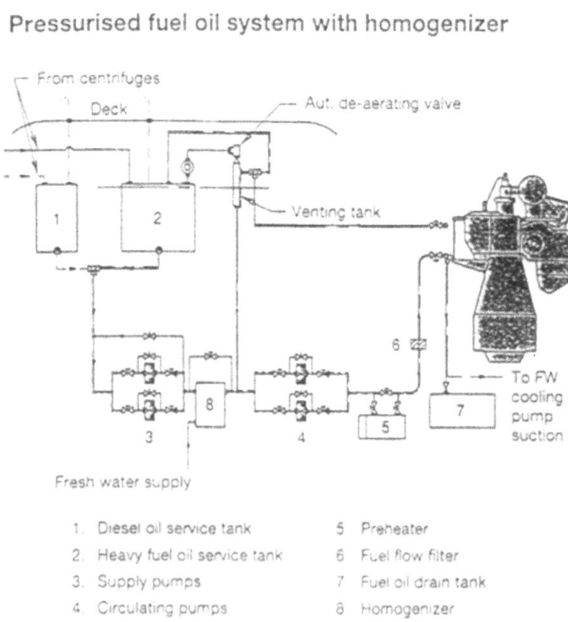

Schematic fuel system with homogenizer.

Fig. 14.19.

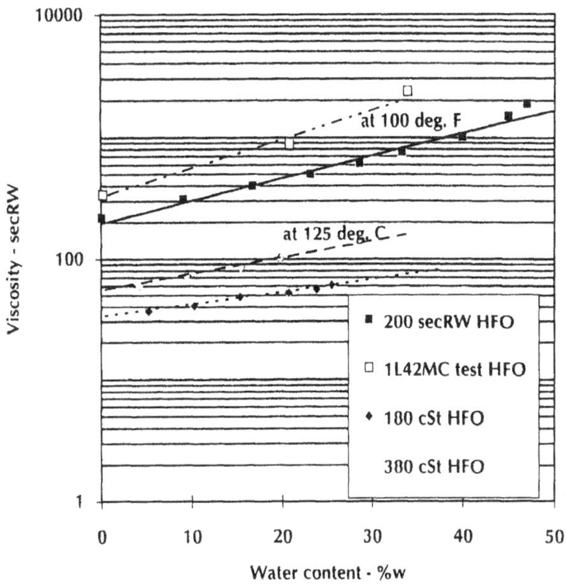

Influence of water content on viscosity of water-in-fuel emulsions.

blocks or cavitation. Emulsions can be created between water and almost any kind of fuel. However, where most heavy fuels emulsify easily, diesel fuels need an emulsifier to stabilize the emulsion.

A summary of the relative NO_x reduction versus actual water content in the water-fuel mixture is shown in Figure 14.20. The figure includes the results of several tests where different FIE layout philosophies were used. A rule of thumb allows 10 percent NO_x reduction for each 10 percent H_2O added. But the actual pump-nozzle design can change this relation drastically, as seen in the spread of the data. Similarly, Figure 14.21 shows the corresponding SFOC penalty for these tests. Therefore, depending on how the FIE was optimized, either a penalty or a neutral SFOC can be foreseen. (It should be kept in mind that zero water content in Figures 14.20 and 14.21 does not necessarily reflect the fuel optimized SFOC, since the FIE may be changed.) The limited effect that water addition has on SFOC is a consequence of the heat-release behavior. Independently of an early maximum heat-release increase or decrease, influenced by the FIE system, the combustion duration becomes shorter with added water, which improves the cycle efficiency.

Similar to the differences reported on SFOC, emissions of CO, HC, and particulates (or soot) vary greatly in the literature also between four-stroke and two-stroke engines (see the earlier H_2O-injection section). The discussed "micro-explosions" in the early literature, depending on emulsion quality, injection system, and combustion conditions (mixing, wall influence, etc.), may affect the SFOC as well as emission formation. Therefore, to obtain a true picture of the emission performance for a specific engine, it is necessary to optimize each engine type.

Fig. 14.20.

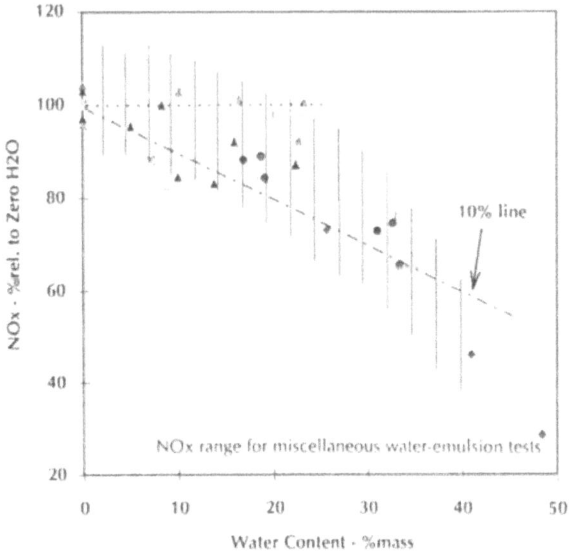

Influence of water on NO_x emission.

Fig. 14.21.

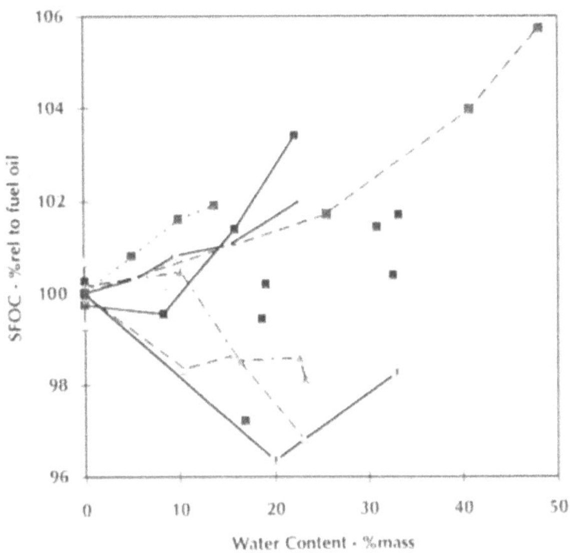

Influence of water on specific fuel-oil consumption (SFOC).

The concern for the particulate emission also requires attention to the cleanliness of the added water.

14.4.2.2 Gas Engines

As described earlier (see Section 14.3.6), a "clean fuel" has a potential benefit to the environment by lowering both NO_x and soot emissions.

Several four-stroke engines have had great success by converting standard DO-operated engines to gas for use in cogeneration plants or for propulsion of LNG carriers to utilize the boil-off gas as a main fuel [10]. However, it is necessary to adapt a direct-injection system in order to obtain the efficiencies of today's standard for DI diesel engines.

Also large two-stroke, slow-speed engines are today available with high-pressure gas injection (some 250 bar) either for power plants or for engine propulsion. A description of the first high-pressure large-bore, gas-injection engine, a 12K80MC-GI-S, is given in Fukuda et al. [65]. The injection system was based on a development for a medium-speed engine with high-pressure gas injection that now is operated in a cogeneration plant.

A schematic of the internal and external systems needed for dual-fuel operation is shown in Figure 14.22. The gas-injection valve is operated by the control-oil pump system by the high-pressure control oil. The normal fuel-injection pump supplies pilot oil in the dual-fuel operation mode. The pilot amount of low-sulfur

Fig. 14.22.

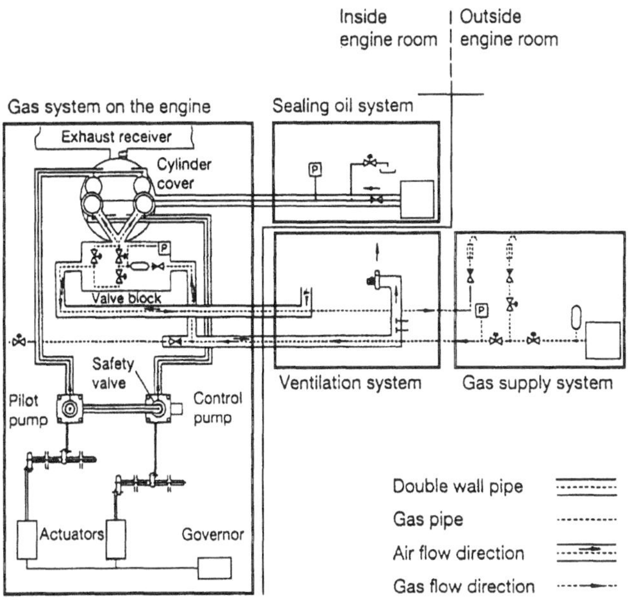

Schematic diagram of internal and external systems for dual-fuel operation [65].

DO corresponds to 8 percent of the total supplied energy at 100 percent load. A safety valve ensures that gas (at a pressure of some 250 bar) only is injected, when the pilot-oil injection has been started to ensure ignition of the gas. A ventilated double-wall, gas-piping system prevents gas outflow to the machinery spaces in case of a ruptured gas pipe.

The combustion and performance characteristics are very similar to the standard diesel version. But, because of the more homogeneous mixture of gas and air compared to the fuel droplets and air in a standard diesel combustion, a reduction of up to 20 percent is seen in the NO_x emission. And at the same time extremely low emissions of SO_x and particulates are seen. For the 12K80MC-GI-S engine operated on natural gas, 3.0 ppm SO_x and 0.033 g/kWh particulate matter were found.

As suggested earlier, because of the very clean combustion from the gas engine, EGR may be a strong method to reduce the NO_x emission further. Due to the very strict regulation that applied for this power plant, a large SCR-DeNO_x system was installed with a final oxidizing-catalyst layer to eliminate also the remaining CO and HC emissions (see the following sections). After the SCR system, NO_x was reduced to less than 10 ppm.

14.4.3 Selective Catalytic Reduction (SCR)

The selective catalytic reduction process is one of the most commonly used methods for the reduction of nitrogen oxides for conventional power plants and industries [66–67] especially when a high degree of reduction is required (in excess of 90 percent). The SCR process has been used for exhaust gas aftertreatment for large two-stroke engines commercially since 1989 [18, 68–69].

The low-speed diesel engines utilize a very high air-excess ratio; therefore, the automotive-type catalysts cannot be used. Furthermore, metals and high sulfur content in heavy fuel oil would cause automotive catalysts to deteriorate after a short period of operation.

The reduction of NO_x takes place through a reaction with ammonia (NH_3) producing N_2 and H_2O. Ammonia or urea is injected into the exhaust gas before it enters the catalytic converter. The exhaust gas should have a temperature of between 320 and 400°C, depending on the SO_x content, to ensure a good efficiency and a nonplugging catalyst [66]. A schematic layout for a SCR system is shown in Figure 14.23 [69]. The nitrogen oxides, which primarily consists of NO and NO_2, are converted according to the following reaction schemes (using NH_3):

$$4NO + 4NH_3 + O_2 = 4N_2 + 6H_2O \tag{14.5}$$

$$6NO_2 + 8NH_3 = 7N_2 + 12H_2O \tag{14.6}$$

For urea an additional decomposition reaction is used that converts urea to NH_3:

$$CO(NH_2)_2 + H_2O = CO_2 + 2NH_3 \tag{14.7}$$

Fig. 14.23.

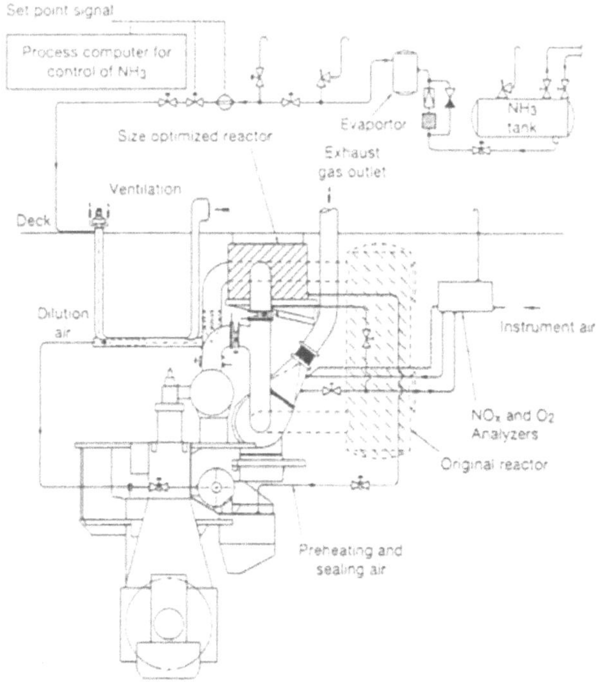

Schematic layout of SCR system for large two-stroke engine [69].

or [66, 70]:

$$CO(NH_2)_2 + H_2O = HNCO + NH_3 \tag{14.8}$$

$$HNCO + H_2O = CO_2 + NH_3 \tag{14.9}$$

The ammonia feed can either be liquid, water-free ammonia under pressure, or it can be a 25 percent aqueous solution at atmospheric pressure. NH_3 is a combustible gas and, therefore, supplied through a double-walled pipe system with appropriate venting and NH_3 leak detection [18]. Urea are nontoxic crystals that need to be dissolved in water (from 61 to 108 g in 100 g H_2O is possible depending on temperature).

The NH_3 is diluted with pressurized air from the scavenge air receiver and mixed in a static mixer before injected into the exhaust gas in a second static mixer. A homogeneous mixture is necessary to obtain high efficiency of the process and to avoid slip of NH_3 from the reactor.

The SCR reactor usually contains several layers of catalyst. The catalyst volume and, consequently, the size of the reactor depend on the activity of the catalyst, the desired degree of NO_x reduction, the NO_x concentration, the exhaust gas pressure, and the acceptable NH_3 slip. The amount of catalyst is usually expressed by its space velocity (abbreviated NHSV), which is defined as the number of normal cubic meters of exhaust gas that are treated per cubic meter of catalyst per hour.

The degree of NO_x removal depends of the amount of ammonia added. Figure 14.24 shows an example of how the NO_x reduction and the slip vary with the NH_3/NO_x ratio for two different catalyst volumes (NHSV) [71]. Both NO_x reduction and NH_3 slip increase with an increasing NH_3/NO_x ratio. The same NO_x reduction can be obtained with half the catalyst volume (10,000 Nm^3/m^3h), if the NH_3/NO_x ratio is raised a few percent. However, the NH_3 slip increases considerably, usually a slip of 5 to 10 ppm will be accepted.

The development of the method was carried out in a pilot test on a large two-stroke engine using a heavy fuel with 2.5 percent sulfur [71]. But SO_2 (and HCl) in the exhaust gas decreases the activity of the catalyst at temperatures less than 320°C and eventually forms ammonium sulfates that will plug the channels in the catalyst:

$$2NH_3 + SO_3 + H_2O = (NH_4)_2SO_4 \tag{14.10}$$

However, the catalyst regenerates when the NH_3 feed is stopped while the exhaust gas is kept flowing. The minimum temperature decreases with decreasing sulfur content. The catalyst activity also deteriorates with overheating above 400 to 450°C.

Due to the catalyst sensitivity with the gas temperature, the SCR reactor in two-stroke applications is installed between the engine and the turbocharger. This presents a restriction in the engine room layout, but also makes the system less

Fig. 14.24.

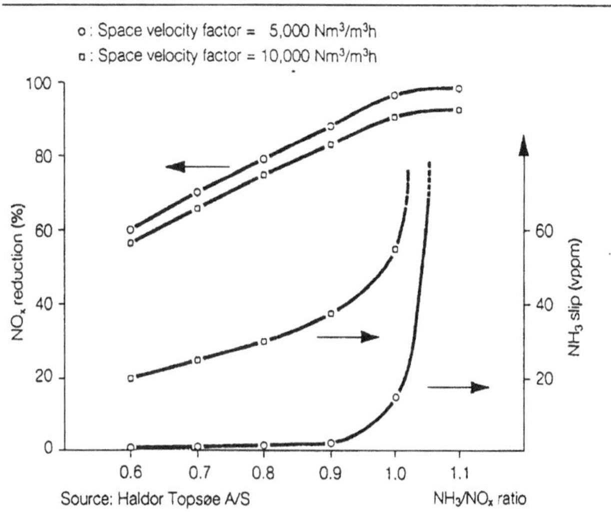

Calculated NH_3 slip and NO_x reduction as function of NH_3/NO_x ratio for two given space velocities [71].

bulky than it would have been due to the "high" pressure before the turbocharger. Two possible SCR positions are indicated in Figure 14.23. Using a special valve arrangement, the reactor can be by-passed during start-up or at low-load operation (less than 20 percent to 25 percent loads), when the catalyst is too cold. In the case of the large gas engine (see Section 14.4.2.2 [65]), a gas burner was installed between the engine and the catalyst to heat up the catalyst during load-up after a cold start.

Today four marine engines are using the SCR reduction technique [68–69] when the engines are operated in California waters. Due to a sulfur restriction in the same area, the engines in this area are operated on a low-sulfur fuel (less than 0.05 percent S). The NO_x emission is at full load reduced from approximately 1300 ppm to less than 130 ppm.

A durability study was initiated in a pilot installation on board a ship [72] because of the concern for the operation of the catalyst, when the engine is operated on HFO. The inlet temperature for the pilot system is between 320 and 340°C, the approximate NO_x concentration is 1300 ppm at 15 percent O_2 (and at around 85 percent engine load), and with approximately 400 ppm SO_2 in the exhaust gas. At a NH_3/NO_x molar ratio of approximately 1.0, a 90 percent reduction of NO_x is obtained with a very small NH_3 slip. The catalyst has been in operation for more than 6000 hours (time of paper, 1995) without clogging or deterioration of the catalyst. The deposits of dust on the catalyst seem to change with the exhaust gas temperature, decreasing to less than half at high temperatures.

The long-term results are similar to the original test on the Faroe Islands [71], where the catalyst was in operation for 4000 hours without any observed deactivation or plugging of the catalyst. In these tests the heavy fuel contained 2.5 percent sulfur and 0.04 percent ash, and the exhaust gas had a NO_x level of around 1200 ppm, a pressure of 2.6 bar, and the inlet temperature was 340 to 390°C.

Along with the reduction of NO_x, the catalyst may also oxidize particulates and other volatile compounds. Reductions of CO and HC by 98 percent and for particulates by 50 percent have been claimed, but the available space for the SCR or actual HFO conditions are not necessarily taken into account. Unfortunately, real test results from the two-stroke engines are still scarce. HC in the pilot plant tests was reduced by more than 60 percent, but data for dust were not included. Four-stroke tests [66, 73–75] show mixed results, HC generally decreases (up to 80 percent to 90 percent), CO varies from decrease to increase (from 130 percent to 40 percent of reference), and PM usually decreases (maximum 45 percent has been reported). The results from gas engines have not been included. The task of the SCR is in these cases regarded as much easier because of the "clean" exhaust (much less particulates and SO_2).

Finally, the SCR technology has no direct effect on the specific fuel consumption. An improvement may be obtained due to the required high-efficiency turbocharger, but the installation has to be paid for and the consumption of ammonia (7 to 8 g/kWh depending on conditions) and catalyst need to be added to the running costs.

14.5

EXHAUST EMISSION CONTROL TECHNOLOGIES—REDUCTION OF OTHER POLLUTANTS

14.5.1 HC Emissions

As discussed in Section 14.3.2 several sources exists for HC formation and, as Tables 14.4 and 14.6 show, several parameters affect HC formation. However, a more detailed analysis of these sources is necessary in order to devise possible design changes to affect the exhaust HC emission. Since very few papers are published for large engines, the small DI literature [16, 76–82] has been used as a starting point. However, each of the possible sources need a critical evaluation, not only due to the large differences reported for the small engines, but also keeping in mind the large differences in design and operational conditions between small and large engines. Possible sources are:

1. fuel-air mixing: overmixing; undermixing
2. quenching: bulk; wall
3. crevices: nozzle sac; piston/ring
4. fuel and lube oil

During the ignition delay period, the spray propagates and mixes (entrains) with fresh air in the combustion chamber. Some of the fuel in the fringes of this spray will mix beyond the lean flammability limit. The mixing process depends on the injection conditions (nozzle design and injection pressure) and on the air motion conditions (swirl and squish effects). Therefore, several design factors influence this mixing process. However, it is possible to estimate the trend of a design change with reasonable assumptions. Correlations have been obtained with "simple" spray models that estimate the mixing process [78–79] in order to evaluate the possible influence on the HC contribution. The short ignition delays, usually around 1 to 4°. CA in well-optimized engines, combined with the slow initial injection rate for standard fuel pumps favor these engines and should ensure only a small HC contribution from overmixing (or the too-lean-to-burn mixture).

Also, undermixing is for large engines not likely in well fuel-optimized engines, except during transients or possibly at low loads, where conventional injection systems are working at their limits. Opening or, more important, closing pressures are important parameters for the low-load behavior.

Bulk quenching should also be limited due to the slow-turning engines and the ample time for the combustion process. The degree of NO_x reduction, though, may affect this statement depending on the required timing retard. As mentioned earlier, excessive timing retard is not likely due to the strong effect on fuel consumption (see Figure 14.10). More likely, as seen in Section 14.4.1.3, the low-NO_x nozzle layout is used, which affects the mixing process by limiting the oxygen access in the high-temperature regions and by shifting the combustion reaction to later in the combustion process. The low-NO_X nozzles also influence fuel consumption as well as other emissions, depending on how strongly the influence affects the combustion duration.

Figure 14.25 shows typical trade-off behaviors between different emissions (CO, HC, NO_X, and PM) and between SFOC and NO_X for three different engines caused by certain changes in the nozzle design and in the injection conditions. All engines are operated at 100 percent loads. Engine A shows the influence of five different nozzles, two of the nozzles labeled low-NO_X nozzles. More than a 20 percent reduction in NO_X emission is seen for these two nozzles, but at the same time giving a remarkable increase in HC, whereas only a slight increase in PM and CO. Engine B shows the strong influence of a mini-sac nozzle on HC and PM (see definition later), which had a slight effect on NO_X also. Finally, engine C shows the influence of two different nozzles (same spray direction and number of nozzle holes, but a 35 percent total hole-area change) at two different injection pressure levels. As indicated, mini-sac nozzle and nozzle hole area change (at

Fig. 14.25.

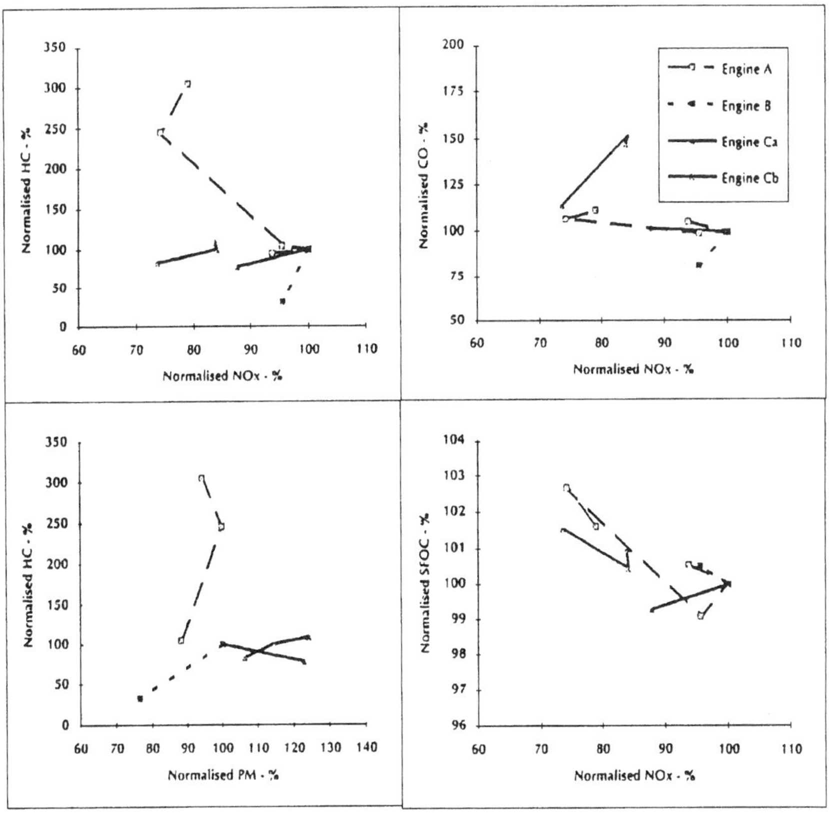

Trade-off behaviors for HC, CO, and SFOC versus NO_X, and HC versus PM—influence of nozzle specifications (data from three engines).

constant pressure) causes marked improvements for some emission components, whereas the spray direction more likely is a simple trade-off between emission components.

Wall quenching is difficult to estimate. The heat load is high, but a great effort is directed at keeping the sprays off the walls both to even out and to limit heat loads to the different engine components that make up the combustion chamber walls. The usual steep HC increase following wall wetting in small engines is not recognized in large engines, which points to the absence of wall wetting.

Crevices are known as important sources of HC emission for the small gasoline engines. The design of the combustion chamber as well as the piston-ring area is therefore important. For diesel engines some of these crevices are of less importance, but two special volumes exist, the injector nozzle sac volume and the clearance volume in the piston-ring-topland design. Both sources could also be looked at from a different point of view. The outflow from the nozzle sac simulates late injection or secondary injections that are likely candidates for bulk quenching effects. Similarly, the flow from the ring-pack also enters the combustion chamber late in the cycle, but additionally, lube-oil effects are probably the main concern for the ring area. To make a recognizable contribution, the crevice volumes need a certain size related to the combustion chamber and the contained masses should be comparable with the injected fuel mass, in the same relation as for small engines.

Figure 14.26 shows a comparison between three nozzle designs: a standard nozzle, a slide-valve design, and a mini-sac nozzle. The cross sections clearly indicate the differences in the possible contained fuel volumes. The slide-valve

Fig. 14.26.

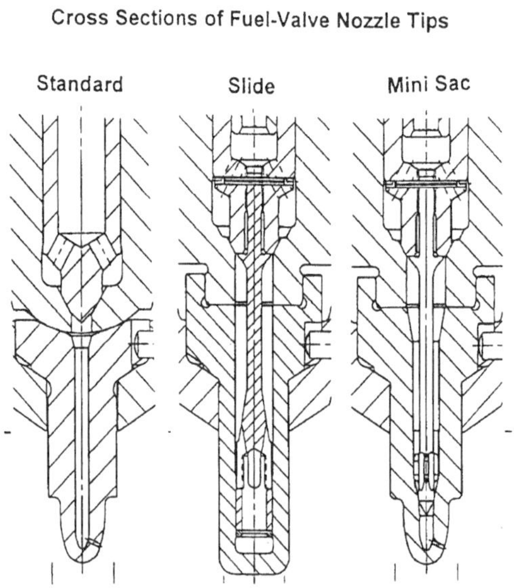

Cross sections of fuel-nozzle tips for three different nozzle designs.

design is equivalent to a VCO design for small engines, whereas the mini-sac design only limits the sac volume. The mini-sac design is a less expensive design compared to the slide valve. Of course, the actual sac volumes of the different nozzles vary with the engine type. Furthermore, depending on temperature level and fuel type, the sac volume may be only partially filled up with liquid or fuel vapor at the time, where the volume is expanded into the cylinder [83]. A slightly leaking valve or secondary injections will worsen the condition. It is important to consider the upstream flow conditions for the nozzle hole design in order to evaluate the actual spray directions, which can affect the actual NO_x performance (as seen in Figure 14.25).

The sac-volume influence on HC and PM as function of engine load is shown in Figure 14.27 for three different engines all operated on diesel fuels (engine B with 0.17 percent sulfur, engine C with 0.05 percent sulfur, and engine D with 0.42 percent sulfur, but using a dust particulate method and therefore not included in the PM figure). The result shows a clear indication of the importance of the sac volume. HC and PM are for some conditions reduced to one third of the standard values. The reduction of the PM for a specific engine depends on the original SOF fraction of the particulates, which changes depending on fuel specification, engine type, and load-speed conditions (see also Section 14.5.4).

Full-load (MCR) HC data from Figure 14.27 are in Figure 14.28 compared to total normalized sac-volume data (sac volume times number of valves divided by the compression volume). The data from [78] Figure 20 and [83] Figure 24 are added in Figure 14.28 in an attempt to relate the results with the results from the small DI engines. The comparison points to two main deductions: Even with the existing scatter, the sac volume is for all engine sizes a serious source for HC emission. An extrapolation to zero sac volume points to major differences in HC sources or the influence of the sources for large two-stroke engines and small DI engines. The load influence, of course, contained in the data from [78] is not relevant for the sac-volume influence; the trend for the sac-volume change is the same.

Finally, the influence of lube oil is important. Large engines are lubricated through feed holes into the cylinder liner in an attempt to lubricate the ring area directly. The lube oil is introduced approximately halfway down the liner, distributed around the periphery, and transported up along the liner by the ring pack. The lube amount corresponds to roughly 1.0 g/kWh at full load and usually increases at low loads. Lube oil has two routes to exit the engine, either through the exhaust gas as unburned HC and particulates, or drained from the scavenging box.

Based on the PM versus lube feed-rate investigation [56, 69], slightly less than half of the lube-oil increase showed up as increased PM (see Figure 14.31 in the PM Section 14.5.3). Also, measurements show that the actual lube-oil film thickness in the upper part of the liner, exposed during the combustion phase, is only a few microns (μm) [84]. Therefore, the lube oil is not likely to participate in the combustion process and will remain unburned and, consequently, will appear as HC and PM. However, assuming that 50 percent of the lube oil disappears (e.g.,

Fig. 14.27.

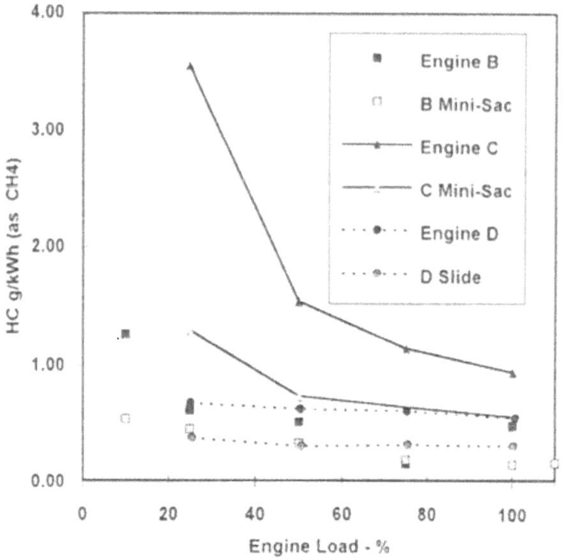

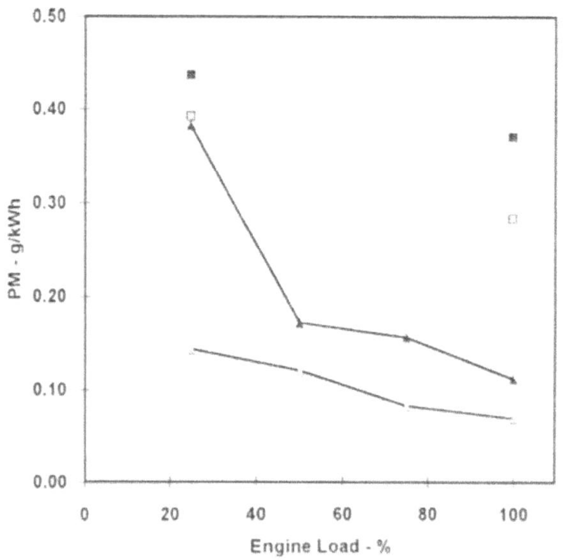

Influence of sac volume on HC and PM emissions versus load for three different engines.

Fig. 14.28.

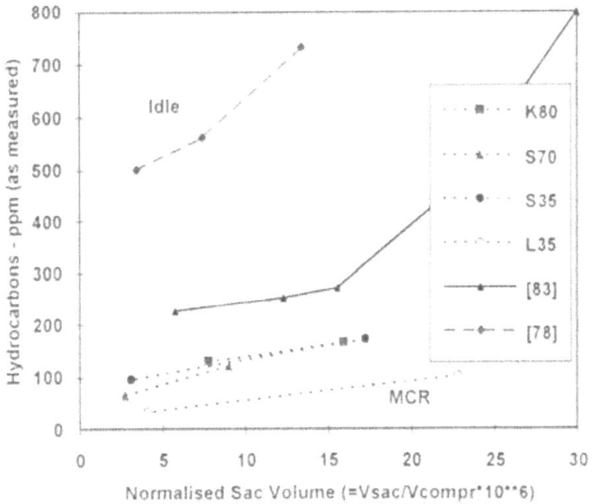

Comparison of normalized sac-volume influence on HC emission for different engines.

to the scavenge box), still 0.5 g/kWh can contribute to HC and PM in the exhaust. But because of high boiling ranges, the large molecules predominantly condense on PM and contribute to the SOF.

As discussed earlier (see Section 14.3.6), fuel and lube have an important effect on emissions. Although it is not the intention to discuss this effect in the present chapter, a few examples are included to illustrate the importance. Two studies have been performed to investigate the combined effect of fuel and lube oil on HC and PM emissions [20, 34]. The first study was on a one-cylinder test engine, and the other on a ten-cylinder power plant engine in connection with a liner-condition evaluation for a new fuel type.

Figure 14.29 shows HC and PM, respectively, as function of engine load for the two engines. For engine E, a standard diesel oil (DO) and standard heavy fuel oil (HFO) were tested with the reference lube oil (MG) and a fully synthetic lube oil (XRN). HC and PM are normalized using the DO-MG condition at 100 percent load. The actual values are not important, but the differences in HC and PM show the possible effect of a change in fuel or lube-oil type. The influence of HFO is already common knowledge, but also the lube-oil influence requires attention for emission-controlled engines. "High" indicates that the test was performed with twice the normal lube-oil feed rate. Engine F was tested using a standard HFO (with 1.93 percent S) and a low-sulfur fuel, BIA (with 0.84 percent S) with two different lube oils: RPT, a standard 70 TBN lube oil, and RPN, a special 40 TBN lube oil. Again differences in both HC and PM of more than 25 percent is found for certain engine conditions, not only due to change of fuel but also due to change of lube oil.

Fig. 14.29.

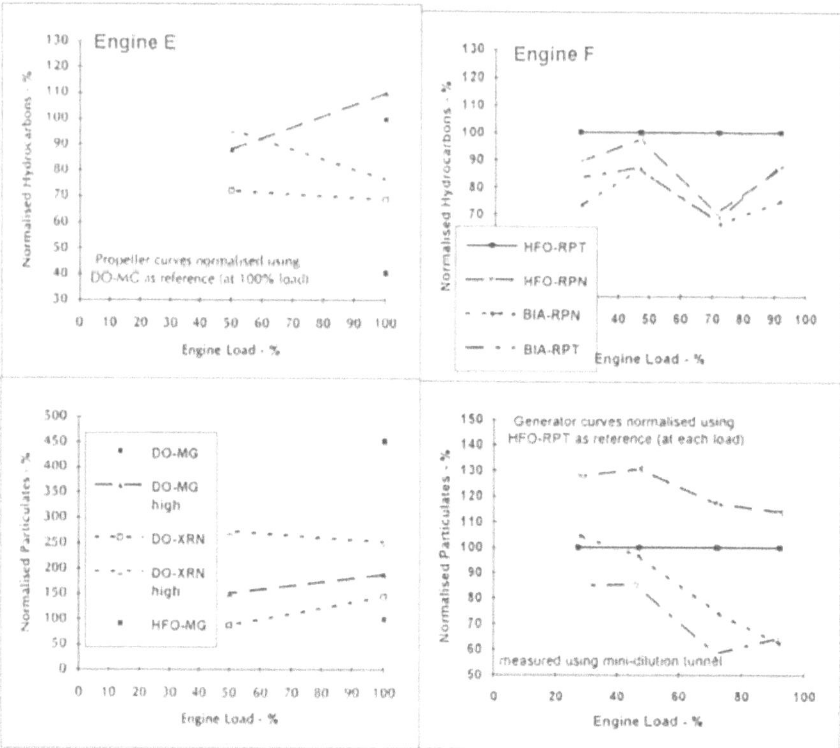

Influence of fuel and lube-oil specifications on HC and PM emissions for two different engines.

14.5.2 CO Emissions

The CO emission has up to now not received much attention with regard to reduction techniques. The main reason for this, at least for the large engines, is the usual low level of CO emission (see Figure 14.7) in relation to the requirements. As mentioned earlier, the low CO level is primarily caused by the high excess-air ratio in the large turbocharged engines. However, in cases where certain internal NO_X reduction techniques were introduced, an increase in CO was seen (see Figure 14.25).

The methods to reduce CO have not yet been explored in detail for the large engines, but methods that will affect HC through improved oxidation may also improve (reduce) the CO emission (i.e., some of the parameters that improve the fuel-air mixing process). Each parameter, though, will need additional investigation to ensure the proper effect.

CO, of course, can be reduced also through catalytic oxidation of the exhaust gas, as mentioned in Section 14.4.3 [67] and discussed further in Section 14.5.4.2.

The method has been used on one gas-fired, multifueled diesel engine only because of the very strict regulation for this engine [65]. Catalysts are still regarded with caution for large two-stroke engines operated on HFO.

14.5.3 Particulate Matter

Because of the diverse collection of material that defines the particulates (see Section 14.3.5), several parameters may be used as basis for the PM reduction methods. But, since several sources of PM are closely connected to the HC sources, and since HC in itself affects the SOF PM fraction directly, a starting point to reduce PM can be taken in the HC reduction technology. Because of this tight connection between HC and PM, several comments have already been given in the previous section on HC emission (14.5.1). Three different types of reduction methods exist [25, 29, 44–45, 85–87]:

1. engine design methods (internal methods)
2. fuel and lube-oil effects
3. aftertreatment methods (will be discussed in the next section)

The engine design involves the injection system as well as the piston ring-pack design. The injection system can be used to control the carbon (soot) fraction of the particulates and to control the SOF fraction through the HC condensation. The ring-pack design controls mainly the SOF fractions of the lube-oil contribution, since only a minor fraction of the lube oil will participate in the combustion.

Soot formation is caused primarily by insufficient fuel-air mixing, which can be improved through fuel-pump and nozzle design by optimizing the injection rate versus the air swirl (high fuel pressure, small nozzle holes, spray directions), as shown in Figure 14.30. The upper plot in Figure 14.30 shows total PM versus load and the lower plot shows the trade-off between PM and NO_x from a special injection study on low-sulfur DO. NO_x increased for increasing injection pressure (except for one test condition) as expected, whereas the influence on PM varied at different loads. The smaller nozzle improved (reduced) both NO_x and PM (except at 50 percent load). Therefore, both PM and NO_x emission need to be part of the optimization. A conceptual path for high-temperature combustion to reduce particulates (soot) without increasing NO_x is proposed in Kamimoto and Bae [88].

Improved oxidation (by continued mixing) throughout the cycle can reduce the already built-up soot. The primary goal is to limit the soot formation rather than to improve the later oxidation (on account of the formation limitation), since the formation process dominates in-cylinder oxidation [29]. To improve oxidation and avoid later HC condensation on the soot conglomerates, the spray should be kept off any of the cold walls, as discussed in the HC section. However, in large engines soot is visible predominantly at low-load operation and during acceleration

Fig. 14.30.

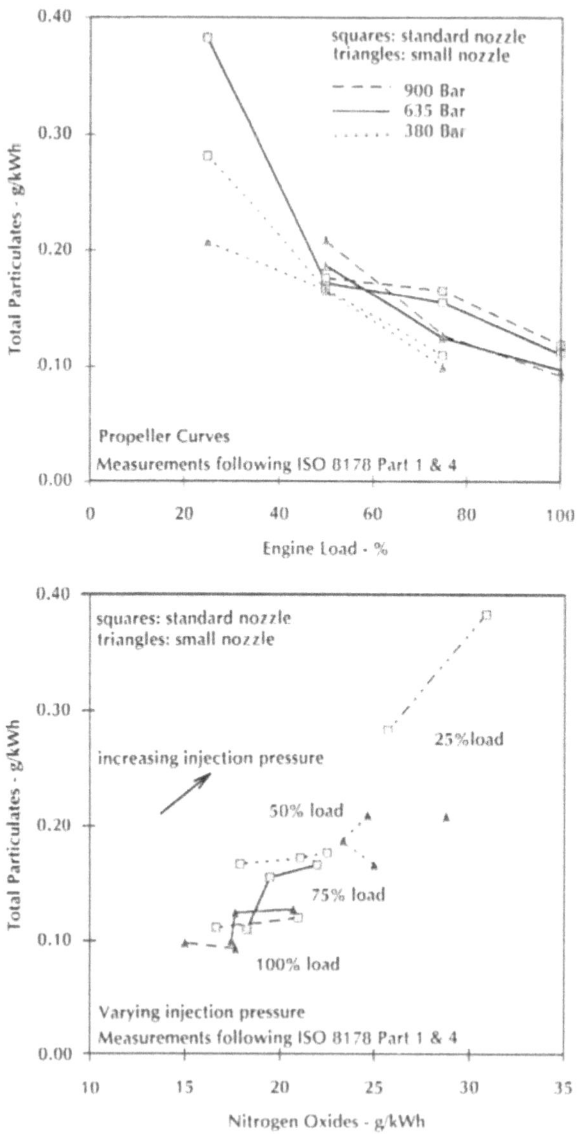

Influence of injection pressure and nozzle specification on total PM versus engine load (upper plot) and on PM versus NO_x (lower plot).

due to lack of air and undermixing, and again at very high loads due to overfueling of the combustion system.

A different wall effect is discussed in Kittelson et al. [89], where a considerable fraction (20 percent to 45 percent for a specific engine case depending on load) of the exhausted soot (carbon) is postulated to originate from stored material on the chamber walls. The deposited particulates is released (or reentrained) during the "violent" expansion, when the exhaust valve is opened. However, most of the increase in the PM concentration during the blowdown period (when the exhaust valve opens) was found to be SOF, and soot increased only by 20 percent. The SOF originated either from the nozzle-sac volume or from lube-oil on the cylinder walls. The soot deposits on the walls have a limited oxidation rate, and the deposition should therefore be restricted as much as possible in order to control the exhaust PM emission.

The SOF fraction originating from the fuel (usually less than 10 percent to 25 percent of the total SOF, based on small engine experience, the major part being lube-oil related [44, 90]) is reduced in general by an improved HC reduction, the main source being the nozzle-sac volume as discussed in the HC section (the influence on the total PM was shown in Figure 14.27 for two different engines).

The piston ring-pack design together with the cylinder liner (affected by ring profiles, topland areas, liner roughness and roundness, temperature levels, and materials) are important for the lube-oil contribution to the SOF [25, 87, 90]. The lube-oil conditions in the large engines operated on HFO are very different from the small engines running on DO, both regarding engine load and combustion environment. However, the main emission problem is the same: to limit the lube-oil consumption as much as possible in order to limit HC as well as PM emissions.

The influence of the lube-oil dosage (or feed rate) on PM (dust, following EPA method 5, see Table 14.2) is shown in Figure 14.31 for a 7L90GSCA stationary engine operated on HFO [18, 56, 69]. Also, the influence of the water quality is added in the figure, when the engine is operated on a water- in-fuel emulsion (see Section 14.4.2.1) with approximately 24 percent water content. A remarkable difference in the particulate-ash content is observed for the engine, when standard plant water is used. The results infer a similar direct effect of any unburnable ash content (metals, impurities) in fuel or lube oil.

In the case of the two-stroke engine operated on HFO, the general liner condition has the higher priority to ensure the reliability of the engine. Several improvements for the ring-liner conditions have been introduced, lately, higher "top" rings, controlled pressure drop across the top piston ring using calibrated leakage slots (CPR rings), and reduced cooling of the cylinder frame [91]. The cylinder-liner "flame ring" or "scraper ring" is also being introduced on the large engines to test the effect on liner cleanliness and secondly on emission performance. A well-running liner with minimum deposit formation on the piston, liner, and ring area also has the lower lube-oil consumption. However, the formulation of the lube oil with respect to the used HFO is still the key factor in this optimization.

Fig. 14.31.

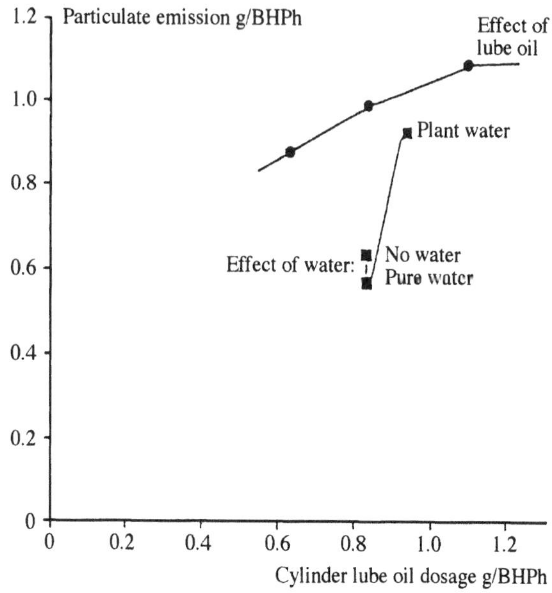

Particulate emissions (dust) as a function of cylinder lube-oil feed rate and of water quality in a 24 percent water-in-fuel emulsion [18].

Direct fuel and lube-oil effects on the particulates are important. A huge effort has been ongoing throughout the 1980s and 1990s to investigate fuel and lube-oil effects for low-emission small diesel engines [30–32, 50–51, 54–55, 86, 92–96]. A major problem is the sulfur contribution to the particulates [30, 32, 51, 54, 96], which in the United States has been the direct cause of lowering the fuel-sulfur content now to 0.05 percent, similar to the low-sulfur diesel available in Europe. But also indirect effects are being considered. Increasing the cetane number could limit the unburned HC and, therefore, reduce the contribution to SOF. Discussions are still taking place on the influence of the aromatic content or the density [31, 50, 94–95], but again, the United States is limiting the aromatic content in the diesel fuel to improve the HC and particulate emissions. Also the volatility (usually defined through the 90 percent end-boiling point, T90) shows influence on PM [30–31, 55]. However, authors find different parameters being more or less important in the analyses, partly because of different engines, operating conditions, or test procedures.

For the large two-stroke engines operated on HFO, with a sulfur limit of 5 percent, no direct restrictions on aromatic content (carbon residue of maximum 22 percent), and a density up to 1, the combustion process shows no deterioration, but emissions of course vary, and especially PM is drastically influenced [20–21, 34, 93]. The specific PM component fractions were shown in Figure 14.8 and the influence of fuel sulfur in Figure 14.9. It is clear that PM can be reduced substantially by lowering the sulfur content in the HFO. Also, the dust content measured by a hot-filter method (JIS Z8808 see Table 14.2) has shown a positive correlation with the sulfur content [93].

An example of the influence of different fuels and lube oils on HC and PM emissions was shown in Figure 14.29 for two different two-stroke engines. The fuel-PM effect was anticipated, but the result of the different lube oils also indicates that a careful evaluation is necessary for lube oils used on low- emission engines. Unfortunately, it is very difficult to optimize fuel and lube oil with respect to emissions due to the wide specification for HFO and in many cases due to unknown fuel qualities.

14.5.4 Exhaust Gas Aftertreatment

In order to reduce the exhaust gas emissions beyond the levels obtained through internal means, special treatment of the exhaust gas is necessary. Similar to the introduction of catalysts on small automotive engines, special catalysts can be used on the large two-stroke engines, when designed properly for the specific requirements (emission levels, excess-air ratio and temperature levels in the exhaust gas, and fuel specifications).

The aftertreatment methods consists of :

1. reducing catalysts (SCR) for NO_X reduction
2. wet or dry desulfurization
3. traps or filters for PM (including electrostatic devices)
4. oxidation catalysts or thermal reactors for HC and CO

Section 14.4.3 discussed the SCR catalyst for NO_X reduction; this section includes a few comments on the reduction of SO_X and PM, where systems are available for two-stroke engines. Finally, a few words will be given on special oxidation catalysts, which have started to become available for certain applications. However, several of the methods are still on an experimental basis and, therefore, will require several running hours to accumulate to prove long-term durability and efficiency.

Because of the large differences in specifications of the fuels used in small and large diesel engines (DO versus HFO), the gas aftertreatment for the small engines is different than for the large two-stroke engines. A typical oxidation catalyst for low-sulfur DO cannot be used for HFO, but an oxidation catalyst was added after the $DeNO_X$ catalyst for a large two-stroke gas engine, as mentioned in Section 14.4.3.

14.5.4.1 Desulfurization and Filtration

Different methods for "flue gas" desulfurization are used on conventional coal and oil-fired power plants, but, due to cost and space requirements, the methods have not been used on board ships. Due to the high excess-air ratio and the different material on the particulates, the exhaust gas from a diesel engine is very different from the exhaust in a conventional plant and, therefore, needs to be handled differently.

Either wet (scrubber) or dry techniques are available [97], but the active process is almost the same. The exhaust gas is either mixed or sprayed with a chemical compound (containing, for example, calcium or sodium) that reacts with the sulfur oxides (SO_2 and SO_3) to produce different sulfates. A few examples on possible chemicals and reactions are given in the following:
using calcium carbonate (lime milk):

$$CaCO_3 + SO_3 => CaSO_4 + CO_2 \tag{14.11}$$

$$CaCO_3 + SO_2 + 0.5O_2 => CaSO_4 + CO_2 \tag{14.12}$$

or lime:

$$Ca(OH)_2 + SO_3 => CaSO_4 + H_2O \tag{14.13}$$

$$Ca(OH)_2 + SO_2 + 0.5O_2 => CaSO_4 + H_2O \tag{14.14}$$

or using sodium bicarbonate:

$$2NaHCO_3 + SO_3 => Na_2SO_4 + 2CO_2 + H_2O \tag{14.15}$$

$$2NaHCO_3 + SO_2 + 0.5O_2 => Na_2SO_4 + 2CO_2 + H_2O \tag{14.16}$$

The product is later removed from the exhaust gas either by washing or by filtering the waste material out (using, for example, baghouse or ceramic filters). The removal efficiency depends on different parameters for the method (e.g., proper mixing, available mixing, and reaction time, or solubility for the wet systems). The process economy can be improved, if the active material can be reprocessed and recycled (e.g., Na_2SO_4) or, if the waste product can be utilized somewhere else. But both methods will add to the complication of the plant and require more space. Also the disposal of the different materials may create problems on ships or islands in sensitive areas either environmentally or in handling.

Both methods are affected by the particulate content in the exhaust gas, but as long as this can be handled by the method, the benefit may be a simultaneous reduction of the exhaust PM.

A few diesel engine power plants today are using dry filtering of the exhaust gas using the baghouse method or electrostatic precipitators to remove particulates, but the method seems too bulky for marine use. The ceramic-trap technology from the small engines has not yet produced a well-proven system. And, due to the use of low-sulfur fuels and the introduction of oxidation catalysts, by which the present regulation can be fulfilled for the small engines, the traps have been postponed again for some years.

The main problem for the many trap devices is the frequent regeneration process that is necessary to clean the filter for deposited soot. In this connection, the type of particulates, the ash, and the metal content from fuel and lube oil add to the severity of the problem not only for the filter itself but also by contaminating the waste product. To ease this filter problem, electrostatic devices are introduced to increase the particle size and allow a simple collection device to take the particles out before the filter. The filter, therefore, will require less frequent regenerations [99].

14.5.4.2 Oxidation Catalysts

Oxidation catalysts have been introduced on small diesel engines in order to reduce HC, CO, and particulates [99]. However, the influence on PM, SOF, and soot in the catalyst is somewhat controversial [100]. Different conditions of the exhaust gas (e.g., temperature and SO_x), influence the processes in the catalyst. The reduction of SOF requires a certain gas temperature, but an overly high temperature increases the SO_2 to SO_3 conversion and, consequently, increases the formation of sulfates. The optimum temperature window was found to between 200 and 350°C [99]. Also the carbon (soot) content of the particulates will oxidize in the presence of O_2 in the exhaust depending on the actual temperature. The oxidation rate is reported negligible at 300°C but significant at 400°C [100]. Different additives added to the fuel or coated on the catalyst have been investigated in connection with lowering the catalyst light-off temperature.

The fuel-sulfur content is wanted reduced not only because of the direct influence on PM sulfates, but also because of an SO_2 influence on the catalyst conversion efficiency for HC and CO oxidation. The HC conversion deteriorates especially at low exhaust temperatures. As the fuel and lube-oil ash content influences the particulate trap regeneration process, the ash also affects the oxidation catalyst. Therefore, low-sulfur fuels and low-additive lube oils are likely for the small engines to ensure good conversion efficiencies and long lifetimes for the catalysts [99].

Catalytic reduction for the large two-stroke engines is still in the testing phase. And before a solution has been found to the problems mentioned earlier, the method will not be introduced. The exception will be on "clean" fuels, where a reduction of HC and CO for some reason is required. A catalytic oxidation catalyst can be installed after the $DeNO_x$ catalyst [67], as mentioned in the 12K80MC-GI-S example in the gas engine (Section 14.4.2.2). More than 95 percent of the CO can be removed and a similar removal can be achieved for HC, except for methane.

Also NH_3 is oxidized in the catalyst, eliminating an eventual slip of NH_3 from the $DeNO_x$ catalyst. However, the oxidation catalyst promotes the simultaneous oxidation of NO to NO_2, which may result in an undesired visible brownish color of the plume from the stack. Furthermore, strong-smelling aldehydes as well as polycyclic aromatic hydrocarbons (PAH) may be reduced by the oxidation catalyst.

Finally, an integration of all of the foregoing methods are being tested for off-highway use [101]. This development involves a combination of an SCR catalyst followed by an oxidation catalyst layer and a knitted fiberglass trap regenerated by either a fuel additive or a catalyst coating.

REFERENCES

1. MEPC, "Prevention of air pollution from ships—Draft Technical Code on Emission of Nitrogen Oxides from Marine Diesel Engines," MEPC 39/6/1, November 15, 1996.
2. IMO Sub-committee on bulk chemicals (BCH) correspondence group on regional control options for NO_x and SO_x, "A regional approach of air pollution by ships," first correspondence paper, December 1992.
3. Koefoed, J.H., "The North Sea Conferences," Institute of Marine Engineers, Paper 4, IMAS 96, October 1996.
4. U.S. Environmental Protection Agency, "Reducing Pollution from Marine Engines—Information on the Marine Engine Rulemaking," EPA Air Docket (LE-131) No. A-92-28, June 1995.
5. U.S. Environmental Protection Agency, "Control of Air Pollution: Emission Standards for New Gasoline Spark-Ignition and Diesel Compression-Ignition Marine Engines." Supplemental Notice of Proposed Rulemaking (SNPRM), *Federal Register*, Vol. 61, no. 26, February 1996.
6. U.S. Environmental Protection Agency, "Federal implementation plans (FIPs) for California-Sacramento, Ventura and South Coast Areas," 40 CFR Parts 52 to 81, *Federal Register*, February 1994.
7. Sjøfartsverket, "Detailed proposal for environmentally differentiated state navigation fees," Stockholm 1997-04-01 (in Swedish.)
8. The World Bank Environment Department, "Industrial Pollution Prevention and Abatement Handbook," preliminary version, June 1995.
9. The World Bank Environment Department, "Pollution Prevention and Abatement Handbook—Part III, Engine-Driven Power Plants," preliminary version, August 1996, revised version, November 1996 and March 1997.
10. Paro, D., "Proposal for a low emissions ship," MotorShip, the 17th Annual Marine Propulsion Conference—Shipping and the Environment: Competitors or Companions, March 1995, Amsterdam.
11. Lustgarten, G.A., "Economics of Gas Turbine and Diesel Applications." Panel session at the 18th CIMAC Conference, Tianjin 1989.
12. Gros, S., "Medium Speed Diesels in Cruise Vessels and Passenger Ferries—Wartsilas Diesel's emission technology," add to The MotorShip, pp. 18–21, December 1993.
13. Fleisch, T.H., "More Dimethyl Ether: Case Is Building for DME as Clean Diesel Fuel." *Diesel Progress Engine & Drives*, pp. 42–45, October 1995.
14. CIMAC, "Exhaust Emission Measurements—Recommendations for Reciprocating Engines and Gas Turbines," No. 12 1991.
15. Kalmbach, S., and J. Schmølling, "Technical Manual for Cleaning the Air—TA Luft, February 1986," Erich Schmidt Verlag, 2nd printing 1986 (in German).
16. Heywood, J.B., *Internal Combustion Engine Fundamentals*, McGraw-Hill Series in Mechanical Engineering. New York: McGraw-Hill, Inc. 1988.
17. Turns, S.R. *An Introduction to Combustion—Concepts and Applications*, McGraw-Hill Series in Mechanical Engineering, McGraw-Hill, Inc. 1996.
18. MAN B&W Diesel A/S, "Emission Control Two-Stroke Low-Speed Diesel Engines." Publication P. 331-96.12, December 1996.
19. Nakano, K., et al., "NO_x reduction test results on a two-stroke low speed diesel engine," 21st CIMAC Conference, Paper D 20, Interlaken 1995.
20. Schmidt-Sørensen, J., and P.S. Pedersen, "The MC Engine: Design for reliability and low maintenance costs," 18th CIMAC Conference, Tianjin 1989.
21. Grøne, O., "The multi-fuel diesel engine." Paper presented at a seminar on unconventional propulsion systems, Institute of Marine Engineers, March 1990, London.

22. Behrens, R., and K. Groth, "Problems caused by burning heavy fuels in diesel engines." Diesel Engine Combustion Chamber Materials for Heavy Fuel Operation, Paper 3, Marine Management 1990.
23. Teetz, C., "Comments on the reduction of corrosion in diesel engines," *VDI Forschungsheft*, no. 626/1984 (in German).
24. Schramm, J., et al., "Modelling of Corrosion of Cylinder Liner in Diesel Engines Caused by Sulphur in the Diesel Fuel," SAE Paper 940818, 1994.
25. Masters, K.J., "Lubricating oil analysis—what is it all about." *Transactions of The Institute of Diesel and Gas Turbine Engineers*, Publication 489, December 1995.
26. Wagner, H.G., "Soot Formation—An Overview," in *Particulate Carbon Formation During Combustion* (D.C. Siegla and G.W. Smith, eds.). New York: Plenum Press, 1981.
27. International Organization for Standardization (ISO), "Reciprocating internal combustion engines—Exhaust emission measurements," ISO 8178 Part 1, "Test bed measurements of gaseous and particulate emissions," September 1995. Part 2, "Measurement of gaseous and particulate exhaust emissions at site," August 1996. Part 4, "Test cycles for different engine applications," September 1995.
28. MacDonald, J. Scott, et al., "Experimental Measurements of the Independent Effects of Dilution Ratio and Filter Temperature on Diesel Exhaust Particulate Samples," SAE Paper 800185, 1980.
29. Johnson, J.H., et al., "A Review of Diesel Particulate Control Technology and Emissions Effects—1992 Horning Memorial Award Lecture," SAE Paper 940233 (SP-1020), 1994.
30. Wall, J.C., and S.K. Hoekman, "Fuel Composition Effects on Heavy-Duty Diesel Particulate Emissions," SAE Paper 841364, 1984.
31. Lange, W., et al., "Potential of Diesel Fuels to Reduce Particulate and NO_x Emissions in Trucks." *Motortechnische Zeitschrift (MTZ)* 54, no. 10, 1993 (in German).
32. Wall, J.C., et al., "Fuel Sulfur Reduction for Control of Diesel Particulate Emissions," SAE Paper 872139, 1987.
33. Henningsen, S., "Influence of the Fuel Injection Equipment on NO_x Emissions and Particulates on a Large Heavy-Duty Two-Stroke Diesel Engine Operating on Water-in- Fuel Emulsion," SAE Paper 941783, September 1994.
34. Henningsen, S., "Fuel and Lube Effects on Exhaust Emissions—An Investigation on Cylinder Liner Conditions," MAN B&W Diesel A/S, unpublished data, 1996.
35. U.S. Environmental Protection Agency (EPA), "Code of federal regulations," CFR 40, Part 60, App. A, Edition 7-1-1989.
36. Verein Deutscher Ingenieure (VDI), "Particulate Measurements," VDI 2066 2nd page, June 1981 (in German).
37. Japanese Industrial Standard, "Methods of Measuring Dust Concentration in Flue Gas," JIS Z8808, 1986.
38. Stoll, H., and H. Bauer, "Smoke Measurements in Diesel Engines" (Bosch smoke test, ETD 020.00), *Motortechnische Zeitschrift (MTZ)* 18, no. 5, May 1957 (in German).
39. Homan, H.S., "Conversion Factors among Smoke Measurements," SAE Paper 850267, 1985.
40. International Organization for Standardization, "Reciprocating internal combustion compression-ignition engines—Apparatus for measurement of the opacity and for determination of the light absorption coefficient of exhaust gas," ISO CD 11614.
41. British Standard, "Ringelmann Smoke Chart," BS 2742:1969.
42. International Organization for Standardization (ISO), "Reciprocating internal combustion engines—Exhaust emission measurements," ISO 8178 Part 9, "Test cycles and test procedures for test bed measurement of exhaust gas smoke emissions from RIC engines operating under transient conditions," ISO/TC 70/SC 8/WG1, No. 51, February 20, 1997 (in preparation).
43. Black, F., and L. High, "Methodology for Determining Particulate and Gaseous Diesel Hydrocarbon Emissions," SAE Paper 790422, 1979.
44. Cartellieri, W., and P. Tritthart, "Particulate Analysis of Light Duty Diesel Engines (IDI & DI) with Particular Reference to the Lube Oil Particulate Fraction," SAE Paper 840418, 1984.

45. Mayer, W.J., et al., "The Contribution of Engine Oil to Diesel Exhaust Particulate Emissions," SAE Paper 800256 (P-86), 1980.
46. Halsall, R., et al., "An Improved Method for Determining the Hydrocarbon Fraction of Diesel Particulates by Vacuum Oven Sublimation," SAE Paper 872136, 1987.
47. Abbass, M.K., et al., "Comparison of methods for the determination of the SOF of diesel particulates and development of a rapid TGA method for the estimation of the unburnt fuel and lubrication oil fractions," presented at the I.Mech. E Seminar, "Experimental methods in engine research and development '91," London, December 1991.
48. Elamir, I.E., et al., "Determination of diesel engine lubricating oil consumption through analysis of the calcium in diesel particulates," presented at the I. Mech. E Seminar, "Experimental methods in engine research and development 91," London, December 1991.
49. Williams, P.T., et al., "The Role of Lubricating Oil in Diesel Particulate and Particulate PAH Emissions," SAE Paper 872084, 1987.
50. Sienicki, E.J., et al., "Diesel Fuel Aromatic and Cetane Number Effects on Combustion and Emissions From a Prototype 1991 Diesel Engine," SAE Paper 902172, 1990.
51. Opris, C.N., et al., "The Effects of Fuel Sulfur Concentration on Regulated and Unregulated Heavy-Duty Diesel Emissions," SAE Paper 930730, 1993.
52. Williams, P.T., et al., "Diesel Particulate Emissions: The Role of Unburned Fuel," *Combustion and Flame*, 75, pp. 1–24, 1989.
53. Lloyds Register, "Marine Exhaust Emissions Research Programme—Phase II, Transient Emission Trails," Publication from Lloyds Register Engineering Services, London, 1993.
54. Dexter, S.G., "Residual fuels—combustion, emissions, and particulates," presented at the I. Mech. E Seminar, "Diesel fuel injection systems," Birmingham, October 1989.
55. Ryan T.W., et al., "The Effects of Fuel Properties and Composition on Diesel Engine Exhaust Emissions—A Review," SAE Paper 810953, 1981.
56. Henningsen, S., and D. Tsalapatis, "Emission Control of Two-Stroke Marine Diesel Engines," Paper presented in CCMS Workshop on Air Pollution from Marine Engines, Athens, January 1994.
57. Pedersen, P.S., "Emission Control of Large-Bore Two-Stroke Engines," FISITA 96, Paper in Proceedings from the XXVI Congress, June 1996, Praha.
58. Collin, L.Th., "A New Compression Ignition Combustion System to Greatly Lower the NO_x Production Without Sacrificing the Fuel Consumption," SAE Paper 960847, 1996.
59. Fleischer, F., "NO_x reduction—a technical challenge for marine diesel engine manufacturers," Paper 13 presented at IMAS 96, "Shipping and the Environment, Is Compromise Inevitable?" Institute of Marine Engineers, October 1996.
60. Wollenweider, J., et al., "Residual fuels in emission-controlled diesel engines—Background, developments and operational results," 21st CIMAC Conference, Paper D69, Interlaken 1995.
61. Niven, H., "Impact of emission legislation on large engine design and operation." Institution of Diesel and Gas Turbines, Publication 475, August 1993.
62. Boot, P., "NO_x-reduction of diesel engines by combustion air conditioning," 20th CIMAC Conference, Paper D67, London 1993.
63. Velji, A., et al., "Diesel engines with water injection passes future NO_x and smoke limits." *Motortechnische Zeitschrift (MTZ)* 57, no. 7/8, pp. 400–407, 1996 (in German).
64. Miyano, H., et al., "The ship test for low-NO_x by stratified fuel-water injection system," 21st CIMAC Conference, Paper D15, Interlaken, 1995.
65. Fukuda, T., et al., "Development of the World's First Large-Bore Gas-Injection Engine," 21st CIMAC Congress, Paper D51, Interlaken, 1995.
66. Søndergaard, K., "Selective catalytic reduction system for the removal of NO_x: The answer to maximizing emissions reductions," MotorShip, the 17th Annual Marine Propulsion Conference—Shipping and the Environment: Competitors or Companions, March 1995, Amsterdam.

67. Morsing, P., and K. Søndergaard, "Catalytic reduction of NO_x, CO and HC emissions from gas engine co-generation plants," Haldor Topsøe A/S, Denmark, Publication May 1990.
68. McMahon, K.R., "Development and operation of deNO_x controlled ships by USS-Posco industries," USS-Posco Industries, Pittsburg, California, Publication 1994.
69. Aabo, K., "Designing and Operating Diesels for Reduced Emissions—MAN B&W Two- Stroke Diesel Engines," Paper presented at Propulsion '93, New Orleans, November 1993.
70. Wollenweider, J., et al., "Diesel emissions control research on the Mobil Explorer engine," 20th CIMAC Congress, Paper D52, London, May 1993.
71. Schoubye, P., et al., "Reduction of NO_x emissions from large diesel engines," 17th CIMAC Congress, Paper D31, Warsaw, June 1987.
72. Sonoda, K., et al., "Research and development of marine deNO_x system," MARIENV '95, pp. 785–792, 1995.
73. Wright, A.A., and B.L. Burlingham, "The SCR option." Institute of Marine Engineers, IMAS 96, Paper 14, October 1996.
74. Cooper, D., "Emission measurements onboard ships—IVL work activities," Paper presented in CCMS Workshop on Air Pollution from Marine Engines, Athens, January 1994.
75. Götmalm, Ö.A., and P.B. Holström, "Retrofit of a catalytic converter." Institute of Marine Engineers, *Transactions of IMarE*, Vol. 108, Part 1, pp. 15–25, 1995.
76. Greeves, G., et al., "Origins of Hydrocarbon Emissions from Diesel Engines," SAE Paper 770259, 1977.
77. Yu, R.C., et al., "Sources of Hydrocarbon Emissions from Direct Injection Diesel Engines," SAE Paper 800048, 1980.
78. Yu, R.C., et al., "The Effect of Mixing Rate, End of Injection, and Sac Volume on Hydrocarbon Emissions from a D. I. Diesel Engine," SAE Paper 831294, 1983.
79. Henningsen, S., "Hydrocarbon Emissions from the Ignition-Delay Period in a Direct-Injection Diesel Engine," SAE Paper 841381, 1984.
80. Henningsen, S., "Hydrocarbons in the Squish Area of A DI Diesel Engine—An In-Cylinder Sampling Study," ASME Paper 86-ICE-3, 1986.
81. Gill, A.P., "Design Choices For 1990's Low Emission Diesel Engines," SAE Paper 880350, 1988.
82. Matsui, Y., and K. Sugihara, "Sources of Hydrocarbon Emissions from a Small Direct Injection Diesel Engine," SAE Paper 871613, 1987.
83. Andoh, H., and K. Shiraishi, "Influence on Injection and Combustion Phenomena by Elimination of Hole Nozzle Sac Volume," SAE Paper 860416, 1986.
84. Moore, S., "The Measurement of Piston Ring Oil Film Thickness in a Two-Stroke Marine Diesel Engine," 20th CIMAC Conference Paper, London, 1993.
85. Greeves, G., and C.H.T. Wang, "Origins of Diesel Particulate Mass Emission," SAE Paper 810260, 1981.
86. Cooke, V.B., and W.R. Vistillo, "Lubrication of Low Emissions Diesel Engines, Part 1 and Part 2," SAE Paper 900814, 1990.
87. Jacobs, R.J., and K. Westbrooke, "Aspects of Influencing Oil Consumption in Diesel Engines for Low Emissions," SAE Paper 900587, 1990.
88. Kamimoto, T., and M.-H. Bae, "High Combustion Temperature for the Reduction of Particulate in Diesel Engines," SAE Paper 880423, 1988.
89. Kittelson, D.B., et al., "Particulate Emissions from Diesel Engines: Influence of In-Cylinder Surface," SAE Paper 900645, 1990.
90. Essig, G., et al., "Diesel Engine Emissions Reduction—The Benefits of Low Oil Consumption Design," SAE Paper 900591, 1990.
91. Egeberg, C.E., et al., "Further Development of the MC Engine Family," 21st CIMAC Congress, Paper D50, Interlaken, 1995.
92. Andrews, G.E., and L.A. Charalambous, "An Organic Diesel Fuel Additive for the Reduction of Particulate Emissions," SAE Paper 912334, 1991.

93. Shiode, K., et al., "Influence of Fuel Properties on Combustion and Emissions of Medium Speed Marine Diesel Engine," 21st CIMAC Conference, Paper D31, Interlaken, 1995.
94. Ullman, T.L., et al., "Effects of Fuel Aromatics, Cetane Number, and Cetane Improver on Emissions from a 1991 Prototype Heavy-Duty Diesel Engine," SAE Paper 902171, 1990.
95. McMillan, M.L., and R. Halsall, "Fuel Effects on Combustion and Emissions in a Direct Injection Diesel Engine," SAE Paper 881650, 1988.
96. Baranescu, R.A., "Influence of Fuel Sulfur on Diesel Particulate Emissions," SAE Paper 881174, 1988.
97. Ives, R., and S.N. Klokk, "Exhaust gas sulphur removal by seawater washing, marine diesel engines," 20th CIMAC Conference, Paper D58, London, 1993.
98. Thimsen, D.P., et al., "The Performance of an Electrostatic Agglomerator as a Diesel Soot Emission Control Device," SAE Paper 900330, 1990.
99. Zelenka, P., et al., "Reduction of Diesel Exhaust Emissions by Using Oxidation Catalysts," SAE Paper 902111, 1990.
100. Johnson, J.E., and D.B. Kittelson, "Physical Factors Affecting Hydrocarbon Oxidation in a Diesel Oxidation Catalyst," SAE Paper 941771, 1994.
101. Hug, H.T., et al., "Off-Highway Exhaust Gas After-Treatment: Combining Urea-SCR, Oxidation Catalysis and Traps," SAE Paper 930363, 1993.

PART V

Spark-Ignition Engines

CHAPTER 15

Introductory Chapter: Fuel Effects

David R. Blackmore
Shell Research and Technology Centre, Shell Research Ltd., Thornton, Chester, United Kingdom

ISBN: 0-12-639855-0
$25.00

15.1 HISTORICAL LANDMARKS

The evolution of fuels for the IC engine has been a continuing story of the matching of fuel design with engine design. Sometimes the one has led the other, at other times the converse.

Thus, in the very early days, the main intent of the engine designer lay in improving the power output and reliability of the gasoline SI engine. The obvious way was to increase the compression ratio of the engine, which was a mechanically feasible route. However, the situation was limited by the phenomenon of knock, and so the constraint of a knock-limited compression ratio was encountered.

In the 1920s, there was a chase to improve the knock resistance of the fuel, and the story of the discovery of lead antiknock additives is well known and well documented. A new satellite industry was born and by the end of the 1920s, octane quality of road fuels in the United States and in Europe had improved dramatically, with a consequent increase in compression ratios and power (and economy) output of road vehicles.

In the last three decades, the issue of emissions has become a requirement of the engine and fuels designer. This requirement has needed to be added to the existing ones for power, economy, and smoothness (driveability). In the case of the gasoline engine, these emissions have consisted of crankcase ventilation, tailpipe exhaust emissions, and evaporative emissions from both fuel tank and carburetor.

Sometimes these emissions requirements and, ultimately, regulations have been met by simple mechanical design features (e.g., positive crankcase ventilation systems), by simple fuel design changes (e.g., reduced volatility to control evaporative emissions, reduced sulphur content of diesel fuel), but more usually by fuel design changes that *enable* mechanical design features to be adopted or optimized.

15.1.1 Lead in Gasoline

By the mid-1960s, lead contents in premium gasoline had risen to high levels (typically 1 g/l, equivalent to four or five octane numbers). Worries were being expressed on two accounts.

Lead alkyl additives were well known to be toxic in their own right, and concerns were increasingly being expressed that the lead emissions entering the ambient air were harmful to public health. It was difficult to prove from medical evidence that there was measurable harm to the mental or physical health of the population, or vulnerable parts thereof, but the situation was felt to be too close for comfort.

It was also increasingly evident that in some cities, air quality was seriously impaired by the exhaust emissions of the motor vehicle, the most notable example being the Los Angeles basin in California. The understanding of photochemical smog had advanced to the point that it was clear that some serious steps were required.

In the late 1960s, the idea of the tailpipe catalytic convertor became a developmental reality, but the lead in the gasoline had another deleterious feature, namely, it irreversibly degraded the performance of these catalysts. No way could be found to develop a lead-resistant catalyst, so it became necessary to remove the lead from gasoline to enable this catalyst technology to progress. Thus, it was that in the early 1970s, the oil companies began to provide unleaded gasoline (of adequately good octane and driveability quality). By 1975 such a grade became mandatory nationwide in the United States. And from there it spread to Europe, Japan, and elsewhere. Today, over 75 percent of the world's gasoline is unleaded.

15.1.2 Additives in Gasoline

Gasoline has been treated with additives for the last 70 years. Initially they were added to control such combustion abnormalities as knock, preignition, or spark plug fouling, all of which had effects on power and driveability of the engine. Other additives were found useful for the prevention of gums during storage, or for the control of carburetor icing.

In the 1960s a new class of detergent additive began to appear. The first additives were aimed at the control of carburetor deposits, so as to preserve the performance of engines in respect of economy and, for the first time, emissions of carbon monoxide (CO). A second generation of detergent additives made their appearance in the 1970s, where strength and dosages were such as to aim for control of deposits in the inlet system and on the inlet valve. Further benefits were demonstrable in terms of emissions and economy.

Yet another development took place in the late 1980s with the advent of increased use of inlet-port fuel injection. Additive formulations were needed to control deposits on the inlet valve in such systems, and particular activity took place in the U.S. market so as to preserve emissions performance at the ever-lower levels demanded by the advancing legislation.

In the 1990s additive development is still moving ahead, with emphasis on two new areas. First, additives are required for the lead-replacement situation, so as to preserve the exhaust valve seat wear protection enjoyed by older vehicles with lead-containing gasolines. This is likely to be a temporary phase, until the old cars with metallurgically "soft" exhaust valve seats disappear from the marketplace. Second, interest is growing in additives to control deposits inside the combustion chamber, and to combat the deleterious effects of such deposits on emissions.

15.1.3 Oxygenates in Gasoline

Normally gasoline has consisted of refined fractions from a barrel of crude hydrocarbon oil. The blends have mainly consisted of a straight-run distillate (naphtha), a catalytically cracked stream, and a catalytically reformed stream. In cases where there have been national constraints on supply, other components obtained from coal, petroleum gases, and biosources have been occasionally used.

A surge of interest in oxygenate sources came in the late 1970s when it was realized that the economics of supply of hydrocarbons was likely to come under threat worldwide. Thus, considerable effort went into defining the technical features of such octane extenders as ethers and the lower alcohols.

Later in the 1980s, it was realized that some of these materials were effective under cold weather conditions to control certain emissions. Thus, fuels containing methyl t-butyl ether (MTBE) were mandated in various U.S. states for use in winter as a measure to control CO emissions in cities.

There has also been an ongoing pressure to adopt the use of alcohol in gasoline, and this has been particularly notable in farming areas of the world where excess grain stocks exist. Thus, gasohol (a mixture of 10 percent ethanol in unleaded gasoline) has been sold in the midwestern states of the United States. Methanol at 15 percent concentration was championed by some, while 85 percent concentration was preferred by others. Flexible-fueled vehicles were designed that operated on fully variable mixtures of methanol and gasoline.

In all these cases, there is a balance to be struck between the cost of the introduction of such oxygenate materials, and the equivalent cost of upgrading the refinery processing against which it has to be compared. The balance of performance behavior in the two cases also needs to be carefully weighed across all the performance and emission features. The verdict is by no means obvious, and indeed depends strongly on the prevailing needs of the economic marketplace and, not least, on the demands of the environmental legislation.

15.1.4 Sulphur in Diesel Fuel

We draw on one further landmark from the diesel fuel area to illustrate the complexity and interrelatedness of environmental control issues.

For many years diesel fuel for both the road truck markets and the recently growing car markets has had a maximum of 0.2 percent sulphur in Europe, higher in some other parts of the world. This sulphur came under environmental scrutiny for two reasons. First, the emission of sulphur (as oxides) in its own right needed to be controlled, even though the automotive diesel fuel contributed a relatively minor 5 percent amount to the total in most industrialized areas. Second, the sulphur in diesel fuel contributes significantly to the particulate emissions by virtue of its oxidation and combination with water to generate sulphates, which bind together with the soot to increase the emissions.

Accordingly, sulphur limits in diesel have recently been reduced in Europe to 0.05 percent, which has meant some considerable technological development in refinery hydrodesulphurization catalysts. This extra and costly refining processing activity necessarily leads to an increase in CO_2 emissions, thus demonstrating an implicit local versus global trade-off. If there were to be a further reduction in diesel fuel sulphur levels to say 0.02 percent, it needs to be appreciated that the benefit in particulate control is not commensurate: The law of diminishing returns applies, and so careful assessment of the cost-effectiveness of such legislation will need to be undertaken.

15.2 RECENT DEVELOPMENTS

In the current decade there have been some important activities that throw light on the role of fuels design on environmental performance.

15.2.1 Auto Oil Process

In the late 1980s, it became evident that the air quality in U.S. cities was not attaining the levels of cleanliness that had been hoped for by the legislators after more than two decades of effort. So yet further rounds of tightening were imminent in the U.S. Clean Air Act, the most striking of which was the call for the introduction of either methanol as a fuel or the adoption of a certain fraction of electric vehicles in California. Other stringent requirements on the durability (now for 100,000 miles) of conventional catalysts on vehicles began to appear.

In partial response to these sorts of pressure, the U.S. auto manufacturers challenged the oil companies in the early 1990s to come to their assistance and to engage on a major collaborative program to measure the effects of fuel property changes on emissions performance of fleets of current and near-future vehicles. A major $40 million bilateral collaborative program was completed on time: The findings showed measurable effects of the fuel property changes, though not unexpectedly, they were small relative to effects of engine design changes. These results were fed into the legislative process, resulting in the identification of CO and ozone as the major pollutants requiring control, and the emergence of the complex model to specify fuel quality. This model operates on the principle of permitting trade-offs of a number of fuel properties simultaneously, thus allowing the refiner maximum flexibility to meet the emissions standards.

A rather similar auto-oil program has taken place in Europe in the mid-1990s, involving a three-way partnership between the oil industry, the motor industry, and the European Commission. The program was designed to link up in advance with air quality standards, thus seeking to define in a rational and cost-effective way the effect of any changes in fuel or engine design on air quality in named and typical

European cities. Thus, a major collaborative program was completed, which combined the European Programme of Emissions and Fuels in Engines (EPEFE) on measured fuel effects on emissions performance in European vehicles with a subsequent air quality program to examine a series of control scenarios on city air. Control features other than fuels were also evaluated, namely inspection and maintenance, traffic management, and others. Overall, the results have shown that there are concerns on NO_x and particulates, interestingly a rather different conclusion than in the United States. These findings are being debated at the political level with the object of defining a set of fuel specifications and vehicle emission limits for adoption in the year 2000, and setting the direction for a further set in the year 2005 (perhaps fully embracing the effects of particulates and stationary sources by then).

15.2.2 City Diesel Fuels

An interesting and instructive situation has occurred recently in Scandinavia. In the search for ultralow sulphur diesel fuels for city use, governments can decide to set up strong tax inducements that make it attractive for oil companies to manufacture and supply such fuels. In the case of Swedish Class 1 fuel, the requirement was for 10 ppm maximum of sulphur and 5 percent maximum of aromatics, together with a restriction on the tail of the fuel. Normally diesel fuel is supplied with a maximum of 0.05 percent sulphur in Europe.

There are some immediate consequences of such an extreme fuel enjoying benefits of extremely low emissions of SO_2. First, the power and economy performance of such a fuel are likely to be diminished. Second, there may be adverse side effects on pump lubricity without the adoption of a new balanced additive package. Third, the supply of such a fuel may mean that refineries, balance is disturbed and that some heavier fuel components will have to be exported to other markets. And fourth, the increased processing severity to achieve low sulphur may need to be bought at the expense of increased CO_2 emissions.

This case provides a very good example of the way that an assessment of a "clean" fuel needs to take into account all aspects, including other primarily environmental ones. It is a matter for a balanced societal decision, which may well be different in different communities. The role of the technical community is to ensure that the full technical picture and its implications are presented in a rational way.

15.2.3 The Developing Engine

Hitherto, we have described the picture as if it is the case of matching fuels to a car population static in its technical requirements. But this is far from the case.

Over the decades of this century, there have been major shifts and developments in gasoline engine design. Not only has there been a gradual evolution of the gasoline engine in terms of its compression ratio and its fuel octane requirement, there have been other notable design changes and innovations:

Overhead cam action
Carburation development
Single-point (throttle-body) fuel injection
Port fuel injection
Four-valve operation
Swirl control systems
Exhaust catalyst development
Lambda sensor mixture strength control
Engine management systems

The process has not stopped and the current prospects are:

On-board diagnostic management system
Variable valve timing
Adaptive transient air-fuel control
Direct gasoline injection

There have also been some new engine design concepts that have been or will be adopted in some measure at least:

Wankel rotary engine
Advanced two-stroke engine
TLEVs, LEVs, ULEVs, and ZEVs for California

There has been a similarly active development in the diesel engine over the decades:

Improved robustness of the engine for power and durability
Injection system developments
- High-pressure, timed, electronic control systems
- Unit pump-injector systems for bigger engines
- Common rail injection for improved fueling match to engine needs

Turbocharging (with intercooling, and with variable geometry)
EGR (with cooling)
Multivalves
High-speed operation for automotive use (IDI to DI shift)
High- and variable-swirl combustion systems

Exhaust catalysts (oxidation, deNO_x)
Exhaust system particulate traps

In general the engine design has become more tuned and adaptive to the performance needs, and these in turn have owed an increasing dependence on the requirements of the emissions legislation.

In the case of every change, there will be a new optimum in fuel appetite, and so the cumulative outcome may be a shifting in the fuel grade structures available in the different marketplaces.

15.3 THE FUTURE

There is bound to be a future for the automotive engine, but will it be a hydrocarbon-fueled IC engine?

15.3.1 Biofuels

There has been much research into biologically grown fuels from special crops, such as Brazil's sugar cane crops that yield ethanol in a fermentation process, or fermented grain in the United States. Another example is the use of rape seed methyl ester for use in diesel fuels. These fuels find a particular application in certain countries and conditions where the economics are right. But in the longer run they are unlikely to command a major share of the fuel market: They would consume unacceptably large areas of land and, additionally, they represent in practice a very poor CO_2 yield.

15.3.2 Gaseous Hydrocarbon Fuels

LPG and CNG have been used to propel road vehicles for decades in particular markets with appropriate needs. They represent a real option, and if they were to be widely adopted across a country, there is no technical reason why they should not be successful. Of course, there are drawbacks, as indeed any variant fuel will have to surmount, but there are some useful advantages, not least are the advantages with respect to CO_2 yield. However, currently there is no escaping the fact that it is intrinsically more costly to contain a pressurized gas than an unpressurized liquid.

15.3.3 Hydrogen

The future for hydrogen as a road fuel has long fascinated researchers and the technical community. There is no serious problem in designing an engine to run on combustible mixtures of hydrogen, nor to control such an engine in practice.

The hydrogen engine intrinsically gains in respect of CO and HC emissions, and its CO_2 yield looks attractive (depending on its sourcing). If hydrogen were to be supplied from an oil or gas source, then the use of the excess carbon would become an important question. The most serious problem is its containment in the fuel tank of a vehicle, and crash safety must be another concern. Distribution of hydrogen to fueling stations will be yet another challenge.

15.3.4 Fuel Cells

There have been many successful attempts to design fuel cells for prototype vehicular use, and this application would be a good outlet for hydrogen, yet further improving its energy efficiency. But containment and distribution still remain a problem. At present the cost of such fuel-cell technology is very high, and much will depend on the development of improved catalyst systems. On-board treatment of hydrocarbon fuel to convert it chemically into an acceptable mixture is another particular possibility. Such fuel-cell powered engines are likely to find particular application in city areas, making this an exciting long-term prospect.

15.3.5 Electric Battery Vehicles

Much interest and development work are currently being spent on electric vehicles, powered by batteries. Four battery systems are in view: advanced lead-acid, sodium-sulphur, nickel-cadmium, and nickel-metal hydride. Each of these systems has its own set of features: Specific power and energy density are the most obvious ones, but durability, replaceability, and recyclability of the batteries are also important. Typically batteries constitute a large proportion of a vehicle's weight (say 30 percent) and over the lifetime of a vehicle, the energy use to generate the electricity is only a little less than the fuel combustion energy of a gasoline or diesel engine.

15.3.6 Hybrid Engines

The concept of hybrid power sources between, for example, battery-storage electric motors and IC engines operating at constant speed or load have been studied and built. Cost and complexity of the control systems have always been a drawback, but recent technical advances may change the picture, and enable such power systems to find application in city-suburban use.

15.4 IN CONCLUSION

The role of the technical community in meeting the environmental challenge has been an exciting one. Enormous strides have been taken, as evidenced by the fact that emission limits have decreased by nearly 99 percent since those early days of the mid-1960s when the first legislation appeared in order to combat the ill effects of high-emitting gas-guzzlers.

And yet, many city centers are still bound by traffic, and there is an ever growing desire to control not only the emissions but the motor car itself.

The solution is in our own hands. The technical community has proved that it possesses the knowledge, the tools, and the skill to move yet further toward a low-pollution world. All that is left to do is to communicate and collaborate, and to generate the collective will.

Perhaps that is the hardest thing of all.

CHAPTER 16

Fuel Effects on Emissions

Yoram Zvirin, Marcel Gutman, and Leonid Tartakovsky
Faculty of Mechanical Engineering, Technion, Haifa, Israel

ISBN: 0-12-639855-0
$25.00

16.1 BACKGROUND

To date, the absolute majority of fuels consumed by internal combustion engines (ICE) are fossil fuels, mainly gasoline and diesel fuel. For big size, stationary, or naval engines various fuel oils are generally used. Automotive fuels are the most important products manufactured and marketed by oil companies becasue large amounts (between 30 percent and 70 percent) of the crude oil run in a refinery is converted into gasoline and diesel fuel [1].

Through the fuels' history, their properties have kept changing because of various reasons, such as crude oil prices, progress in refinery technology, changes in vehicle technology, environmental legislation, and political considerations.

Modern automotive fuels, both gasoline and diesel fuel, must satisfy various requirements, such as: to enable fast refueling; fluently pass from the tank to the engine cylinders; effectively mix with the air; efficiently burn in the cylinders to produce adequate power and minimal amounts of pollutants in a wide range of ambient conditions. The main features required from automotive fuels, following from these demands, are summarized in Table 16.1.

The environmental legislation has become the most important factor affecting requirements of automotive fuels, due to: (a) additional limitations caused by changes in vehicle technology (such as the need of unleaded gasoline for catalyst-equipped vehicles); (b) the growing importance of direct fuel effects (their weighting factor rising sharply as a result of very low emission levels mandated in ecological regulations).

Numerous research works have been performed in order to investigate the fuel composition effects on engine exhaust emissions. Recent comprehensive works are the American Auto/Oil Air Quality Improvement Research Program (AQIRP) [2, 3, 4] and the European Programme on Emissions, Fuels and Engine Technologies (EPEFE) [5, 6]. The former was initiated by three U.S. automakers (GM, Ford, and Chrysler) and 14 petroleum companies, mainly for SI engines. The objective of this cooperative study was to develop data on potential improvements in vehicle emissions and air quality from reformulated gasoline, various other alternative fuels, and developments in vechicle technology. The latter program was aimed at extending benefits achieved from the former to the European conditions

Table 16.1

Main Features Required from Automotive Fuels

Feature	Relationship with engine and vehicle performance
Good combustion quality	Better ignition and combustion qualities lead to better vehicle fuel economy and less emission of pollutants. High octane or cetane numbers are critically important for good combustion quality in SI or CI engines.
Minimized deposit formation	Assists in maintaining engines close to their designed optimal efficiency and relieves the deterioration of performance, fuel economy, and emissions. Deposit control additives are low-cost, widely recognized means for suppressing deposit formation.
High heat of combustion	A smaller fuel quantity needs to be carried in the vehicle tank when its chemical energy content is high.
Suitable latent heat of vaporization	High latent heat of vaporization causes the charge to be cooled and therefore to become denser. However, there is danger of freezing ambient moisture in the carburetor.
Good performance at high and low temperatures	A fractional composition of fuel must enable easy cold start, good driveability, fuel economy, low exhaust and evaporative emissions, and reliable hot restarting without lubricant dilution in a wide range of ambient conditions. Usually, fuels are blended appropriately for both seasonal and geographical variations in temperature.
Materials compatibility	Materials compatibility is essential for preventing corrosion of fuel system components.
Stability	Better fuel stability enables minimizing deposit formation and storing fuel without deterioration for longer periods of time.
Low foaming tendency	Low foaming tendency is relevant for diesel fuels, enabling faster vehicle refueling with lower evaporative emissions.

(fuels, vehicles, and test procedures are quite different in Europe from those in the United States), to study the remaining gaps in the knowledge about fuel/emissions relationships and, finally, to provide the European Commission with the necessary information enabling a strategy to be proposed for vehicles and automotive fuels for the twenty-first century. The EPEFE study involved active participation of 14 vehicle manufacturers, represented by the Association des Constructeurs Europeens d'Automobiles (ACEA), and 11 petroleum companies, represented by the European Petroleum Industry Association (EUROPIA). The scope of EPEFE was restricted to gasoline and diesel fuel with emphasis and priority given to the latter [5].

Effects of different fuel variables on regulated (CO, HC, NO_x, PM) and unregulated (benzene, 1,3-butadiene, aldehydes, PAH, etc.) engine exhaust emissions were investigated in the foregoing and in many other research programs [e.g., 7, 8, 9, 10]. The accumulated knowledge allows main fuel parameters to be defined affecting pollutants emission and fuel/engine/emissions relationships to be revealed with good agreement between different studies.

Table 16.2 summarizes the main gasoline and diesel fuel properties found to have essential effects on engine exhaust emissions [e.g., 5]. Since it is sometimes difficult to separate the effects of some fuel parameters (for example, density and aromatics [5]), there are still some unresolved issues and additional studies

Table 16.2

Main Fuel Properties Affecting Engine Exhaust Emissions

Gasoline	Diesel fuel
Lead content	Sulphur content
Sulphur content	Density
Oxygenates content	Aromatics content
Aromatics content	Cetane number
Benzene content	Distillation characteristics
Olefins content	
RVP	
Distillation characteristics	

are needed in this field. Moreover, even relationships which have already been established must be further investigated in order to validate them for different vehicle technologies, test procedures, and refining processes.

Increasing severity of environmental legislation, together with considerations of long-term security of supply, have led to the rise of activities aimed at developing alternatives to conventional automotive fuels. Some of them, such as hydrogen, natural gas, and others, may provide sharp reductions of engine pollutants emission.

This chapter includes a discussion of the foregoing relationships between automotive fuels composition and engine emissions, and of main trends in fuel specifications developments initiated by severe environmental legislation. Possibilities of exhaust emission reduction by using some alternative fuels are also discussed.

16.2

GASOLINES (SI ENGINES)

16.2.1 Origin, Composition, and Properties

16.2.1.1 Gasoline Origin

As mentioned previously, gasoline is one of the two conventional liquid hydrocarbon fuels widely used today in motor vehicles. Gasoline (other terms sometimes used are *petrol* or *motor spirit*) is a fossil fuel produced from crude oil (technical term is *petroleum*) by a refining process.

The yield of gasoline products from crude oil is greatly dependent on their source. Table 16.3 includes estimates of proportions of distillation products from various crude oil sources. As can be seen from the table, the yield of gasoline fractions (light gasoline and naphtha) can widely range from 25 percent for North African crude to under 2 percent for the South American crude. Therefore, refineries generally need to be much more complex than simple distillation plants in order to accommodate processing of any available crude oil.

Table 16.3

Yield (% wt) of Main Products from Crude Oil by Distillation [11]

	North Africa	North Sea	Middle East	North America	South America
Sulphur	0.1	0.3	2.5	1.0	5.5
Wax	3	9	6	7	2
Light gasoline					
0–70°C	8.9	5.8	4.7	2.4	0.1
Octane No.	73	76	72	75	70
Naphtha					
70–180°C	16.0	11.0	7.9	6.5	1.1
Kerosine					
180–250°C	26.3	18.6	16.4	15.6	4.4
Diesel oil					
250–350°C	18.2	19.1	15.3	19.6	9.6
Cetane No.	55	53	58	45	30
Residue					
350°C+	27.5	36.2	47.2	47.9	76.9

16.2.1.2 Gasoline Composition

Generally, gasoline is a complex mixture of a great number (up to about 400 [1]) of different hydrocarbons. The name implies that these compounds contain carbon and hydrogen only, but many thousands of different combinations are possible depending on how the individual atoms in molecula are arranged. Carbon is a quadrivalent element and can combine with itself by single, double, or triple bonds. The stability of hydrocarbon molecula depends on the strength of chemical bonds and this, in turn, depends on the nature and structure of the various groupings present [1]. Hydrocarbons contained in gasoline belong mainly to the classes of paraffins, cycloparaffins (to a lesser degree), olefins, and aromatics.

Paraffins, or alkalines, (chemical formula C_nH_{2n+2}) are a class of saturated hydrocarbons with only single bonds. There are two types of paraffins in gasoline: normal and isomers (with carbon atoms arranged as a straight chain and branched chain, respectively). The more carbon atoms there are in a hydrocarbon molecula, the more isomers are possible. Generally normal paraffins and isomers are essentially different substances which differ in many properties. For example, the boiling points of normal octane (n-octane) and isooctane are 126°C and 99°C, respectively [1]. The normal, or n-paraffins, usually have low-octane quality compared to isoparaffins with relatively high antiknock performance [12].

Cycloparaffins, or naphthenes, (chemical formula C_nH_{2n}) are a class of hydrocarbons having a cyclic structure. In their simplest form they consist of CH_2 groups arranged in a cycle. Hydrogens attached to carbon atoms can be substituted by methyl or other groups [1]. These products are generally of low-octane quality and require secondary processing in order to enhance their knock resistance [12].

Olefins, or alkenes, (chemical formula C_nH_{2n}) are a class of unsaturated hydrocarbons, containing one or more double bonds. Although olefins have the same general formula as naphthenes, their behavior and characteristics are entirely different [1]. The double bond is a very reactive group, so the oxidation stability of olefins is much lower than that of saturated or aromatic hydrocarbons. Olefins in gasoline affect the emission of 1,3-butadiene, known as a dangerous air toxic. Therefore, olefins content in gasoline has been limited in some national specifications (see Appendix 1: U.S. Federal and Californian Specifications).

Aromatics (chemical formula C_nH_{2n-6}) are a class of hydrocarbons based on the benzenoid ring having three double bonds. The simplest member of this class is benzene (C_6H_6). The benzenoid (aromatic) rings can be fused together in different combinations. These compounds are called polycyclic aromatic hydrocarbons (PAH) or polynuclear aromatics (PNA). Benzene is known as a dangerous product and air toxic. Its content in gasoline is controlled by legislation. Other aromatics are also of concern because they too affect engine exhaust emissions. On the other hand, the aromatics generally have high antiknock performance; thus, they are needed for achieving target values of octane quality.

In addition to hydrocarbons of various classes, gasoline usually contains small amounts of nonhydrocarbon compounds, such as oxygenates, lead, phosphorus, sulphur, water, and others.

Oxygenates are usually added to unleaded gasoline in order to boost its octane quality. The types of oxygenates that are used, their effects on emissions, and restrictions on the use of oxygenates in gasoline are discussed in the following sections.

Lead content in both leaded and unleaded gasoline is highly controlled by legislation because of its high toxicity and poisonous effect on vehicle catalytic converters.

Phosphorus is an additional material that reduces the effectiveness of catalytic converters; therefore, its content in unleaded gasoline is strongly restricted.

Sulphur content in gasoline is limited because of its negative effects on engine exhaust emissions by catalyst deactivation.

Water can be present in a gasoline both in dissolved and free form because of contact with aqueous solutions during gasoline manufacture in the refinery and also due to the usual presence of free water at the bottom of storage tanks. Water may lead to a number of negative effects, such as line blockage, icing of intake system, promotion of corrosion, and so on; therefore, its content in gasoline is usually restricted.

A detailed description of gasoline chemistry can be found in special literature [e.g., 1].

16.2.1.3 Gasoline Properties

Octane Quality In spark-ignition engines the knocking phenomenon is a problem of great concern and gasoline with good octane quality is needed. The use of gasolines of low-knock resistance in high compression ratio engines causes efficiency losses, an increase in pollutants emission, and may lead to catastrophic engine damage under high-load conditions.

Octane number (ON) of a gasoline is defined as the volume percentage of isooctane in a blend with n-heptane (with ON taken to be 100 and 0, respectively), which shows the same antiknock performance as this gasoline when tested in a standard engine under standard conditions. There are two main methods of octane quality rating. These are research octane number (RON) obtained according to ASTM standard D2699 and motor octane number (MON) obtained according to ASTM standard D2700. Both tests are similar and are based on the same laboratory equipment. The main difference between them is the engine operation regimes relating to different driving conditions: RON—mainly urban driving, with engine speed and load relatively low; MON—severe driving conditions with higher engine speed and load.

In real driving, the engine operates most of the time at speeds and loads located between those corresponding to MON and RON. Therefore, the additional parameter of octane quality, known as antiknock index, has gained wide acceptance in the United States and some other countries. The antiknock index of gasoline is defined as an average of its RON and MON:

$$\text{Antiknock index} = (\text{RON} + \text{MON})/2$$

Sometimes, so-called road octane number is used in order to evaluate octane requirements of vehicles on roads. A detailed description of this fuel rating is given in Owen and Coley [1].

An additional important measure of gasoline octane quality is its *sensitivity*, defined as the difference between the RON and MON ratings:

$$\text{Sensitivity} = \text{RON} - \text{MON},$$

which represents the sensitivity of the fuel to changes in the severity of engine operating conditions in terms of knock resistance.

Volatility Gasoline volatility is a measure of its evaporating characteristics. Gasolines with higher volatility evaporate more readily and at lower temperatures; in general, they contain more light and volatile hydrocarbons. The volatility of a gasoline is usually evaluated by the following parameters: distillation performance, Reid vapor pressure (RVP), and vapor lock index (VLI) or vapor-liquid ratio (VLR). The distillation performance, usually evaluated by a test according to ASTM standard D86 [13] is defined in terms of the following parameters:

- percentage of gasoline that is evaporated at a certain temperature, or the temperature at which a certain percentage of gasoline is evaporated, see Figure 16.1;
- distillation residue (i.e., the volume of residue left in a cold flask after the distillation is complete);
- distillation loss, which represents mainly those very light hydrocarbons that are not condensed during the distillation process.

Fig. 16.1.

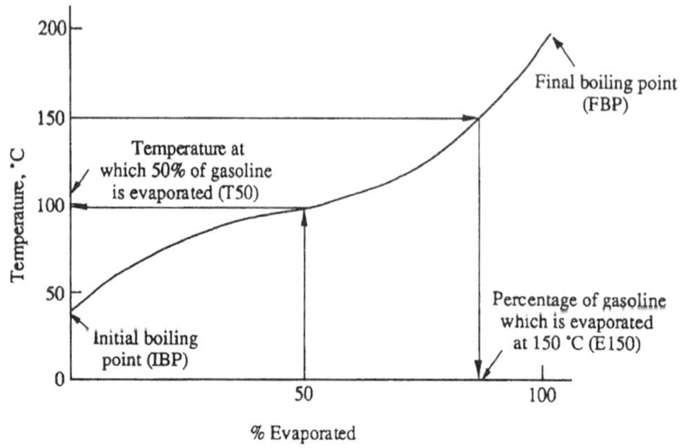

Typical distillation curve of gasoline.

Reid vapor pressure is an important parameter of gasoline volatility and is determined according to ASTM D323 procedure. Higher values of RVP indicate more volatile gasoline.

Vapor lock index values are calculated according to the formula:

$$\text{VLI} = \text{RVP} + n \times \text{E70}$$

where n is a constant. The value of $n = 7$ is widespread, especially in European countries.

VLI is a measure of the likelihood of a gasoline to cause fuel flow irregularities in vehicles on the road, due to the formation of vapor plugs in the engine fuel system (vapor lock). This parameter is very important for driveability and hot startability evaluation of a vehicle at hot ambient conditions. In the United States and some other countries, the term *vapor-liquid ratio* is used instead of VLI. The VLR values are usually evaluated according to ASTM D2533 procedure [13], indicating the volume of vapor formed at atmospheric pressure from a given amount of gasoline at a specified test temperature [1]. The VLR parameter, as also the VLI, is used to define the tendency of gasoline to vaporize in the fuel system of a vehicle.

Oxidation Stability Oxidation stability of a gasoline indicates its suitability for long-term storage and, in part at least, its tendency to form deposits in the engine. Several test methods are used in order to evaluate it. The most commonly used are induction period method, performed according to ASTM D525 procedure, and existent gum test, according to ASTM D381 standard. In the former, the stability is evaluated by oxidation of the gasoline in a closed vessel with oxygen at certain pressure and temperature, by measuring the duration of the induction period. This test is mainly intended for evaluation of gasoline suitability for long-term storage.

The existent gum value is the n-heptane insoluble part of a distillation residue, and it indicates, in part at least, the tendency of a gasoline to cause deposits formation in the engine, fuel filter blockage, and as a result to cause severe driveability problems, and, of course, rise of fuel consumption and pollutants emission.

Some Other Gasoline Properties Corrosivity of gasoline is a problem of great concern because it can lead to damage of fuel system elements, cause filter blockage by corrosion products, and increase wear rates. In addition, dissolved metals such as copper can catalyze oxidation reactions and lead to excessive deposit formation [1].

The corrosivity of gasoline is usually evaluated by the copper strip corrosion test, according to ASTM D130 procedure. There are also some steel corrosion tests outlined in Owen and Coley [1].

Density of gasoline is its mass per unit volume. Usually, gasolines have a density between 0.72 and 0.78 kg/l and it is a function of the composition.

Conductivity of gasoline is a parameter indicating the tendency of gasoline to static electricity build-up, mainly during pumping operations. The higher the fuel conductivity, more rapid is the dissipation of static electricity charge and, hence, there is less risk of an electrical discharge fire. Usually, conductivity values are specified for aviation gasoline and jet fuels [14]. Other gasoline properties, such as heat of combustion, viscosity, appearance, and so on are discussed in Owen and Coley [1].

National requirements of gasoline quality in different countries worldwide are summarized in Appendix 1. This information is mainly based on the CONCAWE data [15]. Average values of some gasoline parameters (typical for Europe and United States) affecting pollutants emission and mentioned earlier in Table 16.2 are given in Table 16.4 [6, 7].

Table 16.4

Typical Values of Gasoline Parameters

Property	European market average gasoline	1990 U.S. industry average gasoline
Sulphur (ppm wt)	300	295
RVP (kPa)	68	57.9
Aromatics (% vol.)	40	34
Benzene (% vol.)	2.3	1.7
Olefins (% vol.)	11	7.7
Oxygenates	0.6% vol. O_2	0.1% vol. ethanol 0.2% vol. MTBE 0.1% vol. TBA
Distillation	E100 = 53% E150 = 84%	T50 = 102°C T80 = 144°C T90 = 163°C

Abbreviations used in this table: RVP (Reid vapor pressure); MTBE (methyl tertiary butyl ether), TBA (*t*-butyl alcohol).

16.2.2 Requirements of Engine Technologies on Fuel Quality

General requirements to automotive fuels, outlined earlier and summarized in Table 16.1, are discussed here in more detail for the case of gasoline use in SI engines.

Pressures to reduce air pollution from motor vehicles have led to a wide range of modifications and innovations in modern engine and vehicle design, such as catalytic converters sharply reducing the tailpipe emissions, evaporative emission control systems preventing discharge of light hydrocarbons into the atmosphere, exhaust gas recirculation (EGR) providing effective means of NO_x emission reduction, and so on. Complex electronic engine management systems, providing precise fuel metering, together with advanced design of combustion chambers, inlet/outlet ports, and so on, enable highly efficient combustion and optimal engine operation to be provided with minimum emissions under a wide range of vehicle operating conditions. The proper engine functioning and maintenance of its performance on the designed level are strongly dependent on fuel quality.

The attempt to further fuel economy improvement (particularly aimed at lowering the emissions of the greenhouse gas CO_2) has led to the development of engines with high compression ratios. The proper operation of such an engine is only possible by using gasoline with high-knock resistance in a wide range of engine operating conditions; that is, gasoline with high-octane quality and low sensitivity is needed.

Motor gasoline must minimize deposit formation in the engine intake system, on intake and exhaust valves, inlet ports, combustion chamber, in carburetor or in injectors, and so on. Deposits lead to a multitude of engine problems and malfunctions, such as octane requirement increase (ORI), abnormal combustion phenomena (surface ignition, preignition), driveability problems, reduction of engine power, increase of fuel consumption and pollutants emission, and so on. One of the least expensive and most effective ways to reduce deposits formation is by using fuel additives (see Section 16.2.3). The use of additives is also an important marketing factor for the modern, highly competitive fuel market. Obviously, the fuel composition significantly affects the tendency of gasoline to form deposits. The use of EGR systems may lead to increased levels of deposits formation in an engine intake system; therefore, the relevant gasoline properties, discussed earlier, are of great importance also here.

The gasoline volatility is an important fuel parameter, with contradictory influence on many engine characteristics, such as cold and hot starting, driveability in a wide range of ambient conditions, engine warm-up time, deposits formation, exhaust and evaporative emissions, and so on. A typical example of volatility effects on SI engine characteristics is illustrated in Figure 16.2 [1].

As can be deduced from Figure 16.2 and the preceding discussion, the optimal gasoline volatility will always be a compromise between different contradictory requirements.

Fig. 16.2.

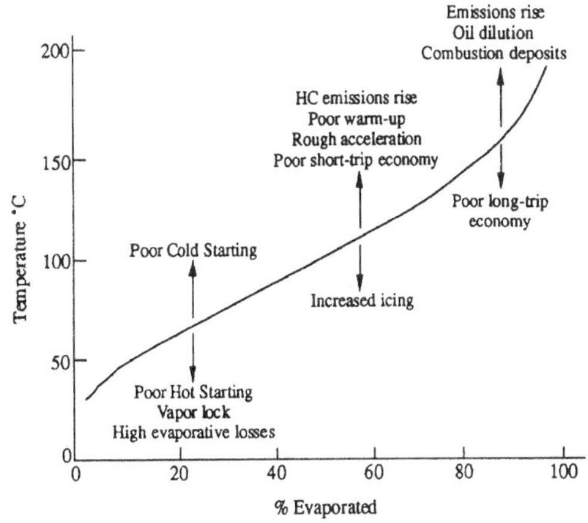

Gasoline volatility effects on vehicle characteristics [1].

Vehicles equipped with catalytic converters may use only unleaded gasolines, free from phosphorous contaminants, because such materials, as mentioned in Section 16.2.1, are catalyst poisons (i.e., sharply reduce their effectiveness). It is important to note here that using unleaded gasoline in old vehicles with "soft" valve seats may lead to severe recession problems of these seats [16]. To overcome them, the use of special fuel additives is needed. At present such additives are already in use in some countries, for example, Austria and Sweden [15, 17]. Recent research programs clearly demonstrate that sulphur also reduces the catalytic converter effectiveness; hence, its content in gasoline must be further restricted.

The environmental legislation is continuously becoming more severe, which has led to additional requirements on gasoline quality, such as reduction of benzene, total aromatics, and olefins content, use of oxygenates, and so on. These requirements are discussed in detail in Section 16.2.4.

16.2.3 Additives to Gasoline

As pointed out in the previous sections, additives to gasoline play an essential role in treatment of fuels aimed at improving their properties in order to meet required specifications and to give them additional competitive benefits. The use of additives enables substantial reduction of engine exhaust emissions in many cases.

Gasoline additives can be classified, according to their functional objectives, to some main groups [1, 18]:

- additives influencing combustion;
- additives protecting fuel systems;
- additives improving lubrication;

- additives improving oxidation stability;
- additives used in gasoline distribution.

Evidently, this classification is quite arbitrary, and other classification approaches are possible too. For example, deposits-control additives (for cleaning both the combustion chamber and the fuel system) may be selected as a separate and important group of additives.

Antiknock additives, anti-ORI additives, antipreignition, antimisfire, and spark-aid additives together with additives that improve fuel distribution between cylinders may be related to the group of additives influencing combustion. Most of the additives of all groups have generally a positive influence on emissions.

The antiknock additives, which have been widely used worldwide, are the lead alkyls tetraethyl lead (TEL) and tetramethyl lead (TML). Over the past two decades, a general reduction in the use of lead compounds has occurred because of two main factors:

- general concern over health effects of the lead;
- the increasing severity of vehicle emissions legislation, which has required the use of catalyst technologies and has resulted in a need for unleaded gasoline to prevent catalyst poisoning.

Extensive research works have been performed over the years for suitable alternatives to lead alkyls as gasoline antiknocks [1, 12]. Organometallic compounds have typically been associated with greatest antiknock activity [12], especially MMT (methyl cyclopentadienyl manganese tricarbonyl), which even was commercialized. However, there are some factors which significantly restrict the development of organometallic compounds and particularly MMT as antiknocks: their high cost, adverse effects on fuel stability and deposits build-up in engines, increased hydrocarbon emissions from catalyst-controlled vehicles, toxicity of manganese emissions, and so on.

Another group of relatively effective antiknock compounds is organic ashless antiknocks, such as N-methylaniline (NMA), amines, N-nitrosamines, iodine, selenium compounds, and so on [18]. None of these compounds were found to be as cost-effective as further crude processing [1].

As mentioned previously, various oxygenates, mainly ethers and alcohols, are widely used at present in unleaded gasolines to ensure required gasoline octane quality. However, their needed quantities are much larger than is common for antiknock additives. Thus, it is more convenient to refer to them as gasoline blending compounds rather than as additives. The main types of ethers, that are used for this purpose are:

- Methyl tertiary butyl ether (MTBE);
- Tertiary amyl methyl ether (TAME);
- Ethyl tertiary butyl ether (ETBE).

The most important alcohol compounds used in gasoline blending, for improving octane quality, and reducing pollutants emission, are:

- Methanol;
- Ethanol;
- Tertiary butyl alcohol or t-Butanol (TBA);
- Isopropyl alcohol or isopropanol;
- Isobutyl alcohol or isobutanol.

There are two major problems limiting the amount of oxygenate which can be added to gasoline intended for use in vehicles designed for conventional hydrocarbon fuel [1]: chemical leaning effect because of the oxygen content, and the adverse effect on vehicle fuel system materials. The effects of oxygenates on engine exhaust emissions are discussed in the next section.

Anti-ORI additives usually operate by removing and/or preventing deposits in the engine combustion chamber and in the ports. Polyetheramines are used for this purpose; they are added to the fuel intermittently and at high treat rates. Detergents in thermally stable carrier oil are used to reduce port deposits, but care must be taken to prevent the adverse effect of such a carrier oil on the formation of deposits in the combustion chamber.

In the case of leaded gasoline, the lead deposits in the combustion chamber may be prevented by using boron compounds such as glycol borates [1]. Halogen-based additives are also effective, but there is great concern about their environmental impact. Indeed, most of the additives marketed now are halogen-free [19].

Antipreignition and antimisfire additives, based on phosphorus-containing compounds, have been used in leaded gasoline. As the use of the latter has declined, so has that of the former.

Spark-aid additives are intended to yield higher spark energy by controlled deposit formation on the plug electrodes. Such additives are usually based on an organic gasoline-soluble potassium or other Group I or II metal compounds.

Additives improving fuel distribution between cylinders operate by forming a low surface energy coating inside of the intake manifold. Such additives are based on a mixture of tallow amines.

The group of additives intended to protect vehicle fuel systems generally includes deposit-control additives, corrosion inhibitors, and anti-icing additives.

The use of *deposit-control additives* is recently becoming more popular, mainly due to ecological concerns. The results of many investigations clearly show that deposit formation in carburetors or fuel injectors (especially port fuel injectors), intake manifolds, ports, and on the valves adversely affect engine performance and, in particular, pollutants emission (see Figure 16.3).

The use of deposit-control additives to gasoline allows engine systems to be kept clean and, therefore, in-service vehicle emissions to be brought as close as possible to the designed levels. In order to ensure fuel system cleanliness, detergent/dispersant additives to gasoline are usually used. These additives are

Fig. 16.3.

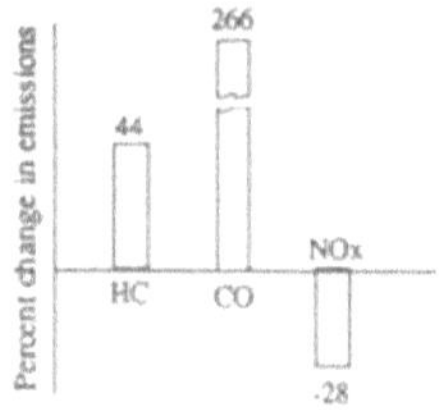

a) injector deposits (injector restriction 23%), [19]

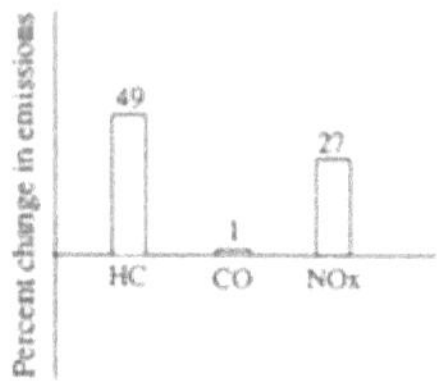

b) intake valve deposits - IVD (IVD rating change from 9-10 to 6), [20]

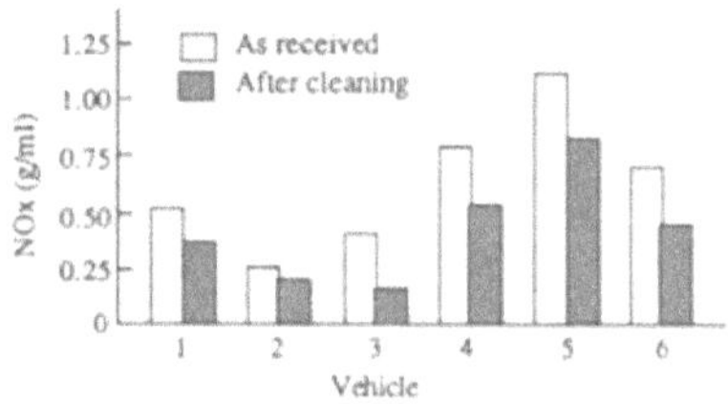

c) intake manifold deposits, reproduced from [20]

Influence of deposits on engine exhaust emissions.

based, as a rule, on polyetheramine, succinimide, or polybuteneamine technologies [1, 18, 21]. From 1995 all gasolines in the United States must contain additives to prevent deposits accumulation in engines and fuel supply systems [15]. In Israel, deposit-control additives in gasoline were mandated from 1994. Some national European specifications (for example, in Sweden and France) also provide for the use of such additives in certain types of gasoline (see Appendix 1).

Corrosion inhibitors are used to prevent deterioration of fuel storage and distribution systems and also to protect vehicle fuel systems from corrosion damage. Another benefit is the reduction of the quantity of corrosion products, which can block filters and nozzles and cause fuel pump wear. A wide range of various compounds are used as corrosion inhibitors, including amine salts of alkenyl succinic acids, alkyl orthophosphoric acids, aryl sulfonic acids, Manich amines, and so on [1, 18].

Anti-icing additives, generally surface-active agents, prevent ice from adhering to the critical parts of the carburetor. Addition of alcohols to the gasoline, which reduces the freezing point of water, is also reliably used to control icing.

Additives for improving lubrication are of limited use with today's highly sophisticated lubricants, although there is some renewed interest in antisludge additives, because of a "black sludge" problem in certain vehicles [1].

Additives against a valve seat recession are sometimes needed for protecting "soft" valve seats of some old vehicles operating on unleaded (or low-leaded) gasoline. As mentioned earlier, in the absence of lead, having a lubrication function, severe problems of valve-seat recession may occur. Such problems lead to performance losses, emissions increase, and, finally, to engine damage. Sodium- and potassium-based compounds are found to be effective as suitable additives against valve-seat recession [17].

Additives improving gasoline oxidation stability may be classified into the subgroups of antioxidants and metal deactivators.

Antioxidants, operate by inhibiting the free radical chain reactions involved in hydrocarbon oxidation [1]. The type and amount of antioxidants needed depend on the gasoline composition and storage demands. The majority of these additives are based on the aromatic diamine and alkylphenol compounds.

Metal deactivators are used to prevent metals present in gasoline (such as copper, for example) from functioning as oxidation reaction catalysts. The most commonly used metal deactivator is N,N-disalicylidene-1,2-propanediamine.

Additives used in gasoline distribution generally include various dyes and markers, drag-reducing agents, and the previously mentioned corrosion inhibitors. Demulsifiers, dehazers, antistatic additives, and sometimes even biocides are occasionally added to a gasoline, depending on storage, handling, and distribution conditions.

16.2.4 Influence of Gasoline Quality on Emissions

As noted earlier in this chapter, there is a strong correlation between gasoline performance and vehicle exhaust emissions. The overall fuel effects on emissions will be discussed henceforth. These are listed in Tables 16.5 and 16.6, which summarize a large amount of present data, including the comprehensive results of the EPEFE and AQIRP programs. These tables are mainly based on those published in [5], with the addition of results concerning the sulphur, aromatics, and volatility effects, which were obtained in the EPEFE and AQIRP programs [22, 23, 24].

Evidently, fuel effects on emissions are dependent on vehicle technology. For example, vehicles equipped with catalytic converters containing palladium (Pd) are generally more sensitive to sulphur content in the gasoline than those with Pt/Rh catalysts [22]. There is an opposite correlation between aromatics reduction and NO_x emissions for catalyst and noncatalyst vehicles.

A comparison of emissions response to fuel quality for various vehicle technologies was performed in the AQIRP program [3, 4, 24]. The main results of this comparison are summarized in Figure 16.4.

As can be seen from this figure, some differences were found in the magnitude of the fuel effects between the 1989 vehicle technology and that of 1983/5

Table 16.5

Summary of Gasoline Parameter Effects on Noncatalyst Cars Emissions [5]

Property	Change	Lead	CO	HC-EXH	HC-EVAP	NO_x	Benzene	Butadiene	Aldehydes
Reduced Lead	0.15→0.08 g/l	↓↓↓	0	0	0	0	0	0	0
Add Oxygenate	0 →2.7% O_2	0	↓↓↓	↓	0– ↑	±0	0	0	↑↑
Reduce Aromatics	40 →25% v/v	0	↓	↓	0	↓	↓↓	0	↑
Reduce Benzene	3 →2% v/v	0	0	0	–0	0	↓↓	0	0
Reduce Olefins	10 →5% v/v	0	±0	↑	–0*	↓	0	↓↓	0
Reduce Sulphur	300 →100 ppm	0	0	0	0	0	0	0	0
Reduce RVP	70 →60 kPa	0	0	±0	↓↓↓	0	0	0	0
Increase E 100	50 →60%	0	+0?	↓?	±0	0	0	0	0
Increase E 150	85 →90%	0	0	↓↓?	0	↑?	0	↓?	↓?

*Some decrease in reactivity.

Note (1): Europia expressed the opinion that the effect of E 100 and aromatics content had not been effectively decoupled. In addition, the effect of olefins of exhaust HC and NO_x were smaller than represented here. Reducing aromatics increases butadiene (replace 0 by ↑).

Note (2): ACEA broadly agreed with the effects as written.

Table 16.6

Summary of Gasoline Parameter Effects on Catalyst Cars' Emissions [5, 22, 23, 24]

Property	Change	Lead	CO	HC-EXH	HC-EVAP	NO_x	Benzene	Butadiene	Aldehydes
Reduced Lead	0.013 → 0.005	↓	↓	↓	0	↓	−0	−0	−0
Add Oxygenate	0 → 2.7% O_2	0	↓↓	↓	0– ↑	+0	0	0	↑↑
Reduce Aromatics	50 → 20% v/v	0	↓↓	↓	0	↑	↓↓↓	+0	+0
Reduce Benzene	3 → 2% v/v	0	0	0	0	0	↓↓	0	0
Reduce Olefins	10 → 5% v/v	0	0	+0	−0*	−0	0	↓↓	0
Reduce Sulphur	380 → 20 ppm	0	↓	↓	0	↓	↓	0	0– ↑***
Reduce RVP	70 → 60 kPa	0	0	−0	↓↓**	0	0	0	0
Increase E 100	35 → 65%	0	↓	↓↓↓	0	↑– ↑↑	0– ↓↓↓	−0?	0– ↓↓
Increase E 150	85 → 90%	0	0– ↑	↓↓	0	↑?	0	↓?	↓?

*Some decrease in reactivity.

**Reduction from a very low level of emission.

***Contradictory results were obtained in EPEFE and AQIRP.

Key

0—no effect	±0 = −2 to 2%	↑ or ↓ = 2–10% effect	↑↑ or ↓↓ = 10–20% effect
↑↑↑ or ↓↓↓	>20% effect	? = Insufficient information	

Fig. 16.4.

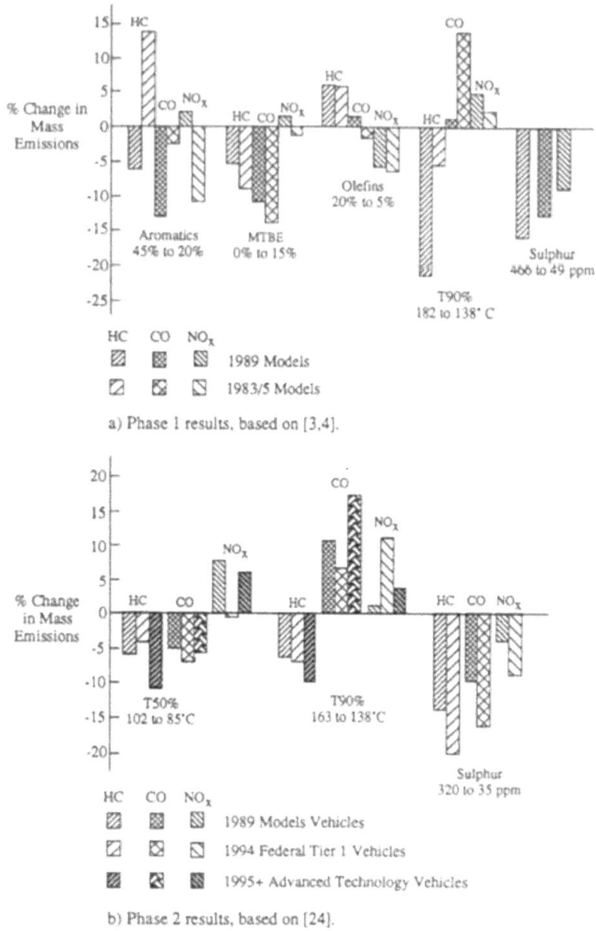

Fuel effects on emissions for various vehicle technologies (AQIRP results).

(Phase 1 results), for example, in their response to changing aromatics content or T90%. On the other hand, the fuel effects on fleets having newer technologies (since 1989) were more uniform, especially for HC and CO emissions. Fuel effects on NO_x emissions were found to be less consistent among the fleets and often less significant [24]. The comparison of fuel effects on emissions for normal-emitting vehicles and high emitters was performed in the framework of the AQIRP program [64]. The results show that most fuel effects (on a relative basis) on exhaust emissions of HC, CO, and NO_x were not distinguishably different in the normal and high emitters tested. Relative fuel effects on toxic emissions and specific reactivity were also found to be similar in the normal- and high-emitting vehicles [64].

It is observed that the AQIRP results indicate for all fleets tested (see Figure 16.4) a negative effect of T90 reduction on carbon monoxide emissions. This fact is not noted in the EPEFE analysis [5]. The data summarized in Tables 16.5, 16.6, and Figure 16.4 are discussed in the following.

Lead Content Lowering of lead content obviously yields a reduction of air pollution by lead, and in the case of unleaded gasoline allows higher catalyst effectiveness and diminished emissions of pollutants.

Oxygenates Content Adding oxygenates leads, both for catalyst and noncatalyst vehicles, to a reduction in CO and HC emissions, but also causes a rise of aldehydes emission, mainly formaldehyde [8, 25], well known as a carcinogenic air toxic with high photochemical activity.

Aromatics Content Aromatics reduction allows reduction of CO, HC, and benzene emissions both for catalyst and noncatalyst vehicles, but it exercises an opposite influence on the NO_x emissions: a reduction of NO_x in noncatalyst vehicles and an increase in NO_x in catalyst cars. The reason for this is the reduced efficiency of NO_x catalyst conversion with low aromatic fuels [23]. Some trend of increase of aldehydes emissions with reduction of aromatics content is noted because partial oxidation of aromatics is not a significant source of aldehydes compared to the oxidation of paraffins.

There is a direct correlation between aromatics content in gasoline and emission of polynuclear aromatic hydrocarbons (PAH), some of which are known as possible carcinogens and others to have mutagenic activity. Evidently, control of these species may contribute to the health of the community. An example of a correlation between fuel aromatics content and PAH emissions for noncatalyst vehicles is given in Table 16.7. As can be seen from the Table, the PAH amount in the exhaust is dependent on the aromatics content, but not necessarily on the PAH concentration in the fuel.

Benzene Content Reduction of benzene content relates directly to benzene emissions in the exhaust gases: The lower is its level in fuel, the lower are its emissions.

Table 16.7

Benz-α-Pyrene (BαP) Emissions [8]

Fuel composition		
Aromatics, % v/v	BαP, μg/l	Emissions of BαP, μg/l
44	0.8	56.8
15	64.7	23.5

Olefins Content Olefins reduction leads to a decrease in the emissions of air-toxic 1,3-butadiene for both catalyst and noncatalyst vehicles. The reduction of light, volatile, and very reactive olefins (such as butenes and pentenes) leads to the improvement of gasoline oxidation stability and to the reduction of ozone formation from evaporative emissions [1, 28].

Sulphur Content The AQIRP and EPEFE studies confirmed the role of fuel sulphur as a deactivator of a vehicle's catalytic converter. Fuel sulphur has the greatest effect on a warmed-up catalyst. Lowering the sulphur content in gasoline leads to reduced emissions of CO, HC, NO_x, and benzene. The emissions response to fuel sulphur content is strongly affected by vehicle technology. As mentioned earlier, vehicles equipped with Pd-based catalytic converters have been found to be more sensitive to fuel sulphur than vehicles with Pt/Rh-based catalysts [22]. Sulphur content is more critical for vehicles with a close-coupled catalytic converter due to its much lower light-off time and, hence, longer operation during a test with a fully warmed-up catalyst. In the European test program, no increase of 1,3-butadiene and aldehydes emissions was found when reducing sulphur content in gasoline, in contrast to the AQIRP results where formaldehyde increase was indicated. Additional research work is needed in order to clarify these results; some possible explanations indicate a difference in vehicles' catalyst technologies and inhibition of formaldehyde formation by increasing sulphur in the fuel, which could occur before the catalyst reaches its operating temperature (in the U.S. FTP, the catalyst operates a greater proportion of the test sequence at fully warmed-up conditions than in the European test) [22].

Vapor Pressure Lowering of the RVP leads to the reduction of evaporative emissions in both noncontrolled and evaporative-emission-controlled vehicles. When RVP of gasoline is lower, fewer light hydrocarbons (such as butane) are contained in it, which leads to a reduction of refueling, evaporative, and running losses of hydrocarbons [26].

Distillation Performance Increase of mid-range volatility (characterized by E100 or T50 distillation points) leads to the reduction of HC emissions for both catalyst and noncatalyst vehicles and CO emissions, mainly for catalyst-equipped cars. From the EPEFE program results, it follows that CO emissions have their lowest value, 50 percent, at E100 [23]. Increase of gasoline mid-range volatility leads to a rise of NO_x emissions. Benzene emissions generally decrease with increasing the mid-range volatility, but the effect becomes weaker at low aromatics content.

The increase of back-end volatility (characterized by E150 or T90 distillation points) contributes toward the reduction of HC emissions, both for noncatalyst and catalyst-equipped vehicles, but leads to some rise in CO and NO_x emissions [24, 27].

Different attempts were made in order to quantify, by equations, the complex relationships between gasoline properties and vehicle exhaust emissions. Such equations have been developed, for example, in the AQIRP program. General principles used in equation development were described in Painter and Rutherford [100]. Regression coefficients and equations for various specific cases can be found in the relevant AQIRP publications [e.g., 3, 101, 102].

The U.S. EPA has developed empirically simple and complex models, which correlate a gasoline's properties to its emissions characteristics. Refiners are required to use these models in order to certify their reformulated gasolines. Until late 1997 use of a simple model is allowed but from January 1998 refiners will be required to use only a complex model for certification. This complex model uses regression equations to calculate changes in emissions of NO_x, total VOCs (volatile organic compounds), and air toxics as a result of fuel property variations. The equations are based on the RVP, distillation parameters, and sulphur, oxygen, aromatics, and olefins content, together with weighting factors defined for old- and new-technology vehicles [15]. The previously mentioned equations of the simple and complex models together with the allowable ranges of fuel properties, weighting factors, and baseline data, as reproduced from the CONCAWE report [15], are given in Appendix 3.

An attempt was made also in EPEFE to quantify the relationships between gasoline properties and emissions by equations based on current data. Such equations [6] are reproduced in Table 16.8. The relationships are not simple and it is not possible to use individual fuel properties alone in order to derive emission factors. However, these equations open new possibilities of predicting the fuel performance

Table 16.8

Auto/Oil Program—Equations of the Fuel/Engine Technologies Responses

Formulas for Gasoline

CO (g/km)

$$[2.459 - 0.05513\ \mathrm{E100} + 0.0005343\ \mathrm{E100}^2 + 0.009226\ \mathrm{Aro} - 0.0003101\ (97-\mathrm{Sulphur})] \times [1 - 0.037\ (\%\ \mathrm{O_2} - 1.75)][1 - 0.008\ (\mathrm{E150} - 90.2)]$$

HC (g/km)

$$[0.1347 + 0.0005489\ \mathrm{Aro} + 25.7\ \mathrm{Aro}\ \exp(-0.2642\ \mathrm{E100}) - 0.0000406\ (97-\mathrm{Sulphur})] \times [1 - 0.004\ (\mathrm{Olef} - 4.97)][1 - 0.022\ (\%\ \mathrm{O_2} - 1.75)][1 - 0.01\ (\mathrm{E150} - 90.2)]$$

NO_x (g/km)

$$[0.1884 - 0.001438\ \mathrm{Aro} + 0.00001959\ \mathrm{E100\ Aro} - 0.00005302\ (97-\mathrm{Sulphur})] \times [1 + 0.004\ (\mathrm{Olef} - 4.97)][1 + 0.001\ (\%\ \mathrm{O_2} - 1.75)][1 + 0.008\ (\mathrm{E150} - 90.2)]$$

Benzene (mg/km)

$$[-0.454 + 5.374\ \mathrm{Fuelbenz} + 0.913 \times (\mathrm{Aro} - \mathrm{Fuelbenz})] \times \mathrm{HC}$$

E100	% vol. evaporated at 100°C (% vol)	**Sulphur**	fuel sulphur content (ppm)
E150	% vol. evaporated at 150°C (% vol)	**Fuelbenz**	fuel benzene content (% vol)
Aro	fuel aromatic content (% vol)	**% O_2**	fuel oxygen content (% wt)
Olef	fuel olefins content (% vol)		

Sources: SAE Paper No. 961076; private communication by M. Hublin, 1997.

Table 16.9

Conversion Factors for Oxygenates Content

To convert from	To	Multiply by
% vol. MTBE	% wt O_2	0.18
% vol. Methanol	% wt O_2	0.528
% vol. Ethanol	% wt O_2	0.366
% vol. TBA	% wt O_2	0.227
% vol. TAME	% wt O_2	0.161

Table 16.10

Change in Exhaust Emissions Due to Sulphur Content Reduction—Comparison of Experimental (EPEFE [22]) and Calculated Data (Sulphur Reduction from 382 to 18 ppm)

	Change in emissions, %	
Pollutant	Experiment	Calculation
HC	−8.1	−9.1
CO	−9.4	−8.7
NO_x	−11.2	−10.1

effects on the emission responses of given emission-control technologies used in motor vehicles. As mentioned by EPEFE [6], the developed equations are valid within a wide range of test procedures, vehicle/engine technologies, and fuel parameters, used in their study, but great care must be taken in extrapolating from these results. The model developed should be validated for fuels and vehicles in production and according to technological evolution processes.

As can be seen from Table 16.8, the oxygenates content is expressed in such formulas as weight percentage of oxygen in the fuel. However, it is sometimes more convenient to use the volumetric content in the fuel of a specific oxygenate (MTBE, ETOH, TBA, etc.). The coefficients for converting from the latter to the former values are summarized, for convenience, in Table 16.9. An example of the comparison between experimental data generated in the EPEFE program [22] and the results of calculations according to the formulas in Table 16.8 is shown in Table 16.10 and Figure 16.5.

As can be seen from the figure and tables, the calculated and measured effects are in quite good agreement. However, a similar comparison with the AQIRP results, performed for studying the effects of back-end volatility on emissions, does not give such a good correlation between the experimental and calculated data. These results are summarized in Table 16.11. As can be seen from this table, the EPEFE formulas do not predict the CO emissions increase as a result

Fig. 16.5.

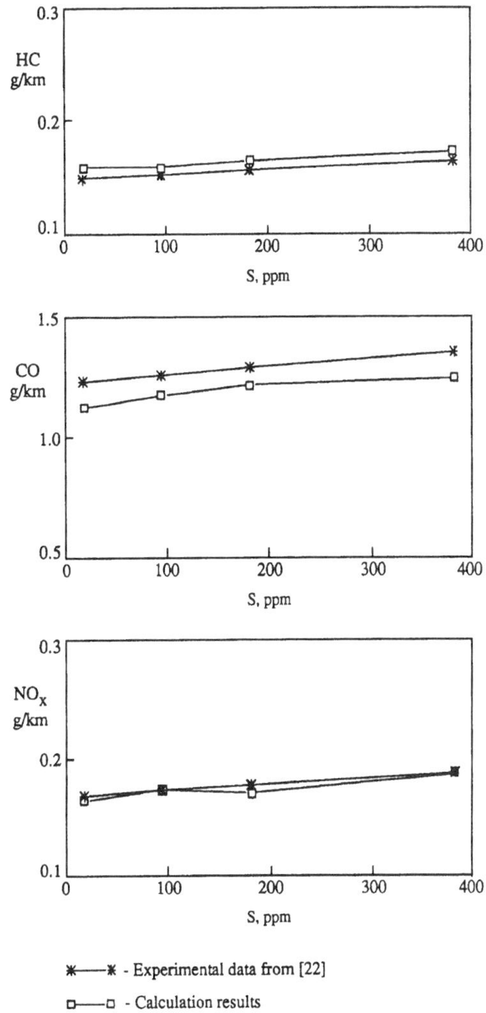

Sulphur effects on exhaust emissions—comparison of experimental and calculated results.

of the back-end volatility increase (see also Tables 16.5, 16.6, and Figure 16.4). Therefore, as mentioned earlier, great care must be taken in using them.

16.2.5 Main Trends in Gasoline Specifications Developments

The main trends of the development in gasoline properties and composition are in accordance with environmental legislation trends and with fuel/emissions responses, as described in the previous section. The ecology-related changes in gasoline specifications in the United States over the last few decades are summarized in the Table 16.12 [28].

Table 16.11

Change in Exhaust Emissions Due to Back-End Volatility Increase—Comparison of Experimental (AQIRP [24]) and Calculated Data (T90 Reduction from 162°C to 137°C)*

Pollutant	Change in emissions, % Experiment (Federal Tier 1 fleet)	Calculation
HC	−6.9	−5.7
CO	6.6	−5.1
NO_x	11.0	3.9

*Distillation parameters, E100 and E150, used in the calculation were approximately estimated based on T50 and T90 data.

As can be seen from the table, the term *reformulated gasoline* (RFG) is used. It was introduced for the first time in the United States in order to distinguish the gasoline formulated for reducing vehicle's emissions from other conventional gasolines. Reformulated gasolines are also sometimes, called, clean gasolines or environmental gasolines. The U.S. Clean Air Act Amendments of 1990 include the requirement for RFG to be sold in major metropolitan areas, which fail to meet ambient ozone and carbon monoxide standards. In general, reformulated gasoline specifications reflect the main trends in gasoline formulations development as outlined in the following.

- *Lead phase-out.* At present, sales of leaded gasoline are totally banned in the United States, Japan, Austria, Sweden, Canada, Brazil, South Korea, Thailand, and elsewhere [15], and this trend is growing rapidly all over the world. In 1994, over 64 percent of the gasoline sold in Europe was unleaded. Furthermore, the lead content in unleaded gasolines is the subject of continuous restriction, for example, the change from the common value of 0.013 g/l to 0.005 g/l or even lower (see Appendix 1). The reduction of lead content in the unleaded gasolines allows, as mentioned earlier, not only to reduce lead emissions, but also to lower catalyst deactivation, and, hence, to reduce emission levels.

- *Addition of oxygenates* to gasoline is recently becoming widespread as a tool for ensuring high octane and for reducing exhaust emissions. However, because of some restrictions (discussed earlier), the use of oxygenates in gasoline is limited, depending on the oxygenate type (as in Europe) or by weight content of the oxygen in the gasoline (as in the United States). Oxygenates limits set out in the European Directive 85/536/EEC are summarized in Table 16.13, reproduced from [15]. The U.S. federal requirements are 2.7 percent O_2 maximum.

Table 16.12

Summary of Ecology-Related Changes in Gasoline Specifications [28]

Year	Agency	Regulation	Purpose
1959	CA	Bromine Number—30 Max. for Southern CA	To limit formation of eye irritants, ozone, and aerosols
1971	CA	Vapor Pressure—9.0 psi Max. summer months	To reduce evaporative hydrocarbon emissions and ozone
1974	US	Unleaded Gasoline Required in Service Stations	To assure proper fuel for exhaust catalyst-equipped vehicles
1976	CA	Sulfur Limit—500 ppm Max.	Reduce sulfur dioxide and sulfur trioxide (sulfate emissions)
1977	CA	Lead Phasedown	To protect public health
1977	US	Manganese Banned Until Waiver Obtained	To prevent increase in hydrocarbon emissions
1977	CA	Manganese Banned	To prevent increase in hydrocarbon emissions
1978	CA	Sulfur Limit—400 ppm Max.	Reduce sulfur dioxide and sulfur trioxide (sulfate emissions)
1980	US	Lead Phasedown	To protect public health
1980	CA	Sulfur Limit—300 ppm Max.	Reduce sulfur dioxide and sulfur trioxide (sulfate emissions)
1981	US	Substantially Similar Rule	To control additive and oxygenate use
1989	US	Vapor Pressure Phase I 10.5/9.5/9.0 psi Max. Summertime	To reduce evaporative hydrocarbon emissions and ozone
1992	US	Vapor Pressure Phase II 9.0/7.8 psi Max. Summertime	To reduce evaporative hydrocarbon emissions and ozone
1992	US	Oxygen Content—2.7 wt % min. Wintertime administered by states	To reduce wintertime carbon monoxide emissions in carbon monoxide nonattainment areas
1992	CA	Vapor Pressure Phase 1 7.8 psi Max. Summertime	To reduce evaporative hydrocarbon emissions and ozone
1992	CA	Deposit Control Additive Requirement	To minimize exhaust emissions caused by carburetor, injector, and intake valve deposits
1992	CA	Lead Banned	To protect public health
1992	CA	Oxygen Content—1.8–2.2 wt % Wintertime	To reduce carbon monoxide wintertime emissions without increasing oxides of nitrogen emissions
1994	CA	Required All Gasoline to be Unleaded	To protect public health and catalysts
1995	US	Deposit Control Additive Requirement	To minimize exhaust emissions caused by carburetor, injector, and intake valve deposits
1995	US	Reformulated Gasoline Simple Model	To reduce ozone in specified and opt in ozone nonattainment areas
		Benzene Limit—1.3 vol % Max. per gallon cap	To reduce toxics

Continues

Table 16.12

(*Continued*)

Year	Agency	Regulation	Purpose
		Oxygen Content—1.5 wt % Min. per gallon cap	Required by CAAA 1990
		Vapor Pressure—7.4/8.3 psi Max. per gallon cap	To reduce evaporative hydrocarbon emissions and ozone
		No Heavy Metals	To protect public health
		Indirect Aromatics Max. of −27 vol %	To reduce toxics
		Sulfur, olefins, and 90% evaporated point<1990 average levels	To prevent increased emissions caused by changes in other fuel properties
1996	US	Lead Banned for Highway Fuel	To protect public health
1996	CA	California Phase 2 Reformulated Gasoline	To achieve maximum cost-effective reductions in criteria and toxic pollutants
		Vapor Pressure—7.00 psi Max.	To reduce evaporative hydrocarbon emissions and ozone
		Sulfur Limit—80 ppm Max. per gallon cap	Reduce sulfur dioxide and sulfur trioxide (sulfate emissions) and minimize temporary deactivation of exhaust catalysts thereby reducing hydrocarbon, carbon monoxide, and oxides of nitrogen emissions
		Benzene Limit—1.2 vol % Max. per gallon cap.	To reduce toxics
		Aromatics Limit—30 vol % Max. per gallon cap	To reduce toxics and hydrocarbon emissions
		Olefins Limit—10.0 vol % Max. per gallon cap	To reduce oxides of nitrogen exhaust emissions and ozone formation from evaporative emissions
		90% Evaporated Point—330°F Max. per gallon cap.	To reduce hydrocarbon exhaust emissions
		50% Evaporated Point—220°F Max. per gallon cap.	To reduce hydrocarbon and carbon monoxide emissions
		Oxygen Content—0.–2.7 wt % Summertime	To reduce carbon monoxide and hydrocarbon emissions without increasing oxides of nitrogen emissions

According to some forecasts [e.g., 29], this value may grow to 3 percent in 2010.

• *Aromatics content reduction.* Because of the need to boost octane, unleaded gasolines typically contain higher concentrations of aromatics [26, 30]. For example, in France premium unleaded gasoline contains an average of about 41 percent by volume aromatics compared to 33.5 percent by volume for leaded grades [31]; similarly, in Great Britain unleaded and leaded gasolines contain, respectively,

Table 16.13

Oxygenates Limits in Gasoline (European Directive 85/536/EEC) [15]

	A (% vol.)	B (% vol.)
Methanol, suitable stabilizing agents must be added[(a)]	3%	3%
Ethanol, stabilizing agents may be necessary[(a)]	5%	5%
Isopropyl alcohol	5%	10%
TBA	7%	7%
Isobutyl alcohol	7%	10%
Ethers containing five or more carbon atoms per molecule[(a)]	10%	15%
Other organic oxygenates defined in Annex section I	7%	10%
Mixture of any organic oxygenates defined[(b)] in Annex section I	2.5% oxygen weight, not exceeding the individual limits fixed above for each component	3.7% oxygen weight, not exceeding the individual limits fixed above for each component

Notes: Member states must permit fuel blends containing levels of oxygenates not exceeding the level set out in column A. If they so desire, they may authorize proportions of oxygenates above these levels. However, if the levels so permitted exceed the limits set out in column B, the dispensing pump must be clearly marked accordingly, in particular to take into account the calorific value of such fuels.

[(a)]In accordance with national specifications or, where these do not exist, industry specifications.

[(b)]Acetone is authorized up to 0.8 percent by volume when it is present as a by-product of the manufacture of certain organic oxygenate compounds.

[(c)]Not all countries permit levels exceeding those in column A even if the pump is labeled.

about 31 percent and 26 percent by volume aromatics. As discussed before, however, aromatics lead to increased vehicle exhaust emissions; therefore, there are environmental pressures to limit and reduce their content in gasoline.

To date, there are no limits on aromatics content in most of the modern gasoline specifications (see Appendix 1). Nonetheless, in California Phase 2 reformulated gasoline requirements for 1996, the aromatics content is already restricted to 25 percent volume maximum [15]. Also, the Swedish specifications for Class 2 environmental gasoline put the so-called *empirical limit* (EL) on aromatics content in this gasoline: (EL = % vol. aromatics/13 + % vol. benzene) [15]. Because the amount of benzene in gasoline is also limited, such a requirement actually limits the aromatics content, but it also allows more flexibility to the refineries. It is possible to forecast, with a high degree of probability, that the aromatics content will be limited and consequently reduced in most of the future gasoline formulations; one of the predicted aromatics limits for the year 2010 is 20 percent by volume [29].

Benzene Content Reduction Benzene is the lightest of the aromatics, known as air toxic, and is present in air environment mainly due to vehicle emissions [8]. The strong correlation between benzene content in fuel and its emissions has been discussed in the previous section. Most of today's unleaded gasoline specifications limit the maximal permissible amount of benzene in gasoline. For example, in Europe (according to CEN standard EN 228), the limit of benzene content is 5 percent by volume. However, for reformulated gasolines marketed now in the United States, it has already been reduced to 1 percent by volume (see Appendix 1). The same trends of the benzene-content reduction in RFG are observed also in the European countries: in Finland, 1 percent by volume; in Sweden, 3 percent by volume. In Italy, according to local agreements between major city authorities and oil companies, the benzene content is limited to 1.8 percent by volume maximum, with a further move to 0.8 percent by volume proposed in Rome [31]. Predictions of future trends indicate the possibility of restricting the benzene content up to 0.6 percent by volume maximum in 2010 [29].

Olefins Content Reduction As discussed in the previous sections, the olefins contained in gasoline mainly contribute to 1,3-butadiene emissions and adversely affect the gasoline oxidation stability. Today, the olefins content is not limited in the European gasoline specifications, but for reformulated gasolines marketed now in California, according to Phase 2 RFG requirements, the amount of total olefins in a gasoline has been restricted to 6 percent by volume maximum (see Appendix 1).

There are considerations for future limits of 5 percent by volume maximum for total olefins and 1 percent by volume maximum for the most volatile and photochemically active olefins (C_4/C_5) [32].

Sulphur content reduction is essential, as has been established, to prevent the deterioration of catalyst effectiveness during the vehicle service life. In Europe, today, (according to EN 228), the maximal permissible sulphur content in unleaded gasoline is 0.05 percent by weight. Allowed sulphur amounts for leaded grades, intended for noncatalyst cars, are generally higher: 0.1 to 0.2 percent by weight maximum (see Appendix 1). However, in some countries the limits for sulphur content in reformulated grades of gasoline have been restricted more severely to values of 0.01 percent by weight or even lower (see Appendix 1). For example, the Phase 2 reformulated gasoline in California may contain only 0.004 percent by weight maximum of sulphur. The most severe restrictions on sulphur content in Europe are in Finland (reformulated gasoline) and Sweden (Class 2—cat.) 0.01 percent by weight.

Vapor Pressure Reduction Since 1989, there is a continuous trend of lowering the maximal RVP values of U.S. gasolines permitted by federal regulations on vapor pressure [28]. Such a trend is common for both winter and summer gasolines.

Today's RVP limit for reformulated gasoline marketed in California is only 48.3 kPa. Some reformulated grades in Europe must also have reduced vapor pressure values (see Appendix 1). For example, the RVP of the Finnish reformulated gasoline must not exceed 70 kPa, while the common value today in European countries with similar climatic conditions is 95 kPa.

Tailoring the Gasoline Distillation The trend in vapor pressure reduction generally corresponds with the reduction of gasoline front-end volatility. At the same time, as can be seen from Tables 16.5 and 16.6, some increase of gasoline mid-range and back-end volatility leads to the reduction of CO and HC emissions. It reflects in the California Phase 2 reformulated gasoline specification, where the T50% and T90% distillation points have been reduced to 99°C and 149°C maximum, respectively, compared to 121°C and 190°C maximum for conventional gasolines. There are forecasts which predict a further rise of mid-range and, to some extent, also back-end volatilities of future gasolines: T50% = 88°C and T90% = 143°C in the year 2010 [29].

The use of deposit-control additives is becoming widespread in modern gasolines. Most gasolines marketed in Europe, both leaded and unleaded, now contain detergent additives [31]. All gasoline sold in the United States from 1995 contains deposit-control additives. It may be assumed, with a high degree of probability, that all future gasolines will contain multifunctional additive packages.

16.3 DIESEL FUELS (CI ENGINES)

16.3.1 Origin, Composition, and Properties

16.3.1.1 Diesel Fuel Origin and Composition

Diesel fuels are complex compounds of hydrocarbon molecules, which generally boil within the temperature range of 150°C to 380°C [1, 33]. They are normally blended from several refinery streams, mostly coming from the primary distillation unit. However, in a conversion refinery, components from other units, like cracking processes, are often used to increase diesel fuel production. In the primary refining unit, distillation takes place at atmospheric pressure, the furnace temperature being set to give maximum distillation without cracking. The quality and quantity of the streams drawn off will be determined both by their boiling range and by the crude boiling used. Tables 16.14 and 16.15 give general indications of how the type of crude oil can influence diesel fuel characteristics and properties [1, 34].

Table 16.14

Influence of Crude Oil Type and Origin on Diesel Fuel Characteristics [1]

Crude oil source	Hydrocarbon type	Cetane number	Sulphur content	Cloud point	Calorific value
United Kingdom/Norway	Paraffinic	High	Low/Medium	High	Low
Denmark	Naphthenic	Moderate	Low	Low	Moderate
Middle East	Paraffinic	High	High	High	Low
Nigeria	Naphthenic	Low	Low	Low	Moderate
Venezuela	Naphthenic	Very Low	Low	Low	High
Mexico	Aromatic	Very Low	Medium	Low	High
Australia	Paraffinic	High	Low	High	Low
Mexico	Paraffinic	High	Low	High	Low

Table 16.15

Influence of Crude Oil Source on Diesel Fuel Properties [34]

	Crude source		
Property	Kuwait crude	Forties crude	Nigerian crude
Density (kg/l at 15°C)	0.8517	0.8558	0.8785
Viscosity (cSt at 40°C)	4.5	4.6	4.5
Cloud point (°C)	4	6	−8
Cold filter plugging point (°C)	−4	1	−11
ATM distillation			
50% distilled at °C	304	294	283
Sulphur (% wt)	1.40	0.19	0.13
Cetane number (ASTM D613)	54.1	52.2	40.9

16.3.1.2 Diesel Fuel Properties

Cetane Number The most universally accepted measure of the ignition quality of diesel fuel is the cetane number (CN) [1, 33]: The higher the number, the easier it is to ignite. The method used to determine the ignition quality in terms of the CN is similar to that used for determining the antiknock quality of gasoline in terms of the octane number. The CN scale is defined by blends of two pure hydrocarbon reference fuels. Cetane (n-hexadecane), a compound with high ignition quality, represents the top of the scale with a cetane number of 100. An isocetane, heptamethyl nonane, a highly branched paraffin which has a very low ignition quality, represents the bottom of the scale with a cetane number of 15. A fuel with the same ignition quality as a mixture of the two reference fuels has a cetane number derived from the equation:

$$\text{CN} = \text{percent n-cetane} + \text{percent heptamethyl nonane}$$

The engine used in CN determination is a standardized single-cylinder, variable compression ratio. The engine, the operating conditions, and the test procedure are specified by ASTM method D613 [1].

The importance of higher cetane numbers is attached to the performance benefits provided in terms of improved cold starting, reduced smoke emission during warm-up, reduced noise, and reduced fuel consumption and exhaust emission. Therefore, there is a general tendency in some countries to increase the cetane number of diesel fuel.

Since the cost of the cetane number determination is quite high, many correlations which predict ignition quality based on the physical properties of diesel fuels have been developed [34–36]. A calculated cetane index (CCI), is often used to estimate it, based on API gravity and the mid-boiling temperature (of 50 percent evaporated), cf. ASTM D976 [34]. It is applicable to straight-run fuels, catalytically cracked stocks, and blends of the two [34]. The use of this index (CCI) is suitable for most diesel fuels and yields values that correspond quite closely to the cetane number. A diesel index is also used, based on the fact that ignition quality is linked to hydrocarbon composition: n-paraffins have high ignition quality and aromatic and naphtenic compounds have low ignition quality [34]. The aniline point (the lowest temperature at which equal volumes of the fuel and aniline become just miscible) is used together with the API gravity to give the diesel index [1, 35, 36]:

$$\text{Diesel index} = \text{aniline point } (^\circ\text{F}) \times \text{API gravity}/100$$

where the API gravity is based on the specific gravity and is calculated by:

$$\text{API gravity, deg} = (141.5/\text{specific gravity at } 60^\circ\text{F} - 131.5).$$

The diesel index usually gives values slightly above the cetane number. It provides a satisfactory indication of ignition quality in many (but not all) cases.

As indicated [35, 36], the formulas known as cetane indices or cetane correlations lead to significant errors in the prediction of the cetane number of a diesel fuel that is very different from the fuels used to derive the index. This stems from the following reasons:

- Relationships between physical properties and chemical structure of the fuel have not yet been developed.
- The physical properties measured commonly (i.e., density, boiling range, aniline point, and viscosity) do not fully describe the fuel chemistry (lack of representation).
- In most predictive equations, commonly measured physical properties have been used randomly without relying on generally accepted physical and/or chemical models.
- The correlations have usually been based on the assumption that aromatic hydrocarbons are low in cetane number, paraffins have high cetane numbers,

Fig. 16.6.

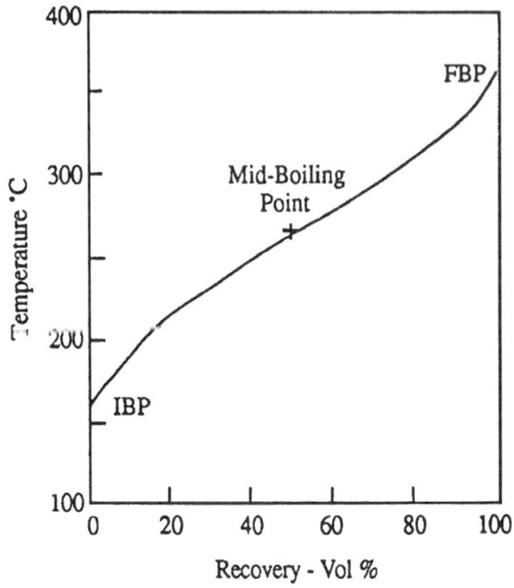

Typical diesel fuel distillation curve [1].

and naphtenes fall somewhere in between. No attempt has been made to represent the degree of branching. This is important since different isomers react differently in regard to cetane numbers just as they do in octane numbers.

Volatility The volatility characteristics of a diesel fuel can be expressed in terms of the temperature at which successive portions are distilled from a sample of the fuel under controlled heating in a standardized apparatus. The distillation method recommended by the standard ASTM D86 is one of the more widely used versions of distillation [1]. Information obtained during the distillation includes initial boiling point (IBP), end point (EP) or final boiling point (FBP), percent of condensate recovered, and percent residue of nonvolatile matter. A typical diesel fuel distillation curve is shown in Figure 16.6.

The volatility (the distillation or boiling range of the fuel) influences many other properties, including density, flash point, autoignition temperature, viscosity, and cetane number. High volatility could cause vapor lock and lower the flash point. The vapor lock can cause engine misfiring or failure to restart after a brief shutdown in hot conditions. However, the higher the volatility, the more easily does complete vaporization of the fuel take place in the combustion chamber. Consequently, high-boiling components may not burn completely, forming engine deposits and increasing smoke levels [1, 33]. Within the range of 350°C to 400°C, however, the effect of low volatility on exhaust emissions is relatively small [33]. The mid-volatility of a diesel fuel has a marked effect on the tendency to smoke, possibly through influence on the injection and mixing of the fuel. However, there is also interest in the 50 percent distillate recovery temperature for the calculation of the cetane index by ASTM D976 [1].

It is emphasized that in practice it is the mix of volatilities that is most important: High-volatility components at the lower end of the curve (Figure 16.6) improve cold starting and warm-up, while low-volatility components at the upper end tend to increase deposits, smoke, and wear [33].

Density The density of diesel fuel can provide useful indications about its composition and performance-related characteristics, such as ignition quality, power, economy, low-temperature properties, and smoking tendency. The density may sometimes be expressed as specific gravity or as API gravity. In units of kg/m^3, the densities of fuels obtained by the different refining processes are approximately as follows [33]:

Straight-run distilled	805 to 870
Hydrocracked gas oil	815 to 840
Thermally cracked gas oil	835 to 875
Catalytically cracked gas oil	930 to 965

Viscosity *of a fluid* indicates its resistance to flow. It is an important property of the diesel fuels because of its influence on the performance of the fuel injection equipment, especially at low temperature, when the increase in viscosity affects the fluidity of the fuel. Increasing viscosity reduces the injector spray cone angle, fuel distribution, and penetration, while increasing the droplet size. It will, therefore, affect optimum injection timing [33]. For a given injector nozzle configuration and a fuel pressure, the viscosity will certainly influence the quantity of fuel injected.

Diesel specifications usually impose an upper limit on viscosity to ensure adequate fuel flow for cold starting and an additional minimum limit is often also specified to guard against loss of power at high temperatures.

Figure 16.7 shows temperature and viscosity characteristics for typical automotive diesel fuel. The figure includes marks of the viscosity range allowed in the British specification for automotive diesel fuel (BS2869), together with the upper and lower limits (cold and hot risk points) for the U.K. climate, as defined by Lucas Diesel Systems.

Diesel Fuel Low-Temperature Characteristics As much as 20 percent of the diesel fuel may consist of relatively heavy paraffinic hydrocarbons, which have limited solubility in the fuels [1]. Paraffins are most likely to deposit out as wax if cooled sufficiently. This is unfortunate because they have high cetane numbers. Wax in a vehicle fuel system is a potential source of problems of operation. The low-temperature properties of the fuel are defined by wax-related tests as follows:

- The cloud point test measures the temperature at which wax becomes visible when the fuel is cooled, according to the method described in ASTM D2500. It has the disadvantage of relying on the tester's judgment.

Fig. 16.7.

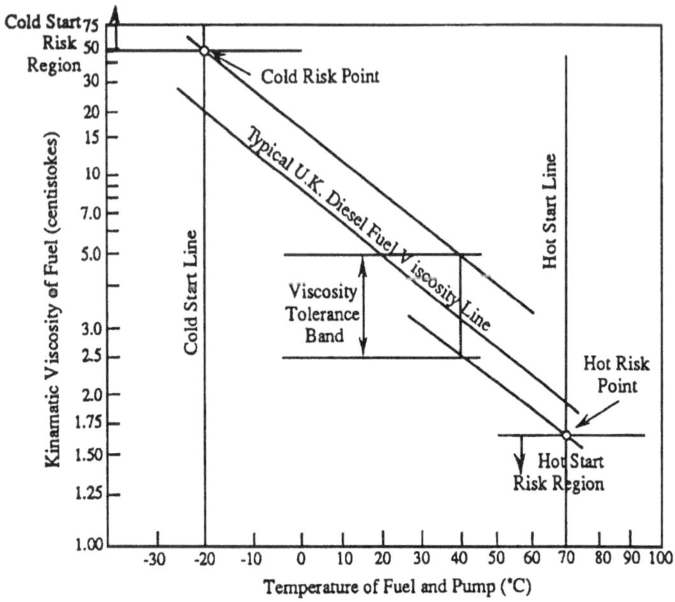

Diesel fuel viscosity limits for the United Kingdom. The upper and lower limits on the 40°C line are those of BS2869, while those at −20°C and 70°C are recommendations by Lucas Diesel Systems [1].

- The wax appearance point is the temperature at which wax starts to come out of the solution (ASTM D3117).
- The pour point test is used to measure the temperature at which the amount of wax out of solution is sufficient to gel the fuel (ASTM D97).
- Other tests, like cold filter plugging test and the low-temperature flow tests, are described in [1].
- Another way to estimate operational performance of a fuel is to combine the cloud point (CP) and the pour point (PP) in an empirical equation to obtain a wax precipitation index (WPI):

$$\mathrm{WPI} = \mathrm{CP} - 1.3\,(\mathrm{CP} - \mathrm{PP} - 1.1)^{0.5}$$

Stability of Diesel Fuel The ability of a diesel fuel to remain unchanged during the period between its manufacture and its eventual use in an engine is obviously a very important quality. The formation of sediment during the long-term storage of diesel fuels, particularly those containing catalytically cracked stocks, has been recognized for many years [37, 38]. Nitrogen and sulphur-containing compounds have always been strongly implicated in the fuel degradation process since these elements tend to become concentrated into fuel sediments. The mechanism by which the oxidation of hydrocarbons progresses occurs in several stages, starting with the chain initiation involving the generation of free radicals. Once a hydrocarbon free radical has been formed, it can combine with oxygen to form a peroxide radical which, in turn, can react with another hydrocarbon molecule, thereby generating other hydrocarbon free radicals and a hydroperoxide. The oxidation process is therefore self-perpetuating. The free radicals can also give rise to polymerization as well as oxidation reactions to form high molecular weight materials. These can deposit in the fuel system. The final stage (chain termination), in the absence of an antioxidant, is a reaction which leads to no free radical products.

More cracked gas oil is being diverted nowdays into diesel fuel. Distillates from cracking operations are more olefinic than those from atmospheric distillation and contain more nitrogen compounds, such as pyrolles and indoles. As a result, they are less stable, being prone to oxidation by free radicals, as explained earlier. This is the reason why oxidation stability limits are being introduced into more diesel fuel specifications.

The ability of a fuel to maintain satisfactory storage stability is dependent upon a series of parameters which control its rate of autoxidation. With diesel and distillate fuels, the end products of these complex autoxidation reactions manifest themselves in two general types, namely, a *soluble gum*, which is essentially nonvolatile, and an *insoluble gum*, which precipitates out in the fuel [38]. It has been postulated that the insoluble gum, usually referred to as *sediment*, is the most harmful product of fuel deterioration in storage.

Although no significant degradation of diesel fuel quality has been observed in conventional storage of locally produced straight-run stocks, formation of degradation products in bus engine fuel systems has been observed by filter clogging and deposit formation [39, 40]. The main cause of these operational problems in the field may be attributed to the fact that diesel fuel, which serves also as an injector coolant, most probably deteriorates at the relatively high temperatures prevailing in part of the fuel tank when cooled and then heated again. The succession of heating and cooling cycles may cause deterioration of thermally unstable diesel fuel.

If oxidation take place, engine operation could be affected due to filter blocking or gummy deposits in the ignition system and on injector nozzles.

There does not yet exist a universally accepted test to predict the stability characteristics of a diesel fuel kept in normal storage facilities for prolonged periods [1]. One accelerated method to test oxidation stability is described in ASTM D2274. The color of the fuel before and after aging can be determined by ASTM D1500. One of the alternative methods for predicting long-term storage stability is ASTM D4625. Another test to evaluate gas oil stability during long-term storage was proposed [40], with a procedure designed to simulate diesel engine operation under field conditions. The test was successfully applied, especially after some improvements [41].

Other alternative fuel oxidation stability test methods, like Esso Test, Du Pont F-21 Test, and so on are described in Owen and Coley [1].

Sulphur Content One known method for reducing the total emission levels of particulate matter (TPM) from diesel engines is to use low-sulphur (<0.05 percent by weight) diesel fuel [6, 42]. During the combustion process, the majority of the sulphur in the fuel is converted to sulphur dioxide (SO_2), which is mostly emitted into the environment where additional chemical reactions may occur, leading to atmospheric pollution. The rest of the SO_2 is oxidized in the oxygen-rich diesel exhaust and forms sulphur trioxide (SO_3). The high diesel exhaust temperature maintains the SO_3 emission in the vapor phase, which has a high affinity toward water. An exothermic reaction leads to formation of sulphuric acid aerosols, which, together with the chemically bonded water, is emitted as particulate matter [42].

Another benefit of decreasing the sulphur level in diesel fuel is the wear reduction. It is known that high sulphur levels in the fuel cause wear in piston rings and cylinder liners. The contribution of sulphur to wear is due to two mechanisms: corrosive wear and deposit formation, which are directly dependent on fuel sulphur levels and engine operating conditions. Most legislation worldwide with respect to diesel fuel quality has focused, until now, on sulphur content.

In the United States, a sulphur limit of 0.05 percent has been adopted for on-highway diesel fuel, and at the same time in Europe for automotive diesel fuel [15]. In 1991 Sweden enacted legislation that resulted in very stringent regulations: The maximum level of sulphur, revised in 1992 for the two classes 1 and 2 diesel fuels, are respectively 0.001 percent and 0.005 percent [15]. To produce fuels with a sulphur content of less than 50 ppm and an aromatics content of 5 percent by volume or less (according to the specifications for the diesel fuel class 1), it is necessary to apply more severe hydro-processing conditions or invoke unconventional hydro-processing technology. Such processing technology will remove polar species from the fuel, which adsorb onto the mating surfaces to form a protective low-friction layer. In fact, there is strong evidence that by severe hydro-treating, the manufactured automotive diesel fuel may increase the risk of premature mechanical failure of certain classes of fuel injection pumps. The events observed for the first time in Sweden, where strict, environmentally driven fuel specifications have been introduced, have confirmed the existence of this risk [43]. The test method to determine the sulphur content is published in ASTM D 2622-82.

Aromatic Content Aromatic components in diesel fuel is a problem of great concern because they tend to contribute to particulate emissions. A maximum content of 10 percent has been imposed by the California Air Resources Board (CARB) and there are suggestions by the Association of European Automobile Constructors (ACEA) for the aromatics content to be included in diesel fuel specifications. This association recommended in 1994 the limitation of the three-ring aromatics content to a maximum 1.0 percent by weight [1]. It is emphasized that aromatics make a major contribution to the lubricity of the fuel, so their removal can give rise to abnormally high rates of wear of the injection pump, a problem which was mentioned earlier [33].

Water and Sediment Content Water cannot be completely eliminated from diesel fuel. The earliest stage at which water can get in is during the manufacturing processes. The main risks of water contamination are during transportation and storage in tanks. The presence of water in storage tanks may encourage growth of fungi or bacteria. Microbial contamination can cause significant problems with the working parts of engines, particularly the fuel system (for example, filter blockage).

Sediments likely to be found in diesel fuel are mainly inorganic in origin: rust, metal particles, and dirt. A certain part can also be organic, from degradation of unstable fuel components, bacterial action at the oil-water interface, or wax from the fuel.

Water and sediments can contribute to filter plugging in the vehicle or in the distribution network and cause problems due to corrosion and wear in the engine and fuel injection systems. A standard test for water and sediment content is by a centrifuge method, ASTM D1796, which is in the ASTM D975 specifications for diesel fuel [1].

Other diesel fuel properties, such as carbon residue, ash content, corrosivity, heating value, and so on are discussed in Owen and Coley [1].

National requirements to diesel fuel quality in different countries worldwide are summarized in Appendix 2. This information is based on standards and CONCAWE data [15]. Average values of some diesel fuel parameters (typical for Europe and the United States), affecting pollutant emissions, are given in Table 16.16.

16.3.2 Requirements of Engine Technologies on Diesel Fuel Quality

Pressure to reduce air pollution from motor vehicles has led to a wide range of modifications and innovations in modern diesel engines. For example, research and development work is devoted worldwide to high-pressure injection systems, with peak pressure at the injection nozzle varying between 1200 and 1800 bar with electronic control; injection timing control as a function of load and speed; using three,

Table 16.16

Typical Values of Average-Market Diesel Fuel Parameters [6, 44]

		United States, Winter 1996		
Property	Europe	East Coast	Midwest	West Coast
Cetane number	51	51	44	49
Density (g/l)	843	846	847	845
Poly-Aro (%)	9			
T95 (°C)	355			
T90 (°C)		314	310	320
Sulphur (ppm)	450	300	400	300

four, and five valves per cylinder in light-duty diesels; using variable-resonance intake systems and common rail systems. Turbochargers with variable geometry, as well as the continuous improvement of EGR use via electronic management systems, will also contribute to easier control of future emission limits [45, 46].

Oxidation catalysts will be used to lower CO, HC, and particulate emissions, and also to improve the odor of diesel exhaust.

In order to decrease the fuel consumption and the emitted quantity of CO_2, the tendency now is to move from indirect-(IDI) to direct-injection (DI) diesels. It is claimed that they are showing 15 percent to 20 percent fuel economy advantages over the comparable IDI.

Even after engines have been optimized regarding emissions and fuel economy, increasingly stringent regulations are unlikely to be met unless fuel quality is maintained. The motor industry around the world has been active in promoting its ideas of appropriate fuel quality. In Europe, ACEA issued its fuels charter in late 1994. Based on the results obtained in the European Programme on Emissions, Fuels and Engine Technologies (EPEFE), a jointly funded project of the European Oil and Automobile Industries, represented respectively by Europia and ACEA, this fuels charter will be changed. From these results it is concluded that the diesel fuel parameters that have the greatest influence on engine emissions are cetane number, sulphur content, density, T_{95} (temperature at which 95 percent fuel is evaporated), and fuel polyaromatic content [6, 47].

U.S. diesel fuel has come under the spotlight from the Engine Manufacturers Association, which is asking for improved quality for 1998. It proposes a huge change in cetane number to a 56 minimum as well as a minimum viscosity of 2.5 cs at 40°C, a big reduction in aromatics, and a lubricity requirement [47].

The quality of diesel fuel can also be improved by the use of additives, whose positive role has led the European motor industry to advocate their use or approve specific high-quality fuels that have been upgraded accordingly. This is documented in the European Standard EN590 and was presented at the 1992 EC Symposium "Auto Emissions" [48]. Additives to improve the quality of diesel fuels are, therefore, beneficial for refineries, customers, and the environment. Like gasoline additives, the use of diesel fuel additives is also an important marketing factor for the modern competitive fuel market.

16.3.3 Diesel Fuel Additives

Additives can enhance various diesel fuel properties. For example, ignition-improver additives are used to increase the cetane number of the fuel. Detergent additives are considered to be of growing importance in controlling the formation of fuel deposits that can have detrimental effects on combustion. The build-up of lacquer and carbonaceous deposits on injector tips can affect the amount of fuel injected and the spray pattern, causing problems of reduced power and higher smoke emission. The cold flow performance of diesel fuel can be adjusted to prevailing climatic conditions by base component selection, but at the expense of cetane quality and an availability penalty. However, by using cold flow additives, the required low-temperature filterability can be obtained at nearly constant cetane number and with less availability constraints.

Diesel fuel additives have the potential to improve the quality of diesel fuels with regard to emissions, noise, engine performance, and customer perception, while offering flexibility in the optimization of refinery production costs.

There are several possible ways to classify diesel fuel additives. In the following, they are divided into two classes:

1. Additives influencing diesel fuel combustion: ignition improvers, combustion improvers, catalysts for regeneration of aftertreatment devices, and detergents.
2. Additives influencing storage and flow (antioxidants, antirust, anticorrosion, antifoam, dehazers, biocides, additives for low-temperature operability, etc.).

16.3.3.1 Additives Influencing Diesel Fuel Combustion

Cetane (ignition) improvers are used to improve the cetane quality of marketed fuel components by reducing the delay between injection and ignition when fuel is sprayed into the combustion chamber. These additives are used in two different ways:

- To increase the cetane number of diesel fuel, which would otherwise fail the specification limits.
- To increase the cetane number of diesel fuel above the specified minimum standard to yield premium-grade products now being marketed by many oil companies in certain locations throughout the world.

The chemicals most commonly used as ignition improvers are nitrates and specifically ethyl-hexyl nitrate (EHN). Certain peroxides have been identified as effective cetane improvers, but there is less experience with the handling of these products in concentrated form in refineries and fuel depots. They are all materials that decompose readily, and at elevated temperatures they generate free radicals that accelerate oxidation of the fuel and initiate combustion.

The response to cetane improvers is dependent on individual fuel charcteristics and on the cetane level [49]. Results presented in Figure 16.8 were obtained from a large sample of fuels and show that, on the average, an improvement of about three numbers was obtained with a treat level of 500 ppm, and with 1000 ppm the gain was five numbers. The results show considerable variation about the average line and it was also found that the lower-cetane fuels had the poorest response. Another important result, seen from the figure, is that the same cetane number

Fig. 16.8.

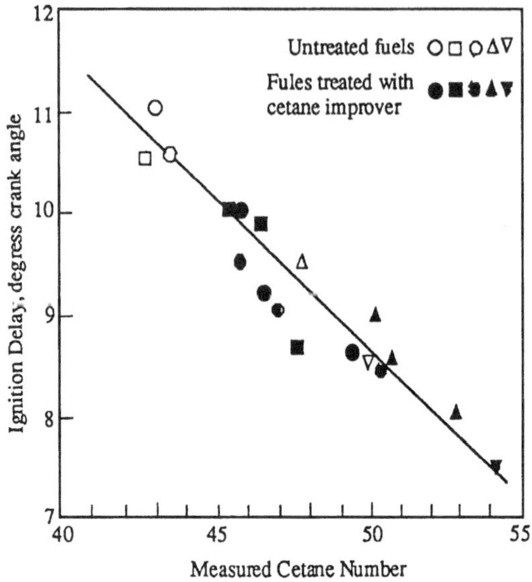

Influence of cetane number on ignition delay [49].

Fig. 16.9.

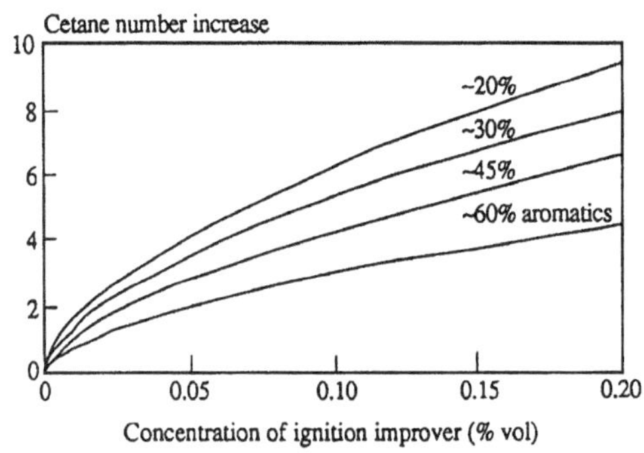

Response of different base fuels to ignition improvers [48].

produces the same ignition delay regardless of whether the former is that of the base fuel or a treated one.

Figure 16.9 shows a very clear monotonic relation between the concentration of the ignition improver EHN and the cetane number for various chemical compositions of the fuel [48]. Unfortunately, parrafins that have a comparatively high cetane number respond best, and aromatics, with their lower intrinsic cetane quality, have a poorer additive response. The graphs in Figure 16.9 show also that increasing the amounts of EHN produce smaller improvements. In practice, for example, the cetane quality of typical European market diesel fuels can be increased by about two to four units using 500 ppm of EHN as shown in Figure 16.10 [48].

Fig. 16.10.

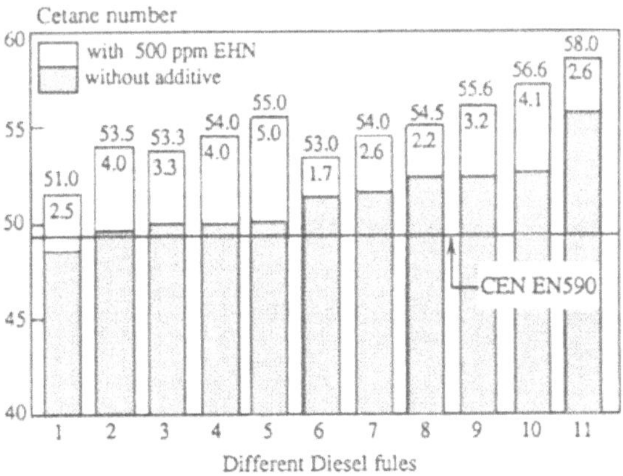

Increasing cetane number by adding ignition improver for different European diesel fuels [48].

Fig. 16.11.

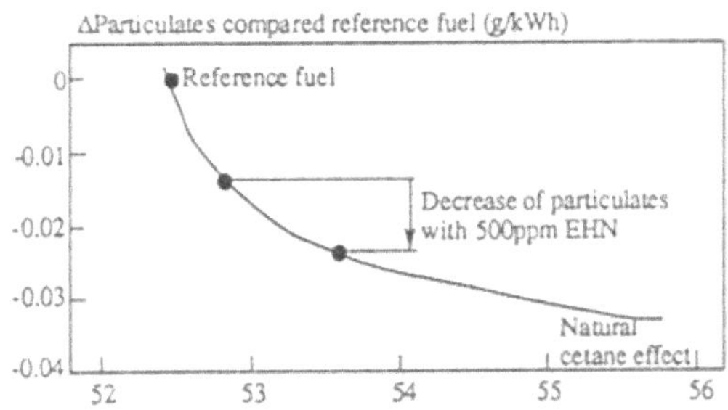

Influence of cetane improvers and of natural cetane effect on particulates reduction of a 6.9 DI engine [48].

A very comprehensive study of cetane response of post-1993 U.S. no. 2 diesel fuels, following the introduction in 1993 of the federally mandated 0.05 percent sulphur on-highway diesel fuel, is presented in Sobotowski [50]. Seven fuels, obtained from the East Coast, Gulf Coast, and Midwest refineries, and emissions-certification fuel, and a low-aromatic California fuel were evaluated. EHN was used as cetane improver in concentration of up to 1 percent by volume. The test results demonstrated relatively low-cetane response of non-California fuels and an exceptionally high response of the low-aromatic California fuel. In the conclusions of this study, an interindustry program was proposed to gain a detailed understanding of diesel fuel composition and property effects on cetane response, and proved a cetane response equation applicable to the range of diesel fuels [50].

Cetane-improved fuels, whether achieved by component blending or by additives, improve cold starting, and reduce idle noise, white smoke, and regulated emissions. Figure 16.11 shows the influence of ignition improver on the decrease of particulates for a heavy-duty engine.

Fig. 16.12.

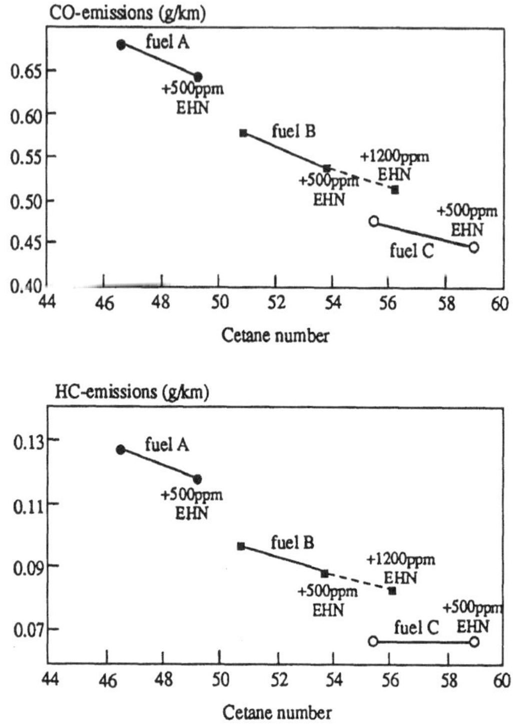

Influence of cetane improvers on HC and CO emissions of 3 IDI diesel [48].

The positive effect of adding ignition improvers to fuels, with different cetane numbers on HC and CO emitted from heavy-duty and passenger-car engines, are shown in Figure 16.12.

Combustion improvers additives for diesel fuels are additives which have a catalytic effect on the combustion process in the engine. The majority of these are organometallic compounds in nature, containing manganese, iron, barium, or calcium. These additives are effective in reducing smoke emissions. During the 1960s and 1970s a range of products containing barium (sometimes together with calcium or iron) were commercialized as smoke-suppressant additives. Despite their obvious effectiveness in reducing smoke from diesel engines, widespread use of smoke-suppressant additives did not develop. Additive cost, together with concern over deposit build-up within engines, the negative effect of additives on particulate emissions, and the high toxicity of barium compounds are all contributory factors [1, 34, 51, 52].

Different studies, carried out recently, have had the objective of using additives for particulate traps regeneration. Diesel soot ignites at 550°C to 650°C. If the diesel engine exhaust gases reach these temperatures during operation, the heat of the exhaust is sufficient to ignite the soot and regenerate the filter. However, diesel exhaust is fairly cool, and typical exhaust gas temperatures in diesel engines are

Table 16.17

Organometallic Compounds Used as Catalytic Additives for Particulate Traps Regeneration [53]

Metal	Organometallic compounds
Ba	carbonate, sulphonate
Ca	carboxylate, naphthenate, sulphonate
Ce	carboxylate, carbonate
Ce + Mn	?
Cu	naphthenate, carboxylate, octoate, pivaloylpinacolonate
CU + others	cetate, naphthenate, octoate + (Ce, Mn, Ni, Pb) organometallic compounds
Fe	acetylacetonate, carboxylate, ferrocene, naphthenate
Li	*t*-butoxy
Mn	corboxylate, methylcyclopentadienyl-Mn-tricarbonyl, oxide, sulphonate
Na	*t*-butoxy
Ni	naphthenate
Pb	naphthenate, tetra-ethyl
Zn	carboxylate

in the range of just 150°C to 350°C. Regeneration temperatures in the low 400s°C have been achieved with catalytic coatings on the particulates trap substrate, while it has been shown that organometallic additive compounds are effective in promoting carbon oxidation by lowering its ignition temperature [34, 53].

Table 16.17 summarizes the organometallic compounds used as fuel additives for catalytic regeneration of particulate traps (aftertreatment devices) in diesel engine exhaust systems.

The exact nature of the organic part of the fuel additive is of limited importance, since after combustion metal oxide or metal sulphate particles remain, which are well distributed within the diesel particulates. The organic part of the additive determines its fuel solubility, and for some metals this solubility is found to be a problem. Several metals have been used in fuel additives, as can be seen in Table 16.17. The concentration of the additives can influence both the activity and the lifetime of the filter, which becomes clogged by ash. As underlined by Neeft et al. [53], data in the literature are ambiguous regarding the relative activity of such additives. Some authors have found copper to be the most active fuel additive; others claim that manganese or its combination with copper is more active.

Many problems still remain to be solved before organometallic additives can be introduced into the market for general application: overly high temperature or thermal stresses during filter regeneration; fuel deposits in combustion chambers or fuel injector nozzles; accumulated ash in the traps, and so on. However, in spite of these problems, it seems that the concept of fuel additives for regeneration of aftertreatment devices provides an interesting and promising route for catalytic removal of soot from diesel exhaust gases.

Detergent Additives An important property of the reformulated diesel fuel (called also premium or low emission) is its detergency, which is essential for maintaining good performance of the engine between periodic servicing. The use of detergent additives, which has become widespread in Europe during the last decade, was prompted by problems of injector nozzle fouling (or coking). Nozzle coking is induced by thermal degradation of fuel and crankcase lubricant components, and worsened by hot combustion gases. This results in slower initial combustion and pressure rise delay in the cylinder, with subsequent increased rate and higher peak pressure [54]. These effects cause increased engine noise, emission of pollutants, and fuel consumption. Detergent/dispersant additives containing surfactants can prevent deposit formation ("keep-clean"), and remove detrimental deposits already formed ("clean-up") in fuel injectors. Thus, they yield and ensure good spray pattern characteristics, and maintain engine performance and pollutants emission at the best levels possible for in-use engines [54–56]. A range of substances is now suitable as detergent additives for diesel fuels: amines, imidazolines, amides, fatty acid succinimides, polyalkylene succinimides, polyalkyl amines, polyether amines, and so on [48, 58]. Evaluating the performance of detergent/dispersant additives is an important aspect in the development of good-quality product for use. Care must be taken when selecting additives in order to avoid any problems created by adverse side effects resulting from their addition to the base fuel [55]. Two recognized test methods are mainly employed now for performance evaluation of detergent additives. In Europe, the test is based on the widely used Peugeot XUD 9 1.91 light-duty IDI diesel engine, and in the United States on the Cummins L-10 engine [54]. The latter is increasingly becoming accepted as the test standard for performance evaluation of detergent additives for controlling nozzle fouling in DI engines. A screening test for evaluating available detergent/dispersant additives to diesel fuel is described in Gutman et al. [58].

The detergent influence on emission performance is illustrated by a test vehicle, which was operated under part-load, city-type driving conditions with and without a detergent in the fuel (Figure 16.13). While the base fuel shows a clear increase in particulates emissions over the test periods, the fuel containing detergent additive maintains nearly constant emission level.

Fig. 16.13.

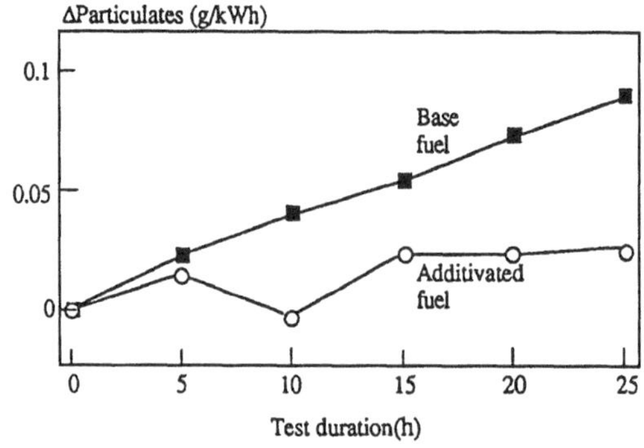

Effect of diesel fuel detergent additive on particulates emissions [48].

16.3.3.2 Additives Influencing Storage and Flow

Stability additives exert a favorable effect on diesel fuel in the course of prolonged storage; on the other hand, relatively little is known about the resistance of various stabilized diesel fuel to thermal stresses [40]. Antioxidants, stabilizers, and metal deactivators are types of additives that are sometimes used in diesel fuels and considered to be prone to oxidative or thermal instability due to the components used in their preparation. The additives work by terminating free radical chain reactions that would result in color degradation and the formation of sediment and insoluble gums. In some countries a fuel might be unacceptable for marketing as automotive diesel fuel if the maximum color specification is exceeded [34]. Additive treatment does not only mean that the refiner has to control the oxidation reactions that occur in uninhibited cracked gas oils. As explained earlier, diesel fuels have to conform to a sulphur specification, and this is achieved by a hydrodesulphurization process. The process also improves fuel stability by removing nitrogen- and oxygen-containing compounds and saturating the more reactive olefinic compounds, which are typically present in catalytically cracked gas oil. When only a mild degree of hydrogen treating is required to meet the sulphur specification, an antioxidant may be added to ensure that the fuel is adequately stabilized [34].

Antioxidants used in diesel fuels are usually hindered phenols that prevent high-temperature, gum-forming reactions.

Stabilizers are amines or other nitrogen-containing basic compounds that prevent sediment formation at ambient temperatures by interfering with acid-base reactions. The effectiveness of additive treatment strongly depends on the dominant fuel characteristics that determine the degradation reactions.

Metal deactivators are sometimes used in conjunction with stability improvers to prevent oxidation reactions from being catalyzed by heavy metal ions, particularly copper, which may be present in trace amounts in the fuel. One of the most commonly used metal deactivator additives is N,N′-disalicylidene-1,2-propanediamine, which works by chelating the dissolved metal to form a noncatalytically active compound [34].

Corrosion inhibitors are often used to prevent corrosion or rusting because it is almost impossible to avoid the presence of water in diesel fuel systems. These additives include esters or amine salts of alkenyl succinic acids, alkyl orthophosphoric acids, alkyl phosphoric acid, and aryl sulfonic acids. Selection of the suitable additive and treat rate is usually determined by using the rusting test ASTM D665 [1].

Biocides are used sometimes for diesel fuel treatment to prevent the growth of bacteria and fungi in the bottom of fuel tanks. The commercial biocides are based on a wide range of chemical types including boron compounds, amines, imines, imidazolines, and others, which need to be soluble in both the fuel and water or in the water phase only [1, 34, 57]. A problem with biocidal treatments is that the bacteria can develop resistance, so the additive type must be changed from time to time [57].

Dehazers and demulsifiers may be used occasionally if the fuel becomes hazy due to the presence of finely dispersed droplets of water. If the haze persists after the normal one or two days' settling time, additive treatment may be necessary to accelerate clearance. Effective dehazer additives include quaternary ammonium salts.

Antifoam additives must reduce the amount of foam build-up in the vehicle tank during filling and must destroy quickly the foam produced. Antifoamants are sometimes added to diesel fuel, often as a component in a multifunctional additive package. Using them enables avoiding the nuisance of stains and unpleasant odor, and reducing the risk of spills polluting the ground and the atmosphere. Typical antifoam additives are silicones having a molecular weight suited to the fuel characteristics [34].

Odor masks and odorants are used because diesel fuel is less volatile than gasoline and the stain and smell of spills will persist. This can be very unpleasant, particularly if clothing is contaminated. Elimination of the smell is undesirable because it facilitates the detections of leaks, so the aim is at modifying it by partial masking with a more acceptable odor. Various products, with a choice of fragrances, are commercially available to satisfy the consumer's preference.

Antistatic additives are added to diesel fuel to avoid the risk of an explosion due to a charge of static electricity building up during fast rates of pumping, as may occur during the filling or the discharge of road tankers. These additives lead to an increase of the fuel conductivity, allowing an electrostatic charge generated during pumping to be dissipated. The treatment is usually with a chromium-based additive [1].

Drag reducers are sometimes used to increase pipeline capacity. They are high molecular weight, oil-soluble polymers, which shear very rapidly and reduce drag. In a pipeline, effectiveness of a flow improver can be expressed in terms of percent drag reduction, defined as:

$$\text{Percent DR} = \frac{p - p_{\mathrm{DR}}}{p} \cdot 100$$

where p is the frictional pressure drop associated with the untreated fluid and p_{DR} is the frictional pressure drop of the fluid containing the drag-reducing polymer [59].

Additives to improve cold weather performance such as cold flow improvers or wax antisettling additives (WASA), were among the first additives used in diesel fuels. However, they are difficult to distribute uniformly throughout the fuel in quantities adequate for them to be efficient. This was the reason for developing new additives for modifying the shape of the wax crystals to enable them to pass the cold filter plugging point test (CFPP) [59]. A number of different materials were found to be effective as CFPP improvers, but all the products in current commercial use are ashless copolymers of ethylene and vinylacetate or other olefin ester copolymers. These additives have no influence on the fuel other than on its low-temperature properties and are compatible with the other types of additives used in automotive diesel fuel. They modify the shapes of the wax crystals, which otherwise are flat platelets, tending to gel together. There are three types of wax crystal modifiers: pour point depressants (PPD), flow improvers, and cloud point depressants (CPD) [1, 33, 34]. Close studies into the effect of wax modifiers have shown that some olefin-ester copolymers appear capable of suppressing wax crystallization by a few degrees [60]. The small compact wax crystals formed in a flow-improved fuel have a greater tendency to settle to the bottom of the fuel tank. This is more of a problem in storage tanks than in vehicle fuel tanks, but wax antisettling additives can nevertheless play a useful part in the avoidance of wax enrichment, as vehicle fuel tanks become empty, especially in very cold climates [34].

Anti-icer additives are used in some countries to prevent ice plugging of fuel lines by lowering the freezing point of small amounts of free water, which may be separated from the fuel. These additives are relatively low molecular weight alcohols or glycols with a strong affinity to water [1].

Lubricity additives are a new class of diesel fuel additives, which have to be used when the diesel fuels lose their own lubricity properties. As explained earlier, the reduction in sulphur content can reduce the lubricating properties of the fuel [1, 47].

16.3.4 Influence of Diesel Fuel Quality on Emissions

Many independent studies have been carried out to assess the effect of fuel properties on diesel engines emissions [42, 53, 61, 62, 63]. A general conclusion from these studies is the clear correlation between certain fuel characteristics and vehicle exhaust emissions. This fact is confirmed by the recent results obtained in the EPEFE research program, which was a very comprehensive investigation of the effects of vehicle/fuel technologies on exhaust emissions [5, 6, 65, 66]. It involved a set of light-duty (LD) vehicles, including engines with indirect-injection combustion chambers (IDI) and engines with direct injection (DI), part of them turbocharged, and the majority intercooled. The heavy-duty (HD) engines were all DI designs and were all turbocharged and intercooled. All of the light-duty vehicles were equipped with exhaust catalysts and all except two with exhaust gas recirculation (EGR) [65].

The test vehicles and engines were all designed to meet the minimum 1996 European emissions standards, as shown in Table 16.18. A matrix for 11 diesel fuels, shown in Figure 16.14, was created in this program to cover the selected variables of density, poly-aromatics, cetane number, and T95. The fuel properties were changed in this study over the range of the EPEFE fuel matrix, that is, density—855 to 828 kg/m^3; poly-aromatics—8 percent to 1 percent by weight; cetane number—50 to 58; T95—370°C to 325°C.

The emissions measured in this study included total hydrocarbons (HC), carbon monoxide (CO), nitrogen oxides (NO_x), and particulates (PM) as well as hydrocarbon speciation and compositional analysis of particulates, tested to the European driving cycle for the year 2000 [66]. The overall fuel effects on emissions,

Table 16.18

European Emission Limits for Diesel Engines [65]

LD limit, g/km		HD limit, g/kWh	
		NO_X	7.0
PM	0.08	PM	0.15
$HC + NO_X$	0.7	HC	1.1
CO	1.0	CO	4.0

Fig. 16.14.

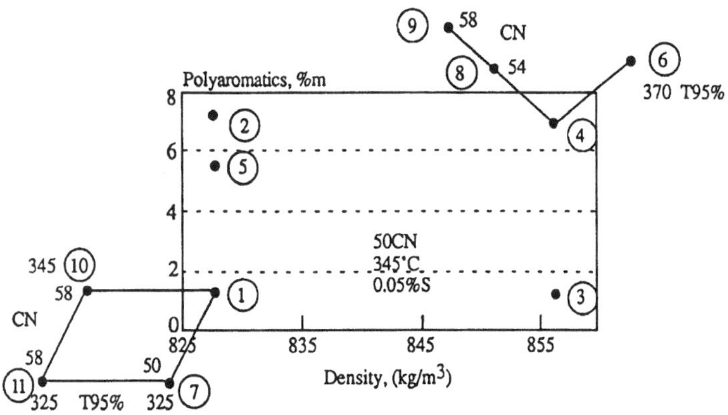

EPEFE diesel fuel matrix for 11 fuels [67].

obtained from the results of a large amount of data, are summarized in Table 16.19. The overall effects on emissions of changing simultaneously all four fuel properties are shown in Figure 16.15 for LD and HD engines.

From the results in Table 16.19 and Figure 16.15, the following important conclusions can be drawn:

- Fuel effects are greater for light-duty vehicles than for heavy-duty engines.
- The variation of diesel fuel quality can lead to a considerable spread in the emissions of CO, HC, and particulates.
- For LD vehicles, the change in fuel properties with regard to the reference point (a hypothetical fuel with density 855 kg/m^3, poly-aromatics 8 percent, CN 50 and T95 370°C) caused a decrease of CO, HC, and PM emissions with a slight increase of NO_X, whereas for HD engines NO_X and PM were decreased with an increase of HC emissions.
- When fuel properties were changed in this way, only PM changed in the same direction for LD and HD engines, but HC and NO_X changed in opposite directions.

Table 16.19

Summary of Estimated Effects of Fuel Property Changes on Diesel Engine/Vehicle Emissions [5]

	Property	Change	CO	HC	NO_x	PM
Light-duty vehicles**	Reduce sulphur	2000 to 500 ppm	0	0	0	↓
	Reduce density*	850 to 820 kg/m^3				
			↓	↓	0	↓↓↓
	Reduce poly-aromatics*	6 to 3% v/v				
	Increase cetane no.	50 to 55	↓↓	↓↓	±0	↓
	Reduce T95	370 to 330°C	−0	−0	−0	−0
Heavy-duty engines**	Reduce sulphur	2000 to 500 ppm	0	0	0	↓↓
	Reduce density*	850 to 820 kg/m^3				
			0	0	↓	↓
	Reduce poly-aromatics*	6 to 3% v/v				
	Increase cetane no.	50 to 55	↓	↓	↓	↓
	Reduce T95	370 to 330°C	−0	−0	−0	−0

Note (1): *Insufficient data are available to reliably separate the influence of these parameters, which will be subject of further study.

**Above effects are not necessarily additive.

Note (2): Europia also requested the following uncertainties to be expressed:

—Light duty replace ↓ by ↓? for HC and CO

—Heavy duty replace ↓ by ↓? for NO_x

—For T95 all zeros should be plus/minus zero

Note (3): ACEA broadly agreed with the effects as written.

Key

0 = no effect	±0 = −2–+2%	↑ or ↓ = 2–10% effect	↑↑ or ↓↓ = 10–20% effect
↑↑↑ or ↓↓↓ =	> 20% effect	? = Insufficient information	

Fig. 16.15.

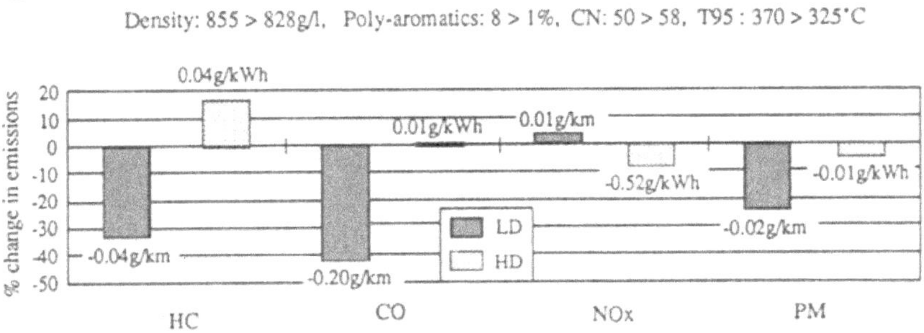

Overall effects on emissions of changing simultaneously fuel properties (two sets of data: LD and HD engines) [65].

Fig. 16.16.

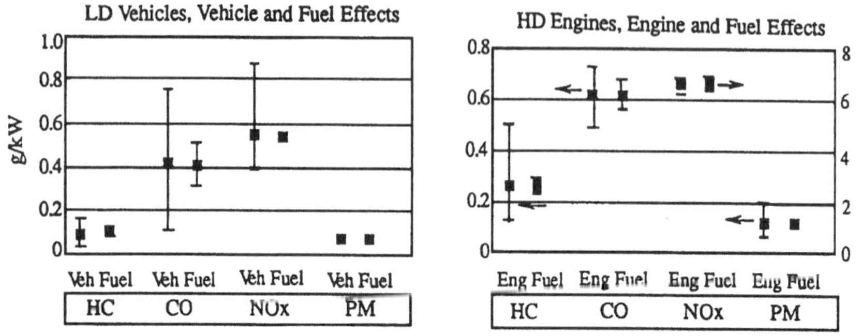

Vehicle and fuel effects on emissions [65].

- The principal fuel parameters affecting vehicle emissions (both light and heavy duty) are sulphur content, cetane number, density, and poly-aromatic content. Distillation characteristics can be considered an additional factor, but their effects appear to be small.

Other conclusions from this research program [5] are:

- Sulphur has little effect on emissions from light-duty vehicles without catalysts but has a larger effect on heavy-duty emissions, especially when an oxidation catalyst is used.
- Emissions from cetane-improved fuels are equivalent to fuels of the same cetane number without additives.

Figure 16.16 shows separately the effects of vehicle and fuel on emissions. In this figure, two marks are shown for each emissions parameter for the two data sets (LD and HD vehicles). The left mark shows the mean and the range of variation for the different test of vehicles and engines, calculated as the average of the results for all tested fuels. The right mark shows the same data for all the tested fuels, based on the average for all vehicles or engines. It is clear from this figure that fuel effects on emissions is strongly dependent on the engine technology.

Figure 16.17 shows the effects of separate fuel changes on HC, CO, PM, and NO_X emissions.

For a better understanding of the relationship between fuel and engine technology, two generations of development, 1992, 1996, and also one heavy-duty engine from the same model line were tested on the EPEFE diesel fuel matrix by a CONCAWE research group. A set of test results concerning, for example, the effects of fuel density change on emissions is shown in Figure 16.18 for the two HD engines. One conclusion of this comparative study is that advanced engine technology has reduced the sensitivity to fuel property changes [67]. The variations observed in Table 16.19 and in Figures 16.15 to 16.18 suggest that

Fig. 16.17.

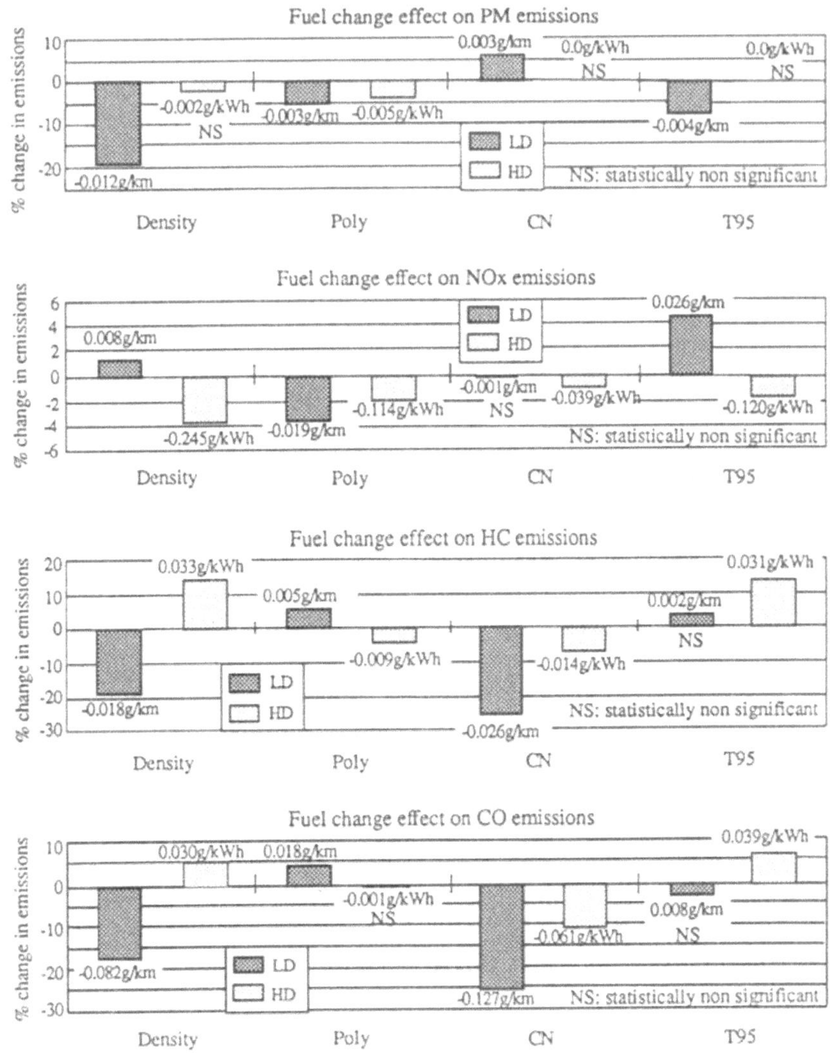

Effects of separate fuel changes on emissions [65].

more remains to be learned about the interaction of vehicle/engine technology and fuels [65].

Based on the experimental data obtained in the EPEFE study and on recent data from the literature which involve parameters not considered in the former, like the effect of sulphur on emissions, quantifying equations were established. These equations [6] are reproduced in Table 16.20. They represent complex relationships between diesel fuel properties and vehicle emissions. These equations

Fig. 16.18.

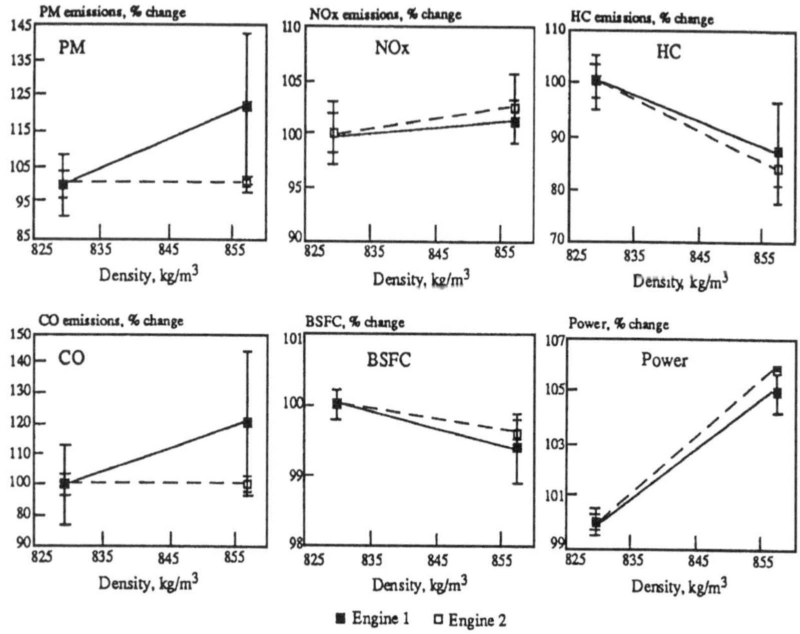

Comparison of engine technology response to fuel density [67].

are not simple and it is not possible to use individual parameters alone to derive emission factors. However, they open new possibilities of predicting the fuel performance effects on the emission responses of given emission-control technologies used in vehicle diesel engines. As underlined in Camarsa et al. [6], these equations were essential for the auto-oil proceedings, since they provided the required input (for both the air quality models and the processes whereby these models were used) for searching for the optimum combination to achieve the air quality objectives for Europe for the period 2000 to 2010. As mentioned by EPEFE [6], the developed equations are valid within a wide range of test procedures, vehicle/engine technologies, and fuel parameters, used in their study, but great care must be taken in extrapolating from these results. According to the developments of vehicle technologies, test procedures, and refining process, it will be necessary to further investigate and validate the conclusions reached.

16.3.5 Main Trends in Diesel Fuel Specification Developments

Environmental legislation pressure will continue to be the major factor in influencing fuel quality. Thus, the main trends of the development in diesel fuel properties and composition are in accordance with environmental legislation trends, and with fuel-emission responses related to the change in engine/vehicle technologies as described in the previous section. The various research programs that were carried

Table 16.20

Auto/Oil Programme—Equations of the Fuel/Engine Technologies Responses

A. Light-duty diesel	
CO (g/km)	$-1.3250726 + 0.003037\ \mathrm{DEN} - 0.0025643\ \mathrm{POLY} - 0.015856\ \mathrm{CN} + 0.0001706\ T_{95}$
HC (g/km)	$-0.293192 + 0.0006759\ \mathrm{DEN} - 0.0007306\ \mathrm{POLY} - 0.0032733\ \mathrm{CN} - 0.000038\ T_{95}$
NO_x (g/km)	$1.0039726 - 0.0003113\ \mathrm{DEN} + 0.0027263\ \mathrm{POLY} - 0.0000883\ \mathrm{CN} - 0.0005805\ T_{95}$
PM (g/km)	$[-0.3879873 + 0.0004677\ \mathrm{DEN} + 0.0004488\ \mathrm{POLY} + 0.0004098\ \mathrm{CN} + 0.0000788\ T_{95}] \times [1 - 0.00016\ (450 - \mathrm{Sulphur})]$
B. Heavy-duty diesel	
CO (g/kWh)	$2.24407 - 0.0011\ \mathrm{DEN} + 0.00007\ \mathrm{POLY} - 0.00768\ \mathrm{CN} - 0.00087\ T_{95}$
HC (g/kWh)	$1.61466 - 0.00123\ \mathrm{DEN} + 0.00133\ \mathrm{POLY} - 0.00181\ \mathrm{CN} - 0.00068\ T_{95}$
NO_x (g/kWh)	$-1.75444 + 0.00906\ \mathrm{DEN} + 0.0163\ \mathrm{POLY} - 0.00493\ \mathrm{CN} + 0.00266\ T_{95}$
PM (g/kWh)	$(0.06959 + 0.00006\ \mathrm{DEN} + 0.00065\ \mathrm{POLY} - 0.00001\ \mathrm{CN})\ [1 - 0.000086 \times (450 - \mathrm{Sulphur})]$

DEN density (g/l) CN Cetane Number Sulphur fuel sulphur content (ppm)

POLY fuel polyaromatic content (%wt)

T_{95} Temperature at which 95% fuel is evaporated

Source: SAE Paper No. 961076

out, especially in United States are California Air Resource Board (CARB), Coordinating Research Council (CRC), Air Pollution Research Commission (APRAC), and so on; in Europe the European Programme on Emissions and Engine Technologies (EPEFE) and others have had the objective to investigate the effects of fuel properties on emissions. These include cetane number, sulphur content, aromatic (or only poly-aromatic) content, density distillation (volatility upper limits), and so on. Based on these studies, new fuel specifications were recommended for the future or have already been implemented [1, 15, 62, 68, 69]. These reformulated products have been given many other names, such as clean diesel, premium diesel, high-purity fuel or low-emission diesel fuel (LEDF). Some of these products are currently available in different countries.

The new specifications would include higher values for cetane numbers, lower sulphur content, a maximum aromatic content (especially polycycle aromatics), lower density value, lower T95 or distillation end point, use of additives to control injector deposits, and so on. In general, reformulated diesel fuel specifications reflect the main trend in the formulation developments of diesel fuels as follows:

Increase of Cetane Number The performance of diesel engines has traditionally been strongly correlated with cetane quality of diesel fuel as a principal parameter. Table 16.21 shows the cetane number specifications for different countries

Table 16.21

Cetane Number Specifications for Different Countries and Sampling Test Results (Winter 1996) [13, 15, 43]

				Winter 1996—quality survey		
				Typical ranges		
Country	Standard	Grade	Cetane number	Minimum	Maximum	Average
Austria	0–Norm EN 590 1.2.94	Winter Fuel Intermediate Summer	49	48.7	50.4	49.7
Belgium and Luxembourg	NBM EN 590	Winter Intermediate Summer	49	47.2	54.4	50.5
Denmark	CEN Diesel	CEN Diesel	49	49.3	53.7	50.8
Finland	Reformulated Diesel	Summer	49			
		Winter	47	48.2	50.8	49.6
France	EN 590:1993	Summer	49			
		Winter Grand Froid		46.5	50.7	49.2
Germany	DIN EN 590:1993	Summer	49			
		Intermediate				
		Winter		47.7	54.7	50.9
Italy	EN 590–93	Summer	49			
		Winter		46.3	54.8	49.3
Japan	JIS K2204 (1992)	Grade 2	45	46.9	59.8	53.2
		Grade 3	45	48.3	54.0	50.3
Spain	EN 590 (1993)	Summer	49			
		Winter		46.7	58.0	51.2
Sweden			50	51.6	53.6	52.6
United Kingdom	BS EN 590:1993		49	47.9	55.2	50.9
United States	ASTM D975 (1994)	Low Sulphur	40	40.8* 39.6** 43.3***	49.0* 49.5** 56.8***	44* 43.9** 48.6***

*U.S. East Coast; **U.S. Midwest; ***U.S. West Coast.

together with sample test results. The engine manufacturers from the United States and Europe continue to request fuels with higher cetane number, so there exist further opportunities for fuel reformulation and for the utilization of cetane number improver. Thus, the European Association of Automobile Constructors (ACEA) and the U.S. Association of Engine Manufacturers are pressing for higher cetane number specifications to enable more stringent emissions targets to be attained. As mentioned in Section 16.3.4, the ACEA fuel charter recommends a minimum cetane number of 53 and the Engine Manufacturers Association proposes a huge change in cetane number to a 56 minimum.

Decrease of Sulphur Content In order to reduce emissions from diesel engines, especially particulates, legislation is being implemented around the world to lower the level of sulphur present in diesel fuels. Table 16.22 shows the allowable maximum sulphur content, for different countries, in comparison to the limits before these new implementations. In the proposal of the Auto Oil European Commission for diesel fuel specification, which is intended to be implemented by the year 2000, the sulphur content will be limited to 350 ppm (see Table 16.23).

Table 16.22

Maximum Allowable Sulphur Content of Automotive Diesel Fuel for Different Countries [14, 15]

		Maximum sulphur content % m/m (max)		
Country	Effective date	Present values		Previous values
European Union	from 01.10.1996		0.05	0.2 (1994)
Austria	from 01.10.1995		0.05	0.15 (from 1986)
Germany	by agreement		0.05	0.2 (1994)
Switzerland	from 01.01.1994		0.05	0.2 (1993)
United States	from 01.10.1993		0.05	0.5 (1992)
Japan	from 01.05.1997		0.05	0.2 (1996)
Israel	from 01.07.1997		0.05	0.2 (1996)
Sweden[(1)]	from 1993	grade EC_1	0.001	0.2 (1990)
		grade EC_2	0.005	
		grade EC_3	0.20	
Denmark[(1)]	from 01.06.1992			
		grade A CEN quality	0.2	
		grade B CEN 0.05% quality	0.05	
		grade C Public Bus Service Ultra Light Diesel	0.05	
Finland[(1)]	from 01.06.1993	Reformulated Diesel	0.005	
		Standard Diesel	0.2	0.2 from 01.01.89

[(1)] A special tax policy for diesel fuels was introduced.

Table 16.23

Vehicle Diesel Fuel Specifications for the Year 2000 [69]

Parameter	Units	Current specification	Proposed year 2000 specification	Predicted market average without proposal	Predicted market average with proposal
Cetane number	number, min.	49	51	51	53
Poly-aromatics	% mass, max.	–	11	9	6
Density	kg/m^3, max.	86–	845	843	835
T95	°C, max.	370	360	355	350
Sulphur	ppm, max.	500	350	450	300

It is possible that further reduction of sulphur content in diesel fuel will be needed by catalysts. As of now, oxidation catalysts are being developed for reducing the SOF (soluble organic fractions) of the total particulate matter (TPM), the vapor-phase HC, and CO emissions from diesel engines. The content of sulphur is related to quantities of SO_4 adsorbed on the catalyst surface and the emitted sulphur compounds. The suppression of SO_4 emissions is important since sulphates have bound water to them and this effect can tend to negate the SOF reduction relative to TPM reduction. This problem can be eliminated if fuel sulphur contents were reduced further from 0.05 percent to 0.01 percent or lower levels [61].

Aromatics Content Limitations Tests on fuels with zero up to about 35 percent total aromatics, typical for currently marketed fuels, show no influence of total aromatics content on particulates emissions, but there were indications of an upturn at higher levels [1]. It is important to note that engines differ in their sensitivity to the influence of aromatic content on particulates emission [1]. Other recent evaluations suggest that while total aromatics may have no influence on particulate emissions, a more significant role may be played by polycyclic aromatics [5, 65].

Limits on aromatics have already been imposed in some countries [1, 15, 62]. For example, the diesel fuel specifications introduced in Sweden in 1991 have limited the maximum aromatics content to 5 percent and 20 percent by volume for Urban Diesel 1 and Urban Diesel 2 respectively [1], and in California limits were introduced by CARB of 10 percent by volume aromatics content and of 1.4 percent by weight (maximum) for polycyclic aromatics from 1993 [15].

In the United States the Engine Manufacturers Association, which is asking for improved quality for 1998, proposes for the new diesel fuel specifications a big reduction in aromatics [47]. The maximum poly-aromatics content is limited to 11 percent mass in the proposed year 2000 specifications for the European Community [69].

Density and T95 The results presented in Hublin et al. [66] indicate that the density and cetane number have the largest effect (in percentage terms) on emissions. Poly-aromatics and T95 have smaller effects. From the specifications for the reformulated and new clean diesel fuel in Sweden, Finland, and Switzerland, it is possible to observe a very slight tendency of reducing the density value. A similar trend is seen from the data in Table 16.23. It can also be observed from this table that there is a tendency to decrease the distillation values (T95) for the new, or future, diesel fuels.

Extended use of diesel fuel additives in the future can be useful for the oil refineries for extra processing and blending as a means to improve the fuel properties (additives to restore the cold properties in case of cutting deeper into the crude oil, ignition improvers for low-cetane material, which is being diverted into automotive diesel fuel, etc.), or for improving the aftermarket properties like detergency, lubricity, and so on.

16.4 ALTERNATIVE FUELS

Continuous increase in the severity of environmental legislation together with considerations of security of energy supply have recently led to significantly increasing interest in alternative fuels. The U.S. Clean Air Act Amendments (CAAA) of 1990 point out the importance of using the "clean alternative fuels," such as methanol and ethanol (and mixtures thereof), reformulated gasoline, natural gas, LPG, electricity and any other fuels which permit vehicles to attain legislated emission standards [15]. Biogas, vegetable oils, ethers, and hydrogen also have been considered as potential fuel alternatives. Reformulated gasoline and its effects on vehicle exhaust emissions have been discussed already in Section 16.2.

Gasoline and diesel fuel may also be produced from synthetic hydrocarbon liquids. The catalytic synthesis of hydrocarbons is performed, generally, from carbon monoxide and hydrogen by the so-called Fischer-Tropsch process [1]. Generally, the sources of CO and H_2 needed for synthesis reactions are coal, natural gas, or methanol. In South Africa, for example, synthetic gasoline has been produced commercially from coal (the SASOL method, see Appendix 1). The process has a low efficiency and is only used where supplies of crude oil are limited [1]. Because of their similar composition, synthetic hydrocarbon fuels have similar effects on vehicle emissions as do conventional ones.

In the following, the previously mentioned alternative fuels and their effects on vehicle exhaust emissions are briefly reviewed.

16.4.1 Alcohols

Two alcohols, methanol and ethanol, are generally considered as alternative motor fuels. Ethanol containing 5 percent water (E95) has been used successfully in the vehicle park of Brazil. In the United States (California in particular), Germany, Japan, Australia, and New Zealand, increasing attention in alcohol fuels is primarily focused on methanol, mainly due to manufacturing cost and supply considerations [1]. As mentioned earlier in Section 16.2, both methanol and ethanol are used together with other oxygenates for blending reformulated gasolines.

Methanol is mainly produced from natural gas, coal, and sometimes from heavy residual oils. Technologies for methanol production from lignite, peat, and biomass are also available. Methanol production from natural gas takes place, generally, near the gas source and not in the consumer's area, in order to avoid the high costs for transport of gas compared to that of easy-to-handle liquid methanol [70]. Ethanol is generally produced by fermentation of biomass feedstocks, mainly sugar cane, wheat, and wood.

The main advantages of using alcohols as alternative motor fuels are:

- Both methanol and ethanol, as mentioned earlier, can be derived from non-crude, sometimes renewable, resources that are much more evenly distributed over the globe, compared to crude oil, of which over 50 percent is located in the Middle East [1]. This fact significantly contributes to the solution of the "security of supply" problem.

- Both methanol and ethanol have excellent antiknock performance, which enables higher compression ratios to be used than are possible with gasoline-fueled engines, so improved engine thermal efficiency can be achieved. Some properties of methanol and ethanol are summarized together with other alternative fuels and compared with gasoline and diesel fuel in Table 16.24.

- Alcohols burn with lower flame temperatures and luminosity, so that the heat losses are lower [1].

- Both methanol and ethanol have very high latent heat of vaporization (see Table 16.24), which allows the achievement of higher volumetric efficiency compared to gasoline-fueled engines due to the increased charge density.

- By using alcohol fuels it is possible to reduce hydrocarbon emissions and to lower the photochemical reactivity of emitted pollutants. An example of alcohol fuel effects on exhaust emissions is shown in Figure 16.19, based on results from the U.S. AQIRP program for methanol and ethanol [74–76]. These were obtained with flexible/variable fuel vehicles (FFV/VFV) for the near-neat methanol (M85) and ethanol (E85) blends with gasoline. Each blend contained 85 percent of alcohol fuel and 15 percent of gasoline. As can be seen from the figure, using methanol or ethanol as alternative fuel leads to a reduction of OMHCE, benzene, and 1,3 butadiene emissions, together with lowering photochemical reactivity of the exhaust gases. At the same time, use of alcohol fuels resulted in increasing NMOG and aldehyde emissions (the latter, quite sharply). There is no clear picture about CO and NO_x emissions, although lower flame temperature may contribute to the reduction of NO_x formation.

The physical and chemical characteristics of alcohols do not make them particularly suitable for compression-ignition engines as a direct replacement for diesel fuel. As can be seen from Table 16.24, the cetane numbers of methanol

Table 16.24

Properties of Various Fuels (Based on [71–73])

Property	Diesel fuel[1]	Gasoline[1]	Methanol	Ethanol	Methane	Propane	Dimethyl ether (DME)	Vegetable oil (RME)[6]	Hydrogen
Density at 15°C, kg/l	0.82–0.86	0.72–0.78	0.79	0.79	0.72[2]	0.51 (liquefied)	0.66	0.88	0.090[2]
Lower Heating Value (LHV), MJ/kg (MJ/ℓ)	42.5 (35.7[3])	44.0 (33.0[4])	19.7 (15.6)	26.8 (21.2)	50.0 (36.0[2])	46.3 (23.6)	27.66 (18.2)	37.2 (32.7)	120 (10.8[2])
Heat of Vaporization[5], KJ/kg	250	305	1110	904	509	426	460 (410 at 20°C)		
Boiling Point, °C	180–360	25–215	65	78	−162	−43	−20	330–340	−253
Flammability Limits (% vol gas in air)	0.6–7.5	0.6–8	5.5–26	3.5–15	5–15	9–9.5	3.4–18		4–77
Octane rating (RON)		91–98	111	108	130	112			
Cetane Number	40–55		5	8			≫55	51–52	

(1) Typical values.

(2) Per nm^3.

(3) For density 0.84 kg/l.

(4) For density 0.75 kg/l.

(5) At 1 atm. and 25°C for liquid fuels; at 1 atm and boiling temperature for gaseous fuels.

(6) RME rapeseed oil methyl ester.

Fig. 16.19.

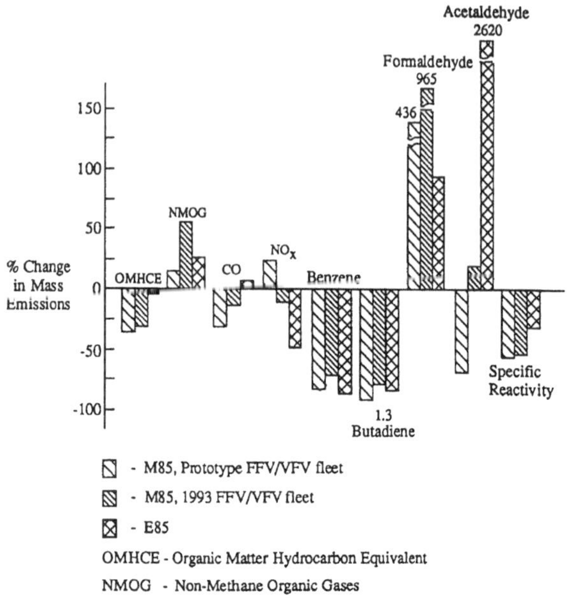

Alcohol fuel effects on emissions (compared to industry average gasoline), AQIRP results [74–76].

and ethanol are 5 and 8, respectively. Use of alcohol as an alternative fuel for the diesel engine is generally possible by one of the following ways [77]: in blended fuels, in engines with forced ignition, and in dual-fuel engines. Blending alcohol with high cetane number fuels is one of the simplest ways of using alcohol fuels, but large amounts of ignition improver and also additives preventing phase separation are generally required. One of the commonly used ignition improvers is the so-called Avocet, containing a nitrate ester dissolved in methanol [78]. Use of neat alcohols is only possible in engines with forced ignition by glow or spark plugs. The forced-ignition method possesses also a good flexibility for a variety of fuels. Dual-fuel engines may be operated with pilot injection or with alcohol carburation (fumigation) into the engine's intake manifold. The pilot injection method allows the ignition of a large amount of alcohol injected into the cylinder by a small ignition dose of diesel fuel. Double-injection nozzles are generally used here. The method of fumigation is beneficial due to its relative simplicity and the potential for application to gaseous as well as liquid fuels. However, knocking or misfiring may occur with this method and it is generally used for relatively low percentages of alcohol in total fuel energy [77, 79]. As follows from various published research works [1, 77, 79], use of alcohols in diesel engines generally contributes toward reduction of particulates and NO_x emissions.

The main shortcomings of alcohols as alternative motor fuels, in addition to the previously mentioned increase in emission of some kinds of pollutants (NMOG, aldehydes, etc.) are:

- Alcohols have a much lower energy content (heating value) than gasoline or diesel fuel (see Table 16.24); therefore, more fuel is needed in order to supply to the engine the same amounts of energy or, in other terms, the volumetric fuel economy will always be lower.

- Alcohols have low vapor pressures and high latent heats of vaporization and, therefore, show poor cold starting and warm-up performance. Near-neat alcohols, containing small amounts of gasoline or other fuels (for example, M85 and E85 fuels) are frequently used in order to overcome this problem.

- Methanol and, to a lesser degree, also ethanol are corrosive to many of the metals, elastomers, and plastic components that are used in conventional fuel systems. The correct choice of suitable materials for the fuel system of alcohol-fueled vehicles is very important and several test programs have been carried out with this aim [1, 80].

- Accelerated cylinder and piston ring wear have been found to occur in neat-alcohol-fueled engines [1]. The possible reasons are washing away of lubricant film during cold starting and/or corrosion due to formic or performic acids formation during combustion.

- Alcohols have lower viscosity relative to diesel fuel and poor lubricity, which would cause increased wear of conventional fuel injection equipment.

- Methanol burns with an almost invisible flame (low luminosity) and so presents somewhat of a safety hazard. Use of near-neat alcohol fuel with the addition of gasoline or other materials allows this problem to be overcome by increasing flame luminosity [1].

16.4.2 Natural Gas

Natural gas is frequently considered as the most promising near-term alternative fuel because of its long-term availability, well-balanced geographic distribution, and clear ecological benefits.

Using natural gas in mobil and stationary engines is not a new concept, several companies for many decades have manufactured gas engines for stationary use such as gas compression and electric power generation [81]. Also, over 1 million natural gas vehicles (NGVs) are operating now on the world's roads. The leading users of NGVs are countries of the former Soviet Union, Italy, Argentina, Australia, New Zealand, United States, and Canada. Table 16.25 includes data about the NGV numbers in different regions worldwide based on the estimate performed for the European THERMIE program [82].

Natural gas, as it is found in the earth, is basically composed of between 88 percent and 96 percent methane with small and varying amounts of nonmethane hydrocarbons, CO_2, water, N_2, H_2S, He, Ar, and other trace gases [1, 81, 82]. In most cases, natural gas from the gas field undergoes an upgrading process. This is necessary for removal of water, H_2S, and excess amounts of higher hydrocarbons

Table 16.25

Regions with Greatest NGV Populations (Based on Data from [82])

Area	Number of vehicles
Former Soviet Union	350,000
Italy	300,000
South America	200,000
Australia	150,000
North America	130,000

Table 16.26

CARB Specification of Natural Gas Used as Automotive Fuel (Based on Data from [15])

Constituent, mole %	Specified value
Methane	88.0 (min)
Ethane	6.0 (max)
C_3 and higher	3.0 (max)
C_6 and higher	0.2 (max)
Hydrogen	0.1 (max)
Carbon monoxide	0.1 (max)
Oxygen	1.0 (max)
Inert gases	1.5–4.5

and inert gases to prevent corrosion damage and condensing problems in pipeline networks, to achieve the required heating value, and to obtain valuable by-products. In order to minimize the negative effect of variations in natural gas composition on vehicle performance and pollutants emission, the California Air Resources Board (CARB) has established special requirements for natural gas sold commercially as vehicle fuel (see Table 16.26).

Natural gas may be stored on board a vehicle either as a compressed natural gas (CNG) in high-pressure cylinders or as a cryogenic liquid. In order to provide vehicles with an acceptable independent range, CNG has to be compressed and stored in tanks on vehicles at pressures of about 200 bar. To liquefy natural gas at atmospheric pressure, its temperature needs to be brought down to about −162°C (see boiling point of methane in Table 16.24). Use of liquefied natural gas (LNG) as an automotive fuel is the subject of some serious safety problems, such as the possibility of excess vapor entering into an enclosed area, which may lead to explosion. There are additional technical challenges, for example, the need to ensure that the liquid phase does not become significantly enriched with nonmethane hydrocarbons compared to initial tank levels [81]. Therefore, almost all NGVs used today are fueled by CNG. Natural gas vehicles operating on CNG may be refueled either in fast-fill mode (refueling times are comparable with those of conventional vehicles) or in slow-fill mode. The latter allows vehicles to be refueled over a period of several hours overnight.

As can be seen from Table 16.24, natural gas has a superior octane quality, which makes it a beneficial fuel for spark-ignition engines. Vehicles using natural gas may be:

- bi-fuel, converted from conventional ones and operating on gas or gasoline at the driver's choice;
- dedicated to the gaseous fuel with forced ignition of fuel-air mixture;
- dual-fuel, in which the gaseous fuel is ignited by injecting a small amount of diesel fuel. Frequently, these vehicles are also bi-fuel because of the possibility to run them on diesel fuel only [83].

The main advantages of natural gas use for fueling vehicles, as partially mentioned earlier, are:

- Natural gas is a primary fossil fuel, widely available worldwide and with well-balanced geographic distribution. Using it as an alternative motor fuel will contribute to the improvement of "security of supply" of energy resources.
- Very high antiknock resistance of natural gas (RON of methane is 130) allows it to be used in dedicated engine concepts with much higher compression ratios as compared to gasoline engines. It is especially beneficial because it opens the possibility of converting heavy-duty diesel engines to high-compression, lean-burn, spark-ignition ones operating on CNG without changing the basic engine design.
- Unlike their gasoline counterparts, natural gas vehicles do not require mixture enrichment for cold starting, so that pollutants emission from NGVs are unaffected by low temperatures.
- Use of natural gas allows drastic reductions to be achieved in pollutants emission compared to either gasoline- or diesel-fueled vehicles. Emissions data from a number of NGVs using modern electronic emission control systems clearly show the ability to meet ULEV emissions standards of CARB [1]. The examples in Figure 16.20 illustrate the changes in emission levels of vehicles operating on natural gas, compared to gasoline or diesel fuel. As can be seen from the figure, there are some superior ecological benefits of NGVs:
    - Low emissions of nonmethane hydrocarbons;
    - Reduction of CO and NO_x emissions, especially substantial for lean-burn, heavy-duty natural gas engines as compared to their diesel counterparts;
    - Extremely low emissions of particulates;

Fig. 16.20.

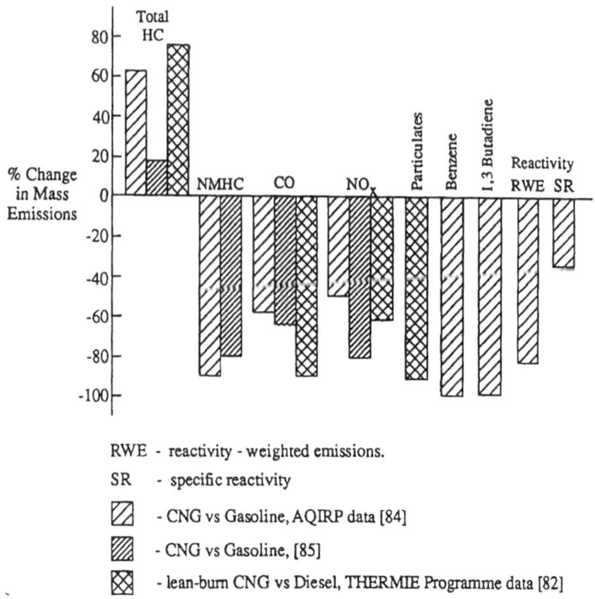

Natural gas effects on emissions.

- Low emissions of air toxics such as benzene and 1,3 butadiene;
- Low photochemical activity of exhaust gases.
- Another environmental benefit of natural gas compared to conventional fuels is zero evaporative emissions due to the sealed fuel system. Also, some reductions in CO_2 emissions have generally been achieved, compared to gasoline engines, due to lower carbon-hydrogen ratios in the natural gas.

- An important advantage of heavy-duty NGVs compared to diesel counterparts is a reported noise reduction (by 3–8 dB(A)) [83, 86].

The *main and principal disadvantages of natural gas* as an automotive fuel lies in the fact that it is a gas. Therefore, special approaches, as mentioned earlier, are needed in order to store it on board a vehicle. Extra weight needed for fuel tanks in buses with CNG compressed to 200 bar, for example, may reach or even exceed 1000 kg in order to provide a reasonable operational range of the vehicle [82].

- Special infrastructure of filling stations must be developed for refueling purposes.

- As can be understood from Figure 16.20, NGVs emit generally much more unburned methane than their gasoline or diesel counterparts. Effective methane oxidation in catalytic converters is a problem because of its high chemical stability. Methane is known as an active greenhouse gas, but NGV's contribution to global warming, even in the scenario of their wide penetration into the market, is considered quite negligible because the great majority of global methane emissions to the atmosphere originate from natural uncontrolled sources [1, 81].

- Like all otto-cycle engines, natural gas ones generally have a fuel economy of about 20 percent lower than that of diesel engines.

From the safety point of view, CNG vehicles are generally not a subject of increased concern and even are expected to be safer than gasoline vehicles [1].

16.4.3 Biogas

Biogas can be produced from a large diversity of sources as long as they contain organic material. Examples of such sources are municipal sewage, agricultural waste, waste dumps, manure, and so on [70]. Depending on the source, the composition of the gas may vary, but the dominant component will always be methane. The conversion of organic mass with biogas formation is generally carried out by a fermentation process. The main and most significant advantage of biogas as an alternative fuel is that it may be produced from renewable energy sources. Because biogas is mainly methane, its effects on exhaust emissions will be similar to those of natural gas.

16.4.4 Liquefied Petroleum Gas

Liquefied petroleum gas (LPG) broadly refers to commercial propane, commercial butane, and mixtures of the two. LPG is generally derived from two sources: natural gas processing and petroleum refining. Produced by either one of them, LPG is essentially a by-product. The name *LPG* is explained by the fact that the gas assumes a liquid state at pressures of 2 to 20 bar; the actual value depends upon the propane-butane ratio [71].

The composition of commercial LPG varies greatly from one country to another. In the United States, automotive LPG generally contains more than 85 percent propane [15, 87]. In Europe, on the other hand, LPG contains nearly equal parts of propane and butane [1, 87]. Table 16.27 illustrates U.S. specifications for automotive LPG.

Over 3 million vehicles are operating on LPG worldwide. The leading users of LPG are Italy, Netherlands, and North America. Some data about the numbers of LPG vehicles in different countries are shown in Table 16.28 [82].

LPG, as also natural gas, is a fuel intended for use in engines with forced ignition (otto-cycle). Generally, any vehicle equipped with an IC engine can be

Table 16.27

U.S. Specifications for Automotive LPG [15]

Property	U.S. ASTM D1835
Propane, % vol.	85 min
Propylene, % vol.	5.0 max
Vapor pressure at 37.8°C, kPa	1430 max
Volatile residue evap. temperature 95%, °C	−38.3 max
or	
Butane and heavier, % vol.	2.5 max
Sulphur, ppm	120 max

Table 16.28

Countries with Greatest LPG Vehicle Populations (Based on Data from [82])

Country or area	Number of vehicles
Italy	1,300,000
Netherlands	610,000
North America	610,000
Japan	300,000
Australia	250,000

converted for operation on LPG, and in most cases spark-ignition gasoline engines will then run in bi-fuel mode (the system can be switched between gasoline and LPG). However, engine optimization for operation on LPG is only possible in dedicated gas engines. A possible problem here is variations in octane quality of LPG as a function of the propane-butane ratio (RON of butane is 94 and it is much lower than that of propane at 112).

The main advantages of LPG as automotive fuel compared to gasoline are:

- Higher-octane quality for potential performance gain.
- Good cold starting and driveability because of its gaseous state.
- Emissions from an LPG engine are substantially lower than those achieved with the gasoline or diesel counterparts. The ecological benefits are mainly similar to those pointed out earlier for NGVs with the exception of higher emissions of NMHC, relative to natural gas [1], and exhaust gases' photochemical reactivity. The latter is mainly due to higher contents in LPG of very reactive olefin compounds. An example of a comparison between different fuel effects on photochemical reactivity (summer smog potential) of exhaust gases is given in Figure 16.21.
- Compared to its diesel counterpart, an LPG engine (similar to NGVs) has a lower noise level [82].

Fig. 16.21.

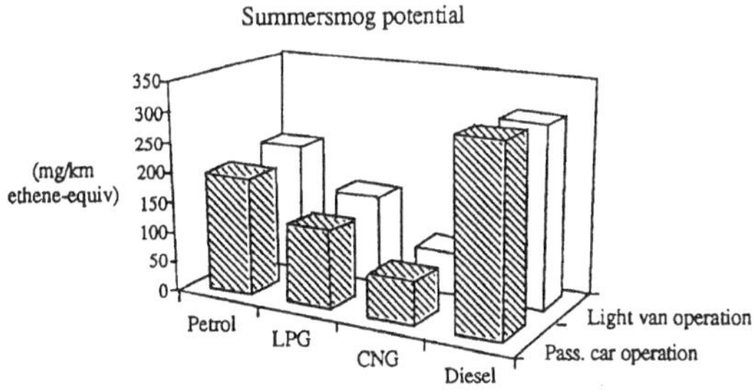

Summer smog potential of the different fuels [88].

The main disadvantages of LPG are:

- Lower fuel energy content (see Table 16.24), which leads to higher volumetric fuel consumption.
- Supplementary safety regulations must be considered, as LPG is stored on board a vehicle under pressure.
- LPG is heavier than air; therefore, use of LPG vehicles is currently the subject of considerable restrictions in some countries, such as the prohibition of parking in confined spaces [82].

16.4.5 Vegetable Oils and Ethers

Vegetable oil is an interesting alternative fuel for compression-ignition engines, mainly because it is produced from renewable energy sources—various sorts of oil-yielding plants. The most common oil plant for vehicle use in Western Europe is rape and the final product produced from it as an automoive fuel is the methyl ester of rapeseed oil (RME). In France 1300 square kilometers of land have been cultivated in 1994 to produce about 130,000 tons of this fuel, frequently named biodiesel [89]. A European Council Directive draft proposal for a specification of vegetable oil methyl esters has been developed, along with the objective of securing a 5 percent market share of total motor fuel consumption for biofuels, of which, as expected, the biodiesel would form the major part [15]. This draft specification, which has been reproduced from CONCAWE Report No. 5/95 [15], is shown in Table 16.29.

Rapeseed oil (or vegetable oil in general), which has been processed to methyl ester, has many characteristics close to those of diesel fuel; for instance, density, viscosity, energy content, and high cetane number (see Table 16.24). From the results of the three-year, large-scale research program, carried out in France [89], and some other publications [e.g., 90], it follows that the use of vegetable oil methyl ester (and particularly, RME) leads to some reduction of hydrocarbons, particulates, and a slight increase of NO_x emissions.

Table 16.29

EU Draft Specification for Vegetable Oil Methyl Ester Diesel Fuel (CONCAWE Data, [15]).

Properties A. Fuel specific properties	Units	Limit	Analytical method
Density at 15°C	g/cm³	0.86–0.90	ISO 3675
Kinematic viscosity at 40°C	mm²/s	3.5–5.0	ISO 3104
Flash point	°C	min. 100	ISO 2719
Cold filter plugging point CFPP	°C	summer max. 0 winter max. < −15	DIN EN 116
Sulphur content	% m/m	max. 0.01	ISO 8754/DIN EN 41
Distillation:			
5% vol. evaporated at	°C	to be indicated	ASTM-1160/ISO 3405
95% vol. evaporated at	°C	to be indicated	
Carbon residue Conradson (10% by vol. residue on distillation at reduced pressure)	% m/m	max. 0.30	ISO 10370
Cetane number	–	min. 49	ISO 5165/DIN 51773
Ash content		max. 0.01	EN 26245
Water content (Karl Fischer)	mg/kg	max. 500	ISO 6296/ASTM D 1744
Particulate matter	g/m³	max. 20	DIN 51419
Copper corrosion (3h/50°C)	corrosion rating	max. 1	ISO 2160
Oxidation stability	g/m³	max. 25	ASTM D 2274
B. Methyl ester specific properties	**Units**		
Acid value	mg KOH/g	max. 0.5	ISO 660
Methanol content	% m/m	max. 0.3	DIN 51413.1
Monoglycerides	% m/m	max. 0.8	GLC
Diglycerides	% m/m		GLC
Triglycerides	% m/m		GLC
Bound glycerine	% m/m	max. 0.2	calculate
Free glycerine	% m/m	max. 0.03	GLC
Total glycerine	% m/m	max. 0.25	calculate
Iodine number	–	max. 115	DIN 53241/IP 84-81
Phosphorous content	mg/kg	max. 10	DGF C-VI 4

Note: Many of the test methods have yet to be finalized.

Reduction of volatile aromatics together with benzene and aldehyde emission rise was also reported [89]. Worsening the volumetric fuel economy is generally proportional to the change in LHV (lower heating value) as shown in Table 16.24.

Results of limited fleet tests, with 30 percent to 100 percent blends of RME and diesel fuel, show no serious side effects and no significant differences in wear rates have been found. However, original elastomers are unsuitable for pure RME use [89]. Deposits formation on certain surfaces has also been reported [89].

Increasing interest has recently been focused on the possibility of using dimethyl ether (DME) as a very promising alternative fuel for compression-ignition engines [72, 91, 92]. DME is manufactured today from methanol in relatively low amounts (about 100,000 ton/year worldwide [91]) mainly for the aerosol industry as a replacement for CFC materials. Haldor Topsoe has developed the process for direct production of DME from synthesis gas [93]. This gas, which is a mixture of H_2, CO, and CO_2 can be generated from various sources, including natural gas, coal, and biomass. The latter allows DME to be qualified as a renewable fuel.

Dimethyl ether (chemical formula CH_3–O–CH_3) is the simplest ether known as environmentally benign. Some relevant properties of DME are listed in Table 16.24. Due to its excellent autoignition quality (cetane number over 55), it is used sometimes as a fuel additive for methanol-operating engines [94, 95]. However, its combustion is possible also as a primary engine fuel.

The main advantages of DME as an alternative automotive fuel are:

- The chemical structure of DME, with its high oxygen content, promises an almost smokeless combustion when burned in a proper manner [72].
- The very high cetane number allows reduction of NO_x emission and noise.

Figure 16.22 shows emissions results that have been obtained by Fleisch et al. [91] with the Navistar T 444E diesel engine operated on pure DME without any exhaust aftertreatment. It must be noted that the optimization for DME of a fuel injection system, combustion chamber, inlet port geometry, and compression

Fig. 16.22.

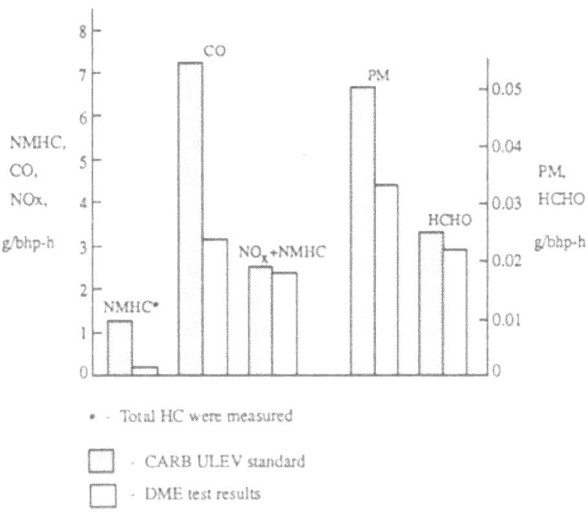

Test results of Navistar diesel engine fueled with DME in comparison with CARB ULEV standards (based on [91] data).

ratio may further lower NO_x and CO emission rates without deterioration of fuel economy.

- Unlike methanol, DME is noncorrosive to metals.
- DME flame luminosity is quite good. It burns with a visible blue flame similar to natural gas. As mentioned earlier, this is an important safety aspect.
- DME provides good engine cold starting.

The main shortcomings of dimethyl ether as an automotive fuel are:

- On board a vehicle, storage and handling of DME need some precautions similar to LPG because the boiling point of DME is too low (−20°C) and it is a gas under normal atmospheric conditions. Special care must be taken to avoid leakage of DME because it could form an explosive mixture with air.
- The energy content and density of DME are significantly lower than those of diesel fuel (see Table 16.24), which cause a proportional increase of volumetric fuel consumption and require changes in fuel injection apparatus, providing an increase of injected fuel volume per stroke.
- Some elastomers are not compatible with this fuel; therefore, careful selection of seal materials is necessary.
- The lubrication performance of DME has not yet been sufficiently investigated, but using a lubricity additive is probably necessary.

16.4.6 Hydrogen

The limited availability of fossil energy sources stimulates the increasing interest in hydrogen (H_2) as a potential alternative automotive fuel. Although the problems of hydrogen production, infrastructure development, refueling, and on-board storage are quite difficult, their solutions seem to be technically feasible in a long-term perspective.

Among the many methods of hydrogen production, stripping it from hydrocarbons (such as natural gas or heavy petroleum oils) is the least expensive large-scale process [70]. The cost of hydrogen produced this way is 3 to 15 times higher than that of natural gas and 1.5 to 9 times that of gasoline [96]. Hydrogen can also be produced by gasification of coal. Production of H_2 by electrolysis of water is used today in some industrial plants [70], but this method is extremely expensive [96]. Production of hydrogen by photoelectrolysis with the use of solar energy is also considered as a potential option.

Storage of hydrogen on board a vehicle remains a major technical challenge. The problem is directly related to the physical properties of hydrogen (see Table 16.24): Very low volumetric energy content and extremely low boiling point

(−253°C). The methods of on-board H_2 storage considered today are as follows:

- Storage of compressed H_2 in pressurized tanks. High pressure (about 300 bar [71]) is required for storage of hydrogen in a gaseous state. This leads to high weight penalties and safety risks.
- Storage of liquefied H_2 in cryogenic tanks. The extremely low temperature of liquefaction (20 K) puts heavy demands on thermal insulation. The process itself is highly energy demanding. It takes almost as much energy to liquefy hydrogen as is contained in the resulting liquid [96]. Environmental heat causes boil-off of the hydrogen from the safety valve, resulting in losses of about 2 percent per day when the vehicle is parked [71]. This venting of hydrogen may also represent a serious safety concern.
- Hydrogen can also be stored in metal hydrides adsorbed on carbon or other materials. The main advantage of this method is that there are no storage losses. The main drawbacks are associated with low hydrogen storage capacity, sensitivity to contamination by impurities in the hydrogen gas, and high cost of materials.
- Hydrogen can be stored as a constituent in a chemical compound (methylcyclohexanol storage is most widely considered). This method of on-board storage requires a catalyst to dehydrate the hydrogenous methylcyclohexane at high temperatures of about 500°C.

Regardless of the on-board storage type, in all current hydrogen-fueled vehicles, it is injected into the intake manifold in gaseous form [71]. The second possibility of liquid H_2 injection directly into the cylinder provides some advantages, such as mixture cooling for low NO_x emissions, reduced heat losses, and no danger of backfires. However, the short injector lifetime means that this type of fuel system for H_2 cannot be a viable option for the near future [71].

The nature of hydrogen, which does not contain carbon and oxidizes during combustion into water, allows ultralow tailpipe exhaust emissions to be obtained from hydrogen vehicles. Only trace amounts of CO, CO_2, and HC, originated from the lubricating oil, may be emitted. Very wide flammability limits of hydrogen in air (see Table 16.24) allow the concept of an ultra-lean-burn engine, low NO_x emissions, and higher thermal efficiency to be realized. An example of regulated pollutant emission from a hydrogen-fueled Mazda vehicle with rotary engine is shown in Figure 16.23 in comparison with the CARB ULEV standards [96].

As can be seen from this figure, levels of emissions which are lower by an order of magnitude than ULEV standards can be achieved by the use of hydrogen as an automotive fuel. As mentioned earlier, almost no CO_2 emissions are produced during hydrogen-fueled vehicle operation. However, from the calculations performed [96], it follows that in total account hydrogen-fueled motor vehicles have no noticeable advantage and in many cases they are worse regarding greenhouse gas emissions, when hydrogen is produced from fossil sources.

Fig. 16.23.

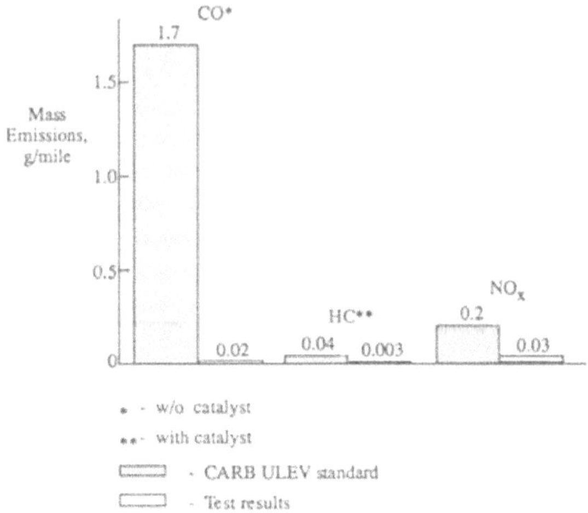

Tailpipe emissions of a hydrogen-fueled Mazda vehicle with rotary engine (based on data from [96]).

16.4.7 Electricity

In connection with the content of the previous subsection, it is important to note that hydrogen can be used as a fuel not only for ICE, but also for fuel cell devices, in which the chemical energy of hydrogen is converted directly to electricity for propulsion of electric vehicles (EV).

The main benefit of fuel cells is their high theoretical efficiency (70 percent to 90 percent [96]) together with zero tailpipe emissions. The main disadvantage is an unsuitably high cost. Development of an effective and reliable on-board source of hydrogen is also a serious technical challenge. Using on-board methanol dissociation is considered today to be a promising method of supplying hydrogen to fuel cells [97]. Because of the previously mentioned shortcomings, automotive fuel cells, as they seem today, are only a long-term possibility. Nonetheless, experiments with EVs powered by fuel cells have recently intensified in the light of California's zero-emission vehicles (ZEV) mandate. One of the latest examples of research and development in this area is the NECAR II vehicle of Daimler-Benz powered by fuel cells [98].

A more realistic candidate for powering a zero-emission vehicle (with today's state-of-the-art technology) is an electric battery. Although battery-powered EVs are not yet competitive with conventional-fuel vehicles (mainly because of the high cost and short driving range), due to the California ZEV mandate, almost all auto makers now devote much effort in order to bring the electric vehicle to the market. Several advanced battery technologies are now in various stages of research and development, the most promising being Ni-Metal Hydrid, Lithium based, Na-

Table 16.30

Comparison of Power and Energy Densities of Some Advanced Batteries for EV (Based on Data from [92, 99])

Battery chemistry	Energy density, Wh/kg	Power density, W/kg
Lead-Acid (conventional)	33–40	100
Nickel-Cadmium	48–60	165–210
Nickel-Metal Hydride	73–80	175
Nickel-Iron	48–50	100
Sodium-Sulphur	100–117	120–150
Nickel-Zinc	65	200
Lithium-Polymer	500*	N/A
Zinc-Air	215	98
Sodium-$NiCl_2$	88–94	75–135
Gasoline	12200	

*Only for test cells; N/A—not available.

$NiCl_2$ and Zn-Air chemistries. Comparison of the power and energy densities of some of the advanced technology batteries and gasoline appears in Table 16.30.

In order to advance the more promising battery technologies, three big U.S. auto makers in cooperation with the U.S. DOE formed the U.S. Advanced Battery Consortium (U.S. ABC) in January 1991 [92]. The prime objectives of the U.S. ABC, as well as other involved organizations for both the mid-term and long-term goals, are to reduce cost, improve performance, and increase the time between charging of the batteries.

REFERENCES

1. Owen, K., and T. Coley, *Automotive Fuels Reference Book.* Society of Automotive Engineers, Inc., 2nd edition, 1995.
2. Burns, V.R., J.D. Benson, W.M. Kreucher, R.M. Reuter, A.M. Hochhauser, and W.J. Koehl, "Description of Auto/Oil Air Quality Improvement Research Program," SAE Paper 912320 (1991).
3. Hochhauser, A.M., J.D. Benson, V. Burns, R.A. Gorse, W.J. Koehl, L.J. Painter, B.H. Rippon, R.M. Reuter, and J.A. Rutherford, "The Effect of Aromatics, MTBE, Olefins and T90 on Mass Exhaust Emissions from Current and Older Vehicles—The Auto/Oil Air Quality Improvement Research Program," SAE Paper 912322 (1991).
4. Benson, J.D., V. Burns, R.A. Gorse, A.M. Hochhauser, W.J. Koehl, L.J. Painter, and R.M. Reuter, "Effects of Gasoline Sulphur Level on Mass Exhaust Emissions—Auto/Oil Air Quality Improvement Research Program," SAE Paper 912323 (1991).
5. MacKinven, R., and M. Hublin, "European Programme on Emissions, Fuels and Engine Technologies—Objectives and Design," SAE Paper 961065 (1996).
6. Camarsa, M., M. Hublin, and R. MacKinven, "Impact of EPEFE Data on the European Auto-Oil Process," SAE Paper 961076 (1996).
7. Mayotte, S.C., V. Rao, C.E. Lindhjem, and M.S. Sklar, "Reformulated Gasoline Effects on Exhaust Emissions: Phase II: Continued Investigation of the Effects of Fuel Oxygenate Content,

Oxygenate Type, Volatility, Sulphur, Olefins and Distillation Parameters," SAE Paper 941974 (1994).

8. McGinty, R.P., and N.P. Dent, "A Review of the Effect of Petrol Composition on Unregulated Motor Vehicle Emissions with Particular Emphasis on Non-Catalyst Vehicles." Accepted for publication in *Environmental Technology*.
9. McCarthy, C.I., W.J. Slodowske, E.J. Sienicki, and R.E. Jass, "Diesel Fuel Property Effects on Exhaust Emissions from a Heavy Duty Diesel Engine that Meets 1994 Emissions Requirements," SAE Paper 922267 (1992).
10. Ryan III, T.W., J. Erwin, R.L. Mason, and D.S. Moulton, "Relationship Between Fuel Properties and Composition and Diesel Engine Combustion Performance and Emissions," SAE Paper 941018 (1994).
11. Garett, T.K., *Automotive Fuels and Fuel Systems, Volume 1: Gasoline*, Pentech Press and Society of Automotive Engineers, Inc., 1991.
12. Searle, G.R., Meeting Octane Quality, The Associated Octel Co., Israeli IP Seminar, 1995.
13. ASTM. *Annual Book of Standards*, Vol. 05.01, 1995.
14. Tartakosky, L., M. Gutman, Y. Kirzhner, Y. Zvirin, and Y. Dayan, *Qualities of Petroleum Products Marketed in Israel Compared to Some Selected Countries*, Report No. EP-03-96 for the Israeli Ministry of Energy and Infrastructure, Technion, Israel 1996.
15. *Motor Vehicle Emission Regulations and Fuel Specifications in Europe and United States, 1995 Update*, CONCAWE Report No. 5/95, Brussels, 1995.
16. Schoonveld, G.A., R.K. Riley, S.P. Thomas, and S. Schiff, "Exhaust Valve Recession with Low-Lead Gasolines," *Automotive Engineering* **95**, no. 11: 72–76 (1987).
17. The Phase Out of Leaded Gasoline. Information list of the Swedish Environmental Protection Agency, October 1990.
18. Gutman, M., Y. Zvirin, and A. Stotter, *Additives to Gasoline*, Report No. EEC-168, Energy Eng. Center, Faculty of Mech. Eng., Technion, 1992.
19. Taniguchi, B.Y., R.J. Peyla, G.M. Parsons, S.K. Hoekman, and D.A. Voss, "Injector Deposits—the Tip of Intake System Deposit Problems," SAE Paper 861534 (1986).
20. Peyle, R.J., "Reformulated Gasoline," *Automotive Engineering* **99**, no. 9: 29–32 (1991).
21. Papachristos, M.J., *Gasoline Quality and Additve Trends in Europe*, The Associated Octel Co., Bletchley, United Kingdom, 1995.
22. Petit, A., J.G. Jeffrey, F.H. Palmer, and R. Steinbrink, "European Programme on Emissions, Fuels and Engine Technologies (EPEFE)—Emissions from Gasoline Sulphur Study," SAE Paper 961071 (1996).
23. Goodfellow, C.L., R.A. Gorse, M.J. Hawkins, and J.S. McArragher, "European Programme on Emissions, Fuels and Engine Technologies (EPEFE)—Gasoline Aromatics/E 100 Study." SAE Paper 961072 (1996).
24. Rutherford, J.A., W.J. Koehl, J.D. Benson, A.M. Hochhauser, J.C. Knepper, W.R. Leppard, L.J. Painter, L.A. Rapp, B. Rippon, and R.M. Reuter, "Effects of Gasoline Properties on Emissions of Current and Future Vehicles—T_{50}, T_{90}, and Sulphur Effects—Auto/Oil Air Quality Improvement Research Program," SAE Paper 952510 (1995).
25. Huynh, N.H., H. Richter, and K.H. Reders, "The Impact of Gasoline Formulation on Engine Performance and Exhaust Emissions," SAE Paper 920297 (1992).
26. Gibbs, L.M., "The Changing Nature of Gasoline," *Automotive Engineering* **102**, no. 1: 99–102 (1994).
27. Wedekind, B., P.J. Bennett, C.L. Goodfellow, J.G. Jeffrey, G.F. Marchesi, and C.R. McDonald, "The Independent Effect of Mid-Range, Back-End Volatility and Aromatics on Emissions from Two European Gasoline Vehicles," SAE Paper 952522 (1995).
28. Gibbs, L.M., "Gasoline Specifications, Regulations and Properties," *Automotive Engineering* **104**, no. 10: 35–40 (1996).
29. Field, S., 1991 Oil Report, Chapter 13: Motor Vehicle Fuels of the Future, *Oil & Gas Journal*, December 9: 17 (1991).

30. Keynes, M., "Global Gasoline, Diesel Quality Continues to Improve," *Oil & Gas Journal*, September 16: 47–48 (1996).
31. *European Gasoline Survey 1994*, Associated Octel Co., Ltd., 1994.
32. "W. European Refiners Face Tough Environmental Rules," *Oil & Gas Journal*, January 25: 24–32 (1993).
33. Garett, T.K., *Automotive Fuels and Fuels Systems. Volume 2: Diesel*, Pentech Press and Society of Automative Engineers Inc., 1994.
34. Owen, K. (editor), *Gasoline and Diesel Fuel Additives*, John Wiley & Sons, 1989.
35. Güldner, O.L., G.H. Burton, and R.B. White, "NRCC Cetane Index-1: An Improved Cetane Number Predictor," SAE Paper 861519 (1986).
36. Ingham, M.G., J.A. Bert, and L.H. Painter, "Improved Predictive Equations for Cetane Number," SAE Paper 860250 (1986).
37. Pedley, J.F., R.W. Jiley, and R.A. Hancock, "Storage Stability of Petroleum Derived Diesel Fuel," *Fuel*, **66**: 1646–1651 (1987).
38. LePera, M.E., and J.G. Sonnenburg, "How Stable Is Diesel in Storage," *Hydrocarbon Processing*, September: 111–115 (1973).
39. Zahavi, E., A.B. Shavit, J. Ben Asher, and N. Por, "Study of Stability of Diesel Fuels under Operational Conditions," Proc. of the 3rd International Conference on Stability and Handling of Liquid Fuels, London: 609–625 (1988).
40. Luria, D., A. Stotter, M. Gutman, and A.B. Shavit, "The Evaluation of Storage Stability of Gas Oil by Engine Test," *Journal of the Institute of Energy* **LIX**: 142–147 (1986).
41. Gutman, M., L. Tartakovsky, Y. Kirzhner, Y. Zvirin, D. Lurie, A. Weiss, and M. Shuftan, Proc. of the 5th International Conference on Stability and Handling of Liquid Fuels, Rotterdam, Vol. 1: 423–431 (1994).
42. Opris, C.N., L.D. Gratz, S.T. Bagley, K.J. Baumgard, D.G. Ledly, and J.H. Johnson, "The Effects of Fuel Sulphur Concentration on Regulated and Unregulated Heavy-Duty Diesel Emission," SAE Paper 930730 (1993).
43. Booth, M., and P.E. Wolveridge, "Severe Hydrotreating of Diesel Can Cause Fuel-Injector Pump Failure," *Oil & Gas Journal*, August 16: 71–76, 1993.
44. Worldwide 1996—Winter Diesel Fuel Quality Survey.
45. Daniels, J., "Can There be a Point of Diminishing Returns," *Automotive Engineer*, August/September: 42–55 (1996).
46. Mullins, P., "Besides Their Excellent Fuel Efficiency, Diesels Have the Potential to Markedly Benefit Urban Air Quality, Ricardo Report," *High Speed Diesels & Drives*, October: 38–40, (1994).
47. *Paramins Post-Industry and Additive News*, Issue 2/96, July 1996.
48. Gairing, M., J.M. Marriott, K.H. Reders, A.A. Reglitzky, and P.E. Wolveridge, "The Effect of Modern Additive Technology on Diesel Fuel Performance," SAE Paper 950252 (1995).
49. Cole, R.D., M.G. Taylor, and R. Rossi, "Additive Solutions to Diesel Combustion Problems," Institution of Mechanical Engineers International Conference on Petroleum Based Fuels and Automotive Applications, London, Paper C310/86 (1986).
50. Sobotowski, R.A., "Investigation of Cetane Response of U.S. Diesel Fuels," SAE Paper 950249 (1995).
51. Miyamoto, N., Z. Hou, A. Harada, H. Ogawa, and T. Murayame, "Characteristic of Diesel Soot Suppression with Soluble Fuel Additives," SAE Paper 871612 (1987).
52. Zeller, H.W., *Effects of Barium-Based Additive on Diesel Exhaust Particulate*, Report of Investigations. 9090, Bureau of Mines. U.S.A., 1987.
53. Neeft, J.P.A., M. Makkee, and J.A. Moolijn, "Review Article: Diesel Particulate Emission Control," *Fuel Processing Technology* **47**: 1–69 (1996).
54. Vincent, M.W., M.J. Papachristos, D. Williams, and J. Burton, "Diesel Fuel Detergent Additive Performance and Assessment," SAE Paper 942010 (1994).

55. Herbstman, S., and K. Virk, "Use of Disprsants/Detergents in Diesel Injector Keep Clean and Clean Up Studies," SAE Paper 912330 (1991).
56. Tartakovsky, L., M. Gutman, Y. Kirzhner, and Y. Zvirin, "A Fleet Test to Study the Effects of Detergent Additives on the Performance of Bus Diesel Engines," Proc. of the 29th International Symposium on Automotive Technology and Automation (ISATA), Florence, Paper 96 EN014 (1996).
57. Gutman, M., Y. Zvirin, and A. Stotter, *Diesel Fuel Additives*, Report No. EC-171, Energy Eng. Center, Faculty of Mechanical Eng., Technion, Haifa, Israel, 1993.
58. Gutman, M., L. Tartakovsky, Y. Kirzhner, and Y. Zvirin, "Development of a Screening Test for Evaluating Detergent/Dispersant Additives to Diesel Fuels," SAE Paper 961184 (1996).
59. Carradine, W.R., "High Performance Flow Improver For Product Lines," *Oil & Gas Journal* **81** no. 32: 92–98 (1983).
60. Damian, B., A. Faure, J. Denis, B. Sillion, P. Claudy, and J.M. Letoffe, "New Additives for Diesels Fuels—Cloud Point Depressant," SAE Paper 861527 (1986).
61. Johnson, J.H., S.T. Bagley, L.D. Gratz, and D.G. Leddy, "A Review of Diesel Particulate Control Technology and Emissions Effects," 1992 Horning Memorial Award Lecture, SAE Paper 940233 (1994).
62. Nikanjam, M., "Development of the First CARB Certified California Alternative Diesel Fuel," SAE Paper 930728 (1993).
63. Miykamoto, N., H. Ogawa, M. Shibuya, K. Arai, and O. Esmilairre," Influence of the Molecular Structure of Hydrocarbon Fuels on Diesel Exhaust Emissions," SAE Paper 940676 (1994).
64. Knepper, J.C., V.R. Burns, W.R. Leppard, W.J. Koehl, R.A. Gorse, Jr., L.A. Rapp, J.D. Benson, A.M. Hochhauser, and R.M. Reuter," Fuel Effects in Auto/Oil High Emitting Vehicles," SAE Paper 930137 (1993).
65. Rickeard, D., R. Bonetto, and M. Signer, "European Programme on Emissions, Fuels and Engine Technologies (EPEFE)—Comparison of Light and Heavy Duty Diesel Studies," SAE Paper 961075 (1996).
66. Hublin, M., P.G. Gadd, D.E. Hall, and K.P. Schindler, "European Programme on Emissions Fuels and Engine Technologies (EPEFE)—Light Duty Diesel Study," SAE Paper 961073 (1996).
67. *Diesel Fuel/Engine Interaction and Effects on Exhaust Emissions*, CONCAWE, Report No. 96/60, Brussels, 1996.
68. Lacey, P.I., and S.R. Westbrook, "Diesel Fuel Lubricity," SAE Paper 950248 (1995).
69. *CONCAWE Review*, Vol. 5, no. 2, 1996.
70. Johansson, A., A. Brandberg, and A. Roth, *The Life of Fuels. Motor Fuels from Source to End Use. An Energy and Emissions System Study of Conventional and Future Options*, Stockholm, 1992.
71. Adler, U. (editor), *Automotive Handbook*, Robert Bosch GmbH, 3rd edition, 1993.
72. Kapus, P., and H. Ofner, "Development of Fuel Injection Equipment and Combustion System for DI Diesels Operated on Dimethyl Ether," SAE Paper 950062 (1995).
73. Heywood, J., *Internal Combustion Engine Fundamentals*, McGraw Hill, 1989.
74. Gorse, R.A., et al. "The Effects of Methanol/Gasoline Blends on Automobile Emissions." SAE Paper 920327, 1992.
75. Burns, V.R., W.J. Koehl, J.D. Benson, R.A. Gorse, and J.A. Rutherford, "Emissions, with Reformulated Gasoline and Methanol Blends in 1992 and 1993 Model Year Vehicles," SAE Paper 941969 (1994).
76. Benson, J.D., J.C. Knepper, R.M. Reuter, W.J. Koehl, W.R. Leppard, B. Rippon, V.R. Burns, L.J. Painter, J.A. Rutherford, A.M. Hochhauser, and L.A. Rapp, "Emissions with E85 and Gasolines in Flexible/Variable Fuel Vehicles—The Auto/Oil Air Quality Improvement Research Program," SAE Paper 952508 (1995).
77. Murayama, T., N. Miyamoto, T. Chikahisa, and H. Ogawa, "Potential and Problems in Alcohol Fuel Combustion Techniques in Diesel Engines," Proc. of the VIII International Symposium on Alcohol Fuels Tokyo: 495–500 (1988).

78. Pettersson, E., and W. Lindberg, "Exhaust Emissions from an Ethanol Fueled Diesel Engine with EGR," Proc. of the 27th International Symposium on Automotive Technology and Automation (ISATA), Aachen, Paper 94EN015 (1994).
79. Ben Zvi. A., M. Gutman, S. Alkalay, and A. Stotter, "Use of Methanol as Additive to Diesel Fuel," Proc. of the 25th Silver Jubilee International Symposium on Automotive Technology and Automation (ISATA), Firenze, Paper 920437 (1992).
80. Gutman, M., A. Stotter, and E. Yanir, "Fuel Line Components Material Compatibility with 90% Commercial Gasoline and 10% Ethanol," Proc. of the VIII International Symposium on Alcohol Fuels, Tokyo: 683–689 (1988).
81. Liss, W.E., and W.H. Thrasher, "Natural Gas as a Stationary Engine and Vehicular Fuel," SAE Paper 912364 (1991).
82. Clean Fuels for Urban Public Transport: LPG and CNG. THERMIE Programme Action No. T85, Performed for European Commission Directorate General for Energy (DG XVII), October 1994.
83. Tartakovsky, L.M., "Comparing Ecological Indices of Diesel and Gas-Diesel Engines," Proc. of the XXIII FISITA Congress, Torino, SAE Paper 905243 (1990).
84. Hochhauser, A.M., V.R. Burns, L.J. Painter, R.M. Reuter, W.J. Koehl, J.C. Knepper, L.A. Rapp, J.A. Rutherford, J.D. Benson, W.R. Leppard, and B.H. Rippon, "Comparison of CNG and Gasoline Vehicle Exhaust Emissions: Mass and Composition—The Auto/Oil Quality Improvement Research Program," SAE Paper 952507 (1995).
85. Hupperich, P., and M. Durnholz, "Exhaust Emissions of Diesel, Gasoline and Natural Gas Fuelled Vehicles," SAE Paper 960857 (1996).
86. Forschen, Planen, Bauen, MAN Ausgabe 1992.
87. Chandler, K., N. Malcosky, R. Motta, P. Norton, K. Kelly, L. Schumacher, and D. Lyons, "Alternative Fuel Transit Bust Evaluation Program Results," SAE Paper 961082 (1996).
88. Hollemans, B., "Perspectives of LPG as Automotive Fuel in the Netherlands." Presented at the Vito-International Symposium Alternative Fuels for Motor Vehicles: Trends and Perspectives, Belgium (1994).
89. Staat, F., and P. Gateau, "The Effects of Rapeseed Oil Methyl Ester on Diesel Engine Performance, Exhaust Emissions and Long-Term Behavior. A Summary of Three Years of Experimentation," SAE Paper 950053 (1995).
90. Scholl, K.W., and S.C. Sorenson, "Combustion of Soybean Oil Methyl Ester in a Direct Injection Diesel Engine," SAE Paper 930934 (1993).
91. Fleisch, T., C. McCarthy, A. Basu, C. Udovich, P. Charbonneau, W. Slodowske, S.E. Mikkelsen, and J. McCandless, "A New Clean Diesel Technology: Demonstration of ULEV Emissions on a Navistar Diesel Engine Fueled with Dimethyl Ether," SAE Paper 950061 (1995).
92. Nichols, R.J., "The Challenges of Change in the Auto-Industry: Why Alternative Fuels?" *Transactions of ASME, Journal of Engineering from Gas Turbines and Power*, **116**: 727–732 (1994).
93. Hansen, J.B., B. Voss, E. Joensen, and I.D. Sigursardottir, "Large Scale Manufacture of Dimethyl Ether—A New Alternative Diesel Fuel from Natural Gas," SAE Paper 950063, 1995.
94. Iwai, N., K. Nagai, H, Yasuda, T. Ayusawa, and K.Y. Kil, "A Study of Cold Startability and Mixture Formation of High-Percentage Methanol Blends," SAE Paper 880044 (1988).
95. Guo, J., T. Chihahisha, T. Murayama, and M. Miyano, "Improvement of Performance and Emissions of a Compression Ignition Methanol Engine with Dimethyl Ether," SAE Paper 941908 (1994)
96. Kukkonen, C.A., and M. Shelef, "Hydrogen as an Alternative Automotive Fuel: 1993 Update," SAE Paper 940766 (1994).
97. Demmler, A., "Daimler-Benz and the No-Emission Vehicle." *Automotive Engineering*, **104** no. 7: 74–75 (1996).

98. "Daimler-Benz Shows Its New NECAR II Fuel Cell Powered Vehicle," *Electric Vehicle Progress*, June 1 (1996).
99. Tartakovsky, L., M. Gutman, Y. Dayan, and Y. Zvirin, *Electric and Hybrid Propulsion for Buses*. Research Report for S. Neaman Institute and "Egged" Co., Technion, Haifa, Israel, 1996.
100. Painter, L.J., and J.A. Rutherford, "Statistical Design and Analysis Methods for the Auto/Oil Air Quality Research Program," SAE Paper 920319 (1992).
101. Rapp, L.A., A.M. Hochhauser, L.J. Painter, J.D. Benson, J.C. Knepper, R.M. Reuter, V.R. Burns, W.J. Koehl, J.A. Rutherford, R.A. Gorse, Jr., and W.R. Leppard, "Effects of Fuel Properties on Mass Exhaust Emissions During Various Modes of Vehicle Operation." SAE Paper 932726 (1993).
102. Koehl, W.J., R.A. Gorse, Jr., J.C. Knepper, L.A. Rapp, J.D. Benson, A.M. Hochhauser, W.R. Leppard, R.M. Reuter, V.R. Burns, L.J. Painter, and J.A. Rutherford, "Effects of Gasoline Sulfur Level on Exhaust Mass and Speciated Emissions: The Question of Linearity—Auto/Oil Air Quality Improvement Program." SAE Paper 932727 (1993).

APPENDIX 1

National Gasoline Specifications (Based on the CONCAWE Data [15])

Country	Standard (date)	Gasoline type	RON min	MON min
Australia	AS 1876 (1994)	Leaded	96	
		Unleaded	P 95 R 91–93	82 82
Austria	O-Norm EN 228	Leaded	Not available	From October 1993
		Unleaded	S 98.0 P 95.0 R 91.0	87.0 85.0 82.5
Belgium	NBM T52-705 1990	Leaded	P 97.5	–
		Unleaded	S 98.0 P 95.0 R 90.0	88.0 85.0 80.0
California (United States)	CARB requirements	Leaded	Not sold from 1.92	
		"Phase 2" reformulated gasoline (1.3.1996)		
Canada	CAN/CGSB-3.5-M87	Leaded	Not sold from 12.90	
		Unleaded Type 1	87.0–90.0 (1)	
		Unleaded Type 2	83.5–87.0 (1)	(2)

(*Continues*)

National Gasoline Specifications (*Continued*)

Country	Standard (date)	Gasoline type	RON min	MON min
Denmark	Danish Petroleum Institute 1/10/90	Leaded	P 98.0	P 88.0
	DS EN 228	Unleaded	S 98.0 P 95.0 R 92.0	88.0 85.0 83.0
Finland		Leaded[1]	P 99.0	P 87.4
	EN 228; 1993	Unleaded (standard)	S 98.0 P 95.0	88.0 85.0
	01/1993	Unleaded (citygasoline)	S 99.0	88.0
	03/1994	Unleaded (reformulated)	P 95.0	85.0
France	NFM 15-005 (1994)	Leaded	P 97.0–99.0	P 86.0
	NF EN 228: 1993	Unleaded	P 95.0	85.0
	Cahier des Charges	Leaded	97.0	86.5
	NFM 15-001 (1994)	Leaded	89–92	
Germany	DIN 51600	Leaded	98.0	88.0
	EN 228: 1993	Unleaded	S 98.0 P 95.0 R 91.0	88.0 85.0 82.5
Greece	Greek Government Gazette (P) 556/93 (R) 581/91	Leaded	P 96–98 R 90	
	EN 228: 1993	Unleaded	P 95	85
Israel	Israeli Standard 90 (1995)	Leaded	S 98 P 96 R 91	86[2] 85[1] 81
		Unleaded	P 95 R 91	85[3] 81[4]
Italy	NC 623-01	Leaded	P 97.0	P 87.0
	UNI-CUNA EN 228: 1993 (1.10.93)	Unleaded	P 95.0	85.0
Japan	JIS K2202 (1991)	Unleaded No. 1	96.0	
		Unleaded No. 2	89.0	
The Netherlands	NEN-EN 228: 1993	Unleaded	P 95	85.0
Portugal	Ministerio da Economia Portaria Feb. '94	Leaded	P 98.0 R 90.0	P 87.0
	NPEN 228 1993	Unleaded	95.0	85.0

(*Continues*)

National Gasoline Specifications (*Continued*)

Country	Standard (date)	Gasoline type	RON min	MON min
South Africa	SABC 299 (1988)	Leaded	97 93 87	
	SABC 1598 (1993)	Unleaded	95 91	85(1) 81(2)
Spain	Royal Decree 1485/1987	Leaded	P 97.0 R 92.0	P 87.0 R 82.0
	EN 228 (1994)(1)	Unleaded	95.0	85.0
Sweden		Leaded	No longer available	
	(1993)	Unleaded (Class 4-SS EN 228)	P 95.0	85.0
		Unleaded (Class 3)		
	(12/1994)	Unleaded (Class 2—Non-Cat.) Unleaded (Class 2—Catalyst) Unleaded (Class 1)	Will be developed	Will be developed
Switzerland	SN 181 161/1 (January 1986)	Leaded	P 98.0	88.0
	SN EN 228: 1993	Unleaded	P 95.0	85.0
United Kingdom	BS 4040 (Amended 1.10.94)	Leaded 4 star Leaded 3 star Leaded 2 star	97.0 93.0 90.0	86.0 82.0 80.0
	BS EN 228: 1993	Unleaded	S 98.0 P 95.0	87.0 85.0
United States	ASTM D4814-94d with	Leaded	No longer available	
	EPA requirements (40 CFR Part 80)	Unleaded	87(2)	82
	EPA regulation based on CAAA	Reformulated gasoline (1.1.95)(3) Phase I (1/1/2000)(4) Phase II		

(*Continues*)

National Gasoline Specifications (*Continued*)

Country	RVP kPa	VLI	Distillation						
			E70 % vol	E100 % vol	E180 % vol min	10%, °C max	50%, °C max	90%, °C max	FBP °C max
Australia									
Austria	Not available								
	S 35–70	<950	15–45	40–65	>85	–	–	–	215
	I mixtures								
	W 55–90	<1150	15–47	43–70	>85	–	–	–	
Belgium	45–95	–	15–45	40–70	90	–	–	–	215
	S 45–80	<1050	15–45	40–65	>85	–	–	–	–
	I mixtures								
	W 60–95	<1200	15–47	43–70	>85	–	–	–	–
California (United States)	Not available								
	48.3 max[(1)]						99/93 (1)	149/143 (1)	
Canada	Not available								
	A 79 max					35–65	70–120	190	
	B 86 max					60	70–117	190	
	C 97 max					55	70–113	185	
	D 69–107					50	70–110	185	
Denmark	S 45–80	700–1100	15	44–68	90	–	–	–	210
	I 58–90	800–1200	20	45–70					
	W 70–95	900–1300	25	45–72					
	S 45–80	<1050	15–45	40–65	>85	–	–	–	215
	I 58–90	<1150	15–47	43–70	>85				
	W 70–95	<1200	15–47	43–70	>85				
Finland	S 60–80	<1050	20–40	40–63	90	–	–	–	210
	I 70–90	<1150	23–43	43–66					
	W 85–100	<1250	25–45	45–68					
	S 60–80	<1050	–	40–63	–	–	–	–	215
	I 70–90	<1150		43–66					
	W 85–100	<1250	–	45–68					
	S 60–70	<1000	20–40	43–63	91				
	I 70–80	<1100	23–43	45–66					
	W 80–90	<1200	25–45	48–70					
France	S 45–79	≤900	10–47	40–70	>85	–	–	210	215
	I 50–86	≤1000							
	W 55–99	≤1150							
	S 35–70	<900	15–45	40–65	>85	–	–	–	–
	I 45–80	<1000	15–45	40–65	>85				
	W 55–90	<1150	15–47	43–70	>85				
	S 35–70	<850	S ≤ 40						
	I 45–80[(1)]	<1000	I ≤ 43						
	W 55–90	<1150	W ≤ 45						

(*Continues*)

National Gasoline Specifications (*Continued*)

Country	RVP kPa	VLI	Distillation E70 % vol	E100 % vol	E180 % vol min	10%, °C max	50%, °C max	90%, °C max	FBP °C max
	S 45–79	≤900							
	I 50–86	≤1000	10–47	40–70	>85	–	–	210	215
	W 55–99	≤1150							
Germany	S 45–70		15–40	42–65	≥90	–	–	–	215
	W 60–90		20–45	45–70	≥90	–	–	–	215
	S 35–70	<950	15–45	40–65	>85	–	–	–	215
	I mixtures								
	W 55–90	<1150	15–47	43–70	>85				
Greece	S 62 max		10 min	30–65	85	–	–	–	215
	W 80 max								
	S 65 max								
	W 80 max								
	S 35–70	900	15–45	40–65	85	–	–	–	215
	W 45–80	1000	–	–	85	–	–	–	215
Israel	45–66	≤ 910				75	125	180	
						73	125	180	215
						70	125	180	
	45–66	≤910				73	125	180	215
Italy	S 40–74		10–45	30–70	85	–	–	–	220
	S 35–70	<900	15–45	40–65	>85	–	–	–	215
	I 45–80	<1000	15–45	40–65	>85	–	–	–	215
	W 55–90	<1100	15–47	43–70	>85				
Japan	44–78					70	125	180	220
The Netherlands	S 40–80	<1050	15–45	40–65	>85				215
	I mixutres								
	W 60–95	<1200	15–47	43–70	<85				
Portugal	S 35–70		15–45	40–65	>85				215
	I 45–80								
	W60–95		15–47	43–70					
	S 35–70	<900	15–45	40–65					
	I 45–80	<1050	15–45	40–65	>85				215
	W 60–95	<1200	15–47	43–70					
South Africa	≤75	20[1]				65	77–115	185	215
		S ≤ 950				65	77–115	185	215
		W ≤ 1000							
		S ≤ 890							
		W ≤ 940							
Spain	S 48–64		10–45	30–70	80				210
	W 55–78								
			15–45	40–65	>85				215

(*Continues*)

National Gasoline Specifications (*Continued*)

Country	RVP kPa	VLI	Distillation E70 % vol	E100 % vol	E180 % vol min	10%, °C max	50%, °C max	90%, °C max	FBP °C max
Sweden			No longer available						
	S 45–80 W 60–95	<1050 <1200	15–45 15–47	40–65 43–70	85				215
	S 45–75 W 65–95			43 min 45 min					215
	S 45–70 W 65–95			47 min 50 min					200
	S 45–70 W 65–95			47 min 50 min					205
			Will be developed						
Switzerland	S 45–70 W 60–90	<990 <1230	15–42 20–47	40 min 42 min	85 85				215 215
	S 35–70 I mixtures W 60–95	<950 <1200	15–45 15–47	40–65 43–70	>85 >85				215 215
United Kingdom	S 45–80 I 103 max W 60–115	<1050 <1240 <1360	15–45	40–65	90				220
	S 45–80 W 65–100	<1050 <1250	15–45 20–50	40–65 43–70	>85 >85				215 215
United States			No longer available						
	AA 54 max A 62 max	20 max(3)				70 70	77–121 77–121	190 190	225 225
	South 50 max North 56 max							167(2)	

Country	Residue % vol max	Oxidation stability, minutes min	Existent gum, mg/100 ml max	Copper corrosion 3 h at 50°C max	Density at 15°C, kg/m^3	Sulphur, % wt max
Australia		240	4	1		0.2
		240	4	1		0.05
Austria	Not sold since October 1993					
	2	360	5	1	735–780 720–770	0.05
Belgium	2.0	–	5	1	720–770	0.10
	2.0				730–780 725–775 720–770	0.10

(*Continues*)

National Gasoline Specifications (*Continued*)

Country	Residue % vol max	Oxidation stability, minutes min	Existent gum, mg/100 ml max	Copper corrosion 3 h at 50°C max	Density at 15°C, kg/m^3	Sulphur, % wt max
California (United States)	Not sold since January 1992					
						0.004/0.003(1)
Canada	Not sold since December 1990					
		240	5	1		0.15
Denmark	2	480	4	1	730–770	0.10
	2	360	5	1	730–780 730–770 710–750	0.05
Finland	2	530	5	1	725–770	0.10
	2	360	5	1	725–780	0.05
						0.04 0.01
France	2		10	1B	720–770	0.15
	2	360	5	1	730–780	0.05
	2		10	1B	700–750	0.20
Germany	2		5	≤1	730–780	0.10
	2	360	5	1	725–780	0.05
Greece	2	360	4	1	720–770 700–750	0.10 0.15
	2	360	5	1	725–780	0.05
Israel	1.5	240	5	1	715–780	0.15
	1.5	360	5	1	715–780	0.05(5)
Italy	2		8	1	725–770	0.20
		360	5	1	725–780	0.05
Japan	2	240	5	1	≤783	
The Netherlands	2	360	5	1	725–780	0.05
Portugal	2	360	5	1	P 720–770 R 710–760	0.10
	2	360	5	1	725–780	0.05
South Africa	2	240	4.0	1		0.15
	2	360	4	1	710–785	0.10
Spain	2	240	5	1B	P 720–780 R 710–760	0.13
	2	360	5	1	725–780	0.05
Sweden	Not available					
	2	360	5	1	725–775	0.05
						0.1
						0.03
						0.01
	Will be developed					

(*Continues*)

National Gasoline Specifications (*Continued*)

Country	Residue % vol max	Oxidation stability, minutes min	Existent gum, mg/100 ml max	Copper corrosion 3h at 50°C max	Density at 15°C, kg/m^3	Sulphur, % wt max
Switzerland	–	240	5	1	–	0.10
	2	360	5	1	725–780	0.05
United Kingdom	2	240	5	1	–	0.20
	2	360	5	1	725–780	0.05
United States	Not sold since January 1995					
	2	240	5	1		0.10
						0.0338$^{(2)}$

Country	Lead content, g/l max	Benzene content, % vol max	Aromatics content, % vol max	Olefins content, % vol max	Phosphorus content, g/l max	Silicon content, g/l max
Australia		5				
	0.013	5			0.0013	
Austria	Not sold since October 1993					
	0.013	3.0			See column "Fuel Additives"	
Belgium	0.15					
California	Not sold since January 1992					
		1.0/0.0.8$^{(1)}$	25/22$^{(1)}$	6.0/4.0$^{(1)}$		
Canada	Not sold since December 1990					
	0.013				0.0013	
Denmark	0.15					
	0.013	5.0			See column "Fuel Additives"	
Finland	0.15					
	0.013	5.0			See column "Fuel Additives"	
		3.0 1.0				
France	0.15	5.0				
	0.013	5.0			See column "Fuel additives"	
	0.08–0.15					0.002
	0.15	5.0				

(*Continues*)

National Gasoline Specifications (*Continued*)

Country	Lead content, g/l max	Benzene content, % vol max	Aromatics content, % vol max	Olefins content, % vol max	Phosphorus content, g/l max	Silicon content, g/l max
Germany	0.15					
	0.013	5.0			See column "Fuel additives"	
Greece	0.15					
	0.013	5.0			See column "Fuel additives"	
Israel	0.15	5				
	0.013	5	To be reported		0.0013	
Italy	0.15					
	0.013	3.0			See column "Fuel Additives"	
Japan	0.001–0.02					
The Netherlands	0.013	5.0			See column "Fuel Additives"	
Portugal	0.40					
	0.013	5.0			See column "Fuel Additives"	
South Africa	0.4					
	0.013					
Spain	0.15					
	0.013	5.0			See column "Fuel Additives"	
Sweden	Not available					
	0.013	5.0			See column "Fuel Additives"	
	0.013	5.0			nil$^{(1)}$	
	0.005	3.0	6$^{(4)}$		0.002	
	0.005	3.0	5.5$^{(4)}$		nil$^{(1)}$	
	Will be developed					
Switzerland	0.15					
	0.013	5.0			See column "Fuel Additives"	

(*Continues*)

National Gasoline Specifications (*Continued*)

Country	Lead content, g/l max	Benzene content, % vol max	Aromatics content, % vol max	Olefins content, % vol max	Phosphorus content, g/l max	Silicon content, g/l max
United Kingdom	0.15					
	0.013	5.0			See column "Fuel Additives"	
United States	Not sold since January 1995					
	0.013				0.0013	
	(1)	1.0-batch basis 0.95-average basis	28.6(2)	10.8(2)		

Country	Oxygenates content	Deposits control	Use of additives
Australia	Adding to gasoline is allowed		The use of additives is allowed.
Austria	Not sold since October 1993		
	According to the Directive 85/536/EEC		The use of additives is allowed but no phosphorous-containing compounds.
Belgium	According to the Directive 85/536/EEC		The use of additives is allowed but no phosphorous-containing compounds.
California (United States)	Not sold since January 1992		
	Oxygen 1.8–2.2% mass	PFI Injector Cleanliness (Chrysler 2.2 l engine). Intake Valve Cleanliness (BMW 318i engine).	Deposit-control additives must be used.
Canada	MTBE ≤ 11% vol.		The use of additives intended to improve a gasoline performance is allowed.
Denmark			
	According to the Directive 85/536/EEC		The use of additives is allowed but no phosphorous-containing compounds.
Finland			
	According to the Directive 85/536/EEC		The use of additives is allowed but no phosphorous-containing compounds.
	2.0–2.7% wt O_2		The use of additives is allowed but no phosphorous-containing compounds.

(*Continues*)

National Gasoline Specifications (*Continued*)

Country	Oxygenates content	Deposits control	Use of additives
France	Methanol ≤ 3% vol. Ethanol ≤ 5% vol. Ethers ≤ 15% vol.		
	According to the Directive 85/536/EEC		The use of additives is allowed but no phosphorous-containing compounds.
	No alcohols and no ketones Ethers ≤ 15% vol.	Carburetor Cleanliness[3] (CEC F-03-T-81): Merit > 8. Injector Cleanliness[4] (GFC-TAE-1-87): flow rate losses < 4%. Intake Valve Cleanliness (CEC F-04-A-87)[5]: Merit ≥ 9. (CEC F-05-A-94)[6]: Merit ≥ 9.	
	Methanol ≤ 3% vol. Ethanol ≤ 5% vol. Ethers ≤ 15% vol.		
Germany	Methanol ≤ 3% vol. Ethanol ≤ 5% vol. Ethers: ≤ 15% vol.		
	According to the Directive 85/536/EEC		The use of additives is allowed but no phosphorous-containing compounds.
Greece			
	According to the Directive 85/536/EEC		The use of additives is allowed but no phosphorous-containing compounds.
Israel	MTBE ≤ 10% vol.		
	Methanol ≤ 3% vol. Ethanol ≤ 5% vol. Ethers ≤ 15% vol. Oxygen ≤ 2.8% wt.	Carburetor cleanliness (CEC F-03-T-81): Merit > 8. Injector cleanliness (Peugeot 205 GTI test): Flow rate losses < 4%. Intake Valve Cleanliness (CEC F-04-A-87): Deposits reduction ≥ 50%	Deposit-control additives must be used.
Italy			
	According to the Directive 85/536/EEC		The use of additives is allowed but no phosphorous-containing compounds.
Japan			
The Netherlands	According to the Directive 85/536/EEC		The use of additives is allowed but no phosphorous-containing compounds.

(*Continues*)

National Gasoline Specifications (*Continued*)

Country	Oxygenates content	Deposits control	Use of additives
Portugal			
	According to the Directive 85/536/EEC		The use of additives is allowed but not phosphorous-containing compounds.
South Africa	The 93 RON grade manufactured by the SASOl oil-from-coal process contains 8–12% alcohols (mainly ethanol).		
	Oxygen $\leq$ 2.8% mass[3] Oxygen $\leq$ 3.7% mass[3]		
Spain			
	According to the Directive 85/536/EEC		The use of additives is allowed but no phosphorous-containing compounds.
Sweden	Not available		
	According to the Directive 85/536/EEC		The use of additives is allowed but no phosphorous-containing compounds.
	Oxygen $\leq$ 2% wt.	Intake Valve Cleanliness[2] (CEC F-05-T-92): Merit $\geq$ 9. Injector cleanliness[3] (GFC-TAE-1-87): flow rate losses $\leq$ 4%.	Additives must not contain ash-forming constituents.
	Will be developed		
Switzerland			
	According to the Directive 85/536/EEC		The use of additives is allowed but no phosphorous-containing compounds.
United Kingdom			
	According to the Directive 85/536/EEC		The use of additives is allowed but no phosphorous-containing compounds.
United States	Not sold since January 1995		
	Oxygen $\leq$ 2% mass Oxygen $\leq$ 2.7% mass for fuels containing aliphatic ethers or/and alcohols (not methanol)		
	Oxygen 2–2.7% mass	PFI Injector cleanliness (Chrysler 2.2 l engine): flow rate losses < 5%. Intake Valve Cleanliness (BMW 318i engine).	From January 1, 1995 all gasolines in the United States must contain deposit-control additives.

(*Continues*)

National Gasoline Specifications (*Continued*)

Country	Remarks
Australia	
Austria	Not sold since October 1993
California (United States)	Not sold since January 1992
	(1) Batch basis/average basis
Canada	Not sold since December 1990
	Manganese content ≤ 18 mg Mn/l (1) Antiknock Index = (RON + MON)/2 (2) MON must not be lower from Antiknock Index at more than 5 units.
Denmark	
Finland	(1) According to [6], leaded gasoline is already not available
France	
	(1) Equivalent vapor pressure at 37.8°C according to "Grabner method." (2) MTBE ≤ 15% vol; ETBE ≤ 15% vol., with residual alcohols ≤ 1% vol. (3) Renault R5 test. (4) Peugeot 205 GTI test. (5) Opel Kadett test. (6) Mercedes M102E test.
Germany	
Greece	
Israel	(1) Allowed 84 until 1/4/1997. (2) Allowed 84 until 1/2/1996 and 85 until 1/4/1997; (3) Allowed 84 until 1/4/1997. (4) Allowed 80 until 1/4/1997. (5) Allowed 0.10 until 1/4/1996.
Italy	
Japan	
The Netherlands	
Portugal	
South Africa	(1) Vapor/Liquid/Ratio at 325 K/Pa (max) calculated according to ASTM D2533 (1) MON = 87 for blends with alcohol content > 0.2% vol. (2) MON = 83 for blends with alcohol content > 0.2% vol. Every alcohol blended with gasoline must contain minimum 85% mass of ethanol with the balance, which is mainly iso- and n-propanol. (3) It is possible to use ethers with 5 or more atoms of carbon in the molecula.

(*Continues*)

National Gasoline Specifications (*Continued*)

Country	Remarks
Spain	(1) Had to be put into force from winter 1994
Sweden	Not available
	(1) Negligible. (2) Mercedes M-102E test. (3) Peugeot 205GTI test. (4) Empirical limit = $\frac{\text{Aromatics, \%vol.}}{13}$ + Benzene, % vol. (max).
	Will be developed
Switzerland	
United Kingdom	
United States	Not sold since January 1995
	(1) This requirement is not official, excluding number of states, where it is already put into force. (2) Antiknock Index = (RON + MON)/2. (3) Vapor/Liquid Ratio according to ASTM D2533.
	(1) Heavy Metals—none without an EPA waiver. (2) Statutory Baseline Parameters—1990 average quality. Emissions reduction relative to 1990 baseline average quality: (3) NO_x—1.5% min; Toxics—16.5% min. (4) NO_x = − 6.8% min; Toxics—21.5% min.

APPENDIX 2

National Specifications for Automotive Diesel Fuel (Based on the CONCAWE Data [15])

Country	National standard reference (date)	Grade	Cetane number (min.)	Cetane index (min.)	Flash point PM°C (min.)	Sulphur content % m/m (max.)
Australia	AS 3570 1989	Summer Winter	45	–	(1)	0.5
Austria	O Norm EN 590 (1.2.94)	Winter Intermediate Summer	49	46	55	(1)0.05
Belgium	NBN EN 590	Winter Intermediate Summer	49	46	55	0.2
California (United States)	CARB Diesel 1.10.1993		48		54	0.05

(*Continues*)

National Specifications for Automotive Diesel Fuel (*Continued*)

Country	National standard reference (date)	Grade	Cetane number (min.)	Cetane index (min.)	Flash point PM°C (min.)	Sulphur content % m/m (max.)
Canada	CAN/CGSB 3.6-M90	A	40		40	0.30
		B	40		40	0.30
Denmark		CEN Diesel	49	46	55	0.2
		CEN Low sulphur	49	46	55	0.05
		Bus Diesel	50	47		0.05
Finland	SF-EN590	C	49	46	55	0.2
		1	47	46	55	
		3	45	43	55	
		4	45	43	55	
	Reformulated Diesel: 1.7.93	Summer	49	49	56	0.005
		Winter	47	47	56	0.005
France	EN590:1993	Summer Winter	49	46	(3) 55	0.2
		Grand Froid	50	49		0.15
Germany	DIN EN590:1993	Summer Intermediate Winter	49	46	55	0.2
Greece	EN 590:1993	Summer Winter	49	46	55	0.3
Ireland	IS EN:590 1993	Summer Winter	50	50	55	0.3
Israel	Israel Standard No. 107 1995	Winter Summer	50	48	66	0.20
Italy	UNI-CUNA EN 590:93 1.10.93	Summer Winter	49	46	55	0.2
Japan	JIS K2204 1992	Special No. 1	(2)	50	50	(4) 0.20
		No. 1	(2)	50	50	0.20
		No. 2	(2)	45	50	0.20
		No. 3	(2)	45	45	0.20
		Special No. 3	(2)	45	45	0.20
The Netherlands		Winter Intermediate Summer				0.2
Portugal	EN 590 1993	Summer Winter	49	46	55	0.3
South Africa	SABS 342 1994		45		55	0.55
Spain	EN 590 1993	Summer Winter	49	46	55	0.3

(*Continues*)

National Specifications for Automotive Diesel Fuel (*Continued*)

Country	National standard reference (date)	Grade	Cetane number (min.)	Cetane index (min.)	Flash point PM°C (min.)	Sulphur content % m/m (max.)
Sweden	SS 155435 13.03.91	Urban Diesel 1 Summer (TD1 Grade)—Winter	50	50	56	0.001
		Urban Diesel 2 Summer (TD2 Grade)—Winter	47	47	56	0.005
		Normal Summer Diesel D 10	49	46	56	0.20
		Winter Diesel 1 D 26	47	46	56	0.20
		Winter Diesel 2 D32	46	46	56	0.20
		Winter Diesel 3 D36	45	43	56	0.20
Switzerland	SN	Summer (01.05–30.09)	49	46	55	0.05
	EN 590	All year	47	46	55	0.05
United Kingdom	B S EN 590:1930		49	46	55	0.2
United States	ASTM D975 (1994)	Low Sulphur No. 1-D-A	40	40(4)	38	0.05
		Low Sulphur No. 2-D-A	40	40(4)	52	0.05
		No. 1-D-A	40	–	38	0.50
		No. 2-D-A	40	–	52	0.50

Country	CFPP °C Max.	Density at 15° kg/m³	Kinematic viscosity mm²/sec at 20°C	Kinematic viscosity mm²/sec at 40°C
Australia	+12; +3; +1; −7;	820–870		1.9–5.5
Austria	−20 −15 +5	820–860		2.0–4.5
Belgium	−15 −5 0	820–860		2.0–4.5
California (United States)		830–860		2.0–4.1
Canada				1.3–3.6 1.7–4.1
Denmark		820–860 820–860 820–855		2.0–4.5 2.0–4.5
Finland	−5	820–860		2.0–4.5
	−26	800–845		1.5–4.0
	−38	800–840		1.4–4.0
	−44	800–840		1.2–4.0
	−15	820–850		2.0–3.5
	−34	800–830		1.4–2.6

(Continues)

National Specifications for Automotive Diesel Fuel (*Continued*)

Country	CFPP °C Max.	Density at 15°C kg/m³	Kinematic viscosity mm²/sec at 20°C	Kinematic viscosity mm²/sec at 40°C
France	0 −15 −20	820–860		2.0–4.5 2.0–4.5
Germany	0 −10 −20	820–860		2.0–4.5
Greece	+5 −5	820–860		2.0–4.5
Ireland	–	820–860		2.0–4.5
Israel	+9 −2	820–860		2.5–6.0
Italy	0 −10	820–860		2.0–4.5
Japan	– −1 −5 −12 −19			2.7 min(3) 2.7 min 2.5 min 2.0 min 1.7 min
Netherlands	−15 −5 0			
Portugal	0 −6	820–860		2.0–4.5
South Africa	−4			2.2–5.3
Spain	0 −10	820–860		2.0–4.5
Sweden	−10(S)(1); −26(W) −10(S); −26(W) −10 −26 −32 −38	800–820 800–820 820–860 800–845 800–840 800–840		1.2–4.0 1.2–4.0 2.0–4.5 1.5–4.0 1.5–4.0 1.4–4.0
Switzerland	−10 −20	820–860 800–845		2.0–4.5 1.5–4.0
United Kingdom	− 15	820–860		2.0–4.5
United States				1.3–2.4 1.9–4.1 1.3–2.4 1.9–4.1

(*Continues*)

National Specifications for Automotive Diesel Fuel (*Continued*)

Country	Distillation (°C) IBP min.	10% min.	50% max.	65% min.	85% min.	90% max.	95% min.	FBP min.	Aromatics content % v.v max.
Australia						357			
Austria				250	350		370		
Belgium				250	350		370		
Californ. (United States)	170–215	205–255	245–295			290–320		305–350	10[1]
Canada						290			
						360			
Denmark				250	350		370		
				250	350		370		
							325		
Finland				250	350		370		
		180					340		
		180					340		
		180					340		
							350		
							310		
France				250	350		370		
				250	350		370		
				250	350		370		30[4]
Germany				250	350		370		
Greece				250	350		370		
Ireland				250	350		370		
Israel						357			
Italy				250	350		370		
Japan					360				
					360				
					350				
					330[1]				
					330				
The Netherlands					350		370		
				250					
Portugal				250	350		370		
South Africa						362			
Spain				250	350		370		
Sweden	180						285	300	5[2]
	180						295		20[3]
				250	350		370		
		180					340		
		180					340		
		180					340		

(*Continues*)

National Specifications for Automotive Diesel Fuel (*Continued*)

Country	Distillation (°C) IBP min.	10% min.	50% max.	65% min.	85% min.	90% max.	95% min.	FBP min.	Aromatics content % v.v max.
Switzerland		180 180(max)		250	350		370 340		
United Kingdom				250	350		370		
United States						228 338(1) 288 338(1)			35(4) 35(4)

Country	Oxidation stability mg/l max.	Copper corrosion 3 h at 50°C max.	Ash % m/m	Water content % m/m (max.)	Cloud point °C max.	Pour point	Fuel additives
Australia	25	2(2)	0.01	0.05/vol.	+15; +6; +4; −4 + 7; 0; −1; −4;		May include aditives.
Austria	25	1	0.01	0.02			Fuel additives use is allowed to improve the fuel quality.
Belgium	25	1	0.01	0.02			Fuel additives use is allowed to improve the fuel quality.
California (United States)							
Canada		1(1)	0.01	0.05(2)/vol			
Denmark	25 25	1 1	0.01 0.01	0.02 0.02			
Finland					– −16 −28 −34 −5 −29		
France	25 15	1 5%(2)	0.01	0.02	– −8		Fuel additives use is allowed to improve the fuel quality. (Detergents, antifoams, etc.)
Germany	25	1	0.01	0.02			Fuel additives use is allowed to improve the fuel quality.

(*Continues*)

National Specifications for Automotive Diesel Fuel (*Continued*)

Country	Oxidation stability mg/l max.	Copper corrosion 3 h at 50°C max.	Ash % m/m	Water content % m/m (max.)	Cloud point °C max.	Pour point	Fuel additives
Greece	25	1	0.01	0.02	–		Fuel additives use is allowed to improve the fuel quality.
Ireland	25	1	0.01	0.02			Fuel additives use is allowed to improve the fuel quality.
Israel	25	1(1)	0.01	0.05/vol.		+5 −6	It is possible to use additives to decrease pour point temperature or to improve performance.
Italy	25	1	0.01	0.02			Fuel additives use is allowed to improve the fuel quality.
Japan						+5 −25 −7.5 −20 −30	
The Netherlands							
Portugal	25	1	0.01	0.02			Fuel additives use is allowed to improve the fuel quality.
South Africa		1(1)	0.01	0.05/vol.			
Spain	25	1	0.01	0.02			Fuel additives use is allowed to improve the fuel quality.
Sweden	25 25 25 25 25 25	1 1 1 1 1 1	0.01 0.01 0.01 0.01 0.01 0.01	0.01 0.01 0.02 0.02 0.02 0.02	(w)-16(s)0(1) (w)-16(s)0 −16 −22 −28		The fuel may include any additives that are chiefly intended for high-speed diesel engines, i.e., engines running at not less than 16 r/s under full load.

(*Continues*)

National Specifications for Automotive Diesel Fuel (*Continued*)

Country	Oxidation stability mg/l max.	Copper corrosion 3 h at 50°C max.	Ash % m/m	Water content % m/m (max.)	Cloud point °C max.	Pour point	Fuel additives
Switzerland	25	1	0.01	0.02			Fuel additives use is allowed to improve the fuel quality.
United Kingdom	25	1	0.01	0.01	0.02		Additives are allowed to improve the fuel quality.
United States		3 3 3	0.01 0.01 0.01	(2)0.05/vol. 0.05/vol. 0.05/vol.	(3)		Stability and biocide additives are allowed. On May 27, 1994 EPA finalized regulations which contain test requirements for fuel additives.

Country	Remarks
Australia	(1) In accordance with the law (2) 3 hours at 100°C
Austria	(1) From 10.01.1995 (Regulation No. 123 —Ministry of Environmental 03.01.1992
Belgium	
California (United States)	(1) PAH—1.4% m/m (max.)
Canada	(1) 3 hours at 100°C (2) Water content and sediment
Denmark	
Finland	
France	(1) 820–855 with Cetane Index ≥ 50 and Cetane Number ≥ 51.; (2) ASTM D655-83 Instruction A; (3) AFNOR T60103 (this for >52°C PM); (4) It is possible to limit the content of PAH (to be defined)
Germany	
Greece	Greek Government Gazette 336/94
Ireland	
Israel	(1) 3 hours at 100°C
Italy	
Japan	(1) If fuel viscosity ≥ 4.7 cSt, T90 must be ≥350°C (2) Cetane number can replace Cetane Index (3) at 30°C (4) 0.05 from May 1997

(*Continues*)

National Specifications for Automotive Diesel Fuel (*Continued*)

Country	Remarks
The Netherlands	
Portugal	
South Africa	(1) 3 hours at 100°C
Spain	
Sweden	(1) (s) summer; (w) winter (2) PAH—0.02% v/v (max.) (3) PAH—0.1% v/v (max.)
Switzerland	
United Kingdom	
United States	(1) 282°C min. (2) Water and sediment content (3) Winter fuel properties must be settled by agreement between refiners and customers. (4) It is necessary to fulfill one of these requirements.

APPENDIX 3: U.S. EPA MODELS FOR CALCULATION OF FUEL EFFECTS ON EXHAUST EMISSIONS

Table 7.5

Formulas for Toxics Emissions, Simple Model[1] (reproduced from CONCAWE Report 5/95, Section 7)

Table 7.5.1

Calculation of Total Toxics Reductions, Simple Model

$$\text{Total Toxics Reduction \%, Summer 1} = \frac{100\{53.2-(\text{toxics emissions, summer1, mg/mile})\}}{53.2}$$

$$\text{Total Toxics Reduction \%, Summer 2} = \frac{100\{52.1-(\text{toxics emissions, summer2, mg/mile})\}}{52.1}$$

$$\text{Total Toxics Reduction \%, Winter} = \frac{100\{55.5-(\text{toxics emissions, winter, mg/mile})\}}{55.5}$$

The toxics emissions for the summer regions and winter are given in Table 7.7.2.

Table 7.5.2

Calculation of Toxic Emissions, Summer, Regions 1 and 2, mg/mile[4]

Toxic emissions =
exhaust benzene + evaporative benzene + running loss benzene + refueling benzene + formaldehyde + acetaldehyde + 1, 3−butadiene + polycyclic organic matter

Exhaust benzene =

$$\left\{1.884 + 0.949(\text{benzene}) + 0.113\,\frac{(\text{aromatics} - \text{benzene})}{100}\right\} \times 1000[\text{exhVOC}]$$

Evaporative benzene emissions = hot soak benzene + diurnal benzene emissions

(*Continues*)

Table 7.5.2 *(Continued)*

Hot soak benzene emissions =

$$(\text{benzene}) \times 0.679[\text{evapVOC}] \times 1000 \times \left\{1.4448 - \frac{0.0684(\text{MTBE})}{2.0} - \frac{0.080274(\text{RVP})}{100}\right\}$$

Diurnal benzene emissions =

$$(\text{benzene}) \times 0.321[\text{evapVOC}] \times 1000 \times \left\{1.3758 - \frac{0.0579(\text{MTBE})}{2.0}\right\} - \frac{0.080274(\text{RVP})}{100}\Big\}$$

Running loss benzene emissions =

$$(\text{benzene}) \times [\text{runlossVOC}] \times 1000 \times \left\{1.4448 - \frac{0.0684(\text{MTBE})}{2.0} - \frac{0.080274(\text{RVP})}{100}\right\}$$

Refueling benzene emissions =

$$(\text{benzene}) \times [\text{refuelVOC}] \times 1000 \times \left\{1.3972 - \frac{0.0591(\text{MTBE})}{2.0} \quad \frac{0.081507(\text{RVP})}{100}\right\}$$

1,3-Butadiene emissions = 5.56 × [exhVOC]

Polycyclic organic emissions = 3.15 × [exhVOC]

Formaldehyde(2,3) =

$$12.56[\text{exhVOC}] \times \left\{1 + \frac{0.421(\text{MTBE}+\text{TAME})}{2.7} + \frac{0.358(\text{EtOH})}{3.55} + \frac{0.137(\text{ETBE}+\text{ETAE})}{2.7}\right\}$$

Acetaldehyde(2,3) =

$$8.91[\text{exhVOC}] \times \left\{1 + \frac{0.078(\text{MTBE}+\text{TAME})}{2.7} + \frac{0.865(\text{EtOH})}{3.55} + \frac{0.867(\text{ETBE}+\text{ETAE})}{2.7}\right\}$$

The formulas for the calculation of VOC emissions for summer, regions 1 and 2, and winter are given in Table 7.5.4.

Table 7.6

Complex Model Formulas

Table 7.6.1

Calculation of NO_x Emissions, Complex Model

$$NO_x \text{ emissions, mg/mile} = NO_x(b) + \frac{NO_x(b) \times Y_{NO_x}(t)}{100}$$

where,

$NO_x(b)$ = baseline exhaust NO_x emissions, mg/mile (see Table 7.7.1)

$$Y_{NO_x}(t) = \text{performance of target fuel} = \left\{w_1 \times \frac{\exp.\, n_1(t)}{\exp.\, n_1(b)} + w_2 \times \frac{\exp.\, n_2(t)}{\exp.\, n_2(b)} - 1\right\} \times 100$$

(% change from baseline)

w_1, w_2 = weighting factors for normal and higher emitters (see Table 7.7.2)
$n_1(t)$, $n_2(t)$ = NO_x equations for normal and higher emitters, target fuel(1)
$n_1(b)$, $n_2(b)$ = NO_x equations for normal and higher emitters, base fuel(1)

$$n_1 = 0.0018571(\text{oxygen}) + 0.0006921(\text{sulphur}) + 0.0090744(\text{RVP}) + 0.0009310(\text{E200}) + 0.0008460(\text{E300}) + 0.0083632(\text{aromatics}) - 0.002774(\text{olefins}) - 6.63 \times 10^{-7}(\text{sulphur})^2 - 0.000119(\text{aromatics})^2 + 0.0003665(\text{olefins})^2$$

$$n_2 = 0.000252(\text{sulphur}) - 0.00913(\text{oxygen}) - 0.01397(\text{RVP}) + 0.000931(\text{E200}) - 0.00401(\text{E300}) + 0.007097(\text{aromatics}) - 0.00276(\text{olefins}) + 0.0003665(\text{olefins})^2 - 7.995 \times 10^{-5}(\text{aromatics})^2$$

(1)For winter NO_x emissions, RVP for both the baseline and target fuels is taken as 8.7 psi.

Table 7.6.2

Calculation of Exhaust Volatile Organic Compounds Emissions, Complex Model

Exhaust VOC Emissions, mg/mile = $VOC(b) + \frac{VOC(b) \times Y_{VOC}(t)}{100}$, where

VOC(b) = baseline exhaust VOC emissions, mg/mile (see Table 7.7.1)

$Y_{VOC}(t)$ = % change VOC from baseline = $\left\{ w_1 \times \frac{\exp. v_1(t)}{\exp. v_1(b)} + w_2 \times \frac{\exp. v_2(t)}{\exp. v_2(b)} - 1 \right\} \times 100$

w_1, w_2 = weighting factors for normal and higher emitters (see Table 7.7.2)
$v_1(t), v_2(t)$ = VOC equations for normal and higher emitters, target fuel[(1)]
$v_1(b), v_2(b)$ = VOC equations for normal and higher emitters, base fuel[(1)]

$$v_1 = 0.0005219(\text{sulphur}) - 0.003641(\text{oxygen}) + 0.0289749(\text{RVP}) - 0.01447(\text{E200}) + 0.0001072(\text{E200})^2 - 0.068624(\text{E300}) + 0.0004087(\text{E300})^2 + 0.0323712(\text{aromatics}) - 0.002858(\text{olefins}) - 0.0003481(\text{aromatics} \times \text{E300})$$

$$v_2 = 0.043295(\text{RVP}) - 0.003626(\text{oxygen}) - 0.000054(\text{sulphur}) - 0.013504(\text{E200}) - 0.062327(\text{E300}) + 0.0282042(\text{aromatics}) - 0.002858(\text{olefins}) + 0.000106(\text{E200})^2 + 0.000408(\text{E300})^2 - 0.000287(\text{aromatics})(\text{E300})$$

(1) For winter exhaust VOC emissions, RVP for both the baseline and target fuels is 8.7 psi.

(2) For fuels with E200 and E300 exceeding the upper limit given in Table 7.7.6 the values are taken as equal to the upper limits.

(3) For fuels with E200, E300, and aromatics outside the limits given in Table 7.9.6 the value of $Y_{voc}(t)$ is modified by an additional term as given in *Federal Register*, vol. 59, no. 32, pp. 7820–21.

Table 7.6.3

Nonexhaust VOC Emissions

Nonexhaust VOC emissions (g/mile)	Phase I (1995–1999)		Phase II (2000 and after)	
	Region I	Region II	Region I	Region II
Diurnal VOC	$0.00736(RVP)^2 - 0.0790(RVP) + 0.2553$	$0.006818(RVP)^2 - 0.07682(RVP) + 0.2610$	$0.007385(RVP)^2 - 0.08981(RVP) + 0.3158$	$0.004775(RVP)^2 - 0.05872(RVP) + 0.21306$
Hot soak VOC	$0.01557(RVP)^2 - 0.1671(RVP) + 0.5399$	$0.014421(RVP)^2 - 0.16248(RVP) + 0.5520$	$0.006654(RVP)^2 - 0.08009(RVP) + 0.2846$	$0.006078(RVP)^2 - 0.07474(RVP) + 0.27117$
Running loss VOC	$0.00279(RVP)^2 - 0.1096(RVP) - 0.7340$	$0.016255(RVP)^2 - 0.1306(RVP) + 0.2963$	$0.017768(RVP)^2 - 0.18746(RVP) + 0.6146$	$0.016169(RVP)^2 - 0.17206(RVP) + 0.56724$
Refueling VOC	$0.006668(RVP) - 0.0180$	$0.006668(RVP) - 0.0180$	$0.0004767(RVP) + 0.011859$	$0.004767(RVP) + 0.011859$
Total nonexhaust VOC[(1)]	$0.02572(RVP)^2 - 0.349032(RVP) + 0.0432$	$0.037494(RVP)^2 - 0.363232(RVP) + 1.0913$	$0.031807(RVP)^2 - 0.3568833(RVP) + 1.226859$	$0.027022(RVP)^2 - 0.300753(RVP) + 1.063329$

(1) Total nonexhaust VOC is sum of diurnal, hot soak, running loss, and refueling VOC emissions.

Table 7.6.4

Calculation of Winter Nonexhaust VOC Emissions, Complex Model

Total nonexhaust VOC emissions shall be set at zero under winter conditions.

Table 7.6.5

Calculation of Change in Total VOC Emissions, Complex Model[(1)]

Change in total VOC emissions (%)	Formula (g/mile)	
	Phase I	Phase II
Summer, Region 1	$100 \times \frac{(\text{Total VOC} - 1.306)}{1.306}$	$100 \times \frac{(\text{Total VOC} - 1.4663)}{1.4663}$
Summer, Region 2	$100 \times \frac{(\text{Total VOC} - 1.215}{1.215}$	$100 \times \frac{(\text{Total VOC} - 1.3991)}{1.3391}$
Winter	$100 \times \frac{(\text{Total VOC} - 0.660)}{0.660}$	$100 \times \frac{(\text{Total VOC} - 1.341)}{1.341}$

[(1)]Total VOC emissions, g/mile = [exhaust VOC (mg/mile)]/1000 + nonexhaust VOC (g/mile).

Table 7.6.6

Calculation of Toxics Emissions

Summer toxics emissions = exhaust benzene + formaldehyde + acetaldehyde + 1,3-butadiene + polycyclic organic matter + nonexhaust benzene emissions

Exhaust benzene, formaldehyde, acetaldehyde, and 1,3-butadiene emissions are calculated using an equation of the type:
Emissions = $\frac{E(b)+E(b)\times YE(t)}{100}$ where,

E(b) = baseline emission, mg/mile (see Table 7.9.5)

$YE(t) = \left\{(W1 \times \frac{\exp E1(t)}{\exp E1(b)} + W2 \times \frac{\exp E2(t)}{\exp E2(b)} - 1\right\}$

W1, W2 = weighting factors for normal and higher emitters (see Table 7.9.2)
E1(t), E2(t) = equations for normal and higher emitters, target fuel
E1(b), E2(b) = equations for normal and higher emitters, base fuel

E1 for exhaust benzene, b1 = 0.0006197(sulphur) − 0.003376(E200) + 0.0265500(aromatics) + 0.2223900(benzene)

E2 for exhaust benzene, b2 = 0.0003370(sulphur) + 0.0112510(E300) + 0.0118820(aromatics) + 0.2223180(benzene) − 0.096047(oxygen)
E1 for formaldehyde, f1 = 0.0462131(MTBE) − 0.010226(E300) − 0.007166(aromatics)
E2 for formaldehyde, f2 = 0.0462131(MTBE) − 0.010226(E300) − 0.007166(aromtics) − 0.031352(olefins)
E1 for acetaldehyde, a1 = 0.0002631(sulphur) + 0.0397860(RVP) − 0.012172(E300) − 0.005525(aromatics) − 0.009594(MTBE) + 0.3165800(ETBE) + 0.2492500(ethanol)
E2 for acetaldehyde, a2 = 0.0002627(sulphur) − 0.012157(E300) − 0.005548(aromatics) − 0.055980(MTBE) + 0.3164665(ETBE) + 0.2493259(ethanol)

(Continues)

Table 7.6.6 (*Continued*)

E1 for 1,3 – butadiene, d1 =	0.0001552(sulphur) – 0.007253(E200) – 0.014866(E300) – 0.004005(aromatics) + 0.0282350(olefins)
E2 for 1,3-butadiene, d2 =	0.0436960(olefins) – 0.060771(oxygenate) – 0.007311(E200) – 0.008058(E300) – 0.004005(aromatics)
Polycyclic organic matter emissions =	0.003355 × VOC exhaust emissions
Nonexhaust benzene emissions =	diurnal + hot soak + running loss + refueling benzene emissions, where:
Hot soak benzene emissions =	10 × (benzene) × (hot soak VOC emissions)[(1)] × {1.4448 – 0.0342(MTBE) – 0.080274(RVP)}
Diurnal benzene emissions =	10 × (benzene) × (diurnal VOC emissions)[(1)] × {1.3758 – 0.0290(MTBE) – 0.080274(RVP)}
Running loss benzene emissions =	10 × (benzene) × (running loss VOC emissions)[(1)] × {1.4448 – 0.0342(MTBE) – 0.080274(RVP)}
Benzene refueling emissions =	10 × (benzene) × (refuelling VOC emissions)[(1)] × {1.3972 – 0.0296(MTBE) – 0.081507(RVP)}

[(1)] As calculated for appropriate phase and region in Table 7.6.3.

Table 7.7

Baseline Data for Calculating Complex Equations

Table 7.7.1

Baseline Exhaust Emissions

Exhaust pollutant (mg/mile)	Phase I (1995–1999)		Phase II (2000 and after)	
	Summer	Winter	Summer	Winter
VOC	446	660	907	1341
NO_x	660	750	1340	1540
Benzene	26.10	37.57	53.54	77.62
Acetaldehyde	2.19	3.57	4.44	7.25
Formaldehyde	4.85	7.73	9.70	15.34
1,3-Butadiene	4.31	7.27	9.38	15.84
Polyorganic Matter	1.50	2.21	3.04	4.50

Table 7.7.2

Weighting Factors for Normal and Higher Emitters

	Phase I (1995–1999)		Phase II (2000 and after)	
	VOC + Toxics	NO_x	VOC + Toxics	NO_x
Normal Emitters (w_1)	0.52	0.82	0.444	0.738
Higher Emitters (w_2)	0.48	0.18	0.556	0.262

Table 7.7.3

Baseline Fuel Properties and Acceptable Range for Models

Baseline fuel properties			Limits of simple model	Acceptable range (complex)	
Fuel Property	Summer	Winter		Reformulated	Conventional
Oxygen (% W)	0.0	0.0	0–3.5	0.00–3.70	0.00–3.70
Sulphur (ppm)	339	338		0.0–500.0	0.0–100.0
RVP(psi)	8.7	11.5	6.6–9.0	6.4–10.0	6.4–11.0
E200 (% v)	41.0	50.0		30.0–70.0	30.0–70.0
E300 (% v)	83.0	83.0		70.0–100.0	70.0–100.0
Aromatics (v/v %)	32.0	26.4	10–45	0.0–50.0	00.0–55.0
Olefins (v/v %)	9.2	11.9		0.00–25.0	0.00–30.0
Benzene (v/v %)	1.53	1.64	0–2.5	0.0–2.0	0.0–4.9

Table 7.7.4

Baseline Nonexhaust Emissions

Nonexhaust pollutant (mg/mile)	Phase I (1995–1999)		Phase II (2000 and after)	
	Region 1	Region 2	Region 1	Region 2
VOC	860.48	769.10	559.31	492.07
Benzene	9.66	8.63	6.24	5.50

Table 7.7.5

Total Baseline Total VOC, NO_x, and Toxics Emissions

Pollutant (mg/mile)	Phase I (1995–1999)		Phase II (2000 and after)	
	Region I	Region II	Region I	Region II
Summer				
NO_x	660.0	660.0	1340.0	1340.0
VOC	1306.5	1215.1	1466.3	1399.1
Toxics	48.61	47.58	86.34	85.61
Winter				
NO_x	750.0	750.0	1540.0	1540.0
VOC	660.0	660.0	1341.0	1341.0
Toxics	58.36	58.36	120.55	120.55

Table 7.7.6

Allowable Ranges of E200, E300, and Aromatics for Exhaust VOC Equations

	Phase I (1995–1999)		Phase II (2000 and after)	
Fuel parameter	Lower limit	Upper limit	Lower limit	Upper limit
E200, % v	33.00	65.83	33.00	65.52
E300, % v	72.00	80.32 + {0.390 × (aromatics)}	72.00	79.75 + {0.385 × (aromatics)}
Aromatics, % v	18.00	46.00	18.00	46.00

Table 7.7.7

Allowable Ranges of Sulphur, Olefins, Aromatics, and E300 for NO_x Equations

	Phase I (1995–1999)		Phase II (2000 and after)	
Fuel parameter	Lower limit	Upper limit	Lower limit	Upper limit
Sulphur, ppm	10.0	450.0	10.0	450.0
E300, % v	70.0	95.0	70.0	95.0
Olefins, % v	3.77	19.0	3.77	19.0
Aromatics, % v	18.0	36.2	18.0	36.8

Index

D

R

S

V

W

90182
9 780126 398557
ISBN 0-12-639855-0
6 08628 39855 6

MIX
Papier | Fördert
gute Waldnutzung
FSC® C083411

Kolibri 360 GmbH
Ferdinand-Jühlke-Straße 7
99095 Erfurt, Deutschland
produktsicherheit@kolibri360.de